ADVANCES IN MATERIALS AND PAVEMENT PERFORMANCE PREDICTION II

CONTRIBUTIONS TO THE 2ND INTERNATIONAL CONFERENCE ON ADVANCES IN MATERIALS AND PAVEMENT PERFORMANCE PREDICTION (AM3P 2020), 27–29 MAY, 2020, SAN ANTONIO, TX, USA

Advances in Materials and Pavement Performance Prediction II

Editors

A. Kumar
Department of Civil Engineering and Geosciences, Delft University of Technology, The Netherlands

A.T. Papagiannakis
Department of Civil and Environmental Engineering, The University of Texas at San Antonio, USA

A. Bhasin
Department of Civil, Architectural and Environmental Engineering, The University of Texas at Austin, USA

D. Little
Department of Civil Engineering, Texas A&M University, USA

CRC Press
Taylor & Francis Group
Boca Raton London New York Leiden

CRC Press is an imprint of the
Taylor & Francis Group, an **informa** business

A BALKEMA BOOK

CRC Press/Balkema is an imprint of the Taylor & Francis Group, an informa business

© 2021 Taylor & Francis Group, London, UK

Typeset by MPS Limited, Chennai, India

Library of Congress Cataloging-in-Publication Data
Applied for

Published by: CRC Press/Balkema
            Schipholweg 107C, 2316 XC Leiden, The Netherlands
            e-mail: Pub.NL@taylorandfrancis.com
            www.routledge.com – www.taylorandfrancis.com

ISBN: 978-0-367-46169-0 (Hbk)
ISBN: 978-1-003-02736-2 (eBook)
DOI: 10.1201/9781003027362
https://doi.org/10.1201/9781003027362

Advances in Materials and Pavement Performance Prediction II – Kumar et al. (eds)
© *2021 Taylor & Francis Group, London, ISBN 978-0-367-46169-0*

Table of contents

Structures

Mixes

Moderators: Yong-Rak Kim (Texas A&M University), Shane Underwood (North Carolina State University), Zhen Leng (Hong Kong Poly. University) and Eshan Dave (University of New Hampshire)

Binders

Keynote Speakers

Dr. Eyad Masad is a professor in the Zachry Department of Civil Engineering at Texas A&M University and a professor in the Mechanical Engineering Program at Texas A&M at Qatar. He is also the Executive Director of Global Initiatives in the Texas A&M Engineering Experiment Station. He received his BSc from the University of Jordan and his MSc and PhD from Washington State University. Dr. Masad is a fellow of the American Society of Civil Engineers (ASCE) and a fellow of the American Association for the Advancement of Science (AAAS). He is the recipient of the James Laurie Prize for 2019 from ASCE. Dr. Masad's research focuses on multiscale characterization of pavement materials, computational modeling, analysis and design of pavement systems.

Multiscale Characterization and Computational Modeling of Asphaltic Materials: Opportunities and Challenges

The past two decades have witnessed significant advances in computational modeling of asphaltic materials. This was paralleled with the development of sophisticated instruments and methods for the characterization of material properties at various scales. These advances have shown great potential and supported the design and construction of sustainable pavements. However, we have not realized their full benefits due to scientific and practical limitations. The presentation will give a critical review of advances in multiscale characterization and computational modeling: what did they help to accomplish and where did they fall short? This will be followed by sharing ideas for overcoming their current limitations in order to support the design of innovative materials and delivering sustainable pavements.

Dr. Kim is the Jimmy D. Clark Distinguished University Professor and Alumni Association Distinguished Graduate Professor in the Department of Civil, Construction, and Environmental Engineering at North Carolina State University and Changjiang Scholar in the Department of Materials Science and Engineering at Chang'an University in China. He has over 30 years of experience in both the laboratory and field aspects of the performance evaluation of asphalt materials and pavements. His research is recognized by over $18 million research funding. He is a Fellow of the ASCE and the Korean Academy of Science and Technology. He holds a PhD from Texas A&M University.

What is Asphalt Concrete Telling Us? A 37-Year Story

This keynote presentation summarizes Dr. Kim's findings on the different ways that asphalt concrete behaves. Dr. Kim will describe the various mechanisms that are critical in modeling the behavior of asphalt concrete, along with experimental data and models. The presentation will demonstrate how the fundamental modeling of asphalt concrete can be applied to practice and will end with a few thoughts on ways to become a successful researcher in pavement engineering.

Since 2011, Markus Oeser is Professor and Director of the Institute of Highway Engineering, at the RWTH Aachen University. Since 2015, he is the Dean of the Faculty of Civil Engineering. Previously, he was a lecturer at the Dep. of Geotechnics, Road Construction and Transport. at the University of New South Wales in Sydney. After studying at the TU Dresden at the Institute for Urban Construction and Road Construction, he worked as a research associate at the TU Dresden and received his doctorate in 2004. His habilitation took place in 2010 at the Institute of Geotechnical Engineering, Road Construction and Transportation of the University of New South Wales.

Study of Asphalt Compaction and Asphalt Performance based on Numerical Simulation and Advanced Measurement Techniques

A high-quality compaction of the asphalt mixture is of great importance for the proper design and construction of high performance asphalt pavements. In order to improve the performance of asphalt, the compaction process of the hot asphalt mix is numerically simulated. The flow of material during the compaction under the paver screed is simulated by discrete element method (DEM). In addition, advanced measurement techniques were applied to monitor the movement of granular material during paver compaction, the movement or kinematic properties of material during paving and the contact force condition of paving materials can be detected. After the paving, the asphalt performance is simulated by finite element method (FEM) using the generated micro-structure of the asphalt mixtures from X-ray Computer Tomography (X-ray CT). Different compaction methods and material properties of the asphalt mixture components are applied to investigate their influence on the performance of the asphalt mixtures. This study will help generating the currently missing theoretical framework to optimize in-situ compaction of asphalt pavement. Meanwhile, it can also be used for guiding the construction of better asphalt compaction equipment (pavers, rollers) and thus ensure the transfer of the acquired basic knowledge into application-oriented research and development.

General topics (Rigid pavements/Geotechnics/Big data)
Moderators: Lev Khazanovich (University of Pittsburgh), Erol Tutumluer (University of Illinois-Urbana Champaign), Katherine Petros (FHWA), Wynand Steyn (University of Pretoria) and Sandra Erkens (TU Delft)

Field performance experience of Sulfur Extended Asphalt (SEA) pavement in Saudi Arabia

Ali Mahdi Alyami & L.S. Toyogon
Saudi Arabian Oil Company (Saudi Aramco), Dhahran, Eastern Province, Saudi Arabia

ABSTRACT: Elemental sulfur is a material that is available in abundant quantities in its natural form as well as an industrial waste product. Sulfur is a viable alternative and sustainable binding agent that can be used as a construction material in large scale. One example is sulfur extended asphalt (SEA) that replaces approximately 30% of asphalt binder by weight in a typical asphalt mixture. Since the introduction of sulfur material in the 1970s, several different technologies for the use of SEA have been developed. A detailed review of these technologies can be found in the literature (Sakib et al. 2019). While most of the work was done in the 1970s-80s, Saudi Aramco has continued the investigating the use of elemental sulfur in pelletized form (to reduce dust and related problems) as a bitumen extender for the sulfur extended asphalt (SEA). In 2006, three roads in the Kingdom of Saudi Arabia (KSA) were built using this approach: Khuraniyah access road, Shedgum–Hofuf road, and Dhahran–Jubail expressway. These sections were evaluated after three years and found to be in satisfactory condition. The main goal of this paper is to present a summary of the materials design, evaluation, and construction practices as well as performance from three projects that utilized significant amounts of SEA in pavement construction in 2017: Wasea Bulk Plant, the Jizan Economic City Mega-Project, and the Fadhili Gas Plant Program. Although these sections have been in place for 2 years, it is important to highlight that these projects were access roads leading in and out of major plants, and therefore experienced a high volume of heavy traffic. Therefore, with the exception of the influence of long-term oxidative aging, these sections serve as an excellent indicator for the performance of SEA mixes.

1 RAW MATERIAL PROPERTIES

The raw materials for SEA are generally the same as conventional asphalt, with the addition of solid elemental sulfur in pellet (granule) form. Following is a review of the raw materials, main characteristics, and requirements.

1.1 Asphalt

The asphalt cement used for the SEA was unmodified and conformed to the requirements of Saudi Aramco Code (SAES-Q-006), which requires the cement to be with a penetration grade 60/70, conforming to Saudi Aramco Product Specification A-970 (W.A 2016).

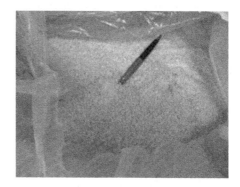

Figure 1. Sulfur pellets.

1.2 Sulfur

Sulfur used for the SEA was pelletized elemental sulfur that conforms to the requirements of Saudi Aramco Product Specification A-102. Figure 1 below illustrates sulfur pellets as produced and available in KSA. Product compliance certificates provided by the supplier are usually furnished with each site delivery (Mohammed 2010).

1.3 Mineral aggregate

The mineral aggregate for the SEA mix was dense, hard, durable, crushed aggregate with a particle size distribution. It should be noted that SEA mixes do not use hydrated lime and/or Portland cement, or fly ash as a mineral admixture.

2 PAVEMENT MIX DESIGN & LABORATORY RESULTS

2.1 Preparing the new mix design for the field pavement (Ali 2018)

The mix designs were developed by means of trial mix testing by an independent testing agency approved by Saudi Aramco. Two mixes were developed, one was to be used for the bituminous base course (BBC) pavement and a second one for bituminous wearing course (BWC). The mix design methodology followed was Marshall Design Procedure for Mixes incorporating Sulfur Pellets.

The resulting BBC mix design used an aggregate gradation Class B with a nominal maximum size of 25mm (1-inch) and well distributed gradation. The binder content was 5.6% by weight, binder consisting of 70% bitumen, and 30% elemental sulfur pellets, total voids on the mix were 4.9 %, while voids in mineral aggregate (VMA) were 12.1%, Marshall stability and flow resulted in 2094 kg and 2.58mm, respectively. In addition, after running many tests for different sets by a certified civil engineering laboratory, the optimum asphalt content was determined to be 5.6 %, for the subject BBC mix design. It is important to recognize that due to the difference in the specific gravity of sulfur and bitumen, an optimum of 5.6% by weight of mix when bitumen-sulfur is used as the binder results in the same volume as approximately 4.5% by weight of mix when conventional bitumen is used as the binder. Simply put, typically when SEA is used the optimum binder content by weight is slightly higher to ensure similar optimum binder content by volume.

The resulting BWC mix design used an aggregate gradation Class B with a nominal maximum size of 19mm (3/4-inch), the binder content was 6.2% by weight, binder consisting of 70% bitumen and 30% elemental sulfur pellets, total voids on the mix were 4.4 %, while voids in mineral aggregate (VMA) were 15.2%, Marshall stability and flow resulted in 2184 kg and 2.77mm, respectively. All of these parameters were within the requirements of Saudi Aramco engineering standards.

3 BATCH PLANT MODIFICATION

3.1 Utilization of sulfur material as part of the hot mix batches

The contractor installed a hot mix asphalt plant dedicated exclusively for this project, and therefore it was easily adjusted and modified to produce SEA without interruption. In fact, single day production switching between conventional and SEA was not permitted nor attempted.

It should be noted that the modifications as presented in this report are the ones chosen and executed by the contractor and are presented only to exemplify one way to accomplish the functions needed

Figure 2.   Location of weighting box.

Figure 3.   The box & pug mill mixer.

for storing, handling, feeding, measuring and mixing sulfur into SEA.

The aggregate and the binder (asphalt + sulfur) were blended together in a twin-shaft pug mill, see Figures 2 and 3. The aggregate was first discharged from the weighting hopper into the pug mill, and briefly mixed (dry-mix time), before the asphalt binder was introduced into the pug mill. Seconds later the sulfur was added for a final mixing of all the ingredients. No physical modifications to the pug mill mixer were necessary for the production of SEA, since the addition of the sulfur is achieved through the weighting hopper immediately above it.

3.2 Sulfur silo

The sulfur silo was used for storing sulfur before introduction into the pug mill. The silo was designed for safe storage, and to avoid contamination. Figure no. 4 shows the sulfur silo used for this project.

3.3 Sulfur conveying system

A sulfur silo feeder was used for conveying sulfur pellets to the weighing scale. Given the form and shape of the sulfur, a forced-air system was selected. This system consisted on a centrifugal air pump that drives

Figure 4. Sulfur silo.

Figure 5. Access road after opening (2017).

Figure 6. Exit of the access road (2019).

air to a closed conduit (8" diameter pipe), which in turn is connected to the controlled outlet gate of the silo. The conduit delivered sulfur pellets directly to the weighing scale and into the pug mill mixer.

4 SEA MIX PRODUCTION CONTROLS

Mix temperature control was essential factor in acquiring the required quality of SEA mix, as well to ensuring the emissions of hazardous gases was below the occupational limits. The temperature of the mix at discharge from the mixer had to be maintained within the temperature range corresponding to a viscosity of 150 to 300 cst for the molten sulfur/asphalt cement binder, but in no case could exceed a temperature of $140\pm5°C$. The aggregate level in the plant hot-bins was maintained approximately 2/3 full during SEA mix production, to minimize fume and odor emission (Mohammed, 2010).

5 COMPLETED SEA PAVEMENT AT FADHILI GAS PLANT PROGRAM, SAUDI ARABIA

Sulfur Extended Asphalt (SEA) pavement was used in the Fadhili Gas Plant Program located in Jubail, Saudi Arabia. This project consisted of the construction of approximately 28 kilometers of four-lane access roads. Roads were designed to withstand heavy traffic loads for the construction of the plant, and subsequent daily truck traffic during operations. SEA concrete mixes for Bitumen Base Course (BBC) and Bitumen Wearing Course (BWC), both Class B were successfully used in the recently completed project. The construction began in October 2016 and ended in June 2017 and the road was open for traffic, as shown in Figure 5.

On July 2019, the road was assessed and there were no noticeable defects, such as cracking or rutting, although it was subjected to a continuous heavy traffic for almost two years. During these two years the pavement experienced air temperatures that varied between approximately 60° C in the summer and approximately 5° C in the winter. Figure 6 shows the current condition of the busiest portion of the road. Table 1 shows the traffic load estimated for this road

Table 1. Road access traffic estimation.

Pavement Categories	Traffic & Load
Access Roadways	Sedans and trucks up to 40 kips axle load at peak hours in 5 days a week with at least 20 vehicles per minute. This is a minimum estimation at the peak stage*.

*Based on above, the total truck passed on this access road exceeded 2.3 Million sedan to truck up to 40 kips axle load in the last two years (2017-2019).

during the first two years of its service. The road in fact is classified as an access road, as per the Saudi Aramco category system.

6 CHALLENGES AND LESSONS LEARNED

6.1 *Elemental sulfur source & supply*

The contractor had to modify the existing batch plant to accommodate the addition of sulfur to produce SEA. The sulfur used in this project was granulated sulfur from Berri Gas Plant, Saudi Arabia as it was the closest source near the site. Note that alternative sources were hundreds of kilometers away.

Although the sulfur sourced from Berri Gas Plant was in compliance with Saudi Aramco product specifications, the prilling process used in this facility produces an increased residual moisture content in the sulfur. This extra moisture resulted in clogging issues in the feed line from the storage to the weighting hopper. Testing was performed on October 2016 to investigate and resolve the moisture issue. As a result, the batching process added a step in which the sulfur was placed outside of the storage facility to sun dry. This extra step exposed the material to contamination from contact with the ground and wind-transported sand. In addition, the added handling operation increased the crunching, and consequently the presence of dust-sized particles.

Berri Gas Plant did not have a mechanism for domestic sales, and therefore a Certificate of Compliance (COC) for the sulfur pellets was not available. Only sulfur truck loading notes from Berri Gas Plant were available for verification, making it difficult to demonstrate quality conformance.

6.2 *Moisture content of elemental sulfur*

Quality personnel were required ensure that the moisture content of elemental sulfur was not more than 3–5%. Thus, it was recommended to conduct a test by the field quality team at every delivery to ensure compliance. Based on the field experience, moisture content requirement is for operational purposes, and does not affect structural strength of the mix. Actual operation observations were noted that when the moisture of elemental sulfur is higher than 3–5%, sulfur is hardly driven by the compressor to the pug mill, and it clogs the pipe conveyor.

6.3 *Sulfur elements size*

Size of pulverized sulfur is recommended to be minimized to less than 2% passing the sieve no. 16. through proper handling and transportation in the field. Physical properties, i.e., particle size distribution, were measured for every delivery of the elemental sulfur, to record the percentage of sulfur fine particles. It is deemed necessary to control the fines for safety purposes, i.e., to mitigate the risk of inhalation of sulfur by exposed field personnel.

6.4 *Sulfur percentage in the mixed asphalt*

As per Saudi Aramco requirements, the binder consisted of 30% elemental sulfur pellets as replacement to the bitumen by weight. Thus, as a quality assurance every batch monitored and recorded compliance to this requirement. Saudi Aramco approved batch plant internal laboratory technician was available full-time, to monitor production and ensure the compliance of the requirements (W.A 2016).

7 CONCLUSION

The experience of the successful completion of Sulfur Extended Asphalt (SEA) pavement works at Fadhili Gas plant program provided a firsthand lesson to the team. The challenges of handling and incorporating elemental sulfur with the asphalt mix design was unraveled. It is also experienced that all personnel in production, placement of sulfur extended asphalt concrete were fully educated of the hazards in handling elemental sulfur.

Three distinct advantages emerge from this study. First, SEA resulted in a pavement that was subjected to heavy construction traffic for a period of over two years and did not result in any perceptible failure in the climatic conditions in the Kingdom of Saudi Arabia. Second, the mix was produced at a relatively lower temperature than conventional asphalt mix, thereby reducing energy consumption and aging during production. Third, SEA was demonstrated as a feasible alternative to prevent bitumen shortage issues while utilizing a substitute and abundant material as a binder, i.e. Sulfur. Based on these advantages, SEA appears to be a significant step in moving towards sustainable and durable materials for pavement construction.

Owing to the aforementioned advantages, SEA is now mandated for all the projects of the Saudi Arabian oil company and it has been recently piloted for government public projects.

REFERENCES

Alyami, Ali., Acero C.E., 2018. *SAER-9037/Successful Sulfur Extended Asphalt Pavement in Fadhili Gas Projects Program*, Saudi Arabia, Dhahran.

Khatri, W.A., 2016. SAES-Q-006/*Asphalt and Sulfur Extended Asphalt Concrete Paving*, Saudi Arabia, Dhahran.

Mehthel, Mohammed., 2010. *SABP-Q-010/Mix Design and Construction of Sulfur Extended Asphalt Concrete*, Saudi Arabia, Dhahran.

Nazmus Sakib, Amit Bhasin, Md. Kamrul Islam, Kaffayatullah Khan & Muhammad Imran Khan. 2019. A review of the evolution of technologies to use sulphur as a pavement construction material. *In: International Journal of Pavement Engineering*.

Organosilane and lignosulfonate: Road subsurface layers stabilisers

D.M. Barbieri & I. Hoff
Norwegian University of Science and Technology, Trondheim, Trøndelag, Norway

S. Adomako
University of Agder, Grimstad, Aust-Agder, Norway

C. Hu
South China University of Technology, Guangzhou, Guangdong, China

ABSTRACT: The "Ferry-free coastal highway route E39" project entails the construction of several long tunnels along the southwestern Norwegian coast, causing the generation of a remarkable quantity of blasted rocks. The aggregates could be used in the road unbound layers close to the place of production to provide a sustainable cost-benefit application. The research investigates two types of stabilising agents adopted to improve the mechanical properties of the crushed rocks, especially the "weak" rocks not fulfilling the requirements specified by the design guidelines. One additive is based on organosilane and the other additive is based on lignosulfonate. Four types of aggregates are investigated in the laboratory by means of repeated load triaxial tests. Resilient modulus is assessed by Hicks & Monismith model and Uzan model. The resistance against permanent deformation is analysed by Coulomb approach. Results prove that both the mechanical parameters are significantly enhanced by the stabilizing agents.

1 INTRODUCTION

Norwegian Public Roads Administration (NPRA) is currently running the "Ferry-free coastal highway route E39" project, which improves the viability along the southwestern Norwegian coast thanks to the creation of an extended tunnelling system (Dunham 2016). This will generate a very large quantity of blasted rocks, which could potentially be used as viable substitutes for natural aggregates in the road unbound layers close to the place of production.

Even if the major part of the rocks has igneous origin and could potentially meet the code requirements, the damage induced by the confined heavy blasting makes the materials weaker. The research investigates how to enhance the mechanical properties by use of stabilising agents.

Two innovative additives are examined: one is based on organosilane and one is based on lignosulfonate. The existing literature has focused on the application of these stabilising agents to clayey and silty soils; therefore, the use of the technologies applied to crushed rocks can broaden their field of exploitation. The current study further analyses the promising results obtained in previous research (Barbieri et al. 2019a, 2020).

Organosilane, here referred to as polymer-based agent (P) as well, is derived from nanoscale technology. It converts the water absorbing silanol groups presented on the silicate-containing surface of the rocks to a 4–6 nm layer of hydrophobic alkyl siloxane resulting in near permanent modifications.

Lignosulfonate, here also referred to as lignin-based agent (L), is a renewable product derived from pulp and paper industry. It is an organic polymer that consists of both hydrophilic and hydrophobic groups, it is non-corrosive, non-toxic and water-soluble.

Four types of aggregates (M1, M2, M3, M4) coming from tunnelling excavations (Barbieri et al. 2019b) are investigated. The pavement design manual N200 (NPRA 2018) sets requirements for the use of crushed rocks in road unbound layers in terms of Los-Angeles (LA) values (CEN 2010) and micro-Deval (MDE) values (CEN 2011). As reported in Figure 1, materials M2 and M3 do not meet the code requirements ("weak rocks") for base courses, differently from materials M1 and M4 ("strong" rocks). M2 and M3 could anyway be used in subbase courses.

2 METHODOLOGY

2.1 *Repeated load triaxial test*

Repeated Load Triaxial Test (RLTT) gives a comprehensive insight into material properties by assessing the stiffness and the resistance to permanent deformation. The behaviour of the tested aggregates is mainly

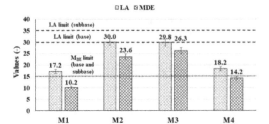

Figure 1. Los Angeles and micro-Deval values of investigated materials.

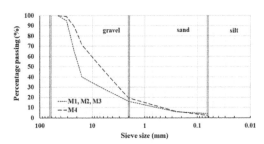

Figure 2. Investigated grain size distributions.

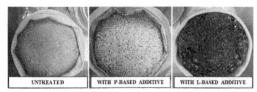

Figure 3. RLTT specimens.

Table 1. Permanent strain rate values defining the material range boundary lines.

Permanent strain rate	Range
$\dot{\varepsilon} < 2.5 \cdot 10^{-8}$	elastic zone
$2.5 \cdot 10^{-8} < \dot{\varepsilon} < 1.0 \cdot 10^{-7}$	elasto-plastic zone
$\dot{\varepsilon} > 1.0 \cdot 10^{-7}$	plastic (failure) zone

connected to the following parameters: stress level, dry density, grading and mineralogy (Lekarp et al. 2000a, 2000b).

RLTT apparatus exerts a uniform confining pressure in all of the directions (σ_3, triaxial or confining stress) by means of pressurised water and an additional vertical dynamic stress (σ_d, deviatoric stress), which is applied according to the chosen sinusoidal pattern and stepwise increases with different levels of σ_3. The RLTT apparatus performs the multi-stage low stress level (MS LSL) loading procedure (CEN 2004): five loading sequences are associated with five different σ_3 values (σ_3 = 20, 45, 70, 100, 150 kPa).

2.2 Sample preparation

M1, M2, M3 gradation corresponds to existing code specification (NPRA 2018), M4 gradation is the one available in the quarry where further tests have been performed (Barbieri et al. 2020), Figure 2.

The additive quantity is chosen after initial trials considering the stabilisation mechanism (Barbieri et al. 2019a). The polymer-based additive is mixed at OMC (w = 5%). The following proportion is used: 36 g for 365 g of water. The lignin-based additive is initially mixed at OMC; the percentage added to the crushed rocks is 1.5% in mass. Lignosulfonate needs a curing time to dry in order to become effective. Each RLTT sample is firstly conditioned at 65°C for 48 hours and then at 22°C for 24 hours before testing to reach w = 1%. Each specimen is compacted and covered by latex membranes (Figure 3). The diameter is 150 mm and the height is approximately 180 mm. Only M2, M3, M4 undergo additive stabilisation.

2.3 Results interpretation

The resilient modulus M_R associated with a change in the dynamic deviatoric stress $\sigma_{d,dyn}$ and constant triaxial stress σ_3 is defined as follows

$$M_R = \frac{\Delta \sigma_{d,dyn}}{\varepsilon_{r,v}} \qquad (1)$$

where $\varepsilon_{r,v}$ is the axial resilient strain. Among the proposed non-linear relationships (Lekarp et al. 2000a), Hicks & Monismith (HM) model is used to interpret experimental results (Hicks & Monismith 1971)

$$M_R = k_{1,HM}\sigma_a \left(\frac{\theta}{\sigma_a}\right)^{k_{2,HM}} \qquad (2)$$

where θ is the bulk stress, σ_a is a reference pressure (100 kPa) and $k_{1,HM}$, $k_{2,HM}$ are regression parameters. Furthermore, Uzan (UZ) model establishes a relationship between three quantities, namely M_R, θ and σ_d (Uzan 1985)

$$M_R = k_{1,UZ}\sigma_a \left(\frac{\theta}{\sigma_a}\right)^{k_{2,UZ}} \left(\frac{\sigma_d}{\sigma_a}\right)^{k_{3,UZ}} \qquad (3)$$

where $k_{1,UZ}$, $k_{2,UZ}$, $k_{3,UZ}$ are regression parameters. Uzan model enables a meaningful representation in a three-dimensional plot.

The permanent deformation is investigated through the Coulomb approach (Hoff et al. 2003). The mobilized angle of friction ρ and the angle of friction at incremental failure ϕ respectively express the degree of mobilized shear strength and the maximum shear strength. They identify three different ranges: elastic, elasto-plastic and plastic (failure). The strain rate $\dot{\varepsilon}$ refers to the development of permanent deformation per load cycle (Table 1).

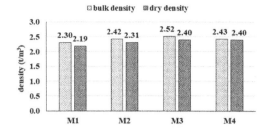

Figure 4. Materials bulk and dry density at w = 1%.

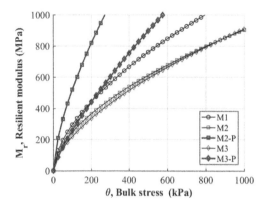

θ, Bulk stress (kPa)

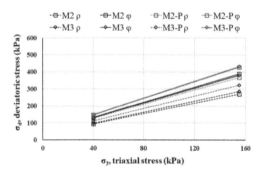

σ₃, triaxial stress (kPa)

Figure 5. Resilient modulus M_R (HM model) and limit angles ρ, ϕ for P-based additive application (M1, M2, M3), w = 5%.

3 RESULTS AND DISCUSSION

Figure 4 compares the bulk and the dry density of untreated materials, i.e. for water content w=1% to enable a uniform comparison.

Figure 5 displays the performance of material M1, M2, M3 with and without the application of P-based additive in terms of resilient modulus M_R and limit angles ρ, ϕ. All the mechanical properties are significantly enhanced, Table 2 reports the values of the regression parameters. The specimens are tested at OMC (w = 5%).

Similarly, Figure 6 illustrates the results connected to the application of the L-based additive to M2, M3. The stabilising agent engenders a significative

Table 2. Regression parameters of M_R HM model and ρ, ϕ limit angles for P-based treatment.

Material	Model parameters			Limit angles	
	$k_{1,HM}$	$k_{2,HM}$	R^2	ρ	ϕ
M1	2994	0.59	0.76	58.4	64.9
M2	2467	0.56	0.62	57.2	65.8
M2-P	5206	0.65	0.58	64.6	67.8
M3	2184	0.62	0.74	58.5	65.3
M3-P	2576	0.78	0.70	61.6	68.0

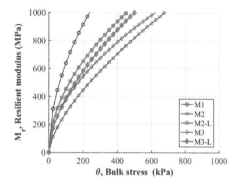

θ, Bulk stress (kPa)

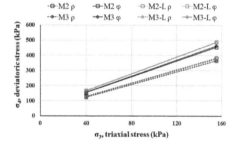

σ₃, triaxial stress (kPa)

Figure 6. Resilient modulus M_R (HM model) and limit angles ρ, ϕ for L-based additive application (M1, M2, M3), w = 1%.

Table 3. Regression parameters of M_R HM model and ρ, ϕ limit angles for L-based treatment.

Material	Model parameters			Limit angles	
	$k_{1,HM}$	$k_{2,HM}$	R^2	ρ	ϕ
M1	6 378	0.52	0.37	62.1	67.3
M2	2 816	0.66	0.86	65.4	68.9
M2-L	4 530	0.52	0.84	64.4	70.3
M3	3 737	0.54	0.69	64.3	69.2
M3-L	3 869	0.59	0.59	65.0	70.2

increase in both the resilient modulus M_R and limit angles ρ, ϕ. Table 3 shows the values of the regression parameters. The samples are tested with water content w = 1%, since the lignosulfonate needs a curing time to dry and attach the material particles.

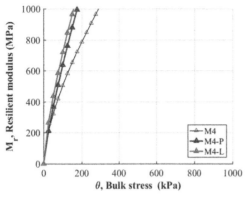

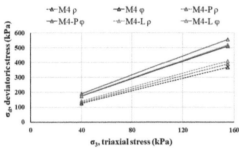

Figure 7. Resilient modulus M_R (HM model) and limit angles ρ, ϕ for P-based, L-based additive application (M4), w = 1%.

Figure 7 displays the results for M4 added with P-based additive or L-based additive at w = 1%. In this case, the P-treated samples undergoes the same conditioning treatment of the L-treated samples. The properties of M4 aggregates, which are already fulfilling code requirements, are further enhanced by the application of the stabilising agents. Table 4 details the values of the regression parameters.

In addition, Uzan model is also adopted to represent the results connected to material M4 (Figure 8): the improvement entailed by the additives can also be clearly seen in the three-dimensional representation. The goodness-of-fit R^2 connected to Uzan model is considerably higher than R^2 connected to Hicks & Monismith model (Table 4, Table 5, Table 6).

4 CONCLUSIONS

The research investigated two stabilizing agents used to improve the mechanical properties of crushed rocks to serve as construction materials in the road base and subbase layers. The additives were based on organosilane and lignosulfonate, respectively. Four types of crushed rocks M1, M2, M3, M4 were tested in the laboratory by means of Repeated Load Triaxial Test.

Based on the rock type, the application of the stabilising agents enhanced the resilient modulus M_R and the mobilized angle of friction ρ and the angle

Table 4. Regression parameters of M_R HM model and ρ, ϕ limit angles for P-based, L-based treatments.

Material	Model parameters			Limit angles	
	$k_{1,HM}$	$k_{2,HM}$	R^2	ρ	ϕ
M4	5046	0.64	0.49	64.6	71.1
M4-P	6378	0.79	0.48	65.7	72.5
M4-L	7194	0.72	0.38	66.8	71.3

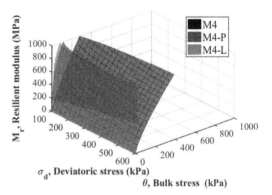

Figure 8. Resilient modulus M_R (UZ model) for P-based, L-based additive application (M4), w = 1%.

Table 5. Regression parameters of M_R UZ model for P-based, L-based treatments.

Material	Model parameters			
	$k_{1,UZ}$	$k_{2,UZ}$	$k_{3,UZ}$	R^2
M4	4 442	0.78	−0.14	0.94
M4-P	9 424	0.37	0.43	0.93
M4-L	10 559	0.31	0.41	0.89

Table 6. Comparison between HM and UZ models.

	HM model			UZ model			
	$k_{1,HM}$	$k_{2,HM}$	R^2	$k_{1,UZ}$	$k_{2,UZ}$	$k_{3,UZ}$	R^2
M2-P	5 206	0.65	0.58	4 813	0.74	−0.08	0.85
M3-P	2 576	0.78	0.70	1 797	1.17	−0.37	0.84
M2-L	4 530	0.52	0.84	4 881	0.45	0.08	0.89
M3-L	3 869	0.59	0.59	3114	0.80	−0.21	0.78

of friction at incremental failure ϕ of the aggregates. Both polymer-based additive and lignin-based additive coat and bond the material particles closely together. The technologies represent non-traditional stabilization approaches to improve mechanical properties of aggregates in pavement subsurface layers.

REFERENCES

Barbieri, D.M., Hoff, I. & Mørk, M.B.E. 2019a, 'Innovative stabilization techniques for weak crushed rocks used in road unbound layers: a laboratory investigation', *Transportation Geotechnics*, vol. 18, pp. 132–41.

Barbieri, D.M., Hoff, I. & Mørk, M.B.E. 2019b, 'Mechanical assessment of crushed rocks derived from tunnelling operations', in W.-C. Cheng, J. Yang & J. Wang (eds),*5th GeoChina International Conference 2018*, Springer, pp. 225–41.

Barbieri, D.M., Hoff, I. & Mørk, M.B.E. 2020, 'Organosilane and lignosulfonate as innovative stabilization techniques for crushed rocks used in road unbound layers', *Transportation Geotechnics*, vol. 22.

Dunham, K.K. 2016, 'Coastal Highway Route E39 - Extreme crossings', *Transportation Research Procedia*, vol. 14, no. 2352, pp. 494–8.

Håndbok N200 vegbygging 2018 (Vegdirektoratet).

Hicks, R.G. & Monismith, C.L. 1971, 'Factors influencing the resilient properties of granular materials', *Highway Research Record*, pp. 15–31.

Hoff, I., Bakløkk, L.J. & Aurstad, J. 2003, 'Influence of laboratory compaction method on unbound granular materials', *6th International Symposium on Pavements Unbound*.

ISO 13286-7 Cyclic load triaxial test for unbound mixtures 2004.

ISO 1097-2 Tests for mechanical and physical properties of aggregates. Part 2: methods for the determination of resistance to fragmentation 2010.

ISO 1097-1 Tests for mechanical and physical properties of aggregates. Part 1: determination of the resistance to wear (micro-Deval) 2011.

Lekarp, F., Isacsson, U. & Dawson, A. 2000a, 'State of the art. I: resilient response of unbound aggregates', *Journal of Transportation Engineering*, vol. 126, no. 1, pp. 66–75.

Lekarp, F., Isacsson, U. & Dawson, A. 2000b, 'State of the art. II: permanent strain response of unbound aggregates', *Journal of Transportation Engineering*, vol. 126, no. 1, pp. 76–83.

Uzan, J. 1985, 'Characterization of granular material', *Transportation Research Record*, no. 1022, pp. 52–9.

Advances in Materials and Pavement Performance Prediction II – Kumar et al. (eds)
© 2021 Taylor & Francis Group, London, ISBN 978-0-367-46169-0

Reconsidering shoulder type for low volume concrete roads

N. Buettner, L. Khazanovich & J. Vandenbossche
University of Pittsburgh, Pittsburgh, PA, USA

ABSTRACT: Structural contribution of a shoulder to the performance of concrete pavements is well recognized. However, it is traditionally quantified through reduction of critical stresses in concrete slabs when the load is applied at the traffic lane/shoulder edge. Oversized vehicles, such as farm equipment, tend to wander across the joint and onto the shoulder of the road. Conventional mechanistic models predict that placement of a portion of a wheel load on an aggregate shoulder causes stresses in the concrete slab to significantly reduce. As a result, such loading scenario is not considered in the design procedures. In this study, the effect of wheel wander on stresses in the concrete slab is reconsidered. Two structural models are developed using the finite element program ISLAB2005. Both tied concrete and aggregate shoulders are considered. The performed analysis provides an insight into the effect of vehicle loads that span across the lane/shoulder joint. Recommended considerations for refinement of the current concrete pavement design practices for low volume concrete roads are provided.

1 INTRODUCTION

Oversized vehicles, such as agricultural vehicles, often travel on low volume concrete roads. Several studies have analyzed the concrete pavement performance effects caused by these vehicles (Ceylan et al. 2015; Lim et al. 2012). While the pavement performance effects caused by oversized vehicles are assessed in the literature, little is established regarding the effects on concrete pavement stresses due to oversized vehicle loading along the pavement shoulder. These vehicles have significant widths and wander partially onto the shoulder of low volume concrete roads to allow other vehicles to pass.

Transverse cracking in concrete pavements is commonly considered a major distress as it triggers pavement rehabilitation or reconstruction. One of the mechanisms of transverse cracking, bottom-up cracking, is caused by excessive tensile stresses at the bottom of a concrete slab due to heavy axle loading. The highest magnitude of concrete slab bottom surface stresses caused by heavy axle loading usually occurs when the load is applied at midslab and adjacent to the lane/shoulder joint. In this case, the maximum stress is also located at midslab of the traffic lane slab and adjacent to the lane/shoulder joint. Many mechanistic-empirical design procedures use this maximum stress for predicting damage. For example, the American Concrete Pavements Association's design program, StreetPave (ACPA 2012), is based on the Portland Cement Association (PCA) design procedure (Packard & Tayabji, 1985). It is commonly used to design low volume concrete roads. In the design procedure, the number of applications until failure for

the concrete pavement is estimated using Equation 1 (Titus-Glover et al. 2005):

$$\log Nf = \left(\frac{-SR^{-10.24} \log(1-P)}{0.0112} \right)^{0.217} \quad (1)$$

where N_f = number of allowable applications until failure; SR = the ratio of concrete slab edge stress due to traffic loadings and PCC flexural strength; P = probability of failure.

The magnitude of the fatigue damage caused by a certain axle type and weight depends on the magnitude of the induced stress and number of load repetitions, as given by the Palmgren-Miner Rule (Miner 1945):

$$FD = \sum \frac{n}{Nf} \quad (2)$$

where FD = cumulative fatigue damage; n = number of applied applications of specific load case and N_f = number of allowable applications until failure of specific load case.

Although the largest stress is induced when the load is applied at the slab edge, a small fraction of the overall traffic travels exactly at the lane/shoulder joint. Therefore, many design procedures, such as AASHTO Mechanistic-Empirical Design Guide (MEPDG) incorporate a traffic wander analysis (ARA 2004). However, usually such an analysis assumes that all the vehicles travel in the design lane. In the past, this assumption was considered to be conservative because conventional mechanistic models predicted that placement a portion of a wheel load on an aggregate shoulder significantly reduced stresses in the concrete slab.

In this study, the effects of oversized vehicles that wander onto the shoulder were analyzed. Both concrete and unpaved shoulders were considered since all can be present on low volume roads. The loading conditions considered include loading a single axle located adjacent to the lane/shoulder joint at midslab. The axle was then moved laterally across the joint a distance of 10 in, with analyses being performed in increments of 2.5 in for a total of 4 analyses being run for each shoulder type.

2 METHODOLOGY

A jointed plain concrete pavement (JPCP) was modeled using the pavement finite element software ISLAB (Khazanovich et al. 2005) as a three-slab system with a shoulder, as shown in Figure 1. The material properties and structural design features used in this analysis are provided in Table 1. As indicated in Table 1, the model was produced with tied concrete shoulders and unpaved (aggregate) shoulders.

Figure 1. Three-slab pavement system considered for finite element analysis.

Table 1. ISLAB inputs for jointed plain concrete pavement model.

Parameter	Input
PCC Slab Thickness	6 in
PCC Modulus of Elasticity	4,000,000 psi
PCC Poisson's Ratio	0.15
PCC Slab Joint Spacing	15 ft
PCC Slab Width	12 ft
Shoulder Type Concrete/Unpaved	
Shoulder Width	8 ft
Base Thickness	6 in
Base Modulus of Elasticity	40,000 psi
Modulus of Subgrade Reaction	200 psi/in
Longitudinal LTE	
Tied Concrete Shoulder	40%
Aggregate Shoulder	1%
Transverse LTE	70%
Nominal Mesh Element Size	6 in x 6 in

A single axle load of 21,000 lbs was applied at midslab. A tire pressure of 80 psi, a tire width of 10 in, and an axle length of 96 in between the wheels was adopted for the single axle. The maximum stress caused by

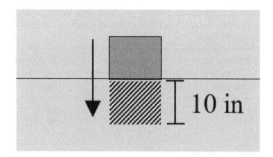

Figure 2. Lateral movement of exterior tire beyond lane/shoulder joint.

the single axle load on the traffic lane slab was determined for various partial loadings of the exterior tire on the shoulder (i.e. 0%, 25%, 50%, 100%), as shown in Figure 2. This partial loading represents progressive wander of an axle onto the shoulder from the initial position depicted in Figure 1.

Two types of interfaces between the pavement surface layer (i.e. traffic lane and shoulder) and the aggregate base layer were considered:

- Unbonded interface model
- Totski model

The unbonded interface model assumes no friction between the surface layer and the base but also assumes that these layers have the same deflection basins. This significantly simplified the subsequent analysis (Ioannides et al. 1992). Because of it, the unbonded interface model has been routinely used in structural modeling of concrete pavements and has been incorporated into MEDPG (ARA 2004).

A more realistic ISLAB2000 Totski model enables analysis of the independent actions of the surface layer and the base (Khazanovich and Ioannides 1994, Khazanovich et al. 2000).

3 RESULTS AND DISCUSSION

3.1 Effect of shoulder type assuming unbonded interface model

Table 2 shows the results of the finite element analysis with the interface between the top surface layer and the base being modeled as unbonded. For an unbonded interface model, it can be observed that a concrete shoulder reduces stresses in the concrete slab compared to the unpaved shoulder when the entire axle load is loaded on the traffic lane slab. When a portion of the exterior tire footprint is loaded on the shoulder, the stresses decrease for both concrete and unpaved shoulders, as shown in Figure 3. For an unbonded interface model, concrete pavements with an unpaved shoulder or a concrete shoulder experience similar fatigue damage if a portion of the tire footprint is placed on the shoulder.

Table 2. Maximum stresses on JPCP with concrete and unpaved shoulders from single axle load assuming unbonded interface model.

% Exterior Tire	Stress (psi)	Stress (psi)
On Shoulder	Concrete	Unpaved
0%	538	600
25%	457	489
50%	354	357
100%	64	6

Table 3. Maximum stresses on JPCP with concrete and unpaved shoulders from single axle load assuming unbonded interface model.

% Exterior Tire	Stress (psi)	Stress (psi)
On Shoulder	Concrete	Unpaved
0%	505	565
25%	441	550
50%	356	503
100%	100	286

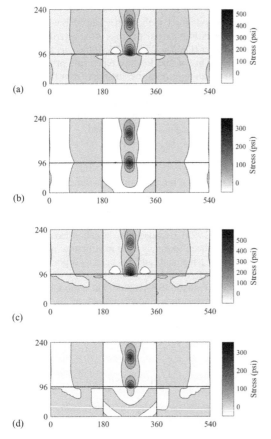

Figure 3. JPCP Stresses assuming an unbonded interface model (dimensions in inches). (a) No shoulder loading – concrete shoulder (b) Half shoulder loading – concrete shoulder (c) No shoulder loading – unpaved shoulder (d) Half shoulder loading – unpaved shoulder.

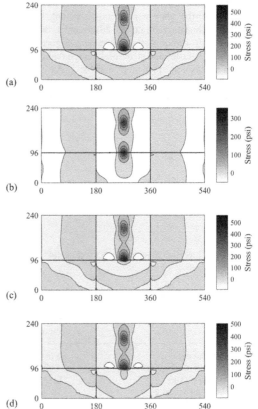

Figure 4. JPCP Stresses assuming a Totski interface model (dimensions in inches). (a) No shoulder loading – concrete shoulder (b) Half shoulder loading – concrete shoulder (c) No shoulder loading – unpaved shoulder (d) Half shoulder loading – unpaved shoulder.

3.2 *Effect of shoulder type assuming totski interface model*

Table 3 displays the results of the finite element analysis assuming a Totski interface model. It can be observed that a concrete shoulder reduces stresses in the traffic lane slab compared to the unpaved shoulder when the entire axle load is loaded on the slab in the driving lane. The effect is even more pronounced when a portion of the exterior tire footprint is loaded on the shoulder, as indicated in Figure 4. While partial loading of the concrete shoulder reduces the critical stress in the traffic lane, a partial loading of an unpaved shoulder (at to 25% of the tire footprint) does not cause a significant reduction in the critical stress. This means that a JPCP with an unpaved shoulder may experience extensive fatigue damage even if a portion of the tire footprint is placed on the shoulder, while fatigue damage for the same wheel path is minor if a concrete shoulder is used. This differs significantly from the results assuming an unbonded interface model, showing that an unbonded model underestimates stresses.

4 CONCLUSIONS

The effects of oversized vehicles that wander onto the shoulder were evaluated in this study. It was shown that a concrete shoulder reduces stresses in the traffic lane slab compared to the unpaved shoulder when the entire axle load is loaded on the traffic lane slab. When a part of the exterior tire footprint is loaded on the shoulder, a concrete shoulder is observed to be more suitable to reduce edge stresses (assuming the Totski interface model). This analysis also shows that the unbonded interface model underestimates the stresses in the traffic lane slab caused by partial loading onto the shoulder. Consideration of oversized vehicle wander onto the shoulder has been demonstrated to be necessary for concrete pavement design procedures for low volume concrete roads.

ACKNOWLEDGEMENTS

This material is based upon work supported by the National Science Foundation Graduate Research Fellowship under Grant No. 1747452. Any opinions, findings, and conclusions or recommendations expressed in this material are those of the authors and do not necessarily reflect the views of the National Science Foundation.

REFERENCES

American Concrete Pavement Association 2012. ACPA StreetPave Software. www.acpa.org/streetpave/Default.aspx.

Applied Research Associates, Inc. 2004. Guide for Mechanistic–Empirical Design of New and Rehabilitated Pavement Structures. *Final Report to NCHRP*. Albuquerque, New Mexico.

Ceylan, H. et al. 2015. Impact of Farm Equipment Loading on Low-Volume Concrete Road Structural Response and Performance. *The Baltic Journal of Road and Bridge Engineering* 10(4): 325–332.

Ioannides A.M. et al. 1992. Structural Evaluation of Base Layers in Concrete Pavement Systems. *Transportation Research Record: Journal of the Transportation Research Record* 1370: 20–28.

Khazanovich L. and Ioannides A. M. 1994. Structural Analysis of Unbonded Overlays Under Wheel and Environmental Loads. *Transportation Research Record: Journal of the Transportation Research Board* 1449: 174–181.

Khazanovich L. et al. 2000. *ISLAB2000 – Finite Element Analysis Program for Rigid and Composite Pavements User's Guide*. ERES Consultants, Champaign, Illinois.

Lim, J. et al. 2012. Effects of Implements of Husbandry (Farm Equipment) on Pavement Performance. *Final Report No. MN/RC 2012*–08. University of Minnesota, Minneapolis, Minnesota.

Miner, M. A. 1945. Cumulative damage in fatigue. *Journal of Applied Mechanics* 12: 149–164.

Packard R. G. and Tayabji S. D. 1985. New PCA thickness design procedure for concrete highway and street pavements. *Proceedings of the 3rd International Conference on Concrete Pavement Design and Rehabilitation*, Purdue University, West Lafayette, Indiana. 225–236.

Titus-Glover, L., et al., 2005. Enhanced Portland Cement Concrete Fatigue Model for StreetPave. *Transportation Research Record: Journal of the Transportation Research Board* 1919: 29–37.

Advances in Materials and Pavement Performance Prediction II – Kumar et al. (eds)
© 2021 Taylor & Francis Group, London, ISBN 978-0-367-46169-0

Aggregate properties affecting adhesive quality in asphalt mixtures

A. Cala & S. Caro
Department of Civil and Environmental Engineering, Universidad de los Andes, Bogotá, Colombia

Y. Rojas-Agramonte
Department of Geosciences, Universidad de los Andes, Bogotá, Colombia

ABSTRACT: Asphalt-aggregate adhesion is a complex phenomenon influenced by properties from both materials. This work aims at stablishing the influence of aggregate's chemistry, mixing temperature and surface texture over the quality of the asphalt-aggregate interface. To accomplish this goal, different asphalt-aggregates combinations were tested with a pull-off debonding test in which these properties of aggregates were changed accordingly. It was found out that SiO_2 content within aggregates promote moisture damage, whereas Fe_2O_3 and CaO contents seems to inhibit it. Low mixing temperatures when performing the initial asphalt-aggregate bonding were also related to low adhesive performance. Finally, aggregate surface texture did not seem to have an effect over the dry quality of the asphalt-aggregate interface but does seem to have a positive effect over its performance under moisture conditions.

1 INTRODUCTION

Hot mix asphalt (HMA) results from mixing hot asphalt binder with aggregates that follow a specific gradation and volumetric proportions. The structural integrity of this material heavily relies on a good adhesive bond between the "soft" binder and the "strong" aggregates. Since a pavement structure is usually subjected to myriad of environmental conditions (e.g. rain, snow, UV-rays) that might alter this bond quality, assessing which material properties generate a better asphalt-aggregate interface is key in the construction of high-quality pavements.

Moisture damage, which has been defined as the loss of cohesive strength within the mastic and/or the degradation of the asphalt-aggregate adhesive bond quality (Kiggundu and Roberts 1988), is one of the most common types of environmental-related distresses of HMAs, and has been identified as one of the main causes of early rehabilitation of asphalt pavements (Caro et al. 2008). Within this type of distress, aggregate chemistry has been shown to be the most contributing factor (Cala et al. 2019; Curtis 1992). Despite this, aggregate chemistry is usually not considered as part of the mix design process of HMAs.

Physical and mechanical properties of binders and aggregates also play an important role in the overall adhesive quality of the asphalt-aggregate system (Bhasin and Little 2007; Ishai and Craus 1977). Aggregate's form, angularity and texture have a direct effect over the total bonding area with the asphalt binder, whereas the binder has to have a low enough viscosity at high temperatures for it to be able to completely coat the aggregate. These shape parameters have been shown to be important within a mixture's performance (Castillo et al. 2018; Masad et al. 2001). Moreover, a mechanical effect is also generated at this interface due to the irregularities at the aggregate's surface; this is more evident in sedimentary aggregates since the binder fills the pores within sediments, creating complex asphalt-aggregate interactions.

This work studies the influence of aggregate's chemistry, mixing temperature and surface texture over the adhesive bond quality and durability of the asphalt-aggregate system, using the experimental methodology proposed by Cala et al. (2019). For the first goal, 6 different low-porosity aggregates (i.e. quartzite, granodiorite, serpentinite, andesitic basalt, andesite and amphibolite) were tested; for the second, 4 different aggregate temperatures were tested for the serpentinite rock; and for the third, a sedimentary sample with the same chemical composition as the quartzite sample but with a different surface texture, was tested.

2 MATERIALS AND METHODS

For all tested samples, an unmodified 60/70 penetration (1/10 mm) asphalt binder was used. Aggregates were obtained from several locations across south and central America, and their chemical composition, obtained through high precision X-Ray fluorescence, are listed in Table 1. The sedimentary quartz-arenite sample used for the surface texture portion of this

Table 1. Chemical composition expressed as weight total percentage of used aggregates.

	Qtz	Grt	Spt	Bas	Ant	Amp
	%	%	%	%	%	%
SiO_2	98.18	65.75	28.49	58.78	55.56	37.97
Al_2O_3	1.09	15.26	23.08	17.29	16.97	12.88
Fe_2O_3	0.09	4.09	7.37	5.47	7.94	12.31
MnO	0.00	0.08	0.48	0.17	0.13	0.22
MgO	0.01	2.45	28.15	1.99	4.07	4.62
CaO	0.06	3.77	0.03	4.45	7.34	18.81
Na_2O	0.00	3.60	0.03	4.21	3.18	1.68
K_2O	0.03	2.68	0.00	4.04	2.16	0.51
TiO_2	0.07	0.53	0.05	0.79	0.94	1.40
P_2O_5	0.01	0.14	0.01	0.04	0.25	0.16
SO_3	0.00	0.01	0.02	0.03	0.02	0.10
Cr_2O_3	0.00	0.01	0.39	0.00	0.01	0.04
NiO	0.00	0.00	0.07	0.00	0.00	0.03

*Qtz = quartzite,   Grt = granodiorite,   Spt = serpentinite, Bas = andesitic basalt, Ant = andesite, Amp = Amphibolite.

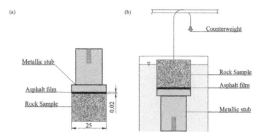

Figure 1. a) Asphalt-aggregate system (AAS), and b) water conditioning process.

Table 2. Average F_{max} for each aggregate per moisture conditioning time (i.e. days submerged in water).

	Qtz	Grt	Spt	Bas	Ant	Amp
Days	N	N	N	N	N	N
0	1340.6	1230.6	1089.3	1669.3	1071.7	1226.5
1	522.7	901.0	849.8	934.9	964.2	1231.7
3	310.6	641.6	785.8	719.7	956.4	1104.3
7	260.4	631.6	654.4	607.7	564.7	1226.4

study, is made of >98% quartz (i.e. SiO_2), similar to the quartzite sample herein used. Surface texture of aggregates was stablished using a Mitutoyo surface roughness tester (Ref: SJ-210).

The adhesion test method is summarized as follows:

1. 1-inch diameter aggregate cores are cut and polished using a grinding machine (Stuers Ref: Accutom-100).
2. The binder is placed in the oven at the desired temperature for 15 minutes, and the aggregate core and a metallic stub are also placed in the oven for at least 30 minutes.
3. With the aid of a modified micrometer, the aggregate core is bonded to a metallic stub by an asphalt film with a thickness of exactly 20 μm, to obtain an asphalt-aggregate system (AAS; Figure 1a).
4. If needed, the system is submerged on ultrapure type 1 water at 22° C for the extent of the conditioning time (Figure 1b).
5. The aggregate core is de-bonded from the metallic stub by a 10 mm/min displacement-controlled condition.

A deeper explanation of the experimental methodology can be found elsewhere (Cala et al. 2019).

The maximum load at failure (F_{max}), the work of fracture (W_f) and the adhesive failure area ($A_\%$) are the results obtained from this test. Due to space constrains, only F_{max} results are presented and analysed in this article.

For the chemical and textural portions of this study, samples were prepared at 150°C, and tested on dry condition and after being submerged during 1, 3 and 7 days in water. For the effect of temperature, the oven was set at 4 different temperatures during the fabrication of the AAS samples: 75°C, 100°C, 125°C and 150°C.

3 RESULTS

For the chemical effect, 3 replicates where fabricated and tested for each aggregate at each conditioning period. For the mixing temperature and surface texture results, only one specimen was tested. Further tests will be conducted in the near future.

3.1 Chemical influence

Table 2 shows the average F_{max} result for each aggregate per conditioning time. It can be seen that the quartzite aggregate (Qtz), which is made almost entirely of SiO_2, is by far the most susceptible to moisture damage, loosing 61% of its F_{max} after just 1 day of conditioning time, whereas for the Amphibolite aggregate (Amp) this loss was virtually 0%. For the dry condition, the Basalt aggregate (Bas) had the best performance (1669 N), whereas the Serpentinite (Spt) and Andesite (Ant) aggregates reached just 65% of that value.

Usually, for design purposes, what is sought after is for material properties not to degrade over time. For this reason, Figure 2 shows the relative loss of F_{max} for each conditioning time with respect to its dry condition. This graph shows the high moisture susceptibility of the Qtz aggregate, and the moisture resistance of the Amp aggregate

A moisture damage parameter per aggregate was arbitrarily defined as the area beneath each of the degradation curves shown in Figure 2. This proxy is higher at a higher moisture susceptibility. For instance, the calculated proxy for Qtz was 4.83, whereas it was

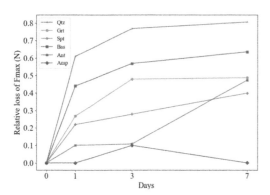

Figure 2. Average F_{max} for each aggregate per moisture conditioning period (i.e. days submerged in water).

only 0.32 for the Amp. After each proxy was calculated, a Pearson correlation coefficient was determined with each compositional oxide from Table 2, to see if there was any linear correlation between the individual oxide components of the aggregates and the moisture damage proxy (i.e. determine if the proxy increased as any oxide within the aggregates content increased). This correlation coefficient takes a value of 1 if there is a perfect positive linear relationship, and a value of -1 if there is a perfect negative linear correlation. The results are shown in Figure 3 (results for TiO_2, P_2O_5, SO_3, Cr_2O_3 and NiO where omitted as they are present only in small quantities within the used aggregates). It can be seen that SiO_2 (0.81) has a strong positive correlation with the proposed proxy, whereas Fe_2O_3 (0.95) and CaO (0.75) have a negative correlation with the index. In other words, aggregates with high contents of SiO_2 are more susceptible to moisture damage, while aggregates with high components of Fe_2O_3 and CaO are more resistant to this damage. This finding supports the chemical susceptibility of SiO_2 to water presented by Hefer et al. (2005).

3.2 Mixing temperature influence

Temperature plays a key role within most adhesives. For the asphalt-aggregate adhesion, the binder has to reach a temperature in which its viscosity is low enough for it to be able to completely 'wet' the aggregate. Moreover, high temperatures also increase the kinetics of the chemical reactions that take place within the active sites of minerals and the polar components from the binder, which are needed for the chemical adhesion between these materials to take place. AAS samples made with Spt were prepared and tested at 4 different temperatures: 75, 100, 125 and 150°C. F_{max} results from these tests are shown in Figure 4. This figure shows that F_{max} values are virtually the same at 125 and 150°C (1090 and 1045 N respectively) meaning that, for this asphalt-aggregate combination, 125°C was enough for the binder to properly coat the aggregate and for the chemical reactions between these two materials to occur. At 100 and 75°C, this

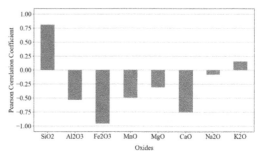

Figure 3. Pearson correlation coefficient between each compositional oxide and the calculated moisture damage proxy.

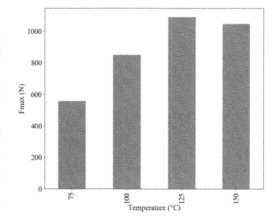

Figure 4. Fmax values at different temperatures for AAS made with Spt.

value decreased by 18 and 46% respectively. It should be noted that at lower temperatures, the bond quality between the metallic stub and the binder is also deteriorated, further lowering the structural integrity of the different AAS tested. Nevertheless, it is clear from these results that temperature plays a key role in obtaining a high-quality asphalt-aggregate interface, as already demonstrated in previous studies.

3.3 Aggregate surface texture influence

To assess the effect of the aggregate surface texture over the F_{max} values, the results obtained from the Qtz were compared with those from a quartz-arenite (Aren) lithology. The former is the metamorphic result of the latter, meaning that they share almost identical chemical composition, which is almost exclusively SiO_2, but the former has little to no porosity. The same testing procedure was used for the preparation of the aggregate cores, meaning that they were levelled and polished to micrometer level tolerances, which was checked with a Mitutoyo micrometer. Since the Aren aggregate is a sedimentary rock, its surface texture after the polishing and levelling was expected to be higher due to the pores in-between sediments. This

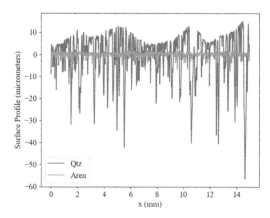

Figure 5. Typical surface texture for the Qtz and Aren aggregates after polishing and levelling.

Table 3. F_{max} results for the Qtz and Aren samples tested at each conditioning period.

	Qtz	Aren	Increment
Days	N	N	%
0	1340.6	1373.32	2.4%
1	522.7	741.6	41.9%
3	310.6	557.32	79.5%
7*	260.4	–	–

* 7 days Aren conditioned sample result have not yet been obtained.

was proven true after obtaining the profile (i.e. surface texture) from the Mitutoyo surface roughness tester (Ref: SJ-210), which is shown on Figure 5 ('x' is the abscise over the surface where the measurements were taken and 'y' the height of the texture profile). The average Ra roughness parameter, which is defined as the average deviation from a flat "zero" plane, for the Qtz and Aren samples tested was 8.250 and 0.797 μm respectively. Table 3 shows the F_{max} results for each conditioning time for the Qtz and Aren samples (data from 3 replicates per conditioning time was obtained for the Qtz samples from section 3.1; no replicates were made for the Aren samples).

Despite that the Ra roughness parameter was more than 10 times bigger in the Aren sample than in the Qtz sample, the former had just a 2.4% F_{max} increment in dry condition. The positive effects of a higher surface texture are evident from the moisture conditioned samples. After 3 days of water conditioning the Aren sample had a 79% higher F_{max} value than its Qtz counterpart, despite of its higher porosity (i.e. water could reach the interface more rapidly). This implies that the effect of the increase surface contact area between the

binder and the aggregate, and the mechanical adhesion effect caused by the increased surface texture of the aggregate, was significant enough to counteract the detrimental effect of a water infiltration at the interface.

4 CONCLUSIONS

Asphalt-aggregate adhesion is a complex phenomenon affected by several different properties from binders and aggregates. From the results herein presented, the following can be concluded:

1. The presence of SiO_2 oxide in the aggregate has a strong correlation with the moisture susceptibility of the asphalt-aggregate interface, whereas the Fe_2O_3 and CaO oxides have a negative correlation with this phenomenon.
2. Low mixing temperatures greatly deteriorate the asphalt-aggregate adhesive quality.
3. Surface texture for the tested aggregate do not seem to have a tangible effect over the dry adhesive quality but it does seem to improve the performance of the asphalt-aggregate interface when subjected to moisture.

REFERENCES

Bhasin, A., and Little, D.N. (2007). Characterization of aggregate surface energy using the universal sorption device. J. Mater. Civ. Eng. 19, 634–641.

Cala, A., Caro, S., Lleras, M., and Rojas-Agramonte, Y. (2019). Impact of the chemical composition of aggregates on the adhesion quality and durability of asphalt-aggregate systems. Constr. Build. Mater. 216, 661–672.

Caro, S., Masad, E., Bhasin, A., and Little, D.N. (2008). Moisture susceptibility of asphalt mixtures, Part 1: mechanisms. Int. J. Pavement Eng. 9, 81–98.

Castillo, D., Caro, S., Darabi, M., and Masad, E. (2018). Influence of aggregate morphology on the mechanical performance of asphalt mixtures. Road Mater. Pavement Des. 19, 972–991.

Curtis, C.W. (1992). Investigation of asphalt-aggregate interactions in asphalt pavements. Am. Chem. Soc. Fuel 37, 1292–1297.

Hefer, A.W., Little, D.N., and Lytton, R.L. (2005). A synthesis of theories and mechanisms of bitumen-aggregate adhesion including recent advances in quantifying the effects of water. J. Assoc. Asph. Paving Technol. 74, 139–196.

Ishai, I., and Craus, J. (1977). Effect of the filler on aggregate-bitumen adhesion properties in bituminous mixtures. In Association of Asphalt Paving Technologists Proc, p.

Kiggundu, B.M., and Roberts, F.L. (1988). Stripping in HMA mixtures: State-of-the-art and critical review of test methods (National Center for Asphalt Technology Auburn, AL, USA).

Masad, E., Olcott, D., White, T., and Tashman, L. (2001). Correlation of fine aggregate imaging shape indices with asphalt mixture performance. Transp. Res. Rec. 1757, 148–156.

Advances in Materials and Pavement Performance Prediction II – Kumar et al. (eds)
© 2021 Taylor & Francis Group, London, ISBN 978-0-367-46169-0

Influence of filler-binder ratio and temperature on the Linear Viscoelastic (LVE) characteristics of asphalt mastics

Mohit Chaudhary, Nikhil Saboo & Ankit Gupta
Department of Civil Engineering, IIT (BHU), Varanasi, India

Michael Steineder & Bernhard Hofko
Institute of Transportation, Vienna University of Technology, Vienna, Austria

ABSTRACT: This study describes an investigation into the Linear Viscoelastic (LVE) limit/range of asphalt mastics prepared with different fillers. Effect of temperature and Filler-Binder (F/B) ratio on the LVE limits of mastics has been studied. The interrelation between Linear Viscoelastic Complex Modulus (G^*_{LVE}) and corresponding LVE limit has been examined. To compare the denouement of the study, the SHRP LVE strain criteria is also included which quantifies the applicability of the criteria to asphalt mastics. The lower LVE limit at high F/B ratio is the antithesis of the LVE limit at higher temperatures. The change in coefficient of determination (R^2) was prominent with change in F/B ratio whereas temperature shows a marginal shift in R^2 value indicating the dominance of F/B in the LVE range. The LVE limits obtained from the study were relatively conservative compared to those from SHRP study showing the unsuitability of applying SHRP criteria directly to mastics.

Keywords: Linear Viscoelastic Complex Modulus, F/B, Mastic, SHRP.

1 INTRODUCTION

Depending on national regulations, particles finer than 63 to 75 micrometers i.e. passing through 0.063 or 0.075 mm sieve are termed as fillers or P200 materials. The asphalt mastic is the mixture of filler and bitumen that fills the interstitial voids between aggregates to enhance the bond strength (Antunes et al. 2016). Many researchers have emphasized on visualizing the asphalt mixture as a mixture comprised with mastic coated aggregates rather than asphalt coated aggregates (Chen et al. 2008). The actual binder which holds the aggregate together is not bitumen but the combination of filler and bitumen popularly called mastic (Hospodka et al. 2018). Therefore, the rheological characterization of mastic is of great importance to improve the performance of asphalt pavements. The characteristics and role of the mastic in an asphalt mix is primarily controlled by the relative amount of filler with respect to the binder content of the mix. The ratio of weight of filler to that of binder which are taken to prepare the mastic is commonly known as Filler-Binder (F/B) ratio.

Many researchers have reported the importance of fillers in the performance of asphalt mixtures. The effect of incorporating divergent fillers to investigate the performance of asphalt mixes prepared with Rice husk ash (Al-Hdabi 2016), Brick Dust (Kuity et al. 2014), Fly ash (Chandra and Choudhary 2013),

Hydrated lime (Miró 2017), has been studied by various researchers.

In the Strategic Highway Research Program (SHRP), the shear stress (τ) and strain LVE limits (Υ) were found to be functions of complex modulus (G^*) defined by following equations (Anderson et al. n.d.; Petersen 1994).

$$\gamma = \frac{12}{G^{*0.29}} \qquad (1)$$

$$\tau = 0.12 G^{*0.71} \qquad (2)$$

Many works have been done on rheological characteristics of asphalt mastics but there is no extensive data available which relates the filler, temperature, F/B ratio and LVE limit. This research intended to narrow this gap by relating various parameters which influence the LVE limits of asphalt mastics.

2 MATERIALS

2.1 Fillers

Several fillers from different origin with varying chemical compositions were selected to explore their effects in asphalt mastics. This research engaged 3 Indian fillers and 3 Austrian fillers i.e. a total of 6 oven dried fillers for the study.

(a)   (b)

(c)   (d)

(e)   (f)

Figure 1. Indian Fillers (a) Red Mud (RM) (b) Marble Dust (MD) (c) Limestone (LS) and Austrian fillers (d) Granite (GR) (e) Basalt (BA) (f) Quartz (QZ).

Table 1. Physical properties of Bitumen.

Properties	Values	Specifications (IS: 73-2013)
Penetration	67 mm	45 mm (min.)
Softening point	58°C	47°C (min.)
Absolute viscosity	3200 poise	2400-3600 poise
Specific gravity	1.0112	0.97-1.02

The physical appearance of all the above mentioned fillers and their origin has been shown in Figure 1.

2.2 Binder

A viscosity graded (VG) asphalt binder, VG 30 was used in the study. 30, in VG 30 indicates the dynamic viscosity of the asphalt binder when heated at 60°C. This binder was collected from a local crude oil refinery in Varanasi, India. Table 1 presents the physical properties of VG 30 bitumen.

2.3 Preparation of asphalt mastics

The Asphalt mastic was prepared in the laboratory by taking required amount of filler corresponding to F/B

ratio in a pan and heating it in a temperature controlled oven for 1 hour at 180°C. The measured quantity of asphalt binder is then heated for 10 minutes at a temperature range of 160°C–180°C. After mixing the binder and filler in the predetermined F/B ratio in a can, it was agitated with the help of a manually operated mixer continuously for 10 minutes to obtain a homogenous mixture.

This study engaged three Filler-Binder (F/B) ratios i.e. 0.5, 1.0, and 1.5. Therefore, 18 cans of asphalt mastics were prepared. The rheological tests were conducted at three temperatures: 10°C, 20°C, and 30°C respectively. The test results presented in this study are an average of three replicates prepared for each temperature at any given F/B ratio.

3 EXPERIMENTAL PLAN

3.1 Linear viscoelastic properties

The viscoelastic behavior of bituminous binder can be assessed with the help of Dynamic Mechanical Analysis (DMA) by using a Dynamic Shear Rheometer (DSR). The complex shear modulus (G^*) and the phase angle (δ) are the primal viscoelastic parameters.

Traditionally, amplitude sweep test is conducted using a DSR to obtain the LVE range of an asphalt binder. The LVE limit/range is defined as the strain level at which the measured complex shear modulus reduces to 95% of its initial value (Saboo 2015). It is noteworthy that the affiliation between stress and strain is influenced by only the loading time (frequency) and temperature if restrained within the LVE region irrespective of the magnitude of stress or strain.

In this study, the LVE limits of the prepared mastic were determined following similar protocol used for asphalt binder. The Amplitude sweep test was conducted on Anton Paar MCR 102 Dynamic Shear Rheometer (DSR) with strain rate varying from 0.01% to 100% taking 25 data points to determine the LVE range of asphalt mastics at a frequency of 10 rad/s. Preparation of sample comprises allowing it to set in the silicon mold followed by sandwiching between the 8 mm diameter parallel plates keeping the testing gap to a value of 2mm.

The excess sample was trimmed with the help of trimming tool and then allowed to equilibrate with the temperature for 10 minutes before starting the test. The test was conducted at intermediate temperatures i.e. 10°C, 20°C, and 30°C on three replicates of each combination.

4 RESULTS AND DISCUSSION

4.1 Variation of LVE limit with temperature and F/b ratio

Figure 2 shows the variation of LVE with the change in F/B ratio as well as temperature of the mastics along

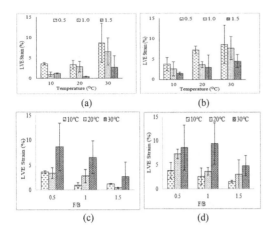

(a)   (b)

(c)   (d)

Figure 2. Variation of LVE range with change in temperature in (a) LS (b) GR and F/B ratio in (c) LS (d) GR.

with error bars. The graphs of one Indian filler (LS) and one Austrian filler (GR) have been shown here for brevity.

It is clear that the LVE limit follows a declining trend with the increase in filler content i.e. F/B ratio at every temperature for both the fillers. It is the expected behavior as high F/B ratio stiffens the mastic resulting in lower LVE limit. It is also noteworthy that the LVE value of high temperatures is always greater than the LVE value of low temperatures at every point.

As the temperature increases, the binder begins to lose its rigidity so did the mastic as a result of which it becomes soft and hence the LVE limit pushed to a higher value with increase in temperature. The same increasing behavior has been observed in both Indian and Austrian filler. But the values corresponding to higher F/B ratios were lower at any respective temperature. Similar trends were observed for other fillers as well. Overall, it may be concluded that the increase in temperature increases the LVE limits while with increase in F/B ratio, the LVE limit decreases.

4.2 Correlation of G*_{LVE} with temperature and F/B ratio

It has been observed that the complex modulus at LVE limit i.e. $G*_{LVE}$ is significantly variable with the different parameters i.e. temperature and F/B. But it is not clear that which parameter is majorly influencing the behavior of mastics. In order to establish the relationship between $G*_{LVE}$ and variables, the inter-correlation of $G*_{LVE}$ with temperature and F/B has been examined and presented in Figure 3.

The correlation has been quantified by the coefficient of determination (R^2) value. The decrease in R^2 value from 0.6712 at 0.5 F/B to 0.5221 at 1.0 F/B clearly indicates the effect of F/B ratio in the LVE behaviour of mastics. In addition to aforementioned values, the R^2 further decreases to 0.246 at 1.5 F/B.

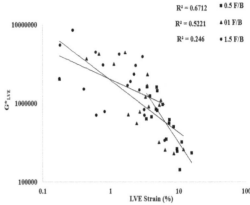

Figure 3. Correlation of $G*_{LVE}$ with F/B ratio.

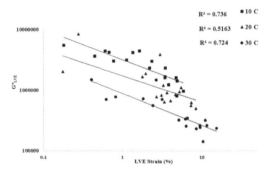

Figure 4. Correlation of $G*_{LVE}$ with temperature.

On the contrary, the change in R^2 value is not very significant with the change in temperature. Therefore, it is evident from Figure 3 and 4 that the F/B has the dominant effect on the LVE limits of asphalt mastics with temperature being the secondary factor.

4.3 Applicability of SHRP binder strain criteria to asphalt mastics

This study investigated the applicability of the SHRP strain criteria directly to the mastics. The LVE limits for asphalt mastics have been plotted against the corresponding $G*_{LVE}$. In addition to that, the LVE limits were also obtained from the Equation (1) to investigate the suitability of SHRP criteria.

The LVE limits obtained from the study are relatively conservative compared to those from SHRP study. It is noteworthy that the neat binder curve and SHRP curve bears a resemblance but mastic curve is totally different from SHRP curve. Therefore, it can be perceived that the SHRP findings cannot be directly applied to the asphalt Mastics. This attribution maybe the outcome of stiffening effect as filler binder interaction provided by the fillers in the mastics. Figure 5 shows the comparative analysis of SHRP curve with neat binder and mastics.

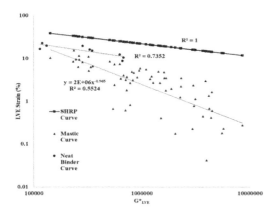

Figure 5. Comparative analysis of SHRP curve with neat binder and mastics.

5 CONCLUSION

This paper focuses on assessing the influence of different Indian and Austrian fillers on the Linear Viscoelastic range of asphalt mastics. The LVE limits were pushed to a higher value at higher temperatures due to loss of rigidity. The increase in F/B ratio resulted in mastic stiffening which in turn decreases the LVE range. The Filler to Bitumen ratio was found to be the momentous factor which influences the change in LVE limits. On the contrary, effect of temperature was not highly pronounced in the LVE behavior of asphalt mastics. Another prominent upshot which can be deduced from the study that the equations developed by the SHRP to correlate the LVE strain with complex modulus cannot be directly applied to asphalt mastics.

ACKNOWLEDGEMENT

This work is a part of Indo-Austrian bilateral project. The authors would like to thank Department of Science and Technology (DST), India and Austrian Agency for International cooperation in Education and Research (OeAD-GmbH) for their support.

REFERENCES

Al-Hdabi 2016. Laboratory investigation on the properties of asphalt concrete mixture with Rice Husk Ash as filler. *Construction and Building Materials.*

Anderson, D.C. & H.B. 1994. *Binder Characterization and Evaluation Volume 3: Physical Characterization.*

Antunes, A.F., L.Q., & R.M. 2016. Effect of the chemical composition of fillers in the filler-bitumen interaction. *Construction & Building Materials,* 85–91.

Chandra, R.C. 2013. Performance Characteristics of Bituminous Concrete with Industrial Wastes as Filler.

Chen, P. K. & P.L. 2008. Experimental and theoretical characterization of the engineering behavior of bitumen mixed with mineral filler. *Materials and Structures/Materiaux et Constructions* 41:1015–1024.

Hospodka, B. H., & R. B. 2018. Introducing a new specimen shape to assess the fatigue performance of asphalt mastic by dynamic shear rheometer testing. *Materials and Structures/Materiaux et Constructions* 51. Springer Netherlands: 1–11.

Kuity, S.J. & A.D. 2014. Laboratory investigation on volume proportioning scheme of mineral fillers in asphalt mixture. *Construction and Building Materials.*

Miró, A.M. 2017. Effect of filler nature and content on the bituminous mastic behaviour under cyclic loads. *Construction and Building Materials* 132: 33–42.

Petersen, J.C. 1994. Binder characterization and evaluation: test methods.

Saboo, P.K. 2015. Optimum Blending Requirements for EVA Modified Binder. *International Journal of Pavement Research and Technology* 8: 172–178.

Advances in Materials and Pavement Performance Prediction II – Kumar et al. (eds)
© 2021 Taylor & Francis Group, London, ISBN 978-0-367-46169-0

Nanomaterial to resist moisture damage of pure siliceous aggregates

H. Chakravarty & S. Sinha
National Institute of Technology Patna, Patna, Bihar, India

ABSTRACT: Moisture damage of flexible pavements hinders construction of durable pavements. Siliceous aggregates are found to aggravate the damage. Hydrated Lime (HL) fillers provide resistance to such damage. In this study, purely siliceous aggregates were considered, and the increment of resistance to moisture damage was studied using nano hydrated lime (NHL). NHL was synthesized using planetary ball milling. A comparative study was carried out on the improvement of moisture damage resistance using both HL and NHL particles. It was observed that although mix fails for unmodified mixes, addition of HL and NHL particles reduce the moisture susceptibility and results in development of moisture resistance mixes. Mix prepared with 1% addition of NHL showed best results in terms of resisting moisture damage.

1 INTRODUCTION

Bitumen is used to bind the aggregates to construct stable surface and binder courses for flexible pavements. However, these courses suffer from durability issues due to oxidation and moisture damage leading to premature failure of the pavements. This leads to a substantial economic loss. For tropical countries experiencing huge amount of rainfall, moisture damage is considered to be one of the most important cause of failure of flexible pavements. Physical, chemical as well as mineralogical content of aggregates have been observed to influence the aggregate-bitumen bond and in turn affect moisture damage (Transportation Research Board 2003; Yoon & Tarrer 1988; Cui et al. 2014; Zhang et al. 2015; Kumar & Anand 2012). Contemporary literature reveals that aggregates with high silica content negatively influence the bond due to its hydrophilic nature (Tarrer & Wagh 1991; Stuart 1990; Kakar et al. 2015).

Hydrated lime has been widely used to resist moisture damage of the flexible pavements (Lesueur et al. 2013; Little & Epps 2001). It forms insoluble calcium salt coat over the siliceous surfaces inhibiting further reaction with moisture. Moreover, upon deposition of calcium ions, the aggregate surfaces are roughened favoring bitumen adhesion resulting in resistance to moisture damage (Little & Jones 2003; Lesueur et al. 2013). Recent times have seen wide use of nano material to various civil engineering problems due to substantially increased surface area, number of particles in unit weight of materials etc (Parviz 2011). However, use of nano hydrated lime and its influence on resisting moisture damage of purely siliceous aggregates is to be still explored.

2 OBJECTIVE

Sheikhpura is a quarry site in Bihar, one of the most flood affected states in India. Mineralogical analysis of the source reveals that the aggregate is 99% siliceous. This provides with an opportunity to study for influence of purely siliceous aggregates on moisture damage. In this study, an effort has been made to synthesize, characterize and use nano hydrated lime to resist moisture damage of flexible pavements, and a comparative study is drawn with regular sized hydrated lime usage. Thus, bituminous mixes were prepared using HL and NHL (1% and 2%) and results are compared with unmodified mix.

3 METHODOLOGY

These materials were thereafter used to prepare bituminous concrete (BC) with 0%, 1% HL, 2% HL, 1% NHL and 2% NHL respectively and named 0_M, 1HL_M, 2 HL_M, 1NHL_M and 2NHL_M accordingly. The Optimum bitumen content (OBC) was obtained for each case. Tensile Strength Ratio (TSR) and Retained Stability (RS) are used as a measure of moisture damage intensity. The methodology adopted for conducting the experiment is given in Figure 1.

4 MATERIALS

4.1 *Aggregates*

Physical characterization is performed on aggregates obtained from Sheikhpura and the results are given

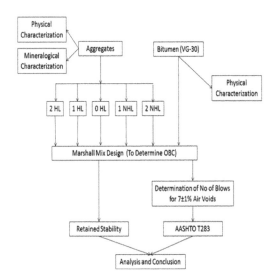

Figure 1. Methodology adopted for the research.

Table 1. Physical characterization of aggregates.

Aggregate impact Value	Loss Angeles Abrasion Value	Flakiness	Water Elongation	Absorption
12.3	17.5	27.5	29.1	0.5

Table 2. Mineralogical characterization of aggregate using XRD.

Quartz Low	Trikalsilikate	Albite Low, Calcian	Vaterite	Zeolite	Anthrosite
99.8	0	0	0.2	0	0

Table 3. Physical characterization of bitumen.

	Penetration at 25°C	Softening Point (°C)	Absolute Viscosity at 60°C (Poises)
Obtained Value	55	49	2578

under in Table 1. X ray diffraction (XRD) is carried out to determine the mineralogical characteristics of the aggregates.

4.2 Bitumen

VG 30 grade bitumen was used from Indian Oil Corporation (IOC), Barauni refinery, India. Tests performed depicted the following results.

4.3 Hydrated Lime (HL)

HL (Purity 90%) has been considered for research. 1% and 2% of HL by weight of aggregate was used as dosage of the additive and subsequently the moisture damage intensity was measured.

4.4 Nano Hydrated Lime (NHL)

NHL was also used in 1% and 2% quantity by weight of aggregate. The NHL used was synthesized using high energy planetary ball milling. The ball powder milling ratio was kept at 10:1 and the rotation speed was optimized to be 300 rpm. 18 hour ball milling thus resulted in producing NHL with particle size of sub 50 nm.

The size was verified by using Scherrer's equation obtained from performing XRD of NHL particles as given under:

$$\beta = \left(\frac{k \times \lambda}{D \times \cos\theta} \right) \qquad (1)$$

Where k is a dimensionless shape factor usually considered as 0.9, λ is the X ray wavelength, β is the line broadening at half the maximum intensity (FWHM) and θ is the Bragg angle

5 EXPERIMENT

For conducting the Marshall mix design, 3 samples were cast at 5.0%, 5.4%, 5.8%, 6.2% and 6.6% respectively for each combinations. Using the requisite stability value, flow value, density value, air void and voids filled with bitumen (VFB) graphs, the optimum bitumen content (OBC) was determined for every combination. The graphs obtained are provided in Figure 2.

The obtained OBC for various mixes are given under:

0_M	1HL_M	2HL_M	1NHL_M	2NHL_M
5.4	5.55	5.6	5.6	5.7

After determination of OBC, 6 samples were cast for each combination to determine the retained stability value. Three (3) samples were kept in water bath at 60°C for 24 hours, and the other set of three samples were kept aside for further testing. These conditioned samples were tested after 24 hours for Marshall Stability followed by the other set of samples at 60°C. The ratio of stability values for conditioned and unconditioned samples is reported as Retained Stability (RS) as given under:

$$RS = \left(\frac{Avgstab_{Cond}}{Avgstab_{UnCond}} \right) \times 100 \qquad (2)$$

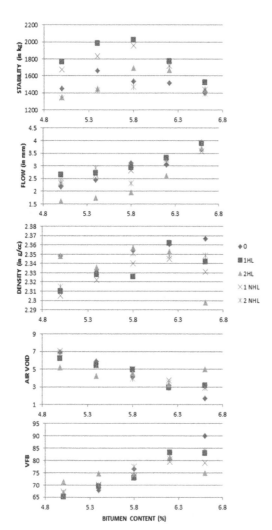

Figure 2. Marshall mix design parameters for OBC determination.

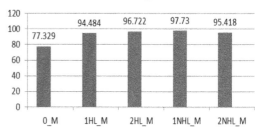

Figure 4. Variation of Tensile Strength Ratio with variation of HL and NHL content.

In order to obtain samples with $7\pm1\%$ air void content, three samples were cast at 20, 25, 30 and 35 blows respectively. The number of blows for acquiring $7\pm1\%$ air void was interpolated to be 33.

Thus, for further testing, six samples were prepared for each combination with 33 numbers of blows with air void in the range 6 to 8%. These samples were divided into two sets, where one set was subjected to moisture conditioning which were termed as wet samples and the other was termed as dry sample. Moisture conditioning was done by initially saturating the samples at 70 to 80% saturation level using vacuum saturation. These samples were wrapped in plastic wrappers to prevent loss of moisture, and were subjected to 24 hours freeze cycle followed by 18 hours of thaw cycle. Thereafter, indirect tensile strength (ITS) was determined at 25°C for each combination by the following equation:

$$ITS = \left(\frac{2 \times P}{\pi \times H \times D} \right) \times 100 \qquad (3)$$

Where, P is the average indirect tensile load at failure, H is the average height of sample and D is the average diameter of sample.

The ratio of the average ITS for wet samples and average ITS for dry sample is the Tensile Strength Ratio (TSR) depicted as:

$$TSR = \left(\frac{ITSwet}{ITSdry} \right) \times 100 \qquad (4)$$

The TSR values obtained for various mixes are given in Figure 4.

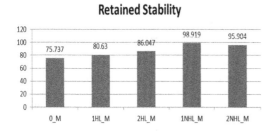

Figure 3. Variation of Retained Stability with variation of HL and NHL content.

Where, $Avgstab_{Cond}$ and $Avgstab_{UnCond}$ are the average stability of conditioned samples and average stability of unconditioned samples at 60°C respectively.

The results obtained are represented in graphical form and is given in Figure 3.

6 ANALYSIS AND CONCLUSIONS

Upon analyzing the experiment data, some interesting results were revealed which are summarized below:

a. Sheikhpura, an aggregate quarry site in Bihar, India was analyzed for its mineralogical composition using XRD. Results revealed the source to be purely siliceous. Literature deems siliceous aggregates to be prone to moisture damage and therefore, this source was considered for further study.

b. Hydrated lime (HL) is an active filler which has been globally accepted and used to resist moisture damage. It was thus considered to check influence on purely siliceous aggregates. Planetary ball milling was performed on these HL fillers to produce nano hydrated lime (NHL), and was used in study.

c. Optimum bitumen content (OBC) was obtained for various mixes containing 0%, 1% HL, 2% HL, 1% NHL and 2% NHL mixes respectively. Gradual increase in OBC was observed with increase in HL filler quantity and fineness (NHL).

d. Retained stability (RS) and AASHTO T283 (TSR) was used to predict moisture sensitivity of the mixes. Without addition of fillers, the mix was unable to produce the requisite resistance against moisture damage. However, with increase in HL and NHL content, resistance increased to satisfactory levels according to AASHTO T283.

e. 1% NHL produced the best results as compared to other mixes in terms of resisting moisture damage. However, given the RS and TSR at 86.07% and 96.72% respectively for 2% HL mix, results are comparable to the values obtained using NHL.

f. Thus, considering results of the experiment, and the time, effort and cost involved in synthesis of NHL, regular sized HL fits better for large scale uses until better technologies evolves to produce NHL at much lesser effort and cost.

REFERENCES

Cui, S., Blackman, B. R., Kinlock, A. J. & Taylor, A., 2014. Durability of Asphalt Mixtures: Effect of Aggregate type and Adhesion Promoters. *International Journal of Adhesion & Adhesives*. 54: 100–111.

Kakar, M.R., Hamzah, M. O. & Valentin, J. 2015. A review on moisture damages of hot and warm mix asphalt and related investigations. *Journal of Cleaner Production*. 99: 39–58.

Kumar, P. & Anand P. 2012. Laboratory Study on Moisture Susceptibility of Dense Graded Mixes. *Journal of Transportation Engineering*. 138: 105–113.

Lesueur, D., Petit, J., & Ritter, H. 2013. The mechanisms of hydrated lime modification of asphalt mixtures : A State - of - the - Art review. *Road Materials and Pavement Design*. 14(1): 1–16.

Little, D. N., & Jones, D. R. 2003. Chemical and Mechanical Processes of Moisture Damage in Hot-Mix Asphalt Pavements. Moisture Sensitivity of Asphalt Pavements. San Diego,California. http://onlinepubs.trb.org/onlinepubs/conf/reports/moisture/00_FRONT.pdf (26 Dec., 2015)

Parviz, A. 2011. Nano Materials in Asphalt and Tar. *Australian Journal of Basic and Applied Sciences*. 5(12): 3270–3273.

Tarrer, A. & Wagh, V. 1991. The Effect of the Physical and Chemical Characteristics of the Aggregate on Bonding. *National Research Council*. Washington, D.C: SHRP-A/UWP-91-510.

Transportation Research Board, 2003. Moisture Sensitivity of Asphalt Pavements, San Deigo, California.: TRB

Stuart, K. 1990. Moisture Damage in Asphalt Mixtures - A State-of-the-Art Report. FHWA975 RD- 90-019. *Federal Highway Administration*. Turner-Fairbank Highway Research Center, 976 McLean, VA.

Yoon, H. & Tarrer, A., 1988. Effect of Aggregate Properties on Stripping. *Transportation Research Records*. pp. 37–43.

Zhang, J., Apeagyei, A., Airey, G. & Grenfell, J., 2015. Influence of Aggregate Mineralogical Composition on water Resistance of Aggregate – Bitumen Adhesion. *International Journal of Adhesion & Adhesives*. 62: 45–54.

Advances in Materials and Pavement Performance Prediction II – Kumar et al. (eds)
© 2021 Taylor & Francis Group, London, ISBN 978-0-367-46169-0

Neural network models for flexible pavement structural evaluation

N. Citir, H. Ceylan & S. Kim
Department of Civil, Construction and Environmental Engineering, Iowa State University, Ames, Iowa, USA

ABSTRACT: In this study, multilayer pavement structure is simplified into one layer of equivalent thickness by using Equivalent Layer Theory (ELT). Artificial Neural Network (ANN)-based pavement structural analysis models were developed to find an equivalent thickness and elastic modulus of the modeled pavement system. The synthetic databases used as inputs in ANN forward and backcalculation models were created using MnLayer, a Layered Elastic Analysis (LEA) program. ANN models were trained to obtain the critical responses at the top and bottom of such layers in pavement systems. The multilayered flexible pavements were subjected to a 20-kip of Falling Weight Deflectometer (FWD) load in a circular area with uniform pressure. ANN models were found to represent a useful alternative approach for not only determining equivalent thickness and modulus but also providing close estimate of deflections of a multilayered flexible pavement system.

1 INTRODUCTION

The State of Iowa has approximately 114,500 miles of public roads, including state primary highways, county roads, city streets, and others. Among these public roads, many Iowa county pavement systems have multilayer pavement structures resulting from multiple cycles of pavement construction and renewal. Such complex pavement structures give Iowa county engineers difficulty in estimating current structural capacities of in-service pavements for developing cost-effective decision-making strategies for managing, maintaining, and rehabilitating county pavement systems.

The mechanical properties of an existing flexible pavement determine its remaining service life, an exceedingly difficult task. For this purpose, material properties of each pavement layer, including layer thickness, Poisson's ratio, and elastic modulus, should be analyzed, but variation in material properties combined with external effects such as traffic loads, climate, etc., makes prediction of pavement responses complex. Pavement responses expressed in terms of stress, strain, and deflection are of interest, particularly for obtaining responses at certain critical locations, such as horizontal tensile strain and stress in the asphalt layer, and vertical compressive strain and stress in the subgrade layer, primary structural failure modes in flexible pavement fatigue cracking and rutting. Deflection is more straight-forward to measure using a Falling Weight Deflectometer (FWD) or to predict using computational software, followed by correlation with the structural performance of pavement layers.

First, to compute pavement responses under specified loading distributed over a circular area, Minnesota Layered Elastic Analysis (MnLayer) has been utilized; it assumes that each layer is linearly elastic, is infinitely wide, and has a constant thickness (Khazanovich & Wang 2007). To simplify a multilayered flexible pavement system as a single-layer pavement, Equivalent Layer Theory (ELT) can be utilized.

Artificial neural networks (ANNs) are known as effective computational tools commonly used in the estimation of layer moduli (Beltran & Romo 2014) and prediction of pavement responses (Ceylan et al. 1999; Ceylan 2002).

In this study, ANNs were used for the backcalculation process of determining equivalent thickness and modulus through training of a database based on synthetic pavement responses, deflection basin specifically, obtained by MnLayer, a linear elastic theory-based software. Although there have been many studies on the backcalculation of layer material properties, there have been few attempts to use ELT theory in backcalculating thickness and elastic modulus together and introducing ANN approaches into this theory to expedite the process and achieve more accurate results. The primary objectives of this paper are thus twofold: 1) to implement ANN-based approaches into an equivalent-layer thickness concept and 2) to propose a fast and robust alternative method developed using ANNs, which is useful for backcalculating layer thickness and elastic modulus of a multilayered flexible pavement system, and for investigating deflection basins in evaluating its structural performance.

2 METHODOLOGY

2.1 Concept of equivalent layer theory

This study used ELT to combine complex pavement structures into a simplified pavement structure of thickness (h_{eq}). ELT, originally proposed by Odemark

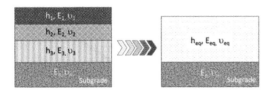

Figure 1. Simplification of a Multilayered Pavement System into a Single Layer Pavement System.

(1949), assumes that the transformed pavement layer has the same stiffness as its original structure. Based on this concept, ELT can be used to convert a multilayer system comprised of multiple layers with different moduli and Poisson's ratios into a single equivalent-layer system in which all layers have the same modulus. Deflections measured on the multilayered pavement system surface need to be matched with deflections calculated for the single equivalent-layer system. Figure 1 depicts a representation of ELT in which it can be seen that pavement layers with thicknesses (h_i), moduli (E_i), and Poisson's ratios (v_i) on sub-grade could be transformed into a single-layer with the equivalent thickness (h_{eq}) on the subgrade. There have been many approximation methods or equations derived from original Odemark's equivalency theory.

2.2 Synthetic database generation

Mechanistic models can be used to mathematically represent pavement physics and compute critical pavement responses (deflections, stresses, and strains) in response to idealized loading and climatic inputs. MnLayer, a software tool based on the Layered Elastic Analysis (LEA) method, uses pavement design options as inputs and their structural responses as outputs in constructing a synthetic database.

LEA models typically assume pavements to be homogenous, isotropic, and linearly elastic, and support calculation of theoretical deflections, stresses, and strains in response to the application of a surface load. In the LEA method, the pavement structure is modeled as a system of horizontal layers with constant properties that allow only elastic deformation within each layer of pavement material.

MnLayer software is used to compute pavement responses since it is a high-performance LEA program requiring comparatively little computational time, as little as 1/20 of other widely used LEA programs; it also provides more accurate results than others (Khazanovich & Wang 2007).

Input datasets representing 10,000 and 1000 different pavement cases comprised of different structural design and mechanical properties were initially generated for ranges defined for a pavement structure characterized by four hot-mixed asphalt (HMA) layers, a subbase and subgrade layer. The dataset was composed of a set of variables that, excluding subgrade-layer thickness, included the thickness, elastic modulus, and

Poisson's ratio for each layer. The pavement's characteristic ranges were based on field investigations, a historical pavement database, and experiences. The elastic modulus of the asphalt layer was assumed to lie between 1,378–13,789 MPa (200,000–2,000,000 psi). For the base layer and subbase, elastic moduli values were assumed to lie between 138–13,789 MPa (20,000–2,000,000 psi) and 103-290 MPa (15,000–42,000 psi), respectively. Poisson's ratios for asphalt concrete, base and subbase were assumed to lie in the range of 0.30–0.40. Thicknesses for asphalt and other layers were between 38–380 mm (1.5–15 in) and 25–450 mm (1–18 in), respectively.

Each input dataset variable was uniformly distributed over the interval [0, 1], then was incorporated into the above ranges, using Equation 2 to find the generated variable.

$$y_i = y_{min} + (y_{max} - y_{min}) \cdot x_i \qquad (1)$$

where y_i = generated variable for the ith case; y_{min} = minimum value of the variable within range; y_{max} = maximum value of the variable within range; x_i = uniformly distributed value of the variable for the ith case.

An output dataset was then created by introducing the input dataset into the MnLayer program. For all pavement structure cases, the multilayered flexible pavements were subjected to a 20-kip of FWD load in a circular area and with uniform pressure.

2.3 Development of ANN models

ANN-based pavement structural analysis models were first developed to find an equivalent thickness and elastic modulus of the modeled pavement system and, followed by computation of pavement responses at specified analysis points. These analysis points, deflection points, were designated on the surface as D_1, D_2, D_3, D_4, D_5, and D_6, with radial locations of 0 (0), 203 (8), 305 (12), 457 (18), 610 (24), and 914 (36) mm (in); and D_{b-a} and D_{t-sg} at the bottom of the asphalt and at the top of the subgrade with radial locations of 0 (0) mm (in), respectively.

In this study, four ANN models were developed using forward, and backcalculation procedures, and in these models, the backpropagation algorithm was utilized. This algorithm generally uses a Levenberg-Marquardt (lm) optimization technique that was chosen in this study since it provided better ANN prediction accuracy than other algorithms based on sensitivity analysis performed by Tarahomi (2019).

In this study, a two-layer neural network (one hidden layer and one output layer) was used. Based on empirically derived rules and trials, the number of hidden layers and neurons in ANNs was selected to be 15 neurons in one hidden layer.

Figure 2 is a schematic demonstration of the overall process of ANN model developments and their independent testing. The F_1- and F_2-models with 19-15-3

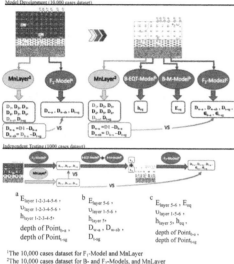

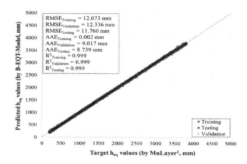

Figure 3. Accuracy Results of ANN-based B-EQT-Model Development by Comparing Target and Predicted h_{eq} Values.

a $E_{layer\ 1-2-3-4-5-6}$,
$v_{layer\ 1-2-3-4-5-6}$,
$h_{layer\ 1-2-3-4-5}$,
depth of Point$_{b-a}$,
depth of Point$_{t-sg}$

b $E_{layer\ 5-6}$,
$v_{layer\ 1-5-6}$,
$h_{layer\ 5}$,
D_{w-a} , D_{w-sb} ,
D_{t-sg}

c $E_{layer\ 5-6}$, E_{eq}
$v_{layer\ 1-5-6}$,
$h_{layer\ 5}$, h_{eq} ,
depth of Point$_{b-a}$,
depth of Point$_{t-sg}$

[1]The 10,000 cases dataset for F$_1$-Model and MnLayer
[2]The 10,000 cases dataset for B- and F$_2$-Models, and MnLayer
[3]The 1000 cases dataset for all Models and MnLayer

Figure 2. Flow Chart of Process of ANN models with MnLayer.

and 10-15-5 architectures, respectively, referring to forward ANN models, are used to compute deflections at analysis points. The deflections obtained by MnLayer and F-models within the asphalt layer (D_{w-a}) and the subbase layer (D_{w-sb}), and at the top of sub-grade layer (D_{t-sg}), were compared. B-EQT- and B-M-models with 14-15-1 architecture refer to backcalculation h_{eq} and equivalent elastic modulus (E_{eq}) ANNs, estimating h_{eq} and E_{eq}, respectively. Inputs for ANN models are listed below Figure 2, and outputs are indicated in red-outlined boxes.

3 RESULTS AND DISCUSSIONS

ANN models developed for backcalculation of layer thickness and elastic modulus, and for prediction of pavement responses (deflections to validate ANN models in this proposed study and strains to calculate fatigue and rutting failure in future studies) were evaluated separately. As stated in Figure 2, two different synthetic datasets were created with 10,000 cases, representing multilayered and single-layer pavement systems, to run MnLayer[1] and Mnlayer[2] so that their outputs were used to develop ANN models. Also, another synthetic dataset created with 1,000 cases was used to run MnLayer[3] so that their outputs were utilized for independent testing of all ANN models. Such independent testing was performed to re-test models using totally untrained data, although model developments (the 10,000 cases) were already comprised of three sets: training, validation, testing (75%, 15%, and 10% of 10,000 cases, respectively).

Table 1 and Figures 3-4 present coefficient of determination (R^2), root mean square error (RMSE), and average absolute error (AAE) as measures of ANN models accuracies in predicting pavement critical responses, equivalent thickness, and elastic modulus. R^2, correlating predicted values with measured values in this study, is a relative measure of fit while RMSE is an absolute measure of fit. The RMSE value shows how close measured values are to the predicted values in its units, but it is sensitive to large outliers in which AAE is less sensitive. A model with higher R^2 (%) and lower RMSE and AAE would achieve greater accuracy.

Table 1 lists all accuracy measures of ANN model development, including training, validation, testing, and independent testing results for F-models for D_{w-a}, D_{w-sb}, and D_{t-sg} and for B-models for equivalent thickness and elasticity modulus. In fact, the accuracy of the F$_2$-model was conditioned by the accuracy of B-models, meaning that independent testing results of B-models were verified by independent testing of F$_2$-model (see in Figure 2).

Figure 3 indicates the accuracy results of training, validation, and testing sets of ANN-based B-EQV-model development. As it is seen, this ANN model developed with outputs of MnLayer[2] predicted the h_{eq} values, which are targets-inputs of MnLayer[2], with the R^2 of over 99%.

Here, the ultimate goal was to match single-layer pavement responses obtained from ANN models with preliminary multilayered pavement responses. Figure 4 presents comparisons of deflections (D_{w-a}, D_{w-sb}, and D_{t-sg}) of the multilayered pavement system computed by MnLayer with deflections of the single-layer pavement system predicted by ANN models. Note that there is no need to use MnLayer for ANN models that compute the deflections seen in Figure 4. In Figure 4, 99.1% of deflections at D_{w-a} computed by MnLayer for a multilayered pavement system can be explained by ANN models with RMSE of 0.0019 mm. Overall, AAEs for deflections are sorted as D_{t-sg}, D_{w-a}, and D_{w-sb} in descending order, being of less than 0.0025 mm. This means that the predicted deflections by ANN models were found to be highly accurate. Figure 4 also shows that when a pavement is subjected to a 20-kip FWD loading, a higher magnitude of deflections are seen at the top of the subgrade, and the change in deflections is under 0.05 mm for the subbase layer.

Table 1. Accuracy Results of ANN Model Developments.

ANNs	Dataset	D_{w-a}	D_{w-sb}	D_{t-sg}
		R^2/RMSE*/AAE*	R^2 / RMSE*/AAE*	R^2 / RMSE* / AAE*
F_1 - Model	Training	0.994/ 1.52E-03 / 2.24E-07	0.863/ 6.72E-04 / 6.03E-08	0.998/ 2.36E-03 / 2.86E-07
	Validation	0.993/ 1.61E-03 / 1.06E-03	0.862/ 6.74E-04 / 4.57E-04	0.998/ 2.66E-03 / 1.80E-03
	Testing	0.995/ 1.45E-03 / 1.01E-03	0.863/ 6.93E-04 / 4.62E-04	0.998/ 2.63E-03 / 1.72E-03
	Ind. Test.	0.994/ 1.61E-03 / 1.07E-03	0.856/ 6.43E-04 / 4.44E-04	0.998/ 2.44E-03 / 1.73E-03
F_2 - Model	Training	0.999/ 6.88E-04 / 3.59E-08	0.941/ 6.20E-04 / 6.72E-08	0.999/ 1.47E-03 / 6.39E-08
	Validation	0.999/ 7.01E-04 / 4.68E-04	0.936/ 6.48E-04 / 3.11E-04	0.999/ 1.61E-03 / 9.20E-04
	Testing	0.999/ 6.96E-04 / 4.80E-04	0.953/ 5.96E-04 / 3.04E-04	0.999/ 1.73E-03 / 9.33E-04 /
	Ind. Test.	0.999/ 6.85E-04 / 4.68E-04	0.942/ 6.01E-04 / 3.07E-04	0.999/ 1.30E-03 / 8.74E-04
B-EQT-Model			h_{eq} (R^2 / RMSE*/AAE*)	
	Training		0.999 / 1.21E+01 / 2.08E−03	
	Validation		0.999 / 1.23E+01 / 9.02E+00	
	Testing		0.999 / 1.18E+01 / 8.74E+00	
B-M-Model			E_{eq} (R^2 / RMSE** / AAE**)	
	Training		0.999 / 2.71E+01 / 2.17E-03	
	Validation		0.999 / 3.12E+01 / 1.95E+01	
	Testing		0.999 / 3.54E+01 / 1.98E+01	

Note: *unit in mm, **unit in MPa

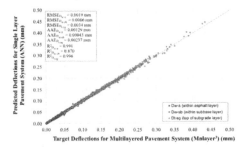

Figure 4. Comparison of Deflections Obtained from ANN Models with MnLayer[3] Solutions.

4 SUMMARY & CONCLUSIONS

In the current study, ANN models were used to back-calculate layer thickness and elastic modulus of a single-layer pavement system converted from a multi-layer pavement system, and to predict deflection basins under a 20-kip FWD load that is uniformly distributed on a circular area. The MnLayer, a LEA-based program, was used to generate a synthetic database that was used as inputs in ANN model developments. To validate the method, deflections of the multilayered pavement system were compared with the deflections of the single-layer pavement system, developed based on the ELT, by using the ANN approach. ANN models developed with R^2 of 0.99 to compute equivalent thickness and modulus, and of more than 0.85 to predict deflections of single-layer pavement systems provided a great performance with R^2 of 0.99 for D_{w-a} and D_{t-sg} and of 0.87 for D_{w-sb} in predicting deflections of multilayered pavement systems. Thus the ANN models proposed in this study can be a useful alternative methodology for computing the critical pavement responses of a single-layer pavement system

transformed from a multilayered pavement system and a suitable tool for use in pavement management practices, specifically to simplify the analysis and design of multilayered pavement systems.

ACKNOWLEDGMENTS

The authors gratefully acknowledge the Iowa Highway Research Board (IHRB), Iowa County Engineers Association Service Bureau (ICEASB), and Iowa Department of Transportation (Iowa DOT) for supporting this study. The contents of this paper reflect the views of the authors who are responsible for the facts and accuracy of the data presented within. The contents do not necessarily reflect the official views and policies of the IHRB, Iowa DOT, ICEASB, and Iowa State University. This paper does not constitute a standard, specification, or regulation.

REFERENCES

Ceylan, H., E. Tutumluer, & E. J. Barenberg. 1999. Artificial Neural Networks for analyzing concrete airfield pavements serving the Boeing B-777 aircraft. *Transportation Research Record* 1684: 110–117.

Ceylan, H. 2002. *Analysis and Design of Concrete Pavement Systems Using Artificial Neural Networks*. Urbana, IL: University of Illinois at Urbana-Champaign.

Khazanovich, L., & Q. Wang. 2007. MnLayer: high-performance layered elastic analysis program. *Transportation Research Record*, 2037(1), 63–75.

Odemark, N. 1949. *Investigations as to the Elastic Properties of Soils and Design of Pavements according to the Theory of Elasticity*. Stockholm, Sweden: Statens Vaeginstitute.

Rezaei Tarahomi, A., 2019. *Computationally Efficient Response Models for Rigid Airfield Pavement Systems Design*. Ames, IA: Iowa State University.

Advances in Materials and Pavement Performance Prediction II – Kumar et al. (eds)
© 2021 Taylor & Francis Group, London, ISBN 978-0-367-46169-0

Sensitivity analysis of simplified viscoelastic continuum damage fatigue model

J. Ding & B.S. Underwood
North Carolina State University, Raleigh, NC, USA

ABSTRACT: This paper investigates the parametric sensitivity in the Simplified Viscoelastic Continuum Damage (S-VECD) fatigue model of asphalt mixture by using model-based and experimental data-based sensitivity analysis. The model-based sensitivity analysis characterizes the change of model response with respect to individual parameters. The experimental data-based sensitivity analysis incorporates the parameter correlation and characterizes the change of model response with respect to correlated parameters. In this study, experimental data from a total of 31 specimens were used to generate parameters through the Markov Chain Monte Carlo (MCMC) technique to account for parameter correlation. Then, the simulated parameters were used to construct the experimental data-based sensitivity indices. The model-based analysis results in the parameter sensitivity ranking as $C_{11} > C_{12} > D_R > \alpha > |E^*|$. The experimental data-based analysis gives the sensitivity ranking as $D_R > C_{11}, C_{12} > |E^*| \approx \alpha$.

1 INTRODUCTION

The fatigue cracking of asphalt pavement has long been problematic. This distress affects the driving experience and ultimately increases the need and frequency of pavement maintenance and rehabilitation. Researchers have been studying to solve this problem by developing models that use fundamental material properties or by developing a simplified test, or both. The simplified viscoelastic continuum damage (S-VECD) fatigue model is one approach and has been shown to be able to predict fatigue life of asphalt mixture under a wide range of loading and environmental conditions. It characterizes the changes in the constitutive relationship as inherent fatigue damage in the material and constructs the damage evolution coupled with linear viscoelasticity theory and a failure criterion.

However, the sophistication of the S-VECD model requires a more comprehensive mathematical approach to analyzing the sensitivity than what is necessary for other, more traditional methods. For example, one common way of doing sensitivity analysis for a certain input variable is to simply change the input variable and construct the variance of the resultant output while the other input variables maintain the same. This method, solely model-based sensitivity analysis, is intuitive and easy to achieve, but not very robust because it does not consider the correlation and interaction between input parameters explicitly. In other words, the sensitivity can be overestimated or underestimated if the parameter correlation exists.

This paper conducts and compares model-based and experimental data-based sensitivity analysis assuming parameters are; (1) independent and perturbed about a nominal value (local sensitivity), (2) independent and uniformly distributed in the variation range (global sensitivity), and (3) correlated and interacted (sensitivity indices from experimental data). Then the optimum sensitivity analysis method is selected, and its interpretation is clarified.

2 BACKGROUND

2.1 *S-VECD model summary*

The S-VECD model consists of the linear viscoelastic behavior, the damage characteristic curve, and failure criterion to predict fatigue life of asphalt mixture. The summary of S-VECD theory is presented here and Underwood et al. (2010) is referred to for more details.

The linear viscoelasticity of the asphalt mixture is characterized through dynamic modulus at different temperature and frequency under linear viscoelastic range, as presented in Equations (1) and (2).

$$\log \left(E' \left(\omega, T \right) \right) = \kappa + \frac{\log \left(\max \left(E' \right) \right) - \kappa}{1 + e^{\delta + \gamma \log(\omega_R)}} \quad (1)$$

$$\log \left(a_T \right) = a_1 T^2 + a_2 T + a_3 \quad (2)$$

where E' = the storage modulus, kPa; ω = angular frequency, rad/s; ω_R = reduced angular frequency, rad/s; a_T = time-temperature shift factor; and κ, δ, γ,

a_1, a_2, and a_3 = model coefficients, obtained from optimization.

The damage characteristic curve is identified from repeated loading experiment at different loading strain amplitudes. It derives from work potential theory, as shown in Equation (3) and (4).

$$\frac{\partial S}{\partial \xi} = \left(-\frac{\partial W^R}{\partial S} \right)^\alpha \qquad (3)$$

$$W^R = \frac{1}{2} C \left(\varepsilon^R \right)^2 \qquad (4)$$

where ξ = the reduced time after applying the t-TS principle; α = the damage growth rate; W^R = the pseudo strain energy; S = the internal state variable; C = pseudo stiffness; and ε^R = pseudo strain.

The pseudo stiffness C and the internal state variable S have a unique relationship that is independent of loading amplitude, loading mode and loading temperature. The graphical expression of this relationship is called the damage characteristic curve. Equation (5) presents the power law relationship between C and S.

$$C(S) = 1 - C_{11} S^{C_{12}} \qquad (5)$$

where C_{11}, C_{12} = model coefficients.

The failure criterion used in the S-VECD model to capture the fatigue failure due to the formation of a macro-cracking is called the D^R criterion, as presented in Equation (6).

$$D^R = \frac{\int_0^{N_f} (1 - C) \, dN}{N_f} = \frac{\text{sum}(1 - C)}{N_f} \qquad (6)$$

where N_f = the number of load cycles at failure.

After the model coefficients are obtained, the number of cycles to failure under a control-strain cyclic fatigue test can be calculated using Equation (7),

$$N_f = \left(D^R \frac{C_{12} + p}{C_{11} \cdot p} \right)^{\frac{p}{C_{12}}} \left(\frac{f_R \cdot 2^\alpha}{p \, (C_{11} C_{12})^\alpha \, |E^*|^{2\alpha} \left(\varepsilon^R \right)^{2\alpha} K_1} \right) \qquad (7)$$

where K_1 = loading shape factor; f_R = the reduced loading frequency, Hz; and $p = 1 - \alpha C_{12} + \alpha$.

2.2 Sensitivity analysis method

The objective of sensitivity analysis is to quantify the relative contributions from individual inputs and parameters to the variation of measured responses. The reasons for sensitivity analysis include: (1) identify whether the model is robust or overly fragile to different parameters; (2) fix insensitive parameters, if any, to simplify the model; (3) determine the parameter ranges that optimally affect the model responses or uncertainties; and (4) instruct experimental design where the most model response sensitivity can be reflected (Smith 2013).

2.2.1 Local sensitivity analysis

Local sensitivity analysis is often achieved by the partial derivative of the response with respect to the individual parameters, $\partial y / \partial q_i$. It focuses on the relative change of the response when parameters or inputs are perturbed at a nominal value. Local sensitivity is often the first step to check if there are any insensitive parameters that can be fixed. This paper uses finite difference approximations technique to construct the local sensitivity analysis for the S-VECD model, as Equation (8),

$$\frac{\partial y}{\partial q_i}(t, q) \approx \frac{y(t, q_i + h_{q_i}, q_{\sim i}) - y(t, q_i, q_{\sim i})}{h_{q_i}} \qquad (8)$$

where $y(t, q)$ = model response; $q_i = i^{th}$ parameter; $q_{\sim i}$ = all the parameters except q_i; t = independent variable; h_{qi} = small increment of q_i.

2.2.2 Global sensitivity analysis

The local sensitivity analysis does not account for the effect of parameter variation since it only captures the output sensitivity adjacent to the nominal value of the parameter. The global sensitivity analysis considers the combinations of parameters throughout the admissible parameter space.

Morris screening method is used in this paper to conduct global sensitivity analysis by averaging over local derivative approximations. The screening methods can rank parameters according to their importance, but are not sufficient to quantify how much more important one parameter is than another.

In Morris screening method, the sensitivity measures for q_i are the sampling mean and variance from r sample points,

$$\mu_i^* = \frac{1}{r} \sum_{j=1}^{r} |d_i^j(q)|, d_i^j = \frac{y(q_i^j + h_{q_i^j}) - y(q_i^j)}{h_{q_i^j}}$$

$$\sigma_i^2 = \frac{1}{r-1} \sum_{j=1}^{r} (d_i^j(q) - \mu_i)^2, \mu_i = \frac{1}{r} \sum_{j=1}^{r} d_i^j(q) \qquad (9)$$

where d_i^j = the elementary effect associated with the ith parameter and jth sample.

2.2.3 Sensitivity analysis using experimental data

The typical local and global sensitivity analysis are focused solely on properties of the model. This section introduces the variance-based sensitivity indices, originated from Sobol decomposition, to analyze the sensitivity using simulated parameter values, as shown in Equations (10) and (11). In this study, experimental data are used to simulate parameter values and the sensitivity indices are computed subsequently. The simulation process used Markov Chain Monte Carlo method is discussed in Section Section 3.2.

$$S_i = \frac{\text{var}[E(Y|q_i)]}{\text{var}(Y)} \qquad (10)$$

$$S_{T_i} = 1 - \frac{\text{var}[E(Y|q_{\sim i})]}{\text{var}(Y)} \qquad (11)$$

where S_i = the first-order sensitivity indices; S_{T_i} = the total sensitivity indices; $E(Y|q_i)$ = the conditional expected value for model response Y; $\text{var}[E(Y|q_i)]$ = the variance for $E(Y|q_i)$.

3 MATERIALS AND METHOD

3.1 Materials and laboratory test

The asphalt mixtures used in this research contain 5.6% of PG 64-22 asphalt by mass and a gradation with a 9.5 mm nominal maximum aggregate size followed by the Superpave requirements. These mixtures were cut and cored to 38 mm × 110 mm (diameter × height) cylinder for further testing. In order to characterize the fatigue properties of asphalt mixture using S-VECD model, two laboratory tests should be performed – dynamic modulus test in accordance with AASHTO TP 133 protocol and cyclic fatigue test in accordance with AASHTO TP 134 protocol. This study used dynamic modulus test results data from 14 specimens of 5 operators and cyclic fatigue test results data from 17 specimens of 5 operators.

3.2 Method

First, the optimum parameter used in N_f prediction, i.e. α, $|E^*|$, C_{11}, C_{12}, and D_R, are constructed via fitting the dynamic modulus and cyclic fatigue test data into linear viscoelasticity model, damage characteristic curve and failure criterion.

Second, the Markov Chain Monte Carlo (MCMC) technique is used to simulate parameter distribution in the S-VECD model from the experimental data. The MCMC method gives the credible interval of the parameters and the prediction interval of the model response. In this study, the prediction interval of the model response is used to simulate the prediction interval of the parameters in the model. More details of the MCMC method application in constructing the parametric distributions of the S-VECD model functions can be found in [Ding et al. 2020]. The input temperature is 10°C and the input frequency is 10 Hz in all the sensitivity analysis.

The local sensitivity analysis is conducted using the optimum parameters as the nominal value.

The global sensitivity analysis is conducted assuming the parameters are independent and uniformly distributed in the ranges of within 10% error (a random but typical assumption when no other information is present) of the optimum parameters, represented as Morris measures μ_{i-1}^* and σ_{i-1}.

The global sensitivity analysis is conducted again assuming the parameters are independent and uniformly distributed in the ranges of the 95% prediction

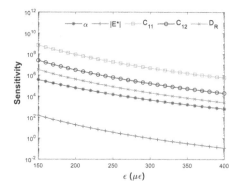

Figure 1. Local sensitivity analysis of different parameters in N_f prediction model.

Table 1. Global sensitivity analysis of different parameters in N_f prediction model.

	μ_{i-1}^*	σ_{i-1}	μ_{i-2}^*	σ_{i-2}
α	4.82×10^9	3.90×10^{10}	8.34×10^5	2.55×10^6
$|E^*|$	2.52×10^6	1.64×10^7	6.32×10^2	2.78×10^3
C_{11}	1.53×10^{13}	1.01×10^{14}	3.25×10^9	1.50×10^9
C_{12}	3.85×10^{11}	2.49×10^{12}	1.25×10^8	5.60×10^8
D_R	7.17×10^{10}	4.54×10^{11}	1.37×10^7	6.02×10^7

interval from MCMC results, represented as Morris measures μ_{i-2}^* and σ_{i-2}.

The sensitivity indices are calculated using all the parameter values generated from the MCMC results which incorporates the parameter correlation.

4 RESULTS AND DISCUSSION

4.1 Local sensitivity analysis results

The local sensitivity results of $\partial y/\partial q_i$ at different input strain magnitude are shown in Figure 1. Note that except D_R, the sensitivity results for α, $|E^*|$, C_{11}, and C_{12} are all negative, i.e. the increase of the parameter results in the decrease in the predicted N_f. The absolute values are presented for comparison.

The results demonstrate that all the parameters in the fatigue prediction model are sensitive. The sensitivity decreases with the increase of input strain magnitude. In addition, the sensitivity contribution ranking is $C_{11} > C_{12} > D_R > \alpha > |E^*|$.

4.2 Global sensitivity analysis results

The global sensitivity analysis results are shown in Table 1.

All the parameters are influential and cannot be fixed. Using the parameter range from assuming 10% error and from MCMC results both give the Morris sensitivity ranking of $C_{11} > C_{12} > D_R > \alpha > |E^*|$ and are in accordance with the local sensitivity results.

34

Table 2. Correlation matrix for parameters in N_f prediction model.

	α	$|E^*|$	C_{11}	C_{12}	D_R
α	1	-0.3982	-0.0224	0.0228	-0.0451
$|E^*|$	-0.3928	1	0.0738	-0.0749	0.041-5
C_{11}	-0.0224	0.0738	1	-0.9923	-0.0151
C_{12}	0.0228	-0.0749	-0.9923	1	0.0144
D_R	-0.0451	0.0415	-0.0151	0.0144	1

Table 3. The sensitivity indices considering correlation between parameters.

	α	$|E^*|$	C_{11}, C_{12}	D_R
S_i	0.0119	0.0191	0.1740	0.7728
S_{Ti}	-0.0043	0.0031	0.2104	0.8082

4.3 Parameter correlation

One common weakness with local and global sensitivity analysis lies in their assumption of independency in parameters, because they solely focus on the properties of the model. In the S-VECD method, the Pearson correlation matrix, a measure of linear correlation between variables, is listed in Table 2. The matrix was constructed using the simulated data from MCMC results representing the prediction interval [Ding et al. 2020]. The high correlation between C_{11} and C_{12} (-0.9923) can possibly lead to different sensitivity results compared to local and global sensitivity analysis.

4.4 Sensitivity indices from experimental data

Due to the non-negligible correlation between C_{11} and C_{12}, they are grouped as damage characteristic curve parameter. The characterized sensitivity indices after grouping are shown in Table 3. The parameters have a different sensitivity ranking from previous local and global sensitivity analysis, that is $D_R > C_{11}, C_{12} > |E^*| \approx \alpha$.

Figure 2 presents a way to interpret the sensitivity indices. In Figure 2 (d), the average value of N_f along a fixed D_R is changing substantially with the change of D_R value. This average value is the conditional expectation $E(N_f|D_R)$ in Equation (10). The variance $\text{var}(E(N_f|D_R))$ quantifies the variability of these average values. The fact that $\text{var}(E(N_f|D_R)) > \text{var}(E(N_f|C_{11},C_{12})) > \text{var}(E(N_f||E^*|)) \approx \text{var}(E(N_f|\alpha))$ results in the first-order sensitivity indices ranking. This indicates that the failure criterion D_R has a higher influence on the variation of N_f prediction and has room to be improved to reduce the uncertainty in S-VECD model compared with α, $|E^*|$, C_{11}, and C_{12}.

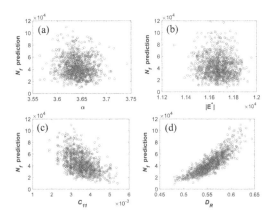

Figure 2. Scatter plots of N_f versus (a) α; (b) $|E^*|$; (c) C_{11}; and (d) D_R constructed using the parameter values simulated from MCMC.

5 CONCLUSION

This paper performs the model-based local and global sensitivity analysis and the experimental data-based sensitivity analysis in S-VECD model.

The local and global sensitivity analysis results show that all the parameters are sensitive and influential. They both give the parameters sensitivity ranking as: $C_{11} > C_{12} > D_R > \alpha > |E^*|$. However, the sensitivity indices computed from simulated parameter values using MCMC results from a total of 31 testing data show the sensitivity ranking as: $D_R > C_{11}, C_{12} > |E^*| \approx \alpha$.

In conclusion, sensitivity analysis without considering the parameter correlation in S-VECD model will overestimate the model fragility with respect to a certain parameter and lead to incorrect results. MCMC technique presents a way in simulating parameter values incorporated with their correlation. Additionally, compared with α, $|E^*|$, C_{11}, and C_{12}, failure criterion D_R has potential to be improved to possibly reduce the uncertainty of N_f prediction in the S-VECD model.

REFERENCES

Ding, J. Wang, Y.D. Gulzar, S. Kim, Y. R. Underwood, B. S. 2020. Uncertainty Quantification (UQ) of Simplified Viscoelastic Continuum Damage Fatigue Model using Bayesian Inference based Markov Chain Monte Carlo (MCMC) Method. *Transportation Research Record: Journal of the Transportation Research Board.* In press.

Smith, R. C. 2013. *Uncertainty Quantification: Theory, Implementation, and Applications* (Vol. 12). Siam.

Underwood, B. S. Kim, Y. R. and Guddati M. N. 2010. Improved Calculation Method of Damage Parameter in Viscoelastic Continuum Damage Model. *International Journal of Pavement Engineering*, Vol. 11, No. 6, 2010, pp. 459–476.

Advances in Materials and Pavement Performance Prediction II – Kumar et al. (eds)
© 2021 Taylor & Francis Group, London, ISBN 978-0-367-46169-0

Development of machine-learning performance prediction models for asphalt mixtures

E. Omer & S. Saadeh
California State University, Long Beach, CA, USA

ABSTRACT: The primary objective of this study is to investigate the effect of in-place air voids (AV), asphalt content (AC), bulk-specific gravity (BSG), and maximum specific gravity of asphalt (G_{mm}) on the fatigue cracking of asphalt mixtures using the Long-Term Pavement Performance (LTPP) database. This study includes 13 sections from different locations covering different mix designs, pavement age between 20-30 years, and two climate zones across the United States. All data were derived from LTPP database. A multiple linear regression, random forest (RF) and support vector machine (SVM) methods between the selected explanatory properties and the fatigue cracking were used for the investigation. A multiple significant linear regression model was developed, and it confirmed the significant relationships between the AC, AV, G_{mm}, percent of aggregate Pass. No. 200, BSG and the fatigue cracking. RF and SVM methods were established using the same data. They validated significant properties and the accuracy of the model.

1 INTRODUCTION

The long-term pavement performance (LTPP) program was established to help understand how and why pavements perform the way they do [1]. Such knowledge will help extend pavement life and save millions of dollars. The LTPP program aims to obtain knowledge of the specific effects of various design features, traffic, environment, materials, construction quality, and maintenance practices on pavement performance. The LTPP collects and stores data from more than 2,500 test sections throughout the United States and Canada and is one of the largest pavement performance experiments. The LTPP database is annually updated and publicly accessible [1].

One of the most widespread and detrimental distresses found in asphalt pavements is fatigue crack-ing. Also known as alligator or long-term cracking, fatigue cracking has numerous negative effects on the short-term and long-term performance of asphalt pavements [2]. If left untreated, this distress can ul-timately lead to complete structural failure of an as-phalt pavement. Typically, fatigue cracking occurs later in life of asphalt pavement and is primarily caused by excessive traffic loading or loads above pavement design strength [2]. However, other fac-tors, includ-ing mix design, may affect the rate of this pavement distress [3–5].

The primary objective of this study is to investigate the effect of in-place AVs, AC, BSG, Gmm, the percent of aggregate passing through the No. 4 sieve, the per-cent of aggregate passing through the No. 200 sieve, bulk-specific gravity and absorption of coarse (CA)

and fine aggregates (FA), and the uncompacted void content of FA on fatigue cracking of asphalt mixtures using the LTPP. Because most models of fatigue crack-ing prediction were based on laboratory results, it is important to use the LTPP for better understanding of pavement behavior. This study result will determine if these properties may affect the rate of fatigue cracking and how this knowledge can be used to create a statis-tically significant model that predicts fatigue cracking behavior in asphalt pavements.

2 METHODOLOGY

The methodology of this study can be summarized as follows:

- Collect data for all properties from LTPP Run a mul-tiple linear regression analysis using Minitab soft-ware between the asphalt pavement properties and fatigue cracking. From this analysis, develop sta-tistically significant linear model to predict fatigue cracking.
- Use the random forest (RF) regression method to validate and show significant factors on fatigue cracking. This method validates the accuracy of the multiple regression models developed to predict fatigue cracking. The RF method was established using commercial software R [6].
- Use a support vector machine (SVM) to test the accuracy of the model by separating the training and test data. The SVM method was established using R software as well [6].

3 DATA COLLECTION

The data used in this analysis accrued from the LTPP InfoPave database. To limit the influence of outside factors on the data, several filters were applied such as: 1) only asphalt pavements between the ages of 20 and 30 years were included because it is widely known that pavement age is a primary influence in fatigue cracking; 2) the analysis was limited to include only pavements in dry, non-freeze and wet zones with an annual temperature range between 0 and 30 degrees Celsius; 3) roadways that have undergone rehabilitation were excluded from this analysis. As result, 13 sections were selected and considered in this study. All the previous discussed properties for each section was collected.

4 DATA ANALYSIS

4.1 Multiple regression result

A multiple regression analysis was completed using Minitab to determine if a statistically significant regression model could be developed to correlate mix design and aggregate properties to percent fatigue cracking in asphalt pavements using LTPP data.

For all models, the research (or alternative) hypothesis was that a significant model could exist between these properties and percent of fatigue cracking. The null hypothesis, therefore, declared that a significant model did not exist between these quantities. To test the hypothesis, a level of significance of 0.05 was used.

An attempt was made to develop a model that incorporated all the mix design and aggregate properties. This iteration of the multiple regression analysis yielded a model, named Model A hereafter, that was statistically significant for the overall model and each of the predictors, shown in Table 1 and Equation 1.

The p-value for all properties was smaller than the designated level of significance (0.05).

$$MEPDG\% = -617.230 - 1.158\%AV$$
$$+6.813AC\% - 33.425BSG + 24.971BSG \text{ of } CA$$
$$+209.690 \text{ BSG of FA} + 16\ 035 \text{ Abs of FA}$$
$$-6.735\% \text{ Passing No. 200} + 2.019 \text{ Uncomp.Voi} \quad (1)$$

Figure 1 verifies the assumption that the residuals are normally distributed because the points follow the straight line. Therefore, the multiple regression model closely fits the presented data. The adjusted R^2 is equal to 97.82%.

4.2 Random Forest

RF regression was used to validate the multiple linear regression method and to show the significant factors on the fatigue cracking of asphalt mixtures. It is also used to investigate if a multi-collinearity problem arose between explanatory properties. RF is a machine

Table 1. Coefficients for model A.

Term	Coef.	SE coef.	t-value	p-value
Constant	617.230	67.760	−9.110	0.0008
% AV	−1.158	0.195	−5.940	0.0040
AC (%)	6.813	0.598	11.380	0.0003
BSG	−33.425	7.018	−4.760	0.0089
BSG of CA	24.971	2.851	8.760	0.0009
BSG of FA	209.690	19.240	10.900	0.0004
Abs of FA	16.035	2.477	6.470	0.0029
% Passing No. 200	−6.735	0.573	−11.740	0.0003
Uncomp. Void	2.019	0.286	7.060	0.0021

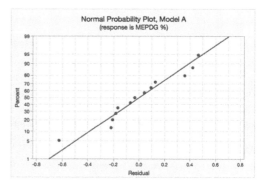

Figure 1. Normal probability plots (Model A).

learning method to classify and regress, introduced by Breiman and Cutler [7]. Many researchers discussed the concept of RF regression [7,8]. This method builds on an ensemble of decision trees from which the prediction of a continuous variable is provided as the average of the predictions of all trees.

In this study, the number of tress in the forest (ntree) is 500 and the number of different descriptors tried at each split (mtry) is 1 [9]. This method was performed using commercial software R [6].

RF method were run using data collected from LTPP. Figure 2 shows RF important properties. From this figure, AbsofFA, BSGofGA, BSGofFA, BSG, and AC properties show a positive increase in mean square-error (%IncMSE), which confirm significant relationships between these properties and fatigue cracking. It appears that AV, uncomp. void and the percent of aggregate passing through the No. 200 sieve have negative IncMSE. This could be due to a multicollinearity problem between AV with another explanatory property. Correlation matrix was obtained to determine the reason for such a result. The correlation matrix showed a high correlation between AV and BSG and between uncomp. void and BSG. This explains why AV, uncomp. void and the percent of aggregate passing through the No. 200 sieve have negative IncMSE. Therefore, in addition to having a

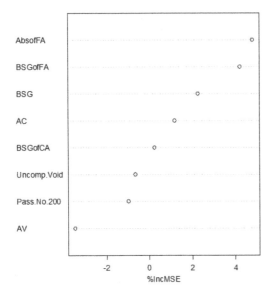

Figure 2.   Variable important measure of the random forest.

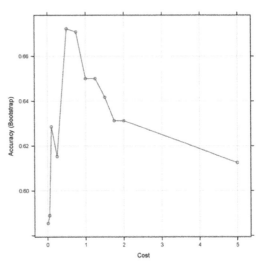

Figure 3.   Accuracy cross validation for linear kernel.

significant p-value and high R^2 for Model A, this confirms the significance of Model A, which can be used for fatigue cracking predication.

4.3   *Support vector machine*

SVM is another machine learning technique was developed by Cortes and Vapnik (10), and it can be used for classification and predictions. In this sec-tion, the fatigue cracking was divided into two groups (low and high), and the threshold value of the fatigue cracking was the average of the fatigue cracking value of the selected sections. It was 5.6%.

The datasets were divided into three parts: 80% of data used to train a machine learning (ML) model, 10% was used for validation, and the remaining 10% of data used to test the accuracy of the model.

The training dataset was used to determine SVM model parameters. A cross validation step was conducted to select the hyper parameters that maxim-ized the accuracy of trained model in validation da-taset.

Two SVM models with linear kernel and non-linear kernel were built and the accuracy in their models were 67% and 83% respectively as shown in figure 3 and 4. The SVM result validated the dataset. It is also confirmed that the fatigue cracking rates is affected significantly by the selected explanatory properties.

5   CONCLUSION

This study examined the effects of in-place AV, AC, BSG, and G_{mm}, the percent of aggregate passing through the No. 4 sieve, and the percent of aggregate passing through the No. 200 sieve on the field fatigue

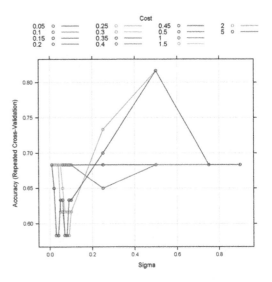

Figure 4.   Accuracy cross validation for no-linear kernel.

cracking of asphalt mixtures. Based on the analysis results, the following conclusions can be drawn:

- Model A has a highest coefficient of determination (R^2), and significant p-value indicating AV, AC, BSG, BSG of CA, BSG of FA, Abs of FA, the percent of aggregate passing through the No. 200 sieve, and uncomp. void significantly impacted fatigue cracking.
- RF confirmed that AC, Gmm, percent of aggregate passing through the No. 200 sieve, and BSG are important properties, and AV, and the aggregate passing through the No. 4 sieve have negative value of (%IncMSE). This result could be due to a correlation between AV with BSG. Therefore, this result validated the accuracy of Models A.

- The SVM result validated the dataset. It is also confirmed that the fatigue cracking rates is affected significantly by the selected explanatory factors.
- Not only traffic loading is the cause of fatigue cracking, but also the mix design and aggregate properties affected the rate of the fatigue cracking.
- It is recommended to revisit this analysis after more data has been added to the LTPP Infopave database.
- It is recommended to investigate how the results of this analysis would change for pavements outside the age range used in this study.

REFERENCES

[1] Elkins GE, Thompson T, Ostrom B, Simpson A, Visintine B (2017) Long-term pavement performance: Information management system user guide (Publication No. FHWA-RD-03-088). McLean, VA: Federal Highway Administration.

[2] Brown ER, Kandhal PS, Roberts FL, Kim YR, Lee DY, Kennedy TW (2009) Hot mix asphalt materials, mixture design and construction (3rd ed.), Lanham, MD: NAPA Research and Education Foundation.

[3] Zelelew H, Senn K, Papagiannakis T (2012) Forensic evaluation of the LTPP specific pavement study projects in Arizona. J. Perform. Constr. Facil. 26(5):668–678. https://doi.org/10.1061/%28ASCE%29CF.1943-5509.00 00288.

[4] Khattak MJ, Peddapati N (2013) Flexible pavement performance in relation to in situ mechanistic and volumetric properties using LTPP data. ISRN Civil Eng. Art 972020. https://doi.org/10.1155/2013/972020.

[5] Haider SW, Chatti K (2009) Effect of design and site factors on fatigue cracking of new flexible pavements in the LTPP SPS-1 experiment. Int. J. Pavement Eng. 10(2):133–147. https://doi.org/10.1080/102984308021 69390.

[6] The R Foundation (2019) The R prject for statistical computing. (accessed January 01,2019) https://www.r-project.org/

[7] Breiman L, Cutler A. (n.d.) Random forests (accessed June 18,2019). https://www.stat.berkeley.edu/~breiman/RandomForests/cc_home.htm.

[8] Svetnik V, Liaw A, Tong C, Culberson JC, Sheridan RP, Feuston BP (2003) Random forest: A classification and regression tool for compound classification and QSAR modelling. J. Chem. Inf. Comput. Sci. 43(6):1947–1958. https://doi.org/10.1021/ci034160g

[9] Geiron A (2017) Hands-on machine learning with Scikit-Learn and TensorFlow: Concepts, tools, and techniques to build intelligent systems. Sebastopol, CA: O'Reilly Media.

[10] Cortes C, Vapnik V (1995) Support-vector Networks. Machine Learning, Vol. 20, No. 3, 00273-297.

Effect of plastic subgrade on full-depth reclamation mix design

A.K. Frye
SME, Kirtland, OH, USA

A.R. Abbas
The University of Akron, Akron, OH, USA

ABSTRACT: This study examined the effect of the amount of subgrade soil and the cement application rate on the mix design of full-depth reclamation (FDR) blends. Four FDR blends containing varying amounts of subgrade soil (0%, 12.1%, 25.6%, and 40.8% by weight) were included in the laboratory testing plan. The FDR blends were mixed with different cement application rates (4%, 6%, and 8%). The test results showed a higher susceptibility to volumetric changes and lower unconfined compressive strength values for the FDR blends containing higher amounts of subgrade. At a cement application rate of 4%, the FDR blend containing 0% subgrade was the only blend that met a minimum unconfined compressive strength of 300 psi, which is commonly used in the mix design of FDR mixtures. While using a higher cement application rate for the other blends resulted in somewhat higher unconfined compressive strength values, none of these blends was able to achieve a minimum unconfined compressive strength of 300 psi, even at a cement application rate of 8%.

1 BACKGROUND

Full-depth reclamation (FDR) is a common pavement recycling technique. In this process, the entire asphalt pavement, including a portion of the base, subbase, and/or subgrade, is pulverized, and the blended material is stabilized, placed, and compacted to form a new base layer, which is then paved using hot mix asphalt. This is generally accomplished through the use of a large, self-propelled reclaimer (for removing and pulverizing the existing pavement), a motor grader (for moving and placing the blended materials), along with equipment for compaction (tamping using a pad-foot, followed by finish rolling).

Chemical stabilization using Portland cement is the most common method used in Ohio for the stabilization of FDR mixtures. The FDR design process begins by reviewing the details of the existing and proposed pavement sections, including the thicknesses of the FDR stabilized layer and the new hot mix asphalt layer, the planned versus existing widths of the roadway, and the planned versus existing elevations. The planned versus existing dimensions of the roadway along with the conditions encountered during sampling are considered in the selection of the proportions of the asphalt, aggregate base, subbase (if present), and subgrade material to be used in the FDR blend(s). In Ohio, the FDR mix design is generally performed using a modified version of ODOT Supplement 1120 (Mixture Design for Chemically Stabilized Soils) to determine the optimum moisture content and the Portland cement application rate to use in the FDR mixture.

This process involves mixing and curing the FDR specimens according to ASTM D1632 (Standard Practice for Making and Curing Soil-Cement Compression and Flexure Test Specimens in the Laboratory) followed by unconfined compressive strength testing according to ASTM D1633 (Standard Test Methods for Compressive Strength of Molded Soil-Cement Cylinders). A minimum unconfined compressive strength in the range of 300 to 400 psi is generally specified for the determination of the Portland cement application rate.

The required application rate for the chemical admixture will depend on the final proportions of the FDR blends, which are determined based on the thickness of the stabilized layer (typically 12 to 16 inches) and the amount of pulverized pavement removed to accommodate the new hot mix asphalt. According to ODOT Supplement 1120, the minimum application rate for Portland cement is 4% of the maximum dry density as determined by the Standard Proctor Test (ASTM D698). However, based on the amount and properties of the subgrade soils incorporated into the FDR blend, the required cement application rate to attain an unconfined compressive strength of 300 psi may exceed 7.5%, which may cause some issues during construction when blending the cement with the FDR materials and may result in shrinkage cracking during curing of the stabilized FDR layer. Because of the required time for sample preparation, curing and testing, any issues discovered after the mix design has been completed may result in significant project delays and would disrupt the construction schedule to complete the additional mix designs with different chemical admixtures and/or higher aggregate proportion for the new FDR blends.

Over the last two decades, several research studies have been conducted to evaluate the effect of the various factors that affect the design of stabilized aggregate bases and FDR mixtures. Guthrie et al. (2002) completed a laboratory study to determine the optimal cement contents for stabilizing aggregate bases. Aggregate gradation was found to be one of the most significant factors in the determination of the required cement application rate, with aggregates having higher contents of fines (particles passing sieve No. 200) resulting in higher optimum moisture contents and requiring higher cement application rates. A follow-up laboratory study was conducted by Guthrie et al. (2007) to determine the influence of reclaimed asphalt pavement (RAP) content on the cement application rate required to achieve a 7-day unconfined compressive strength (UCS) of approximately 400 psi for recycled aggregate bases. The required cement application rate was found to increase with the increase in the RAP content.

Taha et al. (2002) also studied the effects of RAP content and cement application rate on the unconfined compressive strength of stabilized bases and subbases. Several mixtures containing 100%/0%, 90%/10%, 80%/20%, 70%/30%, and 0%/100% RAP/virgin aggregate proportions were included in the laboratory testing plan. The stabilized mixtures were prepared using 0%, 3%, 5%, and 7% Type I Portland cement and were cured for 3, 7, and 28 days in plastic bags at room temperature. The laboratory test results showed an increase in the optimum moisture content, maximum dry density, and unconfined compressive strength with the increase in the amount of virgin aggregates and Portland cement. Longer curing periods were also reported to result in higher unconfined compressive strength. Similar results were reported by Euch Khay et al. (2015) in that the unconfined compressive strength of a stabilized cold in-place recycling (CIR) base decreased with the increase in the amount of RAP for the same cement application rate. It was also reported that acceptable mechanical properties for a stabilized base layer were obtained when using a RAP content of 60% or less. The same trend was also reported by Ghanizadeh et al. (2018) in a laboratory study that evaluated the effect of the amount of RAP and the cement application rate on the unconfined compressive strength for FDR blends prepared using RAP and two different subbase materials having unified soil classifications of SP-SC (poorly-graded sand with clay) and GW-GC (well-graded gravel with clay). It was reported in that study that the GW-GC FDR blend had a significantly higher unconfined compressive strength than the SP-SC FDR blend at a RAP percentage of 20% or less, while the SP-SC FDR blend had a higher unconfined compressive strength than the GW-GC FDR blend at a RAP percentage of 60%. For both FDR blends in that study, an exponential model was used to define the relationship between the unconfined compressive strength and the required cement application rate.

Based on the above discussion, most of the previous studies have focused on the design of FDR blends containing asphalt, base, and subbase materials. However, no studies have been conducted to evaluate the effect of the subgrade soil on the performance of the FDR mixture, especially those containing highly plastic soils that are known to adversely affect the unconfined compressive strength of the resulting FDR mixture.

2 LABORATORY TESTING PLAN

This study was conducted to explore the relationship between the amount of subgrade soil incorporated into an FDR blend and the resulting unconfined compressive strength after Portland cement treatment. One common subgrade soil in northeastern Ohio, consisting of a residual fat clay (with a Unified Soil Classification (USC) of CH – high plasticity clay, liquid limit of 55 and plastic limit of 30) weathered from shale bedrock, was used in the study.

Four different FDR blends containing varying amounts of subgrade soils (0%, 12.1%, 25.6%, and 40.8% by weight) were included in the laboratory testing plan. The FDR blends were selected based on a typical low-volume road or a parking lot in Ohio consisting of 4 inches of hot mix asphalt placed on 8 inches of a dense graded aggregate base underlain by a subgrade consisting of high plasticity clay. The selected material ratios for the four FDR blends used in this study represented an FDR layer thickness of 12 inches with different removal depths (0, 2, 4, and 6 inches) after the pavement section was pre-pulverized to a depth of 12 inches.

To determine the required cement application rates for the four FDR blends that are needed to achieve an unconfined compressive strength of 300 psi, all blends were mixed with the minimum cement application rate of 4% specified in ODOT Supplement 1120. In addition, for Blends 2, 3, and 4 (containing 12.1%, 25.6%, and 40.8% subgrade soil by weight, respectively), two additional cement application rates of 6% and 8% were used (Figure 1).

Mixing and curing of the FDR specimens were conducted in accordance with ASTM D1632 (Standard Practice for Making and Curing Soil-Cement Compression and Flexure Test Specimens in the Laboratory). All specimens were cured in the laboratory for seven days, followed by a 24-hour capillary soak. The weight and dimensions of each specimen were measured before and after capillary soak to evaluate the swelling potential. Specimens were then subjected to unconfined compressive strength testing according to ASTM D1633 (Standard Test Methods for Compressive Strength of Molded Soil-Cement Cylinders). Three replicates were used for this test.

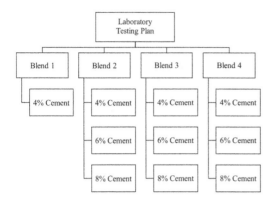

Figure 1. Laboratory testing plan.

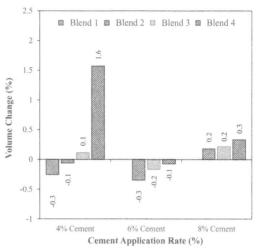

Figure 2. Average volume change from capillary soak (where positive volume change indicates swelling and negative indicates shrinkage).

3 RESULTS AND DISCUSSION

The test results obtained for the four FDR blends at the various cement application rates are presented in Figures 2 and 3. Figure 2 presents the average percentage volume change resulting from the capillary soak, and Figure 3 presents the average unconfined compressive strength. The error bars depicted in Figure 3 represent the variation in the unconfined compressive strength test results from the mean and each error bar is equal to one standard deviation. The dashed line in this figure represents the minimum unconfined compressive strength of 300 psi that is commonly used in the mix design of FDR mixtures.

It can be noticed from Figure 2 that the FDR blends containing higher amounts of subgrade soil showed a higher susceptibility to volumetric changes, with Blend 4 (which contains 40.8% subgrade soil by weight) exhibiting the highest potential for swelling at a cement application rate of 4%. However, the volumetric changes for this blend reduced significantly when the cement application rate was increased from 4% to either 6% or 8%.

The test results in Figure 3 show lower unconfined compressive strength values for the FDR blends containing higher amounts of subgrade, and higher unconfined compressive strength values for FDR specimens with higher cement application rates. Similar to that reported by Ghanizadeh et al. (2018), an exponential decay relationship is noted for most blends between the unconfined compressive strength and the percentage of subgrade incorporated into the FDR blend.

At a cement application rate of 4% (minimum specified by ODOT), Blend 1 (containing 0% subgrade by weight) was the only FDR blend that met a minimum unconfined compressive strength of 300 psi. While using a higher cement application rate for Blends 2, 3, and 4 resulted in somewhat higher unconfined compressive strength values, none of these blends was able to achieve a minimum unconfined compressive strength of 300 psi, even at a cement application rate of 8%. Interestingly, the unconfined compressive

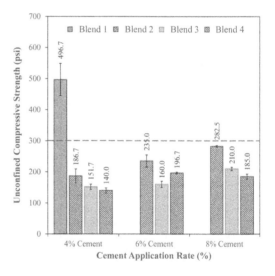

Figure 3. Average unconfined compressive strength.

strength was nearly the same for Blend 4 when the cement application rate was increased from 6% to 8%.

Based on the results presented in Figure 3, it will not be possible to achieve a minimum unconfined compressive strength of 300 psi using the particular subgrade used in this study even by incorporating as little as 12.1% subgrade by weight in the FDR blend and increasing the cement application rate to 8%. However, a small amount of this subgrade might be incorporated into an FDR blend designed to meet a minimum unconfined compressive strength of 200 psi, which is sometimes specified for parking lots and residential roads with very low traffic.

4 CONCLUSIONS

This study examined the effect of incorporating a cohesive subgrade soil consisting of high plasticity clay on the mix design of FDR blends. Four blends containing varying amounts of subgrade soil (0%, 12.1%, 25.6%, and 40.8% by weight) were included in the laboratory testing plan. All FDR blends were mixed with a cement application rate of 4%. Two additional cement application rates of 6% and 8% were used for the FDR blends that contained any amount of subgrade soils. All FDR specimens were cured in the laboratory for seven days, followed by a 24-hour capillary soak.

Below is a summary of the main findings and conclusions of this study:

- Incorporating a cohesive subgrade soil into an FDR blend has a significant impact on the resulting volumetric changes of the blend. However, using a cement application rate higher than 4% may help mitigate these volumetric changes.
- Lower unconfined compressive strength values were obtained for the FDR blends containing higher amounts of subgrade. At a cement application rate of 4%, the FDR blend containing 0% subgrade was the only blend that met a minimum unconfined compressive strength of 300 psi, which is commonly used in the mix design of FDR mixtures. While using a higher cement application rate for the other blends resulted in somewhat higher unconfined compressive strength values, none of these blends was able to achieve a minimum unconfined compressive strength of 300 psi, even at a cement application rate of 8%. This implies that minimizing the percentage of cohesive subgrade in an FDR blend (especially if the subgrade consists of high plasticity clay) should be a goal when evaluating an FDR project.

At present, no predictive models have been established to relate the unconfined compressive strength of an FDR blend to the amount of subgrade incorporated into the blend. Because of the required testing and curing time for quality control/quality assurance

(QC/QA) tests, any issues discovered after the mix design is completed may result in project delays due to the need for additional testing with different chemical admixtures or a higher aggregate proportion to reflect the modified conditions. Therefore, creating a family of curves that relate the unconfined compressive strength to the amount of subgrade soil and cement application rate would give the FDR mix design engineer an estimate of the resulting strength, which can lead to a more proactive decision making process and reduce potential project delays. While a subgrade soil consisting of high plasticity clay was considered in this study, additional research is needed to develop similar relationships for other types of subgrades.

REFERENCES

Euch Khay, S. E., Euch Bin Said, S. E., Loulizi, A. and Neji, J. 2015. Laboratory Investigation of Cement-Treated Reclaimed Asphalt Pavement Material. Journal of Materials in Civil Engineering. Vol. 27(6), 04014192.

Ghanizadeh, A. R., Rahrovan, M., and Bafghi, K. B. 2018. The Effect of Cement and Reclaimed Asphalt Pavement on the Mechanical Properties of Stabilized Base via Full-Depth Reclamation. Construction and Building Materials. Vol. 161, pp. 165–174.

Guthrie, W. S., Sebesta, S., and Scullion, T. 2002. Selecting Optimum Cement Contents for Stabilizing Aggregate Base Materials. Publication FHWA/TX-05/7-4920-2. Texas Department of Transportation.

Guthrie, W. S., Brown, A. V., and Eggett, D. L. 2007. Cement Stabilization of Aggregate Base Material Blended with Reclaimed Asphalt Pavement. Transportation Research Record: Journal of the Transportation Research Board. Record No. 2026, pp. 47–53.

Ohio Department of Transportation (ODOT) Supplement 1120. 2011. Mixture Design for Chemically Stabilized Soils.

Taha, R., Al-Harthy, A., Al-Shamsi, K., and Al-Zubeidi, M. 2002. Cement Stabilization of Reclaimed Asphalt Pavement Aggregate for Road Bases and Subbases. Journal of Materials in Civil Engineering, Vol. 14(3), pp. 239–245.

Advances in Materials and Pavement Performance Prediction II – Kumar et al. (eds)
© 2021 Taylor & Francis Group, London, ISBN 978-0-367-46169-0

Chemo-mechanical properties of synthesized tricalcium silicate

Shayan Gholami & Yong-Rak Kim
Zachry Department of Civil & Environmental Engineering, Texas A&M University, College Station, TX, USA

Hani Alanazi
Department of Civil and Environmental Engineering, Majmaah University, Al-Majmaah, Saudi Arabia

ABSTRACT: Tricalcium silicate based binder provides a simple matrix with only two hydration products: calcium silicate hydrate (C-S-H) and calcium hydroxide (CH). This feature is attractive to researchers who like to more accurately assess the Portland cement-based binders' phases separately. However, different purity than ordinary Portland cement clinkers could alter the properties of hydrated gels. This paper compared the nano-mechanical and chemical properties of synthesized tricalcium silicate (C3S) paste with ordinary Portland cement (OPC) paste using nanoindentation and energy dispersive X-ray spectroscopy (EDS) method. The deconvolution statistical method was used to further analyze the nanoindentation test results and compare with the EDS chemical mapping. Results indicated that C3S paste has a stiffer and more homogeneous (as the C3S paste consists only high-density C-S-H and CH) micromechanical properties than OPC although the calcium to silica ratio of C3S paste is in the range of OPC binders. In addition, the C3S paste seems to have lower porosity than OPC paste due to the lack of observation of the low elastic moduli.

1 INTRODUCTION

A typical Portland cement clinker contains about 50-65% of Alite which is an impure tri-calcium silicate (C3S) phase. Alite's (alongside Belite) hydration reactions play an important role in concrete strength development (Poulsen et al. 2009; Thomas et al. 2001). The hydration reaction products are calcium silicate hydrate (C-S-H) and calcium hydroxide (CH) gels which C = CaO, S = SiO_2, H = H_2O (Cuesta et al. 2018). Since the C3S phase is the main phase in the un-hydrated cement composition, various studies have been conducted on its properties whereupon the pure synthesized powder was hydrated (or carbonated) which led to a paste consisted of C-S-H and CH gels (Huan et al. 2007; Kjellsen & Harald 2004; Kjellsen et al. 2007; Grech et al. 2013). Although the obtained C-S-H and CH gels have similarities to the C-S-H and CH found in Portland cement-based materials, the impurities (i.e., the presence of Alumina and Iron at kiln) and different crystal structures (i.e., different spacing of the atoms in the crystals) could result in different properties (Maki & Kato 1982; MIT Concrete Sustainability 2013). In order to understand the fundamental characteristic of the two primary hydration products (C-S-H and CH) in OPC, this study investigated nanomechanical and chemical properties between hydration products resulting from pure tri-calcium silicate and Alite found in Portland cement clinker. The findings can be used to optimize the performance of Portland cement based concrete pavement within tailoring early

strength hydration reactions. Nanoindentation can provide nanomechanical properties in various types of heterogeneous materials (Constantinides et al. 2003; Nimeĕek 2009), it was thus employed in this study, and the nanomechanical properties were mapped with chemical characteristics detected by energy dispersive X-ray spectroscopy (EDS) method.

2 METHODOLOGY

2.1 Materials and specimen preparation

Two different sets of samples were prepared. Pure synthesized tri-calcium silicate powder as well as Type I Portland cement (OPC) were mixed with water via the water to cement ratio of 0.5 and 0.4, respectively, to achieve the final dimensions of 10 mm x 10 mm cross-section and 10 mm thickness. All the faces of the samples were sealed by the epoxy except the top surface that was ground and polished to reach a nanometer-level roughness. The water to cement ratio in C3S paste was employed greater than typical OPC paste due to the lack of flowability of the mortar. Alite and Belite have C-S-H gel with the same properties, so the OPC sample results can be compared to the C3S sample properly. After ample curing, samples were demolded and cut into appropriate sizes for the nanoindentation and SEM/EDS test.

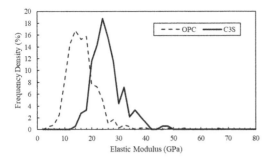

Figure 1. Frequency distribution of the indentation modulus for the OPC and C3S paste specimen.

2.2 Testing procedure

A Hysitron Triboindenter with the Berkovich tip was used to perform nanoindentation tests. The loading function was defined with three steps: 10 seconds loading, 5 seconds holding, and finally 10 seconds unloading (with the same rate as loading). The holding period was considered to eliminate any creep effects. The maximum load was set to 2,000 μN. The reduced modulus which represents elastic properties of both indenter tip and test specimen was determined. Then, the specimen's elastic modulus was calculated based on the following equation:

$$\frac{1}{E_r} = \frac{1 - v_s^2}{E_s} + \frac{1 - v_i^2}{E_i} \qquad (1)$$

The v and E represent Poisson's ratio and elastic modulus while the subscripts of r, s, and i denote reduced, specimen, and indenter, respectively. The elastic modulus and the Poisson's ratio of the diamond indenter are equal to 1,140 GPa and 0.07, respectively. The indentation hardness value was also calculated per the following equation:

$$H = P_{max}/A_c \qquad (2)$$

where the P_{max} is the peak load and A_c is the contact area.

FEI Helios NanoLab 660 (high vacuum SEM with EDX) was used to conduct the EDS chemical analysis on each specimen.

3 RESULTS

3.1 Nanoindentation test

More than 400 grid indents were performed on three different random areas on each sample set. The results were brought together in Figure 1 to compare the frequency occurrence of elastic modulus.

The indentation hardness against the elastic modulus of C3S is shown in Figure 2 with the mean and standard deviation of 1.43 and 0.51 GPa, respectively. The hardness mean is relatively smaller than OPC binders.

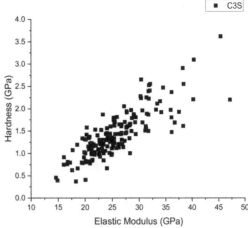

Figure 2. Hardness versus elastic modulus scattered plot for C3S paste.

3.2 Energy dispersive X-ray spectroscopy (EDS)

Chemical mapping (Figure 3) was conducted on the C3S sample to determine the distribution of C-S-H and CH gels in the indented area. It could be seen that the calcium ions could be found all over the surface, while silica ions were absent in some locations where are perhaps related to the locations of CH gel. Also, different C-S-H color contrasts could be related to different densities of C-S-H gel (i.e., low density, high-density, etc.)

4 DISCUSSION

The deconvolution statistical method was applied to the nanoindentation test results of the elastic moduli to statistically distinguish phases assumed to follow Gaussian distribution. Deconvolution results are displayed in Figures 4 and 5. In the C3S paste, it can be illustrated that at least two phases exist. The calculated peaks based on the deconvolution method were 23.17 and 32.76 GPa. Based on the previous studies, the first peak could be associated with the high-density C-S-H gel and the second peak could be related to the ultra-high-density C-S-H gel or CH (Fu et al. 2018; Mendoza et al. 2015). It is not easy to fully distinguish the different phases since the presence of a highly disordered microstructure consists of various layers of calcium, oxygen, and silicon dioxide attached by additional calcium ions (Shahrin 2018). However, based on the area beneath the deconvolution fitted peaks, the ratio of peak 1 to peak 2 could be estimated as 2.24. The interesting question would be whether the second phase consists of only CH or a combination of the ultra-high density of C-S-H and CH. If it only comprises of CH and by assuming that the first phase consists only C-S-H, the ratio of peak 1 to peak 2 can be translated to silica to calcium ratio of 0.69. It should

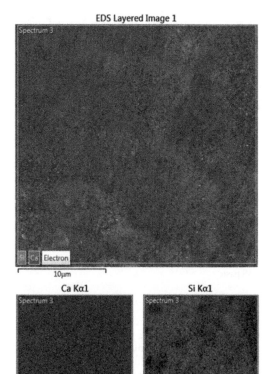

Figure 3. Chemical mapping on the C3S sample.

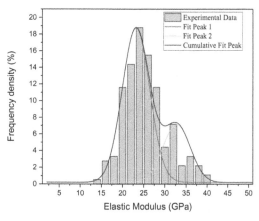

Figure 4. Deconvolution of nanoindentation data for C3S paste.

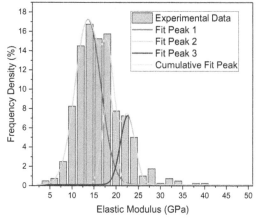

Figure 5. Deconvolution of nanoindentation data for OPC paste.

be noted that there is an overlap area between the fitted peaks and the calculated ratio is just an approximation. Beside the deconvolution method, the EDS chemical mapping was employed to understand the microstructure better. It could be seen that the calcium ions were scattered all over the specimen surface while the silica ions were covering approximately 65% (+/-10%) of the surface (based on the image analysis) which supports the deconvolution ratio. Therefore, the hypothesis that the second phase mainly consists of CH appears true.

Another significant observation from the C3S paste is the absence of elastic modulus less than 10 GPa. This is different from OPC paste whose elastic modulus (Figure 5) started from lower values. This could be related to the lower porosity of C3S paste than OPC.

The OPC paste deconvolution results also display the existence of three phases with elastic modulus peaks of 13.63, 19.33, 22.62 GPa. These phases are probably associated with different densities of C-S-H gel. Also, the elastic modulus associated with CH phase was more distinguishable from C3S than OPC.

Generally, by comparing the elastic moduli histogram between OPC and C3S, it can be indicated that C3S paste has a stiffer and more homogeneous (as the C3S paste consists only high-density C-S-H and CH) micromechanical properties than OPC although

the calcium to silica ratio of C3S paste is in the range of OPC binders (Kunther et al. 2017; Shahrin 2018).

5 CONCLUSION

In this study, an experimental program was executed to examine the chemo-mechanical properties of tricalcium silicate-based material. C3S hydration results in a simple matrix with only two hydration products: C-S-H and CH gels. This study compared the mechanical and chemical properties of resultant gels from C3S with the counterparts of ordinary Portland cement-based binders. Based on test results and relevant statistical analyses, the following conclusions can be drawn:

1. The deconvolution analysis illustrated that the C3S paste consisted of two main elastic modulus peaks which were approximately 23 and 33 GPa with the relative ratio of 2.24. The first peak was associated

with high-density C-S-H and the second one was related to CH gel. Employment of EDS chemical analysis demonstrated the presence of silica about 65% (+/−10%) of the surface.

2. C3S paste has a stiffer and more homogeneous (as the C3S paste consists only two high-density C-S-H and CH) micromechanical properties than OPC although the calcium to silica ratio of C3S paste is in the range of OPC binders.

3. Based on the nanoindentation results, the porosity was higher in the OPC paste than the C3S. The water to cement ratios, as well as the reactivity of synthesized C3S powder, could be the main reasons.

4. For the future work, the effects of water to cement ratio, aging, or carbonation on the C3S microstructure can be investigated for a better understating of the C3S based binders.

REFERENCES

Constantinides, G., Ulm, F.J. & Vliet, K.V. 2003. On the Use of Nanoindentation for Cementitious Materials. *Materials and Structures* 36 (3): 191–96.

Cuesta, A., Zea-Garcia, J.D., Londono-Zuluaga, D., Angeles, G., Santacruz, I., Vallcorba, O., Dapiaggi, M., Sanfélix, S.G. & Aranda, M.A. 2018. Multiscale understanding of tricalcium silicate hydration reactions. *Scientific reports*, 8(1): 1-11.

F Fu, J., Bernard, F., Kamali-Bernard, S. & Cornen, M. 2018. November. Statistical analysis of mechanical properties for main cement phases by nanoindentation technique. In *IOP Conference Series: Materials Science and Engineering* (Vol. 439, No. 4, p. 042018).

Grech, L., Mallia, B. & Camilleri, J. 2013. Investigation of the physical properties of tricalcium silicate cement-based root-end filling materials. *Dental Materials*, 29(2): e20–e28.

Huan, Z. & Chang, J. 2007. Novel tricalcium silicate/monocalcium phosphate monohydrate composite bone cement. *Journal of Biomedical Materials Research Part B: Applied Biomaterials: An Official Journal of The Society for Biomaterials, The Japanese Society for Biomaterials, and The Australian Society for Biomaterials and the Korean Society for Biomaterials*, 82(2): 352–359.

Kjellsen, K.O., & B.L. 2007. Microstructure of Tricalcium Silicate and Portland Cement Systems at Middle Periods of Hydration-Development of Hadley Grains. *Cement and Concrete Research* 37: 13–20.

Kjellsen, K.O., & Harald J. 2004. Revisiting the Microstructure of Hydrated Tricalcium Silicate - A Comparison to Portland Cement." *Cement and Concrete Composites* 26 (8): 947–56.

Kunther, W., Ferreiro, S. & Skibsted, J. 2017. Influence of the Ca/Si ratio on the compressive strength of cementitious calcium–silicate–hydrate binders. *Journal of Materials Chemistry A*, 5(33): 17401–17412

Maki, I., & Kato, K. 1982. Phase Identification of Alite in Portland Cement Clinker. *Cement and Concrete Research*: 93–100.

Mendoza, O., Giraldo, C., Camargo Jr, S.S. & Tobón, J.I. 2015. Structural and nano-mechanical properties of Calcium Silicate Hydrate (CSH) formed from alite hydration in the presence of sodium and potassium hydroxide. *Cement and Concrete Research*, 74, pp. 88–94.

MIT Concrete Sustainability. 2013. Improving Concrete Sustainability Through Alite and Belite Reactivity.

Nìmeèek, J. 2009. Nanoindentation of Heterogeneous Structural Materials. *Czech technical university in Prague.*

Poulsen, S.L., Kocaba, V., Le Saoût, G., Jakobsen, H.J., Scrivener, K.L. & Skibsted, J. 2009. Improved quantification of alite and belite in anhydrous Portland cements by 29Si MAS NMR: effects of paramagnetic ions. *Solid state nuclear magnetic resonance*, 36(1): 32–44.

Shahrin, R. 2018. Nanoindentation Investigation of FIB-Milled Microstructures to Assess Failure Properties of Cement Paste at Microscale.

Thomas, J.J., FitzGerald, S.A., Neumann, D.A. & Livingston, R.A. 2001. State of water in hydrating tricalcium silicate and portland cement pastes as measured by quasi-elastic neutron scattering. *Journal of the American Ceramic Society*, 84(8): 1811–1816.

Advances in Materials and Pavement Performance Prediction II – Kumar et al. (eds)
© 2021 Taylor & Francis Group, London, ISBN 978-0-367-46169-0

Soil stabilization using cement and polyelectrolyte complexes

N. Hariharan, J. Huang, J.G. Saavedra, D.N. Little & S. Sukhishvili
Texas A&M University, College Station, Texas, USA

A.K. Palanisamy
Wageningen University of Research, Gelderland, The Netherlands

E. Masad
Texas A&M University, Doha, Qatar

ABSTRACT: There are frequent concerns related to the mechanical endurance of subgrade soils chemically stabilized with cement especially during moisture intrusion. In this study we evaluate the structural features of polyelectrolyte complexes and propose a working mechanism to consider them as both an alternative and an additive to cement to stabilize subgrade soils. An experimental evaluation of the mechanical properties is conducted to support the proposed mechanism and validate the impact of polyelectrolyte complexes towards increasing the fracture toughness and moisture resistance of limestone screens without compromising strength gain thereby enhancing their overall geomechanical durability.

1 INTRODUCTION

Developing efficient methods to chemically stabilize subgrade soils remains a demanding, unsolved problem in geotechnical and materials engineering. This is because the mineralogical complexity of soils dictates that certain stabilizers will not react with the candidate soil, may not be able to be effectively applied or mixed with the candidate soil, or may not be economical (Makusa 2013). Chemical soil stabilization typically accomplishes one or a combination of three tasks: achieve threshold strength, moderate or eliminate volumetric instability, or provide both strength and volumetric stability (Gallage et al. 2012). Traditional stabilizers, such as portland cement or lime whose reaction mechanisms with the soil minerals are well understood, can usually be selected to increase strength and stability of the soils (Little 1995), but sometimes lead to poor fracture toughness of the chemically stabilized soils (Freeman and Little 2002). This limitation of traditional soil stabilizers has been overcome by the use of polymers (Onyejekwe & Ghataora 2015). More recently, polyampholyte terpolymers comprising of a polyacrylamide-based hydrocarbon chain were reported to achieve equivalent strength and twice the fracture toughness on subgrade soils compared to cement (Iyengar et al. 2012; Zhang et al. 2017). However, the polyacrylamide-based polymers showed high sensitivity to moisture that would compromise their efficacy in most environments. On the other hand, polymer complexes are relativley hydrophobic due to strong ion-pairing interaction and posses excellent toughness in hydrated form (Wang & Schlenoff 2014). By taking advantage of the facile complexation strategy, in this study, we investigated the use of polyelectrolyte complexes (PECs) separately and in combination with portland cement to enhance the geomechanical durability of pavement sublayers composed of limestone screens (LS). The engineering properties measured to assess the geomechanical durability included unconfined compressive strength (UCS), fracture toughness (K_{IC}) and resilient modulus (M_r) with emphasis on promoting mechanical endurance when subjected to diffusion of moisture within the soil matrix. The durability assessment performed in this study did not consider the mechanical and physicochemical impacts on the soils caused due to climatic cycles (wet-dry and freeze-thaw cycles) and weathering.

2 OBJECTIVES

The specific objectives of this study are to:

A. Understand the structural features of the PECs that contribute towards enhancing the geomechanical durability of soils and soil minerals indicated as a measure of strength, stiffness and fracture toughness.

B. Hypothesize the working mechanism of the PECs and explain the composition and synthesis of the PECs used for stabilizing LS.

C. Experimentally illustrate the synergistic benefits of combining cement and PECs to improve the mechanical properties of LS.

3 STRUCTURAL FEATURES OF POLYELECTROLYTE COMPLEXES

PECs are versatile materials and can be synthesized quickly using low cost commercially available polymers. Molecular weights (chain lengths) and the molar ratios of the individual polymers dictate the overall charge and hydrophobicity of the PEC complex. The extent of hydrophobicity in PECs can be easily controlled by choosing different chain lengths of the individual components whereas complex synthetic techniques are required to achieve the same properties in case of copolymers (Reihs et al. 2004). In addition, depending on the soil type, grain size and mineralogy, the chain length of PECs can be altered to bind neighboring soil particles and improve soil stability. The core of the resulting PEC structure where the anionic-cationic polymer merging occurs is charge neutral and hydrophobic, leaving the corona and hence the overall complex positively or negatively charged depending on the composition of the PEC. The charged polymer corona can strongly interact with the broken bonds associated with the edges of various soil minerals. We hypothesize that this interaction will stabilize the soil matrix both mechanically and volumetrically. The hydrophobic central part of the PEC acts as a barrier to reduce the moisture susceptibility of the soils and introduce stability to pavement subgrades that are subjected to constant wetting-drying cycles. In addition, negatively charged PECs, in conjunction with a divalent cation source, such as calcium from lime or cement, can further densify the soil matrix by flocculating the fines through a cation bridging mechanism (Theng 2013) resulting in improved soil durability.

4 SYNTHESIS AND WORKING MECHANISM OF POLYELECTROLYTE COMPLEXES

PECs can be prepared either by direct mixing of two oppositely charged polymers or their dilute aqueous solutions. The solubility of the PECs depends on the nature and molar charge ratios of the individual components. Insoluble complexes are formed when the polymer chains are completely neutralized by the oppositely charged polymer chains. The complete neutralization at stoichiometric ratio (1:1) creates hydrophobic domains that precipitate out of solution. By controlling the charge ratio such that PECs have an excess of positive or negative charge, the hydrophobic domains are stabilized in aqueous solution by the water-soluble free ends of the polymer chains. Figure 1 shows the scheme of preparing water soluble anionic PECs. Cationic PECs can be prepared in a similar way starting from a long chain polycation.

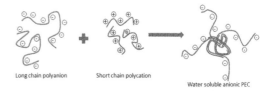

Long chain polyanion   Short chain polycation

Water soluble anionic PEC

Figure 1.   Schematic illustration of anionic PECs.

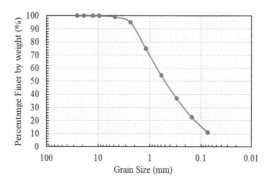

Figure 2.   Gradation curve of the LS.

Edge chemistry and pH play an important role in determining the bonding of various soil minerals with the PECs. For instance, clay minerals such as kaolinite and smectite have an overall negative charge but the edges of these minerals are often composed of broken bonds which become sources of pH dependent charge leaving open the possibility of site-specific chemical interactions. The charged PEC corona has strong possibilities of interacting with these broken bonds, the extent of which would ultimately depend on the composition of the chosen PEC.

5 MATERIALS

5.1 Limestone screens

The LS used in this study were non-plastic and obtained from Knife River Corporation, Texas, USA. Figure 2 depicts the gradation curve of the LS.

5.2 Composition and synthesis of PECs

For the purpose of investigating the geomechanical stability of LS, polystyrene sulfonate (PSS) with average molecular weight, M_w 1,000,000 g/mol, was used as the long chained polyanion and poly-diallyl-dimethylammonium chloride (PDADMAC) with average molecular weight, M_w 250,000 g/mol, was used as the short chained polycation. In this PEC system, the sulfonated styrene forms irreversible complexes with polycations resulting in the PEC being more hydrophobic with restricted segmental mobility. The polymers were individually dissolved in water and the short chained polycation solution was added to the polyanion solution with vigorous stirring to prepare the PEC at a stoichiometric ratio (polyanion to polycation) of 1:0.4 (0.4 M) resulting in an anionic PEC. An anionic PEC was chosen in order to understand the effects of cationic bridging promoted by cement between the PEC and LS both of which have an overall negative charge.

6 EXPERIMENTAL PROGRAM AND METHODS

The mechanical tests performed in this study included the unconfined compressive strength (UCS) test, semi-circular bend (SCB) test to measure K_{IC} and the resilient modulus (M_r) test. All tests were performed on two duplicate samples after dry curing at 40°C for 14 days and wrapped in plastic with about 10 mL of free water for hydration (Little & Nair 2009). The composition of the five samples prepared for mechanical testing are as follows:

Sample 1: Untreated LS
Sample 2: LS + 4% cement
Sample 3: LS + 2% PEC
Sample 4: LS + 4% PEC
Sample 5: LS + 2% PEC + 4% cement

The OMC of the five samples were determined to be 11%, 13.4%, 12%, 13.5% and 14.2% respectively. Type I cement was used to prepare both sample 2 and sample 5. Sample 5 was prepared by homogeneously mixing LS with cement and water first to promote cement hydration prior to PEC addition.

6.1 Unconfined compressive strength test

The UCS of the samples were determined in accordance with ASTM 1633-17 using a sample size of 101.6 mm (4") in diameter and 116.8 mm (4.6") height compacted at their respective OMC.

6.2 Semi-circular bend test

The SCB test was used to measure the plane strain fracture toughness, K_{IC} of the samples in accordance with ASTM D698-12. The samples were fabricated at their respective OMC in a 150 mm-diameter cylinder mold and were compacted to a thickness of 60 mm. The cylindrical samples were cut into two half-circular specimens and a 15-mm notch was introduced (Zhang et al. 2017).

The approach used to calculate K_{IC} was based on linear elastic fracture mechanics (LEFM). K_{IC} was obtained in units of MPa·M$^{0.5}$ at the critical load, P_c using Equations 1 and 2.

$$K_{IC} = \frac{P_c}{2rt} Y_I \sqrt{\pi a} \qquad (1)$$

Where P_c is the applied load in MN, r is the specimen radius in m, t is the specimen thickness in m, a is

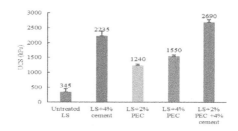

Figure 3. Impact of PEC and cement treatment on UCS of LS.

the notch length in m and Y_I is the normalized stress intensity factor for the specimen used.

$$Y_I = 4.782 + 1.219(\frac{a}{r}) + 0.063 \exp\left[7.045(\frac{a}{r})\right] \quad (2)$$

6.3 Resilient modulus test

The M_r of the samples were determined in accordance with the AASHTO T-307 procedure. The M_r was measured over three moisture contents; OMC-2%, OMC and OMC+2% to assess the moisture sensitivity of the samples. The fluctuation in M_r at various moisture contents was plotted using Equation 3 (AASHTO 2008).

$$\log \frac{M_r}{M_{ropt}} = a + \frac{b - a}{1 + \exp\left[\ln \frac{-b}{a} + k_m \left(S - S_{opt}\right)\right]} \quad (3)$$

Where M_r/M_{ropt} is the ratio of M_r at a given degree of saturation (S) to that at OMC (S_{opt}), a and b are respectively the minimum and maximum values of log (M_r/M_{ropt}) and k_m is a regression parameter.

7 RESULTS AND DISCUSSION

The experimental test results obtained from the UCS, SCB and M_r tests are presented separately in Figures 3–5 respectively. The results shown Figures 3 and 4 represent the average values of the two samples tested along with the standard error. The M_r trends with moisture contents were similar for both samples hence a single data set is shown in Figure 5.

The UCS results in Figure 3 indicate that an optimum dose of cement was more effective in improving the strength of the LS compared to PEC. However, the UCS of the 2% PEC treated sample was nearly quadrupled compared to the control untreated sample and the strength increased further when the PEC dose was increased to 4%. The UCS of the samples treated with 4% cement and 2% PEC comfortably outperformed the other samples. The UCS results suggest that PEC contributes towards increasing the strength of LS and could possibly reduce the cement demand of the LS samples.

The K_{IC} of all treated stabilized samples was found to be significantly higher than the untreated LS

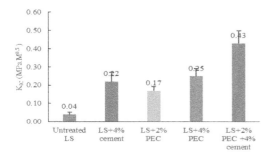

Figure 4. Comparison of K_{IC} of LS samples before and after PEC and cement treatments.

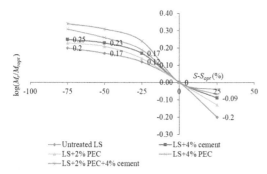

Figure 5. Comparison of the sensitivity of M_r to moisture fluctuations before and after PEC and cement treatments.

sample. A closer examination of Figure 4 indicates three interesting trends; K_{IC} increased by about 30% upon increasing the dosage of PEC from 2% to 4%, 4% PEC addition resulted in slightly higher K_{IC} values compared to 4% cement treatment and the combined use of 4% cement with 2% PEC nearly doubled the K_{IC} of the LS.

The average M_r of samples 1 through 5 at OMC were found to be 55 MPa, 170 MPa, 110 MPa, 144 MPa and 225 MPa respectively. There is evidence that a 4% dose of PEC is optimum for the LS based on Figure 5 wherein the sensitivity of M_r to moisture changes in the cement treated samples is sandwiched between the samples treated with 2% PEC and 4% PEC. Nevertheless, a combination of both treatments provides the best response over a range of moisture contents further emphasizing the impact of PECs on improving the moisture resistance and overall geomechanical durability of the LS tested.

8 CONCLUSIONS

The findings of this work demonstrate that PECs can be a viable alternative to chemically stabilize subgrade soils of varying mineralogy and improve the mechanical endurance of cement stabilized soils when subjected to moisture intrusion. The synergistic benefits of PECs and cement improved the fracture toughness and moisture resistance of the LS investigated while

concomitantly contributing to strength gain thereby enhancing the overall geo-mechanical durability of the material. The polar tails of the PECs appear to promote chemical interaction with the edges and surface of the soil minerals while the charge neutral core is hydrophobic and protects the soil matrix from moisture intrusion. The versatile structural features of the PECs offer possibilities to engineer cost-effective solutions specific to the mineralogy of the soils encountered.

REFERENCES

AASHTO. 2008. *Mechanistic Empirical Pavement Design Guide: A Manual of Practice*. Washington, D.C.

AASHTO T-307. 1999. *Standard Method of Test for Determining the Resilient Modulus of Soils and Aggregate Materials*.

ASTM Standard D698-12. 2012. *Standard Test Methods for Laboratory Compaction Characteristics of Soil Using Standard Effort*. West Conshohocken, PA.

ASTM D1633-17. 2017. *Standard Test Methods for Compressive Strength of Molded Soil-Cement Cylinders*. West Conshohocken, PA.

Freeman, T.J. & Little, D.N. 2002. Maintenance strategies for pavements with chemically stabilized layers (No. FHWA/TX-01/1722-6,). Texas Transportation Institute, Texas A & M University System.

Gallage, C., Cochrane, M. & Ramanujam, J. 2012. Effects of lime content and amelioration period in double lime application on the strength of lime treated expansive sub-grade soils. In *Proceedings of the 2nd International Conference on Transportation Geotechniques* (pp. 99–104).

Iyengar, S.R., Masad, E., Rodriguez, A.K., Bazzi, H.S., Little, D. & Hanley, H.J. 2012. Pavement subgrade stabilization using polymers: Characterization and performance. *Journal of Materials in Civil Engineering*, 25(4), pp. 472–483.

Little, D.N. 1995. *Stabilization of pavement subgrades and base courses with lime*. Dubuque: Kendal/Hunt Publishing Company.

Little, D. N. & Nair, S. 2009. Recommended Practice for Stabilization of Subgrade Soils and Base Materials. NCHRP Web-Only Document 144, Contractor's Final Task Report for NCHRP Project 20-07. *Transportation Research Board, National Research Council*, Washington, DC.

Makusa, G.P. 2013. *Soil stabilization methods and materials in engineering practice: State of the art review*. Luleå tekniska universitet.

Onyejekwe, S. & Ghataora, G.S. 2015. Soil stabilization using proprietary liquid chemical stabilizers: Sulphonated oil and a polymer. *Bulletin of Engineering Geology and the Environment*, 74(2), pp. 651–665.

Reihs, T., Müller, M. & Lunkwitz, K. 2004. Preparation and adsorption of refined polyelectrolyte complex nanoparticles. *Journal of colloid and interface science*, 271(1), pp. 69–79.

Theng, B.K.G. 2013. *Formation and properties of clay-polymer complexes*. Vol.4. Elsevier.

Zhang, J., Little, D.N., Grajales, J., You, T. & Kim, Y.R. 2017. Use of Semicircular Bending Test and Cohesive Zone Modeling to Evaluate Fracture Resistance of Stabilized Soils. *Transportation Research Record*, 2657(1), pp. 67–77.

Pavement performance prediction using mixture performance-volumetric relationship

J. Jeong, A. Ghanbari & Y.R. Kim
North Carolina State University, Raleigh, NC, USA

ABSTRACT: This paper describes the relationship between pavement performance (based on fatigue damage and rutting) and the volumetric information of mixtures used in asphalt field projects. This study, as part of shadow projects for the Federal Highway Administration's Asphalt Mixture Performance-Related Specifications deployment research project, focuses on (1) a procedure to calibrate the performance-volumetric relationship (PVR) and (2) predicting pavement performance under field conditions using the developed PVR. In addition, the performance of as-designed and as-constructed mixture samples is predicted and compared.

1 INTRODUCTION

An important task for state highway agencies is to ensure that pavements are properly constructed so that they will perform well over their intended service life. However, because current quality assurance (QA) specifications do not have factors that relate directly to pavement performance, guaranteeing adequate performance is difficult, even when contractors meet their own QC requirements. Also, properly compensating contractors for their work in a rational way is difficult because pavement performance cannot be anticipated. In the case of the expiration of a contractor's warranty, the maintenance tasks for a prematurely failed pavement would become the sole responsibility of the agency. Given that budget concerns are critical for both agencies and contractors, finding an efficient way to predict pavement performance is important for all parties, including taxpayers. Today, most quality-related specifications are based on volumetric characteristics because such characteristics are easily quantifiable, relatively repeatable, and have some impact on pavement longevity. Therefore, if a relationship between pavement performance and volumetric characteristics can be found, then pavement performance could be predicted without the need for performance tests. Having this relationship, the performance of as-constructed pavements could be predicted from the currently used acceptable quality characteristics (AQCs) and compared to the predicted performance of the as-designed pavement to adjust payments to contractors.

Note that most of this paper's data were generated from the Maine Department of Transportation (MaineDOT) shadow project for the Federal Highway Administration's Asphalt Mixture Performance-Related Specifications deployment research project.

2 OBJECTIVES

The objectives of this paper are to (1) develop a performance-volumetric relationship (PVR) for field mixtures and (2) predict and compare the performance of as-designed and as-constructed pavements.

3 PERFORMANCE-VOLUMETRIC RELATIONSHIP DEVELOPMENT

3.1 Pavement project selection

This study required field mixtures with enough natural sample-to-sample variability to observe performance differences as well as volumetric changes; therefore, the materials were selected from a sufficiently large project.

3.2 Material sampling and acceptable quality characteristics

For the material acquisition process, eleven asphalt mixture samples that had been produced based on a job mix formula (JMF) were acquired from eleven different truck loads in the same paving project to ensure naturally-occurring sample-to-sample variability. Volumetric tests were conducted using these eleven samples to obtain AQCs, which included mixture's maximum specific gravity (G_{mm}), aggregate's bulk specific gravity (G_{sb}), binder content, and the in-place air void contents of each sample.

3.3 Volumetric conditions of mixture samples

As a first step to building the PVRs, the volumetric information for the eleven samples was plotted in the in-place voids in mineral aggregate (VMA) and voids filled with asphalt (VFA) domain. Also, the MaineDOT's mixture specifications were converted

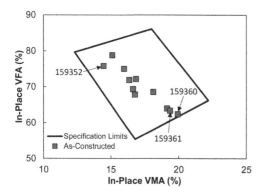

Figure 1. Calculated IP-VMA and IP-VFA for specification limits of eleven samples.

to the same domain. This volumetric information was calculated using Equations 1 and 2.

$$VMA_{IP} = 100 - \frac{\%G_{mm} \times G_{mm} \times P_s}{G_{sb}} \qquad (1)$$

$$VFA_{IP} = 100 \times \left(\frac{VMA_{IP} - V_{a,IP}}{VMA_{IP}}\right) \qquad (2)$$

where VMA_{IP} is in-place VMA, $\%G_{mm}$ is the as-constructed compaction level, G_{mm} is the theoretical maximum density of the asphalt mixture, P_s is the aggregate content, G_{sb} is the bulk specific gravity of the aggregate, VFA_{IP} is the in-place VFA, and $V_{a,IP}$ is the measured in-place air void content.

Figure 1 presents the calculated IP-VMA and IP-VFA of the eleven samples and the specification limits.

3.4 Selection of samples to calibrate the performance-volumetric relationship

The underlying concept of the PVR is that the performance of an asphalt mixture under any volumetric conditions can be predicted by testing the asphalt mixture at only a few selected volumetric conditions and developing the relationship between the performance and the selected volumetric conditions. A previous study [1] suggests that four volumetric conditions located at the widest points within the volumetric condition range (hereinafter called *four corners*) are sufficient to develop the PVR for a given mixture.

As eleven plant-produced samples were used in this study, the only way to change the volumetric conditions in the lab was to target the high and low air void contents to calibrate the four corners. As shown in Figure 1, three samples (159352, 159360, and 159361) out of the eleven samples were selected that were located the furthest away from each other and compacted to high and low densities to construct the four corners in the volumetric domain. The determined PVR conditions reflect that sample 159352 was targeted at two different air void contents, 7.5% and 3.5%, denoted as *A* and *B*, and 2.5% and 7.5% were targeted for samples

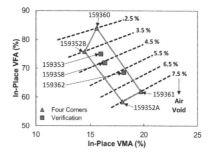

Figure 2. Determined PVR calibration and verification conditions.

159360 and 159361, respectively. Three additional samples (159353, 159358, and 159362) were chosen and compacted at the in-place air voids (4.0%, 4.6%, and 5.7%, respectively) to verify the PVR function. The determined four corners and verification samples are plotted in Figure 2.

3.5 Performance tests

Three performance tests for the four corners and verification samples were conducted using an Asphalt Mixture Performance Tester (AMPT). For the dynamic modulus tests (AASHTO TP 132), three small test specimens (D 38 mm x H 110 mm) obtained from a gyratory-compacted specimen were tested at 4°C, 20°C, and 40°C. For the fatigue tests (AASHTO TP 133), three small test specimens obtained from a gyratory-compacted specimen were tested at 15°C. For the stress sweep rutting tests (AASHTO TP 134), four large test specimens (D 100 mm x H 150 mm) were tested at 20°C for the low-temperature tests and at 45°C for the high-temperature tests. Two large specimens were tested at each temperature. The test specimens were fabricated according to AASHTO R 83 for the large and AASHTO PP 99 for the small specimens.

3.6 FlexMAT™ and FlexPAVE™ Software

3.6.1 FlexMAT™

FlexMAT™ performs complex analysis algorithms to generate dynamic modulus mastercurves, calibrate the simplified viscoelastic damage (S-VECD) model and the shift model for rutting predictions, and ultimately generate output files, which then can be used in the pavement performance simulation software, FlexPAVE™.

3.6.2 FlexPAVE™

FlexPAVE™ employs VECD theory to account for the effects of loading rate and temperature on pavement response and distress mechanisms. It also allows the simulations of pavement structures that consist of asphalt concrete and unbound materials. Project location, traffic conditions, and design vehicle configurations can be assigned for the given project [2]. The output from FlexPAVE™ simulations is pavement

Table 1. Inputs for FlexPAVE™ simulations.

Pavement Structure	100 mm HMA pavement (50 mm wearing course + 50 mm base course) 300 mm aggregate base type B
Climate	Bangor, Maine
Traffic	10 M * ESALs for 20 years

* ESAL is equivalent single-axle load.

Table 2. Volumetric information and FlexPAVE™ simulation results.

	Volumetric Conditions				Performance at 20 years	
	Fatigue test		Rutting test			
Sample	IP-VMA	IP-VFA	IP-VMA	IP-VFA	% damage	Rut depth* (mm)
159352A	17.7	59.4	18.0	53.8	14.0	1.6
159352B	14.0	79.0	14.3	76.8	10.0	1.0
159360	15.6	84.2	15.3	86.0	10.5	0.9
159361	19.8	61.7	19.8	61.6	13.7	1.6
159353	15.7	72.2	15.1	75.7	12.2	1.3
159358	16.5	71.4	16.1	73.4	10.3	1.3
159362	18.2	68.3	18.3	67.9	11.9	1.6

* Rut depth for only asphalt concrete surface.

performance, which is provided in the form of the percentage of damage ('% damage') and rut depth (mm) over the design life of the pavement.

3.7 AMPT results

In order to simulate real field conditions as much as possible, the input values used for FlexPAVE™ were derived from a field project plan that is summarized in Table 1. Note that the project was a mill-and-fill project where the top 50 mm of the existing asphalt layer was replaced by the study mixture. Because fatigue damage in a relatively thin pavement starts mostly at the bottom of the asphalt layer, changing the mixture for the top 50 mm of the asphalt layer would not yield information about the effects of changes in the mixture volumetric conditions on pavement performance. Therefore, the entire 100 mm of the asphalt layer was modeled by FlexPAVE™ for this study. Table 2 presents the volumetric information and simulated pavement performance results.

3.8 Performance-volumetric relationship development

The PVR function was calibrated based on the results of the performance simulations and the volumetric information from the four corners. A linear function was employed for the PVR development based on findings of Wang et al. (2019). Two PVR equations for '%

Table 3. Performance-volumetric relationship coefficients for % damage and rut depth.

Performance	a	b	d	R^2
%Damage (f)	0.317	−0.11	14.5	0.97
Rut depth (r)	0.065	−0.0158	1.30	0.99

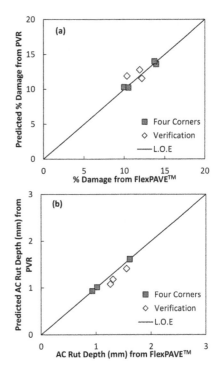

Figure 3. Comparison FlexPAVE™ and PVR results: (a) % damage and (b) asphalt surface rut depth.

damage' and rut depth were developed, as shown in Equations (3) and (4), respectively. Table 3 presents the coefficients used for both equations.

$$\%Damage = a_f \times VMA_{IP} + b_f \times VFA_{IP} + d_f \qquad (3)$$

$$RutDepth(mm) = a_r \times VMA_{IP} + b_r \times VFA_{IP} + d_r \quad (4)$$

where $a_f, b_f, d_f, a_r, b_r,$ and d_r are regression constants.

Note that the PVR is a unique function for the combination of project-specific mixture, climate, and traffic conditions. Therefore, the PVR coefficients for the same mixture would be different for different climate regions or traffic conditions.

3.9 Performance-volumetric relationship verification

The developed PVRs were verified using the three additional samples; Figure 3 presents the results.

Table 4. Comparison of performance between as-designed and as-constructed samples as predicted by performance-volumetric relationship.

Sample	In-place AV (%)	% Damage		Rut depth (mm)	
		Pred. from PVR	Compare to as-designed	Pred. from PVR	Compare to as-designed
As-designed	5.2	12.4	–	1.34	–
159353	4.0	11.3	9.0% D*	1.15	15.0% D*
159358	4.6	11.8	5.0% D*	1.23	8.9% D*
159362	5.7	12.7	2.6% I*	1.40	3.9% I*

* D: Decrease, I: Increase.

The largest error between the simulated and predicted performance was around 15 percent. With the exception of a few samples, the error for the most of the samples was less than 10 percent. These findings indicate that the developed PVRs work well to predict pavement performance without the need for performance tests, which saves a significant amount of both materials and time.

4 PERFORMANCE COMPARISON OF AS-DESIGNED AND AS-CONSTRUCTED PAVEMENTS

4.1 Selection of as-designed conditions

The performance of pavements under as-designed conditions was predicted using the developed PVRs for comparison with the performance of pavements predicted under as-constructed conditions. Capturing as-designed field air void contents is difficult because the initial in-place air void percentages are typically higher than the design air void percentages due to the effects of traffic compaction. Therefore, this study defined as-designed conditions by employing mixture characteristics such as the G_{mm}, Ps, and G_{sb} from a mixture's design and used the averaged in-place air void content of the eleven samples to calculate the in-place VMA and VFA contents.

4.2 Performance comparison of as-designed and as constructed pavements

The three samples used for verification purposes also were employed as the as-constructed samples and their measured in-place air void contents were used to calculate the IP-VMA and VFA. The performance of the as-designed and as-constructed samples was predicted using the PVR; Table 4 presents this comparison.

Based on the predicted pavement performances, two (159353, 159358) of the three as-constructed samples showed 9% and 5% decreases in '% damage' and 15% and 8.9% decreases in rut depth (mm) compared to the as-designed samples respectively, because these two samples' in-place air void percentages were lower than the as-designed conditions. The other sample (159362) was predicted to have 2.6% and 3.9% increases in '% damage' and rut depth, respectively, due to the higher field air void percentage than the as-designed condition. These results show that the PVR can generate reasonable trends of pavement performance in terms of field air void contents.

5 CONCLUSIONS

- PVRs were developed using the 'four corners' method and were able to predict pavement performance at other volumetric conditions. On average, the errors were 5.5% and 4.9% for fatigue damage and rutting, respectively, as obtained from PVRs compared to the simulated performance results derived from FlexPAVE™.
- The predicted performance obtained using the PVR for the as-designed and as-constructed samples showed reasonable trends in terms of in-place air void percentages.

REFERENCES

Kim, Y. R., Guddati, M. N., Choi, Y. T., Kim, D., Norouzi, A., Wang, Y., Keshavarzi, B., Ashouri, M., and Wargo, A. D. 2020. *Development of Asphalt Mixture Performance-Related Specifications*. In press, Final Report for FHWA DTFH61-08-H-00005 project, in press.

Superpave Mix Design: Superpave Series No. 2 (SP-2). 1996. Asphalt Institute.

Wang, Y. D., Ghanbari, A., Underwood, B. S., and Kim, Y. R. 2019. Development of a Performance-Volumetric Relationship for Asphalt Mixtures. *Journal of the Transportation Research Board*, DOI: 10.1177/0361198119845364

Advances in Materials and Pavement Performance Prediction II – Kumar et al. (eds)
© 2021 Taylor & Francis Group, London, ISBN 978-0-367-46169-0

Effect of nominal maximum aggregate size on fatigue damage in concrete

S.R. Kasu
Indian Institute of Technology Kharagpur, India

S. Gunda
CVR College of Engineering, Hyderabad, India

N. Mitra & A.R. Muppireddy
Indian Institute of Technology Kharagpur, India

ABSTRACT: Fatigue damage of the cement concrete pavements is progressive under vehicular loads and these pavements are designed based on cumulative fatigue damage principle where stress level and material characteristics are considered. From the literature, contradictory results on aggregate size with fracture parameters are reported. An attempt is made in this study, to understand the influence of nominal maximum aggregate size (NMAS) on flexural fatigue of concrete specimens without a notch. The main objective is to understand the effect of concrete mixtures made with two NMAS having similar strength range on stiffness change. The results indicated that the concrete made with NMAS of 10 mm had higher the fatigue life and thus loss in stiffness is less compared to the mix with NMAS of 20 mm. The damage model is used to predict the stiffness degradation and found that the model is excellent with a very low root mean square error.

1 INTRODUCTION

Fatigue damage of plain concrete slab used in the pavement systems is progressive with the number of vehicular loads. The cumulative fatigue damage (CFD) concept has been deployed in mechanistic-empirical pavement design guides (AASHTO 1993, IRC 58 2015) to account for the contribution of the expected wheel load in the design. The stress level is found to be the most influencing variable in all the original and modified fatigue models (Smith & Roesler 2004). To keep stress levels well below the flexural strength of concrete, a number of researchers have been trying to change the proportion of ingredients to improve ultimate flexural strength of concrete (Akcay et al. 2012, Beygi et al. 2014, Sireger et al. 2017), and incorporate the fibrous materials into the concrete (Banthia & Sheng 1990, Singh & Kaushik 2003, Lawler et al. 2005).

On the other hand, studies are being carried out on the improvement of fracture energy and the toughness of the concrete material (Nallathambi et al. 1984, Rocco & Elices 2009, Beygi et al. 2014). It is believed that the higher fracture energy of the material would have a higher fatigue life on notched specimens, which may not be true in all the cases especially under dynamic loading conditions. Since crack initiation and growth from the material inherent flaw, the overall damage progress and then failures are abrupt. Similarly, the strain evolution and the corresponding stiffness degradation of the material in the plain concrete slab of pavements are due to repetitive application of loads.

Further, the concrete is heterogeneous and thereby its behavior is complex. The damage from a microscopic random flaw which may be attributed to pores and weaker transition zone can be well explained by the damage evolution model. The strain evolution and or stiffness degradation are the material characteristics used in the development of the damage model. In this study stiffness degradation i.e. loss of stiffness is considered to be the measure of damage per cycle as a function of the number of cycles.

2 OBJECTIVES

Evaluation of change in fracture energy of concrete with the variable size of the aggregate under static loading conditions on notched specimens of concrete is the most common method. However, it is appropriate to conduct repeated cyclic tests to simulate field conditions. Keeping this as an objective of the present study, it is the proposed to conduct the test on concrete mixtures prepared with two nominal maximum aggregate sizes to understand the influence on flexural fatigue behavior of concrete and also to fit a damage model for the mixtures incorporating material properties like stiffness degradation and number of cycles.

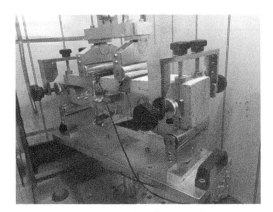

Figure 1. Experimental setup (Kasu et al. 2019).

Table 1. Concrete Mixture Proportions (kg/m³).

Variable	S10	S20
Water	206.86	202.86
Cement	292.43	288.01
Fly ash	157.47	155.33
Sand	834.91	824.20
10 mm/20 mm	853.48	874.98
Superplasticizer	0.85	0.77

3   EXPERIMENTAL PLAN

The flexural static test was conducted on concrete mixtures in accordance with ASTM C78. Figure 1 shows the experimental setup used in this study. For each mixture, a set of five specimens prepared and tested using a 30 kN capacity, closed-loop servo-hydraulic controlled machine under four-point bending condition. Under dynamic loading, 90 specimens were tested with the same test setup. Two vertical spring-loaded linear variable differential transducers (LVDTs) were used to measure the mid deflection of the specimen for every cycle until the specimen fails. The LVDTs with a span range of 10 mm were used.

4   MATERIALS

Two concrete mixtures were studied, referred to as concrete S10 and concrete S20 made with 10 mm and 20 mm nominal maximum aggregate sizes respectively. The concrete mix proportions by weight are presented in Table 1. The beam specimens of 500 mm × 100 mm × 100 mm were cast in the laboratory. The description of the mixing ingredients can be found in the author's previous work (Kasu et al. 2019). The prepared specimens were demolded after 24 hours and cured in a water tank. At 28 days samples were removed from the water and dried prior to conducting static and dynamic testing.

5   EXPERIMENTAL RESULTS AND ANALYSIS

5.1   Static bending tests

The average failure loads and flexural strength at failure obtained from the static bending tests are tabulated in Table 2. The flexural strength of both the mixtures observed to be similar.

5.2   Fatigue bending tests

The maximum stress to be applied in dynamic testing usually ranges between 55 to 95% of ultimate flexural strength. In this study, the maximum stress levels were selected as 80, 85 and 90% of their strengths to represent the heavy and super-heavy wheel loads. The tests were carried out at 2, 5 and 10Hz frequencies to represent vehicles moving at variable speeds. The minimum stress level about 2% kept constant which is equivalent to the seating load of 0.20 kN. The terminal conditions of the test were defined as (i) complete rupture (ii) 50% of initial stiffness and (iii) 1 million cycles. The output data captured as the number of cycles and stiffness per each cycle. It was observed that the specimens of the S10 mix sustain more cycles compared to the S20 mix from SN curves (Stress and number of cycles). Stress levels were found to be more influential on fatigue lives compared to the frequency of loading (Kasu et al. 2019). Therefore, for the present study, the data has been analyzed for the data pooled as per the stress levels and presented here.

5.3   Loss of stiffness

Stiffness of the material is determined from the maximum applied stress and strain. Under constant stress regime, the stiffness per each cycle was found to be decreased. The stiffness degradation is the property of the material obtained from the mid-deflection during cyclic loading. The degradation in stiffness of plain concrete beams is the absolute expression for the structural damage, which depicts the degree of damage i.e. the rate of deterioration. With the advent of continuous data acquisition system facility, the stiffness was recorded for every cycle throughout the testing program. The normalized stiffness, number of cycles, and the corresponding strain are the commonly used variables for describing the fatigue damage.

In this study, the damage parameter used is based on the number of cycles and normalized stiffness × cycles (NSC) as per ASTM D7460. Equation 1 is used to calculate the same.

$$NSC = \frac{E_i - N_i}{E_0 * N_0} \tag{1}$$

Where E_0 is the initial stiffness, E_i is the stiffness at the ith cycle; N_0 is the cycle at which the initial stiffness is determined; N_i is the cycles; $i = 1$ to n.

Figure 2 shows the typical stiffness degradation of concrete flexural fatigue tests under various loading

Table 2. Flexural strengths of concrete mixtures.

Mixture	Avg. failure load, N (Std. dev)	Avg. flexural strength, MPa (Std. dev)	Flexural stresses selected for different stress levels, MPa		
			80%	85%	90%
S10	9452 (564)	4.254 (0.25)	3.403	3.616	3.828
S20	9309 (213)	4.189 (0.10)	3.351	3.561	3.770

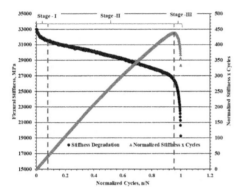

Figure 2. Loss of stiffness with the number of cycles.

Table 3. Loss of stiffness at various stages of cyclic loading for different mixtures.

Mixture	Stiffness reduction to the percentage of Initial Stiffness (X_i) at		Loss of Stiffness in tertiary Stage, %
	Secondary to Tertiary stage (X_1)	Failure (X_2)	
S10	79.16	62.94	16.22
S20	68.15	53.79	14.35

frequencies and stress levels. It was found that the stiffness degradation follows an accumulated three-stage process. The reduction in stiffness of the specimens is relatively large in the first stage and third stage from the start of fatigue loading and when it approaches the failure respectively, but in the second stage, the damage is linearly progressive. In this total process, the damage exhibits a three-stage 'S' shaped curve. This curve built-in with two inflection points, one is at the transition from first to second, and the other is at the transition from second to third stages.

For any quasi-brittle materials like concrete with the change in its toughness by change of ingredients is likely to have an extended life. To identify such behavioral changes in the materials, further analysis is carried out by pooling the test data of all the specimens to identify the stiffness reduction point (second inflection) at which the degradation curve changes its direction from second to the third stage. The loss of stiffness at that point from the plots between NSC and cycles is 79.16 and 68.15% (on average) of the initial stiffness in S10 and S20 mixtures respectively. On the other hand, the stiffness reduction at failure was found to be 62.94 and 53.79% (on average) of initial stiffness for both the mixtures (See Table 3).

5.4 Damage evolution concerning stiffness and number of cycles

The damage in structural components with time and number of cycles is a progressive phenomenon. It may be attributed to the localized microcracking process

zone. At the macro level, it would be difficult to understand how the damage evolution takes place as the concrete made up of a variety of ingredients. It can be well understood by the damage function D(n) for relative stiffness of the material (Chen et al. 2017, Chandrappa & Biligiri 2019) is as shown in Equation 2.

$$D(n) = \left[1 - \left(\frac{\ln(E_n)}{\ln(E_0)} \right) \right] \quad (2)$$

Where E_n is the stiffness at N^{th} cycle; E_0 is the initial stiffness.

The distribution with damage function in reference to relative cycles is

$$D(n) = A \left[\ln \left(1 - \frac{N_F}{n} \right) \right]^{\frac{1}{B}} \quad (3)$$

Where N_F is the number of cycles to failure; A is the scale and B is the shape parameter.

Equating both the functions (Equation 2 and 3) and after rearrangement, the relation will become Equation 4.

$$\left(\frac{\ln(E_n)}{\ln(E_0)} \right) = 1 - A \left[\ln \left(1 - \frac{N_F}{n} \right) \right]^{\frac{1}{B}} \quad (4)$$

The final relationship represents the relative stiffness can be obtained as Equation 5.

$$\ln(E_n) = \ln(E_0) \left\{ 1 - A \left[\ln \left(1 - \frac{N_F}{n} \right) \right]^{\frac{1}{B}} \right\} \quad (5)$$

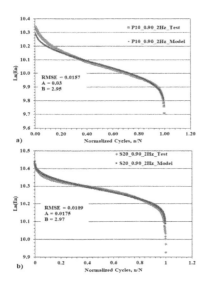

Figure 3. Validation of the damage model for (a) S10 and (b) S20 mixtures.

Using equation (5), the stiffness at a specific cycle is obtained and plotted in Figure 3 along with the test data for a specimen for both S10 and S20 concrete mixtures operated at 2Hz and 90% stress level. As seen from Figure 3, the used model suits well, compared with the experimental data, showing excellent fit with very low root mean square error (RMSE).

6 CONCLUSIONS

In the present study, the effect of concrete mixtures made with two NMAS (S10 and S20) of similar strength range on stiffness change was evaluated experimentally. The tests were conducted under constant stress amplitude with a haversine loading pattern. The damage of the specimens was discussed with stiffness degradation and their stage transformations. The point of inflection, where the stiffness degradation changes its phase from secondary to the tertiary stage was found from the plots between normalized stiffness * cycles (NSC) vs. the number of cycles. The transition points separate the stiffness degradation to understand the enhanced material behavior in the tertiary stage. This helps in the estimation of featured life in the tertiary stage. It can be observed from experimental results that a higher number of cycles required to cause failure and lower losses of stiffness for S10 mix compared to S20 mix. This might be due to larger aggregate size which results in a lower surface area for the development of gel bonds, which is responsible for the lower fatigue strength of concrete. The damage evolution model was used to understand how stiffness degradation evolves with the influence of nominal aggregate size in concrete. The model predictions were compared with the experimental results and a good correlation between the model and experimental data was found.

REFERENCES

AASHTO (1993). "Guide for Design of Pavement Structures." American Association of State Highway and Transportation Officials.

Akcay, B., Agar-Ozbek, A. S., Bayramov, F., Atahan, H. N., Sengul, C., & Tasdemir, M. A. (2012). Interpretation of aggregate volume fraction effects on fracture behavior of concrete. Construction and Building Materials, 28(1), 437–443.

ASTM D7460-10. Standard test method for determining fatigue failure of compacted asphalt concrete subjected to repeated flexural bending, ASTM International, West Conshohocken, PA 19428-2959, USA.

ASTM, Standard Test Method for Flexural Strength of Concrete (Using Simple Beam with Third-Point Loading), ASTM C78/C78M-15, West Conshohocken, PA 19428-2959, USA, 2010.

Banthia, N., and Sheng, J. 1990. "Micro-reinforced cementitious materials." Mater. Res. Soc. Symp. Proc., 211, 25–32.

Beygi, M. H., Kazemi, M. T., Amiri, J. V., Nikbin, I. M., Rabbanifar, S., & Rahmani, E. (2014). Evaluation of the effect of maximum aggregate size on fracture behavior of self-compacting concrete. Construction and Building Materials, 55, 202-211.

Chandrappa, A. K., & Biligiri, K. P. (2019). Effect of pore structure on fatigue of pervious concrete. Road Materials and Pavement Design, 20(7), 1525–1547.

Chen, X., Bu, J., Fan, X., Lu, J., & Xu, L. (2017). Effect of loading frequency and stress level on low cycle fatigue behavior of plain concrete in direct tension. Construction and Building Materials, 133, 367-375.

IRC 58-2015, Guidelines for the Design of Plain Jointed Rigid Pavements for Highways, Fourth Edition, Indian Roads Congress, New Delhi, India, 2015.

Kasu, S.R., Deb, S., Mitra, N., Muppireddy, A.R., & Kusam, S.R. (2019). Influence of aggregate size on flexural fatigue response of concrete. Construction and Building Materials, 229.

Lawler, J. S., Zampini, D., & Shah, S. P. (2005). Microfiber and macrofiber hybrid fiber-reinforced concrete. Journal of Materials in Civil Engineering, 17(5), 595–604.

Nallathambi, P., Karihaloo, B. L., & Heaton, B. S. (1984). Effect of specimen and crack sizes, water/cement ratio and coarse aggregate texture upon fracture toughness of concrete. Magazine of concrete research, 36(129), 227–236.

Rocco, C. G., & Elices, M. (2009). Effect of aggregate shape on the mechanical properties of a simple concrete. Engineering fracture mechanics, 76(2), 286–298.

Singh, S. P., & Kaushik, S. K. (2003). Fatigue strength of steel fibre reinforced concrete in flexure. Cement and Concrete Composites, 25(7), 779–786.

Siregar, A. P. N., Rafiq, M. I., & Mulheron, M. (2017). Experimental investigation of the effects of aggregate size distribution on the fracture behaviour of high strength concrete. Construction and Building Materials, 150, 252–259.

Smith, K. D., & Roesler, J. R. (2004). Review of fatigue models for concrete airfield pavement design. In Airfield Pavements: Challenges and New Technologies (pp. 231–258).

Effect of uncertainty in dynamic modulus on performance prediction

H.A. Kassem
Beirut Arab University, Beirut, Lebanon

G.R. Chehab
American University of Beirut, Beirut, Lebanon

ABSTRACT: Pavement design methods have been developed from the purely empirical 1993AASHTO Guide towards the more realistic *Mechanistic-Empirical Pavement Design Guide* (MEPDG). The ME Design is more accurate and realistic as it overcomes the limitations of the empirical pavement design methods through the incorporation of the principles of material mechanics. Still, it lacks a robust quantification of the reliability of the resulting pavement design. ME Design recognizes that pavement performance is governed by a large amount of uncertainty and variability related to design, construction, traffic loading, and climatic conditions over the expected design life. However, ME Design provides an analytical solution that incorporates reliability uniformly for all pavement types allowing the design of a pavement within a desired level of reliability. This study aims at studying the effect of the uncertainty in the dynamic modulus ($|E^*|$) of asphalt concrete (AC) on the performance prediction using MEPDG. Monte Carlo Simulations are used to model the uncertainty in the $|E^*|$ mastercurve of various AC mixtures. Simulated $|E^*|$ mastercurves are used as input for MEPDG to predict rutting and fatigue cracking for various pavement structures under different loading speeds and climatic conditions resulting in a suit of 15,000 MEPDG runs. A sensitivity analysis is carried out to check how the uncertainty in $|E^*|$ will be forward propagated to predict the performance of pavements. Such results are utilized to assess how uncertainties in material properties, presented by $|E^*|$, need to be incorporated in quality assurance practices of pavement construction.

1 INTRODUCTION

In parallel with the developments in the paving industry, practitioners and researchers have continuously improved methods and specifications for assessing the quality of asphalt concrete (AC) during production and pavement construction. The quality of as-produced and as-constructed AC mixes affect the performance of pavements over its intended service life (Killingsworth 2004). Mixes with unacceptable quality might be prone to high levels of distresses under traffic loading and environmental influences at early stages after the pavement's construction.

Pavement design methods has been developed from the purely empirical 1993AASHTO Guide for Design of Pavement Structures towards the more realistic *Mechanistic-Empirical Pavement Design Guide* (MEPDG). This latter has been continuously improved and developed into the most recent *AASHTOWare Pavement ME Design*. The aim of the ME Design is to mechanistically calculate the stresses, strains, and deflections in pavements due to traffic loading and changes in temperature and correlate them empirically to pavement distresses using transfer functions. The utilization of the ME Design in design can optimize the

pavement performance by minimizing rutting, fatigue cracking, thermal cracking, and roughness during the pavement's design life.

The implementation of ME Design is considered more accurate and realistic as it overcomes the limitations of the empirical pavement design methods through the incorporation of the principles of material mechanics. Still, it lacks a robust quantification of the reliability of the resulting pavement design.

ME Design recognizes that pavement performance is governed by a large amount of uncertainty and variability related to design, construction, traffic loading, and climatic conditions over the expected design life. In addition, significant variabilities are exhibited along the length of the pavements. However, ME Design provides an analytical solution that incorporates reliability uniformly for all pavement types which allows designing a pavement within a desired level of reliability. A robust analysis is required to account for the uncertainty associated with each input parameter for any pavement project in order to improve the accuracy of reliabilitybased pavement designs.

In ME Design, the dynamic modulus $|E^*|$ serves as a key parameter in characterizing the asphalt mix properties. $|E^*|$ is defined as the absolute value of

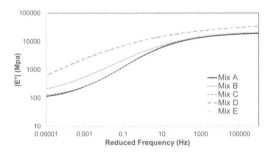

Figure 1. Average |E*| mastercurve for the investigated mixtures.

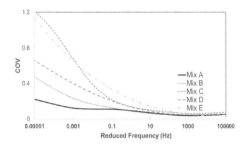

Figure 2. Variation of COV of |E*| as a function of reduced frequency at 20° C for mixes in the study.

the complex modulus, a complex quantity that represents the stress-strain relationship under a continuous sinusoidal/haversine loading. It constitutes a material characteristic that has been utilized for the determination of a mix's ability to resist fracture and permanent deformation under defined conditions (Kahil et al. 2015, and Kassem et al. 2016; Witczak 2007). Thus, quantifying the variability of |E*| is important due to its influence on the predicted performance of asphalt pavements (Kassem 2018). Thus, the major objective of this paper is to study the effect of the uncertainty in |E*| of AC on the performance prediction using MEPDG.

2 UNCERTAINTY IN DYNAMIC MODULUS

For the quantification and modeling of the inherent uncertainty in |E*|, the study by Kassem et al. aimed at collecting |E*| data for mixes that involved at least 8 test replicates. This number was selected as a practical lower bound to the number of tests that would allow for meaningful statistical analyses to be conducted. The data of five of these mixes will be utilized in this research as summarized in Table 1.

The average |E*| mastercurves obtained for each of the mixes is shown in Figure 1. Each mastercurve is determined by fitting the well-known sigmoidal function to the measured |E*| data.

Monte Carlo Simulations consisting of 500,000 realizations of each parameter in the sigmoidal function were conducted to provide realistic estimates of 500,000 realizations |E*| mastercurves. This is used for estimating the uncertainty in |E*| mastercurves as a function of reduced frequency. The uncertainty in |E*| is presented by the coefficient of variation (COV), ration of standard deviation to mean, as shown in Figure 2.

3 MECHANISTIC EMPIRICAL DESIGN RUNS

The plan of this study, as shown in Table 2, entails design runs using ME Design software aiming at the evaluation of the effect of the inherent variability in |E*| on the predicted rut depth and bottom-up fatigue cracking under different climatic and traffic conditions.

The runs were subsequently executed where the fatigue cracking COV and rutting COV were calculated at 2.5, 5, 7.5 and 10 years to monitor the evolution of the results at different stages of the design life of the pavement.

The plan incorporates cold and hot climates with three different traffic speeds because these factors play significant role in defining the regions of the |E*| mastercurves that are required for the prediction of rutting and fatigue cracking. For example, a low reduced frequency is required for the case of hot climate and a slow speed which implied the incorporate of |E*| with higher variability as compared to the case of cold weather and fast speed. For analysis purposes, the effective temperature and effective frequency are calculated for each scenario. The effective temperature presents a unique pavement temperature that is most critical within a pavement structure for a particular distress (Jeong 2014). For each scenario, the effective frequencies and temperatures for both rutting and fatigue cracking are calculated using the average |E*| mastercurve of each mix (El-Basyouny & Jeong 2013). For this purpose, the major climatic parameters used are collected from climatic databases as the following:

- **Chicago:**
- Mean Annual Average Temperature (MAAT) = 50.8°F
- Standard Deviation of Monthly Mean Air Temperature (σ_{MMAT}) = 11.75°F
- Mean Annual Wind Speed = 10.3 mph
- Mean Annual Sunshine = 54%
- Mean Cumulative Rainfall Depth = 27.5 inch
- **Phoenix:**
- MAAT = 75.3°F
- σ_{MMAT} = 15.21°F
- Mean Annual Wind Speed = 6.3 mph
- Mean Annual Sunshine = 85%
- Mean Cumulative Rainfall Depth = 6.3 inch

4 RESULTS

The results of the ME Design runs are used to assess the variability in the predicted rut depth and fatigue cracking percentage in the case of the pavement structure

Table 1. Description of mixes used in this study.

Designation of Mixes	Mix A Coarse graded 9.5mm HMA	Mix B 12.5mm SMA	Mix C Fine graded 25.0mm HMA	Mix D 19mm HMA	Mix E 25mm mix HMA
Asphalt binder	PG 64-22	PG 76-22	PG 64-22	PG 76-16	PG 76-16
Asphalt content, %	5.5	6.5	4.7	4.0	3.8
E* testing conditions	Temp.: 4, 20, & 40°C Freq.: 10, 1, 0.1, and 0.01 Hz			Temp.: 4, 20, & 40°C Freq.: 20, 10, 1, 0.1, & 0.01 Hz	
Number of E* replicates	24	24	24	38	11
Source of Data	Data published in the appendices of NCHRP 9-29 Report 702:			Tested replicates from a paving project	

Table 2. Summary of ME design run scenarios.

Mix Type	Design Life (Years)	Traffic Profile	Traffic Level (AADTT)	Structure	Climate	AC Layer Thickness (in) Speed = 15 mph	Speed = 40 mph	Speed = 60 mph	Number of ME Design Runs
Mix A				AC Layer	Chicago	7.37	6.99	6.83	
					Phoenix	8.16	7.79	7.62	
Mix B				15" Base Layer - Crushed Stone 30000 psi	Chicago	5.99	5.81	5.75	6 combinations of speed and
		Example level 3, Default			Phoenix	6.59	6.32	6.21	climate per mix
Mix C	10		10,000		Chicago	7.45	7.05	6.89	* 5 mixes * 250
					Phoenix	8.27	7.92	7.75	|E*| Realization =
Mix D				Subgrade - A-4 20000 psi	Chicago	4.33	4.22	4.18	7,500 runs
					Phoenix	4.84	4.65	4.55	
Mix E					Chicago	4.54	4.43	4.38	
					Phoenix	4.99	4.84	4.76	

analyzed for each case of the six combinations of climatic conditions and traffic speeds.

The COV in fatigue cracking, as presented in Figures 3, shows that almost it varies between 0.1 and 0.38 for the mixes and pavement conditions in study. This variability is solely due to the forward propagation of variability in |E*| showing that such uncertainty can not be disregarded and thus needed to be accounted for assessing the reliability of pavement performance. The mean of fatigue cracking is 15% for all cases because it is the selected criteria for pavement design. Although it is the same, the uncertainty is not the same but affected by the mixture type. It can be observed that mixes C and E possess the highest fatigue cracking uncertainty which is a reflection of the highest uncertainty in the |E*| mastercurves of these two mixtures. For the same mix, there is an effect for the traffic speed and climatic condition on the uncertainty in fatigue cracking but there is not a clear trend how these factors affect the COV of fatigue cracking. Therefore, to simplify the effect of climatic condition and traffic speed the analysis is conducting by checking the relationship between the obtained COV of the predicted fatigue cracking the COV of effective |E*| for fatigue cracking obtained for each combination of asphalt concrete sample/mix and pavement conditions (basically climate and speed) as shown in Figure 4.

The results show that the variability is dependent on the analysis case represented by effective E* calculated for each of rutting and fatigue cracking. Figure 4 shows a strong positive correlation between the analyzed COVs where it is noticed that as the COV of

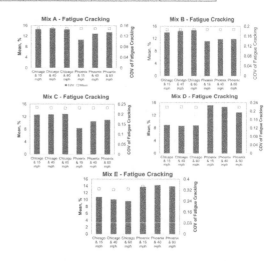

Figure 3. Fatigue cracking results (Mean and COV) for all mixes for each of the six scenarios.

effective |E*| increases, the COV of predicted fatigue cracking increases with the same trend. This factor combined the effect of the mixture type, climatic condition, and speed and directly showing its effect in performance prediction.

Similar to the case of fatigue cracking, the COV in the predicted rut depth is shown in plots of Figure 5. In general, the rut depth shows a COV in the range of 0.08 to 0.25. The COV in the rut depth is the higher for hotter climates (Phonix versus Chicago), higher for

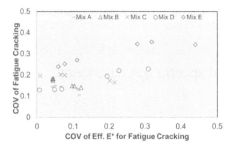

Figure 4. COV of Fatigue Cracking as a function of COV of effective $|E*|$ for fatigue cracking.

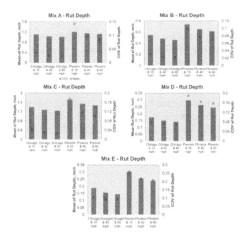

Figure 5. Rut depth prediction results (Mean and COV) for all mixes for each of the six scenarios.

slower speeds, and higher for mixes with high COV of $|E*|$ mastercurve. The effect of hot climate and slow speed is significant in the case rut depth because the uncertainty in $|E*|$ increases as the reduced frequency decreases (increase in temperature and/or slow speed) as shown in Figure 2.

The results of the COV of rut depth as a function of the COV of the corresponding effective $|E*|$ of rutting are shown in Figure 6. A positive correlation is shown where the COV of rutting increases as that of effective $|E*|$ increases. It is shown that as the COV in $|E*|$ increase between 0.05 and 0.65, the COV in rut depth increases from 0.05 up to 0.3. Such results show that the uncertainty in rut depth is important to be taken into consideration because it possesses significant values that affects the probability of pavement unsatisfactory performance. This will be addressed in future research as a further development of the work presented in this paper.

5 CONCLUSIONS

Based on the results of this study, the following conclusions can be drawn:

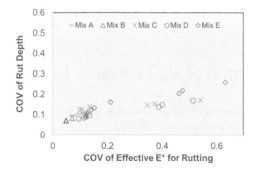

Figure 6. COV of Rut Depth as a function of COV of effective $|E*|$ for rutting.

- Forward propagation of uncertainties in material properties, especially that of $|E*|$ is essential for quantifying the uncertainties in predicted pavement performance which should be incorporated in quality assurance practices in the paving industry.
- The uncertainty in pavement performance is significant for both rutting and fatigue but of a higher magnitude in the case of fatigue especially in hot climatic conditions, slow traffic, and mixes with higher nominal maximum aggregate size.

REFERENCES

Killingsworth, B. M. (2004). NCHRP Research Results Digest 291: Quality Characteristics for Use with Performance Related Specifications for Hot Mix Asphalt. Transportation Research Board of the National Academies, Washington, DC.

Witczak, M. W. (2007). Specification criteria for simple performance tests for rutting (Vol. 1). Transportation Research Board.

Kahil, N. S., Najjar, S. S., & Chehab, G. (2015). Probabilistic Modeling of Dynamic Modulus Master Curves for Hot-Mix Asphalt Mixtures. In Transportation Research Board 94th Annual Meeting (No. 15-4392).

Kassem, H. A., Najjar, S. S., & Chehab, G. R. (2016), Probabilistic Modeling of the Inherent Variability in the Dynamic Modulus Master Curve of Asohalt Concrete. Accepted for publication in: Transportation Research Record: Journal of the Transportation Research Board.

Kassem, H.A., 2018. Probabilistic characterization of the viscoelastoplastic behavior of asphalt-aggregate mixtures (Doctoral dissertation).

Jeong, M. G., McCarthy, L. M., & Mensching, D. J. (2014). Stochastic Estimation of the In-Place Dynamic Modulus for Asphalt Concrete Pavements. Journal of Materials in Civil Engineering, 27(6), 04014181.

El-Basyouny, M., & Jeong, M. (2010). Probabilistic performance-related specifications methodology based on mechanistic-empirical pavement design guide. Transportation Research Record: Journal of the Transportation Research Board, (2151), 93–102.

Advances in Materials and Pavement Performance Prediction II – Kumar et al. (eds)
© 2021 Taylor & Francis Group, London, ISBN 978-0-367-46169-0

Impact of WIM systematic bias on axle load spectra – A case study

Muhamad Munum Masud & Syed Waqar Haider
Michigan State University, USA

Olga Selezneva & Dean J. Wolf
Applied Research Associates, Inc., USA

ABSTRACT: Weigh-in-motion (WIM) is a primary technology used for monitoring and collecting vehicle weights and axle loads on roadways. One way to evaluate the quality of WIM measurements is to analyze axle load spectra over time. Class 9 single-axle (SA) normalized axle load spectra (NALS) can be modeled as a single normal or log-normal distribution with a mean value corresponding to the NALS's peak load frequency value (bell-shaped distribution). The changes in the location of the peak of this distribution can be related to the changes in mean error. Similarly, tandem axle (TA) NALS could be modeled by using a mixture of two normal or lognormal distributions (i.e., the bi-modal distribution). This paper presents a case study for WIM site where the WIM sensor exhibited a drift within a year after calibration. The loading data from single and tandem axles were analyzed to quantify the decline in WIM accuracy and precision. The results presented in the paper can be used as a guideline to select calibration frequency for the WIM sites.

1 INTRODUCTION

Weigh-in-motion (WIM) is a primary technology used for monitoring and collecting vehicle weights and axle loads on roadways. State and other highway agencies collect WIM data for many reasons, including highway planning, pavement and bridge design, freight movement studies, motor vehicle enforcement screening, and vehicle size and weight regulatory studies. Therefore, with so many potential uses, the data collected must be accurate and consistent.

One way to evaluate WIM measurement errors is by using the data collected immediately before and after equipment calibration. The limitation of this approach is that the data represent a snapshot in time and may not be representative of a long-term WIM site performance. Moreover, performing more frequent calibrations need more resources and budget and is not the best approach to study the quality and consistency of WIM data. Consequently, an alternative method is required to characterize temporal changes in WIM data consistency.

This paper investigates other ways of inferring WIM data accuracy and consistency over time. One approach is to relate errors in WIM data to the attributes of the NALS for Class 9 vehicles. There are several advantages in using axle weight data for Class 9 trucks. Class 9 is a recommended class for WIM calibration and validation per ASTM E1318-09. Class 9 typically is the only vehicle class that has

supporting data for computation of WIM precision and bias statistics because ASTM E1318 specifies this truck as a recommended calibration/validation test truck. Class 9 has a stable and well understood gross vehicle weight (GVW) and axle weight distribution that helps in identifying and analyzing changes in WIM data over time. For most roads, Class 9 is the most frequently observed heavy commercial vehicle type for the long-haul. The exceptions are load-restricted roads or secondary roads that have a large percentage of small, lightweight service trucks. Typically, these are recreational, urban, or suburban roads with stop-and-go traffic not conducive to WIM measurements, and thus do not represent recommended WIM site locations.

This paper presents a case study of a WIM site located in Nevada, where the pre- and post-calibration data, and axle load spectra data were well documented. This WIM site is part of the Long- Term Pavement Performance (LTPP) Specific Pavement Studies (SPS) Experiment-10 and equipped with Quartz Piezo (QP) sensors (FHWA 2018; Selezneva and Wolf 2017). The main objective of this paper is to study the shift pattern in single and tandem axle load spectra [i.e., abrupt, seasonal, or systematic changes in bias and standard deviation (SD)] following the calibration event. The single and tandem axle loadings between two consecutive calibration events were analyzed at length to quantify the deterioration in WIM system performance.

2 WIM MEASUREMENT ERROR DESCRIPTION

The target analogy in Figure 1 is a practical way of understanding how accuracy can be quantified by measurement bias. Consistency or precision is related to the repeatability of a process. The variability of repeat measurements can characterize precision under carefully controlled conditions. Figure 1 also illustrates that it is possible to be consistent (or precise, as applied to target shooting) without being accurate or accurate without being consistent (low precision). Ideally, we would like a measurement process to be both accurate and consistent.

3 METHODOLOGY

3.1 *Pre- and post-calibration accuracy data*

Table 1 presents the accuracy data for the Nevada WIM site, extracted from the previous validation and a day before the next calibration. The table provides the mean error and SD for GVW, steering and single axles and tandems for pre- and post-validations.

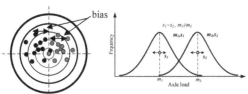

(a) Negative bias for the black data set and positive bias for the red data set, similar precision for both data sets

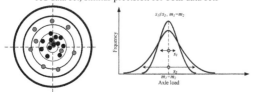

(b) Good (black) and poor (red) precision with no bias

Figure 1. Target analogy for understanding precision and bias.

Table 1. Weight validation history.

| Date | Mean Error and 2SD | | |
	[1]GVW	SA	TA
28-Nov-18 (Post-calibration)	−0.5 ± 4.0	0.0 ± 5.8	−0.7 ± 6.0
13-Aug-19 (Pre-calibration)	13.7 ± 3.2	9.8 ± 4.8	14.6 ± 4.8

[1]ASTM Type-1 accuracy limits ±10%, ±15%, ±20% for GVW, TA, and SA, respectively.

From a comparison between the reported most recent validation of this equipment on November 27, 2018, and the current pre-calibration data set, a steady calibration drift and a positive bias were observed for this site.

3.2 *Site and equipment related factors*

This site was installed on October 26, 2018, by International Road Dynamics (IRD). Before the pre-validation test truck runs, a physical inspection of all WIM and support services equipment was conducted, and no deficiencies were reported. During the on-site pavement evaluation, there were no pavement distresses observed that may affect the accuracies of the WIM system. While some rutting of the pavement was observed, the sensors appeared to be flush with the pavement surface. Visual observation of the trucks as they approach, traverse, and leave the sensor area did not indicate any adverse dynamics that would affect the accuracy of the WIM system. The trucks appear to track down the center of the lane (Associates 2019).

3.3 *Monthly NALS for single and tandem axles*

Monthly NALS developed at the one-month interval after the calibration event are useful for investigating changes in WIM data characteristics between calibration events. The analyses of NALS over time were conducted separately for single and tandem axles of class 9 trucks. Figure 2 shows the single and tandem

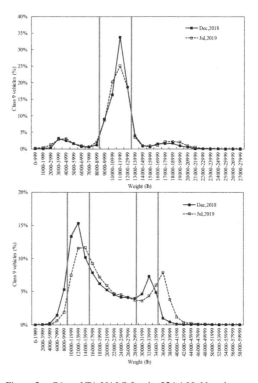

Figure 2. SA and TA NALS for site 32AA00, Nevada.

axle NALS for this site, respectively. The red lines in the figures illustrate the typical ranges for peak loads. It is evident from the figures that the peak load (i.e., mean) and the spread of distribution (SD) were changing over time for single and tandem axle load spectra.

3.4 Axle load spectra attributes

The mean and variance of a single axle NALS were determined by using a discrete distribution by using Eq. (1) and Eq. (2).

$$\mu_x = \sum_x x \times P(X = x) \tag{1}$$

$$\sigma_X^2 = \sum_x (x - \mu_x)^2 P(X = x) \tag{2}$$

Where x is the midpoint of load bin and $P(X = x)$ is observed the normalized relative frequency of x.

For the tandem axle, typically, two peak loads are observed in a NALS. A mixture of statistical distributions, to characterize the predominantly bimodal axle load spectra was considered (Haider and Harichandran 2007; Haider and Harichandran 2009). It was shown that two or more normal probability density functions (PDFs) could be added with appropriate weight factors to obtain the PDF of the combined distribution as shown by Eq. (3):

$$f^* = \sum_i^n p_i f_i \tag{3}$$

where $f^* = $ PDF of combined distribution, $p_i = $ proportions (weight factors) for each normal PDF, and $f_i = $ PDFs for each normal distribution.

For a bimodal mixed normal distribution containing two normal PDFs, the two-weight factors are complementary (i.e., $p_2 = 1 - p_1$), as shown in Figure 3. The bimodal shape of axle spectra can be adequately

captured by using a combination of two normal distributions:

$$f(x; \mu_1, \sigma_2, \mu_2, \sigma_2, p_1) =$$
$$\left(p_1 \frac{1}{\sigma_1 \sqrt{2\pi}} e^{-(x-\mu_1)^2/2\sigma_1^2} + p_2 \frac{1}{\sigma_2 \sqrt{2\pi}} e^{-(x-\mu_2)^2/2\sigma_2^2} \right) \tag{4}$$

Where $\mu_1 = $ the average of empty or partially loaded axle loads, $\sigma_1 = $ the standard deviation of empty or partially loaded axle loads, $\mu_2 = $ the average of fully loaded axle loads, and $\sigma_2 = $ the standard deviation of fully loaded axle loads.

The following statistical parameters were used to analyze differences in SA and TA NALS over time:

- The absolute differences in Peak Load (PL) values were computed based on the data to analyze potential calibration drift or measurement bias overtime for the first 30 days after calibration, and the data collected at the one-month interval (i) after calibration event: $\Delta PL = PL_i - PL_{30}$
- The second Peak Load (PL2) values were computed based on the data for the first 30 days after calibration as a reference to analyze potential calibration drift or measurement bias over time, and data collected one-month interval (i) after calibration: $\Delta PL2 = PL_i - PL_{30}$

Additionally, the differences in tandem axle unloaded peak (peak 1), mean (peak load) for the bins greater than 26,000 lbs., overall mean, and SD was also analyzed.

3.5 Significant differences criteria for NALS consistency

(a) If $>=5\%$ for single NALS (or $>=500$ lb.), then there is a practical difference (measurement bias $>5\%$) between the peak loads for the reference month and ith month.
(b) If $>=5\%$ for tandem NALS second peak or Mean2 (m_2) (or $>=1,500$ lb), then there is a practical difference (measurement bias $>5\%$) between the peak loads for the reference month and ith month.

4 DATA ANALYSIS AD RESULTS

4.1 Single axle load spectra for class 9 trucks

There are minor changes in single axle mean over time, but the trends are not consistent. This is expected since the single axle (or front axle) mainly representing the load of the truck engine and does not vary considerably. Most of these data are within 5% bias or 500 lb load threshold.

4.2 Tandem axle load spectra for class 9 trucks

Figures 4(a) to 4(f) present the actual and relative difference in TA ALS attributes. It is clear from the

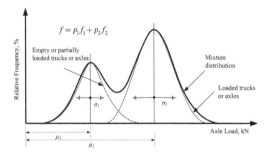

Figure 3. Tandem axle load spectra modeling using bimodal mixed normal distributions.

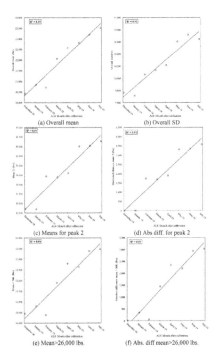

(a) Overall mean   (b) Overall SD

(c) Means for peak 2   (d) Abs diff. for peak 2

(e) Mean>26,000 lbs.   (f) Abs. diff mean>26,000 lbs.

Figure 4. The difference in TA load spectra over time.

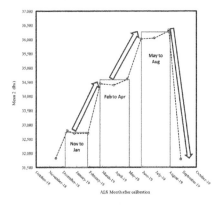

Figure 5. The difference in TA load spectra over time.

results that there is a systematic increase (overestimation of loads) in different TA ALS attributes.

Figure 5 shows the post-calibration changes in the loaded peak over time. It can be seen from the results that there is a systematic and seasonal pattern in load changes from 1st Nov to 31st Jan, 1st Feb to 30th Apr, and from 1st May to 31st Aug. Finally, the re-calibration has eliminated the bias and the peak loads are back to normal range (reduction from 36,324 lbs. to 31,789 lbs.).

5 DISCUSSION

The data analysis showed that this site went out calibration within three to four months. Most of the

records for loaded peak failed the significantly different criteria (Bias >5 % or load >=1500 lbs.) after February 2019. The WIM site consistently overestimated loads for the tandem axles, and the same finding was also confirmed from the positive bias determined in pre-calibration data collected a day before the next calibration event (see Table 1).

Previous research has shown that a 5% negative bias will translate to a 3 to 5-year underestimation in the predicted life (based on cracking) for flexible pavements. While reliability tolerances in performance prediction models can compensate for some of this discrepancy in predicted life, the underestimation in predicted life for flexible pavements is more critical (Haider et al. 2011).

6 CONCLUSIONS

Based on the results of these data analyses, the following are the conclusions:

- Axle loading data available in LTPP is a reliable source to study the effectiveness of calibration and the consistency in WIM data.
- The QP sensor at this site should be calibrated every four to six months to get more accurate data.
- The WIM site may be studied further to evaluate other factors related to pavement conditions and surface roughness.

ACKNOWLEDGMENT

The authors would like to acknowledge the National Cooperative Highway Research Program (NCHRP) for funding the study.

REFERENCES

Associates, A. R. (2019). "WIM System Field Calibration and Validation Summary Report-Nevada SPS-10."
FHWA (2018). "WIM Pocket Guide." Federal Highway Administration, Washington DC.
Haider, S. W., and Harichandran, R. S. (2007). "Quantifying the Effects of Truck Weights on Axle Load Spectra of Single and Tandem Axle Configurations." *the Fifth International Conference on Maintenance and Rehabilitation of Pavements and Technological Control*, 73–78.
Haider, S. W., and Harichandran, R. S. (2009). "Effect of Axle Load Spectra Characteristics on Rigid and Flexible Pavement Performance." *ASCE Journal of Transportation Engineering*, In Press.
Haider, S. W., Harichandran, R. S., and Dwaikat, M. B. (2011). "Impact of Systematic Axle Load Measurement Error on Pavement Design Using Mechanistic-Empirical Pavement Design Guide." *Journal of Transportation Engineering*, 138(3), 381–386.
Selezneva, O., and Wolf, D. (2017). "Successful Practices in Weigh-in-Motion Data Quality with WIM Guidebook Vol 1."

Advances in Materials and Pavement Performance Prediction II – Kumar et al. (eds)
© 2021 Taylor & Francis Group, London, ISBN 978-0-367-46169-0

Sensitivity analysis of AASHTOWare PMED to climatic inputs and depth of water table

K.J. Msechu
University of Tennessee at Chattanooga, Tennessee, USA

M. Onyango
UC Foundation Associate Professor, University of Tennessee at Chattanooga, Tennessee, USA

W. Wu
Associate Professor, University of Tennessee at Chattanooga, Tennessee, USA

S. Udeh
TDOT Pavement Design Manager, Tennessee Department of Transportation, USA

ABSTRACT: Sensitivity analysis is conducted on AASHTOWare Pavement Mechanistic Empirical Design (PMED) by using a Design of Experiment method that considers 2^k factorial design, an unbiased method that analyses the effect of climatic inputs and depth of water table to flexible pavement distress predictions. This study used three LTPP sites located in Tennessee; these sites are of Functional Classes 1, 2 and 7. From the analysis of climatic inputs and depth of water table, temperature was observed to be the most sensitive climatic input followed by Wind speed and the depth of water table. Percent sunshine has a less significant influence on the distress predictions while relative humidity was observed to have a negligible effect to the flexible pavement distress predictions.

1 INTRODUCTION

Pavement design methods developed over the years consider the performance and behavior of pavements under different conditions. Most of the existing pavement design methods are highly based on Empirical approaches. Mechanistic-Empirical Pavement Design approach is more preferred because of its inclusion of different factors that affect the behavior or performance of pavements. The Mechanistic-Empirical pavement design approach considers inputs such as traffic, material properties, climatic condition and the level of water table.

In the efforts of transitioning from the empirical AASHTO 1993 Guide for Pavement Design, AASHTO developed a mechanistic empirical pavement design guide and software namely AASHTOWare Pavement Mechanistic-Empirical Design (AASHTOWare PMED) which in cooperates different models for pavement design and distress predictions.

Data used in PMED are classified according to their sources. Level 1 data are those obtained from actual measurement and are regarded as of the highest quality. Level 2 data are those that are products of regression from provided data and also termed as Regional data; while Level 3 data are those obtained as a representation of a general occurrence of a measurement, they are usually term as National data and are found as the default data in the PMED software (AASHTO 2008).

The long-term performance of pavement is highly effected by traffic, pavement materials and climatic conditions. This paper analyses the effects of climatic inputs and water table depth on flexible pavements. PMED uses the Enhanced Integrated climatic model (EICM) to bridge the gap between climatic conditions and pavement performance. The EICM requires hourly climatic data that can be obtained from weather stations (Khazanovich et al. 2013; Oh et al. 2006; Zaghloul et al. 2006).

Several sensitivity analyses performed using PMED establish relationships between the climatic inputs and the pavement distresses predictions. Various researches on climatic conditions have reported temperature to be the most sensitive climatic input in the prediction of flexible pavement distresses where relative humidity and precipitation were the least sensitive and almost had negligible effects (Li et al. 2013; Schwartz 2015; Yang et al. 2017). Wind speed has been observed to have more effect than percent sunshine on performance predictions (Ahmed et al. 2005).

Rutting and IRI increases with increase in temperature and percent sunshine. Wind speed increases thermal cracking, AC rutting, total rutting and IRI whereas percent sunshine increases thermal cracking,

Table 1. Input levels for design factors.

	High (+)	Low (−)
Temperature (^oF)	110	32
Wind Speed (miles/hour)	60	0
Sunshine (%)	100	0
Relative humidity (%)	100	0
Water table (ft.)	100	0

top-down cracking and bottom-up cracking (Yang et al. 2017). The most sensitive inputs are temperature, wind speed and percent sunshine (Li et al. 2013; Yang et al. 2017).

This paper is focused on the use of a Design of Experiment (DOE) procedures to show the behavior of flexible pavements under different climatic conditions and at different water table depths in the State of Tennessee using available data and sections from Long Term Pavement Performance (LTPP) website (Infopave).

2 METHODOLOGY

This study uses the design of experiment method (DOE) on AASHTOWare PMED Version 2.5.5 to observe the behavior of the system to changes in climatic factors from their maximum and minimum values. Statistical analysis is done by using a full model 2^k factorial design, which favors unbiased and equal testing of each factor affecting the system (Douglasc 2009).

Table 1 shows the high (+) and low (−) values that are used for each climatic and water table depth inputs. In this study precipitation inputs are not considered due to the insensitivity of the PMED to precipitation inputs at low and high values (Schwartz 2015; Yang et al. 2017).

Using five factors for sensitivity analysis with a 2^k factorial design, the 2^k factorial design (where k is the number of factors) yields a total of $2^5 = 32$ runs for analysis.

Hourly Climatic Data (HCD) are created for each of the 32 runs whereby the data generated contains 24-hour climatic data over a period of 36.5 years (greater than the recommended 20-year design period). The 32 HCD files created serve as climatic stations each with a unique combination of climatic factors at different levels, example station 1 contains low values for all climatic inputs in its HCD (24 hours over 36.5 years) file. Table 2 shows the combination of levels for all the 32 Climatic Stations inputs and water table depth as used in the analysis.

From the tables:

A: Temperature.
B: Wind Speed.
C: Percent Sunshine.
D: Relative Humidity.
E: Water Table Depth

Table 2. Climatic input level combinations for HCD stations.

STATIONS	A	B	C	D	E
1	−	−	−	−	−
2	+	−	−	−	−
3	−	+	−	−	−
4	+	+	−	−	−
5	−	−	+	−	−
6	+	−	+	−	−
7	−	+	+	−	−
8	+	+	+	−	−
9	−	−	−	+	−
10	+	−	−	+	−
11	−	+	−	+	−
12	+	+	−	+	−
13	−	−	+	+	−
14	+	−	+	+	−
15	−	+	+	+	−
16	+	+	+	+	−
17	−	−	−	−	+
18	+	−	−	−	+
19	−	+	−	−	+
20	+	+	−	−	+
21	−	−	+	−	+
22	+	−	+	−	+
23	−	+	+	−	+
24	+	+	+	−	+
25	−	−	−	+	+
26	+	−	−	+	+
27	−	+	−	+	+
28	+	+	−	+	+
29	−	−	+	+	+
30	+	−	+	+	+
31	−	+	+	+	+
32	+	+	+	+	+

3 CASE STUDY

This study used, Level 1 materials properties and traffic inputs from three (3) LTPP sites located in Tennessee (Region 1). The LTPP sites included are; 47–3108 (Functional Class 1 (FC1) Rural Principal Arterial – Interstate I-75, with an AADTT of 4078 in year 2014), 47–1028 (Function Class 2 (FC2) Rural Principal Arterial – Other, with an AADTT of 383 in year 1999) and 47–3104 (Functional Class 7 (FC7) Rural Major Collector, with an AADTT of 6 in year 1986). All the distress predictions of the 3 LTPP sites used a 20-year design period and 60, 45, and 40 mph design speeds for FC1, FC2 and FC7 respectively, whereas the vehicle class distribution, growth rates, growth function, axle distribution, axle per truck and Monthly adjacent factors where as those reported by LTPP (Level 1 inputs).

Level 1 materials properties used from LTPP sites data included effective binder content, air voids, unit weight and binder type for asphalt layers, resilient modulus, liquid limit, plastic index, maximum dry density, optimum moisture content and specific gravity for non-stabilized granular base and subgrade layers.

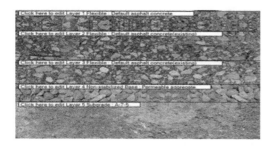

Figure 1. Flexible pavement structure LTPP Site 47–3108.

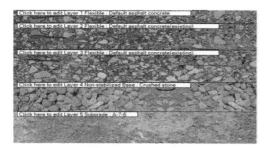

Figure 2. Flexible pavement structure LTPP Site 47–1028.

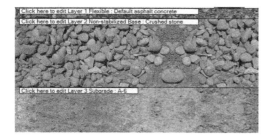

Figure 3. Flexible pavement structure LTPP Site 47–3104.

Figure 1 through 3 show the flexible pavement structures for the three (3) LTPP sites. Figure 1 shows the flexible pavement structure of the rehabilitated LTPP site 47–3108 with 2.7 in. and 5.5 in. asphalt concrete layers, 6.7 in. asphalt concrete treated base layer, 6.1 in. crushed stone subbase and A-7-6 subgrade.

Figure 2 shows a rehabilitated flexible pavement structure of LTPP site 47–1028 with 4.3 in. and 6.2 in. asphalt concrete layers, 5.1 in. asphalt concrete treated base, 3.8 in. crushed stone subbase and A-7-5 subgrade.

And Figure 3 shows the flexible pavement structure of LTPP site 47–3104 with 1.3 in. asphalt concrete layer, 8.7 in. crushed stone base later and a A-6 subgrade.

In this study AASHTOWare PMED version 2.5.5; AC Cracking – Bottom Up, AC Cracking – Top Down and rutting coefficients where locally calibrated. All sites had similar calibrations for cracking coefficients

but differed in rutting coefficients. Site 47–3104 and 47–1028 had similar rutting coefficient for that corresponded to those of Tennessee Region 1 State routes while site 47–3108 rutting coefficients are those for Tennessee Interstate highway (Gong et al. 2017).

4 RESULTS

ANOVA was used to assess the relationships between predicted flexible pavement distresses to climatic inputs and water table depth by analyzing the calculated statistical values (P value and F Value or the sum of squares). The following are the results obtained from the analysis of each of the 3 LTPP sites to changes in climatic and water table depth inputs.

4.1 LTPP SITE 47–3108

4.1.1 Permanent deformation – AC only
The results showed that permanent deformation of Asphalt Concrete surface predictions are sensitive to temperature (A), wind speed (B) and the interaction effect of temperature and wind speed (AB). Temperature appears to be the most sensitive climatic input.

4.1.2 Permanent deformation – Total pavement
From the analysis temperature (A) and wind speed (B) appear to be the most sensitive climatic inputs affecting the permanent deformation of the total pavement predictions.

4.1.3 Top down cracking
Top Down Cracking predictions are observed to be sensitive to temperature (A), the interaction effect of temperature and water table depth (AE), water table depth (E), the interaction effect of wind speed and water table depth (BE), the interaction effect of temperature, wind speed and water table depth (ABE), the interaction effect of temperature and wind speed (AB) and wind speed (B) in the listed order.

4.1.4 Bottom up cracking
Bottom up cracking predictions are relatively sensitive to wind speed (B), temperature (A), interaction effect of temperature and wind speed (AB), water table depth (E), the interaction effect wind speed and water table depth (BE), the interaction effect of temperature and water table depth (AE) and the interaction effects of temperature, wind speed and water table depth (ABE).

4.1.5 Total Fatigue Cracking (Bottom Up + Reflective Cracking)
Temperature (A) was the most sensitive climatic input on the total fatigue cracking predictions. Other inputs that with little effects are the interaction effect of temperature and water table depth (AE), water table depth (E), wind speed (B) and interaction effect of temperature and wind speed (AB) in that order.

4.1.6 Terminal IRI

Temperature (A) was the most sensitive inputs for terminal IRI predictions, where other inputs have less effects, Wind speed (B), the interaction effect of temperature and water table depth (AE), water table depth (E), and interaction effect of temperature and wind speed (AB) in that order.

4.2 LTPP SITE 47–1028

4.2.1 Total transverse cracking

Total transverse cracking predictions are observed to be sensitive to temperature (A), wind speed (B) and the interaction of temperature and wind speed (AB).

4.2.2 Permanent deformation – AC only

The results showed that permanent deformation of Asphalt Concrete surface predictions are sensitive to temperature (A), wind speed (B) and the interaction effect of temperature and wind speed (AB) in that order. Temperature appears to be the most sensitive climatic input. Percent sunshine (C) and the interaction effect of wind speed and percent sunshine (BC) show a little effect to Asphalt Concrete permanent deformation predictions.

4.2.3 Permanent deformation – total pavement

From the analysis temperature (A) and wind speed (B) appear to be the most sensitive climatic inputs affecting the permanent deformation of the total pavement predictions. Water table depth (E) and interaction effect of temperature and water table depth (AE) show little effect to total pavement permanent deformation predictions.

4.2.4 Top down cracking

Top Down Cracking predictions are observed to be sensitive to temperature (A), wind speed (B) and the interaction effect of temperature and wind speed (AB). Percent sunshine (C) and its interactions to wind speed (BC) and temperature (AC) have small effects to top down cracking predictions.

4.2.5 Bottom up cracking

Bottom up cracking predictions are observed to be sensitive to temperature (A), wind speed (B) and the interaction effect of temperature and wind speed (AB). The interactions effects of temperature and percent sunshine (AC), temperature and water table depth (AE) have a small effect on the predictions along with the water table depth (E).

4.2.6 Total fatigue cracking (bottom up + reflective cracking)

Temperature (A) was the most sensitive climatic input on the total fatigue cracking predictions followed by wind speed (B) and the interaction effect of temperature and wind speed (AB). Other inputs that have little effects to the predictions include; the interaction effect of temperature and water table depth (AE) and water table depth (E).

4.2.7 Terminal IRI

Temperature (A) was the most sensitive inputs for terminal IRI predictions followed by wind speed (B). Other inputs that have less effects to the predictions include; the interaction effect of temperature and water table depth (AE), water table depth (E) and percent sunshine (C).

4.3 LTPP SITE 47–3104

4.3.1 Thermal cracking

From the analysis thermal cracking predictions are observed to only to sensitive to temperature (A) inputs.

4.3.2 Permanent deformation – AC only

The results showed that permanent deformation of Asphalt Concrete surface predictions are sensitive to temperature (A) and wind speed (B). Other small effects to the predictions are observed with water table depth (E), the interaction effect of temperature and depth of water table depth (AE), the interaction effect of temperature and wind speed (AB), percent sunshine (C) and the interaction effect of wind speed and percent sunshine (BC) in that order.

4.3.3 Permanent deformation – total pavement

From the analysis water table depth (E) and the interaction effect of temperature and water table depth (AE) were the most sensitive inputs to the predictions of total pavement permanent deformation. Wind speed (B), temperature (A) and interaction effect of temperature and wind speed (AE) show little effect to total pavement permanent deformation predictions.

4.3.4 Top down cracking

Top Down Cracking predictions are observed to be sensitive to water table depth (E) and the interaction effect of temperature and water table depth (AE), temperature (A), the interaction effect of temperature and wind speed (AB), and wind speed (B) in that order.

4.3.5 Bottom up cracking

Bottom up cracking predictions are relatively sensitive to temperature (A), wind speed (B), percent sunshine (C) and all there interactions (AB, AC, BC and ABC).

4.3.6 Terminal IRI

Terminal IRI predictions are observed to be highly sensitive to water table depth (E) and the interaction effect of temperature and water table depth (AE). Other inputs that effects the terminal IRI predictions are; wind speed (B), the interaction effect of temperature and wind speed (AB), temperature (A) and percent sunshine (C).

5 CONCLUSION

The LTPP site 47–3108 with an AADTT of 4078 and structure as shown on Figure 1, show that distress predictions are highly affected by temperature whereby temperature (A) was the most sensitive climatic input for almost all predictions except bottom up cracking where it was approximately close to the most sensitive inputs. Wind speed (B) was the most sensitive climatic input to bottom up cracking predictions and highly affecting both permanent deformation of asphalt concrete surface and total pavement. The interaction of temperature and wind speed (AB) mostly affected permanent deformation of Asphalt concrete surface, bottom up cracking and had little effects on the predictions of top down cracking, total fatigue cracking and terminal IRI. Water table depth (E) and its interaction to temperature (AE) affected top down cracking, total fatigue cracking and terminal IRI, water table depth (E) and its interaction to wind speed (BE) effects bottom down cracking while the interaction of temperature, wind speed and water table depth affected bottom up cracking and top down cracking predictions.

The LTPP site 47–1028 with an AADTT of 383 and structure as shown in Figure 2 shows that temperature (A) is the most sensitive climatic input followed by wind speed (B) for all distress predictions. The interaction effect of temperature and wind speed (AB) highly affected all distress predictions following temperature and wind but did not affect total transverse cracking, total pavement permanent deformation and terminal IRI predictions. Water table depth (E) and its interaction to temperature (AE) affected; total pavement permanent deformation, bottom up cracking, total fatigue cracking and terminal IRI predictions. Percent sunshine (C) affected asphalt concrete permanent deformation, top down cracking and terminal IRI. The interaction of percent sunshine with wind speed (BC) affected Asphalt concrete permanent deformation and top down cracking while its interaction with temperature (AC) affected top down cracking and bottom up cracking.

The LTPP site 47–3104 with an ADTT of 6 and structure as shown of Figure 3, shows temperature (A) the most sensitive input to thermal cracking, Asphalt concrete surface permanent deformation and bottom up cracking. Water table depth (E) was the most sensitive input for total pavement permanent deformation and terminal IRI predictions. The interaction between temperature and depth of water table (AE) affected both permanent deformations, top down cracking and terminal IRI. Wind speed (B) affected both permanent deformations, terminal IRI, top down and bottom up cracking, its interactions with temperature (AB) affected asphalt concrete permanent deformation, top down and bottom up cracking and terminal IRI while its interaction with water table depth (AE) affected both permanent deformations, top down cracking and terminal IRI. Percent sunshine (C) affected asphalt concrete permanent deformation, bottom up cracking and terminal IRI while its interaction with wind speed (BC) affected asphalt concrete permanent deformation and bottom up cracking and its interaction with temperature (AC) affected bottom up cracking.

From the three (3) LTPP sites with different flexible pavement structures and traffic volumes given above, it can be concluded that temperature is the most sensitive climatic input, followed by wind speed and the depth of water table along with their interactions. Percent Sunshine is observed to have some influence on the distress predictions but not at a high significance while relative humidity has negligible effects to the distress predictions.

From the sensitivity analysis, the importance of obtaining quality data for pavement design can be seen. Poor quality inputs especially of those with high sensitivity will lead to over or under design of a road pavement structure which directly affects the cost and/or safety of projects and the constructed roads.

REFERENCES

AASHTO, A. 2008. Mechanistic-empirical pavement design guide: A manual of practice. *AAoSHaT Officials, Editor.*

Ahmed, Z., Marukic, I., Zaghloul, S. & Vitillo, N. 2005. Validation of enhanced integrated climatic model predictions with New Jersey seasonal monitoring data. *Transportation research record,* 1913, 148–161.

Douglasc, M. 2009. Design and analysis of experiments. Douglas C. Montgomery. Wiley, London.

Gong, H., Huang, B., Shu, X. & Udeh, S. 2017. Local calibration of the fatigue cracking models in the mechanistic-empirical pavement design guide for Tennessee. *Road Materials and Pavement Design,* 18, 130–138.

Khazanovich, L., Balbo, J. T., Johanneck, L., Lederle, R., Marasteanu, M., Saxena, P., Tompkins, D., Vancura, M., Watson, M. & Harvey, J. 2013. Design and construction guidelines for thermally insulated concrete pavements.

Li, R., Schwartz, C. & Forman, B. Sensitivity of predicted pavement performance to climate characteristics. 2013 Airfield and Highway Pavement Conf, 2013.

Oh, J., Ryu, D., Fernando, E. G. & Lytton, R. L. 2006. Estimation of expected moisture contents for pavements by environmental and soil characteristics. *Transportation research record,* 1967, 134–147.

Schwartz, C. W. 2015. *Evaluation of LTPP Climatic Data for Use in Mechanistic-Empirical Pavement Design Guide (MEPDG) Calibration and Other Pavement Analysis,* US Department of Transportation, Federal Highway Administration, Research . . . .

Yang, X., You, Z., Hiller, J. & Watkins, D. 2017. Sensitivity of flexible pavement design to Michigan's climatic inputs using pavement ME design. *International Journal of Pavement Engineering,* 18, 622–632.

Zaghloul, S., Ayed, A., Halim, A. A. E., Vitillo, N. & Sauber, R. 2006. Investigations of Environmental and Traffic Impacts on Mechanistic–Empirical Pavement Design Guide Predictions. *Transportation research record,* 1967, 148–159.

Advances in Materials and Pavement Performance Prediction II – Kumar et al. (eds)
© 2021 Taylor & Francis Group, London, ISBN 978-0-367-46169-0

Electrokinetic treatment of clay soil – A baseline case study introduction

Nikiforos G. Pavlatos
Section of Pavement Engineering, Faculty of Civil Engineering and Geosciences, Delft University of Technology, Delft, The Netherlands

Athanasios (Tom) Scarpas/Skarpas
Department of Civil Infrastructure and Environmental Engineering, Khalifa University of Science and Technology, Abu Dhabi, United Arab Emirates

Lambert J.M. Houben
Section of Pavement Engineering, Faculty of Civil Engineering and Geosciences, Delft University of Technology, Delft, The Netherlands

ABSTRACT: Although electrokinetic phenomena were relatively recently discovered, they have come a long way and even dominate certain sectors. However, they have rarely been utilized in the discipline of civil engineering. While the limited civil engineering applications are accompanied by inherent uncertainty and changing field conditions, laboratory applications that mitigate those challenges often suffer from conditions that differ substantially from real life engineering applications, thus making questionable the validity of extrapolating results from the laboratory to on-site projects. The work presented here aims to bridge that gap by conducting a laboratory experiment that reproduces – as closely as possible – the conditions that can be expected in civil engineering projects. To that end an experimental set-up is constructed that allows for electrokinetic treatment of clay soil and even though it facilitates constant monitoring of the desired parameters, it does not deviate substantially from how electrokinetic treatment would look like in an actual project.

1 INTRODUCTION

Electrokinetic treatment of clay soil involves the application of electric current to the clay soil and its subsequent effects. As a result of the electric current application a net pore water movement is expected to develop (often called eletro-osmotic flow or electro-osmosis) and movement of charged particles might also occur (described as electrophoresis). Electrokinetic phenomena in general involve liquid containing ions and charged solid surfaces, and they were first discovered more than 200 years ago (Biscombe 2017; ;autheror 1801; Porret 1816; Reuss 1809). They have come a long way since their discovery; electrokinetic principles are a very common approach in liquid actuation and particle manipulation in microscale analysis systems (Bilitewski et al. 2003; Chang and Yeo 2010; Squires and Quake 2005).

Despite their immense advance in other sectors, electrokinetic phenomena have seldom been used in civil engineering applications. The few cases of engineering applications fall under one of the two main categories; either the adoption of electro-osmotic flow to achieve consolidation via dewatering (e.g. Indraratna et al. 2015), or the usage of electro-osmotic flow to remove contaminants from polluted soil (e.g. Acar and Alshawabkeh 1993; Acar and Alshawabkeh 1996). The field applications change the conditions of the system

with time (e.g. initially saturated clay soil loses moisture) and as it is inherent for field applications limited control and high uncertainty can be expected. On the other hand, applications under laboratory conditions that provide higher control and lower uncertainty due to a number of reasons cannot always safely be extrapolated to field applications.

Reported laboratory applications often suffer from one or more of the following limitations: a) time scale of the laboratory testing is much smaller than the time required for the field application (e.g. in Beddiar et al. 2005 only a time period of 10 hours is examined), b) dimensions of the laboratory experiment are considerably smaller than those of the field application (e.g. in Moayedi et al. 2012 a 150mm long soil specimen is treated), and c) the boundary conditions applied to the laboratory experiment are different than the conditions prevailing in the field (e.g. in Airoldi et al. 2009 the experimental cell fully encloses the treated soil and the absence of an open surface means that any pressure changes can have disproportional impact, whereas the compaction direction differs by 90 degrees when compared to the direction of the compaction taking place in field applications).

In this work, we consider electrokinetic treatment of clay soil as a baseline case (i.e. no pollutant or any other substance is transferred out from/delivered in the treated soil, water level at anode and cathode is maintained constant avoiding any water accumulation or dewatering, soil top surface and anodic/cathodic chambers are open as expected in field applications avoiding any air entrapment and pressure build-ups). The electrokinetic treatment is continuous over a period of few months, and the characteristic length of the experimental apparatus has an order of magnitude of 1 meter. The compaction is performed in the vertical direction like it always occurs in the field. Even ordinary tap water is used throughout the experiment instead of e.g. demineralized water as this is far more likely to be the case in an engineering application.

Electrokinetic treatment of clay soil has the potential to be exploited in order to perform in situ soil stabilization (and not per se by means of dewatering). This idea is very attractive in cases where existing structures are built on top of the clay soil, as the structures remain completely intact. Such a typical case of existing structure can be found in pavement engineering, where the pavements are built on top of a natural subgrade. The motivation behind examining a base line case as a first step is to build up a better understanding of the underlying mechanisms and get a feeling about the feasibility of the method – especially the time requirements associated with it.

2 EXPERIMENTAL SET-UP

In this study, two large titanium plates connected to a power supply are used as anodic and cathodic electrodes (anode and cathode). Wet clay soil is placed in the largest middle part of the set-up in four layers; each layer is compacted and has a height of ~30 mm after compaction. The volume of treated clay soil is confined at two opposite sides from the walls of a HDPE (high density polyethylene) container while two paint-coated perforated stainless steel plates covered with sponge cloth restrain the remaining two sides of the clay soil volume. The bottom part of the treated volume is in contact with the HDPE container, while the top part is open and in contact with the room atmosphere (a very loose plastic cover was thrown on top during the treatment in order to limit the evaporation).

The two areas that are created at each side of the set-up between the perforated plates, the anode, and the cathode, form the anodic and the cathodic chamber respectively. A system of pumps and capacitive proximity sensors is designed in order to constantly keep the water height in both chambers at 100 mm; tap water is added in the anodic chamber and excess water is removed from the cathodic chamber. A total of 19 passive electrodes (thin rods of titanium) is distributed inside the setup before any clay soil or water is added. These passive electrodes are fixed in position by means of small PVC blocks that are glued at the bottom of the HDPE container and have a hole drilled through their center for placing the titanium rods. The 19 passive electrodes along with the anode and the cathode (21 electrodes in total) are connected to a control panel that allows for measuring electric

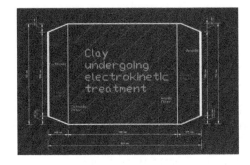

Figure 1. Top view of the set-up.

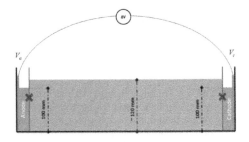

Figure 2. Side view of the set-up.

potential differences between the different locations. A multimeter which is added at the circuit of: power supply – anode – anodic chamber – treated clay soil – cathodic chamber – cathode – power supply, allows for measuring the electric current that flows through the circuit. The power supply is designed to provide a constant potential difference as an output for the entire duration of the treatment, while the electric current is free to fluctuate. In Figure 1 a top view sketch of the set-up is presented, while in Figure 2 a side view sketch of the set-up is depicted.

3 RESULTS

The clay soil underwent electrokinetic treatment for 127 days. Throughout the treatment a constant potential difference of 40V is provided by the power supply, while the electric current is measured periodically at infrequent intervals. At the same intervals, the amount of excess water pumped out of the cathodic chamber was measured. By dividing the mass of the excess water by the time period between the latest (n) and the one before the latest (n-1) measurement, a mass flow rate can be calculated (this mass flow rate does not account for any water lost due to evaporation). Furthermore, potential differences between all the 21 electrodes are measured at the same intervals. From the time evolution of those potential differences, a view of the time changing electric potential field is feasible. If the experiment is considered as an 1D case and we interpolate from the measuring points, it is possible to calculate the potential difference between the anodic chamber – clay soil interface and the clay soil – cathodic

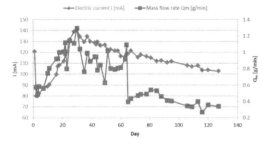

Electric current I, Mass Flow Rate Q_m

Figure 3.   Electric current I; Mass flow rate Q_m.

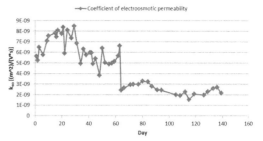

k_{eo} Coefficient of electroosmotic permeability

Figure 5.   Coefficient of electroosmotic permeability k_{eo}.

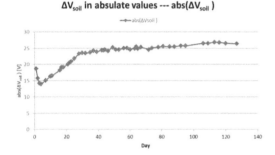

ΔV_{soil} in absulate values --- abs(ΔV_{soil})

Figure 4.   Electric potential difference between the edges of the treated volume of clay soil – absolute values plotted.

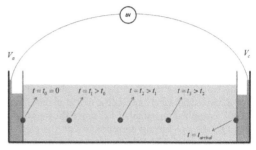

Figure 6.   Concept of arrival time.

chamber interface (the two points marked with purple x in Figure 2). In Figure 3 the measured electric current I (left axis) and the calculated mass flow rate Q_m (right axis) are plotted for the 127 days that the experiment was running, while in Figure 4 the absolute value of the electric potential difference between the two ends of the treated clay soil (assuming 1D case) is depicted for the same period of time.

It is possible to macroscopically describe the electroosmotic flow by means of the coefficient of electroosmotic permeability k_{eo}. The coefficient of electroosmotic permeability k_{eo} is analogous to the coefficient of hydraulic permeability k_h, but the electric potential gradient dV/dx drives the flow instead of the hydraulic gradient dh/dx. The combined flow (in case of soil surface with negative charge – which is true for this study) is described by Equation 1 (Hausmann 1990).

$$u = -k_h \frac{dh}{dx} - k_{eo} \frac{dV}{dx} \qquad (1)$$

$$Q_v = u \cdot A \Leftrightarrow u = \frac{Q_v}{A} \qquad (2)$$

$$Q_v = \frac{Q_m}{\rho_w} \qquad (3)$$

$$Q_{m,av} = \frac{\sum (Q_{m,i} \cdot \Delta t_i)}{\sum \Delta t_i} \qquad (4)$$

In Equation 1, u is the superficial velocity (ignoring e.g. porosity and tortuosity) and can be calculated by dividing the volumetric flow rate Q_v by the cross-sectional area A (Equation 2), where the volumetric flow rate Q_v is calculated by dividing the mass flow rate Q_m by the water density ρ_w (Equation 3). Since in this study the anodic and cathodic chambers have an open top and the water column is maintained in both chambers at a height of 100 mm (Figure 2), the $-k_h \cdot dh/dx$ term in Equation 1 is neglected. With either knowing, having measured or calculated everything else, the coefficient of electroosmotic permeability k_{eo} throughout the experiment is calculated and plotted in Figure 5.

So far everything has been considered in a macroscopical framework, which – even though very useful – cannot be used for estimating traveling time. We define the concept of arrival time as the time required for a water volume in the anodic chamber that enters the clay soil volume at $t = t_0 = 0$ to exit the clay soil volume and enter the cathodic chamber at $t = t_{arrival}$; this concept is depicted in Figure 6. In order to calculate the arrival time, the measured mass flow rate (right axis in Figure 3) that applies to different time intervals needs to be transformed to a time weighted average mass flow rate that describes the flow from $t = t_0 = 0$ until the time of interest. This is done by means of Equation 4 and is plotted in Figure 7. Assuming an average mass flow rate of 0.9 g/min and knowing all necessary items like e.g. the porosity (calculations not included in this work) one can calculate the interstitial volumetric flux, and given the dimensions of the

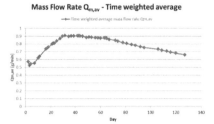

Figure 7. Time weighted average mass flow rate $Q_{m,av}$.

treated clay soil the arrival time can be determined. An arrival time of ~29 days was calculated and according to Figure 7 the initial assumption of an average mass flow rate of 0.9 g/min requires no update.

4 CONCLUSIONS

In this work a baseline electrokinetic treatment of clay soil is carried out where an electric potential difference of 40 V is applied for a total of 127 days. The mass flow rate Q_m shows a strong correlation with the electric current I (Figure 3). They both increase during the first days and after a few weeks they both start dropping, while the day on which they both peak coincides. Out of the 40 V electric potential difference provided by the power supply, the portion that is applied on the treated clay soil is not constant but changes with time (Figure 4); even on the most favorable periods, out of the 40 V provided by the power supply not much more than 25 V are applied to the soil, thus resulting in inefficient energy consumption. Further research work exploring the origin of this hefty "electric potential loss" can result in considerably improved efficiency in future applications.

The calculated coefficient of electroosmotic permeability changes substantially with time. Even though the range is rather extensive from 9.0e-9 m²/(s·V) to 1.0e-9 m²/(s·V), all these values are well within the typical reported range for k_{eo} (Mitchell & Soga 2005). Since the anodic chamber is always kept at a constant water level (no dewatering) and no other chemical substances are added in the stream or the soil, it becomes evident that the treatment itself affects the coefficient of electroosmotic permeability k_{eo}. This aspect has been reported in some studies (Shang 1997), and it becomes clear that the common practice of calculating the coefficient of electroosmotic permeability from a short-term experiment and handling it as a material property that remains constant can lead to substantial errors.

The calculated arrival time of 29 days is encouraging for the potential of electrokinetic treatment in the discipline of civil engineering, since construction or other similar works take place over periods of several weeks and months if not years. Even though the time scale requirements are deemed as very satisfactory, research in shortening the time requirements is always welcome. Shortening the time requirements can be achieved by making the process more efficient, or mitigating the existing inefficiencies that have already been identified. Furthermore, electric potential differences higher than 40 V are likely to speed up the phenomena.

REFERENCES

Acar Yalcin B., Alshawabkeh Akram N., 1993. Principles of electrokinetic remediation. *Environmental Science & Technology*, Vol. 27, Issue 13, 2638–2647.

Acar Yalcin B., Alshawabkeh Akram N., 1996. Electrokinetic remediation. I: pilot-scale tests with lead-spiked kaolinite. *Journal of Geotechnical Engineering*, Vol. 122, 173–185.

Airoldi Fabio, Jommi Christina, Musso Guido, Paglino Elena, 2009. Influence of calcite on the electrokinetic treatment of a natural clay. *Journal of Applied Electrochemistry*, Vol. 39, 2227–2237.

Beddiar Karin, Fen-Chong Teddy, Dupas André, Berthaus Yves, Dangla Patrick, 2005. Role of pH in electroosmosis: experimental study on NaCl-water saturated kaolinite. *Transport in Porous Media*, Vol. 61, 93–107.

Bilitewski Ursula, Genrich Meike, Kadow Sabine, Mersal Gaber, 2003. Biochemical analysis with microfluidic systems. *Analytical and Bioanalytical Chemistry*, Vol. 337, Issue 3, October, 556–569.

Biscombe Christian J. C., 2017. The discovery of electrokinetic phenomena: setting the record straight. *Angewandte Chemie International Edition*, Vol. 56, Issue 29, 8338–8340.

Chang Hsueh-Chia, Yeo Leslie Y., 2010. *Electrokinetically driven microfluidics and nanofluidics*. New York, United States: Cambridge University Press.

Gautherot N., 1801. Mémoire sur le galvanisme. *Annales de Chimie, ou Recueil de Mémoires Concernant la Chimie et les Arts Qui en Dépendent, et Spécialement la Pharmacie*, Vol. 39, 203–210.

Hausmann R. Manfred, 1990. *Engineering principles of ground modification*. Singapore: McGraw – Hill.

Indraratna Buddhima, Chu Jian, Rujikiatkamjorn Cholachat, 2015. *Ground Improvement Case Histories: Chemical, Electrokinetic, Thermal and Bioengineering Methods*. Oxford, United Kingdom: Butterworth Heinemann.

Mitchell K. James, Soga Kenichi, 2005. *Fundamentals of soil behavior*. New Jersey, United States: John Wiley & Sons.

Moayedi H., Kazemian S., Huat B. B. K., Vakili A. H., 2012. Electro-strengthening of highly organic soil using environmentally friendly admixtures. *International Conference on Sustainable Design, Engineering and Construction (ICSDEC2012)*, Fort Worth, Texas, United States, 449–456.

Porret R. Jr., 1816. Curious galvanic experiments. *Annals of Philosophy; or, Magazine of Chemistry, Mineralogy, Mechanics, Natural History, Agriculture, and the Arts*, Vol. 8, 74–76.

Reuss F. F., 1809. Notice sur un nouvel effect de l'électricité galvanique. *Memoires de la Société Impériale des Naturalistes de Moscou*, Vol. 2, April, 327–337.

Shang J. Q., 1997. Zeta potential and electroosmotic permeability of clay soils. *Canadian Geotechnical Journal*, Vol. 34, 627–631.

Squires Todd M., Quake Stephen R., 2005. Microfluids: Fluid physics at the nanoliter. *Reviews of Modern Physics*, Vol. 77, July, 977–1026.

Advances in Materials and Pavement Performance Prediction II – Kumar et al. (eds)
© 2021 Taylor & Francis Group, London, ISBN 978-0-367-46169-0

High strength and good durability of stabilized dolomite fines

Issam Qamhia & Erol Tutumluer
Department of Civil and Environmental Engineering, University of Illinois at Urbana-Champaign, IL, USA

Heather Shoup
Illinois Department of Transportation, Springfield, IL, USA

Debakanta Mishra
School of Civil and Environmental Engineering, Oklahoma State University, OK, USA

Hasan Ozer
School of Sustainable Engineering and the Built Environment, Arizona State University, AZ, USA

ABSTRACT: Dolomite and limestone are two most widely quarried carbonate aggregates in the State of Illinois available for use as pavement materials. In particular, dolomites are believed to have mineralogical compositions that can provide improved durability and functionality for certain pavement applications requiring better quality materials. This paper presents two case studies where dolomitic aggregates were utilized in closely-monitored field test sections and showed superior performance when compared to limestone aggregates. The first section is for dolomitic aggregate Quarry By-products (QB) used as cement-stabilized base/subbase in low to medium volume roads, while the second case is for dolomite used as an unbound material for a gravel (unsurfaced) road application. The performances of pavement sections with dolomite aggregate and their durability aspects are compared to those of limestone sections. Preliminary findings indicate that dolomitic fines contributed to improved durability and/or higher strength gain due to delayed reactions associated with carbonate cementation governed by 'dissolution precipitation' reactions when exposed to temperature and moisture changes induced by freeze-thaw cycles.

1 INTRODUCTION

In general, little has been done to understand the effect of aggregate mineralogy on long-term pavement performance for providing a strong pavement foundation through the use of unbound and chemically stabilized bound base/subbase layers of aggregates. This paper provides two case studies with preliminary results that contrast both the performance and durability trends of field test sections having dolomite and limestone fines in pavement base/subbase courses. The two case studies, referred to as Case 1 and Case 2, were conducted at the Illinois Center for Transportation (ICT) established at the University of Illinois at Urbana-Champaign (UIUC). Both the dolomite and limestone fines, typically passing No. 40 sieve materials on which Atterberg limit tests are conducted, were utilized in pavement test sections trafficked under accelerated pavement testing and exposed to harsh weather effects over the winter.

2 CASE 1: QUARRY BY-PRODUCT (QB) FINES

2.1 Background

As part of a recent project at UIUC, the ICT R27-168 (Qamhia et al., 2018; 2019), Accelerated Pavement Testing (APT) was used to evaluate seven different material combinations of cement- or fly-ash-stabilized Quarry By-product (QB) materials or a combination of 70% QB and 30% fractionated course recycled aggregates used in base and subbase layers under flexible pavements. Low surface rutting, low Falling Weight Deflectometer (FWD) deflections, and low pressure on top of the subgrade were measured for all sections with stabilized QB layers (Qamhia, 2019). A follow-on research project, the ICT R27-SP38 (Qamhia et al., 2019a), investigated the wet-dry and freeze-thaw durability of these seven material combinations by collecting field samples and conducting laboratory durability tests.

This paper will focus on material combinations from three of the test sections with QB materials from two different sources denoted herein as QB2 (dolomite) and QB3 (limestone). The samples are named after the field test sections from which they were extracted: C2S4 (QB2 and 3% cement as a base layer), C3S1 (QB3 with 3% cement as a base layer), and C3S2 (QB2 with 3% cement as a subbase layer). C2S4 and C3S2 are essentially the same material combinations but were tested in the field for different applications (i.e. base vs. subbase). A full suite of durability testing and characterization results for the other material combinations can be found elsewhere (Qamhia et al., 2019a).

2.2 Extraction of field samples

AASHTO T 136 freeze-thaw and AASHTO T 135 wet-dry standard test procedures were followed. Ideally, cylindrical test samples with dimensions conforming to the size of the standard Proctor mold were to be used. However in an earlier ICT-R27-168 project, several attempts made to extract intact cylindrical cores of the stabilized base/subbase layers were not successful. The materials eroded in the presence of water introduced by the coring process. The fine fragments produced from dry coring ended up clogging the coring bit thus creating high friction and preventing the recovery of fully intact cores.

Alternatively, large blocks of intact stabilized QB base/subbase sections were extracted using a mini excavator. The extracted materials were stored for further handling. Following the extraction and storage of intact blocks of field samples, the samples were cut into cuboid prisms using a large saw-cutting equipment normally used for cutting hard rocks. A dry saw-cutting procedure was adopted to ensure samples did not disintegrate due to the presence of water. The samples were shaped into cuboids (prisms) with a square cross section having a 71 mm (2.8 in.) side length and a 122 mm (4.6 in.) height. Sample extraction and preparation are presented in Figure 1(a–c).

The as-received dry densities and the relative densities of all field-extracted samples are presented in Table 1. The dry densities were calculated based on the as-received density and the as-received moisture content of each sample, which were measured at Illinois Department of Transportation (IDOT) Central Bureau of Materials (CBM). Table 1 lists lower field densities compared to the maximum dry densities for samples extracted from C3S1 with QB3. These low densities could contribute to the rather poor wet-dry and freeze-thaw durability trends for QB3, to be discussed later. However, note that other test sections with QB2 mixed with fractionated coarse aggregates also had low relative densities yet satisfactory durability. The results from these sections are presented elsewhere (Qamhia et al., 2019a) As it will be discussed later, chemical composition is the main factor for the

(a) Sample extraction   (b) Saw-cutting samples

(d) Wet-dry durability testing

(c) Durability samples   (e) Freeze-thaw durability testing

Figure 1. Extraction, preparation, and testing of wet-dry and freeze-thaw durability samples collected from field sections.

Table 1. As-received dry densities and relative densities of field-extracted samples.

Sample	Durability Test	MDD (kN/m³)	Avg. dry density (kN/m³)	Field Relative Density (%)
C2S4	wet-dry	21.6	20.8	96.3
	freeze-thaw		20.7	95.8
C3S1	wet-dry	20.4	17.5	85.8
	freeze-thaw		17	83.3
C3S2	wet-dry	21.6	20.8	96.3
	freeze-thaw		20.4	94.4

observed durability trends for samples with QB2 and QB3 extracted from the field.

2.3 Testing samples for durability

Testing of wet-dry durability samples was conducted in accordance to the AASHTO T 135 standard procedure. After each cycle, measurements of moisture change, volume change, and soil-cement loss by a brushing method were taken. The standard method 'B' in the AASHTO T 135 specification was employed. The testing procedure involved wetting of the seven-day cured samples for five hours at a room temperature of $21 \pm 2°C$ ($70 \pm 3°F$), followed by oven-drying for 42 hours at a temperature of $71 \pm 3°C$ ($160 \pm 5°F$). Sample weight was recorded after each step. Samples were then tested by brushing all areas with two firm strokes using a wire scratch brush and applying a brushing force approximately 13 N (3 lbf). The wetting and drying steps were repeated for a total of 12 cycles for each tested specimen. Wet-dry durability testing is illustrated in Figure 1(d).

Testing of freeze-thaw durability samples was conducted according to the AASHTO T 136 standard. Similarly, this procedure involves using a brushing method to determine the percentage of soil-cement loss produced after each cycle of freezing and thawing. The standard method 'B' was also used. The testing procedure involves the placement of samples in a freezing cabinet in air for 24 hours at a temperature of $-23°C$ ($-10°F$), followed by thawing in a moist room for 23 hours at 100% relative humidity and a temperature of $23 \pm 2°C$ ($73.5 \pm 3.5°F$). The samples were brushed and tested in the exact same manner as the wet-dry samples. After 12 cycles of testing, the samples were dried to a constant weight at a temperature of $110 \pm 5°C$ ($230 \pm 9°F$). Freeze-thaw sample testing is shown in Figure 1(e).

Figure 2 presents the wet-dry durability test results. All laboratory samples survived the 12 cycles of wet-dry durability testing. IDOT Standard Specifications for Road and Bridge Construction specifies that the loss in weight/mass shall be less than 10% after 12 cycles of wetting and drying (IDOT, 2016). Samples with QB2, namely C2S4, and C3S2, had less than 2% soil-cement loss after 12 cycles of wet-dry durability. On the other hand, C3S1 with cement-stabilized QB3 had significantly higher soil-cement loss. Given both materials had the same gradation and cement content, this is a clear indication that mineralogy and chemical composition of QB could affect wet-dry durability. Also, note that samples extracted from C3S1 had low relative densities compared to other cement-stabilized samples.

Figure 3 presents the freeze-thaw durability test results. Samples with QB2 (dolomite) survived 12 cycles of testing and accumulated significantly lower soil-cement losses than samples with QB3 (limestone). Note that samples with QB3 completely disintegrated after 7-8 cycles. Similarly, the type of QB had a clear effect on the freeze-thaw durability.

2.4 Chemical compositions of QB2 and QB3

The type, i.e. origin and chemical composition, of QB can have a significant impact on the wet-dry and freeze-thaw durability. Table 2 presents the chemical composition of QB2 and QB3 materials determined by X-Ray Fluorescence (XRF) test. QB2 with more dolomitic fines, i.e. higher magnesium oxide (MgO), could develop higher strength gain over time and thus contributed to better long-term durability (Hou et al., 2019). Since the grain size distributions are quite similar for both QB2 and QB3, the chemical composition is the main factor that can relate to the obvious discrepancies in durability performance trends. QB2 has significantly higher percentages of MgO (33.4–36.7%) compared to the 9.5–11% MgO in QB3. Significantly higher strength gain and improved durability observed

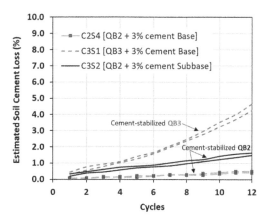

Figure 2. Estimated soil-cement loss for wet-dry durability testing of field-extracted samples.

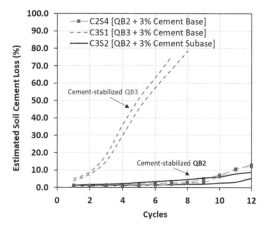

Figure 3. Estimated soil-cement loss for freeze-thaw durability testing of field-extracted samples.

Table 2. Chemical composition of QB2 and QB3.

Quarry	Crushing Stage	Percent by Weight (%)		
		CaO	MgO	Others
QB2	Primary	54.7	36.7	8.2
	Secondary	48.5	33.4	17.4
	Tertiary	50.4	34.2	14.5
QB3	Primary	58.7	11.0	29.5
	Secondary	71.4	10.1	17.9
	Tertiary	71.4	9.5	18.4

with time of QB2 can be linked to better cementing reactions with MgO content. Nevertheless, the effect of QB chemical composition on short- and long-term durability needs to be further investigated by considering a larger dataset to investigate any statistical significance of chemical composition on durability.

3 CASE 2: AGGREGATE BASES WITH DOLOMITE FINES

3.1 *Background*

As part of the ICT R27-182 project, unbound aggregate pavement sections were constructed and evaluated for rutting performance through an APT effort. Among the base course materials tested were limestone and dolomite aggregates with high percentages of nonplastic fines and uncrushed river gravel (Mishra, 2012; Mishra et al., 2013). Test Cells with crushed dolomite aggregates had higher backcalculated layer moduli and low surface rutting after having been exposed to freeze-thaw cycles over the winter. A follow-up laboratory study was conducted to better assess the reasons behind this improved performance.

One plausible mechanism contributing to the strength gain within the crushed dolomite layer could be 'carbonate cementation' within the fines fraction (Graves, 1987). Significant strength gain in high carbonate base course materials upon soaking has been reported in literature due to cementation of the fines fraction through dissolution and precipitation of Calcium Carbonate ($CaCO_3$) according to Equation 1 (Graves, 1987):

$$CO_2 + H_2O + CaCO_3 = Ca^{2+} + H^+ + HCO_3^-$$
$$+ HCO_3^{2-} \qquad (1)$$

To further investigate the hypothesis regarding strength gain by the crushed dolomite material with high fines upon extended periods of soaking and exposure to freeze-thaw cycles, unconfined compressive strength tests as per ASTM D 2166 were conducted on the aggregate fines (material finer than 0.075 mm, or passing No. 200 sieve). Two identical cylindrical specimens, 71 mm (2.8 in.) in diameter and 142 mm (5.6 in.) in height, were prepared at a target moisture content of 14%. Note that the target moisture content of 14% was selected through iterative trial-and-error approach to arrive at the minimum moisture content where the compacted specimen exhibited cohesive characteristics (Mishra, 2012). One of the compacted specimens was immediately tested for the unconfined compressive strength, whereas the other was exposed to 24 hours of freezing ($-16°C$ or $3°F$) followed by 24 hours of thawing ($16°C$ or $61°F$) as specified in ASTM D 6035. Care was taken to ensure that no moisture was lost by the specimen during the freeze-thaw period.

Figure 4 compares the unconfined compressive strength (Q_u) values for the two specimens described above. As shown in the figure, subjecting the specimen to one freeze-thaw cycle (24-hours of freezing followed by 24-hours of thawing) resulted in an increase in the unconfined compressive strength value from 224 kPa (32.5 psi) to 581 kPa (84.2 psi). This preliminary investigation supported the possibility of strength gain by the carbonate fines in the crushed dolomite

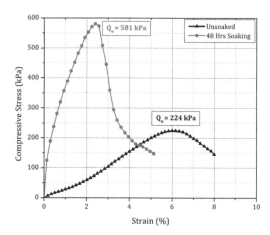

Figure 4. Effect of freeze-thaw cycles on the unconned compressive strength of carbonate (dolomite) fines.

material upon exposure to extended periods of soaking and freeze-thaw cycles. Further investigation of this phenomenon is required to verify the mechanism responsible for such strength gain.

4 CONCLUSIONS

This paper presented a preliminary study investigating the effect of chemical composition or mineralogy of fines in aggregate materials on mechanical properties and durability. Based on the preliminary results, it can be inferred that chemical composition is a major factor contributing to durability and strength gain. In particular, dolomite aggregate fines show better long-term durability and strength gain through cementation when compared to limestone fines both in both stabilized/bound and unbound base/subbase applications. Further studies on the effect of chemical composition on durability can help to further validate the more durable and improved performance of dolomite aggregates for constructing long lasting and sustainable pavement layers.

REFERENCES

Graves, R.E., 1987. Strength developed from carbonate cementation in silica/carbonate systems as influenced by cement-particle mineralogy. M.S. Thesis, University of Florida.

Hou, W., Qamhia, I., Mwumvaneza, V., Tutumluer, E. and Ozer, H., 2019. Engineering Characteristics and Stabilization Performance of Aggregate Quarry By-Products from Different Sources and Crushing Stages. Frontiers in Built Environment, 5, p. 130.

Illinois Department of Transportation (IDOT), 2016. Standard specifications for road and bridge construction. Springfield, IL: IDOT.

Mishra, D., 2012. Aggregate characteristics affecting response and performance of unsurfaced pavements on weak subgrades. Doctoral Dissertation. University of Illinois at Urbana-Champaign.

Mishra, D. and Tutumluer, E., 2013. Field performance evaluations of Illinois aggregates for subgrade replacement and subbase—Phase II. Illinois Center for Transportation/Illinois Department of Transportation.

Qamhia, I., Tutumluer, E. and Ozer, H., 2018. Field Performance Evaluation of Sustainable Aggregate By-product Applications. Illinois Center for Transportation/Illinois Department of Transportation.

Qamhia, I.I., Tutumluer, E., Ozer, H., Shoup, H., Beshears, S. and Trepanier, J., 2019. Evaluation of chemically stabilized quarry byproduct applications in base and subbase layers through accelerated pavement testing. Transportation Research Record, 2673(3), pp. 259–270.

Qamhia, I., Tutumluer, E., Ozer, H. and Boler, H., 2019a. Durability Aspects of Stabilized Quarry By-product Pavement Base and Subbase Applications. Illinois Center for Transportation/Illinois Department of Transportation.

Qamhia, I.I., 2019. Sustainable pavement applications utilizing quarry by-products and recycled/nontraditional aggregate materials. Doctoral Dissertation. University of Illinois at Urbana-Champaign.

Advances in Materials and Pavement Performance Prediction II – Kumar et al. (eds)
© 2021 Taylor & Francis Group, London, ISBN 978-0-367-46169-0

On uncertainty in asphalt binder unit response mastercurves

Aswathy Rema & Aravind Krishna Swamy
Department of Civil Engineering, Indian Institute of Technology, New Delhi, India

ABSTRACT: Various temperature shift factor approaches are used to account for time-temperature equivalencies while constructing master curves. Even with good quality control measures, significant scatter in dynamic shear modulus ($|G^*|$), phase angle (Φ) data is observed. This, in turn, results in uncertainty in finalized relaxation modulus, G(t) and creep compliance, J(t) master curves when interconversion techniques are used. This uncertainty could be ascribed to factors like material testing practices, human errors, experimental errors, numerical approximations etc. This study presents a comprehensive approach to evaluate the uncertainty that is propagated to J(t) and G(t) values during master curve construction and interconversion process. Two shift factor construction methods, i.e, asymmetric Kaelble shift factor method and symmetric free shifting approach are considered during the master curve construction process. The uncertainty parameters such as Normalised Uncertainty Range (NUR) indicated that asymmetric-Kaelble method results in lower uncertainty when compared to symmetric-free shifting approach.

Keywords: Uncertainty quantification; Uncertainty propagation; Viscoelasticity; Interconversion technique; Mastercurves

1   INTRODUCTION

The viscoelastic nature of bitumen complicates the response and performance prediction of Asphalt Concrete (AC) to a large extent. Viscoelastic properties like shear creep compliance J(t) and shear relaxation modulus G(t) are determined using actual physical testing or interconversion approaches suggested by various researchers (Park and Scaphery 1999). Due to the practical difficulty in conducting experiments (applying input within a short period of time and record corresponding response over longer duration), high-speed data acquisition system with large memory capacity, interconversion process is preferred over experimental determination. Under such circumstances, researchers determine dynamic shear modulus ($|G^*|$) and phase angle (Φ) by conducting frequency-temperature sweep tests using a dynamic shear rheometer. These measured values are further used to construct corresponding storage modulus (G*) master curves. From G' mastercurves the unit response functions (i.e. J(t) and G(t)) are determined using interconversion methods (Park and Scaphery 1999).

As bitumen is a temperature-dependent material, the viscoelastic response is generally described through master curve using time-temperature superposition principles (Mateos and Soares 2015). This principle finds equivalencies by horizontal shifting of individual isotherms until a smooth function is obtained between various loading frequencies and test temperature. This shifting is attained through the determination of the temperature shift factor. There are various shift factor construction approaches such as Kaelble method, Arrhenius method, WLF method, free shifting approach, etc (Pellinen et al. 2004; Rowe and Sharrock 2011). As no functional form is associated with temperature shift factor vs temperature, both sigmoidal coefficients and the temperature shift factors are considered as unknown parameters in free shifting approach. This study considers the Kaelble method and free shifting shift factor methods (for temperature shift factors) along with asymmetric and symmetric sigmoidal functions (Rema and Swamy 2019).

Due to practical limitations (like time, resource), very few samples are often tested for its properties. Even with good quality control during testing and analysis protocols, it is seen that there is significant scatter in the finalized master curves constructed. Rema and Swamy (2019) attributed this scatter to the flexibility offered by individual coefficients within the master curve construction approach. To overcome the limitation of small sample size, the resampling method like Bootstrapping was used in this study. Bootstrapping is a kind of resampling method that estimates distribution parameters by using the resampled data from the initial population. This resampled dataset is generated mostly in thousands through the replacement method. Bootstrapping offers advantages like quicker resampling, faster computation on present-day computers, and distribution independent approach (Rema and Swamy 2019). Further, to address the scatter in data, uncertainty quantification techniques can be used (Cullen and Frey 1999). Uncertainty Quantification (UQ) is a broad interdisciplinary area of research which includes majorly, advanced computational mathematics and statistics. The focus of UQ methods is to estimate the likelihood or probability of a parameter of interest, knowing the fact that there is uncertainty in the whole modeling process.

2 METHODOLOGY

For the demonstration of this uncertainty evaluation and propagation framework, nine asphalt binder samples (VG 30) taken from the same container were short term aged separately. Further, $|G^*|$ and Φ values were measured over a range of different temperature and loading frequencies using a dynamic shear rheometer. Further, J(t) and G(t) master curves were constructed through approximate interconversion technique using storage modulus (G^*) master curves. During this master curve construction, asymmetric-Kaelble shift factor and symmetric-free shifting approach were considered. Subsequently, J(t) and G(t) values for all nine samples were obtained using back-calculation at various reduced time locations. Considering these values as the initial population, the Bootstrap simulation was used to generate 10000 samples (at every reduced time location). For evaluating uncertainty, Normalized Uncertainty Range (NUR) was used (Rema and Swamy 2019). Additionally, the uncertainty in the input parameters was propagated to the unit response

functions through Monte Carlo simulation coupled with Latin Hypercube Sampling. The uncertainty propagated to the outcome was again evaluated using NUR.

3 RESULTS AND DISCUSSION

Since symmetric-free shifting and asymmetric-Kaelble shifting approach provided extremes on uncertainty bounds, both approaches were evaluated in this work. The variation of mean values of unit response master curves at various reduced time locations using the two shift factor approaches are compared in Figure 1. As expected, with increasing reduced time, J(t) and G(t) showed increasing and decreasing trend, respectively. At a particular reduced time, the marginal difference in mean values can be found between both shifting approaches.

G′ J(t) and G(t) values were bootstrapped 10000 times (at each time location) and the NUR was calculated for all the three parameters. The mean and NUR calculated for storage modulus values using both shift factor methods are compared in Figure 2. Even though marginal differences in mean values are observed, significant variation is observed with NUR values at all reduced time. In general, NUR with asymmetric Kaelble method was lower when compared to the free shifting method for storage modulus values. Lower NUR was observed at intermediate reduced time values when compared to extremes.

Similarly, the NUR values obtained for the unit response functions using approximate interconversion and propagation approach are compared in Figure 3. The results indicate that in almost all cases, the asymmetric Kaelble method is associated with lower uncertainty when compared to free shifting approach. Also, NUR values obtained through propagation were higher when compared to interconversion method.

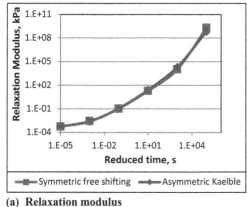

(a) Relaxation modulus

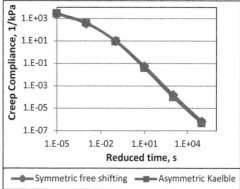

(b) creep compliance

Figure 1. Variation of mean values with reduced time.

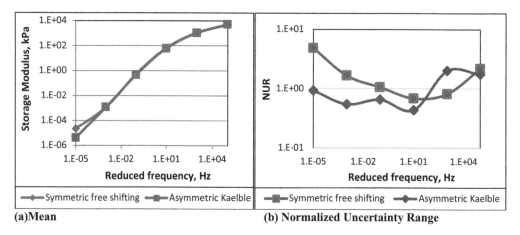

(a)Mean   (b) Normalized Uncertainty Range

Figure 2. Variation of storage modulus with reduced frequency.

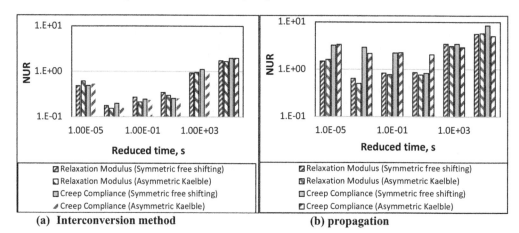

(a) Interconversion method   (b) propagation

Figure 3. Variation of NUR with reduced time.

4 CONCLUSION

This study evaluated the effect of master curve construction methods on uncertainty associated with unit response functions. The result shows that the NUR for uncertainty propagation was higher due to the numerical approximations involved, simulations involved and the variability in the input factors. The lower NUR in case of asymmetric Kaelble method is due to the fact that this method gives due prominence to flexibility for free shifting of individual isotherms while limiting overfitting. The drawbacks of symmetric free shifting were the lack of flexibility and often overfitting, eventually resulting in higher uncertainty. Moreover, NUR reflected that more uncertainty exists at the lower and higher reduced time compared to intermediate reduced time locations.

REFERENCES

Cullen, A.C., and Frey, H.C. 1999. Use of Probabilistic Techniques in Exposure Assessment: A Handbook for Dealing with Variability and Uncertainty in Models and Inputs, Springer US, New York.

Mateos, A., and Soares, J.B. 2015. Validation of a Complex Modulus predictive equation on the basis of Spanish asphalt concrete mixtures", *Journal of Material de Construction*, 65(317).

Park, S.W., and Scaphery, R.A. 1999. Methods of interconversion between linear viscoelastic material functions part I- A numerical method based on Prony series, *International Journal of Solids and Structures*, 36, 1653–1675.

Pellinen, T., Witczak, M., and Bonaquist, R. 2004. Asphalt Mix Master Curve Construction Using Sigmoidal Fitting Function with Non-Linear Least Squares Optimization, *Recent Advances in Materials Characterization and Modeling of pavement Systems, ASCE*, 83–101.

Rema, A., and Swamy, A.K. 2019. Effect of Construction Methodology on Uncertainty in Asphalt Concrete Master Curve, *Journal of Transportation Engineering Part B-Pavements, ASCE*, 145(3), 04019021:1–12

Rowe, G.M., and Sharrock, M.J. 2011. Alternate Shift Factor Relationship for Describing Temperature Dependency of Viscoelastic Behavior of Asphalt Materials. *Transportation Research Record: Journal of the Transportation Research Board*, 2207, 125–135

Advances in Materials and Pavement Performance Prediction II – Kumar et al. (eds)
© 2021 Taylor & Francis Group, London, ISBN 978-0-367-46169-0

Towards the application of drones for smart pavement maintenance

J.D. Rodriguez, R. Balieu & N. Kringos
KTH Royal Institute of Technology, Stockholm, Sweden

ABSTRACT: Drones have benefited multiple sectors because of their simplicity, low cost, and adaptability. However, the use of this technology for pavement monitoring is not well extended. The main goal of the present research is evaluating the application of Unmanned Aerial Systems (UAS) for pavement monitoring, by means of case study in the Norvik Port, Sweden. The study presents different aspects and issues that should be considered while implementing a UAS. The main results of the work show the improvement opportunities, successes, capability and feasibility of the UAS selected by the Norvik operator to capture the different defects that can occur in the pavement of the port. As a conclusion, it was suggested that UAS are a viable tool for monitoring defects in the pavement. However, the precision, accuracy, quality and relevancy of the data are influenced by the rigor and quality control applied during the implementation process.

1 INTRODUCTION AND MOTIVATION

One of the critical activities in pavement asset management is the collection of information related to the conditions of the pavements. These programs require periodic and systematic inventory and condition surveys of all roadway assets to determine the roadway's level of service (LOS), set maintenance priorities and make trade-off decisions. (Hart & Gharaibeh, 2011)

Today, most of the inventory and survey work is done manually in the field by professionals. Although the process of data collection, in this case, is standardized by norms like the ASTM D6433 (ASTM, 2018), the most recurrent critic of this kind of operation is the subjectivity of the inspectors. To improve some of these aspects, road agencies around the world have developed vehicles with several mounted sensors and data processing tools at the vehicle to automatize the data collection work. Such methods are, however, not exempt from criticism either. For example, Zhang *et al.* (2016) believes that such practices are expensive, time-consuming, tedious, require specialized staff regularly, and can exhibit a high degree of variability, "thereby causing inconsistencies in surveyed data over space and across the evaluation." (Zhang *et al.*, 2016)

In recent years, there has been an increased interest in the use of drones or Unmanned Aerial Vehicles (UAV) for multiple purposes. Literature shows an agreement regarding the main advantages of UAS, its flexibility and the low cost of the systems. For instance, Zakeri *et al.* (2016) mention that "the advantages of UAV systems are their low cost, fast speed, high maneuverability and high safety for collecting images." The initial cost for a UAS system, like the one used in Norvik, can be in the order of 12,000 USD, while the cost of a vehicle for surveying is far superior.

Running costs are similar as the main component is the labor involved in taking the information. In terms of resolution, a good UAS can reach resolutions of more than 0,2 mm/pixel close to the ones achieved by modern survey vehicles at a fraction of the cost. Thus, UAV can be seen as an alternative for the automatic collection of information of pavement distresses.

The present study aims to evaluate the application of UAS as an alternative for the surveying of pavement structures by making use of a roadmap in which the various decisions and options that need and can be considered are discussed in detail. The roadmap structure is applied to the case study of a new cargo port that is currently under construction in Sweden and which will be utilizing advanced smart technology for its pavement maintenance.

2 CASE STUDY FOR THE NORVIK PORT

The case study focuses on the Norvik port in Sweden. The port is being built as a response to the necessity in the Stockholm region, Sweden, to attend bigger ships and the projected increment in the maritime cargo traffic in the Baltic. (Stockholm Hamn AB, 2016). The container terminal will open spring 2020 and the ro-ro side in the fall of 2020. The port aims to become one of the most energy-efficient cargo ports. Therefore, the smart maintenance of its infrastructure can contribute significantly to this, among others, due to reduced downtime and a reduction of ad-hoc maintenance actions. The study uses a roadmap to assess the UAS applicability in the port infrastructure. According to Rodriguez (2019), the implementation of UAS can be divided into four stages. Figure 1 gives a brief overview of each stage. This study uses this roadmap to

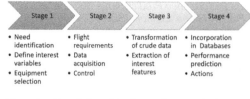

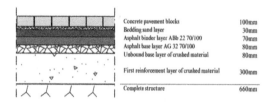

Figure 1. Description of the four main stages in the use of UAS for pavement maintenance. Adapted from (Rodriguez, 2019).

Concrete pavement blocks	100mm
Bedding sand layer	30mm
Asphalt binder layer ABb 22 70/100	70mm
Asphalt base layer AG 32 70/100	80mm
Unbound base layer of crushed material	80mm
First reinforcement layer of crushed material	300mm
Complete structure	660mm

Figure 2. Typical pavement structure of the container area. (Stockholm Hamnar AB, 2018).

assess the UAS applicability in the port infrastructure by following each one of the stages described below.

2.1 Stage 1: Need identification and equipment selection

Among the different paved areas in the port, the pavement in the container storage area was selected for further research. This structure consists of an unbound granular base, two bitumen layers and pavement concrete blocks on the top (Stockholm Hamnar, 2018). Figure 2 presents the typical pavement structure of the container storage area. The main issues that might affect this area consist of differential settlements, deformations and cracking in the top blocks due to the action of concentrated loads for long periods. To detect these distresses with the UAV camera it is required to measure cracking areas, vertical and lateral displacements in the top pavement blocks

2.1.1 Experimental setup

The sample consisted of several pavement blocks laid out in a finished surface. The control consisted of pavement stone samples evenly placed as in the finished pavement found in the port. Once the control sample was captured, layout 2 shown in Figure 3 (a) was arranged to measure the data capture capability of the UAV regarding elevation differences among the blocks. Two blocks were uniformly raised 1.5 cm and 2.5 cm for this setup. This layout will be named as "Layout 2."

Layout 3 was placed to evaluate the capability to capture lateral movements in the pavement concrete blocks. For this layout, the blocks were separated 1 cm, 2 cm, and 3 cm. Figure 3 (b) shows the layout used for this sample.

The UAV model used for the test is a DJI Inspire 2 with a ZENMUSE X5S camera in its UAV solution.

Figure 3. Pavement blocks layout. (a) LEFT: Layout 2 Height difference. (b) RIGHT: Layout 3 block lateral displacement.

It is a 20.8 MP camera with a 17,3 mm focal distance from the same manufacturer as the drone. This camera provides a ground sample distance of about 3 cm at a flight altitude of 120 m. A full list of specifications is available at the manufactures website.

2.2 Stage 2: Data acquisition

For data acquisition, it was necessary to account for several factors that might affect the correct development of the test. Blurriness in the captured images was one of the most concerning effects generated by the wind, while the limited amount of batteries and limited time to run the flights were another concerning issues in the flight planning.

For the case study, the imagery was not tied to ground control points due to the small sample size. However, the actual positioning of UAVs is vital in the maintenance and inspection of infrastructure. Geo-localized information enables the construction of databases in a Pavement Management System (PMS) to compare the progression of the deterioration of an area at different times.

The critical parameter that was manipulated in the experiment was the resolution of the imagery. This parameter can be controlled indirectly through the flight elevation of the UAV. Usually, at lower flight height, it is possible to obtain better resolution imagery. Besides flight patterns, speed, or altitude, current planning software solution uses to include options to select the desired imagery resolution for the most widely used camera sensors. The programmed flight parameters corresponded to 0.5, 1, and 2 cm pixel resolutions with a photo overlap of 70% with flight elevations of approximately 25 m, 45 m and 90 m, respectively; other settings were left automatically to be decided by the software within the recommendable limits.

2.3 Stage 3: Postprocessing

The next step is to process and extract the interest features from the field data. First, the data were categorized and its quality controlled for each flight.

The first stage of the postprocessing was completed with the remaining data. The Digital Elevation

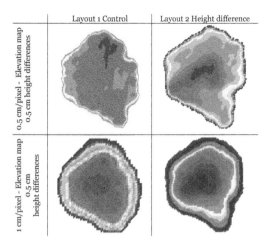

Figure 4. Sample of pavement blocks elevation maps for the control and modified layout for 2 different image resolutions.

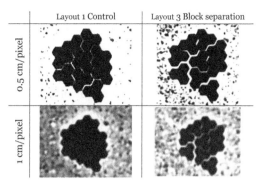

Figure 5. Pavement blocks modified orthophoto imagery for the control and modified layout for 2 different image resolutions.

Models (DEM) and orthophoto mosaics were created from the pictures with the help of specialized software as a first output for data extraction. However, the output data from the processing software itself might not be enough to retrieve the distress data easily. The use of color elevation maps created from DEM facilitates the recognition of the defects at plain sight or with machine learning algorithms. (Jensfelt, pers. comm., 2019). This meant a significant simplification in the identification of distresses manually or by machine learning algorithms. Figure 4 shows the results obtained after converting the DEM into elevation maps for the control sample and Layout 2, for a resolution of 0.5 cm/pixel and 1 cm/pixel.

A different technique is necessary for the detection of the lateral displacements for layout 3. For this purpose, several methods can be applied to UAS data. Currently, most studied techniques are the use of AI or ML. Those techniques work by applying several transformations or filters to images until the features of interest are recognizable and measurable clearly. As a means to evaluate the usability of the data, several filters were applied to the Orthophoto mosaics with image editing software. Figure 5 shows the results obtained after applying the filters for the control and Layout 3, for a resolution of 0.5 cm/pixel and 1 cm/pixel

2.4 Stage 4: Data usage

For the data obtained in the Norvik port, the main interest was to evaluate the viability of the UAV collected data to identify different defect signs with sufficient detail. In general, if the data allowed the identification of the height difference in the pavement blocks or if the lateral movements of the blocks were adequately noticeable. The case study covered the first steps of this process until the stage in which the information outputs that could be potentially used within a database.

3 RESULTS AND DISCUSSION

Figure 4 presents the results for the post-processing involving the production of elevation maps. In the elevation maps, it is possible to distinguish some differences between the control and Layout 2. For the elevation created from the highest resolution imagery (0.5 cm/pixel), the shape of the two blocks that were raised is differentiable in the image. However, when the elevation map is created from a lesser resolution data (1 cm/pixel), the shape starts to become less defined, adopting a cone-like shape instead.

About the lateral displacements, Figure 5 shows that with the 0.5 cm /pixel control and modified set shows that it is possible to identify different separation widths with the help of UAV imagery. The image results show that while the imagery resolution worsens, the identifiable features are less. The 1 cm/pixel sets, the normal separations among blocks are no longer recognizable. However, for layout 3, it is still possible to identify the different separation distances in the blocks.

From Figures 4 and 5, it becomes clear that it was possible to detect differences in the block regarding its elevation and lateral placement. So, the systematic measurements of the pavement surface via drones would enable early detection of a broad spectrum of defects in the pavement. Settlements rutting and freeze-thaw cycles could be fully identified in its initial stages by looking at displacements, defects, and differences in the elevation of the pavement blocks.

As expected, when the resolution of the imagery becomes worse, the data that can be extracted from a set of images is minor. However, the flights showed that with adequate conditions and sufficient quality, significant defects in the pavement could be captured with resolutions of 2 cm/pixel, which in practice means the coverage of larger extensions with a single flight. ML algorithms can identify more features than the human eye in several fields (Ouerhani, Bur & Hügli, 2006). Therefore, with the application of those techniques, it is possible to extract more detailed features with lesser quality data.

Nevertheless, it is still possible to find shortcomings in the data. For instance, the cone-like shapes in the 1cm/pixel data (Figure 4) are the result of missing information between several points and the interpolation from the software. The boundary effect is equally important in the layouts presented. Due to construction restrictions, a limited amount of pavement stones was used in the experiment. The small number of blocks created a boundary effect that affected the final DEM. A larger amount of blocks can be used to correct this issue.

Although it is viable to use UAS for pavement monitoring, there are still elements that must be addressed to ensure that a whole range of defects in the pavement is covered with the UAV surveying. The present work did not include, for example, the deterioration in the pavement blocks. No cracking or chemical deterioration was evident in the blocks since pavement is newly laid. However, among imagery, it is possible to address these concerns through the use of cameras or other types of sensors and later develop automated learning algorithms to automatize the process.

4 CONCLUSIONS

The main focus of the present study was to evaluate the possibilities and how UAV data collection can be implemented for infrastructure pavement maintenance purposes. The results showed that the used UAS possessed the capacity to collect information in which small distresses of the pavement are identifiable. That capacity is related to the quality of the data taken infield, which, in the case of the present study and type of sensor, was represented by the imagery resolution.

Imagery resolution was manipulated indirectly through the flight height producing different information sets. The 0.5 cm/pixel data sets originated results in which the defects were identifiable at plain sight. While the resolution worsened, the identification of the defects in the imagery became more difficult. Still, 0.5 cm/pixel data required more time and resources for its processing compared with lesser quality data. When looking at the whole process, data collection from UAV mounted sensors display a collection of trade-off parameters that must be adequately selected to produce the required results. As a global conclusion, it is possible to mention that UAV data has the potential to be successfully incorporated in the operative and maintenance decisions for pavement. However, at a network level, being aware of the limitations and potential of the technology, adapting the UAS components to needs and constraints, and selecting the best methods of processing the collected data is vital to ensure that the information is relevant and has the adequate quality level to be incorporated furtherly in the decision making.

The long-term objective of the Norvik port project is to deliver utilizable data to the Port Authorities to involve it in the decision-making system. This through the calibration of defect prediction models with actual data collected by the UAV and other sources. Those models will forecast how damage will be progressing in the pavement and what will be the real effect of the maintenance on it. Afterward, multiple scenarios can be modeled to select the best maintenance alternative with the technical and budgetary restrictions given by the port administrators.

REFERENCES

ASTM (2018) 'Standard Practice for Roads and Parking Lots Pavement Condition Index Surveys', pp. 1–5. doi: 10.1520/G0044-99R13.Copyright.

Hart, W. S. & Gharaibeh, N. G. (2011) 'Use of Micro Unmanned Aerial Vehicles in Roadside Condition Surveys', *Transportation and Development Institute Congress 2011.* (Proceedings), 2, pp. 80–92. doi: 10.1061/41167(398)9.

Ouerhani, N., Bur, A. & Hügli, H. (2006) 'Linear vs. Nonlinear Feature Combination for Saliency Computation: A Comparison with Human Vision BT – Pattern Recognition', in Franke, K. et al. (eds). Berlin, Heidelberg: Springer Berlin Heidelberg, pp. 314–323.

Rodriguez, J. D. (2019) *Towards the application of UAS for road maintenance at the Norvik Port, TRITA-ABE-MBT NV – 19557.* Available at: http://kth.diva-portal.org/smash/get/diva2:1329881/FULLTEXT01.pdf.

Stockholm Hamn AB (2016) *Norvik Hamn Projektdirektiv.*

Stockholm Hamnar (2018) 'Stockholm Norvik Hamn. Mark Hamnplan'.

Zakeri, H., Nejad, F. M. & Fahimifar, A. (2016) 'Rahbin: A quadcopter unmanned aerial vehicle based on a systematic image processing approach toward an automated asphalt pavement inspection', *Automation in Construction*, 72, pp. 211–235. doi: https://doi.org/10.1016/j.autcon.2016.09.002.

Zhang, S. *et al.* (2016) 'Characterizing Pavement Surface Distress Conditions with Hyper-Spatial Resolution Natural Color Aerial Photography', *Remote Sensing*, 8(5). doi: 10.3390/rs8050392.

Advances in Materials and Pavement Performance Prediction II – Kumar et al. (eds)
© 2021 Taylor & Francis Group, London, ISBN 978-0-367-46169-0

Characterization of road fleet weight using Big Data tools

C.L.S. Romeiro Jr., H.F. Grimm, L.A.T. Brito, M.R. Garcez & L.F. Heller
Pavement Laboratory of Federal University of Rio Grande do Sul, Porto Alegre/RS, Brazil

ABSTRACT: This paper investigated 2 years of WIM datasets from a Brazilian highway to compare the results obtained by a large dataset versus simplifications usually taken by Brazilian pavement designers about the traffic parameters, especially the ESAL. Since the database contained more than 3.4 million entries, it was key to use tools for Big Data analysis; Python programming with the Pandas, NumPy and Matplotlib libraries were employed for this purpose. In addition, to generate axle loads spectra from the dataset, ESAL per year were calculated on different scenarios of the fleet : (i) considering each axle weight contribution (maximum use of dataset); (ii) considering the weight means of each axle type (medium use of dataset); (iii) considering the 100% of the axles are operating on the Brazilian legal limits (minimum use of dataset). The results highlighted the adequateness of the tools used in the analysis and pointed out the importance of considering all available entries to characterize the fleet in order to improve traditional premises in pavement design.

1 INTRODUCTION

Data Science techniques are spread across a lot of fields once data collection has become more efficient and systematic due to technological advances, resulting in the so-called Big Data concept.

The increase of data availability, however, creates a new challenge: how to optimize information extraction from large datasets. Computational tools commonly used to manage smaller amounts of data fail to perform in large datasets. Appropriate tools for this end usually involve computational programming languages. Still, the use of these languages is interesting to make it possible to search for trends and patterns in the data with data science techniques (e.g., machine learning algorithms). Python programming language, for example, is used mainly for data science analytics, with many libraries already developed to handle large data volume such as Pandas, NumPy, Matplotlib, TensorFlow.

On pavement studies, traffic information plays an important role, mainly on pavement design methods. Development of Weigh-In-Motion technology has brought the Big Data concept on the field, especially on highways with high traffic volume. Now it is possible to view and analyze vehicle volume, and their loads collected continuously over long periods (e.g., one year, two years), allowing the characteristics of each entry to be accounted for, rather than using a statistical approach.

Many studies focused on data collected from WIM systems have only computed axles load spectra (input for AASHTOWare Pavement Design) and search for patterns on data (Jasim et al. 2019; Li et al. 2016; Lu et al. 2009; Hyun et al. 2015; Sayyady et al. 2012), dismissing pattern recognition regarding fleet characteristics.

Load spectra generated with such large databases are very useful, also, to calculate the Equivalent Single Axle Load (ESAL) of the fleet, which still is an essential input to some pavements design method, such as the Brazilian design guide.

This paper aimed to develop a notebook on Python language to analyze the data collected on a Low Speed Weigh in Motion located on a Brazilian highway. In addition to showing the utility of the programming tool for Big Data analysis and visualization aid, the research carried out looked into the impact of using the data available on ESAL calculation.

2 CASE STUDY AND METHODOLOGY

Dataset available for this study was provided by a Low Speed Weigh in Motion ($\sim$20 km/h) system located on the BR-116 highway between the cities of Rio de Janeiro and São Paulo in Brazil (see Figure 1) for the years of 2016 and 2017. The system operated 24/7 in one direction, totaling 731 days of collected data.

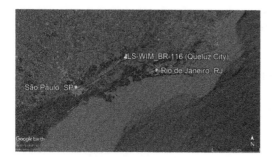

Figure 1. Localization of LS-WIM on BR-116 Highway.

Table 1. Number of entries of the dataset before data cleaning (raw data).

Year	Lines	Columns
2016	1,714,815	9
2017	1,719,052	9

Dataset (see the number of entries in Table 1) was received from the LS-WIM operators on a .csv file extension. A Jupyter Notebook using Python language with the support of Pandas and NumPy libraries was created to manage and analyze data. Also, the Matplolib library was used to provide better visualization of data through some plots.

Dataset lines correspond to each vehicle that passed on the LS-WIM system, while the columns refer to a series of information about these vehicles, such as control ID, vehicle category class, date and hour of the reading, and the weight of each axle group.

Before any characterization analysis, the algorithm routine cleared the dataset from errors to avoid distortion in the study.

The main errors types detected were: Error I – LS-WIM system unable to classify vehicle, yielding to an "ERR" output on the "Brazilian Category Type". Error II – inconsistencies on the axle group weight reading, such as groups with zeroed total weight.

Table 2. Number of lines (entries) with errors detected.

Year	Error I	Error II
2016	31,831 (1.86%)	97 (<0.01%)
2017	30,608 (1.78%)	–

Once the dataset did not provide axle type arrangement (single, dual, or tridem), it was implemented a code to relate it with the category type.

With the information in place, it was possible to visualize the axle load spectra per axle type, incorporating an analysis of the difference of the load spectra of 2016 and 2017. Yet, it was obtained the total number of each axle type with occurrence on the dataset.

Table 3. Actual limits of axles weight on Brazilian regulation.

	Legal Limits (kg)
Single Axles (single wheel)	6,000
Single Axles (dual wheels)	10,000
Tandem Axles	17,000
Tridem Axles	25,500

Besides the data characterization, the study evaluated the impact of three different consideration rules regarding the fleet weight on ESAL calculation:

i. Total Dataset: ESAL calculated from each axle weight on dataset individually (maximum use of dataset);
ii. Mean of Dataset: ESAL calculated from the mean of axles weight on dataset (medium use of dataset);
iii. All Axles Loaded: ESAL calculated considering 100% of the axles operating on the Brazilian legal limits (presented in Table 3) (minimum use of dataset).

3 RESULTS

The axles load spectra per year are represented in the histograms illustrated in Figure 2.

There is a clear difference towards a heavier fleet during the year 2016. Also, the axle's weight means, presented in Table 4, show higher values for 2016.

The total number of occurrences per axle type (see Table 5) identified the highest presence of Single Axles with Single Wheels in the fleet for both consecutive years.

The ESALs calculated per year using three different considerations about the fleet weight are summarized in Figure 3.

The "All Axles Loaded" scenario was the only one that presented higher ESAL in 2017. It is likely the single parameter that varies between years is the number of vehicles (or axles) on the fleet, which was higher in 2017. It is essential to point out the considerable difference between this scenario and the "Total Dataset". That is, if this scenario was considered on a pavement design (e.g., AASHTO 1993 Method), required layers thicknesses would be greater than necessary.

Contrasting, the "Mean of Dataset" scenario presented an ESAL relatively lower than the one obtained with the individual contribution of each axle weight data available ("Total Dataset"). In 2016, the difference between these two scenarios was 31.5%, and in 2017 it was 33.8%. This consideration of a pavement design method that uses the ESAL concept would lead to an insufficient structure.

The ESAL obtained on the "Total Dataset" scenario confirmed that the fleet of 2016 presented higher loads than 2017, which was observed visually on the histograms of Figure 2.

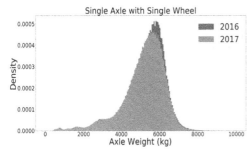

Single Axle with Single Wheel

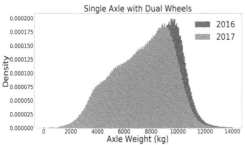

Single Axle with Dual Wheels

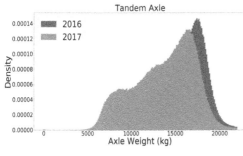

Tandem Axle

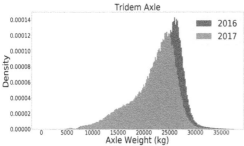

Tridem Axle

Figure 2.   Axles load spectra of 2016 and 2017.

Table 4.   Axles weight means and standard deviation

| | 2016 | | 2017 | |
	μ (kg)	σ (kg)	μ (kg)	σ (kg)
Single Axles (single wheel)	5245	1115	5188	1194
Single Axles (dual wheels)	7650	2337	7330	2206
Tandem Axles	14461	3748	13735	3537
Tridem Axles	23035	4747	21804	4531

Table 5.   Number of occurrences of each axle type.

	2016	2017
Single Axles (single wheel)	1,687,318	1,689,869
Single Axles (dual wheels)	1,266,843	1,286,508
Tandem Axles	1,215,420	1,230,442
Tridem Axles	486,739	490,175

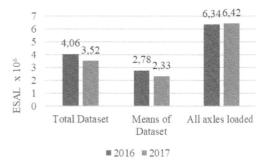

Figure 3.   ESAL (millions) obtained per year.

4   CONCLUSIONS

This paper aimed to present how useful Big Data tools can be in the context of analyzing data provided by LS-WIM systems. A large amount of data (more than 3.4 million entries) was accounted for thanks to the use of a proper tool, which was Pandas, NumPy, and Matplotlib libraries on Python language through the Jupyter Notebooks.

Data cleaning and characterization was performed to generate the axles load spectra. Besides that, these tools were useful to determine specific parameters (e.g. ESAL) using all available data.

Since it was available data of only two years, it was not possible to evaluate if there was a statistically significant difference.

ESAL calculated with simplifications and assumptions about the axles load can be very distant from the ESAL obtained from axles load spectra, with consequences on pavement design results.

For future works, it will be incorporated analysis using machine learning techniques to identify patterns on data that can provide better predictions about the traffic loads behavior.

REFERENCES

Hyun, K. (Kate), Hernandez, S., Tok, A., & Ritchie, S. G. 2015. Truck Body Configuration Volume and Weight Distribution. *Transportation Research Record: Journal of the Transportation Research Board, 2478*(1), 103–112.

Jasim, A. F., Wang, H., & Bennert, T. 2019. Evaluation of Clustered Traffic Inputs for Mechanistic-Empirical Pavement Design: Case Study in New Jersey. *Transportation Research Record: Journal of the Transportation Research Board.* 7

Li, J. Q., Wang, K. C. P., & Lou, J. 2016. Impact of time coverage of traffic data collection on Pavement ME Design. *International Journal of Pavement Research and Technology*, *9*(1), 1–13.

Lu, Q., Zhang, Y., & Harvey, J. T. 2009. Estimation of Truck Traffic Inputs for Mechanistic–Empirical Pavement Design in California. *Transportation Research Record: Journal of the Transportation Research Board*, *2095*(1), 62–72.

Sayyady, F., Stone, J. R., List, G. F., Jadoun, F. M., Kim, Y. R., & Sajjadi, S. 2012. Axle Load Distribution for Mechanistic–Empirical Pavement Design in North Carolina. *Transportation Research Record: Journal of the Transportation Research Board*, *2256*(1), 159–168.

Advances in Materials and Pavement Performance Prediction II – Kumar et al. (eds)
© 2021 Taylor & Francis Group, London, ISBN 978-0-367-46169-0

Short continuously reinforced concrete pavement structural model

L.S. Salles & L. Khazanovich
Department of Civil and Environmental Engineering, University of Pittsburgh, Pittsburgh, USA

J.T. Balbo
Department of Transportation Engineering, University of Sao Paulo, Sao Paulo, Brazil

ABSTRACT: Short continuously reinforced concrete pavement (CRCP) have been proposed as a solution for bus stops and terminals. Experimental sections designed with 50 m long concrete slabs, short in comparison to traditional CRCP, showed a unique cracking behavior. From a structural point of view, only a continuous model without cracks or joints was able to match field stresses in the short CRCP. However, since transverse cracks are visible at the slab surface, the continuous model may not be the most ideal structural model for short CRCP. Concerning this, partially developed cracks based on fracture mechanics were incorporated in a new model for this structure. Results show that, in contrast to the cracked model currently used for traditional CRCP, both the continuous and the partial crack models accurately match field deflection basis. However, both models also show different critical stresses under negative thermal differentials.

1 INTRODUCTION

Implementation of bus lanes in large urban areas can improve traffic by increasing average bus velocity which makes public transportation more competitive when compared to private transportation. However, constant rehabilitation and maintenance of bus lanes, especially near bus stops, can become a costly issue for highly urbanized areas. Therefore, the search for more durable and less problematic pavement structures becomes paramount for public transportation infrastructure.

Continuously reinforced concrete pavement (CRCP) has been proposed as a potential solution for bus corridors stops and terminals. CRCP is a concrete pavement without contraction joints. The major difference between CRCP and Jointed Plain Concrete Pavement (JPCP) is that the former has a high reinforcement percentage placed above the slab's half-height. The major role of this reinforcement is to keep inevitable transverse shrinkage cracks tight. In this way, cracks are imperceptible to the user and also provide highly satisfactory load transfer efficiency (LTE) by means of aggregate interlock. Therefore, in contrast to JPCP, there is not a CRCP crack control or inducement because of the steel structure designed to maintain CRCP slabs' structural and functional integrity. Several studies point out that the CRCP's major advantages are its durability and low maintenance needs, making it a suitable solution for bus corridors (Dossey & Hudson 1994; Gharaibeh et al. 1999; Tayabji et al. 1995; Won 2011).

In search of a better understanding of this structure as a solution for bus stops and terminals, four CRCP sections were constructed at the University of São Paulo (USP) campus in 2010. The CRCP sections were designed with only 50 m of length, which is short in comparison to traditional CRCP. The experimental short CRCP showed a unique crack pattern with a long cracking process resulting in fewer cracks than expected. It was assumed that the short structure length along with the lack of anchorage at the slabs longitudinal edge made the crack development slower, preventing cracks from appearing at the slab surface. Non-destructive ultrasonic testing also indicated that cracks were not fully developed to the slab bottom even after 5 years of construction (Salles 2017). However, despite the cracking behavior, falling weight deflectometer (FWD) and dynamic load testing results indicated typical CRCP performance, i.e., high crack LTE, low deflections (save for the longitudinal edges), and low stresses (Salles et al. 2015).

Salles et al. (2019) numerically simulated a dynamic load test performed on the experimental short CRCP with concrete pavement finite element software ISLAB2005. Only a continuous (no cracks or joints) 50 m long structural model was able to match field stresses for the short CRCP. However, since the transverse cracks are visible at the slab surface, the continuous model may not be the best representation of the short CRCP structure. Based on this, the objective of the present paper is to introduce and validate a new model with partial cracks not fully developed to the slab bottom.

1.1 Traditional CRCP structural model

Traditionally, CRCP is modeled as a cracked (full-depth cracks) slab with transverse cracks simulated as joints using an aggregate interlock model. The base layer is considered to have an unbounded interface with the top concrete layer, both presenting the same deflection profile. The cracks are considered to be propagated through the base layer and the aggregate interlock stiffness is computed based on crack width and crack shear wear damage. Since the the Mechanistic Empirical Pavement Design Guide (MEPDG) shows that crack width is a function of crack spacing, crack spacing significantly affects the predicted CRCP stresses induced by traffic and environmental loading. MEPDG computes concrete critical stresses located at the slab top using with a finite element model with small crack spacing. The design procedure uses these stresses to predict punchouts which are considered to be the major structural distress in CRCP (ARA 2003).

2 SHORT CRCP STRUCTURAL MODELS

2.1 General layout and slab transition

Table 1 presents material parameters and properties used for simulating the short CRCP concrete slab, asphalt base, and the adjacent asphalt pavement (Figure 1). Transition joints between the asphalt pavement and the CRCP slabs were simulated as a low LTE (10%) aggregate interlock joints.

Table 1. Short CRCP model general parameters.

Concrete Slab		
Concrete Density	kg/m3	2,400
Coef. of Thermal Expansion	°C-1	$8 \times 10\text{-}6$
Modulus of Elasticity	MPa	30,000
Poisson's Ratio		0.15
Asphalt Base and Adjacent Asphalt Pavement		
Base Thickness	mm	60
Asphalt Density	kg/m3	2,200
Resilient Modulus	MPa	3,500
Poisson's Ratio		0.35
Subgrade		
Modulus of Sub. Reaction	MPa/m	60

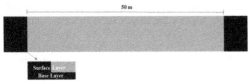

Figure 1. Model layout and transition joint.

In order to simulate the short CRCP as a bus stop in an asphalt pavement (Figure 1), instead of an unbounded interface between slab and base, the Totski model was adopted. The Totski model enables analysis of the independent bending of the CRCP and base layer. An interlayer spring layer placed between the concrete slab and base layer accommodates the direct compression of the base. If the Totski model is used for the base layer, ISLAB2005 allows the user to specify different joint properties in the base and in the top layer. To model continuous behavior of the base, joints in the base were assumed to be rigid, i.e. with perfect shear and moment load transfer.

2.2 Modeling of partial cracks

Salles et al. (2019), modeling short CRCP with ISLAB2005, revealed that the aggregate interlock model for CRCP cracks resulted in much worse correspondence with measured stresses than ignoring cracks altogether (continuous model). Therefore, there is a need to find a more realistic model for describing the structural behavior of short CRCPs.

In this study it was hypothesized that tight CRCP cracks have an ability to transfer moment in addition to shear transfer. To investigate this hypothesis, a partial depth crack model was used to characterize crack behavior in short CRCPs. This model was first used for analysis of pavement structures by Roesler and Khazanovich (1997). The model is based on the relationship between crack depth and line spring stiffness derived by Rice using a linear fracture mechanics approach. The model was later implemented into the ISLAB2005 finite element program.

When the crack depth ratio is 0, i.e. no crack, the crack is simulated with conventional aggregate inter-lock with infinite shear and flexural load transfer efficiency. Conversely, when the crack is in full depth (crack depth ratio = 1), the moment transfer efficiency is zero. For a partial depth crack the moment transfer efficiency is between these two extremes. The higher the crack depth the lower the moment transfer is.

In this study, the surface transverse cracks were assumed to be developed to or from the depth of the longitudinal reinforcement (around 80 mm). So for a 250 mm thick slab, the crack depth ratio would be 0.32.

2.3 Model validation

Three different models – continuous, cracked and partial crack – (illustrated in Figure 2) were used to simulate FWD tests performed at the experimental short CRCP in April 2014. FWD tests were carried out with a 60 kN load on a 30 cm diameter plate. Additional information on these and other FWD tests can be found elsewhere (Salles & Balbo 2016). Slab thickness at the location where the FWD tests were conducted was determined by a non-destructive ultrasonic test.

For each crack model mentioned above, backcalculation was performed for FWD deflection basins

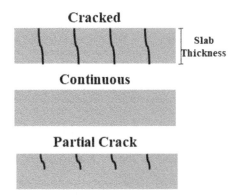

Figure 2. Cross sections of short CRCP structural models.

collected at two locations. By varying concrete modulus of elasticity and modulus of subgrade reaction, the discrepancy between the measured and theoretical deflections was minimized. Figure 3 presents FWD deflection basins (field data) and corresponding theoretical deflection basins for the three models. Transverse cracks in the cracked and partial crack models were simulated based on cracks visible at the slab surface from the time of the FWD test. For the cracked model, cracks were simulated as high LTE joints (90%) using the aggregate interlock model currently applied by the AASHTO PavementME design procedure. The partial crack model was simulated with crack depth ratio of 0.36 (Figure 2a) and 0.35 (Figure 2b).

As can be seen, the continuous slab and the partial crack models match the structure deflection basin accurately (the sum of square error between field and theoretical deflections of all sensors was below 1 for both models) while the cracked model fails to even capture the deflection behavior, i. e. smaller deflections away from the load. For the FWD test presented in Figure 3a, the load was applied 80 cm from the transverse crack; in this way, sensors spaced 90 cm and 120 cm were in adjacent slab panel. The deflection basin of the cracked model indicates an increase in deflection in the loaded panel reaching a maximum level at the crack. For Figure 3b, the crack was around 30 cm from the FWD load and most of the remaining sensors were on the unloaded slab. Again, the cracked model shows the influence of a fully developed crack without moment transfer on the deflections. Results corroborate the findings of Salles et al. (2009) indicating a significant moment transfer between adjacent panels for the short CRCP that is currently ignored by traditional CRCP design and evaluation.

2.4 Critical stresses on the continuous and partial crack models

Figure 4 shows critical top and bottom stresses for a 250 mm thick concrete slab under different thermal differentials. Both models use parameters presented in

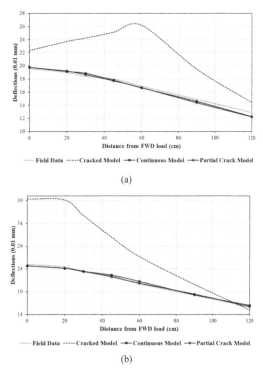

(a)

(b)

Figure 3. Field deflections matching with three different structural models.

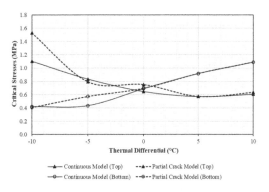

Figure 4. Critical stresses for continuous and partial crack models.

Table 1. A single axle load with 80 kN load was placed at the slab corner. For the partial crack model, a 50 m long concrete slab was simulated with constant crack spacing set at 1.25 m. The crack depth ratio was 0.32.

The models present similar results for critical bottom stresses induced by positive thermal gradients. However, when moderate to high (smaller than −5°C) negative thermal gradients are simulated, the stresses at the slab top are significantly different with the partial cracked model presenting more critical stresses. Figure 5 shows the stress distribution on the slab top for both models under a 10°C thermal gradient. The critical stresses for both models are located at the

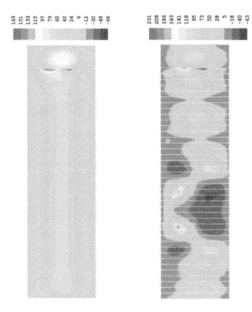

Figure 5. Stress distribution for continuous and partial crack model under a negative thermal differential.

corner where the SA was applied. It is worth mentioning that the continuous model was validated for top stresses due to positive thermal differentials (Salles et al. 2019).

3 CONCLUSIONS

A novel structural model for short CRCP analysis was proposed by incorporating a fracture mechanics-based partial depth crack model. The partial depth crack model predicts deflection basins similar to those measured using the Falling Weight Deflectometer. The aggregate interlock crack model traditionally used for modeling CRCP behavior failed to match FWD deflection basins. Results of this study indicate that a significant amount of moment transfer between panels currently not taken into consideration for design and analysis of traditional CRCPs.

REFERENCES

Dossey, T & Hudson, W. R. 1994. Distress as function of age in continuously reinforced concrete pavements: models developed for Texas pavement management information system. *Transportation Research Record,* 1455: 159–165.

Tayabji, S. D., Stephanos, P. J. & Zollinger, D. G. 1995. Nationwide field investigation of continuously reinforced concrete pavements. *Transportation Research Record,* 1482: 7–18.

Gharaibeh, N. G., Darter, M. I. & Heckel, L. B. 1999. Field performance of continuously reinforced concrete pavement in Illinois. *Transportation Research Record,* 1684: 44–50.

Won, M. C. 2011. Continuously reinforced concrete pavement: identification of distress mechanisms and improvement of mechanistic-empirical design procedures. *Transportation Research Record,* 2226: 51–59.

Salles, L. S. 2017. *Short Continuously Reinforced Concrete Pavement Design Recommendations Based on Non-Destructive Ultrasonic Data and Stress Simulation.* Doctorate Thesis. University of Sao Paulo.

Salles, L. S., Balbo, J. T. & Pereira, D. S. 2015. Nondestructive Performance Evaluation of Experimental Short Continuously Reinforced Concrete Pavement. *International Journal of Pavement Research and Technology,* 8: 221–232.

Salles, L. S., Balbo, J. T. & Pereira, D. S. 2015. Nondestructive Performance Evaluation of Experimental Short Continuously Reinforced Concrete Pavement. *International Journal of Pavement Research and Technology,* 8: 221–232.

Salles, L. S., Khazanovich, L. & Balbo, J. T. 2019. Structural analysis of transverse cracks in short continuously reinforced concrete pavements, *International Journal of Pavement Engineering,* DOI: 10.1080/10298436.2019.1570194

ARA. 2003. Guide for Mechanistic-Empirical Design of New and Rehabilitated Pavement Structures. Report *NCHRP 1-37A,* Applied Research Associates, Inc., Albuquerque, NM.

Salles, L. S. & Balbo, J. T. 2016. Experimental continuously reinforced concrete pavement parameterization using non-destructive methods. Ibracon Structures and Materials Journal, 9: 263–274.

Roesler, J. R., & Khazanovich, L. 1997. Finite-Element Analysis of Portland Cement Concrete Pavements with Cracks. Transportation Research Record, 1568(1), 1–9. DOI: 10.3141/1568-01

Advances in Materials and Pavement Performance Prediction II – Kumar et al. (eds)
© 2021 Taylor & Francis Group, London, ISBN 978-0-367-46169-0

Digital twinning of asphalt pavement surfacings using Visual Simultaneous Localization and Mapping

W.J.vdM. Steyn & A. Broekman
Department of Civil Engineering, University of Pretoria, Pretoria, South Africa

G.J. Jordaan
University of Pretoria, Pretoria, South Africa
Jordaan Professional Services (Pty) Ltd, Pretoria, South Africa

ABSTRACT: Quantification of pavement surfacing characteristics relates performance indicators pertaining to the mean texture depth, coarseness and loss of aggregates. Traditional measurement methods require skilled personnel and methodical execution of established methods to indirectly infer these characteristics. High accuracy Visual Simultaneous Localization and Mapping (VSLAM) techniques has seen widespread adoption in areas of robotics and visual odometry. Advances in hardware miniaturization, digital processing of optical imagery and reconstruction has seen the development of commercial scanners that provide mobile platforms and accuracies comparable to that of traditional, laboratory-grade laser scanners. Incorporating state-of-the-art, non-destructive digitization technology, together with well-established investigative methods, aims to improve both the fidelity and speed of surface data acquisition. The pipeline for the creation of a digital twin is simple in its execution, producing sub-millimeter accuracy and dense reconstructions over a wide range of dimensional scales. The twin model enables not only calculation of texture and profile characteristics, but also describes the orientation and curvature of particles along with changes in surface coarseness over time. The digitization process ties in directly with the continued drive toward digitization as part of improving pavement rehabilitation using predictive analytics and machine learning, improving existing preventative maintenance strategies.

Keywords: photogrammetry, pavement surfacing, digital twin, vSLAM

1 INTRODUCTION

Surface characteristics are defined by measuring the geometric properties of the pavement surface. The Mean Texture Depth (MTD) and Mean Profile Depth (MPD) represent two widely adopted metrics to define the surface. The MTD measures the macrotexture using a volumetric approach and is also referred to as the sand patch test (ASTM, 2006) whereas the MPD calculates the average profile depth over a 100 mm baseline (ASTM, 2003). The successful execution of these measurements requires calibrated instrumentation and trained personnel, only yielding a small amount of information compared to the complexity and time involved.

In the context of pavement engineering, the velocity and volume of Big Data (Núñez et al., 2014) is proving increasingly significant. Velocity for the increasing frequency of inspections that are becoming possible using digital solutions, and the volume of generated data over spatial-temporal domains.

Digital reconstruction techniques using optical systems are providing viable alternatives to investigate surfacing characteristics producing a "digital twin" (a virtual replica of the road surface in this case) of the physical asset.

2 RECONSTRUCTION METHODS

Reconstruction of 3D geometry from projections (2D photographic images) is an inherently ill-posed problem. One of the most successful reconstruction methods to date has been that of photogrammetry. Structure from Motion (SfM) and Clustering Views for Multi-View Stereo (CMVS) techniques infer the pose estimation and reproduces a reconstructed cloud point respectively from a set of photographs (Paixão et al., 2018). Photogrammetry is steadily replacing laser scanning for research applications. For the measurement of loaded tractor tire footprints on soft soil (Kenarsari et al., 2017), the laser scanner's accuracy and resolution were reported as $40\,\mu m$ and $50\,\mu m$ respectively compared to the standard deviation of the photogrammetry process measuring $100\,\mu m$. Measurement of pavement surface texture has been replicated using laser-based approaches with comparative accuracy (Sengoz et al., 2012).

Where photogrammetry is considered an offline process with processing proceeding after data acquisition, SLAM depends on real-time mapping of the operating environment by estimation the camera motion and pose (Taketomi et al., 2017). After the initialization phase, feature matching and tracking algorithms are fused to estimate the camera pose relative to the reconstructed map that is updated in real-time. State-of-the-art hardware is capable of sampling 300 000 points per second with accuracies of ± 30 mm up to 100 m (Jones et al., 2019).

3 DIGITIZATION PIPELINE

The creation of a digital twin is divided into three primary steps: sample preparation, data acquisition and digital processing.

3.1 Sample preparation

A representative sample area measuring approximately 300 mm on each side is marked off and cleaned of any loose debris and loose aggregates. Two coatings of white spray paint are applied, from all four cardinal directions at a 45° angle, to the surface and allowed to dry. A lightly colored surface is required for registration by the scanner. Reflective markers are randomly placed on planar surfaces of the aggregates with a spacing of 50 mm to 75 mm (Figure 1) for a total of around 30 markers. The reflective markers are required for accurate pose estimation due to the lack of features from the pavement surface. After a period of 6 weeks (with traffic), 30 per cent of the markers were still adhered to the aggregate). Scale markers in the shape of arrows were 3D printed and placed around the perimeter of the scanning area. These markers assist in tracking during scanning surface in addition to providing scale for the geometry, with the edges measuring either 10 mm or 20 mm on a side.

3.2 Data acquisition

The commercially available EinScan Pro was used for scanning different seals near Meyerton, located on the outskirts of Johannesburg. Both handheld and fixed tripod methods can be used with the latter offering twice the accuracy (50 μm) compared to the former (100 μm). The fixed scanning method was employed owing to the higher accuracy (resolution dn repeatability) and ease of use. Prior to scanning the samples, the calibration procedure was followed using the included calibration plate.

The aggregate's spherical morphology for most tests required the use of different scanning angles relative to the horizontal through adjustment of the scanner's tripod mount. 16 scans for each of the 3 pitch angles were acquired for a total of 48 scans per sample, starting from the perimeter to assist in pose estimation and tracking. The scanner projects a target area prior to

Figure 1. Prepared asphalt surface with reflective and scale markers.

Figure 2. Fixed mode scanning of a pavement surface sample using the EinScan system.

each scan that assists the operator to maintain uniform coverage (Figure 2). The auxiliary color camera was not utilized for the scanning process. With each scan, new datapoints are added to the model and displayed in real-time on the computer screen. A mobile power supply unit provided reliable power to the laptop computer and scanner for the duration of the data acquisition process and can be supplemented with the vehicle's inverter if required. All scans were conducted after sunset to reduce the glare produced by sunlight.

3.3 Digital processing

After the scanning process is complete, overlapping points among different scans are discarded and the resultant point cloud is meshed. The final model is exported to scale in the standard STL file format, without any post-processing options applied (decimation and smoothing). A typical scan will consist of approximately 5 million points. The digitization and

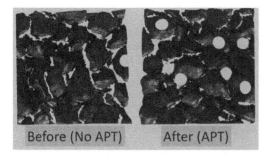

Before (No APT)   After (APT)

Figure 3. Digitized twins of 19 mm seals before (right) and after (left) accelerated testing.

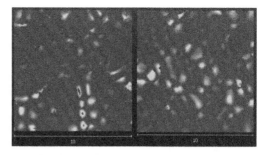

Figure 4. Curvature ($\lambda = 9.5$ mm) intensity map before (left) and after (right) accelerated testing.

export process for a sample can be completed within 20 minutes.

The open-source Blender animation and modelling software suite is used to obtain a fixed sample size. A Boolean cut is performed on a representative section where few markers are present using a "die" cube measuring 100 mm on each side. The intersection between the coplanar cube and geometry produce the sample measuring exactly 100 cm^2 that is subsequently reported in a STL file format. Figure 3 illustrates two digital twins (measuring 100 cm^2) of a 19 mm seal before and after testing with an accelerated pavement tester (APT).

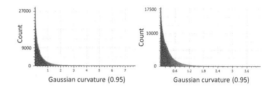

Figure 5. Curvature ($\lambda = 9.5$ mm) distribution before (left) and after (right) accelerated testing.

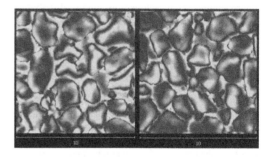

Figure 6. Coarseness ($\lambda = 9.5$ mm) intensity map before (left) and after (right) accelerated testing.

4 ANALYSIS

CloudCompare, the open source 3D point cloud and mesh processing software, is used to perform analytics on the standardized sample. Statistical tools require points instead of a mesh; hence 1 million random, interpolated points are sampled from the mesh surface, yielding a point density of 100 points per mm^2.

4.1 *MTD and MPD*

The MTD and MPD can be accomplished using the digitized twin model. With a predefined surface area, the volume required to fill the crevices can be calculated by fitting the surface within the smallest possible cubic volume. This volume divided by the surface area directly approximates the MTD. Similarly, the mean profile average depth can be calculated to determine the MPD measurement using either cross sections or on any line within the entire sample area.

4.2 *Expanded metrics*

The density of points for the digital twin enables additional analysis pertaining to the particle size, orientation, coarseness over different wavelengths and curvature. The included curvature measurement fits a best fitting quadratic about each of the points, with

the resulting distribution inferring particle distribution and reorientation. Figure 4 illustrates the curvature before and after 20 000 load cycles for a 19 mm seal. The kernel size of 9.5 mm is chosen as half the nominal aggregate size to selectively filter the distribution. Figure 5 both highlights the change in the curvature distribution as the particles re-orientate under the influence of traffic and serve as the color intensity map for Figure 4. These distributions highlight the significant suppression of strong curvature features with particles preferring co-planar orientations with the wheel surface.

Where the MTD and MPD provide an effective or mean value pertaining to the texture and profile respectively, the distribution of the coarseness can be defined using the included coarseness tool. The coarseness is calculated by fitting a plane through a selection of points. The number of points is again defined by the user as the kernel size. Using the same 19 mm seal sample with a kernel size of 9.5 mm, Figure 6 illustrates the change in distribution of the coarseness before and after the load application. Prior to traffic, the particles assume a somewhat random orientation with irregular surface features parallel to the riding surface and wide separation distances in-between particles. After the application of 20 000 wheel load cycles, the particles were reconfigured into a much tighter matrix with substantially smaller separation distances, in certain cases developing contact with one another. Most of the particles are arranged with planar parallel with the riding surface (particles with low coarseness), with others protruding from the surface that are constrained from reorientation by the neighboring particles.

5 CONCLUSIONS

With the rapid emergence of the 4th Industrial Revolution the importance of preventative maintenance strategies, measurement and investigative measurements should not be overlooked nor discarded. Instead, modern technologies provide innovative ways to improve the efficacy and accuracy of measurements whilst addressing the need to utilize Big Data in a meaningful and practical way. Decreasing costs and improvements in data acquisition, storage, distribution and utilization will accelerate the adoption of digitization methods. The results presented illustrate the unique benefits of acquiring digital twins of pavement surfacings in unprecedented detail and new metrics that describe its characteristics over different points in time, distilling large amounts of data into usable information. This technique was shown to enable a digitized and objective measurement of the surface coarseness of a pavement over a larger area, providing an improved understanding of the potential effects of such coarseness on tire-pavement interaction.

ACKNOWLEDGEMENTS

The Department of Civil Engineering and its staff is acknowledged for supporting the research conducted by the Chair in Railway Engineering at the University of Pretoria.

REFERENCES

American Society for Testing Materials (ASTM). 2006. Standard test method for measuring pavement macro-texture depth using a volumetric technique, ASTM E 965-96, Pennsylvania, USA.

American Society for Testing Materials (ASTM). 2003. Standard practice for calculating pavement macro-texture mean profile depth, ASTM E 1845-01, Pennsylvania, USA.

Jones, E., Sofonia, J., Canales, C., Hrabar, S. & Kendoul, F. 2019. Advances and applications for automated drones in underground mining operations. *Proceedings of the Ninth International Conference on Deep and High Stress Mining*, The Southern Africa Institute of Mining and Metallurgy, Johannesburg, pp. 323–334.

Kenarsari, A.E., Vitton, S.J. & Bears, J.E. 2017. Creating 3D Models of Tractor Tire Footprints using Close-range Digital Photogrammetry. *Journal of Terramechanics*, Volume 74, pp. 1–11.

Núñez, A., Hendriks, J., Li, Z., De Schutter, B. & Dollevoet, R. 2014. Facilitating Maintenance Decisions on the Dutch Railways Using Big Data: The ABA Case Study. *IEEE International Conference on Big Data*, Washington, DC, USA, pp. 48–53.

Paixão, A., Resende, R., Fortunato, E. 2018. Photogrammetry for Digital Reconstruction of Railway Ballast Particles – A Cost-efficient Method. Construction and Building Materials, Volume 191, pp. 963–976.

Sengoz, B., Topal, A. & Tanyel, S. 2012. Comparison of Pavement Surface Texture Determination by Sand Patch Test and 3D Laser Scanning. *Periodica Polytechnica Civil Engineering*, pp. 73–78.

Taketomi, T., Uchiyama, H., Ikeda, S. 2017. Visual SLAM Algorithms: A Survey From 2010 to 2018. *IPSJ Transactions on Computer Vision and Applications*, Issue 9, Article 16.

Advances in Materials and Pavement Performance Prediction II – Kumar et al. (eds)
© 2021 Taylor & Francis Group, London, ISBN 978-0-367-46169-0

Influence of nanoclay in viscosity graded asphalt binder at different test temperatures

Mayank Sukhija & Nikhil Saboo
Department of Civil Engineering, IIT (BHU), Varanasi, India

ABSTRACT: Nanotechnology has been progressively insinuating into the research field of asphalt binder modification. However, studies on various characteristics of nanoclay modified binder including resistance to ageing, temperature susceptibility and rheological performance, are still obscure. This study focuses on studying the efficacy of nanoclay modification on the properties of viscosity graded (VG 30) asphalt binder. High shear mixer was employed for the modification of different dosage (2–6%) of nanoclay to counteract the above mentioned performance characteristics. Conventional tests along with the rheological performance were evaluated for nanoclay modified and unmodified binders.

Master curves at 20°C and 60°C obtained from the frequency sweep (FS) test data revealed that improvement in rheological properties is more obvious at lower frequencies. Different Indices were evaluated analyzing FS data across different temperatures. A satisfactory and translucent level of modification was observed from Modification Index. Alongside, decrease in ageing index confirms the resistance to ageing potential of nanoclay modified asphalt binder. A noticeable change in black curve was observed for 6% NC modified asphalt binder while 2% and 4% NC modified asphalt binder was found to be less significant. However, the improvement with 2% and 4% NC inclusion was still higher than VG 30. From the fabrication of nanoclay modified asphalt binder to the rheological characterization process, this study confirms that nanoclay holds great prospective in alleviating the performance of VG 30 binder at all the test temperatures.

Keywords: Nanotechnology, Ageing, Temperature Susceptibility, Complex Modulus, Phase Angle, Black Curve.

1   INTRODUCTION

In the past few decades, increase in traffic intensity and loading condition coupled with new axle configuration and higher tire pressure have exasperated the severity of the conditions which leads to permanent deformation (Airey 2002). Modification of the asphalt binder is one of the best alternative to counteract such problems and improve the life of asphalt mixes (Saboo & Kumar 2016).

Nanomaterials has been gaining ground among the researchers and engineers for its inclusion in asphalt pavements. Nanotechnology provides a new asphalt material at a micro-level that has the ability to enhance the mechanical properties at macro-level (Mills-Beale & You 2011). Nanoclay, a silicate layered structure, is considered as one of the nanomaterial, capable of improving the field performance of asphalt pavements (Jahromi & Khodaii 2009). However, due to the presence of inherent electrostatic charge, nanoclay demands a higher rate of dispersion. Varun & Gehlot (2018) reported that nanoclay have a tendency

to cling with each other at lower rate of dispersion, resulting in small lumps. Ghile (2006) performed an experimental work and suggests that nanoclay modification influence some performance characteristics of asphalt binder along with asphalt mixture.

As far as ageing is considered, it is one of the ideal cause of failure in asphalt pavements. Asphalt binder at high temperature exists as a thin layer over the surface of aggregates and hence, leads to oxidation and volatilization (Singh et al. 2017). Ashish et al. (2017) stated that nanoclay modification has been found to reduce the ageing potential and it plays a vital role in enhancing the rheological characteristics of asphalt binder under a specified temperature range. Moreover, the study done by Ashish et al. (2017) claims that nanoclay addition turns the binder towards less temperature susceptible zone compared with neat asphalt binder.

This study is an attempt to differentiate the binder on the basis of modification level, ageing potential as well as temperature susceptibility using different performance predictor and analysis.

Table 1. Physical Properties of Asphalt Binder.

Processing Variables	VG30	2%NC	4%NC	6%NC
Penetration, dmm	62	39	36	33
Softening Point, °C	48	57	59	62
Viscosity 60°C, Poises	2704	3010	3154	3280
High Temperature PG Grade	PG 70	PG 76	PG 76	PG 76
True Fail Temperature, °C	74.3	76.5	77.5	80.4

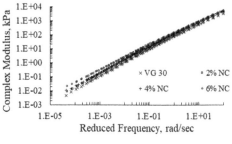

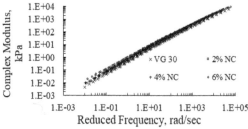

Figure 1. Master Curves at (a) 20°C and (b) 60°C.

2 MATERIALS AND PHYSICAL CHARACTERIZATION

Viscosity graded (VG) asphalt binder, VG 30, was used in the present study which was further modified with 2%, 4% and 6% nanoclay by weight of asphalt binder.

For the preparation of modified asphalt binder, the predefined dosage of nanoclay were intercalated into 200 gm of preheated VG 30 asphalt binder. Blending of nanoclay was done employing a high shear mixer (4000 rpm, 40 minutes). In addition to this, approx. 50 gm of binders were subjected to short term ageing process in a thin film oven following AASHTO T179 (AASHTO 2011). Table 1 represents the physical properties of VG 30 and nanoclay modified asphalt binder (2%, 4% and 6%) along with high temperature performance grading (PG). It is perceived that nanoclay modification increases the softening point as well as viscosity along with reduction in penetration value. Likewise, high temperature PG grade also shifts from PG 70 to PG 76.

3 EXPERIMENTAL PLAN

Performance characteristics including Modification Index, Ageing Index, Temperature Susceptibility were evaluated after conducting rheolCgical investigations such as linear viscoelastic range (LVE), performance grading (PG) of the binder and frequency sweep (FS) test at a temperature range of 10°C–70°C using dynamic shear rheometer (DSR).

FS test was conducted at various frequency range from 0.1 rad/sec to 100 rad/sec at 10°C–70°C. Master curves of complex modulus were plotted at selected reference temperatures of 20°C and 60°C which is similar to the normal field temperature at which distresses such as fatigue and rutting respectively becomes dominant.

In this paper, FS data was quantified using different indices including Modification Index and Ageing Index. The determination of Modification Index and Ageing Index were done utilizing equation 1 and equation 2 respectively.

To assess the suitability of nanoclay modified asphalt binder, temperature susceptibility of binders were compared by determining different rheological parameters at varying temperatures. Furthermore, Black curve, a plot between phase angle and complex modulus, was constructed for different binders to differentiate the binders on the basis of rheology.

$$\text{Modification Index} = \frac{(\text{Complex Modulus})_{\text{modified}}}{(\text{Complex Modulus})_{\text{unmodified}}} \quad (1)$$

$$\text{Ageing Index} = \frac{(\text{Complex Modulus})_{\text{aged}}}{(\text{Complex Modulus})_{\text{unaged}}} \quad (2)$$

4 RESULTS AND DISCUSSION

Amplitude Sweep data indicated that if the strain level is kept below 0.1% for all the asphalt binders at different temperatures, the range of linear viscoelasticity can be achieved. Therefore, in the present study, strain value was fixed at 0.1% for testing of asphalt binders.

4.1 Data obtained from master curves

Complex Modulus master curve at a reference temperature of 20°C and 60°C plotted from the data of FS test is portrayed in Figure 1 (a,b). It can be observed that nanoclay addition has a pragmatic influence on the performance characteristics of VG 30 binder. However, the improvement in performance, at both the reference temperature, is predominant at lower frequency range.

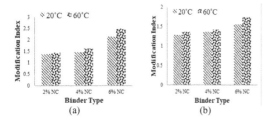

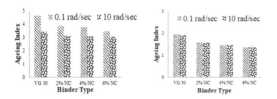

Figure 2. Modification Index at (a) 0.1 rad/sec and (b) 10 rad/sec.

Figure 3. Ageing Index at (a) 20°C and (b) 60°C.

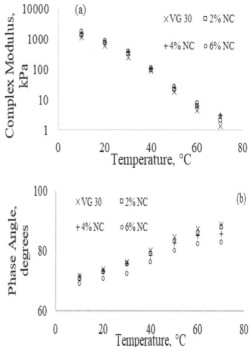

Figure 4. Variation with temperature (a) Complex Modulus and (b) Phase Angle.

4.2 Evaluation of modification index

Figure 2 (a,b) presents the Modification Index on the basis of complex modulus obtained at two different frequency values i.e. 0.1 rad/sec and 10 rad/sec respectively. For reference, Modification Index of VG 30 (0% NC) was kept equals to 1. It is noteworthy that with the increase in dosage of nanoclay, the modification level increases. However, the increase being more dominant at lower frequency region. In addition, a satisfactory and clear visualization of modification is observed from Modification Index at lower frequency. The outcomes of Modification Index were found to be congruous with the results of master curve plotted at the reference temperatures (Figure 1). Similar trends were seen at all the other temperature, therefore, not presented here for brevity.

4.3 Determination of ageing index

Effect of ageing on the rheological behavior of asphalt binder containing nanoclay was evaluated by comparing Ageing Index at the reference temperatures as shown in Figure 3 (a,b). In general, Ageing Index is affected by change in frequency and temperature. However, results reveals that, VG 30 binder is highly influenced by nanoclay dosage. The binder without nanoclay is found to be highly susceptible to ageing than the one modified with nanoclay. Similar trends were observed at other temperatures and are not shown here. Such behaviour can be attributed to the plate like structure of nanoclay which retards the oxidation and reduces the hardening process of asphalt binder. It is expected that nanoclay modification reduce the deformation and rutting distresses occurred in asphaltic pavements.

4.4 Temperature Susceptibility

Figure 4a shows the variation of complex modulus with respect to temperature at 10 rad/sec (80 km/hr). Generally, with change in temperature, higher the variation in the properties of asphalt binder higher is the temperature susceptibility. It can be seen that complex modulus of VG 30 binder decreases significantly with increase in temperature. Decrease in slope from 0.116 for VG 30 to 0.112 for 6% NC proves that the change is less significant with increase in nanoclay dosage in VG 30 binder. Trends obtained at different frequencies were found to be similar and hence not presented here. This states that, inclusion of nanoclay improves the temperature susceptibility of VG 30 binder.

Likewise, phase angle is considered to be highly sensitive to the modification and change in chemical structure of asphalt binder than complex modulus. Therefore, an attempt has been made in this paper to determine the temperature susceptibility of asphalt binder using phase angle values at 10 rad/sec. Figure 4b represents the variation of phase angle with temperature. It is observed that nanoclay modified binder has lower phase angle at all the specified temperature range (10°C–70°C) than VG 30, indicating improvement in elastic response of modified asphalt binder. Additionally, the temperature susceptibility was quantified using a linear curve fit and it is found that the slope increases from 0.261 (6% NC) to

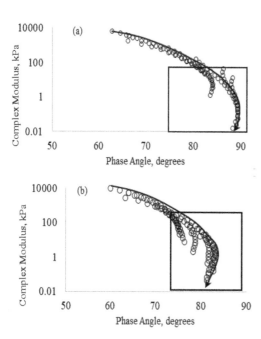

Figure 5. Black Curves (a) VG 30 and (b) 6% NC.

nanoclay modified binder at a temperature range of 10°C–70°C.

5 CONCLUSION

This paper examined the efficacy of nanoclay for VG 30 asphalt binders with the help of different performance analysis. In respect of master curves, it was observed that incorporation of nanoclay improves the performance at lower frequency range while the improvement at higher frequency range is marginal. A better and significant level of modification was observed from Modification Index at lower frequency and found to be congruous with the results of master curve plotted at reference temperatures. As far as ageing potential is considered, increase in hardness, due to ageing, decreases with the intercalation of nanoclay. Additionally, black curve shows the shift towards elastic response for nanoclay modified binders, indicating better temperature susceptibility and higher level of modification. The results confirms that incorporation of nanoclay plays a vital role in enhancing the performance characteristics of VG 30 binder at intermediate to high temperature zone.

0.309 for VG 30 binder. This increase in slope indicates higher temperature susceptibility for the base binder. Phase angle at higher temperature approaches to 90°, representing a shift towards viscous behavior. This change of shift is comparatively less for nanoclay modified asphalt binder. In addition, difference in variation of phase angle at lower temperatures is marginal compared to higher temperatures. In respect to the present discussion it is presumed that improved elastic behavior due to the inclusion of nanoclay might leads to better performance of VG 30 binder.

4.5 Discussion on black curve

Figure 5(a,b) shows the black curve of VG 30 and 6% NC asphalt binders respectively. A noticeable change was not observed in 2% and 4% NC modified asphalt binder, therefore, not shown here for brevity. In line with the above statement, 6% NC inclusion significantly influences the behavior of asphalt binder, indicating two phase structure change. Shift of black curve is represented by arrow inside the callout. It is found that the inner end of the black curve shifts towards lower phase angle i.e. from 88° in VG 30 to 80° for 6% NC, representing increase in elastic response. This divulge that nanoclay modified binder is less temperature susceptible. The obtained results can be tailed with the variation of complex modulus and phase angle with temperature (Figure 4) and proves the efficacy of

REFERENCES

AASHTO. 2011. "Standard Method of Test for Effect of Heat and Air on Asphalt Materials (Thin-Film Oven Test), T179."

Airey, Gordon D. 2002. "Rheological Evaluation of Ethylene Vinyl Acetate Polymer Modified Bitumens." *Construction and Building Materials* 16 (8). Elsevier: 473–487.

Ashish, Prabin Kumar, Dharamveer Singh, and Siva Bohm. 2017. "Investigation on Influence of Nanoclay Addition on Rheological Performance of Asphalt Binder." *Road Materials and Pavement Design* 18 (5): 1007–1026.

Ghile, Daniel Beyene. 2006. "Effects of Nanoclay Modification on Rheology of Bitumen and on Performance of Asphalt Mixtures, *MASTERS OF SCIENCE IN*,"

Jahromi, Saeed Ghaffarpour, and Ali Khodaii. 2009. "Effects of Nanoclay on Rheological Properties of Bitumen Binder." *Construction and Building Materials* 23 (8): 2894–2904.

Mills-Beale, Julian, and Zhanping You. 2011. "Nanoclay-Modified Asphalt Binder Systems." *Nanotechnology in Civil Infrastructure*, 257–270.

Saboo, Nikhil, and Praveen Kumar. 2016. "Optimum Blending Requirements for EVA Modified Binder." *Transportation Research Procedia* 17 (January). Elsevier: 98–106.

Singh, Bhupendra, Nikhil Saboo, and Praveen Kumar. 2017. "Use of Fourier Transform Infrared Spectroscopy to Study Ageing Characteristics of Asphalt Binders" 6466 (November). Taylor & Francis.

Varun, and Tarun Gehlot. 2018. "Use of Nano-Clay in Asphalt Binder Modification." *IOSR Journal of Mechanical and Civil Engineering (IOSR-JMCE) e-ISSN* 15 (1): 25–30.

Microwave heating simulation of asphalt pavements

H. Wang, P. Apostolidis, H. Zhang, X. Liu & S. Erkens
Section of Pavement Engineering, Faculty of Civil Engineering and Geosciences,
Delft University of Technology, Delft, The Netherlands

A. Scarpas
Department of Civil Infrastructure and Environmental Engineering,
Khalifa University of Science and Technology, Abu Dhabi, United Arab Emirates
Section of Pavement Engineering, Faculty of Civil Engineering and Geosciences,
Delft University of Technology, Delft, The Netherlands

ABSTRACT: Microwave heating is a promising heating technology for the maintenance, recycling and deicing of pavement structures. Many experimental studies have been conducted to investigate the microwave heating properties of asphalt mixtures in the laboratory. However, very few studies investigated the application of microwave heating on asphalt pavements. This study aims to simulate microwave heating of paving materials using the finite element method. Results show that the developed three-dimensional model, which couples the physics of electromagnetic waves and heat transfer, shows a great potential for optimizing the design of microwave heating prototypes for pavement applications.

1 INTRODUCTION

Microwave heating has been widely applied in various industrial fields, such as food and construction materials processing. Microwave has the potential to provide rapid, uniform, high efficient, safe and environment-friendly heating technology of materials (Jones et al. 2002; Metaxas & Meredith 2008). Due to the above advantages of microwave heating, there have been increased interests in utilizing microwave heating in the paving industry. Specifically, three main applications in pavement engineering: (i) pavement maintenance, such as crack healing in asphalt, pothole patching; (ii) recycling of the old pavement materials (heating of reclaimed asphalt pavement using a microwave tunnel); and (iii) snow melting or deicing (Wang et al. 2020).

In the conventional heating methods, such as hot-air heating and infrared heating, energy is transferred from the surfaces of the material to the internal by convection, conduction and radiation (Metaxas & Meredith 2008). In contrast, microwave heating is achieved by molecular excitation inside the material without relying on the temperature gradient. Therefore, microwave heating is a direct energy conversion process rather than heat transfer from external heat sources (Wang et al. 2019). This fundamental difference in transferring energy endows microwave heating many exclusive advantages, such as no air emissions or liquid pollutants, speed heating, volumetric heating, selective heating, easier to control and isolation of

risk conditions, strict control of programmed heating, etc. (Benedetto & Calvi 2013). Although microwave heating technology was tried and some prototype equipment was developed for paving materials production (Benedetto & Calvi 2013; Eliot 2013; Jeppson 1986) and pavement maintenance (Al-Qhaly & Terrel 1988; Bosisio et al. 1974; Terrel & Al-Qhaly 1987) applications, it is still not commercially used in this field at the present time, mainly due to the high operating costs. Therefore, this study aims to design a microwave heating system for paving materials and pavement structures through finite element method (FEM). The effects of microwave power, operating frequency, and moving speed on the heating efficiency of (asphalt) pavements were investigated.

2 FINITE ELEMENT MODEL OF MICROWAVE SYSTEM

2.1 *Multiphysics governing equations*

Microwave heating involves electromagnetic waves and heat transfer phenomena. To simulate the electro-magneto-thermal phenomena in a real-time system, the COMSOL Multiphysics software has been utilized for modelling microwave heating in pavements made from asphalt.

Electromagnetic analysis of a medium corresponding to a paving material, such as asphalt mixture, involves solving Maxwell's equations subject to certain boundary conditions. These equations can be formulated in differential form, which can be handled by FEM

$$\nabla \times \mathbf{H} = \mathbf{J} + \frac{\partial \mathbf{D}}{\partial t} \tag{1a}$$

$$\nabla \times \mathbf{E} = -\frac{\partial \mathbf{B}}{\partial t} \tag{1b}$$

$$\nabla \cdot \mathbf{D} = \rho_e \tag{1c}$$

$$\nabla \cdot \mathbf{B} = 0 \tag{1d}$$

To apply the Maxwell equations, the constitutive relations describing the macroscopic properties of asphalt mixture need to be determined. For linear materials, the polarization is directly proportional to the electric field; the magnetization is directly proportional to the magnetic field. Assuming asphalt mixture is an isotropic and linear material, the constitutive equations can be written as

$$\mathbf{J} = \sigma \mathbf{E} \tag{2a}$$

$$\mathbf{D} = \varepsilon \mathbf{E} \tag{2b}$$

$$\mathbf{B} = \mu \mathbf{H} \tag{2c}$$

where $\mathbf{H}$ is the magnetic field intensity; $\mathbf{J}$ is the electric current density; $\mathbf{D}$ is the electric displacement or electric flux density; $\mathbf{E}$ is the electric field intensity; $\mathbf{B}$ is the magnetic flux density; ρ_e is the electric charge density; σ is the material electrical conductivity; ε is the material permittivity; μ is the material permeability.

Applied microwave energy is converted into power based on the electric field distribution at a particular location. The absorbed power term is considered a source term in heat transfer equations to calculate transient temperature profile. The equation governing diffusion of heat into continua is as

$$\rho C_p \frac{\partial T}{\partial t} = \nabla \cdot (K \nabla T) + Q_e \tag{3}$$

where ρ is the density; C_p is the specific heat at constant pressure; k is the thermal conductivity; T is the temperature at time t; Q_e and is the internal heat source (absorbed power). The surface of the material exchanges heat with surrounding air by convection expressed as

$$-\mathbf{n} \cdot \mathbf{q} = h(T - T_a) \tag{4}$$

where $\mathbf{q}$ is the conductive heat flux; h is the surface convective coefficient; $\mathbf{n}$ is the normal vector on the boundary; T is the transient temperature and T_a is the ambient temperature.

The electro-magneto-thermal phenomenon often encountered in microwave heating is usually solved in a coupled manner. The distributed heat source, which includes resistive heating (ohmic heating) and magnetic losses in **Eq. 5** (Kopyt & Celuch 2007), is computed in a stationary, frequency-domain electromagnetic analysis. Then a transient heat transfer

simulation showing how the heat redistributes in the asphalt pavement was followed

$$Q_e = Q_{rh} + Q_{ml} \tag{5a}$$

$$Q_{rh} = \frac{1}{2}\text{Re}(\mathbf{J} \cdot \mathbf{E}) \tag{5b}$$

$$Q_{ml} = \frac{1}{2}\text{Re}(i\omega \mathbf{B} \cdot \mathbf{H}) \tag{5c}$$

where Q_{rh} is the resistive heating of dielectric material; Q_{ml} is the magnetic loss of magnetic material interacting with the magnetic field component of microwave. Re() is the real part of the variable.

2.2 Model definition

The microwave heating unit is a metallic box connected to a microwave source via a rectangular waveguide. The dimensions of the heating unit are 0.3 m (length) × 0.15 m (width) × 0.05 m (height). The waveguide is made of aluminum. To reduce surface losses, the inside walls are coated with copper, a high-conductivity metal. The applied impedance boundary condition on these walls ensures the small resistive metals losses get accounted for. As can be seen in Figure 1, there are two rectangular ports in the heating unit. Only Port 1 is excited by a transverse electric (TE) wave. The TE$_{10}$ mode was chosen at an arbitrary trial frequency of 1 GHz. The thickness of asphalt pavement (2 m × 2 m) is set as 0.2 m. The asphalt pavement layer is modeled as a dielectric material having electrical conductivity of $\sigma = 3.85 \times 10^{-7}$ S/m, relative permeability of $\mu = 1.03$, and a relative permittivity of $\varepsilon_r = 5.68$, with a loss tangent of $\delta = 0.176$. The thermal conductivity is k = 1.446 W/(m·K). Furthermore, the density is 2632 kg/m^3 and the specific heat is 756.5 J/(kg·K).

To ensure a relatively high level of simulation accuracy and a reasonable computation time, the domains of the three-dimensional model were meshed with different element sizes. The maximum mesh size in the

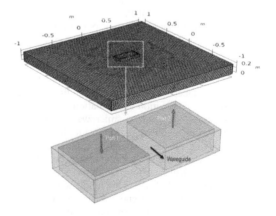

Figure 1. Model of asphalt pavement and microwave heating unit above pavement surface.

air domain in the microwave heating unit should be smaller than 0.2 wavelengths. For the asphalt pavement layer, the mesh size was scaled by the inverse of the square root of the relative dielectric constant. In this study, different input power values (1000 kW, 2000 kW, 4000 kW, and 8000 kW) were assigned to the port for the parametric analysis.

3 FINITE ELEMENT SIMULATIONS

Sensitivity analyses were conducted, given the importance to identify the main operational factors that influenced the efficiency of microwave heating. The thermal field distribution of asphalt pavement after 300 s microwave heating with the supplied power of 4000 kW is presented in Figure 2. Figure 2b shows the temperature distribution of pavement surface underneath the microwave ports. It shows a special heating pattern which is related to the electromagnetic wave shape. Based on the temperature distribution, it is recommended to move the microwave unit horizontally to achieve uniform heating and avoid repeated heating. Figure 2c shows the temperature distribution of the selected cross section (indicated by the dashed line) along the depth. It clearly shows the microwave energy attenuated gradually along the depth, resulting in a temperature gradient. The temperature distribution exhibits a pattern of wave propagation in the longitudinal direction.

To quantitatively analyse the effect of supplied microwave power on heating efficiency, point temperature evaluation underneath the center of Port 1 was conducted. Temperature evolution with time with different input powers was presented in Figure 3. As expected, a higher heating efficiency can be achieved when applying higher supplied power. When the supplied power increased to 4000 kW, surface temperature

of asphalt pavement can reach approximately 200°C after 300 s, which is sufficient for pavement maintenance and rehabilitation, etc. However, when applying an input power of 8000 kW, the surface temperature is extremely high after 300 s of heating, which will burn the material. One may argue that applying 8000 kW can reach the required temperature in a very short time. However, this leaves very limited time for the operational works for either maintenance or rehabilitation.

As pointed out earlier, microwave energy attenuated along the depth of asphalt pavement. It is important to know how the temperature evolves with the depth of pavement structure. Figure 4 shows the temperature variation with pavement depth at the location of centre of Port 1. From pavement bottom to surface, the temperature generally shows an increasing trend with clear wavelike fluctuations. The fluctuations are more prominent at a higher supplied power. The formation of temperature fluctuation is because of the harmonic nature of electromagnetic wave propagation. The peak and valley of temperature fluctuation correspond to the peak and valley of propagating wave considering the wavelength of applied microwave is 0.075 m according to Eq. 6. Based on the analyses from Figure 4, there are two points worthy to be noticed: (1) gradient heating along pavement thickness can be achieved by microwave to effectively control the heating depth; (2) particular temperature fluctuations can be realized through adjusting the microwave frequency to satisfy peculiar functions (e.g., heating certain layers of multilayer asphalt pavement structure).

$$\lambda = \frac{c}{f} \tag{6}$$

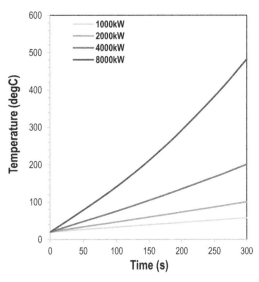

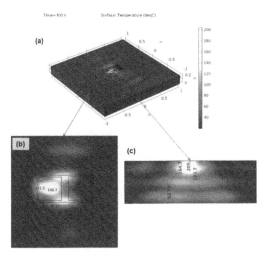

Figure 2. Thermal field distribution of asphalt pavement after 300 s heating (a) entirety (b) top surface (c) cross section.

Figure 3. Temperature evolution with time with different input power values.

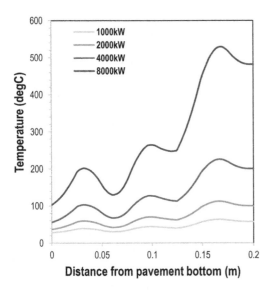

Figure 4. Temperature development of asphalt pavement along the depth after 300 s heating.

where λ is the wavelength, is the speed of light, and f is the operational frequency.

The heat transfer phenomena and the temperature profile along the thickness of studied medium as shown in Figure 4 coincide with the results obtained elsewhere (Bosisio et al. 1974; Sun 2014).

4 CONCLUSIONS

Based on the study results, the three-dimensional finite element method for microwave heating shows a great potential for optimizing the design of microwave heating prototypes for asphalt pavement applications. Effects of operational parameters on heating efficiency are simulated. A supplied power of 4000 kW at 1 GHz was expected to achieve sufficient heating temperatures without generating excessively high temperatures. Gradient and selective heating of asphalt pavement can be achieved by microwave heating.

For future studies, optimization of the design of microwave heating unit needs to be done to achieve a more homogenous thermal field. Practically, a moving microwave system should be added to the current mode to investigate the effects of moving speed on heating efficiency. In addition, effects of electromagneto-thermal properties of asphalt pavements can be examined.

REFERENCES

Al-Qhaly, A.A., & R.L. Terrel. 1988. Effect of microwave heating on adhesion and moisture damage of asphalt mixtures. *Transportation Research Record* (1171), 27–36.

Benedetto, A., & A Calvi. 2013. A pilot study on microwave heating for production and recycling of road pavement materials. *Construction and Building Materials 44*, 351–359.

Bosisio, R.G., et al. 1974. Asphalt road maintenance with a mobile microwave power unit. *Journal of Microwave Power 9(4)*, pp. 381–386.

Eliot, M. 2013. *Microwave processing unit for pavement recycling and asphalt pavement production*. WO 2013/166489 A1.

Jeppson, M.R. 1986. *Microwave method and apparatus for heating loose paving materials*. US 4,619,550.

Jones, D.A., et al. 2002. Microwave heating applications in environmental engineering-a review. *Resources, Conservation and Recycling 34*(2), 75–90.

Kopyt, P., & M. Celuch. 2007. Coupled electromagnetic-thermodynamic simulations of microwave heating problems using the Fdtd algorithm. *Journal of Microwave Power and Electromagnetic Energy, 41*(4), 18–29.

Metaxas, A.C., & R.J. Meredith. 2008. *Industrial microwave heating*. The Institution of Engineering and Technology: IET Power and Energy Series 4.

Sun, T. 2014. Key models of heat and mass transfer of asphalt mixtures based on microwave heating. *Drying Technology 32(13)*, pp. 1568–74.

Terrel, R.T., & A. Al-Qhaly. 1987. Microwave heating of asphalt paving materials. *Association of Asphalt Paving Technologists 56*, pp. 454–491.

Wang, H., et al. 2019. Laboratory and numerical investigation of microwave heating properties of asphalt mixture. *Materials 12*(1).

Wang, H., et al. 2020. Accelerated healing in asphalt concrete via laboratory microwave heating. *Journal of Testing and Evaluation 48*(2).

Advances in Materials and Pavement Performance Prediction II – Kumar et al. (eds)
© 2021 Taylor & Francis Group, London, ISBN 978-0-367-46169-0

Simulation of granular material in Distinct Element Model based on real particle shape

C. Wang, P. Liu & M. Oeser
Institute of Highway Engineering, RWTH Aachen University, Aachen, Germany

X. Zhou
Department of Civil and Environmental Engineering, Michigan Technological University, Houghton, USA

H. Wang
School of Highway, Chang'an University, Xi'an, China

ABSTRACT: An innovative approach was developed in this research to generate aggregates based on real shape through DEM. The shape indexes of aggregates were captured by Aggregate Image Measuring System (AIMS). The output was then processed by MATLAB to obtain the edge points of particles. The edge points were used to generate aggregate in DEM. Based on the former procedures, the aggregates morphological database that record shape indexes were established. The generation of DE models was optimized in the study to approach more realistic model geometries. Clump-based models with real morphologies were compared with laboratory tests and conventional ball-based models in the repose angle test. The clump-based model fitted better with the laboratory tests compared with ball-based models. Finally, a pavement compaction model was established. The influence of aggregate gradation and the movement of asphalt mixture during the compaction were investigated.

1 BACKGROUND OF STUDY ON GRANULAR MATERIAL

Traffic safety can be attributed to several factors in terms of environment, driver navigation, and pavement condition (Wilson 2006). The quality and durability of asphalt pavement, which mainly depend on the framework and material of the asphalt mixture are fundamental to sustainable transportation. The morphological features of aggregate can remarkably influence the frame structure, which consequently affects the general performance of asphalt pavement such as texture and friction (Maerz 2003; Xie et al. 2019).

Aggregates are generally divided into coarse- and fine grain particles determined by their sizes, which both have different effects on the functional performance of road surface (Maerz 2003; Wang et al. 2017; Wang et al. 2020). As for internal structure, coarse aggregates are closely related to the interlocking of the pavement framework (Chen et al. 2005; Pan et al. 2006; Rao et al. 2007). Besides, the shape parameters of aggregate have been proved as important input factors in different pavement models, including FEM and DEM (Abbas et al. 2007; Liu & You 2011; Masad et al. 2007). In the fields of asphalt mixture, the coarse aggregates have a high mass fraction. Thus, the properties of coarse aggregate such as shape properties are predominant factors in the construction procedure and

the quality of pavement. (Arasan et al. 2011; Singh et al. 2012).

DEM were carried out in 1970s. The method was called Distinct Element Method at first to distinguish from the popularly used Finite Element Method (FEM) (Cundall & Strack 1979, 1983; Liu & You 2011; Liu et al. 2019). DEM was designed to simulate mechanical process of granular materials. This simulation has the access to provide an innovative approach to enhance the understanding of building materials' properties (Bardet & Huang 1992; Liu et al. 2019; Olsson et al. 2019).

In DEM simulation, bulk materials are treated as an assembly of 2D disks or 3D spheres (Cundall & Strack 1979, 1983), or else as clumps of these shapes made by rigidly connecting and overlapping multiple disks or spheres (Bardet & Huang 1992, Bardet 1994; Favier et al. 1999, 2002; Potyondy & Cundall 2004; Oda et al. 1983). In real granular systems, the effect of rolling resistance has been regarded to represent the steric effect due to surface roughness or non-sphericity about the contact point (Calvetti et al. 1997; Iwashita & Oda 2002; Misra & Jiang 2002; Zhou & B.D 2005). Normally the rolling resistance contact model is based on the linear model, that incorporates a torque acting on the contacting pieces to counteract rolling motion. However, the virtual "angularity" imposed on particles cannot discriminate the interlocking conditions of bulk material with different particle sphericity. Therefore, the particle with real shape in DEM has more practical contribution to simulate the mechanical behavior of particles. In addition, it should be emphasized that the particle flow in DEM simulation with real shape can guarantee the efficiency in model repetition, and no need for revalidation

2 RESEARCH STRATEGIES

In this study the realistic shape of the aggregate was generated and applied into DEM simulation. First, the image of particle samples was obtained using AIMS, and the shape indexes of particle (were also captured with the imaging process analysis. Secondly, these indexes of particles were imported into simulation with shape-converted point coordinates of particle through MATLAB. The physical properties of the aggregates were calculated based on their contour by bubble pack calculation. The template of particle model was then reconstructed in DEM and utilized to generate certain amount of bulk materials. Coming to the following stage is to simulate the mechanical behavior of granular material in the DEM simulation with real particle shape information. In the end, the laboratory test related to the simulation was conducted to validate the model parameters.

3 DISCRETE ELEMENT MODEL GENERATION BASED ON AIMS APPLICATION

3.1 *Characterization on morphological properties of coarse aggregates*

The Aggregate Imaging System (AIMS) is utilized to capture the images of particles, and can be used to analyze the aggregate morphological property with a wide range of particle types and sizes (Ortiz & Mahmoud 2014). The application of AIMS II is able to characterize the surface on both macro- and micro- texture scales.

Masad et al. revealed that AIMS II is able to offer a precise, highly reproducible morphological description of the aggregate (Masad 2005). In this study, AIMS II was used to obtain angularity parameters of particles.

The change in the gradient angularity on aggregate boundary is related to the sharpness of the corners of its 2D images, which can be quantitatively described as the gradient angularity (GA) of a particle (Wang et al. 2017). The GA index calculates the inclination of gradient vectors on particle boundary points from the horizontal axis in an image.

The images of the aggregates were obtained and converted by image processing technology to get binary image. During the digital image acquisition the edge detection and image measurement was finished in MATLAB.

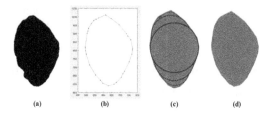

Figure 1. (a) 2D particle image; (b) Particle edge points acquired by MATLAB; (c) Particle reconstruction in PFC2D; (d) Finalized clump particles compare with the particle image that marked as red background.

Edge detection is a sort of image processing technology used to discriminate and obtain the boundaries of objectives within an image. In general, the edge detection algorithms include Sobel, Canny, Prewitt, Roberts, and fuzzy logic approaches. In this research the Canny algorithm was adopted utilizing MATLAB for particle image segmentation and data traction, since it can trace the shadow of particle in binary images and also detect gradient magnitude.

3.2 *Model establishment*

There are several steps to obtain the index of aggregates and input them into DEM: (1) image acquisition: the digital information of the aggregates were captured using AIMS; (2) Edge point identification and export: the contour recognition of particles was finished by MATLAB, the export datum of particles were then imported into DEM; (3) modular calibration: the template of particles was generated in DEM according to the imported data, the physical properties of aggregates were calculated; (4) The mineral and morphological information of each particle was recorded as its identification index, and a gene database of aggregates was developed. Figure 1 has shown the procedure to capture information of a single particle and then import into DEM simulation.

The database of aggregate utilized for simulation import was then developed based on the acquisition of aggregate through AIMS and MATLAB processing. The detailed indexes of particles are recorded with the particle into the database like the fingerprint of an aggregate. Every particle in the database can be called into simulation due to each index as required.

4 DISCRETE ELEMENT MODELLING AND LAB VALIDATION

4.1 *Contact model*

A contact model describes how elements behave when they come into contact with each other. The interaction between bulk material and equipment will be defined after the model has been generated. The geometry contact between particles and paver can be simplified into

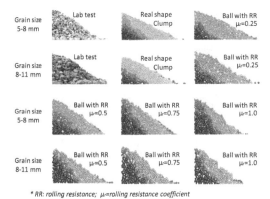

Grain size 5-8 mm | Lab test | Real shape Clump | Ball with RR μ_r=0.25

Grain size 8-11 mm | Lab test | Real shape Clump | Ball with RR μ_r=0.25

Grain size 5-8 mm | Ball with RR μ_r=0.5 | Ball with RR μ_r=0.75 | Ball with RR μ_r=1.0

Grain size 8-11 mm | Ball with RR μ_r=0.5 | Ball with RR μ_r=0.75 | Ball with RR μ_r=1.0

* RR: rolling resistance; μ_r=rolling resistance coefficient

Figure 2. Comparison analysis of repose angle from lab test and simulation.

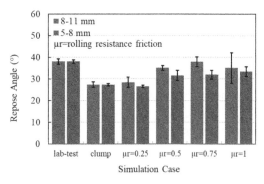

Figure 3. Comparison between the average experimental angle of repose and the simulated ones obtained using different rolling friction coefficient.

the contact between particles with walls in simulation, which can improve the calculation efficiencies as well as guarantee the accuracy of simulation. The material model used in the PFC is set as granular material without cohesive behavior. The surface friction of aggregate was obtained via frictional angle test, while the stiffness of aggregate was measured according to ASTM D6758-18 (ASTM. 2007).

4.2 *Comparison analysis in simulation*

Measuring the repose angle of aggregates in simulation was also conducted to compare and validate the model. The particles are generated according to designed gradation and then fall into the hollow cylinder with gravity until the cylinder was full. The cylinder was slowly raised vertically to allow the particles to naturally fall into a conical pile. The repose angle was then measured until the particles move at equilibrium state.

Figure 2 compared the results of repose angle from laboratory test and simulation.

5 RESULTS AND DISCUSSIONS

The simulation results of single-sized spherical particles and clumps were carried out separately. Simultaneously, the experimental tests of granule repose angle were carried out as well. Figure 2 illustrated the comparison between lab test and simulation. Based on the measurement data in laboratory, the variation in particle grain size does not have significant effect on the repose angle of particles. The data from Figure 3 indicated that for the variation in repose angle between two grain sizes, the simulated results of clumps with real particles shapes match the lab results relatively well, and their results errors are both within 5%. However, the value of repose angle from simulation is less than that from laboratory test, the value of the gap is approximately 10°. The reason why this variation exists can be attributed to the influence of the form

dimension. On the other hand, for the clumps in simulation, it is hard to simulate the sharp edges and corners of real aggregates, which can be the reason of the gap value as well. In addition, the simulation results of spherical elements with rolling resistance performs obvious difference between the size of 5-8 mm and 8–11 mm, which is different from the results obtain from experimental measurement. The results from this tests depicted that the repose angle of larger spherical particles is significantly larger than which of smaller one regardless of the rolling resistance coefficient. From this comparison analysis, it can be concluded that the simulation results are consistent with the one obtained from the laboratory test.

6 CONCLUSIONS

Based on the results of experiments and simulation, the main conclusions of this study can be summarized as follows:

An innovative approach was developed to generate aggregates that based on their real morphological properties via DEM simulation.

In DEM simulation, the results of repose angle measurement have improved that the aggregates with real shape has a better simulation result than which of spherical particles with rolling resistance.

The value of repose angle from simulation is less than that from lab test, the value of gap is approximately 10°. The reason why this variation exists can be attributed to the influence of the form dimension, and also because of the limitation for clumps.

More experiments are supposed to be conducted in terms of both laboratory and field test to further validate the calculation accuracy of the model.

ACKNOWLEDGMENTS

This research is supported by the research project of the National Natural Science Foundation of China (NSFC) (No. 51878063, 51578075), and is based on a part of the research project carried out at the request of the German Research Foundation (DFG), under research project No. OE 514/1-2 (FOR2089). The authors also appreciate the support from China Scholarship Council (Grant No. 201706560034, Grant No. 201706560033). The authors are solely responsible for the content.

REFERENCES

Abbas, A., Masad, E., Papagiannakis, T., and Harman, T., 2007. Micromechanical Modeling of the Viscoelastic Behavior of Asphalt Mixtures Using the Discrete-Element Method. *International Journal of Geomechanics*.

Arasan, S., Yenera, E., Hattatoglu, F., Hinislioglua, S., and Akbuluta, S., 2011. Correlation between Shape of Aggregate and Mechanical Properties of Asphalt Concrete. *Road Materials and Pavement Design*.

ASTM. Standard test method for measuring stiffness and apparent modulus of soil and soil-aggregate in-place by an electro-mechanical method.

Bardet, J.P., 1994. Observations on the effects of particle rotations on the failure of idealized granular materials. *Mechanics of Materials*.

Bardet, J.P. and Huang, Q., 1992. Numerical modeling of micropolar effects in idealized granular materials. *Mechanics of granular materials and powder systems*, 37 (1), 85–92.

Calvetti, F., Combe, G., and Lanier, J., 1997. Experimental micromechanical analysis of a 2D granular material: Relation between structure evolution and loading path. *Mechanics of Cohesive-Frictional Materials*.

Chen, J.-S., Chang, M.K., and Lin, K.Y., 2005. Influence of coarse aggregate shape on the strength of asphalt concrete mixtures. *Journal of the Eastern Asia Society for Transportation Studies*.

Cundall, P.A. and Strack, O.D.L., 1979. Discrete Numerical Model for Granualr Assemblies. *Geotechnique*.

Cundall, P.A. and Strack, O.D.L., 1983. Modeling of microscopic mechanisms in granular material. *In*: *Studies in Applied Mechanics*.

Favier, J.F., Abbaspour-Fard, M.H., and Kremmer, M., 2002. Modeling Nonspherical Particles Using Multisphere Discrete Elements. *Journal of Engineering Mechanics*.

Favier, J.F., Abbaspour-Fard, M.H., Kremmer, M., and Raji, A.O., 1999. Shape representation of axi-symmetrical, non-spherical particles in discrete element simulation using multi-element model particles. *Engineering Computations (Swansea, Wales)*.

Iwashita, K. and Oda, M., 2002. Rolling Resistance at Contacts in Simulation of Shear Band Development by DEM. *Journal of Engineering Mechanics*.

Liu, Y. and You, Z., 2011. Discrete-Element Modeling: Impacts of Aggregate Sphericity, Orientation, and Angularity on Creep Stiffness of Idealized Asphalt Mixtures. *Journal of Engineering Mechanics*.

Liu, Y., Zhou, X., You, Z., Ma, B., and Gong, F., 2019. Determining Aggregate Grain Size Using Discrete-Element Models of Sieve Analysis. *International Journal of Geomechanics*, 19 (4), 04019014.

Maerz, N.H., 2003. Technical and Computational Aspects of the Measurement of Aggregate Shape by Digital Image Analysis. *Journal of Computing in Civil Engineering*.

Masad, E., Muhunthan, B., Shashidhar, N., and Harman, T., 2007. Quantifying Laboratory Compaction Effects on the Internal Structure of Asphalt Concrete. *Transportation Research Record: Journal of the Transportation Research Board*.

Masad, E.A., 2005. *Aggregate Imaging System (AIMS): Basics and Applications*. Austin, Texas 78763–5080.

Misra, A. and Jiang, H., 2002. Measured kinematic fields in the biaxial shear of granular materials. *Computers and Geotechnics*.

Oda, M., Konishi, J., and Nemat-Nasser, S., 1983. Experimental micromechanical evaluation of the strength of granular materials: Effects of particle rolling. *In*: *Studies in Applied Mechanics*.

Olsson, E., Jelagin, D., and Partl, M.N., 2019. New discrete element framework for modelling asphalt compaction. *Road Materials and Pavement Design*, (July), 1–13.

Ortiz, E.M. and Mahmoud, E., 2014. Experimental procedure for evaluation of coarse aggregate polishing resistance. *Transportation Geotechnics*.

Pan, T., Tutumluer, E., and Carpenter, S.H., 2006. Effect of Coarse Aggregate Morphology on Permanent Deformation Behavior of Hot Mix Asphalt. *Journal of Transportation Engineering*.

Potyondy, D.O. and Cundall, P.A., 2004. A bonded-particle model for rock. *International Journal of Rock Mechanics and Mining Sciences*.

Rao, C., Tutumluer, E., and Kim, I.T., 2007. Quantification of Coarse Aggregate Angularity Based on Image Analysis. *Transportation Research Record: Journal of the Transportation Research Board*.

Singh, D., Zaman, M., and Commuri, S., 2012. Inclusion of aggregate angularity, texture, and form in estimating dynamic modulus of asphalt mixes. *Road Materials and Pavement Design*.

Wang, H., Wang, D., Liu, P., Hu, J., Schulze, C., and Oeser, M., 2017. Development of morphological properties of road surfacing aggregates during the polishing process. *International Journal of Pavement Engineering*.

Wang, C., Wang, H., Oeser, M. and Mohd Hasan, M.R., 2020. Investigation on the morphological and mineralogical properties of coarse aggregates under VSI crushing operation. *International Journal of Pavement Engineering*, pp.1–14.

Wilson, D.J., 2006. An Analysis of the Seasonal and Short-Term Variation of Road Pavement Skid Resistance, (May), 13, 39–50, 60.

Xie, X., Wang, C., Wang, D., Fan, Q., and Oeser, M., 2019. Evaluation of Polishing Behavior of Fine Aggregates Using an Accelerated Polishing Machine with Real Tires. *Journal of Transportation Engineering, Part B: Pavements*, 145 (2), 04019015.

Zhou, Y.C. and B.D, W., 2005. Rolling friction in the dynamic simulation of sandpile formation. Physica A: Statistical Mechanics and its Applications.

Advances in Materials and Pavement Performance Prediction II – Kumar et al. (eds)
© 2021 Taylor & Francis Group, London, ISBN 978-0-367-46169-0

Application and evaluation of novasurfacing technology

X.J. Zhang, W. Hong, S.L. Song, J.H. Luan, Q. Liu & P.J. Zhang
Gansu Henglu Traffic Survey and Design Institute Co., Ltd, Lanzhou, Gansu, China

ABSTRACT: In order to solve the problems of poor adhesion and durability at conventional micro surfaces, The process of spraying modified emulsified asphalt, paving and mixing super-abrasive layer containing glass fiber, novasurfacing are used. The SCB semi-circular bending test compares the performance comparison between the novasurfacing and the conventional micro-surfacing. The coefficients of friction, flatness, and PCI are used to compare the performance changes before and after the repair of the road section. Use the net annual value method to evaluate its economic benefits. The results show that the supervised abrasion layer has advantages in both performance and economic benefits.

1 INTRODUCTION

With the rapid increase in traffic volume and the severe existence of overload, The surface structure of the asphalt pavement constructed in the early period decreased in depth and friction coefficient. the pavement was severely damaged, such as rut pits and cracks, which jeopardized driving safety.

Gansu Province, China is located in an arid desert area. Large temperature difference, high UV intensity, large vehicle traffic and other factors cause roads to be prone to damage, and driving safety is at risk. In order to improve the road driving quality and the anti-skid performance of the pavement, postpone the overhaul and repair time of the pavement. Fast maintenance technology with convenient construct-ion and little interference to traffic is urgently needed.

The novasuefacing is a super abrasive layer. Special equipment is needed to spray modified emulsified asphalt binder and glass fiber in asphalt mixture. A new abrasion layer is formed after rolling. It can effectively improve the rut resistance and abrasion resistance of the sidewalk, and extend its service life, compared to the traditional micro surface. Currently, three types of fibers are mainly used in pavement engineering: cellulose fibers, polyester fibers and mineral fibers (Mcdaniel et al. 2009). Celauro added basalt fibers to the micro-surfacing and studied its effects on resistance to permanent deformation resistance and surface texture. The results show that it is beneficial to introduce basalt fibers to fight rutting (Celauro et al. 2018). Karine Krummenauer study on the incorporation of chromium-tanned leather residue to the cold asphalt micro-surface layer was performed using the dry mix process. Reach shows that the micro-surface layer lanes with added fiber wear less than the lanes without fiber (Krummenauer et al. 2009).

Micro-surfacing pavement repair and maintenance technology was used in Germany as early as the 1960s and 1970s, and then quickly spread in Europe and the United States (Erwin et al. 2009). In recent years, although many scholars in China have carried out a large number of application tests, there has been little comprehensive evaluation of the performance gap between novasurfacing and conventional micro-surfacing. In this paper, we investigate the application of a novasurfacing, compare the performance of the novasurfacing with ordinary micro-surfacing, and track and evaluate its road use effect one year later. It provides theoretical basis and technical support for the promotion and application of the novasurfacing.

2 MATERIAL CHARACTERISTICS

The main materials used in the novasurfacing mixture are polymer modified asphalt emulsion, fiber, mineral aggregate, filler, water and additives. Their quality directly affects the pavement performance of the mixture.

2.1 Asphalt material

The asphalt selected for the novasurfacing is specially modified emulsified asphalt for spraying. The indexes, technical requirements and test results of emulsified asphalt are shown in Table 1.

2.2 Aggregate

The coarse aggregate should be basalt, and the fine aggregate is made of artificial sand with appropriate gradation. The material properties meet the technical

Table 1. Basic properties of the emulsified asphalt.

Test item	Unit	Technical requirements	Inspection results	Inspection method
Demulsification speed		Slow cracking	Slow cracking	T66017
Particle charge		Cation	Cation	T0653
Residue on sieve	(%)	≤0.1	0.07	T0652
Engel viscosity E25	(Ev)	3 ~ 15	5.74	T0622
Asphalt standard viscometer C25 ~ 3 (S)	s	12 ~ 60	33.6	T0621
Evaporation residue				
Solid content	(%)	≥62	63.0	T0651
Penetration (100 g, 25°, 5 s)	(0.1 mm)	40 ~ 90	68.0	T0604
Softening point	(°C)	≥57	59.0	T0606
Density (5°, 5 cm/min)	(cm)	≥20	26.7	T0605
Storage stability	(%)	<1	0.7	T0655

Table 2. Basic properties of the aggregate.

Material	Pilot projects	Unit	Skills requirement	Test results	Experiment method
	Stone crushing value	%	≤26	12.6	T0316
	Loss of Wear Loss	%	≤28	13.5	T0317
Coarse aggregate	Stone polishing value	PN	≥42	53	T0321
	Sturdy	%	≤12	8.4	T0314
	Needle-like content	%	≤12	6.2	T0312
Fine aggregate	Sturdy	%	≤12	14	T0340
Mineral powder	Sand equivalent	%	≥65	66	T0334

Table 3. Basic properties of the aggregate.

Test items	Skills requirement	Test results	Experiment method
Apparent relative density (g/cm³)	≥2.50	2.73	T0352-2000
Water content (%)	≤1	0.17	T0103-1993
Exterior	No agglomeration	No agglomeration	Visual inspection
Hydrophilic coefficient	<1	0.51	T0353-2000

requirements of JTG E42-2005. The test results are shown in Table 2.

2.3 Mineral powder

The mineral powder obtained by grinding limestone alkaline stone is used. The test results of the material are shown in Table 3.

2.4 Fiber

Glass fiber is an important material for the novasurfacing. Non-twist roving-type glass fiber is used. The fiber cutting length is controlled at 5–10 mm.

3 TEST RESULTS AND ANALYSIS

3.1 Wet wheel wear value

The wet wheel abrasion test is used to determine the abrasion resistance of the emulsified asphalt slurry sealant mixture after molding, according to the test specification JTG E20-2011. The results are shown in Table 4. Compared with ordinary micro-surfacing, the 6-day wet-wheel abrasion value of the novasurfacing is smaller and the wear resistance is better.

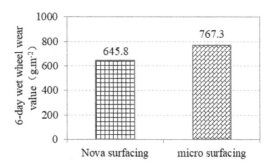

Figure 1. Compare the 6-day wet-wheel abrasion value results of novasurfacing and micro surfacing.

3.2 Shear resistance

A 45° oblique shear test was used to evaluate the adhesion between the novasurfacing and the road surface.

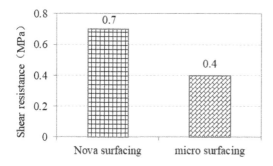

Figure 2. Compare the shear resistance results of novasurfacing and micro surfacing.

Table 4. Compare the crack resistance results of novasurfacing and micro surfacing.

Mixture type	Num	Deflection (mm)	Force Value (kN)	Break Energy (J/m^2)
Novo surfacing mixture	1	3.1	1.57	1520.94
	2	3.3	1.46	1505.63
	3	3.5	1.61	1760.94
	4	3.4	1.38	1466.25
	Ave	/	/	1563.44
Micro surfacing mixture	1	2.5	1.34	1046.88
	2	2.1	1.38	905.63
	3	2.9	1.52	1377.50
	4	2.8	1.21	1058.75
	Ave	/	/	1097.19

After the specimen was cured until it was completely hardened, the core was removed and the composite specimen was subjected to a shear test. The tests show that the failure forms of the specimens are all relatively moved between the layers. Compared with the microsurface, the bond strength of the supervised abrasion layer and the original road surface increased by 75%. The values of the test results are shown in Figure 2.

3.3 Crack resistance

The SCB semi-circular bending test was used to evaluate the crack resistance of the super-abrasive layer and the micro-surfacing mixture. The test results are shown in Table 4. The fracture energy of the super-abrasive layer mixture is significantly higher than that of the micro-surfacing. Compared with the micro-surfacing, the fracture energy can be increased by about 42%.

3.4 Friction performance

The pendulum friction meter was used to test the friction coefficient of the road surface. The test results are shown in Figure 3.

The results show that the road surface friction coefficient increases from 49 to 79 on the original road

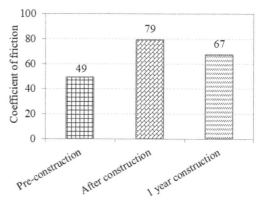

Figure 3. Pavement coefficient of friction results before and after construction and one year later.

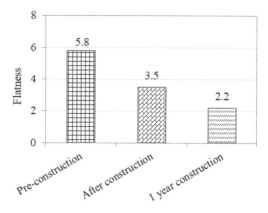

Figure 4. Pavement flatness test results before and after construction and one year later.

surface, indicating that the novasurfacing can greatly improve the friction coefficient of the original road surface and reduce driving danger. One year after opening, the average value is 67, which can provide good anti-skid performance services.

3.5 Flatness

A three-meter ruler was used to detect the flatness of the carriageway surface. The test results are shown in the Figure 4. The results show that the novasurfacing can effectively improve the smoothness of the road surface, and the traffic lane shows that the smoothness is 2.2 mm, which is a good performance.

3.6 Pavement condition index

The Pavement Condition Index (PCI) is a numerical index between 0 and 100 which is used to indicate the general condition of a pavement. used in transportation o standardized by (American Society for Testing and Materials).

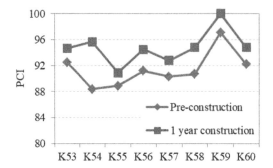

Figure 5. Pavement PCI results before constru-ction and one year later.

The surface damage of the test section of the nova-surfacing one year after opening to traffic was investigated. The investigation found that the main pavement diseases were cracks, including transverse cracks and partial network cracks. According to Figure 5, the average value of the original pavement was 91.4 and 94.8 one year after opening, which still has obvious advantages over the original pavement before construction.

3.7 Economic benefits

Due to the different service life of different maintenance technologies, in order to compare the economic benefits of ordinary micro-surfacing and super-viscous wear layers, this article uses the net annual value method to calculate the average annual maintenance cost during the service life for comparison. According to the maintenance time Ti as the base year (that is, $T_1 = 1$), the total present value of the maintenance costs within the service life is averaged to the base year to calculate the net annual value, in which the service life of the ordinary microsurface and NovaSurfacing Calculated according to 3 years and 5 years ($T_2 = 3, 5$).

There are two methods to calculate the maintenance cost: net present value (NPV) or net annual value (NAV). The formula is as follows:

$$NPV = \sum_{T_1=1}^{n} \frac{c}{(1+d)^{T_1}} \quad (1)$$

$$NAV = NPV \times \left[\frac{d \times (1+d)^{T_2}}{(1+d)^{T_2} - 1} \right] \quad (2)$$

In the formula:

NPV is the total present value of various maintenance costs during the analysis period, yuan;

NAV is the average annual maintenance cost of the analysis period, yuan;

D is interest rate, % (Calculated at 0.5%);

C is maintenance costs, yuan;

T_1 is maintenance and maintenance time, year;

T_2 is the time after the indicator reaches the level before maintenance, year.

After calculation, the net annual value at the micro-surfacing is 7.26 yuan, and the net annual value of the novasurfacing is 6.83 yuan, as shown in Figure 2. It can be seen that the net annual value of the novasurfacing technology is 6.3% lower than that of ordinary micro-surfacing.

4 CONCLUSION

This article uses a variety of experimental methods to compare the performance comparison between the novasurfacing and the conventional microsurface. The results show that the novasurfacing has several advantages. The specific conclusions are as follows:

1. The wet wheel abrasion value indicates that the fibers help to enhance the abrasion resistance and durability of the repaired pavement, and the shear resistance performance indicates that the super-adhesive abrasion layer has better adhesion characteristics with the original road mask, which effectively solves the problem of chipping on the repaired road phenomenon.
2. The SCB semi-circular bending test shows that the novasurfacing can significantly improve the crack resistance at the micro-surfacing and reduce the reflection cracks. the PCI value is also confirmed.
3. The novasurfacing effectively improves the friction coefficient and flatness of the original road surface, which is conducive to driving comfort and safety.
4. The net annual value indicates that the novasurfacing has economic benefits.

ACKNOWLEDGMENTS

The research work reported in this paper was supported by the science and technology program of Gansu Province (18YF1GA034)

REFERENCES

Celauro, C. Pratìco, F.G. 2018. Asphalt mixtures modified with basalt fibres for surface courses. Construction and Building Materials, 170:245–253.

Erwin, T. Tighe, S. L. 2008. Safety Effect of Preventive Maintenance: A Case Study of Microsurfacing// Transportation Research Record: Journal of the Transportation Research Board.

Mcdaniel, R. S. 2015. Fiber Additives in Asphalt Mixtures. Nchrp Synthesis of Highway Practice.

Krummenauer, K. Jairo Joséde Oliveira Andrade. 2009. Incorporation of chromium-tanned leather residue to asphalt micro-surface layer. Construction & Building Materials, 23(1):574–581.

Advances in Materials and Pavement Performance Prediction II – Kumar et al. (eds)
© 2021 Taylor & Francis Group, London, ISBN 978-0-367-46169-0

Application of phase change self-regulating temperature material in asphalt pavement

X.J. Zhang, S.L. Song, P. Yang & Y.Z. Yu
Gansu Henglu Traffic Survey and Design Institute Co., Ltd, Lanzhou, Gansu, China

H.L. Xu & W.D. Liu
Beijing Qintian Technology Group Co., Ltd

ABSTRACT: With the global warming and the frequent occurrence of extreme weather, many roads and urban roads are often damaged by ice and snow under the low temperature, rain, snow and freezing climate. The problem of ice and snow pavement has been puzzling the road maintenance department. In this paper, from the point of view of active control of the temperature of asphalt concrete pavement, the mix design and mixture performance of phase change self regulating asphalt mixture are briefly described, and the effect of engineering application is tracked and monitored. The results show that the phase change material can improve the ability of asphalt mixture to deal with the change of environment temperature, keep the asphalt mixture in a good temperature range for a long time, not only play the role of snow melting and ice melting, but also avoid the early diseases such as low temperature cracks, effectively improve the safety of highway driving, and extend the service life of pavement.

1 INTRODUCTION

Asphalt mixture is a typical temperature sensitive material, a large number of studies show that environmental factors, especially temperature significantly affect the performance of asphalt pavement (Cheng 2018; Ming 2015). In order to reduce the asphalt pavement diseases caused by temperature, we need to explore the technical measures to improve the adaptability of asphalt concrete pavement to the environmental conditions(Ma 201; Wei 201) from the new angle of actively improving the temperature condition of asphalt concrete pavement, so as to reduce the frost and ice formation of the pavement and its impact in the cold rain and snow freezing weather.

Based on the existing research of phase change materials, this paper focuses on the way of adding phase change materials into asphalt mixture through theoretical analysis and indoor test, and systematically studies the mixture proportion design and actual temperature regulating effect combined with the ac-tual engineering application.

2 MECHANISM AND TECHNICAL ADVANTAGES OF PHASE CHANGE AND TEMPERATURE REGULATION OF ASPHALT MIXTURE

Through chemical chain grafting, the long carbon chain in the asphalt is sheared and the hydroxyl of DTC road phase change material is shield, which makes the asphalt increase the softening point and be more stable at high temperatur; reduce the embrittlement point and be more flexible at low temperature. DTC road phase change temperature regulating material essentially improves the performance of as-phalt mixture, just like putting on the "space suit" for the road, so that it can store energy, conduct phase change temperature regulation at different temperatures, intelligently regulate the temperature of as-phalt pavement, avoid the temperature sensitivity of general asphalt mixture road surface, overcome the impact of environmental temperature, and prone to high-temperature rutting, low-temperature cracking and other damage phenomena The temperature fatigue life of asphalt mixture is extended by 54%, the adaptability of its environment temperature is improved, and the service life of road is nearly doubled. When the environment temperature changes dramatically, black ice is produced on the road. The self regulating asphalt mixture uses solar energy to store heat through phase change, reduce the temperature difference between the road surface and the environment temperature, so as to eliminate the harm of black ice.

In addition, the construction is simple and easy. Before mixing in the asphalt concrete mixing station, put the phase change temperature regulating agent beside the feeding port. Dry mix the mineral aggregate and fiber first, spray the asphalt for 10s, and put the phase change temperature regulator into the mixing plant at one time from the observation port of the mixer without prolonging the mixing time. The material has no special requirements for asphalt content, heating temperature of mineral aggregate and asphalt, and delivery temperature of asphalt mixture, does not affect the traditional process, and is easy to add and use. There is no pollution to the external environment, which provides us with a new way to improve road traffic safety.

3 MIX DESIGN OF ASPHALT MIXTURE

3.1 Mix design

The mixing amount of phase change materials for road temperature regulation shall be based on the total mass of asphalt mixture, which shall be determined according to the climate conditions of the project area. The mixing range shall not be more than 0.45% or less than 0.30%, and the mixing proportion shall be adjusted by gradient of 0.05%. The addition of road temperature regulating phase change material does not affect the conventional mix design of as-phalt mixture.

3.2 Performance verification of adding phase change material mixture

On the basis of ordinary road materials, phase change temperature regulating materials are new road materials which can change the performance of asphalt mixture by adding related materials. We have tested the splitting strength, freeze-thaw splitting, stability and low-temperature crack resistance of the self-adjusting temperature asphalt mixture. The test results show that each index of the self-adjusting temperature asphalt mixture under the reasonable mixing amount meets the requirements of the specification, all the performances are within the controlled range, and the application is not affected.

Among them, the results of freeze-thaw splitting test of self regulating asphalt mixture with different content are shown in Table 1, in which the splitting tensile strength of the test piece without freeze-thaw cycle is expressed by RT1, the splitting tensile strength of the test piece undergoing freeze-thaw cycle is expressed by RT2, and the splitting strength ratio of freeze-thaw is expressed by TSR. The ratio of freeze-thaw

Table 1. Freeze thaw splitting test results of self regulating temperature asphalt mixture

Gradatio	Phase change material content/%	R_{T1}/MPa	R_{T2}/MPa	TSR/%
	0	1.212	1.134	93.6
	0.1	0.986	0.937	95.0
sup-13	0.3	0.793	0.771	97.2
	0.5	0.621	0.586	94.3
	0.7	0.641	0.578	90.1

splitting strength of self regulating temperature asphalt mixture increases first and then decreases with the increase of phase change material content, and the ratio of freeze-thaw splitting strength of self regulating temperature asphalt mixture with different content meets the requiremen value of 80% «technical code for construction of highway asphalt pavemen» (JTG F40-2004).

4 EVALUATION OF ENGINEERING APPLICATION EFFECT

4.1 Engineering survey

In the project, sk2123 + 910-sk2125 + 000 section of G30 Lianhuo expressway is selected as the test section, with a total length of 1.09 km. This road section is located in Hexi corridor. It is cold and long in winter. The ice and snow damage on the road surface is serious, and part of the road section is in continuous downhill section, which makes the harm of ice and snow on the road more serious and seriously affects the driving safety. The original asphalt surface course shall be milled for 5 cm first, then the cracks shall be treated after cleaning, and the subgrade settlement shall be treated. The 1 cm modified asphalt stress absorption layer shall be made, and then the 4cm high-performance modified asphalt concrete sup-13 (mixed with 0.35% phase change material) shall be paved.

4.2 Application effect in summer

The temperature of the ordinary road section and the test section of adding phase change material is monitored by infrared thermal imager. The results show that the temperature of the road section with phase change material is about 7° lower than that of the ordinary road section. The test results are shown in Figure 1 and Figure 2

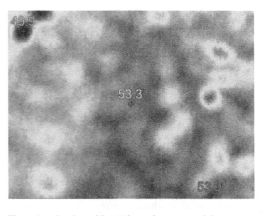

Figure 1. Section without phase change material.

4.3 Application effect in winter

The temperature of the test road and the ordinary road in winter was measured by infrared thermal imager. The results show that the temperature of the section with phase change material is 4.5° higher than that of the ordinary section. The test results are shown in Figure 3 and Figure 4.

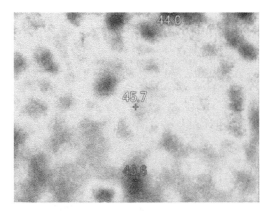

Figure 2. Adding phase change materials.

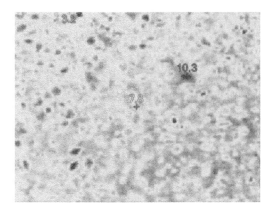

Figure 3. Section without phase change material.

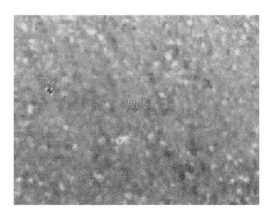

Figure 4. Adding phase change materials.

By using infrared thermal imager to compare the temperature of the road section with phase change material and the ordinary road section in winter, the temperature difference between the two roads can be directly reflected (red is the road section with phase change material added, blue is the ordinary road section). The test results are shown in Figure 5.

This test compares the test road section with ordinary asphalt pavement in full time. The results show that the temperature of the road section with phase change material is 2.5°–7.2° higher than that of ordinary road section in the same environment. Especially when the temperature of ordinary pavement is less than −10°, the temperature of the pavement with phase change material is within −5°, which improves the adaptability of asphalt pavement to extreme low temperature and large temperature difference, and the effect of temperature regulation is obvious. The curve of road temperature with time is shown in Figure 6.

The road surface temperature released a lot of latent heat under the low temperature environment in winter, which can effectively eliminate the black ice, freezing rain ice layer and frozen layer. Through the field observation, it is found that the road surface of ordinary road section is bright and slippery, while the road section with phase change material does not appear the phenomenon of road icing. The driving safety is improved. The comparison of pavement icing effect is shown in Figure 7.

Figure 5. Comparison of overall temperature of pavement in winter.

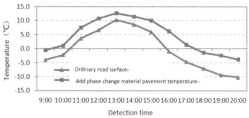

Figure 6. Comparison of road temperature in different periods.

Figure 7. Snow melting comparison between ordinary road section and DTC road section.

5 CONCLUSION

In this paper, Based on the research of phase change materials for road use, combined with the actual engineering application, this paper focuses on the way of adding phase change materials into asphalt mixture, and track and monitor the self regulating effect of asphalt mixture test section. The results show that adding phase change materials can improve the adaptability of asphalt road to extreme low temperature, large temperature difference and other climatic environment, that is, to carry out extreme temperature "Peak cutting and valley filling" can effectively eliminate the black ice on the road and improve the driving safety of asphalt pavement in winter! To avoid the temperature sensitivity of general asphalt pavement, to overcome the high temperature rutting, low temperature cracking and other damage caused by the environmental temperature, and to extend the temperature fatigue life of asphalt mixture.

ACKNOWLEDGMENTS

The research work reported in this paper was supported by the science and technology program of Gansu Province (18YF1GA034).

REFERENCES

Cheng Y C, Bi H P, Ma G R, et al. 2018. Pavement performance of nano materials-basalt fiber compound modified asphalt binder. *Jilin Daxue Xuebao (Gongxueban)/Journal of Jilin University (Engineering and Technology Edition),* 48(2):46–465.

Ming R, Xiao-Ling Z, Ming-Ming C, et al. 2015. Analysis of Rutting Resistance Performance of Asphalt Pavement Structure in High Temperature Area. *Journal of Wuhan University of Technology.*

Ma, Biao, Li, Jun, Liu, Ren Wei,. 2011. Study on Road Performance of Phase-Change Temperature-Adjusting Asphalt Mixture. *Advanced Materials Research* 287–290:97–981.

Wei S, Biao M , Xue-Yan Z, et al. 2018. Temperature responses of asphalt mixture physical and finite element models constructed with phase change material. *Construction and Building Material* 178:52–541.

Advances in Materials and Pavement Performance Prediction II – Kumar et al. (eds)
© 2021 Taylor & Francis Group, London, ISBN 978-0-367-46169-0

Curing and adhesive characteristics of single-component polyurethane binders

M. Zhang, K. Zhong & M. Sun
Research Institute of Highway Ministry of Transport, Beijing, China,
Key Laboratory of Transport Industry of Road Structure and Material, Beijing, China

X. Wang
Chongqing Jiaotong University, Chongqing, China

Z. Sun
Delft University of Technology, Delft, The Netherlands

ABSTRACT: Polyurethane is a relatively new type of binder for pavements. In this paper, curing and adhesion characteristics of two single-component polyurethane (PU) binders were studied. And chemical compositional, pull-out and shear analyses were performed to evaluate the curing performance of two PU binders. Meanwhile, the strength of PU were also adopted to represent adhesion characteristics (the binding capacity to the aggregate). With the increase of natural curing conditions, the results show that the isocyanate index of PU-I and PU-II decreases, while the Urea Index have the opposite trend. Both the isocyanate index and urea index of PU are stable after 4 days natural curing, indicating that PU are nearly fully cured after 4 days curing. As for pull-off and shear strength, PU-I and PU-II have better binding capacity to stone and rubber. Mean-while, the binding capacity of stone-PU–rubber system is the worst.

1 INTRODUCTION

Over the service of asphalt pavements, stiffening and embrittlement of asphalt binders is occurred due to environmental factors such as oxygen, temperature, ultraviolet radiation and moisture. Mainly because of the degradation of asphalt binders, asphalt pavements often suffer from defects such as thermal and fatigue cracking, and in view of these, it is of great importance the exploration of alternative binders for pavement structures. (Hou et al. 2018; Ryms et al. 2017).

Polyurethane contains polyurethane groups (-NHCOO-) of molecular chain, which has excellent properties such as high mechanical strength, toughness, ductility and oxidation resistance. (Demirel et al. 2019; Li et al. 2019; Swinton et al. 2016). Since polyurethane has been widely used in developing various synthetic polymers for engineering application. (Anbarlouie et al. 2018; Boutar et al. 2018; McCreath et al. 2018).

Due to the unique properties of PU, its application in the transportation industry is attracting more and more attention. Some scholars have proposed to use PU instead of traditional asphalt binder (Sun 2016). However, the reaction process of PU is really complex, which is susceptible to many environmental factors such as humidity and temperature. And the adhesion capacity of the PU is the key factor to determine whether the material can be widely used (Adams et al. 2014; Cong et al. 2019). In this paper, curing performance and adhesion property of single-component polyurethane were studied. FTIR, pull-out and shear tests were taken, in order to study the curing performance of different PU. On the basis of curing process, the adhesion property of PU to basalt stone and rubber is also studied.

2 MATERIALS AND PREPARATION

There were two kinds of polyurethane adopted in this research, which were named as PU-I and PU-II. The raw material of single-polyurethane is isocyanate, which is chemically modified and made by adding catalyst, cross-linking agent, chain extender, foaming agent and other additives. The isocyanate monomer of PU-I is toluene diisocyanate (TDI), and the isocyanate monomer of PU-II is diphenyl-methane-diisocyanate (MDI). Basalt and rubber cube were used to simulate aggregates in mixtures.

3 CURING OF SINGLE-COMPOMENT POLYURETHANE BINDER

Single-component polyurethane is the production of reaction between isocyanate (-NCO) and moisture (H_2O) in the air. In order to study the curing

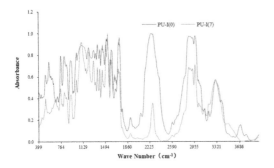

Figure 1. FTIR spectra of PU-I initially and at the end of cure.

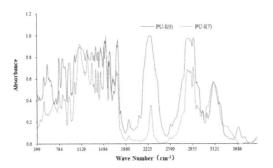

Figure 2. FTIR spectra of PU-II initially and at the end of cure.

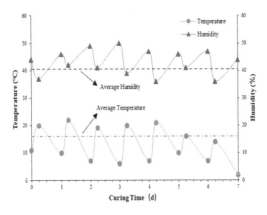

Figure 3. Temperature and humidity data of the 7 curing day.

performance of PU-I, PU-II, FTIR was adopted to measure the absorbance of key functional groups in single-component polyurethane after different curing time (Figure 1–2). Furthermore, the maximum curing time for single-component polyurethane was 7 days, and the temperature and humidity data of the 7 curing days (average temperature: 16.°; average humidity: 40.9%) were shown in Figure 3.

The changes of isocyanate, urea group and other major functional groups are mainly analyzed by Peak

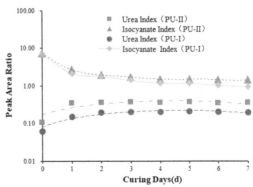

Figure 4. FTIR Results of Single-component Polyurethane after Different Curing Time.

area ratio method. Specifically, the benzene group ($1599\,cm^{-1}$) was adopted as the reference group because it would not participate in reaction for polyurea curing, and the wavenumber for isocyanate group and urea group are $2265\,cm^{-1}$ and $1642\,cm^{-1}$ respectively Moreover, Isocyanate Index and Urea Index were calculated by peak area ratio method according to the equation (1) and equation (2) (Chen 201; Carrera et al. 2010).

$$\text{Isocyanate Index} = \frac{A_{-NCO}}{A_{Benzene}} \quad (1)$$

$$\text{Urea Index} = \frac{A_{-NH-CO-NH-}}{A_{Benzen}} \quad (2)$$

The isocyanate index of PU-I and PU-II are named, According to Figure 4, The Isocyanate Index of PU-I and PU-II decreases with the increase of curing time, while the Urea Index have the opposite trend. Meanwhile, both the Isocyanate Index and Urea Index of PU-I and PU-II are stable after 4 days curing, indicating that PU-I and PU-II are nearly fully cured after 4 days curing.

4 EFFECT OF CURING TIME ON TENSILE AND SHEAR STRENGTH

In order to simulate the bonding state between the binder and the aggregate in the mixture, the specimens of stone/rubber-polyurethane-stone/rubber adhesion system were prepared (4 cm * 4 cm, 10 cm * 10 cm). The pull-off test (pull-off rate: 50 mm/min) was adopted to measure the pull-off strength of different adhesion systems and use AB glue to hold the two ends of the adhesive system to block. The adhesive properties of PU-I, PU-II on rubber and stone at different curing times were studied (Figure 5). The 45 degree shear test (shear rate: 10 mm/min) was adopted to measure the shear strength of different adhesion systems.

Figure 5. Pull-off test of different Single-component polyurethane.

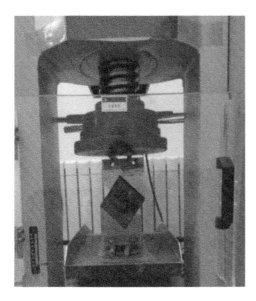

Figure 6. Shear test of different Single-compoent Polyurethane.

The adhesive properties of PU-I, PU-II on rubber and stone at different curing times were studied (Figure 6).

According to Figure 7, there is a gradual growth trend for pull-off strength. Comparing the adhesion property of PU-I, PU-II to stone and rubber, PU-I has better adhesion property to rubber, PU-II has better bond property to stone. According to three kind of adhesive systems, the pull-off strength between stone and stone is the highest, followed by rubber and stone, rubber and rubber is the lowest.

According to Figure 8, The shear strength increases gradually in 2∼7 days and is less than the pull-off strength. The adhesive property of PU-I to stone is better. And the adhesive characteristic of PU-I, PU-II to rubber material are similar. According to three kind of adhesive systems, the shear strength of rubber-PU-rubber system is the highest, and stone-PU-rubber is the lowest.

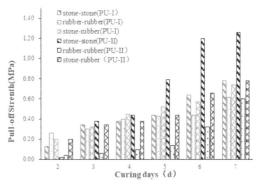

Figure 7. Comparison of pull-off strength of three adhesion systems of polyurethane with different components.

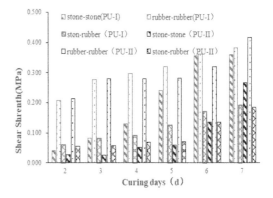

Figure 8. Comparison of shear strength of three adhesion systems of polyurethane with different components.

5 CONCLUSION

(1) Both the isocyanate index and urea index of PU-I and PU-II are stable after 4 days curing, indicating that PU-I and PU-II are nearly fully cured after 4 days curing. Yet figures 5 and 7 show that adhesive strength continues to increase. The possibility that cure continues in a manner that has less of an effect on isocyanate index and urea index may be considered.

(2) PU-I and PU-II have better adhesion property to stone and rubber. The pull-off and shear strength of the polyurethane adhesion system from 2 to 7 days is roughly increasing, and the shear strength is less than the pull-off strength.

(3) According to three kind of adhesive systems, the shear strength of stone-PU-rubber is the lowest, and the pull-off strength between rubber and rubber is the lowest. The rubber material is a special elastomer, the deformation ability of compression is higher than that of tension, which is needed to considered.

ACKNOWLEDGMENT

The authors would like to appreciate the research funding by the Fundamental Research Funds for the Central Research Institute (Grant No. 2019-0121). This financial support is gratefully acknowledged

REFERENCES

Adams, J.M., Kim, Y.R. 2014. Mean profile depth analysis of field and laboratory traffic-loaded chip seal surface treatments. *International Journal of Pavement Engineering* 15(7):64–656.

Anbarlouie, M., Mahdikhani, M., Maleki, A., 2018. The contribution of encapsulated polyurethane adhesive in improving the static torsional resistances of selfhealing concrete beam comparing bonded FRP technique, *Constr Build. Mater*. 191: 904–911.

Boutar, Y., Naïmi, S., Mezlini, S., et al., 2018. Fatigue resistance of an aluminium onecomponent polyurethane adhesive joint for the automotive industry: effect of surface roughness and adhesive thickness. *Int. J. Adhes. Adhes* 83:14–152.

Carrera, V., Partal, P., Garcí-Morales, M., Gallegos, C., Pérez-Lepe, A., 2010. Effect of processing on the rheological properties of poly-urethane/urea bituminous product s. *Fuel Processing Technology* 91(9)

Cong, L., Yang F., Guo G., Ren, M., Shi J., Tan, L., 2019. The use of polyurethane for asphalt pavement engineering applications: A state-of-the-art review. *Construction and Building Materials,* 225(3).

Demirel, S., Tuna, B.E., 2019. Evaluation of the cyclic fatigue performance of polyurethane foam in different density and category, *Polym Test*. 76:14–153.

Hou, X., Xiao, F., Wang, J., Amirkhanian, S., 2018. Identification of asphalt aging characterization by spectrophotometry technique. *Fuel* 226 23–239.

Li P., Guo Y., Zhou M., et al, 2019. Response of an-isotropic polyurethane foam to compression at different loading angles and strain rates. *Int. J. Impact Eng.* 127: 154–168.

McCreath, S., Boinard, P., Boinard, E., et al., 2018. High clarity poly (caprolactone diol)-based polyurethane adhesives for polycarbonate lamination: effect of isocyanate and chain-extender. *Int. J. Adhes. Adhes* 86: 84–97.

Ryms, M., Denda, H., Jaskuła, P., 2017. Thermal stabilization and permanent deformation resistance of LWA/PCM-modified asphalt road surfaces. *Constr. Build. Mater* 142: 328–341.

Swinton, M.C., Maref, W., Bomberg, M.T., et al., 2006. In situ performance evaluation of spray polyurethane foam in the exterior insulation basement system (EIBS), *Build. Environ* 41: 187–1880.

Sun, M., 2016. Research on performance of polyurethane porous elastic pavement mixture. *Southeast University, UDC:625.7/8.*

Chen, J., Ma,X., Wang, H., Xie, P., 2018. Experimental study on anti-icing and deicing performance of polyurethane concrete as road surface layer [J]. *Construction and Building Materials* 161:59–605

Structures

Moderators: Karim Chatti (Michigan State University), Elie Hajj (University of Nevada Reno), Ghassan Chehab (American University of Beirut) and Emin Kutay (Michigan State University)

Advances in Materials and Pavement Performance Prediction II – Kumar et al. (eds)
© 2021 Taylor & Francis Group, London, ISBN 978-0-367-46169-0

Evaluation of in-service ravelled asphalt pavements in Kuwait

T. Ahmed
Australian College of Kuwait, Kuwait, Kuwait

E.Y. Hajj, A. Warrag & M. Piratheepan
University of Nevada, Reno, USA

ABSTRACT: Evaluation of ravelled asphalt pavements was conducted in this study as a step toward an effort to investigate the possible causes of ravelling of asphalt pavements in the arid climatic region of Kuwait. Pavement distress surveys along with sand patch tests which were conducted for selected sections to identify the existed distresses and to estimate their pavement condition indices (PCIs). Several cores were extracted from the pavement sections for further laboratory evaluation. Pavement condition surveys and sand patch test results showed that all pavement sections, including the recently maintained sections within the last 3 years, were severely deteriorated due to the existence of ravelling. Sections' PCIs ranked from "Fair" to "Serious", indicating significantly poor functional performance. Aggregate sieve analyses and as-built volumetric properties results showed slight deviations from the specified requirements by the state of Kuwait, which can be related to the ravelling problem.

1 INTRODUCTION

1.1 Background

Ravelling can be defined as the deterioration of pavement due to the loss of aggregate particles from the asphalt pavement surface as because of the loss of adhesion between the aggregate and the asphalt binder. The existence of ravelling in the asphalt pavements significantly contributes to the development of other distresses such as, fatigue cracking and potholes formation, which can lead to premature and early failure of pavement. This problem gets worse in the presence of severe climatic and loading conditions such as high temperature and heavy traffic.

In arid areas like Kuwait, ravelling has been the primary cause of poor performance in many of the asphalt pavements. According to the Köppen climate classification, arid, or desert, climate is known to have extreme dry heating conditions all year long with moderate winters that last only for a short time. (Peel et al. 2007). Therefore, Kuwaiti asphalt pavement mix design procedure did not emphasis on the moisture sensitivity of the produced asphalt mixes due to the fact that there is low annual rainfall in the region, which varies from 75 to 150 millimeters and lasts for a very short time. In 2018, the meteorological department in the state of Kuwait reported that rainfall in Kuwait in November 2018 exceeded twice that of the year 1997. November 2018 recorded 261 mm, while it was 114 mm in 1997.

Because of this unexpected heavy rainfall, a significant number of asphalt pavements suffered from severe ravelling and stripping problems. Therefore, local transportation authorities in Kuwait are now trying to find an effective solution to this problem.

The development of ravelling in an asphalt pavement is influenced by several factors (Karol et al. 2016; Liantong et al. 2009; Miradi 2009) such as;

– Quality control (QC) and quality assurance (QA) deficiencies during construction, such as high moisture content in asphalt mixture or aggregates, improper production or compaction temperatures, insufficient compaction efforts, and the use of contaminated materials especially dusty aggregates.
– Asphalt mixture design characteristics such as low asphalt binder content, high air voids, high percentage of coarse aggregates, etc.
– Concentrated stresses under heavy traffic, chains, or studded tires.
– Extreme weather conditions, such as the number of very hot or cold days during the year, especially in the presence of moisture.
– Aging of asphalt, as the properties of binding material deteriorate over time.
– Or combination of the above.

Several research studies linked asphalt pavement ravelling to the stripping resulted from moisture damage. Stripping can be a result from either the adhesion failure between the asphalt binder and the aggregate particles, or the cohesion failure within the asphalt mastic itself (asphalt binder mixed with fine aggregate) (Liantong et al. 2009).

Understanding the mechanism of failure is a very essential step in determining the correct repair strategy for the affected asphalt pavement. Therefore, a comprehensive performance evaluation of the ravelled pavement should be undertaken before applying any treatment method. The performance evaluation should include both, laboratory and field studies, and the repair strategy should cover the design, construction, and monitoring practices of an asphalt pavement in order to prevent future premature failures due to ravelling.

1.2 Research objectives

This research study aimed to conduct an end-of-life evaluation of ravelled asphalt pavement in Kuwaiti arid climatic region. This work is part of an ongoing study to identify the main causes of ravelling distress and to improve the in-service performance.

2 RESEARCH METHODOLOGY

In order to achieve the research study objectives, five existing ravelled pavement sections were selected and evaluated accordingly.

Mainly, there are four road function classes in the state of Kuwait (Aljassar et al. 2007; MPW 2012b). These classes include;

1) Expressways and Freeways (EF): EF includes motorways and expressways that are major through routes for traffic.
2) Arterial Roads (AR): AR includes through traffic routes that are usually of lower design standard than EF.
3) Collector Roads (CR): CR includes roadways that are used to distribute local traffic through a district or to serve a place of importance within a local community.
4) Local Roads (LR): LR provides direct access to a local residential or a commercial area. They are used for short journeys only.

The five pavement sections of this study were selected to cover several in-service ages and functional classes in the state of Kuwait. The evaluation of these sections included the following two tasks;

Task I: Pavement condition surveys and sand patch tests: In this task, pavement condition surveys and sand patch tests were conducted for the selected pavement sections. Pavement conduction surveys were performed using visual inspections and distress measurements according to ASTM D6433-18 standards (ASTM 2018b). Sand patch tests were done according to ASTM E965–15 standards to assess ravelling and flushing (ASTM 2018c). An increase in texture is associated with ravelling; a decrease in texture can be an indication of flushing.

Task II: Field cores collection and evaluation: For this research study, the quality control/quality assurance (QC/QA) data were not available, therefore the research team focused on measuring the as-built properties of the selected sections. Since, ravelling

happens at the surface layer, the as-built properties of the top asphalt layer of the selected sections were measured and then were compared against the Kuwaiti roads and highways specifications (Kuwait MPW 2012a). A total of twelve cores were collected from each pavement section. Six cores were obtained from within wheel paths, and others were obtained from between wheel paths. Cores have been taken within the wheel paths were used to evaluate the density of top asphalt surfacing materials, while cores from between the wheel paths were obtained for asphalt mixture and binder testing.

The obtained cores were investigated to determine their volumetric properties such as bulk (G_{mb}) and maximum theoretical (G_{mm}) specific gravities, percentage of air voids (AV), asphalt binder content (AC), and aggregate gradations.

3 RESULTS AND ANALYSIS

3.1 Pavement condition surveys and sand patch tests' results

The summary of the sections' characteristics including their condition surveys and sand patch tests' results are shown in Table 1.

Sand patch test was used to determine the average macrotexture depth of the asphalt pavement surface. The test was done according the ASTM E965–15 (ASTM 2018c). The average depths of the pavement surfaces were calculated using the following equation.

$$MTD = 4V/\pi D^2 \tag{1}$$

where, $MTD =$ Mean texture depth of pavement macrotexture (mm).
$D =$ Average diameter of sand patch circle (mm).
$V =$ Volume of sand used ($69,350 \, mm^3$).

The higher the MTD value, the higher the surface roughness will be, which can be associated with Weathering/Ravelling distress. Normally, the coarse texture of a pavement surface ranges from 0.5 to 50 mm. Kuwaiti standards specifies a minimum of 0.8 mm for texture depth of newly constructed bituminous wearing course layers (MPW 2012a). As mentioned earlier, due to a lack of QC and inspection data during construction for the selected projects, the research team assumed that the base value for texture depth at all locations to be 0.8 mm. The difference between the current texture depth and the newly constructed project texture depth, MTD_{diff}, was calculated as follow:

$$MTD_{diff} = MTD - 0.8 \tag{2}$$

Based on the pavement condition surveys data and the sand patch test results, the presence of ravelling lowered the pavement rating for sections 1, 3, 4 and 5, giving the fact that these sections have been periodically maintained every 3–5 years. An exception is given to section 2, which has not been maintained for almost 10 years. Figure 1 shows samples of the observed distresses.

Table 1. Summary of pavement condition surveys and sand patch tests results.

Property	Section 1	Section 2	Section 3	Section 4	Section 5
	Locations Characteristics				
Functional Class	EF	CR	AR	CR	LR
AADT[1] (veh/day)	135,000	40,000	50,000	40,000	2,500
Last Maintenance	3–5 years	>5 years	3–5 years	5 years	5 years
	5 cm (2 inch) Mill and Overlay with Dense Graded Mix				
Mix Design Traffic Level	Heavy (>10^6 ESALs)				
Measuring Parameters	Sand Patch Test Results				
D (mm)	190	170	165	150	170
MTD (mm)	2.45	3.06	3.24	3.92	3.06
MTD$_{diff}$ (mm)	1.65	2.26	2.44	3.12	2.26
Roughness Level	Medium	Medium	Medium	Medium	Medium
Distress Type	Distress Severity[2], and Density[3]				
Long. & Transverse Cracking	Low 4%	Low 7%	–	Medium 6.5%	Medium 7%
Potholes		Medium 2%	–	–	High 2%
Rutting	High 35%	–	Medium 75%	–	–
Weathering/Ravelling	Medium 87%	Medium 80%	Low 20%	Medium 20%	High 87%
PCI	18	42	62	48	58
Rating	Serious	Very Poor	Fair	Poor	Fair

[1]*AADT (veh/day) = Annual Average Daily Traffic*
[2]*Distress Severity = Low, Medium, or High*
[3]*Distress Density = 100 × (Total Distress Quantity/Sample Area)*

Section 1: Rutting   Sections 2 & 3: Raveling

Section 4: Transverse Crack   Section 5: Pothole

Figure 1. Examples of observed distresses on sections.

Figure 2. Cores sampling and preparation process.

3.2 Field cores collection and evaluation results

In this task, the volumetric properties of the top AC surface layer were determined. In order to obtain the gradation of the field cores, an extraction and recovery process of the asphalt binder and aggregates were conducted in accordance with ASTM D2172/D2172M-11(ASTM 2018a). Figure 2 shows cores sampling and preparation process. Table 2 summarizes the measured volumetric properties of the collected core samples with cells highlighted in yellow indicating a property that did not meet the respective specification. All the volumetric and aggregate gradation properties were obtained with accordance with the latest AASHTO standards (AASHTO 2019).

Based on the measured volumetric properties, that section 1 exhibited lower in-place air voids with a 4.7% asphalt binder content; indicating excessive

Table 2. Summary of measured properties for collected Core Samples.

Volumetric Properties	G_{mm}	G_{mb}	AV (%)	AC (%)
Section 1	2.63	2.55	3.0	4.7
Section 2	2.44	2.32	4.8	4.9
Section 3	2.53	2.39	5.4	4.3
Section 4	2.45	2.28	7.0	4.1
Section 5	2.58	2.38	7.8	3.9
Kuwaiti Specs	–	–	4–6	4–5.5

AC (%) = Asphalt Content by Weight of Aggregate

compaction during and after construction. This section was maintained within the last 3 to 5 years, and it already exhibits a high severity rutting. Sections 2 and 3 met the required specifications; however, they still

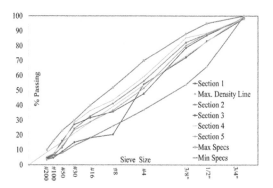

Figure 3. Recovered aggregate gradation analyses.

showed signs of ravelling associated with low field performance. Sections 4 and 5 exhibited higher in-place air voids along with relatively low asphalt binder contents; indicating a dry asphalt mixture that contributed to its low field performance.

Figure 3 shows the recovered aggregate gradation analyses.

As shown in Figure 3, the recovered aggregate gradations of the five sections showed slight differences form Kuwaiti specifications for surface layer aggregate gradation. This can be attributed to the loss of some aggregate particles from the surface due to continuous ravelling.

4 CONCLUSIONS, AND FINDINGS,

An end-of-life evaluation of selected ravelled pavements was completed in this research study as a step forward towards the overall effort to investigate and identify possible causes of asphalt pavement ravelling in Kuwait arid climatic region. Five pavement sections were identified and selected for further examination and evaluation.

Pavement distress surveys and sand patch test results showed that all pavement sections were significantly deteriorated due to ravelling and exhibited very low ratings given their relatively short in-service ages. The aggregate sieve analyses and as-built volumetric properties results showed slightly small to medium deviations from the Kuwaiti specifications, which could be the result of the loss of surface particles due to the ravelling problem.

Based on work done so far, it can be concluded that the ravelling could be resulted from a combination of the extreme weather conditions of Kuwait arid region, poor asphalt mixture design characteristics and concentrated stresses under heavy traffic. This can be seen from the pavement condition survey results and the as-built volumetric properties of the pavement sections which lead to the sections poor performance with respect to their design traffic levels and respective in-service ages.

Further laboratory evaluation of the collected cores is currently being conducted including moisture damage and rutting sensitivities, extracted asphalt binder penetration and performance grading (PG).

Final conclusions and recommendations will be made after combining the findings from this research study with the results from the undergoing laboratory evaluation of the collected cores.

ACKNOWLEDGMENTS

The activities presented in the paper is part of the research grant provided by Kuwait Foundation for the Advancement of Science (KFAS).

REFERENCES

American Association of State Highway and Transportation Officials (AASHTO). (2019). Standard Specifications of Transportation Materials and Methods of Sampling and Testing.

Aljassar, Ahmad H.,Ali, Mohammed A., and Al-Saleh, Omar I. 2007. Traffic Factors for Design of Roadway Geometrics and Pavement Structures in Kuwait. Kuwait journal of science and engineering, Vol 34 (2B), PP. 17–33.

ASTM D2172/D2172M. (2018a). Standard Test Methods for Quantitative Extraction of Asphalt Binder from Asphalt Mixtures. American Association of State Highway and Transportation Officials.

ASTM D6433–18. (2018b). Standard practice for roads and parking lots pavement condition index surveys. American Society for Testing and Materials.

ASTM E965-15. (2018c). Standard Test Method for Measuring Pavement Macrotexture Depth Using a Volumetric Technique. American Society for Testing and Materials.

Karol R. Opara, Marek Skakuj, Markus Stöckner. (2016). Factors affecting ravelling of motorway pavements-A field experiment with new additives to the deicing brine. Construction and Building Materials, Vol 113, pp 174–187.

Kuwait Ministry of Public Works (MPW), Roads Administration. (2012a). General Specifications for Kuwait Roads and Highways. State of Kuwait.

Kuwait Ministry of Public Works (MPW), Roads Administration. (2012b). Design Manual for Roads and Bridges. State of Kuwait.

Liantong Mo, M. Huurman, Shaopeng Wu and A.A.A. Molenaar. (2009). Fatigue Characteristics and Damage Model Development of Bitumen-Stone Adhesive Zones. Transportation Research Board 88th Meeting, Washington DC.

Little, D. N. Jr.; Jones, D. R. IV. (2003). Chemical and Mechanical Processes of Moisture Damage in Hot-Mix Asphalt Pavements. National Seminar, TRB Committee on Bituminous–Aggregate Combinations to Meet Surface Requirements Transportation Research Board (TRB), San Diego, CA, pp. 37–74.

Miradi, Maryam. (2009). Knowledge discovery and pavement performance: intelligent data mining. Ph.D. thesis, Faculty of Civil Engineering and Geosciences, Delft University of Technology.

Peel, M., Finlayson, B. and McMahon, T. (2007). Updated world map of the Köppen-Geiger climate classification. Hydrology and Earth System Sciences Discussions, [online] 4(2), pp.439–473.

Advances in Materials and Pavement Performance Prediction II – Kumar et al. (eds)
© 2021 Taylor & Francis Group, London, ISBN 978-0-367-46169-0

Calibration/validation of the Texas mechanistic-empirical flexible pavement design method

E. Alrashydah, A.E. Masad & A.T. Papagiannakis
University of Texas at San Antonio, San Antonio, TX, USA

ABSTRACT: This paper summarizes a study dealing with the calibration/validation of the Texas mechanistic-empirical (TxME) flexible pavement design approach. In doing so, Texas-specific data were used, sourced from DSS, a Texas DOT maintained pavement performance database. Calibration coefficients were established for the layer rutting, bottom-up fatigue cracking and transverse (thermal) cracking models. For each distress, analysis of variance was conducted to establish the factors that affect prediction errors. Calibration factors were estimated through different optimization techniques such as the generalized reduced gradient (GRG) and curve fitting algorithms. The calibration coefficients resulted in significant quality of fit improvements compared to those obtained using the default calibration factors.

1 INTRODUCTION

The Texas mechanistic-empirical (TxME) flexible pavement design model was developed under Texas Department of Transportation (TxDOT) projects (Hu et al. 2011; Hu et al. 2012; Zhou et al. 2010). It is intended as a tool for predicting flexible pavement distresses over time, given material, environmental and traffic data input. It consists of PC software with a graphical user interface that allows specifying pavement type, pavement layer properties, site location, and traffic spectra. Location allows the user to select environmental data from Texas weather stations near the design site. Traffic is characterized in terms of load spectra consisting of average annual daily truck traffic (AADTT), vehicle classification distributions, monthly adjustment factors (MAFs) and load frequency distributions by axle configuration. Material properties can be input as either fundamental properties (e.g., asphalt concrete master curves, soil-water characteristic curves) or index properties (e.g., resilient moduli, plasticity indices). Traffic input can be customized with site-specific data or approximated using software defaults.

2 OBJECTIVE

The objective of this paper is to summarize an effort undertaken to calibrate/validate the damage models of the TxME, using Texas-specific flexible pavement performance data stored in the DSS database (Lubinda 2015). The damage models discussed next include,
layer rutting, bottom-up asphalt concrete cracking and transverse (thermal) cracking.

3 TxME PAVEMENT DAMAGE MODELS

3.1 *Asphalt layer rutting*

Asphalt rutting model adopts the rutting model VESYS (Jordhal & Rauhut 1984) and was modified by expressing the elastic deformation in layer i as the difference in deformations between its top and bottom, denoted by $U_i^+ - U_i^-$ and introducing a correction function f and a calibration factor k_{AC} with a default value of 1.0817

$$RD_{AC} = \sum_1^m k_{AC} f \int_0^{N_1} (U_i^+ - U_i^-) \mu_i N^{-\alpha_i} dN \qquad (1)$$

where, μ_i and α_i are the plastic deformation parameters for asphalt concrete sublayer I and f is correction function for temperature, dynamic modulus and overlay thickness.

3.1.1 *Granular base and subgrade rutting*
The plastic deformation model for granular bases ($RD_{granularbase}$) and subgrades ($RD_{subgrade}$) incorporated into the TxME follows a similar format to the asphalt concrete plastic deformation model in (1)

$$RD_{granularbase} = k_{granularbase} \int \Delta U \mu N^{-\alpha} \qquad (2)$$

$$RD_{subgrade} = k_{subgrade} \int \Delta U \mu N^{-\alpha} \qquad (3)$$

where, U represent elastic deflection and μ and α are material properties. The calibration factors for the granular base and subgrade rutting, $k_{granularbase}$ and $k_{subgrade}$, respectively, have default values of 1.0. The difference between elastic deflections ΔU is estimated for each sub-layers i, as the deflection at the top of the sublayer minus the deflection at its bottom (i.e., $U_i^+ - U_i^-$).

3.2 Asphalt layer fatigue cracking

The bottom-up asphalt layer fatigue cracking damage model is defined in terms of the number of repetitions to failure (5)

$$N_f = k_i N_i + k_p N_p \qquad (4)$$

where N_i and N_p are the number of strain cycles necessary for crack initiation and propagation, respectively and k_i, k_p are the corresponding calibration factors with default values of 1.0. N_i is expressed as a function of the strain level ε

Fatigue damage FD is accumulated using Miner's hypothesis:

$$FD = \sum \frac{n_{i,j,k,l,m}}{N_{i,j,k,l,m}} 100 \qquad (5)$$

where, n_i is the actual number of load cycles applied under conditions i, and, N_i is the number of load cycles to fatigue failure under conditions i. The percent of the pavement surface that experiences fatigue-cracking FC is computed using:

$$FC(\%) = \frac{100}{1 + e^{C \log FD}} \qquad (6)$$

where, C is a calibration factor that has a default value of -7.89. In this form, Equation (6) yields $FC = 50\%$ area cracked when $FD = 1$.

3.3 Thermal (transverse cracking)

Thermal cracking model in TxME design approach is a variation of the low temperature-cracking model described the M-E PDG. Increments in the crack length ΔC are computed using (7) and introducing calibration coefficient k with a default value of 0.5:

$$\Delta C = kA(\Delta K)^n \qquad (7)$$

where, A and n are fracture parameters obtained by the Texas Overlay Tester and ΔK is the stress intensity factor caused by thermal stresses given by:

$$\Delta K = \sigma(0.45 + 1.99C^{0.56}) \qquad (8)$$

where, C is the crack length, computed as the sum of the crack increments ΔC and σ is the daily maximum horizontal stress at the depth of the crack tip.

The relationship used to compute the total thermal cracking amount CA (feet/mile) is given by:

$$CA = \frac{100B}{e^{0.698147}\left(\frac{\rho}{m}\right)^\beta} \qquad (9)$$

where, ρ is the length of time (months) it takes for the sum of the crack increments $\sum \Delta C$ to accumulate to a length equal to the thickness of the asphalt concrete layer h_{AC}, m is the month number, while B and β are calibration constants with default values of 42.24 and 1.1854, respectively. The calibration factor B was selected to reflect cracking failure, that is the month when the sum of the crack increments has reached the thickness of the asphalt concrete layer (i.e., when by definition $\rho = m$), 50% of the maximum possible length of thermal cracks, estimated to be 4,224 feet/mile.

4 PERFORMANCE DATA

77 pavement sections were extracted from THE DSS database (Lubinda 2013). They were divided into the following groups by structural configuration:

A* = not overlaid regardless of base type
D = overlaid with unbound base
E = overlaid with cement stabilized base
F = overlaid with lime stabilized base and
G = Perpetual

5 RUTTING MODEL CALIBRATION

The pavement type, subgrade modulus (M_r) and the Average Annual Daily Truck Traffic (AADTT) were shown to be the main factors affecting TxME predictions based on one-way ANOVA results. In addition, three-way ANOVA was carried out to determine the need to further subdivide these groups with respect to the interaction between the three significant factors namely, pavement type, subgrade modulus, and AADTT. The results showed that AADTT volumes and to a lesser extent the subgrade modulus (M_r) are statistically significant. Given the limited number of sections in some of the pavement groups, it was decided to subdivide them by AADTT level only.

Rutting observations were available only for the pavement surface. Since the layer calibration factors weigh the estimated plastic deformation linearly (Equ. 1 to 3), the Excel®Solver was used to estimate them by minimizing the SSE between monthly observed surface rutting and the sum of the estimated plastic deformations in the AC, base and subsase layers. The resulting calibration coefficients were 0.632, 0.368 and 1.043 for AADTT < 500 and 0.690, 0.100 and 3.604 for AADT > 500, resp. These calibration factors resulted in reductions in bias and standard error (S_e) of 71% and 19%, respectively, compared to those estimated using the default values. The following plots

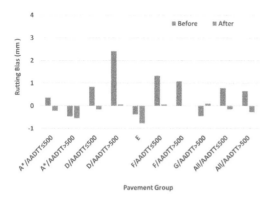

Figure 1. Bias in rutting predicitons before/after calibration.

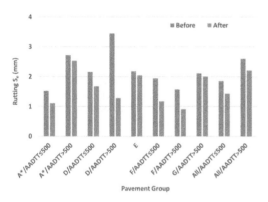

Figure 2. S_e in rutting predicitons before/after calibration.

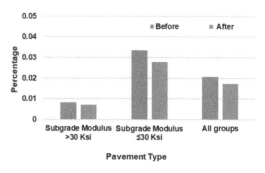

Figure 3. Bias in fatig. crack. predicitons before/after calibration.

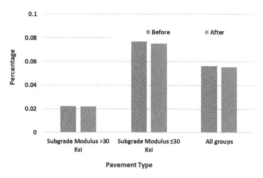

Figure 4. S_e in fatig. cracking predicitons before/after calibration.

show the reduction in bias and S_e for the rutting models based on pavement type and AADTT level.

plots show the reduction in bias and S_e for the fatigue models based on pavement type and AADTT level.

6 FATIGUE CRACKING MODEL CALIBRATION

For the fatigue-cracking model, there are three calibration factors, namely k_i and k_p that weigh the number of cycles for crack initiation and propagation, resp., and C, which defines the transfer function between fatigue damage (FD) and wheel path area cracked (FC). Their corresponding default values are 1.0, 1.0 and -7.89, respectively. For the purpose of calibration, only sections with non-zero cracking observations were usable. This reduced the number of sections used for calibration from 77 to 19. The ANOVA results conducted on the prediction errors before calibration suggest that the only factors that are significant are the base thickness and the subgrade modulus (M_r). In addition, 2-way ANOVA between these two variables suggested that only the subgrade M_r is significant in predicting fatigue cracking. The Excel®Solver and the curve fitting algorithms in MATLAB®were used to find the calibration factors. The recommended calibration coefficients were, 1.0, 1.0 and -4.2415. The following

7 THERMAL CRACKING MODEL CALIBRATION

The TxME thermal cracking calibration coefficients to be determined are k, B and β with default values of 0.5, 42.24 and 1.1854, respectively. The coefficient k weighs linearly the predicted crack increments ΔC in Equation (7) and hence, it affects the rate of their accumulation and the length of time it takes for their sum to reach the thickness of the asphalt concrete layer. By definition, the month m when this takes place, is ρ and hence, at this point the ratio ρ/m equals one. Furthermore, at this point, the Cracked Amount CA becomes 50B. For the default B value of 42.24, 50B equals 2,114 ft, which is the consensus maximum length of transverse thermal cracks expected per mile. As a result, it was decided not to alter B, but focus the calibration in establishing the values of k and β that generate TxME estimates best fitting the observed transverse cracking amounts. For this purpose, only 20 sites with non-zero transverse crack observations were usable.

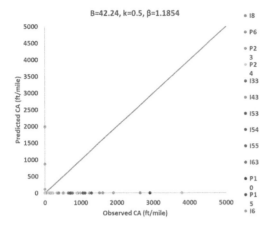

Figure 5. *CA* predictions versus observations before calibration.

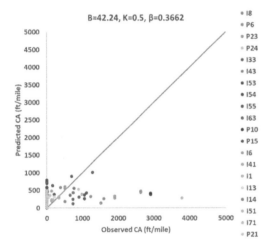

Figure 6. *CA* predictions versus observations after calibration.

Preliminary evaluation of the transverse cracking model suggested that the coefficient k weighing the crack increments needs to be increased, while the exponent β of the *CA* transfer function needs to be decreased from their default values. A multitude of TxME simulations was conducted using the calibration dataset by selecting a wide range of coefficients. In these simulations, k ranged from 0.1 to 170, while β ranged from 1.1854 to 0.1. The resulting standard error S_e in predicting transverse cracking (*CA*)

was plotted versus k and β. A surface was fitted through these prediction errors using a cubic spline interpolation approach. It revealed that in reducing prediction errors, the effect of β is by far more significant than the effect of k. The effect of k was further explored by analyzing its effect on transverse cracking prediction error S_e and bias, while keeping $\beta = 0.3$, which is the minimum value established earlier. The minimum error was obtained for values of $k = 0.1$ and $\beta = 0.3662$ and amounted to 654 ft/mile, which is much lower than the 4,226 ft/mile obtained with the TxME default values. Figure 5 and Figure 6 show the transverse cracking predictions versus the observed before and after calibration, respectively. Overall, calibration reduced significantly the transverse cracking error predictions.

8 CONCLUSIONS

This paper presented an effort to calibrate the pavement damage models in the TxME using as a reference Texas DOT collected pavement performance observations. For rutting and fatigue cracking, it was possible to do so by back-calculating mechanistic response parameters and then fitting the constants through a solver. For the transverse (i.e., thermal) cracking model, a trial and error approach was necessary. This calibration effort resulted in significantly improved pavement performance predictions.

REFERENCES

Hu, S., F. Zhou, and T. Scullion (2012), "Texas M-E Flexible Pavement Design System: Literature Review and Proposed Framework," Technical Report: 0-6622-1, Texas A&M Transportation Institute, College Station, TX

Hu, S., X. Hu, F. Zhou, and T. Scullion (2011). Development, Calibration, and Validation of a New M-E Rutting Model for HMA Overlay Design and Analysis, *Journal of Materials in Civil Engineering*, Vol. 23, pp. 89–99.

Jordahl, P.R. and B. Rauhut, (1984) Flexible Pavement Model VESYS IV-B, FHWA-RD-84-021 Final Rpt., FCP 35E1–041.

Lubinda et al., (2013) Texas Flexible Pavements, and Overlays: Calibration Plans for M-E models and related software, Report FHWA/TX-13/0-6658-P4, TXDOT.

Zhou, F., E. Fernando and T. Scullion (2010), Development, Calibration, and Validation of Performance Prediction Models for the Texas M-E Flexible Pavement Design System Technical Report: 0-5798-2, Texas A&M Transportation Institute, College Station, TX.

Advances in Materials and Pavement Performance Prediction II – Kumar et al. (eds)
© 2021 Taylor & Francis Group, London, ISBN 978-0-367-46169-0

Modified area under pavement profile for the light weight deflectometer measurements

Ahmadudin Burhani, Issam Khoury & Shad Sargand
Department of Civil Engineering, Ohio University, Athens OH, USA

Roger Green
Ohio Research Institute for Transportation and the Environment, Lancaster, OH, USA

ABSTRACT: Non-destructive test equipment such as falling weight deflectometer and light weight deflectometer are commonly used to measure pavement responses by recording the deflection of sensors produced by dropping a known load from a known height. Although they have a similar operating principle, few studies have examined the relationships between the devices, addressing light weight deflectometer testing with multiple geophones applied to layered systems. In this study, falling weight and light weight defelectometer tests were performed to establish a correlation between their measurements, and the resulting relationships were used to devise a new pavement deflection basin parameter called Area under Pavement Profile (AUPP) based on the light weight deflectometer measurements. The devised AUPP can be used to predict tensile strain at the bottom of an asphalt layer without performing back-calculation and/or structural modeling. Three-dimensional Finite Element Model (FEM) was developed in the ABAQUS program to validate the results. Comparison of the FEM and the field data shows close agreement.

1 INTRODUCTION

The falling weight deflectometer (FWD) has been widely used for pavement evaluation in recent decades. The use of FWD appears to be increasing for structural evaluation of existing pavements. It is typically a truck-trailer setup designed to be towed on the road to the measurement location (Fleming 2000, 2002). However, the equipment is expensive to purchase and operate, making it cost prohibitive for many local agencies. It can be difficult or impossible to bring a FWD trailer onto a site where there is restricted access during construction (Burhani 2016; Fleming et al. 2007; Sargand et al. 2016).

An alternative device, the light weight deflectometer (LWD) was developed in 1981 to overcome the accessibility problems of the FWD (Grasmick et al. 2015). It is a more compact and portable unit which uses the same basic operating principle as the FWD to measure the load response of a layered pavement system (Benedetto et al. 2012; Elhakim et al. 2014; Mooney & Miller 2009). Many different types of LWD are commercially available in the market. The Prima 100 LWD, manufactured by the Carl Bro was employed for this study. Previous studies demonstrated that the Prima 100 LWD provides the option to measures vertical displacement at the center as well as at the typical radial offsets of 12 inch (300 mm) and 24 inches (600 mm) from the plate center for unbound materials. The LWD and FWD captured deflections at

the load plate center and at the offset sensors can be used to back-calculation layer elastic moduli (Grasmick 2013; Horak et al. 2008; Senseney & Mooney 2010;. A typical schematic of FWD and LWD for a pavement structure considered in this study is shown in Figure 1.

1.1 *Background*

Numerous studies have explored correlation of LWD measurements with the results from other in-situ testing devices (Alshibli et al. 2005; Fleming et al. 2007; Grasmick et al. 2013, 2015; Horak et al. 2008; Siekmeier et al. 2000; Vennapusa & White 2009). Most of these studies focused on the LWD data at the load plate center. For instance, some common relationships are presented in Table 1, where composite stiffness (E_{LWD}) of the LWD were compared to composite stiffness of FWD (E_{FWD}).

Similarly, many researchers have addressed the theoretical relationships between deflection basin parameters and pavement responses such as: (a) tensile strain at the bottom of asphalt layer for fatigue cracking; (b) vertical compressive strain on the top of the base layer; and (c) vertical compressive strain on the top of the subgrade (Kim & Park 2002). Specifically, the tensile strain (ε_{ac}) at the bottom of the asphalt concrete (AC) layer produced fatigue cracking, and the allowable number of load repetitions is related to the tensile

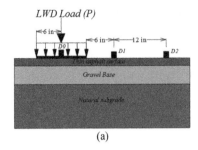

(a)

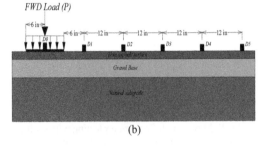

(b)

Figure 1. Schematic of the devices: (a) Prima 100 LWD sensor arrangement, (b) FWD sensor arrangement (1 inch = 25.4 mm).

Table 1. Reported E_{FWD} and E_{LWD} relationships (Grasmick et al. 2015; Shafiee et al. 2013).

Equation (MPa)	(R^2)	Description	Reference
$E_{LWD} = 0.97$ E_{FWD}	0.60	450-mm granular capping over silty soil and clay	(Fleming et al. 2002)
$E_{LWD} = 1.21$ E_{FWD}	0.77	260-mm lime-cement treated clay subgrade	
$E_{LWD} = 1.05$ $E_{FWD} + 4.36$	0.84	Silty Soil	Petersen et al. (2007)
$E_{LWD} = 1.03$ E_{FWD}	0.97	Granular subgrade	(Nazzal et al. 2007)
$E_{LWD} = 1.42$ E_{FWD}_209	0.81	Thin asphalt layer (130–150 mm)	(Steinert et al. 2005)

strain at the bottom of the HMA layer (Ceylan et al. 2005). A research conducted at the University of Illinois by Thompson (1989; Thompson & Elliot, 1995), analyzed the ILLI-PAVE database for conventional flexible pavements to establish relationship between FWD deflection basin parameter called, the Area under Pavement Profile (AUPP), and the tensile strain (ε_{ac}) at the bottom of an asphalt layer. The FWD sensor deflections at radial offset distances of 0, 12, and 24, 36 inches (0, 300, 600, and 900 mm) from the center of the loading plate were used to develop an algorithm. A proposed relationship is presented in Equation 1 and Equation 2.

$$AUPP = \frac{1}{2}(5D_0 - 2D_1 - 2D_2 - D_3) \qquad (1)$$

Where:

$D_0 = $ FWD sensor deflection at the center of the loading plate, mils

$D_1 = $ FWD sensor deflection 12 inches (300 mm) from the center of the loading plate, mils

$D_2 = $ FWD sensor deflection 24 inches (600 mm) from the center of the loading plate, mils,

$D_3 = $ FWD sensor deflection 36 inches (900 mm) from the center of the loading plate, mils.

The (ε_{ac}) for conventional flexible pavements with an aggregate base in term of AUPP, the AC tensile strain can be predicted using Equation 2.

$$Log(\varepsilon_{ac}) = 0.821 * Log(AUPP) + 1.210$$
$$R^2 = 0.973 \text{ and } SEE = 0.0579 \qquad (2)$$

where (ε_{ac}) is the strain (micro-strain) at the bottom of AC layer when AC thickness ranged from 3 to 8 inches (76 to 203 mm), AC modulus ranged from 100 to 1400 ksi (689 to 9652 MPa), and modulus of subgrade ranged from 1 to 12.3 ksi (7 to 85 MPa). This relationship was developed from the ILLIP-PAVE database.

1.2 Objectives

The goal of this study was to investigate the feasibility of employing the LWD for evaluating the response of pavement structure. The main objective was to determine whether the tensile strains calculated from the AUPP upon LWD deflection data can reliably predict the tensile strains calculated from the AUPP upon FWD deflection data. This was achieved by adapting the relationship proposed by Thompson (1989, 1995). The result was validated through a three-dimensional finite element model.

2 RESEARCH DESCRIPTIONS

A field research program was undertaken to investigate the feasibility of employing LWD for evaluating structural capacity of thin layered pavement systems. A total of 66 locations were tested at 21 sections across the state of Ohio. The sections contained thin asphalt layer built on the gravel base and placed over natural subgrade. Dynatest standard FWD test using a 12-inch (300 mm) diameter plate immediately followed by the Prima 100 LWD were employed. The Prima 100 LWD plate diameter was 12-inches (300 mm) and a total dropping mass of 22 lb (10 kg with a drop mass height of 33.5 inch (850 mm), which on average produced a contact pressure of 22 psi (152 kPa). A minimum of three test locations for each test section were tested. For both tests, the deflections data was collected at the load plate center and at radial offsets of 12-inch (300 mm) and 24-inches (600 mm) from the plate center. Maximum deflections were normalized by their respective peak contact stress. Although deflections induced by the FWD were measured using geophones positioned at the drop point and at up to six radial offsets of

Table 2. Characteristics of tested sections (1 inch = 25.4 mm, 1 ksi = 7 MPa).

Tested Section	Layer Thickness			Backcalculated Layer Moduli		
	AC (in)	Base (in)	Subgrade (in)	E_1 (ksi)	E_2 (ksi)	E_3 (ksi)
B.S-C70-11	3.0	5–9	–	1001	326	4
R.-C117-01A	3.0	5–9	–	928	290	6
R.-C117-01b	3.0	5–9	–	652	261	14
r.-c117-03	3.0	8–12	–	800	270	9
M.-C6-14	3.0	8–12	–	780	304	19
W.C.C.-C123-04	3.0	8–12	–	690	240	13
T.B.-C134-12	2.8	6–10	–	890	350	18
F.-C68-08	2.8	6–10	–	650	230	7
M.R.-T269-03	2.8	6–10	–	600	220	7
A.R.-C12-05	2.8	6–10	–	830	360	22
C.-C66-06	2.5	5.5–12	–	841	377	23
U.-C12-03	2.5	5.5–12	–	870	320	10
B.R.-C51-04	2.5	5.5–12	–	825	367	22
B.-C10-02	2.5	5.5–12	–	775	290	17
W.D.-C184-19	2.5	5.5–12	–	770	280	15
P.-C236-03	2.5	5.5–12	–	884	320	19
H.H.-C26-07	2.5	5.5–12	–	912	320	20
K.-C22-14	2.5	5.5–12	–	920	318	21
S.-C3-15	3.0	10–12	–	785	295	15
B.P.-C160-12	3.0	10–12	–	900	340	18
E.S.-C71-08	3.0	10–12	–	775	275	10

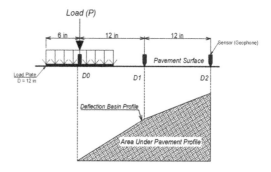

Figure 2. AUPP (LWD 3-sensors) modified deflection basin parameter (1 inch = 24.5 mm).

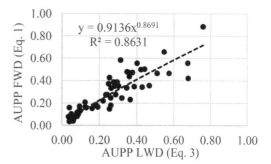

Figure 3. Comparison of AUPP's.

12-inch (300 mm) apart, this study considered FWD measurements corresponding to radial distances equal to those of the Prima 100 LWD.

The deflection data were used to determine the resilient moduli of pavement layers, using Modulus 6.0, back-calculation software. Table 2 summarizes the characteristics and the back-calculated layer moduli for each pavement sections included in this study.

3 MODIFIED AUPP

The FWD and LWD data (deflection data) were considered to investigate whether the AUPP based on the LWD data can be used instead of FWD data to estimate the tensile strain at the bottom of AC layer. The LWD data at radial distances 0, 12, and 24 inches (0, 300, and 600 mm) were considered, and a revised geometric schematic for the device was established. The Prima 100 LWD sensors arrangement along with a revised geometric schematic are shown in Figure 2 and Equation 3 respectively.

$$(AUPP)_{LWD} = \frac{1}{2}(3D_0 - 2D_1 - D_2) \qquad (3)$$

Where:

D_0 = LWD sensor deflection at the center of the loading plate, mils

D_1 = LWD sensor deflection 12 inches (300 mm) from the center of the loading plate, mils

D_2 = LWD sensor deflection 24 inches (600 mm) from the center of the loading plate, mils.

The normalized deflections were used in performing regression analysis to evaluate whether there is a proper relationship between both AUPP's. Thompson's relationship (Eq. 1), which is based on the FWD data, and the devised relationship (Eq. 3), which is upon LWD data were plotted and the result is presented in the Figure 3. As indicated in the Figure 3, regression analysis yielded a nonlinear model with a power function, that gives the best fit based on the best regression coefficient of determination, $R^2 = 0.86$. This result was substituted into Eq. (2), and a modified relationship was developed. The new relationship is presented in Eq. (4), which can be used to predict tensile strains at the bottom of a thin asphalt layer using LWD deflection data.

$$Log(\varepsilon_{AC}) = 0.7135 * Log(AUPP_{LWD}) + 1.178 \qquad (4)$$

Where: ε_{ac} = tensile strain at the bottom of the AC layer.

$AUPP_{LWD}$ = Area under Pavement Profile based on LWD sensor deflection data.

4 FINITE ELEMENT MODEL

A three-dimensional finite element model (3D FEM) was developed in the Abaqus program to simulate the

FWD and LWD measurements to validate the back-calculated layer moduli determined from Modulus 6.0. The model contained three layers, including thin AC surface, base, and natural subgrade. The model dimensions were chosen to cover all sensors of both devices.

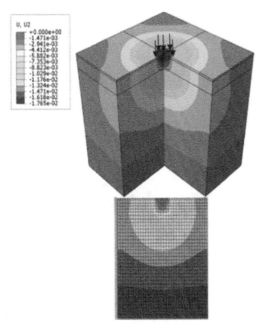

Figure 4. A 3D FEM of flexible pavement layered system for a tested section of B.S-C70-11 (1 inch = 25.4 mm).

All pavement layers were assumed to have linear elastic response to the applied loads. The thickness for each layer was assigned based on field core measurements, and the elastic moduli for all layers were obtained from back-calculation results corresponding to each section, presented in Table 2. The simulated 9,000 lb (40 KN) load was applied to an actual steel circular disc of radius 6 inch (152 mm) which produces stress of 79.6 psi (549 kPa). The contact interaction between adjacent layers was assumed to be fully bonded. Since the applied load presses down on the pavement layers, the vertical boundaries were modeled with roller supports with vertical motion possible. The horizontal boundary at the bottom was modeled with pin supports. The mesh was constructed with eight node linear brick reduced integration elements (C3D8R). A finer mesh was applied within 24 inches (600 mm) offset from the drop point as well as in the entire load plate areas, where were of greatest interest. The 3D FEM was used to predict the responses of pavement; to simulate the tested pavement sections. Back-calculated layers moduli and measure thicknesses were used to characterize each pavement sections. Figure 4 shows a sketch of 3D flexible pavement model.

Deflections were calculated corresponding to the sensor locations, D_0, D_1, and D_2 and compared with measured deflections. Also, maximum tensile strains at the bottom of AC layer were calculated based on Eq. (2) (Thompson Eq.) as well as Eq. (4) (Modified Eq.). The results were compared with those predicted by the 3D FEM, which found in good agreement, as presented in Table 3.

Table 3. Predicted 3D FEM versus Normalized FWD and LWD Deflection data.

Tested Section	FWD Measured Deflections (in)			LWD Measured Deflections (in)			FEM Predicted Deflections (in)			Tensile Strain Eq. (2) ε_{ac} (μs)	Tensile Strain Eq. (4) ε_{ac} (μs)	Tensile Strain FEM ε_{ac} (μs)
	D_0	D_1	D_2	D_0	D_1	D_2	D_0	D_1	D_2			
B.S-C70-11	0.14	0.12	0.09	0.12	0.1	0.08	0.17	0.15	0.10	2.35	2.43	2.40
R.-C117-01A	0.38	0.32	0.23	0.35	0.29	0.22	0.50	0.31	0.27	6.19	5.16	6.20
R.-C117-01B	0.23	0.15	0.09	0.26	0.11	0.09	0.25	0.20	0.05	5.91	5.79	5.30
R.-C117-03	0.19	0.12	0.08	0.23	0.11	0.06	0.23	0.19	0.11	4.95	4.70	5.70
M-C6-14	0.33	0.25	0.15	0.24	0.17	0.10	0.36	0.26	0.17	7.44	6.48	6.30
W.C.C.-C123-04	0.45	0.27	0.14	0.39	0.22	0.11	0.44	0.17	0.13	11.85	10.25	9.80
T.B.-C135-12	0.12	0.11	0.09	0.11	0.08	0.07	0.22	0.20	0.14	1.97	1.74	2.10
F.-C68-08	0.45	0.40	0.30	0.46	0.39	0.29	0.41	0.33	0.23	6.51	7.01	6.90
M.R.-T269-03	0.07	0.03	0.02	0.08	0.02	0.01	0.19	0.10	0.08	2.97	3.17	3.30
A.R.C12-05	0.11	0.08	0.05	0.05	0.03	0.02	0.09	0.04	0.03	2.98	1.95	2.00
C.-C66-06	0.06	0.04	0.04	0.07	0.04	0.03	0.20	0.09	0.01	1.35	1.74	1.50
U.-C12-03	0.05	0.04	0.03	0.07	0.03	0.03	0.12	0.04	0.01	1.59	1.96	2.10
B.R.-C51-04	0.05	0.04	0.03	0.07	0.03	0.03	0.13	0.11	0.02	1.46	1.91	1.60
B.-C10-02	0.30	0.19	0.11	0.30	0.14	0.08	0.43	0.13	0.09	8.16	7.54	7.70
W.D.-C184-19	0.27	0.16	0.12	0.29	0.16	0.15	0.34	0.29	0.08	6.96	7.12	6.90
P.-C236-03	0.26	0.16	0.09	0.24	0.13	0.08	0.22	0.17	0.05	7.02	7.55	8.00
H.H.-C26-07	0.30	0.18	0.09	0.41	0.20	0.11	0.37	0.12	0.07	8.88	8.62	9.20
K.-C22-14	0.39	0.25	0.13	0.29	0.18	0.10	0.36	0.27	0.05	10.17	8.70	9.70
S.-C3-15	0.17	0.14	0.10	0.14	0.10	0.08	0.29	0.11	0.10	3.49	3.35	3.40
B.P.C160-12	0.16	0.12	0.07	0.29	0.13	0.09	0.29	0.24	0.06	4.19	5.74	4.70
E.S.-C71-08	0.23	0.18	0.11	0.40	0.21	0.14	0.31	0.27	0.20	5.23	6.74	5.80

5 CONCLUSION

In this study, an experimental and numerical study was performed to investigate the feasibility of employing LWD for evaluating the response of pavement structure in terms of tensile strain at the bottom of AC layer. The Prima 100 LWD normalized sensor deflections reveal a satisfactory relationship with the benchmark test, FWD test, corresponding to the sensor locations of both tests. This was reflected in the positive relationship revealed between the AUPP's, which yielded a nonlinear model, giving a good correlation, $R^2 = 86$. Upon this, a modified equation for the LWD deflection data was developed by adapting Thompson proposed relationship for FWD, which can be used to predict tensile strains at the bottom of a thin AC layer. The result was validated via a 3D FEM, which predicted the tensile strains at the bottom of AC layer by an average of 11 and 9% for the Thompson and modified equations respectively. Thompson proposed equation generally tends to give similar result as of FEM. The modified equation result was overestimated by an average of 14%. Also, FEM overestimated maximum measured deflections of the FWD and LWD tests for an average of 56% and 51% respectively.

REFERENCES

Alshibli, K. A., Abu-Farsakh, M., & Seyman, E. (2005). Laboratory evaluation of the geogauge and light falling weight deflectometer as construction control tools. Journal of Materials in Civil Engineering, 17(5), 560–569.

American Association of State Highway and Transportation Officials. (1993). AASHTO Guide for Design of Pavement Structures, Washington, D.C.

Benedetto, A., Tosti, F., & Di Domenico, L. (2012). Elliptic model for prediction of deflections induced by a Light Falling Weight Deflectometer. Journal of Terramechanics, 49(1), 1–12.

Burhani, A. (2016). Correlation Study on the Falling Weight Deflectometer and Light Weight Deflectometer for the Local Pavement Systems (Master thesis, Ohio University).

Ceylan, H., Mathews, R., Kota, T., Gopalakrishnan, K., and B. J. Coree. 2005. Rehabilitation of Concrete Pavements Utilizing Rubblization and Crack and Seat Methods. IHRB Project Final Report TR-473, Center for Transportation Research and Education, Ames, Iowa.

Elhakim, A. F., Elbaz, K., & Amer, M. I. (2014). The use of light weight deflectometer for in situ evaluation of sand degree of compaction. HBRC Journal, 10(3), pp 298–307.

Fleming, P.R. (2000) "Small-scale dynamic devices for the measurement of elastic stiffness modulus on pavement foundations," Nondestructive testing of pavements and backcalculation of moduli: third volume, ASTM STP 1375, S.D. Tayabji and E. O. Lukanen, eds., pp. 41–59, West Conshohocken, PA.

Fleming, P. R., Lambert, J. P., Frost, M. W., and Rogers, C. D. 2002. "In-situ assessment of stiffness modulus for highway foundations during construction," 9th Int. Conf. on Asphalt Pavements, Copenhagen, Denmark CD-ROM, 12.

Fleming, P., Frost, M., & Lambert, J. (2007). Review of lightweight deflectometer for routine in situ assessment of pavement material stiffness. Transportation research record: journal of the Transportation Research Board, (2004), 80–87.

Grasmick, J. G. (2013). Using the light weight deflectometer with radial offst sensors on two-layer systems for construction quality control/quality assurance of reclaimed and stabilized materials, M.S. thesis, Colorado School of Mines, Golden, CO

Grasmick, J. G., Mooney, M. A., Senseney, C. T., Surdahl, R. W., & Voth, M. (2015). Comparison of Multiple Sensor Deflection Data from Lightweight and Falling Weight Deflectometer Tests on Layered Soil, Geotechnical Testing Journal, Vol. 38, No. 6, (pp. 851–863).

Horak, E., Maina, J. W., Guiamba, D., & Hartman, A. (2008). Correlation study with the light weight deflectometer in South Africa.

Kim, Y. Richard, and Park, Heemun. (2002). Use of Falling Weight Deflectometer Multi-Load Data for Pavement Strength Estimation. (Report No. FHWA/NC/2002-006 for the North Carolina Department of Transportation). Raleigh NC, June 2002.

Mooney, M. A., & Miller, P. K. (2009). Analysis of lightweight deflectometer test based on in situ stress and strain response. Journal of geotechnical and geoenvironmental engineering, 135(2), 199–208.

Nazzal, M. D., Abu-Farsakh, M. Y., Alshibli, K., & Mohammad, L. (2007). Evaluating the light falling weight deflectometer device for in situ measurement of elastic modulus of pavement layers. Transportation Research Record, 2016(1), 13–22.

Petersen, J. S., Romanoschi, S. A., & Hossain, M. (2007). Development of stiffness-baseD specifications for in-situ embankment compaction quality control (No. K-TRAN: KSU-04-6).

Sargand, S., Green, R., Burhani, A., Alghamdi, H., & Jordan, B. (2016). Investigation of In-Situ Strength of Various Construction/Widening Methods Utilized on Local Roads, (134991).

Senseney, C., & Mooney, M. (2010). Characterization of two-layer soil system using a lightweight deflectometer with radial sensors. Transportation Research Record: Journal of the Transportation Research Board, (2186), 21–28.

Shafiee, M. H., Nassiri, S., & Khan, M. R. H. (2013). Evaluation of New Technologies for Quality Control/Quality Assurance (QC/QA) of Subgrade and Unbound Pavement Layer Moduli, center of transportation engineering and planning, CTEP, University of Alberta.

Siekmeier, J. A., Young, D., & Beberg, D. (2000). Comparison of the dynamic cone penetrometer with other tests during subgrade and granular base characterization in Minnesota. In Nondestructive testing of pavements and backcalculation of moduli: third volume. ASTM International.

Steinert, B. C, Humphrey, D. N., and Kestler, M.A., 2005, "Portable Falling Weight Deflectometer Study," Report No. NETCR52, New England Transportation Consortium, Storrs, CT.

Thompson, M.R.,1989, "ILLI-PAVE Based NDT Analysis Procedures," Nondestructive Testing of Pavements and Backcalculation of Moduli, ASTM STP 1026, A. J. Bush III and G. Y. Baladi, Eds., American Society for Testing and Materials, Philadelphia, pp. 487–501.

Thompson, M. R., and Elliot, R. P., (1985), "ILLI-PAVE-Based Response Algorithms for Design of Conventional Flexible Pavements", Transportation Research Record 1043, TRB, Washington, D. C., pp. 55–57.

Advances in Materials and Pavement Performance Prediction II – Kumar et al. (eds)
© 2021 Taylor & Francis Group, London, ISBN 978-0-367-46169-0

Evaluation of reflective cracking in composite pavement based on different rheological models

Ki Hoon Moon
Korea Expressway Corporation (KEC), South Korea

Augusto Cannone Falchetto
University of Alaska Fairbanks, USA

Hae Won Park
Inha University, South Korea

Di Wang
Technische Universität Braunschweig, Germany

ABSTRACT: Three rheological models including a newly developed formulation based on the current Christensen Anderson and Marateanu (CAM) model, named sigmoidal CAM model (SCM), are used to predict the evolution of reflective cracking in a typical composite pavement structure currently used in South Korea. Three different asphalt mixtures were selected and dynamic modulus tests were performed. The mechanistic-empirical pavement design guide (MEPDG) was then used for evaluating the pavement distress and to estimate the effect of the three different models on such phenomena. It is found that while the CAM model may not be entirely reliable due to its inability in fitting the data in the high-temperature domain, the SCM might result in relatively more conservative pavement design.

Keywords: composite pavement; reflective cracking; dynamic modulus; mechanistic-empirical pavement design guide (MEPDG)

1 INTRODUCTION

Asphalt pavement overlay is a common practice for rehabilitating the existing concrete pavement having more than 15 to 20 years in the expressway network in South Korea (Kim et al. 2019). This is strictly related to the remarkable improvement in the production of stone mastic asphalt (SMA) mixture (Moghaddam et al. 2012; Norambuena-Contreras et al. 2019) and the development of a low-noise porous asphalt (LNPA) (Kim et al. 2019).

After applying an overlay of asphalt material on concrete pavement, reflective cracking (RC) is the most severe distress commonly experienced (Baek & Al-Qadi 2006; Bennert et al. 2009; Wang & Zhong 2019). RC presents similar crack initiation and progress mechanism to the bottom-up (BU) cracking while finding a location in correspondence of the concrete pavement joints underneath the surface (Bennert et al. 2009; Wang & Zhong 2019). As a consequence of RC, the pavement surface exhibits a series of transverse cracking that crosses the entire pavement transverse profile (Baek & Al-Qadi 2006; Bennert

et al. 2009; Wang & Zhong 2019). Two major negative effects can be identified due to the presence of RC: a poor driving experience due to the cracks and serious pavement deterioration occurring not only in the layer consisting of asphalt material but also in the concrete layer due to infiltration of water and moisture.

Several researchers focused on evaluating RC while proposing different solutions for mitigating such distress. This includes the use of geotechnical fabrics and fibers and field testing (Maurer & Malasheskie 1989) forensic analyses (Zhou & Scullion 2005) and interlayer grid reinforcements (Walubita et al 2019). RC was also addressed in several studies based on numerical and FEM simulations (Bennert et al. 2009; Wang & Zhong 2019).

Despite the significant research attention devoted to RC, only limited studies considered the incorporation of the mechanical properties (i.e., rheological characteristics) of asphalt mixtures such as the dynamic modulus within a global evaluation of the pavement performance during its expected service life. It is well known that the dynamic modulus (AASHTO T342 2015) of asphalt mixtures can provide critical information on the material response (Corrales-Azofeifa & Archilla 2018). Depending on the selected mathematical models used for generating master curves, experimental results of dynamic modulus can provide different predictions ultimately highly affecting the evaluation of the pavement performance concerning different distresses such as RCs (Bennert et al. 2009; Corrales-Azofeifa & Archilla 2018).

In this paper, the impact of different mathematical models used for generating master curves of dynamic modulus, on the predicted performance of rehabilitated concrete pavements subjected to asphalt overlay is investigated. First, three different asphalt mixtures widely used in South Korea such as hot mix asphalt (HMA), SMA, and LNPA (MOLIT 2017) are prepared and the dynamic modulus is experimentally measured (AASHTO T342 2015). Then, three different mathematical models, including a newly developed model, are selected to analyze the rheological properties of the tested asphalt mixtures. Finally, the current MEPDG design software (AASHTO 2018), is used to evaluate the effect of different asphalt materials and mathematical models on the prediction of RC in the composite pavement.

2 MATERIAL AND TESTING

A hot mix asphalt (HMA), an SMA and an LNPA mixture commonly used in South Korea are selected for this study (MOLIT 2017). Schematic information on the prepared materials is presented in Table 1.

The dynamic modulus test (DMT) is performed based on the current AASHTO T342 (2015) standard. Three replicates per each mixture set are prepared and tested. Therefore, a total of nine asphalt mixture samples are used. More details on DMT is provided in Figure 1 and Table 2.

3 MASTER CURVES MODELS

The first Model (1) consists of the sigmoidal function by Pellinen et al. (2004):

$$Log\,|E^*(\omega)| = \delta + \frac{\alpha}{1 + e^{\beta + \gamma \cdot (Log\omega_r + Loga_T)}} \quad (1)$$

where $|E^*(\omega)|$ is the dynamic modulus (GPa); δ is the minimum value of the dynamic modulus; α is the value of the fitted dynamic modulus; β and γ are fitting parameters; ω is the frequency (rad/s); ω_r is the reduced frequency (rad/s), and a_T.

The Christensen, Anderson and Marasteanu (CAM) model (1999) is the second master curve expression (i.e., Model 2) adopted:

$$Log|E^*(\omega)|$$
$$= LogE_g - \frac{w}{v} \cdot Log\left[1 + (10^{Log\omega_r + Loga_T - Logt_c})^v\right] \quad (2)$$

Table 1. Asphalt mixtures.

ID	Mix	Asphalt binder	NMAS (mm)	VMA (%)	VFA (%)	Air voids (%)
1	HMA	PG 76-22	13	14.6	72.0	4.4
2	SMA	PG 76-22	13	17.5	79.0	2.4
3	LNPA	PG 82-34	10	31.3	41.6	20.0

* VMA: Voids in the Mineral Aggregates,
VFA: Voids Filled with Asphalt.

Figure 1. Dynamic modulus test.

Table 2. Dynamic modulus testing.

Specimen size	Diameter: 100 mm; Specimen height: 150 mm; Gauge length: 100 mm
# of replicates	Three specimens per mixture type
Temperature	−10, 4.4, 21.1, 37.8 and 54.4°C (21.1°C reference temperature: T_S)
Frequency	0.1, 0.5, 1, 5, 10 and 25 rad/s

where E_g is the Glassy modulus of asphalt mixture (Podolsky et al. 2018), and v, w, and t_c are fitting parameters. It must be remarked that this exponential function tends to present a deviation in the function shape compared to an "S" shaped sigmoidal function.

The third model (i.e., Model 3) consists of a combination of Models 1 and 2 and it is identified as the sigmoidal CAM model (SCM):

$$Log|E(\omega)^*| = LogE_g - \frac{w}{v} \cdot \text{Sigmoidal function}$$

$$Log|E(\omega)^*| = LogE_g + \frac{w}{v} \cdot \frac{1}{\left[1 + e^{z \cdot (Log\omega_r + Loga_T) + t_c}\right]^{\frac{1}{v}}}$$
$$(3)$$

4 DATA ANALYSIS

The master curve of $|E^*(\omega)|$ was obtained by fitting expressions 1, 2 and 3 to the experimental results. Optimization was performed based on the Generalized Reduced Gradient method built-in in Excel. The outcome of the fitting process is presented in Figures 2, 3 and 4 for the three types of mixtures, respectively.

It can be observed that Model 1 (sigmoidal) and Model 3 (SCM: sigmoidal CAM Model) present a

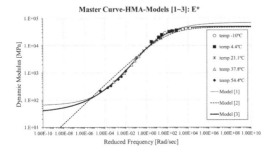

Figure 2. Dynamic modulus master curve of Hot mix asphalt (HMA).

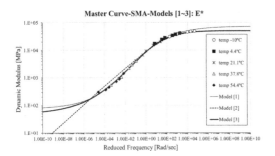

Figure 3. Dynamic modulus master curve of Stone mastic asphalt (SMA).

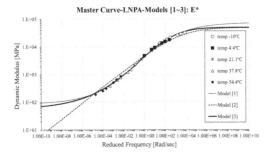

Figure 4. Dynamic modulus master curve of LNPA.

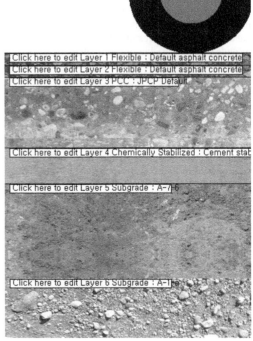

Figure 5. MEPDG pavement structure.

clear symmetric "S" shaped trend. Model 1 shows higher $|E^*(\omega)|$ both at low and high frequency compared to Model 3. In the case of Model 2 (CAM model), similar $|E^*(\omega)|$ prediction compared to Model 3 is observed at high frequency. However, a different trend is exhibited at low frequency where a substantial deviations form models 1 and 3 are experienced. Such a trend suggests that Model 3 (SCM model) can successfully provide a lower bound to $|E^*(\omega)|$ for the sigmoidal model.

Together with visual comparison, a simple t-test with a 5% significance level was adopted to statistically estimate the difference in the fitting capability of the three master curves models. Distinct differences in $|E^*(\omega)|$ can be observed between Model 1 (sigmoidal) and Model 2 (CAM model) at almost all tested frequencies. In addition, Model 1 (sigmoidal)

and Model 3 (SCM model) result in significantly different predictions at low and high frequencies. Within these two models, similar fitting of $|E^*(\omega)|$ is found for a mid-frequency range. When comparing Model 2 (CAM model) and Model 3 (SCM model), statistically distinct values $|E^*(\omega)|$ at low and intermediate frequency are found while the model return equivalent predictions at high frequency. Based on the results of the analysis performed, SCM can be potentially used as an alternative mathematical model in place of the conventional sigmoidal formulation. This is because Model 3 (SCM) presents the characteristics not only of a sigmoidal formulation but also the benefits of the simple expression of typical of the CAM model.

5 PAVEMENT DESIGN ANALYSIS

The current MEPDG design software is used in the present research effort to evaluate the effect of incorporating the three different rheological master curve models on the performance prediction of a composite pavement system (AASHTO 2018). Figure 5 presents the schematic of the adopted pavement structure which is based on the current expressway design guide (e.g., composite pavement design guide written in Korea Expressway Corporation) in South Korea (MOLIT 2017). A set of $|E^*(\omega)|$ data was generated using the three different models analyzed for the following set of frequencies and temperatures: 0.1, 0.5, 1.0, 10 and 25 rad/s and −10, 4.4, 20.1 37.8 and 54.4°C. Then,

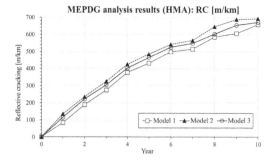

Figure 6. Reflective cracking for Hot mix asphalt (HMA).

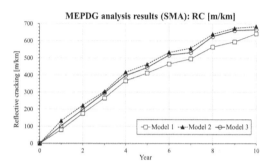

Figure 7. Reflective cracking for Stone mastic asphalt (SMA).

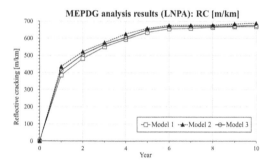

Figure 8. Reflective cracking for LNPA.

these data were used as input in the MEPDG software. This guarantee that comparisons among the three models are incorporated in the pavement design analysis. Figures 6–8 present the results of the MEPDG calculation in terms of RC when the three master curves models investigated in the present study are incorporated in the design simulation.

Distinct upper and lower bounds on reflective cracking are obtained for all tested cases. In the previous section, a higher $|E^*(\omega)|$ could be predicted with Model 1 (sigmoidal) compared to Model 3 (SCM model). This reflects in the actual response of the pavement obtained with the MEPDG design software. It also needs to be mentioned that the results of the MEPDG analysis performed with the CAM Model (Marasteanu & Anderson 1999) might be not

necessarily very robust as a poor prediction of dynamic modulus and phase angle at a higher temperature and low frequency was observed. In view of the RC plots, while the MEPDG output of Model 1 (sigmoidal) and Model 3 (SCM) appears to be relatively close, the latter provides a more conservative prediction of the evolution of the pavement distresses (higher degree of distress).

6 CONCLUSIONS

The newly developed sigmoidal CAM model (SCM) provides lower limits of the dynamic modulus in comparison with the sigmoidal model. In addition, the pavement design analysis performed with SCM results in a higher degree of RC. This may result in a more conservative evaluation of the pavement and of its performance during its service life. While this may eventually lead to the design of more durable pavement with higher service quality when an overlay is adopted, an economic analysis would need to be performed to obtain an overall estimation of the actual benefit of using a different master curve model in the design procedure.

REFERENCES

AASHTO T 342-15 (2015) Standard method of test for determining dynamic modulus of hot mix asphalt concrete mixtures; American Association of State Highway and Transportation Officials: Washington, DC, USA.
AASHTO (2018) User manual of AASHTOWare-Pavement. American Association of State Highway and Transportation Officials: Washington, DC, USA.
Baek, J., Al-Qadi, I.L. (2006) Finite element method modeling of reflective cracking initiation and propagation: Investigation of the effect of steel reinforcement interlayer on retarding reflective cracking in hot-mix asphalt overlay. Transp. Res. Rec. 1949, 32–42.
Bennert, T., Worden, M., Turo, M. (2009) Field and laboratory forensic analysis of reflective cracking on Massachusetts interstate 495. Transp. Res. Rec., 2126, 27–38.
Corrales-Azofeifa, J.P., Archilla, A.R. (2018) Dynamic modulus model of hot mix asphalt: statistical analysis using joint estimation and mixed effects. J. Infrastruct. Syst., 24, 04018012.
Kim, C.H., Kang, H.J., Jang, T.S., Cheon, B.H. (2019) Effectiveness of low noise porous asphalt (LNPA) for reduction of noise compared to conventional pavement types. In Proceedings of the KSNVE Annual Conference Proceedings, Pyeongchang, Korea, 260. (In Korean)
Kim, D.H., Lee, J.M., Moon, K.H., Park, J.S., Suh, Y.C., Jeong, J.H. (2019) Development of remodeling index model to predict priority of large-scale repair works of deteriorated expressway concrete pavements in Korea. KSCE J. Civ. Eng., 23, 2096–2107.
Marasteanu, M.O., Anderson, D.A. (1999) Improved model for bitumen rheological characterization. In Eurobitume Workshop on Performance-Related Properties for Bituminous Binders; European Bitumen Association: Belgium, 133.

Maurer, D.A., Malasheskie, G.J. (1989) Field performance of fabrics and fibers to retard reflective cracking. Geotext. Geomembr., 8, 239–267.

Moghaddam, T.B., Karim, M.R., Syammaun, T. (2012) Dynamic properties of stone mastic asphalt mixtures containing waste plastic bottles. Constr. Build. Mater., 34, 236–242.

MOLIT (2017) Guideline for production and construction of asphalt mixture. Ministry of Land Infrastructure and Transportation: South Korea, Seoul. (In Korean)

Norambuena-Contreras, J., Yalcin, E., Hudson-Griffiths, R., García, A. (2019) Mechanical and self-healing properties of stone mastic asphalt containing encapsulated rejuvenators. J. Mater. Civ. Eng., 31, 04019052.

Pellinen, T.K., Witczak, M.W., Bonaquist, R.F. (2004) Asphalt mix master curve construction using sigmoidal fitting function with non-linear least squares optimization. Recent Adv. Mater. Charact. Model. Pavement Syst., 83–101.

Podolsky, J.H., Williams, R.C., Cochran, E. (2018) Effect of corn and soybean oil derived additives on polymer-modified HMA and WMA master curve construction and dynamic modulus performance. Int. J. Pavement Res. Technol. 11, 541–552.

Walubita, L.F., Mahmoud, E., Lee, S.I., Komba, J.J., Nyamuhokya, T.P. (2019) Grid reinforcement in HMA overlays-a field case study of highway us 59 in Atlanta district of Texas (No. 19-02991). In Proceedings of the Transportation Research Board 98th Annual Meeting.

Wang, X., Zhong, Y. (2019) Reflective crack in semi-rigid base asphalt pavement under temperature-traffic coupled dynamics using XFEM. Constr. Build. Mater., 214, 280–289.

Zhou, F., Scullion, T. (2005) Overlay tester: a simple performance test for thermal reflective cracking. J. Assoc. Asph. Paving Technol., 74, 443–484.

Advances in Materials and Pavement Performance Prediction II – Kumar et al. (eds)
© 2021 Taylor & Francis Group, London, ISBN 978-0-367-46169-0

Monitoring asphalt pavement behavior through in-situ instrumentation

G.B. Colpo, L.A.T. Brito & J.A.P. Ceratti
Federal University of Rio Grande do Sul, Porto Alegre, Brazil

D.M. Mocelin
North Carolina State University, Raleigh, USA

ABSTRACT: This paper presents a study of the mechanical responses from an in-service road section, in terms of strains and vertical stress, comparing data obtained from sensors installed at the bottom of the asphalt layer to the pavement-modeled responses in two different approaches: using a linear elastic analysis software (AEMC) and a viscoelastic analysis software (FlexPAVE™). Data in situ was gathered by means of a customized acquisition system with embedded signal processing. The temperatures measured in-situ were also compared to those estimated by the MERRA (Modern-Era Retrospective Analysis for Research and Applications) climate database, which is embedded in the newest AASHTOWare Pavement-ME version. The results demonstrated high quality of the signal processing capabilities of the data acquisition system developed to the project and returned excellent affinity to the modeled response, thereby demonstrating high potential to serve as a continuous structure monitoring tool.

1 INTRODUCTION

For the design of new pavements and in rehabilitation projects, it is vital to collect and understand information regarding the behavior of pavements under different load and environmental conditions.

Given the difficulties and limitations of calibrating the laboratory-determined performance models, it is essential to simulate the pavement behavior accurately under the actual loading spectra, which can be aided by strain, stress, and temperature information from instrumentation systems in the pavement layers. Such in-situ measurements help better to understand pavement responses to loading and environmental changes, allowing identifying critical differences between laboratory testing and actual field performance (Al-Qadi and Nassar 2003).

Sensors embedded in pavement sections have proven their usefulness in providing measured responses as a basis of comparison. Various researches have instrumented pavement layers using strain gages and pressure cells to measure the dynamic response to load in the form of strains and pressures (Leandri et al. 2013; Priest and Timm 2006; Timm 2016; Timm et al. 2004). This technology is employed in various countries, displaying satisfactory results that often contribute to the validation of pavement design methods (Priest and Timm 2006; Van Deusen et al. 1992).

However, the sensors installation process is tedious and cumbersome and can be carried out during or after the construction of the pavement. The most common locations for installation are at the bottom of the asphalt layer and the top of the subgrade; both points are of interest due to the major distresses' liability – asphalt layer fatigue and permanent subgrade deformation (Loulizi et al. 2001; Timm 2016).

The objective of this study was to investigate the behavior of an asphalt concrete layer subject to real traffic and climate variations. Instrumentation installed in the pavement along with a bespoken data acquisition system allowed for temperature, stress, and strain monitoring. Pavement modeling using linear elastic and viscoelastic analysis software allowed for their comparison of the in situ data while providing data for the proposed system validation.

2 METHODOLOGY

2.1 Field instrumentation

The structure of test section selected consisted of a flexible pavement with an asphalt layer of dense graded asphalt concrete (HMA; 12.5 mm NMAS; 5.2% binder content by weight) with a polymer-modified binder (PG 76–11), 10 cm in thickness, and underlying pavement layers composed by dense-graded aggregate base (15 cm) and granular subbase (30 cm) over a clayey subgrade.

The selected structure is part of a road segment on BR-116, in southern Brazil. The instruments used were installed at the bottom of the asphalt layer (at 10 cm depth) with the following array: six strain gauges (SG) (Figure 1a), two pressure cells (PC) (Figure 1b) and six solid-state thermistors (Figure 1c). One of the pressure cells featured a 1.0 MPa nominal maximum pressure (PC1) and the other 2.5 MPa (PC2); the strain gages were oriented to measure longitudinal and transverse strain, and the thermistors have the temperature range from $-55°C$ to $+150°C$ with $\pm0.5°C$ accuracy. The sensors were covered by a protective layer that also helps to avoid slipping during asphalt mixture compaction. The sensors were selected based on the instrumentation performed by the National Center for Asphalt Technology test tracks (Timm et al. 2004).

The first set (PC2 + 3 SG) was spaced 5 m apart longitudinally from the second set (PC1 + 3 SG) and located under the external wheel path of the outer road lane. Figure 2 shows the sensor layout along the external wheel path.

A custom data acquisition system (DAS) was developed as part of this research. The DAS performs signal conditioning, analog signal filtering, amplification, digitization, and data pre-processing before sending it to an embedded single board computer for storage. The pre-processing implements a high pass filter to remove baseline level variations (caused mainly due to temperature variations) and a trigger algorithm that sends a flag indicating the start of a dynamic event. The event flag sets when the signal variation is higher than a user-defined threshold. The system samples the analog signals at a rate of 900 points per second, per channel. The system was designed to be of low power consumption and capable of handling data streaming to a cloud-based service.

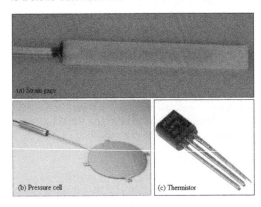

Figure 1.   Sensors used on the pavement.

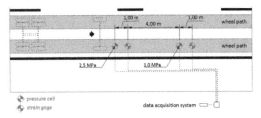

Figure 2.   Sensors installation layout.

2.2   Modeling

The longitudinal strain gage displays alternate contraction and extension due to the moving load, which is a result of different efforts depending on the position of the gauge to the axle, while the transversal strain is always extension, and vertical stress is always compression (Perret 2003).

To validate field results, modeling of the tested structure using a linear elastic analysis software AEMC and a viscoelastic analysis software FlexPAVE^{TM}, provided a baseline comparison to the instrumented section.

The simulations were performed considering a standard axle (single axle – dual tired), with a standard load of 20 kN per wheel, a tire pressure of 560 kPa, wheel spacing of 32.4 cm, and fully bonded layers. The analysis points were under the wheel and in the center of the half axle.

The dynamic modulus tests followed AASHTO R-84 (2017), and the results were fitted to a sigmoidal function as per AASHTO TP-107 (2014). The resilient moduli of the underneath layers were determined by retroanalysis of deflection basin results.

The pavement parameters used in the analysis are in Table 1.

2.3   Temperature conditions

Deformation and pressure responses should be analyzed accounting for pavement temperature due to the high thermal sensitivity of asphalt mixtures.

Temperatures obtained by the sensors installed at the bottom of the asphalt layer (10 cm depth) were extracted from the research dataset for a given period of field collection. In situ temperatures were compared to temperatures obtained from the MERRA database via AASHTOWare Pavement-ME software (NCHRP 2004), after equating to the same depth and location of the sensors.

MERRA climate database is available using AASH-TOWare Pavement-ME software interface. Its database uses a reanalysis model to combine computational models with terrestrial, atmospheric, oceanic, and satellite observations worldwide. MERRA climate data includes temperature, precipitation, wind speed,

Table 1.   Material properties and layers thickness used.

Material	Behavior	Thickness (cm)	Resilient Modulus (MPa)	Poisson's ratio
Asphalt layer	Linear elastic/ viscoelastic	10	Dynamic Modulus	0.30
Granular base	Linear elastic	15	307	0.35
Granular subbase	Linear elastic	30	369	0.35
Subgrade	Linear elastic	–	107	0.40

relative humidity, and cloud coverage provided hourly to any location in the world.

To estimate pavement temperature, the software uses a thermal balance equation shown below (equation 1). Details of this equation for different pavement depths are available in the Guide for Mechanistic-Empirical Design – AASHTO (NCHRP 2004).

$$Q_i - Q_r + Q_a - Q_e \pm Q_c \pm Q_h \pm Q_g = 0 \qquad (1)$$

where Q_i = incoming short wave radiation; Q_r = reflected short wave radiation; Q_a = incoming longwave radiation; Q_e = outgoing longwave radiation; Q_c = convective heat transfer; Q_h = effects of transpiration, condensation, evaporation, and sublimation; Q_g = energy absorbed by the ground.

3 RESULTS AND ANALYSIS

Figure 3 illustrates the comparison obtained for the strain and stress data collected in the field using the instrumentation system, and the FlexPAVE™ and AEMC software. The temperature considered in this analysis, to determine the dynamic modulus of asphalt concrete, was 25°C, the average temperature of the pavement section, and the speed was 30 km/h.

A good agreement between the pulses measured in the field with those simulated by the FlexPAVE™ program is evident. Notwithstanding, the maximum values of strain calculated by AEMC were superior to those measured in the longitudinal and transversal directions. Regarding the vertical pressure, the peak values determined by AEMC and FlexPAVE™ were inferior to the field, as presented in Table 2.

The simulated pavement responses are consistent with field measurements, validating the data acquisition system. Regarding the shape and magnitude of the signal obtained by FlexPAVE™, the transversal and longitudinal strains showed a better correspondence with the field-collected data. The vertical stress shape measured by the pressure cell shows a difference in the unloading path, compared to simulated data (Figure 3c). The shape of the unloading path is related to the relaxation behavior of asphalt materials, and the observed difference may be an inability of the program to describe this viscoelastic phenomenon well.

Strain and stress responses should be analyzed taking into account pavement temperature due to the high thermal sensitivity of asphalt mixtures. Figure 4 shows temperature data, in degrees Celsius, collected by the thermocouples at the bottom of the asphalt layer (10 cm), for a period of approximately 42 hours, in which a large oscillation throughout the day can be observed, with an amplitude of up to a temperature variation of 22°C, which drastically influences the response of the pavement structure.

Figure 5 shows a selection of data collected in the field from 09/30/2016 to 01/10/2016 and those calculated using MERRA climatic data for the same period at a depth of 10 cm in the asphalt layer.

Table 2. Parameters obtained in instrumentation and simulations for similar conditions.

Parameters	Field	FlexPAVE™	AEMC
Longitudinal strain ($\mu\varepsilon$)	121	133	157
Transversal strain ($\mu\varepsilon$)	82	82	104
Vertical stress (kPa)	194	162	164

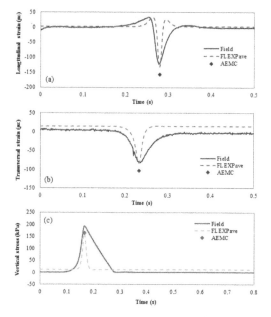

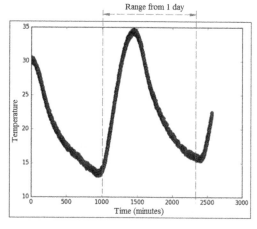

Figure 3. Field-collected signals, FlexPAVE™ simulations results, and the maximum value determined by AEMC for longitudinal strain (a), transversal strain (b), and vertical pressure (c).

Figure 4. Temperature data collected at a depth of 10 cm below asphalt surface.

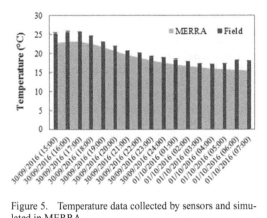

Figure 5. Temperature data collected by sensors and simulated in MERRA.

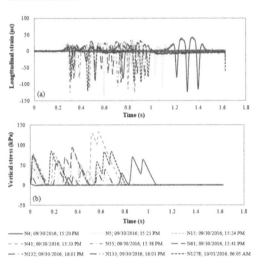

Figure 6. Field-collected signals: longitudinal strain (a), and vertical stress (b).

Temperature data collected from the field were, on average, 10% higher than simulated data using the MERRA climate database, for the presented period. The proximity of both data indicates the validation of the data collection system.

Figure 6 presents the longitudinal strain (a) and the vertical stress, (b) measured for a series of vehicles from 09/30/2016 to 10/01/2016. The temperature at the bottom of the asphalt layer varied from 19°C to 30°C for this data. The legend represents the vehicle numbers, being N1 vehicle number 1, N2 vehicle number 2, and so forth.

The signals' quality can be deemed excellent due to low noise/signal ratio, high-speed response, and resolution. All the different types of axles can be easily identified for every vehicle, as well as the strain and stress peaks. The values showed here corroborate with the values obtained from the analysis software in similar conditions, around 150 με of longitudinal strain and 150 kPa of vertical stress.

4 CONCLUSIONS

A brief list of the findings of this study are as follows:

– The data processing has proven adequate, with a high-quality signal, excellent signal-to-noise ratio, and acquisition rate compatible with the highway operational speed.
– The FlexPAVE^{TM} simulations validated the instrumentation results, as well as the temperature obtained from the MERRA database.
– The implemented system is considered able to monitoring stresses and strains developed on the asphalt layer and to be used as a tool to help practitioners on monitoring and maintenance plans.

REFERENCES

Al-Qadi, I. L., & Nassar, W. N. 2003. Fatigue shift factors to predict HMA performance. *International Journal of Pavement Engineering* 4, pp. 69–75.

AASHTO. 2014. AASHTO TP-107: Standard method of test for determining the damage characteristic curve of asphalt mixtures from direct tension cyclic fatigue tests. Washington.

AASHTO. 2017. AASHTO R-84: Standard practice of developing dynamic modulus master curves for asphalt mixtures using the asphalt mixture performance tester (AMPT). Washington.

Leandri, P., Bacci, R., Di Natale, A., Rocchio, P., & Losa, M. 2013. Appropriate and reliable use of pavement instrumentation on in-service roads. *Airfield and Highway Pavement 2013: Sustainable and Efficient Pavements*, pp. 1424–1433. DOI: 10.1061/9780784413005.120.

Loulizi, A., Al-Qadi, I. L., Lahouar, S., & Freeman, T. E. 2001. Data collection and management of the instrumented smart road flexible pavement sections. *Transportation Research Record, Journal of the Transportation Research Board* 1769, pp. 142–151. DOI: http://dx.doi.org/10.3141/1769-17.

National Cooperative Highway Research Program – NCHRP. 2004. Guide for Mechanistic-Empirical Design of New and Rehabilitated Pavement Structures. *Part 2 – Desing Inputs, Chapter 3 – Environmental effects, Final Report*. Washington.

Perret, J. 2003. Déformations des couches bitumineuses au passage d'une charge de trafic. *Thèse*. École Polytechnique Fédérale de Lausanne, Lausanne.

Priest, A. L, & Timm, D. H. 2006. Methodology and calibration of fatigue transfer functions for mechanistic-empirical flexible pavement design. *National Center for Asphalt Technology*, NCAT, Auburn University, NCAT Report 06-03.

Timm, D. H. 2016. Key Concepts in dynamic signal processing from instrumented pavement sections. *The Roles of Accelerated Pavement Testing in Pavement Sustainability*. DOI:10.1007/978-3-319-42797-3_22.

Timm, D. H., Priest, A. L., & Mcewen, T. V. 2004. Design and instrumentation of the structural pavement experiment at the NCAT Test Track. *National Center for Asphalt Technology*, NCAT, Auburn University, NCAT Report 04-01.

Van Deusen, D. A., Newcomb, D. E., & Labuz, J. F. 1992. A review of instrumentation technology for the Minnesota Road Research Project. *University of Minnesota*.

Advances in Materials and Pavement Performance Prediction II – Kumar et al. (eds)
© 2021 Taylor & Francis Group, London, ISBN 978-0-367-46169-0

Dealing with non uniqueness of calculated asphalt pavements responses

K. Gkyrtis, A. Loizos & C. Plati
National Technical University of Athens, Laboratory of Pavement Engineering, Athens, Greece

ABSTRACT: The availability of several analysis tools and utilized models for Asphalt Concrete (AC) material characterization implies that the calculated pavement responses are non-unique. Potential variations are expected to have an impact on the bearing capacity assessment and the pavement life expectancy estimation. In the present research study, the investigation considers two powerful and worldwide used viscoelastic analysis tools, namely the 3D-Move and the ViscoRoute. Although both programs are similar, there are differences in the formulation and the way the input information is provided in the format required. This study emphasizes on the produced output (in particular the calculated tensile strains at the AC bottom, related to AC fatigue performance) and its potential impact on pavement engineering and decision-making issues, considering locally utilized AC materials.

1 BACKGROUND AND OBJECTIVES

Pavements response and performance prediction is a challenging task for pavement scientists and engineers that are engaged in both pavement design and analysis (Loizos et al. 2018). Individual material characteristics together with the available analysis theories and tools drastically affect pavement evaluation process with a direct financial impact.

Although Multi-Layered Elastic Theory (MLET) is often used for mechanistic analysis and design of asphalt pavements as a common practice, Asphalt Concrete (AC) materials behavior is rather different. The dynamic modulus (E^*) is a fundamental parameter that accounts for the temperature and loading frequency dependency of AC. Several studies have already focused on the development of models and tools for a more realistic AC material characterization (Darabi et al. 2012; Im et al. 2017; Masad et al. 2005; Motamed et al. 2013; Oeser et al. 2008; Olard & Di Benedetto 2003; Pellinen & Witczak 2002; Zelelew & Papagiannakis 2010; Zhang et al. 2019).

Overall, advanced AC computational modeling is essential for an accurate performance prediction of asphalt pavements (Castillo & Caro 2014; Kassem et al. 2018). Further advances in the response analysis include among others the combination of material modeling with complex loading simulations (Liu et al. 2018), or the concept of the domain analysis for determining pavement damage (Gamez et al. 2018).

Nevertheless, the calculated pavement strains may differentiate because of the several analysis tools or utilized models. Potential variations are expected to have an effect on the bearing capacity assessment and the life expectancy estimation. Since pavement instrumentation is not always and everywhere a viable

solution, the selection of an analysis tool for strain calculations affects pavement evaluation in terms of pavement design optimization, Quality Control (QC) after construction as well as long-term field performance assessment.

On this context, this investigation considers two powerful viscoelastic analysis tools, namely the 3D-Move (Siddharthan et al. 2000) and the ViscoRoute (Chabot et al. 2010). Both calculation tools characterize AC in the linear viscoelastic (LVE) region and have been worldwide utilized (e.g. Leandri et al. 2015; Siddharthan et al. 2018). The emphasis of the analysis is put on the calculated tensile strains at the AC bottom, which are related to the one of the two main traffic-induced distress modes of asphalt pavements (Ulloa et al. 2013). From a relevant sensitivity investigation on thin and thick asphalt pavements, differences within computed strains among the two packages fell in the range of 6% for a wide range of moving speeds and AC temperatures (Hajj et al. 2011).

In the following sections, results from a preliminary investigation of an ongoing research experiment are presented and discussed, in respect to field performance assessment of locally utilized AC materials.

2 EXPERIMENTAL PROCESS

In the current study, laboratory results on the dynamic modulus testing from field cores were used in order to proceed with a response analysis process based on the considered analysis tools. In particular, cores from a new asphalt pavement section of a heavy-duty motorway were extracted prior to the construction of the wearing course and were further utilized to meet the

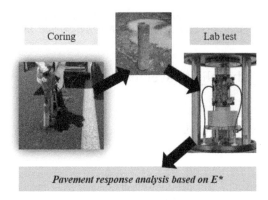

Figure 1. Framework of the experiment.

Table 1. Aggregate gradation for AC materials.

Sieve (mm)	Binder course	Base course
19	100	91
9.5	78	65
4.75	52	44
0.075	4	3.4

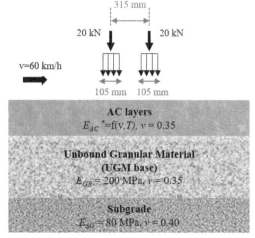

Figure 2. Pavement model for response analysis.

Table 2. Fit of Huet-Sayegh model.

Core	S1	S2	S3	S4	S5	S6
RMPSE %	6.1	6.2	20.6	3.5	7.0	5.4

research objective of this study, which is the assessment of strain variations between the afore-mentioned tools through the consideration of the AC viscoelasticity. A conceptual framework of the experimental process is illustrated in Figure 1.

Both AC binder and base course consist of asphalt mixtures with limestone aggregates. The asphalt material has a penetration index ranging from 35–36PEN and a softening point of 55–56°C. Aggregate gradation is presented in Table 1.

As per the laboratory testing, it included the dynamic modulus test (AASHTO 2001) on six cores. Thirty combinations of five temperatures (4, 15, 20, 25 and 37°C) and six frequencies (25, 10, 5, 1, 0.5 and 0.1 Hz) were used to impose a controlled sinusoidal (haversine) compressive load. Difficulties to perform the test at −10 and 54°C, an issue that has been also mentioned elsewhere (Bennert & Williams 2009; Georgouli & Loizos 2017), led to the selection of the previous temperature spectrum.

To proceed with the response analysis, the experimentally measured dynamic modulus was directly utilized as input in 3D-Move, yet ViscoRoute required the calibration of the Huet-Sayegh model to account for AC viscoelasticity. Model calibration was based on the least squares optimization method.

Thereafter, the calculation of pavement responses was followed. A typical asphalt pavement structure is simulated (as per Figure 2) including AC layers, unbound granular base and a subgrade. AC thicknesses ranged from 17–19 cm based on coring results, whereas a constant thickness of 28 cm was assigned to the base. AC materials were considered as viscoelastic, while base and subgrade materials were assumed to be linear elastic with typically assumed stiffness moduli (Figure 2).

The standard axle-loading configuration (8-ton axle with dual tires) was adopted as per Figure 3, and the critical tensile strains at the AC bottom were calculated at the temperature of 20°C.

3 RESULTS AND DISCUSSION

At first, the Huet-Sayegh model (used in ViscoRoute) was proved to be easily adaptable to the experimentally measured dynamic modulus, with relatively low Root-Mean-Square-Percentage-Errors (RMSPE), apart from one case (S3), which was excluded from further analysis (Table 2). For comparison purpose, the sigmoidal function (used in 3D-Move) exhibited lower RMSPE during calibration. Any variations in the prediction of the E^* master curve could also induce variations in strain calculations.

A typical example of the comparison between the experimental and theoretical E^* (based on the Huet-Sayegh model) is shown in Figure 3. It is noted that for the construction of the master curves a reference temperature of 20°C was considered.

In general, strain profiles were found to follow similar trend between the two calculation tools, with both positive and negative components in the longitudinal strains (Figure 4) and only positive components in the transverse strains (Figure 5). This is in accordance with previous results for the considered axle loading configuration (Chabot et al. 2010).

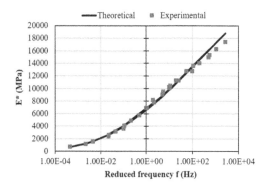

Figure 3. Example of Huet-Sayegh master curve fit for S6.

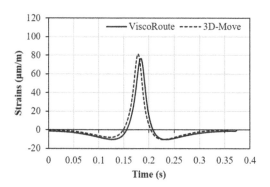

Figure 4. Longitudinal strain profiles for S2.

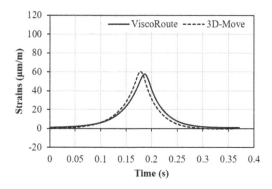

Figure 5. Transverse strain profiles for S2.

Focusing on the maximum strains that are linked with AC fatigue performance, it can be seen from Figure 6 that peak longitudinal strains range from 57–89 μm/m for both analysis tools. However, peak strains are not identical for the two tools. In particular, strains from the 3D-Move are consistently calculated higher than the ViscoRoute considering the available data. Although strain differences fall in the range of around 10%, it is expected that this deviation might be further magnified during the estimation of the allowable traffic repetitions in a fatigue life analysis.

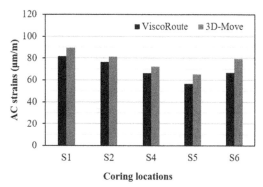

Figure 6. Peak longitudinal AC strains.

Overall, from a pavement evaluation perspective, the selection of a particular calculation tool strongly depends on the target of the analysis, the significance of the pavement section (e.g. heavy-duty motorway versus rural road) and the availability of all the necessary information in respect to both bitumen and mix. For instance, specialized knowledge is needed in the 3D-Move calculation tool for the asphalt binder viscosity data (A-VTS parameters) that requires additional laboratory testing.

4 CONCLUDING REMARKS

This study provided with preliminary implications on the potential variations in the calculated asphalt pavement responses through existing tools. The 3D-Move and ViscoRoute calculations tools were considered and several issues were discussed (e.g. need for model calibration or knowledge of specialized information for bitumen and mix components of AC).

From the preliminary investigation, it was demonstrated that the 3D-Move tool tends to produce higher strains. The differences in the viscoelastic analysis may be significant in some cases, an issue that is expected to have an impact on both pavement design and analysis processes in terms of life expectancy estimation. Further investigation is ongoing in respect to a greater temperature spectrum and different AC materials that are commonly used in road construction activities in order to assess their field performance during both the early pavement's life and its service life.

Overall, pavement engineers should be aware of the capabilities and limitations of several available tools, since in the absence of benchmark strains through instrumentation, calculated responses are non-unique. Nevertheless, selecting an analysis tool can act dominantly in terms of long-term pavement evaluation and management of maintenance strategies by considering principles of pavement mechanics. As such, structural performance assessment needs to be dealt with as rationally as possible in order to reach a reliable and sustainable decision-making during both pavement design and analysis.

REFERENCES

AASHTO T342-11 2001. Standard method of test for determining dynamic modulus of HMA. American Association of State Highway and Transportation Officials, Washington.

Bennert, T. & Williams, S.G. 2009. Precision of AASHTO TP62-07 for use in mechanistic-empirical pavement design guide for flexible pavements. *Transportation Research Record: Journal of the Transportation Research Board* 2127: 115–126.

Castillo, D. & Caro, S. 2014. Probabilistic modeling of air void variability of asphalt mixtures in flexible pavements. *Construction and Building Materials* 61: 138–146.

Chabot, A., Chupin, O., Deloffre, L. & Duhamel, D. 2010. ViscoRoute 2.0: a tool for the simulation of moving load effects on asphalt pavement. *Road Materials & Pavement Design* 11: 227–250.

Darabi, M.K., Abu-Al-Rub, R.K., Masad, E.A. & Little, D.N. 2012. Thermodynamic-based model for coupling temperature-dependent viscoelastic, viscoplastic, and viscodamage constitutive behavior of asphalt mixtures. *International Journal for Numerical and Analytical Methods in Geomechanics* 36: 817–854.

Gamez, A, Hernandez, J.A., Ozer, H. & Al-Qadi, I.L. 2018. Development of domain analysis for determining potential pavement damage. *Journal of Transportation Engineering, Part B: Pavements* 144(3): 04018030(1–12).

Georgouli, K. & Loizos, A. 2017. E* prediction algorithm for pavement quality control assessment. *Paper presented at the 10th International Conference on the Bearing Capacity of Roads, Railways and Airfields (BCRRA)*, June 28–30, 2017, Athens, Greece, pp: 799–805.

Hajj, E.Y., Ulloa, A., Sebaaly, P.E. & Siddharthan, R.V. 2011. Equivalent loading frequencies to simulate asphalt layer pavement responses under dynamic traffic loading. *Transportation Research Board*, AFD80: Committee Meeting.

Im, S., You, T., Ban, H. & Kim, Y-R. 2017. Multiscale testing-analysis of asphaltic materials considering viscoelastic and viscoplastic deformation. *International Journal of Pavement Engineering* 18(9): 783–797.

Kassem, H.A, Chehab, G.R. & Najjar S.S. 2018. Quantification of the inherent uncertainty in the relaxation modulus and creep compliance of asphalt mixes. *Mechanics of Time-Dependent Materials* 22(3): 331–350.

Leandri, P. Losa, M. & Di Natale, A. 2015. Field validation of recycled cold mixes viscoelastic properties. *Construction and Building Materials* 75: 275–282.

Liu, P., Xing, Q., Wang, D., Oeser, M. 2018. Application of Linear Viscoelastic Properties in Semianalytical Finite Element Method with Recursive Time Integration to Analyze Asphalt Pavement Structure. Advances in Civil Engineering, Article ID 9045820: 1–15.

Loizos, A., Cliatt, B., Plati, C. & Gkyrtis, K. 2018. Assessment of variations in asphalt pavement mechanical responses. *Paper presented at the International Conference on the Advances in Materials and Pavement Performance Prediction (AM3P)*, April 16–18, 2018, Doha, Qatar, pp: 351–355.

Masad, E., Tashman, L., Little, D. & Zbib, H. 2005. Viscoelastic modeling of asphalt mixes with the effects of anisotropy, damage and aggregate characteristics. *Mechanics of Materials* 37: 1242–1256.

Motamed, A., Bhasin, A. & Liechti K.M. 2013. Constitutive modeling of the nonlinearly viscoelastic response of asphalt binders; incorporating three-dimensional effects, *Mechanics of Time-Dependent Materials* 17 (1): 83–109.

Oeser, M., Pellinen, T., Scarpas, T. & Kasbergen, C. 2008. Studies on creep and recovery of rheological bodies based upon conventional and fractional formulations and their application on asphalt mixture. *International Journal of Pavement Engineering* 9(5): 373–386.

Olard, F. & Di Benedetto, H. 2003. General "2S2P1D" model and relation between the linear viscoelastic behaviors of bituminous binders and mixes. *Road Materials and Pavement Design* 4(2): 185–224.

Pellinen, T.K. & Witczak, M.W. (2002), Stress dependent master curve construction for dynamic (complex) modulus, *Journal of the Association of Asphalt Paving Technologists* 71: 281–309.

Siddharthan, R.V., Hajj, E.Y. & Nasimifar, M. 2018. Moving deflectometer devices to predict critical pavement responses. *Paper presented at the International Conference on the Advances in Materials and Pavement Performance Prediction (AM3P)*, April 16–18, 2018, Doha, Qatar, pp: 75–78.

Siddharthan, R., Krishnamenon, N., Sebaaly, P.E. 2000. Pavement Response Evaluation Using Finite-Layer Approach. *Transportation Research Record: Journal of the Transportation Research Board* 1709: 43–49.

Ulloa, A., Hajj, E.Y., Siddharthan, R.V. & Sebaaly, P.E. 2013. Equivalent loading frequencies for dynamic analysis of asphalt pavements. *Journal of Materials in Civil Engineering* 25(9): 1162–1170.

Zelelew, H.M. & Papagiannakis, T. 2010. Micromechanical Modeling of Asphalt Concrete Uniaxial Creep Using the Discrete Element Method. *Road Materials and Pavement Design* 11(3): 613–632.

Zhang, R., Sias, J.E., Dave, E.V. & Rahbar-Rastegar, R. 2019. Impact of Aging on the Viscoelastic Properties and Cracking Behavior of Asphalt *Transportation Research Record: Journal of the Transportation Research Board* 2673(3): 406–415.

Advances in Materials and Pavement Performance Prediction II – Kumar et al. (eds)
© 2021 Taylor & Francis Group, London, ISBN 978-0-367-46169-0

Evaluation of emerging performance tests and pavement performance prediction models using the FlexPAVE™ software

Amir Golalipour
Senior Project Manager, Engineering & Software Consultants, Inc.
Turner-Fairbank Highway Research Center, Georgetown Pike, McLean, Virginia, USA

David J. Mensching
Asphalt Materials Research Program Manager, Federal Highway Administration
Turner-Fairbank Highway Research Center, Georgetown Pike, McLean, Virginia, USA

ABSTRACT: This paper outlines the impact of changes in material properties, mixture density levels, and pavement structure on FlexPAVE™ outputs. This study is focused on evaluating the rutting and fatigue performance characteristics of asphalt pavements throughout pavement life cycle using FlexPAVE™ software. A dense grade asphalt mixture from a field project is utilized in this evaluation. The results of this study provide insight on FlexPAVE™ pavement performance predictions for various factors.

1 INTRODUCTION

1.1 Background

Pavement performance prediction models are based on parameters such as climate, traffic, environment, material properties, etc. While all these factors are playing important roles in the performance of pavements, the selected material properties have a significant impact on the prediction model outcomes.

Recently, the simplified viscoelastic continuum damage (S-VECD) and the shift models, which characterize the fatigue and rutting behaviors of asphalt concrete, have been utilized in the FlexPAVE™ software to predict pavement performance. However, despite the advantages of these models in providing accurate and advanced material characterization, their pavement performance predictions need to be evaluated using field sections (Wang et al. 2018).

This paper will outline the impact of changes in material properties, mixture density levels, and pavement structure on FlexPAVE™ outputs. This study is focused on evaluating the rutting and fatigue performance characteristics of asphalt pavements throughout pavement life cycle using this software.

2 MATERIALS & METHODS

2.1 Materials

The plant-produced asphalt mixture used in this study was dense graded HMA mix with 9.5 mm nominal maximum aggregate size (NMAS). This mixture was produced in Dilley, Texas. The mix type, based on TX specification, is 344-SP-C, produced using PG 76-22 asphalt binder with no RAP.

Samples of plant-produced asphalt mixture were delivered in buckets. The sampled materials were stored in the lab, split, and quartered using a quartering template according to AASHTO R 47. Afterward, the samples were reduced to appropriate sizes for the evaluation of volumetric properties. The loose asphalt mixture samples were then compacted into cylindrical specimens using a Superpave gyratory compactor (SGC). The volumetric specimens were compacted to the design gyration level (N_{design} = 50 gyrations).

Table 1. Volumetric data summary for production samples.

Sample	Binder specific gravity, (Gb)	V_a (%)	VMA	VFA	P_b%
PMLC	1.036	4.5	17.13	73.73	5.97

The objective of the testing and analysis performed was to characterize performance-related properties of this mixture at different air void levels (4, 7, and 10%). Volumetric data for performance test specimens are summarized in the Table 2.

Table 2. Performance test specimen volumetrics.

Specimen	VMA %	VFA %
4% V_a	16.75	75.40
7% V_a	19.30	63.50
10% V_a	21.81	54.50

2.2 Test procedures

This study is intended to investigate the impact of changes in different asphalt mixture parameters using pavement performance prediction tools. A laboratory evaluation of Superpave volumetrics and material-level performance, using the Asphalt Mixture Performance Tester (AMPT), was conducted. Cylindrical test specimens were characterized in the AMPT for viscoelastic, rutting, and fatigue properties using dynamic modulus ($|E^*|$) (AASHTO TP 132), cyclic fatigue (AASHTO TP 133), and Stress Sweep Rutting (SSR) tests (AASHTO TP 134). Data from these tests are used as inputs and coupled with pavement layers' information for pavement performance analysis.

2.2.1 Performance test specimen fabrication

The performance testing specimens are prepared for the SSR test according to the AASHTO R 83 specification. The amount of material placed in the SGC compaction molds was adjusted in order to achieve the target air void content for PMLC specimens. Considerable care was taken to ensure that the specimens met the dimensional tolerances indicated in AASHTO R 83.

Small-scale specimens were also prepared for dynamic modulus, and cyclic fatigue performance testing according to AASHTO PP 99 specifications. Small specimen dimensional tolerances are indicated below.

Table 3. Summary of small-scale performance specimen dimensional requirements.

Item	Limits
Mean Diameter	36 to 40 mm
Standard Deviation of Diameter	0.5 mm or less
Height	107.5 to 112.5 mm for $|E^*|$ and Fatigue 127.5 to 132.5 mm for
End Flatness	0.5 mm or less
End Perpendicularity	1.0 mm or less

2.2.2 AMPT performance testing

Dynamic modulus of a viscoelastic material (i.e. asphalt mixture) is the absolute value of the complex modulus calculated by dividing the peak-to-peak stress by the peak-to-peak strain. The Dynamic modulus $|E^*|$ tests is performed according to AASHTO TP 132. For each mixture, tests were performed at three different temperatures: 4.0°C, 20.0°C, and 40.0°C and three different frequencies: 0.01, 0.1, 1, and 10 Hz. The lowest frequency of 0.01 Hz was applied only at the highest test temperature of 40.0°C.

The permanent deformation properties of asphalt mixtures can be evaluated using the SSR test – AASHTO TP 134. The test can be used to determine shift model coefficients which can then be used to predict permanent deformation under variable loading, stress, and temperature conditions for an asphalt mixture using four specimens: two at a high temperature

and two at a lower temperature. The temperature for the project SSR tests is taken from the closet weather station, for which the corresponding adjusted high PG temperature is PG 76-22 (98% reliability, 0 mm for the surface and not adjusted for traffic). The SSR test temperatures are 37.5°C and 56.3°C.

The SSR test is a cyclic compression test which applies various levels of deviatoric stress upon the performance specimen over a series of three loading blocks at 200 cycles each. The SSR test is terminated after 600 cycles are complete. The results from the test can be processed using FlexMAT™-Rutting, to determine the shift model coefficients, which can then be utilized to predict rutting performance of asphalt pavements through the FlexPAVE™ pavement analysis program.

One of the fatigue tests that recently has received significant attention in pavement community is the AMPT cyclic fatigue test. This test employs viscoelastic continuum damage principles to characterize a material's fatigue resistance. The characterization is determined by performing controlled actuator displacement, cyclic tension ("pull-pull") testing according to AASHTO TP 133. The researchers at North Carolina State University (NCSU) have developed a Simplified Viscoelastic Continuum Damage (S-VECD) model for characterizing the fatigue properties of asphalt mixtures. The S-VECD model is a mathematically rigorous approach where fundamental material properties are incorporated. The key function is the damage characteristic curve (C versus S) that relates the amount of damage (S) in a specimen to the material integrity or pseudo stiffness (C) (Underwood et al. 2012).

In order to represent asphalt mixture's fatigue cracking resistance in a single parameter, Wang et al. (2018) developed a cracking index property, S_{app}, which is based on concepts of the S-VECD model. S_{app} takes into account different mix fatigue factors including material's modulus and toughness and is a measure of the amount of fatigue damage the material can tolerate under loading. Higher S_{app} values indicate better fatigue resistance of the mixture. S_{app} can be defined as 'the apparent damage capacity' (FHWA-HIF-19-091 Technical Brief).

3 RESULTS & DISCUSSIONS

3.1 Dynamic modulus ($|E^*|$)

Dynamic modulus data showed some changes in stiffness of the mixture as the air void level changed. The most significant differences in $|E^*|$ values were observed at lower frequencies (higher temperatures).

The dynamic modulus test results for an asphalt mixture at various temperatures can be shifted along the frequency axis to form a single dynamic modulus master curve at a desired reference temperature. This procedure takes advantage of the principle of time-temperature superposition and allows a single curve to be used to describe the mixture response at multiple loading frequencies and temperatures. For this project, master curves were developed using the MasterSolver spreadsheet application described in AASHTO PP 61. The reference temperature for the master curves was 20°C. The results of 7% V_a sample is shown as an example for this process.

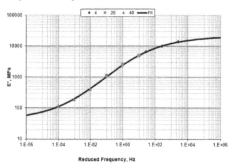

Figure 1. Dynamic modulus (E*) master curve for 7% V_a sample.

The comparison of master curve results confirmed the observation from the dynamic modulus data that lower air void samples exhibited higher stiffness and lower phase angle, which is indicative of more elastic behavior.

3.2 Stress sweep rutting

Average permanent deformation results for SSR tests based on two low temperature and two high temperature stress sweep tests are summarized. Based on the SSR results, air void level increases resulted in a decrease in rutting resistance, as shown in Figure 2.

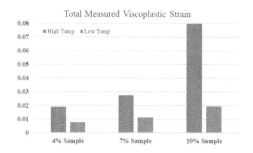

Figure 2. SSR test data showing total measured viscoplastic strain.

3.3 Cyclic fatigue

The AMPT cyclic fatigue testing indicated a difference in the fatigue properties as the air void level changed. These differences were more apparent as the air void level increased to 10%. This trend can be observed from C-S curve as well as S_{app} criteria results. Fundamentally, C-S curve shows the relationship between material integrity and microstructural damage accumulation. A material with a lower, more gradual slope

is more favorable as this corresponds to a lower damage accumulation rate. Overall, based on this figure, the sample with the lowest air void (4%) shows better fatigue behavior as compared to the other mixes.

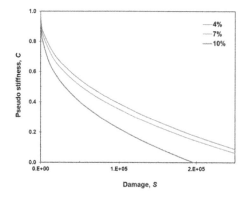

Figure 3. Damage characteristics curve of all the mixtures: pseudo stiffness versus damage.

The curves shown in the above figure may not be sufficient to rank mixtures' fatigue performance, as it does not consider the significant influence of the material's resistance to deformation (Underwood et al. 2006). Under loading conditions where other mechanisms, such as viscoplasticity, begin to contribute significantly, the performance ranking could change. Hence, a more comprehensive method of comparison is needed to characterize fatigue behavior of these materials. Such comparisons are conducted by using cracking index property, S_{app}. The cracking index property for mix production samples with different air void levels is presented in Figure 4. Asphalt mixture with 4% air void exhibits the best fatigue performance (the highest S_{app} value).

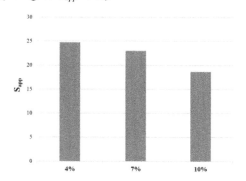

Figure 4. S_{app} criteria of mixtures at different air void levels.

3.4 Pavement performance predictions

FlexPAVE™ is a three-dimensional finite element program that is capable of moving load analysis under realistic climatic conditions. It utilizes the S-VECD model to predict asphalt pavement life. FlexPAVE™ simulations can account for the effects of temperature

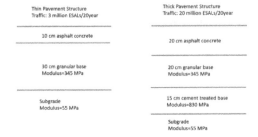

Figure 5. Pavement structures used for FlexPAVE™ analysis.

using the temperature–time history data obtained from the Enhanced Integrated Climatic (EIC) Model. This software also uses a layered viscoelastic model to calculate stress and strains for different pavement layers (Wang et al. 2018).

For this analysis, two different pavement structures are considered (Figure 5). The objective is to evaluate FlexPAVE™ sensitivity to different pavement design.

3.4.1 FlexPAVE™ analysis

Based on the pavement analysis, the impact of density could be very significant in terms of fatigue damage evolution. Below figures clearly show that the pavement with 10% air voids exhibited almost two times the percent damage than that observed in the pavement with 4% air void mixture. The differences are more pronounced for the thicker pavement structure. For both pavement structures, the pavement with 4% showed the best fatigue damage performance.

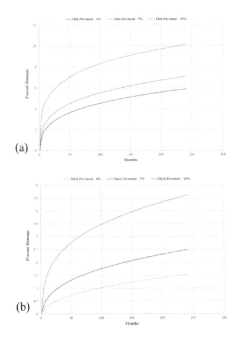

Figure 6. Damage evolution for different pavement structures: (a) Thin structure, and (b) Thick structure.

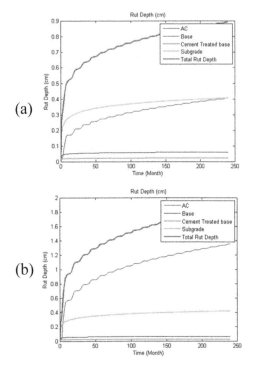

Figure 7. Rut depth for thick pavement structure: (a) 4% V_a, and (b) 10% V_a.

Another feature in FlexPAVE™ software is to predict rut depth for all pavement layers. The rutting results also showed that the pavement section with 10% air void resulted in almost two times more total rut depth. Figure 7 summarizes these results for thick pavement structure. It should be noted that the same trend was observed for the thin pavement structure too.

4 CONCLUSIONS

Based on the findings, the exploratory project demonstrated the identification of the effects of air void level on the stiffness, fatigue and rutting properties of asphalt mixtures. The findings indicate that FlexPAVE™ software shows promise for use as a pavement prediction software.

REFERENCES

FHWA Technical Brief (2019). "Cyclic Fatigue Index Parameter (Sapp) for Asphalt Performance Engineered Mixture Design." The Federal Highway Administration (FHWA) TechBrief, FHWA-HIF-19-091.
Underwood, B. S., Y. R. Kim, and M. N. Guddati. (2006). "Characterization and Performance Prediction of ALF Mixtures Using a Viscoelastoplastic Continuum Damage Model." Journal of Association of Asphalt Paving Technologists, Vol. 75, pp. 577–636.

Underwood, B. S., Baek C., and Y. R. Kim (2012) "Simplified Viscoelastic Continuum Damage Model as Platform for Asphalt Concrete Fatigue Analysis". Journal of the Transportation Research Board, No. 2296, Transportation Research Board of the National Academies, Washington, D.C., 2012, pp. 36–45.

Wang, Y., B. Keshavarzi, and Y. R. Kim (2018). Fatigue Performance Prediction of Asphalt Pavements with Flex-PAVE TM, the S-VECD Model, and D R Failure Criterion. Transportation Research Record: The Journal of the Transportation Research Board.

Advances in Materials and Pavement Performance Prediction II – Kumar et al. (eds)
© *2021 Taylor & Francis Group, London, ISBN 978-0-367-46169-0*

Assessment of reflective cracking performance life for high polymer-modified asphalt overlays

J. Habbouche
Virginia Transportation Research Council, Charlottesville, Virginia, USA

E.Y. Hajj & P.E. Sebaaly
Pavement Engineering & Science Program, University of Nevada, Reno, Nevada, USA

ABSTRACT: Reflective cracking is one of the major type of distresses associated with the use of asphalt concrete (AC) overlays for rehabilitating deteriorated asphalt pavements. This paper briefly describes the research effort completed to evaluate the reflective cracking performance life of high polymer-modified (HP) AC mixtures used as overlays. Sixteen AC mixtures were produced in the laboratory using conventional polymer modified (PMA) and HP asphalt binders. These mixtures were evaluated in terms of their dynamic modulus and resistance to reflective cracking. The engineering property and laboratory performance data were combined into a mechanistic analysis that took into consideration the existing AC pavement condition, mix-specific material properties, traffic condition, and climate. Overall, the HP AC mixes resulted in an increase in both time to reach initial cracking and reflective cracking performance life of the AC overlay when compared to their respective PMA AC overlay mixes.

1 INTRODUCTION

The combined effects of traffic loading and climate will cause asphalt concrete (AC) pavements to deteriorate over time. When major work becomes necessary, AC overlays have been one of the commonly used methods for rehabilitating aged and deteriorated asphalt pavements. Consequently, reflection of cracks from existing pavements becomes one major type of distress influencing the life of an AC overlay. Thus, the long-term performance of many AC overlays will highly depend on their ability to resist reflective cracking, the specific conditions of the existing pavement, and the combination of materials, traffic, and environmental conditions (Habbouche et al. 2017, 2019).

While conventionally polymer-modified asphalt (PMA) mixtures, with 2–3% polymer content, have shown improved long-term performance when used as overlays, it is also believed that AC mixtures with high polymer (HP) content may offer additional advantages for flexible and composite rehabilitated pavements subjected to heavy and/or slow moving traffic loads (Habbouche et al. 2018). This innovative technology involving the use of a new styrene-butadiene-styrene (SBS) polymer structure at a higher rate was introduced without causing production problem such a binder pump clogging or reduced mixture workability.

The objective of this paper is to assess the use of HP AC overlays to mitigate reflective cracking in rehabilitated flexible pavement sections. This was accomplished by evaluating AC mixtures in the laboratory in terms of dynamic modulus, $|E^*|$, and resistance to reflective cracking; and by incorporating the engineering property and performance models into commonly designed pavement structures through comprehensive mechanistic analyses (Habbouche et al. 2019, 2020, in-press).

2 LABORATORY EVALUATION

2.1 *Mix designs*

Table 1 presents some of the mix designs information for all AC mixtures. More details regarding the raw materials and mix designs can be found elsewhere (Habbouche 2019; Habbouche et al. 2019a; Habbouche et al. 2019b). Two asphalt binders, PG 76-22 PMA and HP (PG 88-34 & PG 94-28), were sampled from two different sources (herein labeled as source A and source B). Two aggregates' mineralogy were evaluated under this effort: Southeast Florida limestone (labeled as FL) and Georgia Granite (labeled as GA).

Two aggregate gradations were evaluated from each aggregate source with Nominal Maximum Aggregate Size (NMAS) of 9.5 mm and 12.5 mm. Sixteen mixtures (i.e., 8 PMA and 8 HP) were designed and produced in the laboratory following the Florida Department of Transportation (FDOT) Superpave methodology. Recycled asphalt pavement (RAP)

Table 1. PMA and HP AC Mixtures.

Asphalt Mixture ID	Description[#]	OBC (%)[*]
FL95_PMA(A) FL95_PMA(B)	FL aggregate; 9.5 mm NMAS; PMA binder from source A and source B	6.2
FL95_HP(A) FL95_HP(B)	FL aggregate; 9.5 mm NMAS; HP binder from source A and source B	5.9
FL125_PMA(A) FL125_PMA(B)	FL aggregate; 12.5 mm NMAS; PMA binder from source A and source B	5.5
FL125_HP(A) FL125_HP(B)	FL aggregate; 12.5 mm NMAS; HP binder from source A and source B	5.4
GA95_PMA(A) GA95_PMA(B)	GA aggregate; 9.5 mm NMAS; 20% RAP; PMA binder from source A and source B	4.8
GA95_HP(A) GA95_HP(B)	GA aggregate; 9.5 mm NMAS; HP binder from source A and source B	4.9
GA125_PMA(A) GA125_PMA(B)	GA aggregate; 12.5 mm NMAS; 20% RAP; PMA binder from source A and source B	4.2
GA125_HP(A) GA125_HP(B)	GA aggregate; 12.5 mm NMAS; HP binder from source A and source B	4.9

[#]RAP = reclaimed asphalt pavement
[*]OBC = optimum asphalt binder content for volumetric design.

materials were only used with AC mixtures manufactured using GA aggregates and PMA asphalt binders at a content of 20% by dry weight of aggregate.

2.2 Dynamic modulus

The dynamic modulus |E*| provides an indication on the overall quality of the asphalt mixture. The test was conducted in accordance with AASHTO T 378 using the Asphalt Mixture Performance Tester (AMPT) under various combinations of loading and frequencies. All mixtures were evaluated at the short-term aging condition in accordance with AASHTO R 30. All |E*| test specimens (100 mm diameter by 150 mm height) were compacted to 7 ± 1% air voids. Figure 1 shows the |E*| values for all evaluated AC mixtures at the critical temperature for cracking in Florida (25°C) and at a loading frequency of 10 Hz representing highway travel speed. Overall, the HP AC mixtures showed similar or lower |E*| values when compared with those of the control PMA mixtures at intermediate and high temperatures regardless of the aggregate source, the NMAS of the AC mixture, or the binder source.

2.3 Resistance to reflective cracking

The Texas overlay test (OT) was used to evaluate the mixtures' resistance to reflective cracking in

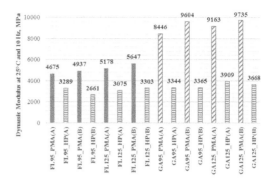

Figure 1. |E*| values at 25°C and 10 Hz.

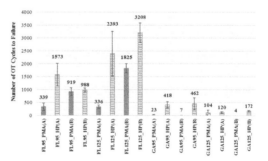

Figure 2. Number of OT cycles to failure at 25°C.

accordance with Tex-248-F procedure. The OT specimens (150 mm long by 75 mm wide and 37.5 mm thick) were only subjected to short-term aging and were compacted to 7 ± 1% air voids. The test is conducted in a controlled displacement mode until failure occurs at a loading rate of one cycle per 10 seconds with a maximum displacement of 0.6350 mm at 25 ± 0.5°C. Figure 2 presents the number of cycles to failure for all evaluated AC mixtures. They were defined as the number of cycles to reach a 93% drop in initial load measured from the first opening cycle.

A power function defined in Equation 1 was used to fit the load reduction curve function of the number of loading cycles to determine the crack propagation rate (CPR) and the crack resistance index (CRI) (Garcia et al. 2016).

$$NL = N^{CPR} = N^{(0.0075*CRI-1)} \qquad (1)$$

where NL = normalized load; CPR = crack propagation rate; and CRI = crack resistance index.

The critical fracture energy (G_c) at the maximum peak load of the first loading cycle is considered as the energy required to initiate crack. The area under the hysteresis loop, limited for the tension phenomena, is considered essential to compute the fracture parameters that characterize the crack initiation stage of the OT. The critical fracture energy is calculated as the ratio of this area and the OT specimen cross-section area. A design interaction graph plotting G_c versus CPR was established (Garcia et al. 2016).

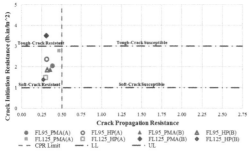

*1 lb.in/in^2 = 17.86 kg.m/m^2

Figure 3. Interaction plot for FL AC mixes.

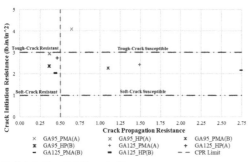

*1 lb.in/in^2 = 17.86 kg.m/m^2

Figure 4. Interaction plot for GA AC mixes.

This interaction plot includes four major parts: tough-crack resistant; tough-crack susceptible; soft-crack resistant; and soft-crack susceptible. Figures 3 and Figure 4 illustrate the cracking resistance interaction plots for AC mixes manufactured using FL and GA aggregates, respectively.

All FL AC mixes showed a *CPR* value lower than 0.5 indicating a good cracking resistance. All FL mixes, except for FL125_PMA(B), showed a G_c value between 1 and 3 indicating a good resistance to crack initiation. All GA PMA AC mixes showed a *CPR* value greater than 0.5 indicating a brittle behavior and a low resistance to crack propagation. These mixes, except for GA95_PMA(A), showed G_c values between 1 and 3 indicating a good resistance to crack initiation. On the other hand, all GA HP mixes show *CPR* values lower than 0.5 and G_c values between 1 and 3 indicating soft-crack resistant mixes.

3 REFLECTIVE CRACKING PERFORMANCE LIFE

3.1 *Fracture parameters A and n*

Various models have been developed to analyze and/or predict reflective cracking. The Texas Transportation Institute (TTI) system consider the Paris' law-based fracture mechanics model expressed in Equation 2

Table 2. Fracture Parameters for PMA and HP AC Mixtures at 25°C.

Mix ID	*A*	*n*
FL95_PMA(A)	9.98E-02	6.60E-01
FL95_PMA(B)	7.15E-02	6.62E-01
FL95_HP(A)	3.81E-03	1.36E+00
FL95_HP(B)	1.71E-02	1.16E+00
FL125_PMA(A)	2.90E-02	1.02E+00
FL125_PMA(B)	5.58E-04	1.46E+00
FL125_HP(A)	2.30E-03	1.49E+00
FL125_HP(B)	6.17E-04	1.93E+00
GA95_PMA(A)	6.14E-01	2.02E-01
GA95_PMA(B)	2.70E-01	5.56E-01
GA95_HP(A)	4.92E-02	8.79E-01
GA95_HP(B)	7.94E-02	7.62E-01
GA125_PMA(A)	6.30E-01	1.11E-01
GA125_PMA(B)	–	–
GA125_HP(A)	2.87E-01	4.48E-01
GA125_HP(B)	2.47E-01	5.44E-01

for the evaluation of reflective cracking propagation (Zhou et al. 2008). The determination of fracture parameters (i.e., *A* & *n*) for the PMA and HP AC mixes (refer to Table 2) requires the determination of the following: (1) *SIF* function of crack length (*c*); (2) *NL* using OT test function of *c*; (3) *NL* using OT test function of *N*; (4) *c* function of *N*; and (5) *SIF* function of *N*. *A* and *n* were computed as the intercept and slope of *SIF* versus *N* curve, respectively.

$$\frac{dc}{dn} = A^* (SIF)^n \tag{2}$$

where *c* = crack length (mm); *N* = number of loading cycles; and *SIF* = stress intensity factor amplitude.

In general, the *n* value is characteristic of the asphalt binder, meanwhile the *A* value is characteristic of the AC mixture itself (i.e., aggregate gradation and asphalt binder). Lower *A* and *n* values were observed for the PMA AC mixes when compared to their corresponding HP AC mixes. It should be mentioned that *A* and *n* values could not be calculated for GA125_PMA(B) mix due to the low number of OT loading cycles to failure (i.e., *N* = 4). Moreover, this mix was the stiffest among all evaluated AC mixes and contains 20% of RAP material. Accordingly, a mechanistic analysis could not be conducted.

3.2 *Mechanistic analysis*

The thickness of the AC overlays for PMA pavements was designed following the FDOT Flexible Pavement Design Manual (FDOT 2016). The thickness of the AC overlay for the HP pavement sections was reduced according to the structural layer coefficient for HP AC mixes in Florida (i.e., 0.54). The structural designs of all PMA and HP rehabilitated pavement sections are summarized in Table 3. FDOT mandates the use of

Table 3. Structural Designs for Rehabilitated Pavement Sections.

FDOT ESAL	Label	PMA AC Overlay (cm)	HP AC Overlay (cm)	Existing PMA AC (cm)	Base: Existing Layer (mm) [M_r (MPa)]	SG M_r (MPa)
Level C (7 million ESAL)	R-C1	8.9	7.6	1.3	30.5 [231.0]	79.3
	R-C2	11.4	9.5	6.4	27.9 [305.4]	37.9
	R-C3	8.9	7.6	1.3	25.4 [305.4]	79.3
Level D (20 million ESAL)	R-D1	10.2	8.3	5.1	30.5 [231.0]	79.3
	R-D2	14.0	11.4	8.9	31.8 [305.4]	37.9
	R-D3	10.2	8.3	5.1	25.4 [305.4]	79.3

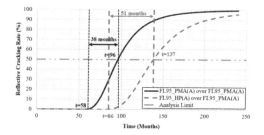

Figure 5. RCR versus time of R-C1 for PMA and HP AC overlays on top of an existing PMA AC layer.

30.5 cm thick stabilized subgrade layer with a M_r of 84.5 MPa.

The Texas Asphalt Concrete Overlay Design and Analysis System (TxACOL) software (Zhou et al. 2008) was used to estimate the reflective cracking rate in the PMA and HP AC overlay. The mechanistic analysis for reflective cracking considered multiple factors such as traffic loading and speed, environment, existing pavement condition, and characteristics of AC overlay material.

In order to simulate the deteriorated condition of an existing AC layer due to fatigue cracking before rehabilitation, a reduction in the stiffness of the existing PMA AC layer was applied. A damaged |E*| master curve was calculated following the approach used in AASHTOWare Pavement ME (ARA 2004) with a damage accumulation of 0.6 representing a fair condition of the existing AC layer over its service life. It should be mentioned that all existing AC layers before rehabilitation were assumed to be made of PMA AC mixes. Only the new AC overlay was considered either as an undamaged PMA or HP AC mix. The reflective cracking analysis criterion was selected to be 50%.

Figure 5 illustrates an example of reflective cracking propagation rate *(RCR)* for pavement section R-C1 for two cases: FL95_PMA(A) AC overlay on top of an existing damaged FL95_PMA(A) AC layer, and FL95_HP(A) AC overlay on top of an existing damaged FL95_PMA(A) AC layer. For the case of the PMA AC overlay, the cracks started to reflect in the overlay (i.e., *RCR* >0%) at an initial time (i.e., $t_{initial}$) of approximately 58 months (4.8 years) after construction. The *RCR* reached its failure criterion (i.e., 50%) after 96 months (8.0 years) ($t_{RCR=50\%}$) from construction. Thus it took 38 months (3.1 years) for the PMA AC overlay to reach failure after initial cracking has occurred.

For the case of HP AC overlay, the cracks started reflecting on top of the AC overlay after 86 months

(7.1 years) from construction. The *RCR* reached its failure criterion after 137 months (11.4 years). Thus, it took 51 months (4.3 years) for the HP AC overlay to reach failure after initial cracking has occurred. In summary, the illustrative example showed that, for the same traffic and environmental conditions, a 3.0 inch of HP AC overlay is expected to perform better than a 3.5 inch PMA AC overlay as demonstrated with the observed 41 month delay in reaching failure criterion. Similar analyses were conducted for all pavement sections. The percent increase in performance life ranged between 42.7% and 366.0% and between 14.3% and 201.8% for traffic level C and D, respectively.

4 SUMMARY AND CONCLUSIONS

Materials shipped from Florida were assessed and used for the development of sixteen AC mixtures using PMA and HP asphalt binders. The sixteen AC mixtures were evaluated in terms of their dynamic modulus and resistance to reflective cracking. The engineering properties and laboratory performance data were combined into a mechanistic analysis to evaluate the reflective cracking performance life of HP AC mixtures used as overlays. The analysis took into consideration the existing pavement condition in terms of damaged modulus for the existing AC layer, mix-specific material properties, traffic condition, and Florida climate. The HP AC overlay mixes resulted in an increase in both time to reach initial cracking and performance life of the AC overlay. Thus, HP AC overlays, if properly designed are expected to exhibit an acceptable or better performance in terms of mitigating reflective cracking when compared to the respective PMA AC overlay mix.

REFERENCES

ARA, Inc. 2004. Guide for Mechanistic-Empirical Design. Transportation Research Board, Washington, D.C.
Florida Department of Transportation. 2016. *Flexible Pavement Design Manual*. Tallahassee.
Florida Department of Transportation. 2018. *Standard Specifications for Road and Bridge Construction*. Tallahassee.

Garcia V., Miramontes A., Garibay J., Abdallah I., & Nazarian S. 2016, *Improved Overlay Tester for Fatigue Cracking Resistance of Asphalt Mixtures*, Center for Transportation Infrastructure Systems, The University of Texas at El Paso, Report No. TxDOT 0-6815-1.

Habbouche, J., Hajj, E.Y., Sebaaly P.E. & Hand A.J.T. 2020. Fatigue-Based Structural Layer Coefficient of High Polymer-Modified Asphalt Mixtures, *Transportation research Record: Journal of the Transportation Research Board*, in-press.

Habbouche, J., Hajj, E.Y., Morian N.M., Sebaaly P.E. & Piratheepan M. 2017. Reflective Cracking Relief Interlayer for Asphalt Pavement Rehabilitation: From Development to Demonstration, *Journal of Road Materials and Pavement Design*, Volume 18: pp. 30–57.

Habbouche, J., Hajj, E.Y., & Sebaaly P.E. 2019a. *Structural Coefficient of High Polymer Modified Asphalt Mixes*, Florida Department of Transportation, https://www.fdot.gov/research, Tallahassee.

Habbouche J. 2019. *Structural Coefficients of High Polymer Modified Asphalt Mixes Based on Mechanistic-Empirical Analyses and Full-Scale Pavement Testing.* Doctoral Dissertation. University of Nevada, Reno, Nevada.

Habbouche, J., Hajj E.Y., Piratheepan M., & Sebaaly P.E. 2019b. Impact of High Polymer Modification on Performance Characteristics of Asphalt Mixtures. *Journal of the Association of Asphalt Pavement Technologists*, in press.

Habbouche, J., Hajj, E. Y., Sebaaly, P. E., and Piratheepan, M. 2018. A Critical Review of High Polymer-Modified Asphalt Binders and Mixtures. *International Journal of Pavement Engineering*, DOI: 10.1080/10298436.2018. 1503273.

Zhou F., Hu S., Hu X., & Scullion T. 2008, *Mechanistic-Empirical Asphalt Overlay Thickness Design and Analysis System*, Texas Transportation Institute (TTI), Texas Department of Transportation, Technical Report 0-5123-3.

Advances in Materials and Pavement Performance Prediction II – Kumar et al. (eds)
© 2021 Taylor & Francis Group, London, ISBN 978-0-367-46169-0

Effect of seasonal and daily FWD measurements on back-calculated parameters for pavements – LTPP SMP study

Syed W. Haider, Hamad B. Muslim & Karim Chatti
Michigan State University, USA

ABSTRACT: Falling weight deflectometer (FWD) test is used to evaluate structural capacity of an existing pavement. However, FWD measurements are influenced by the temporal variations and therefore, consideration of such effect is essential for accurate assessment of pavement structural condition. Long-term Pavement Program (LTPP) Seasonal Monitoring Program (SMP) was designed to evaluate the influence of such temporal variations on pavement structural characteristics. This paper presents typical effects of seasonal and daily variations on pavement deflections and back-calculated moduli. Available FWD deflection measurements and back-calculated moduli, in the LTPP database for different pavement layers for all SMP pavement sections were evaluated. Results showed that HMA moduli decrease from morning to afternoon (i.e., increasing FWD pass number), and for different seasons in a year (i.e., different months). PCC moduli and k-values also exhibited significant variations with increasing FWD pass number. Sub-surface layer material properties for both flexible and rigid pavement sections also showed noticeable variations when measured in different seasons of the year.

1 INTRODUCTION

FWD is widely used to measure pavement deflections, which are analyzed and used to back-calculate moduli for different layers of flexible and rigid pavements. These calculated parameters can help in determining the structural capacity of existing pavements at project and network levels. However, these measurements are prone to be influenced by the temporal (seasonal temperature-moisture, and daily temperature) changes and, therefore, considering such effect can help eliminate errors in assessing existing pavement condition. Hence, there is a need to evaluate the impacts of these temporal variations on FWD measurements. Consequently, quantifying such effects on these measurements will assist in a better understanding of back-calculated layer moduli based on these FWD tests.

The LTPP SMP study is designed to evaluate the influence of temporal changes on pavement structural characteristics due to daily and seasonal variations in temperature and moisture. The LTPP SMP sites are instrumented for measuring subsurface moisture and temperature data to explain the observed temporal changes in FWD measurements. Therefore, such data can be analyzed to seek the full potential of FWD deflection data for investigating the corresponding pavement condition for both flexible and rigid pavements.

2 LTPP DATA AVAILABILITY

The original plan for the SMP experimental study required 48 flexible and 16 rigid pavement sections

as shown in Table 1. However, Table 2 shows currently available data in the LTPP database used for analysis in this paper.

3 FWD DEFLECTION MEASUREMENTS

The primary objective for the data mining was to obtain FWD deflection measurement availability during a day and different seasons for all the SMP pavement sections. Table 3 shows the available FWD measurements by different passes on the same day for all flexible pavement sections in different climates. FWD pass serves to distinguish multiple runs of the same lane number on the same day (Elkins et al. 2018). The

Table 1. The SMP experimental design cells and target number of sections per cell (Rada et al. 1994).

Pavement type	SG	No freeze		Freeze	
		Dry	Wet	Dry	Wet
Flexible-thick AC	Fine	3	3	3	3
(>5-inch surface)	Course	3	3	3	3
Flexible-thin AC	Fine	3	3	3	3
(<5-inch surface)	Course	3	3	3	3
JPCP	Fine	1	1	1	1
	Course	1	1	1	1
JRCP	Fine	1	1	1	1
	Course	1	1	1	1

Note: Numbers indicate the desired number of pavement sections

Table 2. Available pavement sections in the LTPP SMP study.

| Pavement type | SG | No freeze | | Freeze | | |
		Dry	Wet	Dry	Wet	Total
Flexible	Fine	4		3	4	10
HMA>5"	Course	4	6	26	6	38
	Sub Total	**8**	**6**	**29**	**10**	**53**
Flexible	Fine			2	3	38
HMA<5"	Course	1	1	2	5	13
	Sub Total	**1**	**1**	**4**	**8**	**13**
Total (Flexible)		**9**	**7**	**33**	**18**	**67**
JPCP		**2**	**2**	**6**	**6**	**16**

Note: JRCP sections have been excluded.

Table 3. Available FWD deflection basins by FWD pass in the different climatic region—SMP flexible pavements.

FWD Passes	Dry, Freeze	Dry, No Freeze	Wet, Freeze	Wet, No Freeze
1	340 (9)	207 (7)	779 (33)	599 (18)
2	267 (9)	160 (5)	445 (23)	458 (17)
3	230 (9)	139 (5)	349 (14)	318 (17)
4	104 (9)	59 (5)	150 (13)	114 (17)
5	14 (4)	17 (5)	4 (3)	23 (8)

Note: Values in parenthesis show the number of sections. An additional 2, 12, and 21 deflection basins are available in DF, WF, and WNF climates where more than 5 FWD passes are available.

Table 4. Available FWD deflection basins by FWD pass in the different climatic region—SMP rigid pavements (E and k-value).

FWD Passes	Dry, Freeze	Dry, No Freeze	Wet, Freeze	Wet, No Freeze
1	54 (2)	73 (2)	153 (6)	131 (6)
2	43 (2)	59 (2)	105 (5)	85 (4)
3	14 (2)	23 (2)	37 (5)	18 (3)
4	2 (1)	1 (1)	2 (2)	2 (2)

Note: Values in parenthesis show the number of sections

data in the table shows that several flexible pavement sections contain multiple FWD passes on the same day.

Similarly, Table 4 shows the available FWD deflection measurements (i.e., for back-calculation of layer moduli and LTE calculations) for multiple passes on rigid pavement sections in the SMP experiment. The overall availability for flexible and rigid pavements shows that most pavement sections have about three passes during a day.

4 EVALUATION OF FWD DEFLECTION MEASUREMENTS — SMP FLEXIBLE PAVEMENT SECTIONS

Currently, available FWD deflections measurements and back-calculated layer moduli in the LTPP database

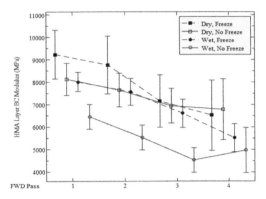

Figure 1. Variations in average HMA layer moduli with FWD pass—SMP flexible pavement with thick HMA sections.

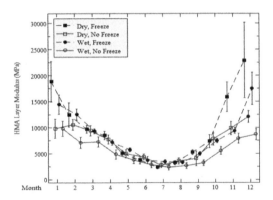

Figure 2. Variations in average HMA layer moduli with FWD testing month—SMP flexible pavement thick HMA sections.

for different materials and pavement structures were evaluated. Figure 1 shows the variation in the average HMA layer moduli, the 95% mean confidence interval, with FWD pass for the different climatic regions. A decrease in average HMA moduli was observed with increasing FWD pass number (i.e., morning to afternoon). However, somewhat mixed trends were observed within different climates. The reason could be different HMA mixtures and field aging of pavement sections.

Overall results show that, on average, the subgrade moduli are not impacted by the FWD pass for pavement sections with different HMA layer thicknesses and located in different climatic zones. It is expected that unbound layer moduli are not much impacted within a day; however, such moduli values may show a significant influence of different seasons (months). As expected, significant variations in HMA layer moduli are observed with higher values in winter months and lower values in summer months (see Figure 2).

Also, substantial differences are observed in the aggregate base modulus for flexible pavements with

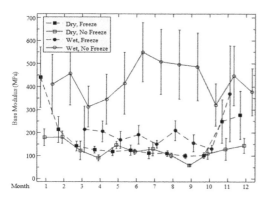

Figure 3. Variations in average aggregate base moduli with FWD testing month—SMP flexible pavements with thin HMA layer.

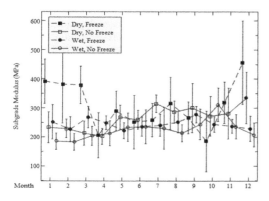

Figure 4. Variations in average subgrade moduli with FWD testing month—SMP flexible pavements with thick HMA layer.

thin and thick HMA layers in different months. The aggregate base layers for flexible pavement sections with thin HMA layers show higher moduli values for winter than summer months, especially in wet climates (see Figure 3). The subgrade layers for flexible pavement sections also exhibited substantial variations in moduli values with the highest values for the dry-no-freeze and lowest for wet-no-freeze climates (see Figure 4).

The results from the FWD measurements and estimated layer moduli on SMP flexible pavements are summarized below:

- Diurnal measurements: It seems that only HMA layer moduli vary with FWD pass in a day. No significant variability is observed for the subgrade moduli. The result implies that multiple FWD measurements during a single day may not be needed for flexible pavements if the HMA moduli can be corrected for different temperatures with a high level of confidence (unless the objective is to back-calculate the dynamic HMA modulus master curve).

- Seasonal measurements: Significant effects of seasonal variations (i.e., between months) on HMA and aggregate base moduli values were observed, especially for pavements with thinner HMA layers. This may imply the need for conducting FWD testing in each season to quantify such variations in material properties. The need for such testing requirements can be minimized if models are available to incorporate such effects of seasonal variability on the material properties. The AASHTOWare Pavement-ME incorporate such strategies for correcting material moduli based on predicted temperatures and moisture changes through the Enhanced Integrated Climate Model (EICM).

5 EVALUATION OF FWD DEFLECTION MEASUREMENTS — SMP RIGID PAVEMENT SECTIONS

Like flexible pavement sections in the SMP experiment, the parameters determined based on FWD deflection testing were summarized to isolate the effect of daily and seasonal measurements on rigid pavement sections. Results indicate an influence of FWD passes on calculated PCC moduli and k-values for all rigid pavement sections. Also, there is an essential influence of diurnal measurements on these calculated material properties. Significant variations can be observed in the material properties in different seasons for different climatic regions (see Figure 5). Such variations in back-calculated material properties are expected since the slab curling/warping is a function of pavement temperature and moisture gradients due to diurnal and seasonal FWD deflection measurements.

The results also confirm that FWD deflection-based LTE values are effected by daily variations of temperature. Figure 6 illustrates the impact of seasonal FWD deflection-based LTE values in different climatic zones for all rigid pavement sections in the SMP experiment. Higher LTE values are expected at higher temperatures due to slab expansion and joint locking while lower LTE value corresponds to slab contraction in lower temperatures. Therefore, it is vital to consider the daily and seasonal temperatures for such measurements for rigid pavements.

The above results from the FWD deflections and estimated parameters on SMP rigid pavements are summarized below:

- Diurnal measurements: It seems that significant variations in the deflection-based layer properties can be anticipated for rigid pavements. This effect is more critical in rigid than in flexible pavements. This observation implies that multiple FWD measurements during a single day may be needed for rigid pavements.

- Seasonal measurement: Significant effects of seasonal variations (i.e., between months) were observed on deflection-based material properties

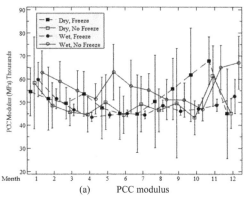

(a)   PCC modulus

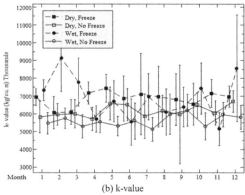

(b) k-value

Figure 5. Variations in average material properties with FWD testing month—SMP rigid pavements.

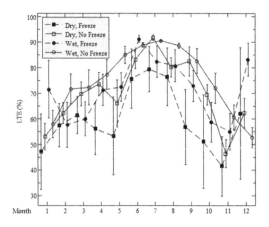

Figure 6. Variations in LTE with FWD testing month—SMP rigid pavements.

for rigid pavements located in all climatic zones. Such findings may imply conducting FWD testing in each season to quantify such variations in material properties. The need for such testing requirements

can be minimized if models are available to incorporate such impact of seasonal variability on the material properties. The AASHTOWare Pavement-ME includes such strategies for correcting material moduli based on predicted temperature gradients and moisture changes through the EICM.

6   SUMMARY

This paper presents an overall insight that how the measured FWD deflections and the corresponding back-calculated parameters for both flexible and rigid pavements are prone to temporal (i.e., seasonal temperature/moisture and daily temperature) variations. For flexible pavements, the asphalt layer modulus decreases as the temperature increases, resulting in more considerable deflections. Similarly, with the increase in FWD pass (i.e., from morning to afternoon), the HMA layer moduli decrease. Therefore, temperature corrections are generally applied to correct the asphalt layer modulus. Sub-surface layers back-calculated moduli also show significant variations with different HMA layer thicknesses and different climatic zones. PCC pavements are also affected by temperature because of slab curling due to the temperature differential between the top and bottom of the slab. Due to these reasons, back-calculated PCC layer moduli show variations. Temperature variations can also affect joint and crack behavior in rigid pavements. As a result, deflection testing conducted at joints and subsequently calculated LTE based on these measured deflections are not the true representative of the current joint condition. Based on these results, it is imperative to investigate such variations to refine current guidelines on FWD field measurements that properly account for daily and seasonal variations in pavement condition, response, and performance.

ACKNOWLEDGMENT

The authors would like to acknowledge the NCHRP for funding this research.

REFERENCES

Elkins, E. E., T. Thompson, B. Ostrom, and B. Visintine (2018). "Long-Term Pavement Performance Information Management System User Guide," Office of Infrastructure Research and Development, Federal Highway Administration, FHWA-RD-03-088 (revision).

Rada, G., G. Elkins, B. Henderson, R. Van Sambeek, & A. Lopez (1994). "Seasonal Monitoring Program: Instrumentation Installation and Data Collection Guidelines," Report No. FHWA-RD-94-110. FHWA, US Department of Transportation.

Advances in Materials and Pavement Performance Prediction II – Kumar et al. (eds)
© 2021 Taylor & Francis Group, London, ISBN 978-0-367-46169-0

Effect of thresholding algorithms on pervious pavement skid resistance

A. Jagadeesh & G.P. Ong
National University of Singapore, Singapore

Y.M. Su
National Kaohsiung University of Science and Technology, Taiwan

ABSTRACT: The primary usage of pervious concrete mixtures in pavement construction is to improve the functional performance of pavements through increasing skid resistance and hydroplaning speeds (by dispersing the hydrodynamic pressure developed at the tire-pavement interface) and increasing the water infiltration and tire-pavement contact area at higher speeds, thereby reducing the wet weather accidents. Past skid resistance simulation models for pervious pavements considered the use of artificial pore grid models, but with today's advanced X-Ray computed tomography (XRCT) technologies, it is now possible to develop realistic pavement models for skid resistance simulation. This paper aims to investigate the effect of thresholding - a crucial step in image segmentation - on skid resistance of pervious concrete pavements, whose pore structures are derived from XRCT. It was found from the analysis presented in this paper that the various thresholding algorithms are found to be either under- or over-estimating the uplift and drag forces, as compared to the discharge-based thresholding algorithm. Errors in skid number due to thresholding are however found to be marginal (with up to 2 SN or 3% error).

1 INTRODUCTION

With recent advancement in the computational resources and modelling techniques, it is now possible to model complex skid resistance simulation models, thereby avoiding the unnecessary time (and potential risk) involved in field experiments. The development of skid resistance simulation model for pervious pavements started off in the literature with the artificial pore grid model (Zhang et al. 2013) and then extended to advanced XRCT based models (Tang et al. 2019).

Zhang et al (2013) developed the skid resistance simulation model for pervious pavements using artificial pore network. The advantages of Zhang's model include shorter computation time and resources because of the use of single shell elements instead of solid elements for modeling tire components. The variables involved in the model includes porosity, pore size and center-to-center distance or spacing between pores. The model was further applied to investigate the effect of pore size and spacing on skid resistance given constant porosity (Zhang et al. 2014). However, it was noted that Zhang et al. (2014)'s study did not consider the possibility of varying pore size and porosity given a constant pore spacing.

Ding & Wang (2018) studied the effect of pervious pavements on hydroplaning using the smooth pavement skid resistance model with water film thickness (WFT) computed from empirical equations. While the

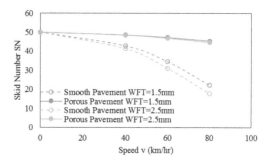

Figure 1. Skid number-speed relationship for smooth and pervious pavements using Zhang et al. (2013)'s model.

model is simple to use, the validation against pervious pavements were not considered in their works. Zhang et al (2013) in fact found that modelling of pores in skid simulation model is important as the skid number (SN) with speed relationship (as shown in Figure 1) are different between smooth and pervious pavement models for various lower water-film thicknesses (WFT).

Recently Tang et al (2019) developed a skid resistance simulation model for pervious pavements using XRCT with consideration of variable water film thicknesses. It is noted that their studies did not consider the effect of thresholding during image segmentation and potential errors in model estimation may occurs as proven by Jagadeesh et al. (2018, 2019a, 2019b) whom recently studied the effect of thresholding on pavement functional parameters, such as permeability, porosity and pore size.

2 METHODOLOGY

2.1 *Materials*

Three different pervious concrete samples of 150 mm diameter and 250 mm height consisting of two single-graded mixtures P1 and P2 with nominal maximum aggregate sizes (NMAS) of 9.5 mm and 12.5 mm, and one dense mix P3 with NMAS of 12.5 mm was produced in the laboratory. Medical XRCT was carried out to obtain the internal pore structure and various global thresholding algorithms were used to divide images of varying grey scale intensities into solid and pore phases. Details on the materials, XRCT and image processing algorithms can be found in Jagadeesh et al. (2019). Three different pervious concrete slabs were casted using the same mixtures and their British Pendulum Number (BPN) were obtained. The skid number at low speed (SN_0) was computed using Equation (1) (Fwa & Chu 2019; Leu & Henry 1978):

$$SN_0 = -35 + 1.32\,BPN \tag{1}$$

2.2 *Image processing*

The freeware ImageJ was used in this paper to obtain the threshold values for various global thresholding algorithms. The Simpleware ScanIP N-2018.03 software was used in the conversion of scanned images to the finite volume meshes. Thresholding of solids and air voids based on the grey scale intensities was carried out using commonly used segmentation algorithms and the discharge-based thresholding algorithm developed by Jagadeesh et al. (2019). Figure 2 compares the raw XRCT slice against the minimum and Yen threshold slices.

The white region in the figure represents the cement and aggregate phase, whereas the black region represents the void phase. The Enhanced Volumetric Marching Cubes (EVoMaC) algorithm (Young et al. 2008) was used to convert the masks into finite volume meshes

2.3 *Skid resistance simulation model development and validation*

A skid resistance simulation model was developed using ANSYS Static Structure, CFX and Fluid Structure Interaction tools. It consists of two sub-models: the tire-pavement sub-model and the fluid sub-model. The tire-pavement sub-model consists of a rigid smooth pavement surface and a single layered tire model (ASTM E524 smooth tire), and the fluid sub-model consists of the pervious pavement pores obtained from XRCT and fluid zone under the tread. Figure 3 shows the tire-pavement and

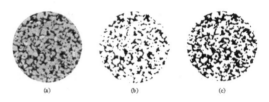

Figure 2. Pervious concrete slice (a) Raw image, (b) Minimum threshold segmented image and (c) Yen threshold segmented image.

(a) Tire-pavement sub-model

(b) Fluid sub-model

Figure 3. Skid resistance simulation model.

fluid sub-models. The fluid sub-model consists of the Navier-Strokes equations with consideration to k-ε turbulence model, and multiphase modelling using the Volume of Fluid method. The footprint area calibration was carried out to obtain the unknown shell thickness properties. More details on the material properties and boundary conditions can be found in Zhang et al. (2014).

Validation for the developed model was carried out using past experimental studies on concrete pavements (Anderson et al. 1998; Rose & Gallaway 1977) and the back-calculation procedure developed by Fwa & Ong (2008), as shown in Figure 4. The skid number SN is obtained as the ratio of horizontal resistive force F_x to the normal wheel load F_z using Equations (2) and (3). It is noted that the uplift and drag forces (F_{uplift} and F_{drag}) are obtained from the simulations and friction coefficient μ as the ratio of SN_0 to 100.

$$SN = (F_x/F_z) \times 100 \tag{2}$$

$$F_x = \mu(F_z - F_{uplift}) + F_{drag} \tag{3}$$

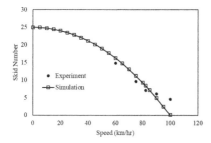

(a) Comparison against Anderson et al. (1998) (WFT=12.5mm)

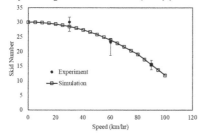

(b) Comparison against Anderson et al. (1998) (WFT=7.62mm)

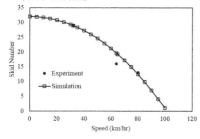

(c) Comparison against Rose and Gallaway (1977) (WFT=7.62mm)

Figure 4. Model validation.

Table 1. Effect of thresholding on uplift force, drag force and skid number.

Sample	Algorithm	F uplift (N)	F drag (N)	SN
P1	Minimum	297.91	−73.34	67.17
	Intermodes	345.44	−77.28	66.36
	Otsu bilevel	369.34	−77.42	65.99
	Otsu trilevel	357.64	−79.21	66.13
	Isodata	373.60	−78.90	65.90
	Kapur	396.08	−80.18	65.53
	Sahoo	408.39	−80.35	65.33
	Yen	419.13	−81.64	65.14
	Li	328.13	−77.44	66.62
	Volumetric	391.23	−81.12	65.58
	Discharge	399.58	−85.07	65.37
P2	Volumetric	372.80	−77.48	36.72
	Discharge	374.65	−79.83	36.65
P3	Volumetric	315.37	−69.32	46.02
	Discharge	314.47	−73.46	45.94

Table 2. Comparison of percentage error for different thresholding algorithms with respect of discharge-based algorithm.

Sample	Algorithm	Percent error		
		F uplift	F drag	SN
P1	Minimum	−25.44	−13.79	2.75
	Intermodes	−13.55	−9.15	1.51
	Otsu bilevel	−7.57	−8.99	0.95
	Otsu trilevel	−10.50	−6.89	1.17
	Isodata	−6.50	−7.25	0.80
	Kapur	−0.88	−5.75	0.24
	Sahoo	2.21	−5.54	−0.06
	Yen	4.89	−4.03	−0.35
	Li	−17.88	−8.97	1.91
	Volumetric	−2.09	−4.63	0.32
	Discharge	0.00	0.00	0.00
P2	Volumetric	−0.49	−2.94	0.18
	Discharge	0.00	0.00	0.00
P3	Volumetric	0.29	−5.63	0.17
	Discharge	0.00	0.00	0.00

3 RESULTS AND DISCUSSION

Table 1 shows the effect of different thresholding algorithms on the uplift force, drag force and skid number while Table 2 computes the percentage error of the thresholds obtained from the various algorithms with respect to discharge-based thresholding algorithm. It can be observed from the tables that as the nominal aggregate maximum size increases, skid number decreases. This is line with the permeability and porosity results of the same samples reported by Jagadeesh et al. (2019a).

Figure 5 further shows the effect of various thresholding algorithms on skid number at a speed of 64 km/h and a water film thickness of 7.62 mm for a 50 mm porous layer thickness. It was found from past studies that an increase in the threshold values results in a significant increase in porosity, permeability (Jagadeesh et al. 2018, 2019a), pore size characteristics (Jagadeesh et al. 2019b) and a poor segmentation of pervious pores may potentially skid numbers from skid resistance models. This is illustrated in Figure 5(a) which shows that an increase in threshold value results in an overall increase in fluid uplift force and fluid drag force. Note that the negative fluid drag force means the fluid force being against the wheel sliding direction.

However, it was found from Figure 5(b) that the effect of thresholding on skid number may in fact be insignificant for a given water film thicknesses with errors of up to 2 SN (or 3%). This is in line with the findings reported by Zhang et al. (2014, 2015).

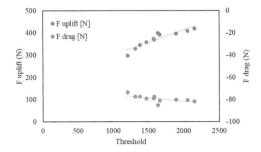

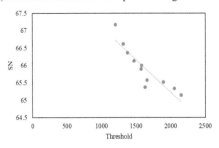

(a) Effect of threshold on fluid uplift and drag forces

(b) Effect of threshold on skid number SN

Figure 5.   Effect of threshold on uplift force, drag force and skid number.

4   CONCLUSION

This study analyses the effect of various threshold segmentation algorithms on the skid resistance characteristics of the pervious pavements using medical XRCT scanning, digital image processing and numerical simulations. First, the development of the skid resistance simulation model was carried out using the fluid-structure interaction tools and validated with the past experimental studies. The experimental British pendulum number was used in the current study to determine the wet-pavement friction coefficients. It was found in this paper that an increase in the threshold value results in the overall increase of fluid uplift and drag forces, resulting in the decrease of skid number. However, the effects on skid number due to thresholding was found to be marginal.

The model developed in this paper can be further extended to include advanced water film thickness models and thermo-mechanical skid resistance simulation models to investigate its effect on accumulated water film thickness, dry friction coefficient, braking distance, and rolling resistance. It can be also extended to the study of skid resistance characteristics of dense-graded pavements. Overall, it is expected that the present research will help in understanding the effect of threshold characteristics in the field of XRCT-based tire-pavement interaction studies.

REFERENCES

Anderson, D.A., Huebner, R.S., Reed, J.R., Warner, J.C. & Henry, J.J., 1998. *Improved surface drainage of pavements* (No. Project 1–29,).

Ding, Y. & Wang, H., 2018. Evaluation of Hydroplaning Risk on Permeable Friction Course using Tire–Water–Pavement Interaction Model. *Transportation Research Record*, 2672(40), pp. 408–417.

Fwa, T.F. & Chu, L., 2019. The concept of pavement skid resistance state. *Road Materials and Pavement Design*, pp. 1–20.

Fwa, T.F. & Ong, G.P., 2008. Wet-pavement hydroplaning risk and skid resistance: analysis. *Journal of Transportation Engineering*, 134(5), pp. 182–190.

Jagadeesh, A., Ong, G.P. & Su, Y.M., 2018, July. Porosity and permeability evaluation of pervious concrete using three-dimensional x-ray computed tomography. In *4th International Conference on Transport Infrastructure (ICTI 2018)*, South Africa.

Jagadeesh, A., Ong, G.P. & Su, Y.M., 2019a. Development of Discharge-Based Thresholding Algorithm for Pervious Concrete Pavement Mixtures. *Journal of Materials in Civil Engineering*, 31(9), p. 04019179.

Jagadeesh, A., Ong, G. & Su, Y.M., 2019b, October. Digital sieving of pervious concrete air voids using X-ray computed tomography. In *11th Asia Pacific Transportation and the Environment Conference (APTE 2018)*. Atlantis Press.

Leu, M.C. & Henry, J.J., 1978. Prediction of skid resistance as a function of speed from pavement texture measurements. *Transportation Research Record*, 666, pp. 7–13.

Rose, J.G. & Gallaway, B.M., 1977. Water depth influence on pavement friction. *Journal of Transportation Engineering*, 103(4).

Tang, T., Anupam, K., Kasbergen, C., Scarpas, A. & Erkens, S., 2019. A finite element study of rain intensity on skid resistance for permeable asphalt concrete mixes. *Construction and Building Materials*, 220, pp. 464–475.

Young, P.G., Beresford-West, T.B.H., Coward, S.R.L., Notarberardino, B., Walker, B. & Abdul-Aziz, A., 2008. An efficient approach to converting three-dimensional image data into highly accurate computational models. *Philosophical Transactions of the Royal Society A: Mathematical, Physical and Engineering Sciences*, 366(1878), pp. 3155–3173.

Zhang, L., Ong, G.P. & Fwa, T.F., 2013. Developing an Analysis Framework to Quantify and Compare Skid Resistance Performance on Porous and Nonporous Pavements. *Transportation Research Record*, 2369(1), pp. 77–86.

Zhang, L., Ong, G.P. & Fwa, T.F., 2014. Influences of pore size on the permeability and skid resistance of porous pavement. In *Sustainability, Eco-Efficiency and Conservation in Transportation Infrastructure Asset Management-Proceedings of the 3rd International Conference on Transportation Infrastructure, ICTI* (Vol. 2014, pp. 621–632).

Zhang, L., Ong, G.P. & Fwa, T.F., 2015. Numerical Study on the Influence of Aggregate Size on Skid Resistance Performance of Porous Pavements. *Asian Transport Studies*, 3(3), pp. 284–297.

Advances in Materials and Pavement Performance Prediction II – Kumar et al. (eds)
© 2021 Taylor & Francis Group, London, ISBN 978-0-367-46169-0

Investigation of cracking performance of asphalt mixtures in Missouri

Behnam Jahangiri, Hamed Majidifard, Punyaslok Rath & William Buttlar
Department of Civil and Environmental Engineering, University of Missouri-Columbia, USA

ABSTRACT: This study investigates the performance of four asphalt mixture types that were recently used on paving projects across the state of Missouri, USA. Three-of-four mixtures contained reclaimed asphalt pavement (RAP) with up to 35% asphalt binder replacement (ABR), while the fourth mixture possessed 33% ABR, entirely via recycled asphalt shingles (RAS). Performance testing including the disk-shaped compact tension (DC(T)) test, along with indirect tension (IDT) creep and strength testing were performed on field cores and plant-produced, lab-compacted (PPLC) samples to evaluate the cracking potential of each of the studied mixtures. Testing results suggested that each of the mixtures have high cracking potential, i.e., exhibit brittle behavior. ILLI-TC software was employed using creep and DC(T) inputs to simulate low temperature cracking potential in the Midwest climate.

1 INTRODUCTION

Decades ago, traditional asphalt mixtures involved simple combinations of virgin asphalt binder and aggregate, designed to meet the load-bearing requirements of the roads and airfields. Accordingly, simple tests such as Marshall Stability and Flow could adequately address the screening and quality control needs for the design and control of most asphalt concrete mixtures. Today, modern, heterogenous, sustainable asphalt mixtures contain a myriad of components, such as reclaimed asphalt pavement (RAP), recycled asphalt shingles (RAS), warm mix and antistripping agents, rejuvenators, ground tire rubber, fibers, and even waste plastic.

Arguably, the DC(T) test is one of the limited standardized methods available to evaluate the fracture energy of asphalt mixtures at low temperatures (ASTM 2013; Dave & Hoplin 2015), and appears to be suitable for the control of thermal and block cracking (Buttlar et al. 2018; Jahangiri et al. 2019). The DC(T) test has been widely used by researchers and agencies alike to determine the low temperature characteristics of asphalt mixtures that contain recycled materials such as RAP and RAS (Behnia et al. 2011). Cascione et al. (2011) used post-consumer RAS in asphalt mixtures used on high-volume road shoulder surfaces, along with RAP, and demonstrated that softer base binders were needed to ensure adequate fracture resistance. Braham et al. examined the effects of higher mixture aging leading to lower DC(T) fracture energy values at long-term aging levels (Braham et al. 2009). Buttlar et al. (2016) reported an increase in DC(T) fracture energy with the use of a softer binder or a polymer-modified binder.

The Indirect Tensile Strength test (IDT) was developed as a part of the SHRP A-005 project by Roque et al. at the Pennsylvania State University to characterize asphalt mixture performance at low temperatures (Roque & Buttlar 1992). The IDT test is performed to ascertain the tensile strength and the creep properties of the asphalt mixture specimen, which are critical factors for thermal cracking characteristics of asphalt mixture (Hill et al. 2013). Behnia et al. (2011) fitted a power-law function onto IDT creep compliance master curves to characterize low temperature behavior of various.

As the literature suggests, the mixture performance is evaluated using various tests in order to mitigate different distress types. In this study, four different sections recently paved across Missouri were selected to be characterized in terms of their high and low temperature mixture performance properties. The asphalt material used to pave these sections contained different amounts of recycled materials and additives. Also, two different sample production procedures, namely, field core and PPLC were considered for each section. These measures helped achieve a better understanding of the effects of mixture contents and sample fabrication process on the subsequent performance test results.

2 METHODOLOGY AND TESTING PROCEDURE

2.1 DC(T) test

The DC(T) test was developed to characterize the fracture behavior of asphalt concrete mixtures at low temperatures. The testing temperature is 10°C warmer than the PG low temperature grade of the mixture, per ASTM D7313-13. The DC(T) test procedure includes conditioning of the fabricated specimen in a temperature-controlled chamber for a minimum of two hours. After conditioning, specimens were suspended on loading pins in the DC(T) apparatus and seated with a small, 0.1 kN pre-load. The test is performed at a constant Crack Mouth Opening Displacement (CMOD) rate, which is controlled by a CMOD clip-on gage mounted at the crack mouth. The CMOD rate specified in ASTM D7313-13 is 0.017 mm/s (1 mm/min).

2.2 *IDT creep and strength tests*

IDT creep and strength tests were performed following AASHTO T-322. To carry out the IDT creep test, three samples were conditioned at three different temperatures including 0, −12 and −24°. Each sample was kept at the testing temperature for 2 hours. The creep loading level was reached in one second, using a closed-loop, servohydraulic universal test machine.

After plotting the creep compliance curves at different temperatures versus time in a log-log space, the curves are shifted horizontally relative to a pre-selected referenced temperature to construct a unique continuous curve, called the master curve. A power-law function is then fitted to the master curve (Buttlar et al. 1998), as follows:

$$D(t) = D_0 + D_1 * t^m \qquad (1)$$

Equation 1 is used to obtain the m-parameter, linked to mixture viscoelastic relaxation characteristics. A generalized Voight-Kelvin model is also fit to the master curve, for use in ILLI-TC simulation software (Buttlar et al. 1998). The IDT strength test is finally performed by applying an increasing ram displacement at a constant rate until tensile failure occurs in the specimen (AASHTO T-322).

3 MATERIALS AND SAMPLE PREPARATION

Modern asphalt mixtures contain a wide range of recycled materials and additives, i.e., RAP, RAS, GTR, etc. Table 1 provides compositional details for of each of the four tested sections in Missouri. All of these sections were constructed in 2016 and were cored soon thereafter (within two weeks after construction), which represents the short-term aging condition. Asphalt binder replacement (ABR) by RAP and RAS, total percentage of asphalt content by mixture mass (P_b), PG of the virgin binder, and nominal maximum aggregate size (NMAS) are summarized Table 1. For convenience, the percentage of ABR, and RAP / RAS contents have been included in the section labels. As an example, MO13_1 (17-17-0) has an ABR of 17%, resulting from 17% replacement by RAP and 0% by RAS.

Table 1. Properties of four sections.

Section	ABR by: RAP	ABR by: RAS	P_b % Binder	Virgin (mm)	NMAS
MO13_1 (17-17-0)	17	0	5.7	PG64-22 H	9.5
US63_1 (35-35-0)	35	0	5.1	PG58-28	12.5
US54_6 (31-31-0)	31	0	5.1	PG58-28	12.5
US54_1 (33-0-33)	0	33	5.2	PG58-28	12.5

4 RESULTS AND DISCUSSION

4.1 *DC(T) Testing Results*

Figure 1 presents the measured DC(T) fracture energy from both field cores and plant mixtures. Considering the desired DC(T) fracture energy threshold of 460 J/m^2 for medium traffic roads, only the US54_1 section was found to pass, albeit only marginally. The US63_1 section was the most susceptible section to the onset of thermal cracking; it likely will show signs of thermal cracking distress within the first few winters after construction. The US54_6 section displayed much higher fracture energy as compared to the US63 section, which had a similar RAP level. Neither section involved grade bumping on the low temperature side to account for the stiffness imparted by the addition of RAP to the mixture. The other observations from DC(T) fracture tests are as follows.

- The MO13_1 (17-17-0) section used the stiffest virgin binder (PG64-22 H) and has the lowest amount of ABR. Since the yielded fracture energy is close to the limit of 460 J/m^2, one grade bump on the low temperature side of the PG is suggested in future designs to more easily reach the desirable fracture energy criterion.
- Comparing US54_6 (31-31-0) and US54_1 (33-0-33) with the same neat binder grade, identical aggregate structure and amount of ABR, it appears that the combination of RAS and WMA agent improved the DC(T) fracture energy.

Comparing the field cores with plant mix DC(T) fracture energies, it can be observed that the field cores consistently exhibited higher fracture energy values as compared to their plant mixed counterparts. Although it is desirable that both the values agree with each other, field construction introduces a number of variables that differ from plant-produced mixtures, which affect the final fracture energy of the mixture in the constructed asphalt pavement.

4.2 *IDT Creep and Strength Testing Results*

IDT testing was performed in order to characterize the linear viscoelastic behavior of the mixtures

172

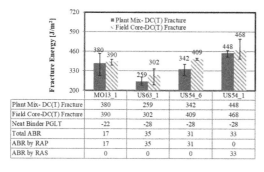

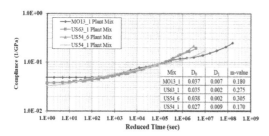

	MO13_1	US63_1	US54_6	US54_1
Plant Mix- DC(T) Fracture	380	259	342	448
Field Core-DC(T) Fracture	390	302	409	468
Neat Binder PGLT	-22	-28	-28	-28
Total ABR	17	35	31	33
ABR by RAP	17	35	31	0
ABR by RAS	0	0	0	33

Figure 1. Comparing DC(T) fracture energy of field cores and plant mixes.

Figure 2. IDT creep compliance master curve of pant mixtures (Reference Temperature: −24).

investigated. Creep compliance of each mixture was calculated at three temperatures as described in section 2. Shift factors were used to horizontally shift the creep compliance curves in time, leading to the final master curves. The lowest testing temperature, −24°, was selected as the reference temperature. A power law function was then fitted on the master curve. Figure 2 shows the creep compliance master curve for each of the sections using a traditional log-log plot. As shown, MO13_1 and US54_1 exhibited the highest and lowest instantaneous creep compliance (D_0), respectively. The stress relaxation capability of each mix can be examined by evaluating the m-value, which is a function of the slope of the creep compliance curve at long loading times. Based on the measured m-values, the MO13_1 and US54_1 sections showed very low stress relaxation ability after aging in the field.

After performing the IDT creep test at three temperatures, the samples were let to relax the recoverable strain overnight and then were broken in IDT strength test. Table 2 contains the IDT strength of the mixtures. As seen, all four mixtures yielded very similar IDT strength. The lowest average strength was recorded by US54_1 while the highest strength was captured by US63_1. Comparing to the DC(T) test, it appears that the lower IDT strength indicates a softer behavior which resulted in higher fracture energy.

4.3 ILLI-TC modeling results

Researchers at the University of Illinois, Urbana-Champaign developed a transverse cracking analysis

tool called Illi-TC as a part of a national pooled funded study on low-temperature cracking. The user first inputs the mixture tensile strength in MPa, calculated from the IDT test (or it can be estimated from the DC(T) test peak load result), the DC(T) fracture energy in J/m^2, creep compliance data from the IDT test, and mixture coefficient of thermal expansion and contraction. The simulation tool creates a finite element mesh depending on the specified geometry (pavement depth) and uses a cohesive zone fracture constitutive model at element boundaries to simulate thermal cracking potential. Field validation of this tool has been presented by Dave et al. (2013). More recently, Dave et al. showed that IlliTC is sensitive to the variation in fracture energy of asphalt mixtures and shows a difference in transverse cracking of asphalt pavements when the fracture energy is varied by values as low as $25 J/m^2$ (Dave & Hoplin 2015)

As Missouri was not included in the climatic data, the intermediate climate in Illinois was considered as the closest region to the Missouri climate and was used to compute the critical cracking events and amount of thermal cracking. The number of critical events indicates the number of nights that the pavement stress reached above 85% of its tensile strength. The inputs for the mixtures used for Illi-TC modeling and the modeling results are shown in Table 2. The Illi-TC results show the combined effect of various mixture factors on the thermal cracking behavior of the investigated sections. The US63_1 section had the lowest number of critical events. This follows, since the stress relaxation (m-value) of this mix was relatively high and the IDT strength value was the highest among the mixtures. This resulted in the lowest number of critical thermal cracking events. However, due to the low level of DC(T) fracture energy, the modeling results predicted a severe amount of cracking (>200m per 500 m of pavement length) in the 5 years of cracking simulation. It is also worth mentioning that the US54_1 section which had the highest fracture energy, was predicted to have 9 critical cooling events in 5 years. This implies that, in addition to fracture energy, the viscoelastic and strength of the asphalt is critically important, as far as low temperature cracking is concerned.

5 FINDINGS AND CONCLUSION

In this study, four asphalt mixtures used to pave different routes across the state of Missouri were investigated with performance testing and thermal cracking simulation. Cracking test results along with IDT creep compliance master curves were implemented to predict the critical cooling events and cracking amounts using the Illi-TC software program. The conclusions drawn from the results of the study are as follows.

- The investigated Missouri mixtures are prone to cracking based on performance property values

Table 2. Input parameters in Illi-TC and resultant critical events and the amount of cracking.

Section	MO13_1	US63_1	US54_6	US54_1
Fracture Energy (J/m^2)	385	263	342	448
Mix CTEC (mm/mm/^{o}C)	3.00E-05	3.00E-05	3.00E-05	3.00E-05
IDT Strength (MPa)	3.9	4.5	4.5	3.9
VMA	16.3	14.1	14.7	14.4
Layer Thickness (cm)	7.5	7.5	7.5	7.5
Critical Events (output)	8	1	13	9
Crackin per 500 m Length	>200 m	>200 m	>200 m	>200 m

obtained. Similarly the Illi-TC software predicted a severe level of thermal cracking for all of the investigated sections. This is consistent with field results for sections containing recycled materials and very little binder bumping to softer virgin binder grades.

- Measures such as using a softer binder system, more robust aggregate structures, and modifiers such as ground tire rubber could be implemented to increase fracture resistance of mixtures containing recycled materials in Missouri.
- The effect of reheating plant produced asphalt mixtures should be addressed in future research, and in forthcoming performance test specifications.
- In addition to DC(T) fracture energy limits, viscoelastic parameters such as creep compliance and m-value, along with thermal cracking simulations, can be used to achieve better low temperature field performance when designing modern, heterogeneous recycled asphalt mixtures.

REFERENCES

ASTM. (2013). ASTM D7313-13: Standard Test Method for Determining Fracture Energy of Asphalt-Aggregate Mixtures Using the Disk-Shaped Compact Tension Geometry. ASTM International Standards, 2–8. https://doi.org/10.1520/D7313

Behnia, B., Dave, E., Ahmed, S., Buttlar, W., & Reis, H. (2011). Effects of Recycled Asphalt Pavement Amounts on Low-Temperature Cracking Performance of Asphalt Mixtures Using Acoustic Emissions. Transportation Research Record: Journal of the Transportation Research Board, 2208, 64–71.

Braham, A. F., Buttlar, W. G., Clyne, T. R., Marasteanu, M. O., & Turos, M. I. (2009). The effect of long-term laboratory aging on asphalt concrete fracture energy. Journal of the Association of Asphalt Paving Technologists, 78.

Buttlar, W., Rath, P., Majidifard, H., Dave, E.V. and Wang, H., (2018). Relating DC (T) Fracture Energy to Field Cracking Observations and Recommended Specification Thresholds for Performance-Engineered Mix Design. Asphalt Mixtures, p. 51.

Buttlar, W. G., Hill, B. C., Wang, H., & Mogawer, W. (2016). Performance space diagram for the evaluation of high- and low-temperature asphalt mixture performance. Road Materials and Pavement Design, 1–23.

Buttlar, W., Roque, R., & Reid, B. (1998). Automated procedure for generation of creep compliance master curve for asphalt mixtures. Transportation Research Record (98), 28–36. https://doi.org/10.3141/1630-04

Cascione, A. A., Williams, R. C., Buttlar, W. G., Ahmed, S., & Hill, B. (2011). Laboratory evaluation of field produced hot mix asphalt containing post-consumer recycled asphalt shingles and fractionated recycled asphalt pavement. AAPT (80), 377–418.

Dave, E. V., Buttlar, W. G., Leon, S. E., Behnia, B., & Paulino, G. H. (2013). IlliTC – low-temperature cracking model for asphalt pavements. Road Materials and Pavement Design, 14 (December), 57–78.

Dave, E. V., & Hoplin, C. (2015). Flexible pavement thermal cracking performance sensitivity to fracture energy variation of asphalt mixtures. Road Materials and Pavement Design, 16(sup1), 423–441.

Hill, B., Behnia, B., Buttlar, W. G., & Reis, H. (2013). Evaluation of Warm Mix Asphalt Mixtures Containing Reclaimed Asphalt Pavement through Mechanical Performance Tests and an Acoustic Emission Approach. Journal of Materials in Civil Engineering, 25 (December), 1887–1897.

Jahangiri, B., Majidifard, H., Meister, J., & Buttlar, W. G. (2019). Performance Evaluation of Asphalt Mixtures with Reclaimed Asphalt Pavement and Recycled Asphalt Shingles in Missouri. Transportation Research Record, 2673(2), 392–403.

Roque, R., & Buttlar, W. G. (1992). The development of a measurement and analysis system to accurately determine asphalt concrete properties using the indirect tensile mode (with discussion). Journal of the Association of Asphalt Paving Technologists, 61.

Advances in Materials and Pavement Performance Prediction II – Kumar et al. (eds)
© 2021 Taylor & Francis Group, London, ISBN 978-0-367-46169-0

New methods of structural assessment on project level

D. Jansen & U. Zander

Bundesanstalt fuer Strassenwesen (BASt), Bergisch Gladbach, Germany

ABSTRACT: The so-called RSOs are the first set of standardized guidelines in Germany to describe a systematic approach to structural assessment of pavements at project level. The target group includes both public road construction authorities and private operators. For asphalt pavements, the procedure is fully described; for concrete pavements, the rules and regulations are currently being drawn up. The method is based on the principles of mechanistically-empirical pavement design and requires testing on drill core samples within predefined structurally homogeneous sections. For the formation of the sections, non-destructive testing methods are used in particular, which will also play a greater role in the assessment in the future. The transfer of the principles described for application at the project level to the network level is a goal that is pursued in ongoing research projects. The paper describes the method.

1 INTRODUCTION

The road network in Germany is one of the densest and most heavily trafficked in the world. The availability of the infrastructure is therefore of paramount importance. Appropriate and economic planning and implementation of maintenance measures is indispensable. However, this also requires methods that prepare the corresponding decisions. In addition to recording and evaluating the surface condition, this also includes substance evaluation, i.e. the prognosis of the residual life. With the new set of rules 'Guidelines for the Assessment of the Structural Substance of the Superstructure' (RSO), a standardized procedure will be available for the first time.

2 THE RSO GUIDELINES

The RSOs are a set of guidelines separated according to the asphalt and concrete construction methods. While the draft of the RSO for asphalt pavements has already been completed and is in an extensive test phase before its official introduction, the RSO for concrete pavements is still in the making. Both sets of guidelines are drawn up by the national committees, supported by BASt research projects and subsequently introduced by the Ministry of Transport. At least the RSO for asphalt pavements is based on the mechanistic-empirical design method, which in turn defines fatigue cracking on the bottom of the asphalt package as a failure criterion.

The RSOs are designed for application at the project or object level and can theoretically also be used to evaluate a construction service in relation to the achievable service life, for example within the framework of public-private partnership (PPP) projects. Use at network level is only theoretically possible, because of the costs due to the necessary sampling and testing effort. R&D projects are currently examining what a modified procedure for application at network level might look like. Details can be found in the last chapter of this article.

The assessment is usually only to be carried out for the most trafficked lane, which is usually the right lane. If there are indications that make an evaluation of one or more lanes appear sensible, proceed accordingly. In principle, the RSO procedure can be used for all road classes and traffic areas.

3 RSO FOR ASPHALT PAVEMENTS

The procedure defined according to RSO for asphalt pavements can be roughly divided into four process steps. First, structurally homogeneous sections are analyzed, followed by sampling using drill core samples in each of the structurally homogeneous sections. The cores are then tested in the laboratory for fatigue behavior, stiffness and bonding. Finally, a model is created from which the residual life is derived.

3.1 *Formation of structurally homogeneous sections*

It can be assumed that road sections of the same construction method, the same age, the same traffic load, with the same materials and other characteristics were similarly damaged or lost in total useful life over the period of use already endured. Thus, a section of

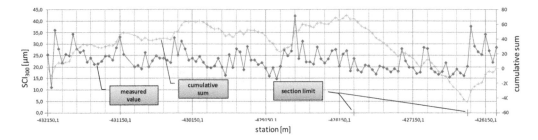

Figure 1. Example of subdividing bearing capacity data into homogenous sections.

the overall project identified as a structurally homogeneous section can then be subjected to the same evaluation over its entire length.

The challenge lies in identifying the structurally homogeneous sections. To anticipate, it should be noted that this process step is both of high importance and requires a high degree of engineering expertise for the evaluation, which means that pure computational and programmatic considerations currently do not lead to a reliable result. The focus of the section formation is particularly on those characteristic values that are related to fatigue crack formation or the size of horizontal strains on the bottom of the asphalt package, since, as mentioned above, the subsequent evaluation procedure is based on this criteria. Thus, for example, surface characteristics are not to be used or only to a limited extent useful for section formation. These should at best be used to determine the cause of failure.

Many of the necessary data can be taken from construction files and pavement databases. However, it is expressly recommended to carry out measurements in-situ. These measurements include bearing capacity measurements and ground penetrating radar (GPR) measurements. These also could be the basis for tracking any changes while repeating the assessment years later. Here it is to be paid attention that suitable measuring parameters are selected. Furthermore, when carrying out measurements, care must be taken to ensure exact and unambiguous location in the lateral as well the transversal direction.

For stationary bearing capacity measurements, like the Falling Weight Deflectometer (FWD), a maximum measuring point distance of 25 m makes sense. Continuously measuring methods such as the Traffic Speed Deflectometer can also be used. The measurements are to be carried out with a load of 50 kN in the right wheel track. For the evaluation, with regard to the section formation, characteristic values are to be selected which are suitable for the failure criterion explained above. These can be, for example, the SCI_{300} value related to the strain at the bottom of the asphalt package and parameters describing the subgrade bearing capacity.

Measurements with the GPR should also be made in the same lane and wheel track. The antenna center frequency should be selected in such a way that the bound superstructure is reproduced in sufficient

quality. Usually 2 GHz antennas should be selected. The scan rate should be at least 10 scans/meter. For the formation of structurally homogeneous sections, the evaluation of the radargrams without drill cores is sufficient. For preliminarily calibration, i.e. a permittivity about seven can be assumed. A more accurate calibration of the GPR data will be necessary at a later date. Therefore, if necessary, representative drill core sampling points must be defined for the GPR evaluation, which can then coincide with the sampling points for the structural evaluation in order to optimize the number of drill cores.

The data collected shall be subdivided into homogeneous sections. The cumulative sum method can be used as an aid here, which indicates possible section changes by significant deviation from the mean value or by a change in the slope of the cumulative curve of deviations, see Figure 1. A t-test significance test is also used for further limitation. However, the procedure should only be understood as an aid, i.e. the results should be checked by an engineer, who also crosschecks with the raw-data as well as the specific local situation. The section boundaries derived in this or other ways are determined for all data sets and then presented in comparative graphical form. Based on this, the engineering definition of the structurally homogeneous sections takes place. An example of this is given in Figure 2.

Among other things, attention must be paid to the sensible combination of similar or very short sections. This also applies, for example, to the case in which a possibly connected section is separated by a bridge structure. For the merging of sections, the data basis in the original, i.e. the value levels and other characteristics, must also be considered.

3.2 Drill core excavation

A minimum number of cores shall be taken per structurally homogeneous section. At least 16 cores shall be taken for the first kilometer of a section and five cores for each additional kilometer.

Sampling may be performed as 'sectional' or 'cross-sectional' sampling, see Figure 3. Section sampling involves random distribution of drill cores throughout the section. The samples are taken in the right wheel track. During cross-sectional sampling, all drill cores

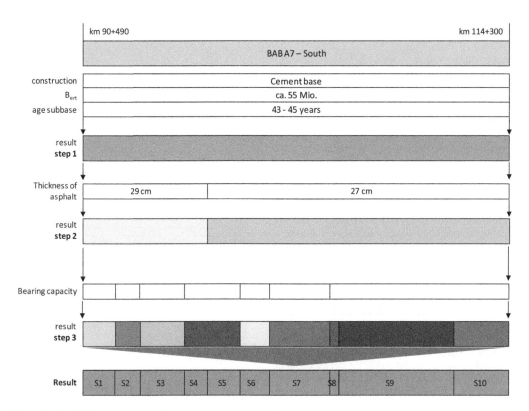

Figure 2. Combining assessment of homogenous.

are taken directly and closely one behind the other or side by side near the right wheel track. Section sampling is preferable to cross-section sampling for statistical reasons, but may be sufficient, for example, in the case of sampling new construction lots.

Drill cores with a diameter of 15 cm must be taken at full depth from the asphalt package. If there is no bond of layers, all layers should be removed if possible. The drill cores must be without cracks, otherwise they must be replaced by further drill cores taken near the cracked one. The experience gained to date shows that careful planning of the drill core extraction points, taking into account both aspects of traffic safety and the results of the non-destructive tests, makes a significant contribution to the achievable daily output and clarification of the overall situation. For example, the extracted drill cores can be used to calibrate the layer thickness of the GPR measurements and to clarify local bearing capacity problems that did not lead to section formation.

Since the structure of the road should not be damaged by the sampling, care must be taken to ensure that the sampling points are closed properly. In the case of sampling of new construction work, sampling may also take place before the wearing course is paved on the binder course. Suitable assumptions for the surface course must then be made for the evaluation or taken from laboratory made samples.

Table 1. Temperatures and frequencies at ITS test.

Stiffness test						
Temperature [°C]	−10	5	20			
Frequency [Hz]	10	3	1	0.3	0.1	10

Fatigue test	
Temperature [°C]	20
Frequency [Hz]	10

3.3 *Laboratory testing*

The temperature-stiffness function is determined for each individual asphalt layer removed by means of indirect dynamic tensile strain (ITS) tests. The same test method is also used to determine the fatigue function for the lowest effective layer or layer on other specimens. This is the layer or layer which is the last to have a complete top layer bond, i.e. which is subjected to fatigue on its underside. The test temperatures, frequencies and sequence can be found in Table 1.

3.4 *Structural assessment*

In order to determine the residual life or the still bearable load changes, the road pavement is first modeled

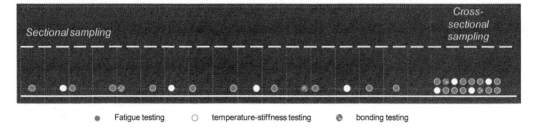

Fatigue testing   ○ temperature-stiffness testing   ⊗ bonding testing

Figure 3. Sectional and cross-sectional sampling.

as a multilayer model. The asphalt package is divided into sub layers according to a given scheme and the other bound and unbound layers are each left as a complete layer. In addition, temperature gradients typical for the location of the road and their relative probability of occurrence for the forecast period are determined. The relative occurrence probability of 26 axle load classes, each with a width of one ton in the range of 1 to 26 tons, is also determined. The methods for this, temperatures and loads, are taken from the corresponding rules and regulations for mechanistic-empirical pavement design, RDO Asphalt.

Suitable assumptions are made for the modulus of deformation and stiffness on the subgrade and the layers above, except for the asphalt package. These are generally 45 MPa for the subgrade, 120 MPa for the top surface of the unbound base course and 2,000 MPa for hydraulically bound base courses. The sub layers of the asphalt package each receive a stiffness derived from the given temperature gradients and the temperature-stiffness function.

The multilayer model is loaded with a top load, r = 150 mm. The size of the load results from the maximum of the respective axle load class. For each case of the superposition of 12 temperature gradients, dependent from the probability of 13 surface temperature classes from −12.5 to 47.5°C and 26 load classes, the maximum strain at the bottom of the decisive layer or layer is calculated with the help of multilayer elastic theory. Using the known fatigue function, the permissible number of load changes is then determined for each calculation case. From the quotient of the predicted number of load cycles and the permissible number of load cycles, a Miner summand is calculated for each load case. The predicted number of load changes, which combined with the traffic load (vehicles per time unit) gives the residual life, must be determined for the case in which the Miner sum is less than or equal to one.

The method can be used both deterministically, see Figure 4, and probabilistically. In the case of a deterministic approach, the calculation procedure described above must be carried out for the section with average stiffnesses and layer thicknesses. When using the probabilistic method, the stiffnesses, fatigue functions and layer thicknesses are to be converted into classes which are assigned to a probability. The calculation shall then

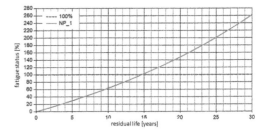

Figure 4. Example for deterministic result.

be performed for all combinations of classes and the failure probability of the section shall be calculated. A section is considered to have failed if the probability of failure exceeds 10% (for motorways).

4 RSO FOR CONCRETE PAVEMENTS

As mentioned above, the RSO concrete is still in the process of being produced. Since the material and failure behavior of concrete differs fundamentally from that of asphalt, other concepts must be chosen. It can also be seen that concrete fails relatively abruptly at the end of its useful life without previous degradation, which makes forecasting difficult. The concept of RSO concrete will therefore be based in particular on the statistical evaluation of material parameters and surface characteristics.

5 SUMMARY AND OUTLOOK

The RSOs will be the first systematic procedure in Germany to assess the structural state of roads at object level. Based on a growing background of experience, the regulations will be further developed. In future, the multilayer method is to be replaced by FEM. A corresponding research project is in progress with the aim of achieving appropriate calculation times while simultaneously applying probabilistics. Furthermore, the first steps towards the application of the RSO principle at network level have been started. This is to be achieved by assuming typical, classified fatigue and stiffness functions. It can be assumed that non-destructive measurements will play a role in the classification.

REFERENCES

FGSV 2009. *Richtlinien für die rechnerische Dimensionierung des Oberbaus von Verkehrsflächen mit Asphaltdeckschicht.* Cologne: FGSV-Verlag.

FGSV 2014. *Richtlinien für die Bewertung der strukturellen Substanz des Oberbaus von Verkehrsflächen in Asphaltbauweise (Draft-not published).*

Advances in Materials and Pavement Performance Prediction II – Kumar et al. (eds)
© 2021 Taylor & Francis Group, London, ISBN 978-0-367-46169-0

Characterizing low-temperature field produced asphalt mix performance

Joyce Kamau, Joseph Podolsky & Chris Williams
Iowa State University, USA

ABSTRACT: The Northern United States and Canada experience winter between 4-6 months of each year and thus are more prone to experience low temperature cracking as the primary distress in their asphalt pavements. This cracking results from a sudden drop in temperature or repeated freeze and thaw cycles, causing thermal stress build-up that exceeds the asphalt pavement's tensile strength. Cracks may allow water infiltration into the pavement, causing moisture-induced damage, which reduces pavement life, and thus maintenance is required; this adds costs to the Department of Transportation (DOT). This research assesses the low-temperature cracking resistance of asphalt mixtures used in the State of Iowa by correlating the low-temperature performance of field-produced mix based on lab specifications. The disk-shaped compact tension (DCT) was used to evaluate low-temperature mixture fracture energy. From this Study, ten mixtures were found to have fracture energies ranging from 265.25 J/m^2 to 470 J/m^2 for the DCT test, where most do not meet the required fracture energy for their specified, designed levels of traffic and the minimum value of 400 J/m^2. Storage of asphalt as loose mixture, aging of mixture and reheating in the laboratory may have caused reduction in fracture resistance. A distress survey is recommended before the specification are revised.

1 INTRODUCTION

Thermal cracking, also known as low-temperature cracking, is one of the major distresses experienced by pavements in cold regions that experience large daily temperature fluctuations (Behzad 2011; Mihai 2004). Thermal stress builds up as the temperature decreases, causing thermal contraction, which results in the development of cracks once the pavement tensile strength is surpassed (Jenny 2017).

To improve thermal cracking performance in the field, characterization of material is important in predicting pavement performance. This has been achieved by characterizing material properties, such as the asphalt cement binder stiffness (Jenny 2017; Marasteanu et al. 2012). Bending Beam Rheometer (BBR) and Dynamic Shear Rheometer (DSR) were introduced in 1990 to characterize low-temperature performance mix. However, pavement performance is influenced by other components within the mix; Generally, most mix contains approximately 95% aggregate and only the rest 5% is the Asphalt binder. The need for asphalt mix test has led to the development of tests such as the Semi-Circular Bend (SCB) test, Illinois Flexibility Index Test (IFIT), and Disc-Shaped Compact Tension (DCT) test (Eshan et al. 2016; Gabriel & Kim 2019). which has been used by most researchers and Departments of Transportation (DOT)s within the Midwestern United States as asphalt mix characterization tests for low-temperature performance.

Fracture energy (Gf), determined by DCT and SCB, is the energy required to initiate and create a unit surface of a crack, indicating the fracture resistance of a mix (Michael et al. 2005; Rivera-Perez et al. 2018). The Gf is affected by aggregate type, test temperature, the addition of recycled material, and air voids of the mix (Marasteanu et al. 2012; Michael et al. 2005; Xinjun 2010). High Gf indicates that the pavement has high tensile strength and thus can dissipate tensile stress build up more easily in the pavement at low temperatures when the pavement is under loading.

The disk-shaped compact tension (DCT) is of Key interest in examining the fracture mechanics of asphalt concrete at low temperature(Wagoner et al. 2005). DCT was recommended during the national pooled study 2 to be used for low temperature cracking specification (Marasteanu et al. 2012).

The current specification for Fracture energy (Gf) measured through Disk-shaped Compact Tension (DCT) (ASTM D7313-07a) test in Iowa is contained in IM510 and pertains to when binder replacement exceeds 30% for mixtures containing only RAP or 25% for mixtures containing recycled asphalt shingles. "The average of two specimens shall meet the minimum fracture energy requirements tested at 10°C warmer than the low climatic temperature (normally specified as the low-temperature performance grade:

Very High Traffic (VT) 690 J/m2
High Traffic (HT) 460 J/m2
Standard Traffic (ST) 400 J/m2"

Table 1. Different mixture properties.

Mix	Year	Binder type	Traffic level	Recycled material % in mix	Binder content
1	2013	PG58-28	high	11% RAP, 4% RAS	5.33
2	2014	PG64-22w/hG	high	9.5% RAP,	5.28
3	2013	PG58-28	high	5% RAP, 4% RAS	5.33
4	2013	PG58-28	high	12% Slag,	4.49
5	2014	PG58-28	high	9.5% RAP,	5.48
6	2018	PG58-28V	Very high	19% RAP,	4.75
7	2018	PG58-34H	high	15% RAP,	5.34
8	2018	PG58-28S	standard	–	5.89
9	2018	PG58-28H	high	15% RAP	5.36
10	2018	PG58-28S	standard	17% RAP	5.02

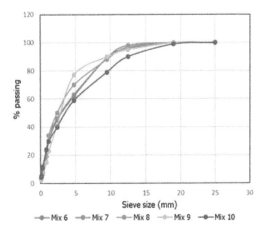

Figure 2. The new design mix gradation.

There is overlap in aggregate gradation between mixtures 1 and 3, and same for mixtures 2 and 5. However, they have different binder content and film thickness. Binder replacement was between 11.9% to 22.7% with 8 values above 15%.

2.2 Methods

Boxes containing ten different field produced asphalt mixtures were obtained during paving of projects. Boxes of the mix were reheated, and laboratory compacted using a gyratory compactor to produce 150mm (6in.) diameter by 50mm height s specimens. Theoretical maximum specific gravity (G_{mm}) was determined using AASHTO T 209-12 and Bulk specific gravity (G_{mb}) using ASTM D2726. Air voids were determined, and specimens with air voids of 7% ± 0.5% were used for subsequent testing. Before testing, specimens were prepared according to ASTM D7313-13. To determine the fracture energy of each mix design, four replicate specimens for each specimen were tested using the disk compact tension (DCT) test at low temperatures.

The DCT test was carried out in accordance with ASTM D7313-13. Each mixture was tested at a temperature 10°C higher than the low temperature of their PG grade i.e., for PG 64-22 could be tested at −12°C. A 1.0mm/min rate of the crack mouth opening displacement (CMOD) was used. Continuous load and CMOD are measured, and fracture energy is computed as the area under the load-CMOD curve.

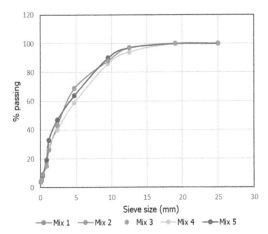

Figure 1. The old design mix gradation.

Iowa Material IM510 Method of design of asphalt mixture.

2 MATERIAL AND METHODS

2.1 Material

Ten mixtures five from the old design and the other five from the Ndesign were used and they had properties as shown in Table 1. The mixture were from five of the six Iowa Transportation department districts and represented the two regions of class 1 projects PG binder as of Iowa reference guide-2016-2.

The material had combined aggregate gradation shown in figure 1 and 2.

For the purpose of this research, the mix was named mix 1 to 5 for the old design mixtures and mix 6 to 10 for the new end design. Nine mixtures contained varying percentages of recycled material (RAP, RAS, and slag), with some containing a combination of two.

3 RESULT AND ANALYSIS

Only three of the mixtures had fracture energy value above minimum fracture energy for standard traffic of 400J/m^2. There are two mixtures that are below 300J/m^2 while the rest are in between 300J/m^2 and 470J/m^2 as shown in Figure 1

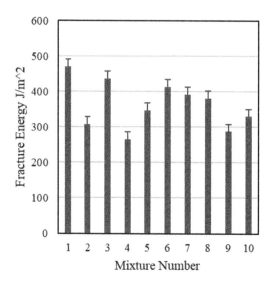

Figure 3. Fracture energies.

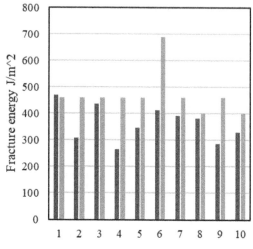

■ Obtained GF   ■ minimum expected value of Gf

Figure 4. comparison of fracture energy obtained to the expected minimum values.

Mix 8 and 10 designed for standard traffic have G_f of 381.25 and 330.50 while they are supposed to have a minimum value of 400J/m². Mix 1 is the only mix that has a G_f value that meets the criteria it was designed for (high traffic 460J/m²). Mixtures 3 and 6 satisfy the standard traffic specification of 400J/m²; however, they do not meet the high traffic standard value they were designed for. Mixtures 2, 4, 5, 7 and 9 do not meet their designed criteria nor the standard traffic. Mixture 4 had the highest percentage of slag and RAP and had the lowest fracture energy. From the results, it should be noted that even mixtures without recycled material did not meet their specified criteria.

The mean fracture energies varied between 10 J/m² to 277.25 J/m² according to set criteria five mixtures differed from the criteria with a difference above 100 J/m² and only 3 mixtures had a difference below 50 J/m². There was no effect of old or new design noted as almost all the mixture failed all the mixtures. The mixtures with the same gradation mix 1 and 3 had fracture energy above the standard traffic level, however, mix 3 did not meet the design criteria level while mix 1 meet the minimum fracture energy level it was designed for.

It is evident there is a problem with the field produced mix as it does not meet the minimum value of fracture energy for their designed traffic levels From the ten mixtures used Slag, RAP and RAS could not be associated with the problem as one mixture containing 11 percent RAP and 4 percent RAS met the criteria for high traffic volume; on the other hand a mixture without recycled material did not meet it's designed criteria of standard traffic volume.

4 CONCLUSION

It can be concluded that there is a problem with field-produced mix as the vast majority of mixtures do not meet the specified fracture energy for low temperature cracking. This indicates that most field-produced mixtures have low crack resistance compared to the set criteria, most of pavement constructed by this mixtures will have poor performance / will fail a few years after exposure to loading.

However, it should be noted that the mixtures used were produced as earlier as in 2013 this may have caused the mixtures to age and hence the low fracture resistance. Additionally, the mixtures had to be heated again in the lab for Theoretical maximum density determination and before compaction. The mixture were stored in the loose nature and this would have accelerated the aging of the binder unlike in a pavement where only limited surface is exposed to air for binder aging.

DCT test may not have been carried out for the lab produced mixture during construction, since the binder replacement was less than 30% for RAP and 25% for RAS thus DCT test was not a requirement according to the specifications. The specification on the need for DCT test should be revised to; when the asphalt binder replacement exceeds 15% from the current value of 30%. A performance study; distress survey is recommended to evaluate the performance of the pavement constructed using these 10 asphalt mixtures as this will direct the Iowa DOT on revising the specifications on low temperature cracking or set a guideline to be followed by the contractors.

REFERENCES

Behzad et al. 2011. Effects of Recycled Asphalt Pavement Amounts on Low-Temperature Cracking Performance of Asphalt Mixtures Using Acoustic Emissions. *Journal of the Transportation Research Board*

Eshan V. et al. 2016. Effects of Mix Design and Fracture Energy on Transverse Cracking Performance of Asphalt Pavements in Minnesota *Journal of the Transportation Research Board*

Gabriel Nsengiyumva and Yong-Rak Kim,2019. Effect of Testing Configuration in Semi-Circular Bending Fracture of Asphalt Mixtures: Experiments and Statistical Analyses *Journal of the Transportation Research Board*

Iowa Material IM510 Method of design of asphalt mixture.

Jenny et al. 2017. low temperature cracking analysis of asphalt binders and mixtures; *Cold Regions Science and Technology.*

M. Marasteanu et al. 2012. Investigation of Low Temperature Cracking in Asphalt Pavements *National Pooled Fund Study – Phase II. Minnesota Department of Transportation technical report.*

Michael P et al. 2005. Investigation of the Fracture Resistance of Hot-Mix Asphalt Concrete Using a Disk-Shaped Compact Tension Test; *Journal of the Transportation Research Board.*

Mihai et al. 2004. Low temperature cracking of asphalt concrete pavements; *Minnesota department of transportation*

J. Rivera-Perez et al. 2018. Impact of Specimen Configuration and Characteristics on Illinois Flexibility Index *Journal of the Transportation Research Board*

Xinjun Li et al. 2010. Factors study in low-Temperature Resistance of Asphalt Concrete. *Journal of Materials in civil Engineering.*

Advances in Materials and Pavement Performance Prediction II – Kumar et al. (eds)
© 2021 Taylor & Francis Group, London, ISBN 978-0-367-46169-0

Investigation of the motion and the flow-field of a wheel tracking device

K. Kavinmathi
Indian Institute of Technology Madras, Chennai, India

ABSTRACT: Wheel tracking device is used to find the rutting susceptibility of asphalt concrete. Typically in most of these devices, like the French rutting tester and the Hamburg Wheel tracking device, the wheel follows a to-and-fro motion while subjecting the sample to a moving load. This motion is unlike what is observed in highways. It is widely known that at higher temperatures, a moving vehicle subjects the pavement to permanent deformation in the longitudinal direction in addition to the permanent deformation in the vertical direction. Thus, both the direction of wheel motion and the longitudinal component of force can affect the flow-field of the tested samples. Hence, these factors must be considered while modelling their mechanical behavior.

In this study, the influence of the direction of the wheel loading and the influence of longitudinal force on the flow-field was examined by conducting a finite element analysis on an asphalt concrete sample. Linear viscoelastic model was used to describe the mechanical behavior of asphalt concrete at 60°C. A sensitivity analysis was conducted with three different coefficients of friction (0, 0.1, and 0.35), and two different modes of wheel motion (to-and-fro motion, and unidirectional motion). Preliminary analysis showed a significant difference in rut depth, for samples subjected to "to-and-fro" wheel motion, with and without longitudinal force.

1 INTRODUCTION

The testing process of a wheel tracking device consists of subjecting an asphalt-concrete sample to repeated loading by a moving wheel. This is performed to replicate the rutting process occurring in field, but at a faster rate (BS-EN:12697-22, 2003). The testing process itself can be time-consuming if one has to test a large number of samples to get repeatable results. To overcome the limitation associated with the testing process, few researchers like Bodin et al. (2013), Tsai et al. (2016), and Wang et al. (2016), have performed the Finite Element Analysis of asphalt-concrete samples that were tested in a wheel tracking device. Among them, Wang et al. (2016) specified the appropriate boundary conditions to replicate the test setup and simulated the testing process by characterizing asphalt concrete with a linear viscoelastic model. On the other hand, Fakhri and Towfigh studied the mechanical response of asphalt-concrete samples by characterizing it as a viscoplastic material.

In practice, the samples used in a wheel tracking device are smooth. On the contrary, the actual surface of the pavement can have undulations and roughness. Thus, when a vehicle-tire moves on such a surface, the loading boundary conditions can vary from what is assumed and used in conventional pavement analysis. Mainly, the longitudinal component of tire load can vary at the interface of vehicle-tire and pavement,

and the amount by which it varies depends on various factors. Among them, the degree of roughness, the speed and maneuver of vehicle play a major role in influencing the friction characteristics at the interface (Khavassefat et al. 2015; Markow et al. 1988). Thus, the longitudinal traction resulting from the interaction of the rolling tire with the pavement surface at a chosen maneuver can affect the pavement's response and the resulting distress of pavement (Yang et al. 2016). Especially distresses such as top-down cracking is mainly caused by the interaction of such load components with the surface of Asphalt concrete (Gideon & Krishnan 2012). But in most of the studies involving pavement analysis, the longitudinal component of traction is generally ignored for the ease of analysis. Thus, to better understand the pavement response, one needs to choose the right type of friction coefficient, which can vary with the speed and the condition of the pavement surface (IRC:73-1980, 1980).

In this study, the performance of asphalt-concrete samples to different loading boundary conditions in a wheel tracking test is studied. Additional conditions were analyzed by choosing various values for the coefficient of friction. This is done to represent the sample surface at different conditions. The sample's behavior is modeled using a commercial finite element software, Comsol (Comsol 2012). Since the study aims at finding the rutting susceptibility of asphalt concrete samples at 60°C, a linear viscoelastic fluid-like model was used for characterizing its behavior.

Table 1. Prony Series Coefficients for Relaxation Modulus at 60°C (Gideon and Krishnan, 2012)

Relaxation Modulus (MPa)	Relaxation Time (s)
867.53 (E_1)	0.069 (ζ_1)
207.48 (E_2)	2.366 (ζ_2)

2 METHODOLOGY

2.1 Structure of asphalt concrete sample

A 3D model of asphalt concrete sample with dimensions - 0.3 m (length) x 0.15 m (width) x 0.05 m (depth) was constructed using the commercially available software package COMSOL 4.4 (Multiphysics 2012) and it is illustrated in Figure 1. The dimensions for the sample were chosen as per the standard, BS-EN:12697-22, 2003. Since the focus of this study was to examine the influence of the longitudinal coefficient of friction on the stress-strain response and the permanent deformation of the asphalt concrete samples, the material model corresponding to asphalt concrete at 60°C is chosen.

2.2 Constitutive equation

In order to precisely capture the viscoelastic response, the asphalt-concrete sample was modeled using a linear viscoelastic model. Particularly, the generalized Maxwell model (with two Maxwell components in parallel) was used for the characterization (Findley & Davis 2013). This model can be described using spring and dashpot analogs as a parallel combination of two Maxwell models. The 1-D constitutive equation for such a model can be expressed as follows:

$$\ddot{\sigma} + \left(\frac{1}{\tau_1} + \frac{1}{\tau_2}\right)\dot{\sigma} + \left(\frac{1}{\tau_1 \tau_2}\right)\sigma = (E_1 + E_2)\ddot{\varepsilon}$$

$$+ \left(\frac{E_1}{\tau_2} + \frac{E_2}{\tau_1}\right)\dot{\varepsilon} \tag{1}$$

Where σ is the stress, ε is the strain, τ_1, τ_2 are the relaxation times and E_1, E_2 are the relaxation moduli for the Maxwell models. The model parameters for the simulation were taken from the study conducted by Gideon and Krishnan (2012) and are shown in Table 1. The Poisson's ratio ν was assumed to be 0.35.

The weak form for the asphalt concrete sample was developed using the method described by Zhang et al. (2015). The 3D constitutive equation for such a model is shown as follows:

$$\sum_{x=1}^{x=2} \sigma_{ij}^x = \sum_{x=1}^{x=2} K^x(\varepsilon_{kk}^e - \varepsilon_{kk}^{x,\mu}) + \sum_{x=1}^{x=2} G^x(\varepsilon_{ij}^\theta - \varepsilon_{ij}^{x,\mu}) \tag{2}$$

$$\tau_x \frac{d\varepsilon_{kk}^{x,\mu}}{dt} + \varepsilon_{kk}^{x,\mu} - \varepsilon_{kk}^e = 0 \tag{3}$$

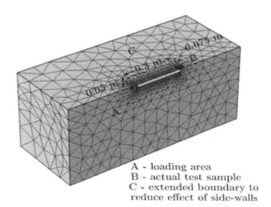

A - loading area
B - actual test sample
C - extended boundary to reduce effect of side-walls

Figure 1. Axisymmetric- Finite Element Model.

$$\tau_x \frac{d\varepsilon_{ij}^{x,\mu}}{dt} + \varepsilon_{ij}^{x,\mu} - \varepsilon_{ij}^e = 0 \tag{4}$$

$$\sigma_{ij,j} = \rho \ddot{u}_j \tag{5}$$

where, σ_{ij}^x is the total stress component of the viscoelastic fluid model τ_x, is the relaxation time for the xth Maxwell component, K^x, G^x are the components of bulk relaxation and shear relaxation modulus for the xth Maxwell component, ε_{kk}^e is the elastic part of the volumetric strain component, ε_{ij}^e is the elastic part of the deviatoric strain component, $\varepsilon_{kk}^{x,\mu}$ is the viscous part of the volumetric strain for the xth Maxwell component, $\varepsilon_{ij}^{x,\mu}$ is the viscous part of the deviatoric strain for the xth Maxwell component, ρ is the density of the material. The above set of equations has a displacement field. 'u_i' as the main dependent variables and the strains ($\varepsilon_{ij}^{x,\mu}$, $\varepsilon_{ij}^{x,\mu}$), is the internal dependent variables that form coupling between Eqn.2, Eqn.3 and Eqn.4. Eqn.2, along with the equilibrium equations (Eqn.5), were modeled in COMSOL. Here, Eqn.3 and Eqn.4 were treated as ODEs.

2.3 Boundary condition

Suitable boundary conditions were chosen for the asphalt-concrete sample to replicate the laboratory conditions. Particularly, the wheel-tracker consists of a wheel with a load of 700 N, and it moves over a loading platform of length 0.3 m. The radius of the wheel contact is assumed to be 0.05 m and the wheel is assumed to make 60 passes per minute. Hence, the speed of the wheel movement is found to be 0.3 m/s.

Using these details, the boundary conditions for the simulation were set. This is inclusive of fixed boundary conditions for the sides and the bottom of the sample structure. The top surface of the sample was treated as a free boundary, except for the loading area. A built-in analytical function called "flc2hs(x, scale)" with C^2 continuity was used to simulate the moving load applied by the wheel-tracker. Further, this function was reformulated in such a way to represent the

185

existing "to-and-fro" motion of the wheel tracker and a hypothetical case involving a "unidirectional" motion exhibited by the wheel tracker. Both these functions (corresponding to the z component of wheel load) are represented by the equation shown below,

$$\sigma_{zz}(x,t) = P[H_2(x - vt_1) - H_2(x - vt_1 - l_1)] \quad (6)$$

where, $\sigma_{zz}(x,t)$ are the moving wheel load function corresponding to a to-and-fro and a unidirectional type of loading, P is the contact pressure of the tire, $H_2(x - v_0t)$ is the heavi-side moving load function which is C^2 continuous, vs the speed of moving wheel, l is the length of the tire contact patch, x,t are the spatial and temporal variables t_1, l_1, are chosen according to the type of motion. To compare the response of a lab-fabricated sample with that obtained from the field, additional load functions that were inclusive of a longitudinal friction coefficient. These functions can account for the surface-roughness of pavements. The different coefficients of longitudinal friction chosen are – 0 (laboratory prepared smooth-surfaced sample), 0.1, and 0.35 (field-collected sample with a rough surface). These factors when multiplied with the vertical component of loading is used as longitudinal loading.

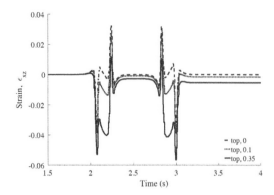

Figure 2. Oscillatory Loading: Variation of xz-component of Strain at the Surface

obtained from a smooth-surfaced sample. On the contrary, the temporal variation of the xz component of strain (Figure 2) shows a substantial difference when the samples with different friction coefficients were compared. The difference in the response obtained from a smooth-surfaced sample and a sample with a friction coefficient of 0.1 is 17.9 %. Similarly, the response obtained from a sample with a friction coefficient of 0.35 exhibited a peak strain, which is higher than the peak strain from a smooth-surfaced sample by 40 %.

3.2 Influence of Direction of Motion of Wheel Load on the Response of Asphalt Concrete Sample:

In the second part of this study, the direction of motion, and its effect on the flow-field is studied. Here, the response from a to-and-fro motion (Figure 2) of the wheel load was compared with the response from the unidirectional movement (Figure 3) of the wheel. Though there is a negligible difference in the magnitude of the measured response, there was a substantial change in the pattern of variation of the response. For instance, a unidirectional loading exhibits an alternating pattern of tension-compression-tension-compression for responses observed within one loading cycle. On the contrary, a to-and-fro type of loading exhibits a compression-tension-tension-compression type of pattern in the response. It is to be noted that both the patterns were observed using responses that ere captured during one loading cycle. In other words, this can be understood as subjecting the sample to two consecutive tensile loadings during each cycle of to-and-fro motion. Thus, the type of failure at the end of several loading cycles can be quite different from what is usually observed. Hence, this change in response pattern also needs to be understood to model the mechanical behavior of asphalt concrete samples.

2.4 Meshing and Time Discretization

'Tetrahedral' mesh elements were used for the spatial discretization of the sample such that the meshes closer to the loading area were refined to get an accurate solution. The quadratic interpolation function was used for the displacement field variables. Based on mesh convergence studies that were conducted previously, the optimum value of the computation time step was fixed at incremental time steps of 0.0001 s (Kavinmathi and Narayan, 2017).

3 RESULTS AND DISCUSSION

3.1 Influence of Longitudinal Friction Coefficient for Rough-surfaced Field-Collected Samples

The influence of the longitudinal friction coefficient on the mechanical response of an asphalt concrete sample can be found by comparing the responses obtained using a moving load with and without longitudinal coefficient of friction. The response of the sample (in terms of the xz and the zz component of strain) is captured at every 0.001 s while a wheel load moves on its surface. The temporal variation of the response was measured for samples that have varying levels of friction coefficient. Particularly, the zz component of strain was found to show negligible variation between the different types of surface. The variation in the response obtained from a sample with a smooth surface and a sample with a surface friction of 0.1 is found to be 10.76 %. Similarly, the zz component of strain (peak) was 14.5 % higher than the response

4 CONCLUSION

The main aim of this study was to carry out a full-fledged Finite Element analysis of asphalt concrete

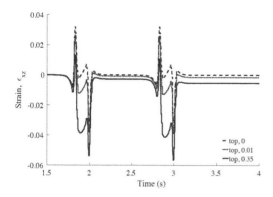

Figure 3. One Directional Loading: Variation of xz-component of Strain at the Surface.

samples by characterizing it as linear viscoelastic fluid material. However, the limitation of this approach is that it can under/overestimate the strain values. Using this investigation, the influence of the longitudinal coefficient of friction and the direction of wheel-motion on the flow-field of sample was evaluated. The following observations were made:

• The strain at the end of one loading cycle was found to vary with the longitudinal coefficient of friction. Particularly, the zz-component and the xz-component of strain showed a increasing trend.
• Though additional observations showed that there is a negligible difference in the magnitude of response between the two types of motion, there was an observed variation in the pattern of the flow-fields. Thus, the investigation from this study indicates that the longitudinal coefficient of friction should be considered in the stress/strain analysis, while one can safely ignore the direction of motion of the wheel.

ACKNOWLEDGMENT

The author acknowledges the funding provided by the Department of Science and Technology, Government of India, through the grant DST/TSG/STS/2011/46. The author also wishes to thank the staff of Pavement Engineering Laboratory, Indian Institute of Technology-Madras, for providing the experimental data and for supporting this research.

REFERENCES

Bodin, D., M. Moffatt, and A. Lim. (2013). Feasibility Study of Using Wheel-Tracking Tests and Finite Element Modelling for Pavement Permanent Deformation Prediction. *Technical report.*

BS EN 12697–26:2003: Bituminous mixtures – Test methods for hot mix asphalt – Part 22: Wheel tracking, *British Standards*

Comsol (2012). Comsol Multiphysics User's Guide. The Heat Transfer Branch, (1215), 709 – 745, *the Heat Transfer Branch.*

Fakhri, M., and A. Towfigh . Evaluation of Rutting Distress for Asphalt Concrete Mixture Using Abaqus Program.

Findley, W. N., and &. F. A. Davis., Creep and relaxation of nonlinear viscoelastic materials.2013.

Gideon, C. S. and J. M. Krishnan., Influence of Horizontal Traction on Top-Down Cracking in Asphalt Pavements. *In 7th RILEM International Conference on Cracking in Pavements. Springer,* Netherlands, 2012.

IRC: 73-1980 Geometric Design Standards for Rural (non-urban) Highways, *Indian Road Congres*, New Delhi, India.

Gideon, C. S. and J. M. Krishnan., Influence of Horizontal Traction on Top-Down Cracking in Asphalt Pavements. *In 71 st RILEM International Conference on Cracking in Pavements. Springer,* Netherlands, 2012.

Kavinmathi, K. and SPA Narayan., Effect of Fluid-Like Characteristics of Bituminous Layers and Inertia on the Rutting Behaviour of Flexible Pavements. *In 71st RILEM International Conference on Advances in Construction Materials and Systems,* India, 2017.

Khavassefat, P., D. Jelagin, & B. Birgisson. 2015. Dynamic Response of Flexible Pavements at Vehicle - Road Interaction. *Road Materials and Pavement Design,* 16(2), 256-276.

Markow, M. J., J. K. Hedrick, B. D. Brademeyer, & E. Abbo. 1988. Analyzing the Interactions between Dynamic Vehicle Loads and Highway Pavements. *Transportation Research Record,* 1196, 161-169.

Tsai, B.W., Coleri, E., Harvey, J.T.& Monismith, C.L. 2016. Evaluation of AASHTO T 324 hamburg-wheel track device test. *Construction and Building Materials*, 114, 248-260.

Wang, P., M. Ke, H. Zhang, F.Wang, and Z. Lu (2016). Performance Experiments and Finite Element Simulation on Basalt Fiber Reinforced Asphalt Mixture. *(Iceta),* 1121–1127.

Yang, W., Tiecheng, S., Yongjie, L.& Chundi, S. 2016. Prediction for Tire-Pavement Contact Stress under Steady-State Conditions based on 3D Finite Element Method. *Journal of Engineering Science & Technology Review,* 9(4).

Zhang, Y., B. Birgisson, and R. L. Lytton (2015). Weak Form Equation Based Finite-Element Modeling of Viscoelastic Asphalt Mixtures. *Journal of Materials in Civil Engineering*, 28(2), 04015115.

Advances in Materials and Pavement Performance Prediction II – Kumar et al. (eds)
© 2021 Taylor & Francis Group, London, ISBN 978-0-367-46169-0

Layered nonlinear cross anisotropic model for pavements with geogrids

M.E. Kutay & M. Hasnat
Michigan State University, East Lansing, MI, USA

E. Levenberg
Technical University of Denmark, Lyngby, Denmark

ABSTRACT: This paper reports the details of a new layered elastic nonlinear cross anisotropic model that can consider the confinement effects of geogrids within pavements. The algorithm, herein called MatLEACANGG, was verified against well-known programs such as ELLEA2 (Layered Elastic-Cross Anisotropic model) and MichPave (Nonlinear Finite Element Model). Influence of the Geogrid on the vertical microstrain of unbound layers was illustrated via runs on four different structures (with and without Geogrid).

1 INTRODUCTION

Geogrids are useful in improving the structural response of unbound layers used in both flexible and rigid pavements (Giroud & Han, 2004). There has been numerous research on the characteristics of the geogrids used in unbound layers(Al-Qadi et al., 2008; Kwon & Tutumluer, 2009; Kwon et al., 2009; Ling & Liu, 2003). Although significant work has been done on understanding the effects of geogrids in pavement performance, models and computationally efficient software that can be used in mechanistic-empirical pavement design approaches are limited. To respond this need, NCHRP 1-50 project was completed (Luo et al., 2017). This project resulted in a set of artificial neural networks (ANNs), which can be used in the Pavement ME software to compute the critical stresses and strains. Unfortunately, the ANNs are not publicly available, therefore, their use is only limited to AASHTOWare applications. The objective of this study was to develop a computationally efficient layered elastic model that can consider the nonlinearity, cross-anisotropy as well as the confining effects of geogrids in unbound layers of the pavements.

2 LAYERED ELASTIC NONLINEAR CROSS ANISOTROPIC MODEL WITH GEOGRIDS (MATLEACANGG)

2.1 Layered elastic cross-anisotropic model

The layered elastic cross-anisotropic model implemented herein is based on the formulations presented by Levenberg et al. (Levenberg et al., 2009). The only difference is that the way the formulations were implemented into an algorithm in MATLAB. The matrix-based methodology presented by Khazanovich and Wang (Khazanovich & Wang, 2007) in implementation of MnLayer was followed herein to develop the program called MatLEACA, which stands for Matlab-based Layered Elastic Analysis with Cross Anisotropy. The main material-related inputs to the MatLEACA are the vertical modulus (E_z), horizontal to vertical modulus ratio ($n_E = E_r/E_z$) two Poisson's ratios (ν_{zr} and ν_{rr}) and shear modulus (G_{rz}) for each layer. The other inputs are the contact radius and pressure of the surface load, as well as the evaluation point coordinates (in r- and z-direction). It is noted that the MatLEACA is a sub algorithm of the MatLEACANGG software described in this paper.

2.2 Inclusion of nonlinearity

The approach followed herein for including the effect of nonlinearity (i.e., stress-dependency) of the modulus of unbound layers is identical to the approach presented by Varma and Kutay (2016). It is well known that the multilayered solutions can only work with constant modulus in radial direction. However, in such solutions, since stress changes in radial direction, the stress-dependent modulus cannot be rigorously considered. On the other hand, approximate solutions can be achieved when a representative stress state within a sublayer is used in the nonlinear modulus equation (Huang, 2004; Zhou, 2000). In this study, the stress state in the center of the sublayer (in z-direction) is calculated at a radial distance r=3.5a, where a=radius of the surface contact load (see Figure 1). Varma and Kutay (2016) showed (via comparison with Nonlinear Finite Element Analysis) that this location is reasonable to compute an 'average' stress-dependent modulus for the given layer. The form of the

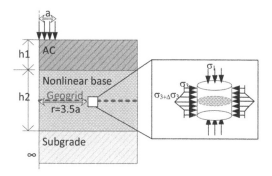

Figure 1. Conceptual illustration of inclusion of Geogrid in the multilayered structural model.

stress-dependent unbound layer modulus used in this study is as follows:

$$M_R = k_1 P_a \left(\frac{I_1}{P_a}\right)^{k_2} \left(\frac{\tau_{oct}}{P_a} + 1\right)^{k_3} \quad (1)$$

where, I_1 first invariant of the stress tensor, τ_{oct} is octahedral shear stress k_1, k_2, k_3 are material-level constants, K_o is the coefficient of earth pressure at rest and P_a is atmospheric pressure (=101.3 kPa).

It should be noted that nonlinearity was incorporated via an iterative process:

- First, an initial set of elastic moduli are assumed for each layer.
- Then, the stresses produced by the MatLEACA using the initial values of moduli are used to evaluate the new modulus using equation (1).
- The iteration is continued until the computed modulus from the stresses predicted by the MatLEACA and input modulus used in the layered solution converges.

It is noted that the ratio of horizontal to vertical modulus is assumed constant. The modulus obtained using the equation (1) is used as the vertical modulus (E_z) and the horizontal modulus (E_r) is computed from the E_z using the ratio n_E (where $n_E = E_r/E_z$). At this stage, the shear modulus (G_{rz}) is computed using the following relationship (Barden, 1963; Levenberg, 2009; Wolf, 1945):

$$G_{rz} = \frac{E_z E_r}{E_z + E_r (1 + 2\nu_{zr})} \quad (2)$$

where G_{rz} is the shear modulus, ν_{zr} = Poisson's ratio, E_z and E_r = vertical and horizontal moduli, respectively. For simplicity, the two Poisson's ratios (ν_{zr} and ν_{rr}) were assumed to be equal to each other, and do not change with stress state.

2.3 Inclusion of Geogrid in unbound layer

In this study, the approach presented in the NCHRP 1-50 report was followed. This model was based on

the model developed by the Yang and Han (Luo et al., 2017). The concept is based on the idea that the Geogrid's main influence on the unbound layer is in the form of a linearly increasing additional confining stress, as illustrated in Figure 1 (Luo et al., 2017). As the vertical stress is applied near the Geogrid, lateral expansion due to plastic and resilient deformation is restrained by the Geogrid. This restraint is applied to the unbound layer in terms of additional confinement $\Delta\sigma_3(z)$) which is assumed to be linear around the zone of influence (which is about ±7.5 cm above and below the Geogrid). The maximum additional confining stress at the level of Geogrid ($\Delta\sigma_3^{max}$) can be calculated by iteratively solving the following equation:

$$\Delta\sigma_3^{max} = \frac{2M}{(1 - \nu_g)\,\delta\alpha}$$
$$\left(\frac{\nu_{13}\sigma_1}{E_z} + (1 - \nu_{33})\frac{\sigma_3 + \Delta\sigma_3^{max}}{E_r} + \varepsilon_{3p}\right) \quad (3)$$

where;

$$\varepsilon_{3p} = 0.85\varepsilon_o e^{-\left(\frac{\rho}{N}\right)^\beta} \left(\sqrt{J_2}\right)^m (\alpha I_1 + K)^n. \quad (4)$$

$$\alpha = \frac{2\sin\phi}{\sqrt{3}\,(3 - \sin\phi)}. \quad (5)$$

$$K = c \cdot \frac{6\cos\phi}{\sqrt{3}\,(3 - \sin\phi)}. \quad (6)$$

where M = Geogrid sheet stiffness (kN/m), ν_g = Poisson's ratio of the Geogrid, δ depth of influence (~15 cm), I_1 = first invariant of the stress tensor, J_2 = Second invariant of the stress deviator tensor, ε_{3p} = plastic strain in lateral direction, $\varepsilon_o \rho \beta m$ and n = plastic strain model coefficients, N = cycles, ϕ, c = friction angle and cohesion of the unbound layer, σ_1 and σ_3 are the vertical and horizontal stresses, respectively. It should be noted that the model above was developed for laboratory specimens with axisymmetric conditions. This model has been extended to the field conditions herein to come up with an approximate/practical solution. The main assumption is that the stress is evaluated close to the centerline of the load where $\sigma_2 \sim \sigma_3$. More rigorous model is probably needed to improve the accuracy.

Once the $\Delta\sigma_3^{max}$ is calculated, it is incorporated into the following formulation to compute a modified vertical modulus;

$$E_{z-mod}(z) = k_1 P_a \left(\frac{I_1 + \Delta\sigma_3(z)}{P_a}\right)^{k_2} \left(\frac{\tau_{oct}}{P_a} + 1\right)^{k_3} \quad (7)$$

where $E_{z-mod}(z)$ = modified vertical modulus (constant until $\delta = 7.5$ cm above the geogrid, increases linearly with depth until the Geogrid, then decreases until $\delta = 15$ cm below the Geogrid, after which it is

Table 1. Case 1 structure for verification of MatLEACANGG.

Layer	h_i (cm)	E_z (MPa)	E_r (MPa)	v_{zr}	v_{rr}	G_{rz} (MPa)
1	10.2	3,445	2,067	0.35	0.3	1,034
2	20.3	207	138	0.45	0.38	48
3	25.4	103	69	0.35	0.4	25
4	50.8	48	34	0.3	0.4	12
5	semi-inf	31,005	27,560	0.2	0.15	11,713

constant), $\Delta\sigma_3$ (z) =additional confining stress, triangularly distributed (see Figure 1). Finally, the effective vertical modulus E_{z-eff}) is computed using the following formula:

$$E_{z-eff} = \frac{1}{h} \int_{z=0}^{h} E_{z-mod}(z)dz. \qquad (8)$$

where h = height of the unbound layer. The E_{z-eff} is essentially an average modulus. The horizontal modulus is computed by simply multiplying by the ratio of horizontal modulus to vertical modulus of the original unbound layer (i.e., $E_{r-eff} = n_E E_{z-eff}$. Finally, the effective shear modulus G_{rz-eff} is computed using the equation (2). The remaining parameters are assumed constant.

3 MODEL VERIFICATION AND ANALYSIS RESULTS

In order to verify the models presented herein, two independent software programs were used. First, the cross-anisotropic implementation of the MatLEACANGG was verified against the ELLEA2 V0.8 program (Levenberg et al., 2009). Table 1 shows the properties of the Case 1 structure used for this purpose. A circular load of 40 kN was applied at the surface. The contact radius was 15 cm. A comparison of (a) vertical surface (z-direction) displacement versus depth along the centerline of the load, (b) surface (z-direction) displacements versus radial coordinate and (c) the stresses are shown in Figure 2. As shown, MatLEACANGG and ELLEA2 agrees well with each other.

The second verification was for implementation of the nonlinear (stress-dependent) response of the unbound layers. The MichPave finite element modeling software was used for this purpose. In the NCHRP 1-50 project, four structures listed in Table 2 were analyzed (Luo et al., 2017). In Table 2, structures of Exp. 1 and Exp. 2 were analyzed using MichPave and MatLEACANGG algorithms (Exp. 3 and Exp. 4 can only by analyzed in MatLEACANGG algorithm because they include Geogrids). Same loading configuration was applied (i.e., 40kN circular load with 15 cm radius).

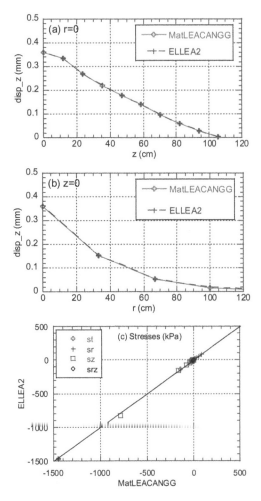

Figure 2. Comparison of ELLEA2 and MatLEACANGG for (a) vertical surface (z-direction) displacement versus depth along the centerline of the load, and (b) surface (z-direction) displacements versus radial coordinate.

Figure 3 shows a comparison of MichPave and MatLEACANGG for (a) vertical microstrain versus depth along the centerline of the load, and (b) radial microstrain at the bottom of asphalt versus radial coordinate. As shown, a very good match is visible, although the nonlinear models of MichPave and MatLEACANGG model are different. The MichPave's nonlinear model is $M_R = k_1 I_1^{k_2}$ whereas MatLEACANGG's nonlinear model is shown in equation (1). It is noted that, the same k1 and k2 values listed in Table 3 was entered into the MichPave, which resulted in similar equivalent modulus for the unbound layer at a given stress state. It is worth noting that the MatLEACANGG runtime was about 0.2 seconds, whereas the MichPave runtime was about 25 seconds for the same structure.

Figure 4 shows results of MatLEACANGG runs on the four structures shown in Table 2, where the

Table 2. Structural properties of the four experiments simulated.

Layer	Exp. 1 h_i(cm)	Exp. 2 h_i(cm)	Exp. 3 h_i(cm)	Exp. 4 h_i(cm)	All Exp. Ez (MPa), ν, n_E(=Er/Ez)
1 (AC)	15	15	15	15	5995, 0.22, 0.99
2 (Unb.)	15 (No GG)	25 (No GG)	15 (GG bot.)	25 (GG ctr.)	See Table 3
3 (Subg.)	Inf	Inf	Inf	Inf	69.9, 0.40, 0.6

Notes: Exp. = experiment, Unb. = unbound layer, Subg. = subgrade layer, GG = Geogrid, GG bot. = geogrid is at the bottom of the layer, GG ctr. = geogrid is at the center of the layer.

Table 3. Nonlinear modulus and plastic strain coefficients of the unbound base layer used in MatLEACANGG model.

k_1	k_2	k_3	n_E (=Er/Ez)	$v_{zr} = v_{rr}$	e_o	ρ
1545.0	0.75	−0.10	0.45	0.3	0.149	72.40

β	m	n	ϕ	c (kPa)	M (kN/m)	v_g
0.25	1.70	-2.16	51.30	20.20	420.0	0.30

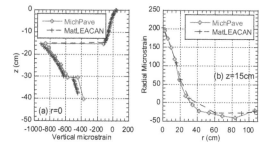

Figure 3. Comparison of MichPave and MatLEACANGG for (a) vertical microstrain versus depth along the centerline of the load, and (b) radial microstrain at the bottom of asphalt versus radial coordinate.

vertical strain within the unbound layer has been plotted. As shown in 4, there is relatively minor reduction in the vertical microstrain at the initial cycle of loading (N = 1), whereas, at N = 10000 cycles, there is significant mobilization of the granular layer due to the existence of the Geogrid and vertical microstrain decreases significantly. This vertical microstrain actually corresponds to the resilient microstrain used in the plastic strain formulations of the MEPDG. Therefore, it has profound influence on the predicted rutting within the unbound layer.

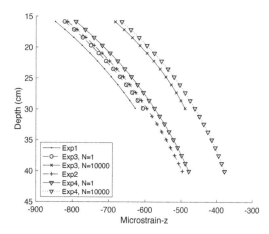

Figure 4. Effect of Geogrid on the vertical microstrain within the unbound layer for four different structures shown in Table 2.

4 CONCLUSIONS

This paper reported the results of a study aimed at developing an efficient layered elastic algorithm to predict the performance of nonlinear anisotropic unbound layers, with and without a Geogrid reinforcement. The algorithm, herein called MatLEACANGG, was verified against well-known programs such as ELLEA2 and MichPave. Influence of the Geogrid on the vertical microstrain of unbound layers was illustrated via runs on four different structures (with and without Geogrid).

REFERENCES

Al-Qadi, I. L., Dessouky, S. H., Kwon, J., & Tutumluer, E. (2008). Geogrid in flexible pavements validated mechanism. *Transportation Research Record*, 2045, 102–109. https://doi.org/10.3141/2045-12

Barden, L. (1963). Stresses and Displacements in a Cross-Anisotropic Soil. *Géotechnique*, 13(3), 198–210. https://doi.org/10.1680/geot.1963.13.3.198

Giroud, J. P., & Han, J. (2004). Design method for geogrid-reinforced unpaved roads. I. Development of design method. *Journal of Geotechnical and Geoenvironmental Engineering*, 130(8), 775–786. https://doi.org/10.1061/(ASCE)1090-0241(2004)130:8(775)

Huang, Y. H. (Yang H. (2004). *Pavement analysis and design*. Pearson/Prentice Hall.

Khazanovich, L., & Wang, Q. (Chuck). (2007). MnLayer: High-Performance Layered Elastic Analysis Program. *Transportation Research Record: Journal of the Transportation Research Board*, 2037(1), 63–75. https://doi.org/10.3141/2037-06

Kwon, J., & Tutumluer, E. (2009). Geogrid base reinforcement with aggregate interlock and modeling of associated stiffness enhancement in mechanistic pavement analysis. *Transportation Research Record*, 2116, 85–95. https://doi.org/10.3141/2116-12

Kwon, J., Tutumluer, E., & Al-Qadi, I. L. (2009). Validated mechanistic model for geogrid base reinforced flexible pavements. *Journal of Transportation Engineering*, *135*(12), 915–926. https://doi.org/10.1061/(ASCE)TE.1943-5436.0000046

Levenberg, E. (2009). Backcalculation of anisotropic pavement properties using time history of embedded gauge readings. *Geotechnical Special Publication*, *190*, 79–85. https://doi.org/10.1061/41042(349)11

Levenberg, E., McDaniel, R., & Olek, J. (2009). *Validation of NCAT Structural Test Track Experiment Using INDOT APT Facility*. https://doi.org/10.5703/1288284314311

Ling, H. I., & Liu, H. (2003). Finite element studies of asphalt concrete pavement reinforced with geogrid. *Journal of Engineering Mechanics*, *129*(7), 801–811. https://doi.org/10.1061/(ASCE)0733-9399(2003)129:7(801)

Luo, R., Gu, F., Luo, X., Lytton, R. L., Hajj, E. Y., Siddharthan, R. v., Elfass, S., Piratheepan, M., & Pournoman, S. (2017). Quantifying the Influence of Geosynthetics on Pavement Performance. In *Quantifying the Influence of Geosynthetics on Pavement Performance*. Transportation Research Board. https://doi.org/10.17226/24841

Wolf, K. (1945). Distribution of stress in a half plane and a half space of anisotropic material. *Zeitschrift Für Angewandte. Hath. Und Mech*, *15*, 249–254.

Zhou, H. (2000). Comparison of Backcalculated and Laboratory Measured Moduli on AC and Granular Base Layer Materials. In S. D. Tayabji & E. O. Lukanen (Eds.), *Nondestructive Testing of Pavements and Backcalculation of Moduli: Third Volume* (pp. 161–172). ASTM International. https://doi.org/10.1520/STP14766S

Advances in Materials and Pavement Performance Prediction II – Kumar et al. (eds)
© 2021 Taylor & Francis Group, London, ISBN 978-0-367-46169-0

Highlights on interesting findings of High Modulus Asphalt Concrete (HMAC/EME) for cold regions in cooperative research project

S. Lamothe, S. Proteau Gervais, C. Neyret, M.M. Boussabnia, D. Perraton & A. Carter
Laboratoire sur les chaussées et matériaux bitumineux (LCMB) – École de technologie supérieure, Canada

B. Pouteau
Centre de Recherche – Eurovia, Mérignac, France

M. Proteau
Centre Technique Amériques (CTA) – Eurovia, Brossard, Canada

H. Di Benedetto
École nationale des travaux publics de l'État (ENTPE), Vaulx en Velin, France

ABSTRACT: This paper presents a methodology to support growth and implementation of a new asphalt mix, such as high modulus asphalt concretes (HMAC) for cold regions. This requires laboratory studies (fatigue, E* and sensitivity) and elaboration of trial sections with several steps. Even though that different fatigue test methods were used in this project (tension-compression and 2 points bending), similar Wohler's law parameters were obtained with referring to a discriminant failure criterion. Overall, the laboratory investigation shows good to excellent thermomechanical properties of HMAC tested as regards stiffness, fatigue, low temperature, and rutting resistances. Moreover, the ongoing sensitivity study shows that HMAC mixes make with bitumen and filler dosage fluctuations in accordance to regulation are not more sensitive than usual HMA.

1 INTRODUCTION

To support the growth and implementation of HMAC for cold regions, a cooperative research project funded in part by the Natural Sciences and Engineering Research Council of Canada (NSERC) was launched on April 1, 2017 for a four-year period. The research project led by the LCMB of *École de technologie supérieure* (ÉTS), with the CTA from Eurovia-Canada and the Research Center from Eurovia-France. Knowing that engineering properties are at the heart of HMAC mix design, two main topics were targeted in the research project: 1) to define thermomechanical properties of HMAC mixes and, 2) to link on-site to laboratory fatigue performance. The aim of the paper is to present the latest development of this project.

2 PROBLEM STATEMENT

The mix design of HMAC is based on the French concept used to produce *"Enrobé à Module Élevé, EME"*. The implementation of HMAC in view of Mechanical-Empirical pavement design (M-EPD) is a complex process. To be able to do that, a number of key issues still require clarification. How can we prescribe those mixes when different tests are used with different parameters to quantify their fatigue resistance? Which calibration coefficients should be used for passing from the lab to the field in order to comply with the empirical part of the M-EPD? Till now, very limited number of field trials has been made to evaluate the performance of those mixes. On the other hand, the sensitivity of this type of asphalt to variations in its composition on its thermomechanical behavior should be evaluated.

3 COOPERATIVE RESEARCH PROJECT

The collaborative research project has two distinct phases: I. Laboratory studies and II. Trial section.

4 LABORATORY STUDIES

The objectives of the laboratory investigation are: 1) to define HMAC fatigue law parameters to apply in the M-EPD approach and, 2) to check thermomechanical mix sensitivity in accordance to allowable production plant variations (bitumen and filler).

4.1 Material

The HMAC is designed with modified bitumen (PG88-28) by using a 0/14mm fully crushed high quality aggregates.

4.2 Laboratory-compacted slabs and samples preparation

Two types of laboratory-compacted slabs were used to produce specimens for thermomechanical tests: 1. Small slabs (SS: $125 \times 180 \times 500$ mm^3) and, 2. Big slabs (BS: $125 \times 400 \times 600$ mm^3). At CTA's facilities, HMAC were mixed at 185°C, cured and compacted at 170°C with a French LPC wheel compactor to make the BS. After a curing time of 14 to 60 days in laboratory conditions, slabs were sawed or cored in the thickness of the slabs, in the direction of the compaction. The specimens were tested after 14 to 60 days.

4.3 Fatigue study

The aim is to correlate the Whöler's law parameters of HMAC using various fatigue test methods. The tension-compression (TC) and the 2 and 4 points bending (2PB, 4PB) tests were used. The 4PB results are currently under analysis and are not shown here. Two failure criteria, $N_{f50\%}$ and $N_{fII/III}$, were considered to fix the fatigue life N_f of the specimens. Wöhler parameters from TC and 2PB tests were compared.

4.3.1 Failure criteria

The classical criterion is based on 50% modulus ($|E^*|$) decrease, $N_{f50\%}$. The criterion $N_{fII/III}$ was calculated from the mean value of: 1. Criterion from modulus evolution, N_{fE*_N}; 2. Criterion from phase angle evolution, $N_{f\phi E\_N}$; and 3. Criterion from Black domain evolution, $N_{fE*\_\phi E}$. The $N_{fII/III}$ approach allows taking to account the evolution of the rigidity of the material (N_{fE*_N}) as well as the viscous behavior ($N_{f\phi E\_N}$ and $N_{fE*\_\phi E}$) during fatigue testing.

4.3.2 Experimental device and test principle

The 2PB is a non-homogeneous and standardized test commonly used in Europe (EN 12697-24). The test applies a continuous sinusoidal waveform displacement at the top of a trapezoidal specimen.

Uniaxial homogeneous TC test is used in Quebec (CTA and LCMB), in France (ENTPE) and USA (U. of New Hampshire). The axial strain is measured with three (3) extensometers located at 120° around the sample. The sinusoidal signal of strain, $\varepsilon_1(t)$, and stress, $\sigma_1(t)$, were fitted to the experimental data and used to calculate the norm of the modulus, $|E^*|$, and phase angle, ϕ_E, of the complex modulus (E^*) in the axial direction.

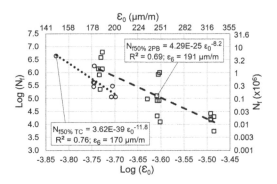

Figure 1. $N_{f50\%}$ data and Wöhler curves for TC and 2PB tests (10°C and 25 Hz).

4.3.3 Experimental program

Twenty-seven (27) specimens, coming from BSs, were tested in fatigue, with 8 cylindrical samples (TC: Ø75×h150 mm^2; LCMB) and 19 trapezoidal shape samples (2PB: b25×e25×B70×h250 mm^4; Eurovia-Fr). All tests were done at 10°C and 25Hz.

4.3.4 Test results

Figure 1 shows the Wöhler curves obtained for TC and 2PB tests based on $N_{f50\%}$. For the same strain amplitudes, 2PB tests give a higher fatigue life values than those obtained from TC tests. The strain value to fail at a million cycles, ε_6, is 191 μm/m for 2PB tests and 170 μm/m for TC tests. The fatigue slope of the Wöhler' law, a_2, from TC tests is higher than the one from 2PB tests (11.8 and 8.2, respectively). It is clear that fatigue life value is sensitive to the test method used when $N_{f50\%}$ failure criterion is used. Di Benedetto et al. (2004) reported higher N_f values from 2PB tests compared to TC tests. Several studies point out that the $N_{f50\%}$ criterion does not adequately describe the fatigue behavior of bituminous mixtures, especially when high polymer content is used (Perraton et al. 2018; Tapsoba et al. 2013).

Figure 2 shows the Wöhler curves obtained for TC and 2PB tests based on $N_{fII/III}$ failure criterion. Overall, the ε_6 are quite similar for both tests, and vary from 180 μm/m for TC to 186 μm/m for 2PB. In comparison to $N_{f50\%}$, slope value for TC tests drop from 11.8 to 7.4 and does not change for 2PB tests (8.2 and 8.1). Although 2PB still gives a higher life duration, the differences are less significant based on $N_{fII/III}$. Those results will serve to develop a unified bituminous fatigue characterization, regardless of the test methods used. Clearly, this assertion will have to be validated further with other fatigue tests.

4.4 Sensitivity study

The objective of the sensitivity study is to better anticipate HMAC thermomechanical changes with bitumen and filler dosage fluctuations.

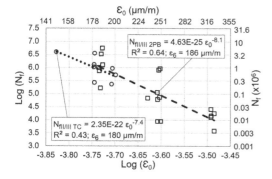

Figure 2. $N_{fII/III}$ data and Wöhler curves for TC and 2PB tests (10°C and 25 Hz).

4.4.1 Experimental program

From the control mix (M0), four bituminous mixes (M1 to M4) were tested by varying amounts (% mass) of bitumen (%b ± 0.25) and filler (%f ± 1.25). Variations were fixed in accordance to most critical cases that may occur in plant and allowed by quality standard. The compactibility, stiffness, low temperature, and rutting resistance of mixes were evaluated. Fatigue tests are still ongoing.

4.4.2 Test results

Table 1 presented the summary of the test results obtained for the thermomechanical properties of HMAC. Results show that a variation of the bitumen content (%b) does not significantly affect the stiffness ($|E^*_{(10 °C-10 Hz)}|$), the rutting resistance (rut$_{30,000;100,000cycles}$), and the low temperature cracking resistance (θ_{break_avg}) of mixes. In addition, the results show that the filler content (%f: particles <80 µm) does not have a significant impact on the measured properties. HMAC studied appears very promising from a manufacturing point of view since it does not require more attention during the manufacturing process than conventional hot mix asphalt (HMA). Moreover, the HMAC show good compaction under Superpave Gyratory Compactor (SGC) with an air voids content $V_a \leq 5\%$ at $N_{Design:\ 100\ gir.}$, exceptional resistance to rutting at 60°C (rut$_{100,000\ cycles} < 2.5\%$), low temperature cracking ($\theta_{break\_avg} \leq -27°C$), and high rigidity $|E^*_{(10°C-10\ Hz)}| > 16,000$ MPa.

5 TRIAL SECTION

Laboratory investigations only give a partial appreciation of a product's performance. Thus, field trials must be done. Phase II has two main objectives. Firstly, to quantify the cost-performance benefits of HMAC. To do that, we must clearly define the mechanics of the pavement design tools to input the HMAC properties for calculations. Secondly, to evaluate the on-site performance of structures used in cold regions incorporating the HMAC as asphalt base layer. Throughout

Table 1. Summary of the thermomechanical properties obtained for EME in sensibility study.

Parameters	Unit	Mixes M1	M2	M0	M3	M4
Filler (f)	%	−1.25	−1.25	0	+1.25	+1.25
Bitumen (b)	%	−0.25	+0.25	0	−0.25	+0.25
$V_{a\_SGC;10gir.}$	%	13	12	9	11	10
$V_{a\_SGC;100gir.}$	%	5	5	4	5	4
$V_{a\_SGC;200gir.}$	%	4	5	4	5	3
Rut$_{30,000cycles}$	%	N/M	N/M	1.9	1.1	1.2
Rut$_{100,000cycles}$	%	N/M	N/M	2.1	1.3	1.2
$\sigma^a_{break\_average}$	kPa	6.6	7.3	6.1	5.9	6.5
$\sigma_{break\_stand.deviation}$	kPa	0.3	1.2	0.6	0.9	0.7
$\theta_{break\_average}$	°C	−31	−28	−28	−27	−31
$\theta_{break\_stand.deviation}$	°C	2	3	1	5	3
$|E^*_{(10°C-10Hz)}|$	MPa	16,400	17,100	16,500	16,900	16,400
$E^b_{00\_2S2P1D}$	MPa	150	150	170	170	210
$E^{a,b}_{0\_2S2P1D}$	MPa	41,200	41,100	42,800	41,200	41,200
h^b_{2S2P1D}	−	0.400	0.400	0.435	0.400	0.435
δ^b_{2S2P1D}	−	2.45	2.40	2.45	2.40	2.45
$\tau^b_{E\_2S2P1D}$	−	0.2	0.2	0.2	0.1	0.2
$C^b_{1\_WLF}$	−	22.82	17.33	21.71	18.27	26.50
$C^b_{2\_WLF}$	−	174.07	142.64	164.04	157.80	204.73

[a]Samples from SS given a higher V_a content than those of BS. For this reason, some values have been adjusted, with a 95% confidence interval, to allow comparison (V_a fixed to 3%).
[b]The reference temperature is 15°C, k = 0.135 and $\beta = 9,000$ for 2S2P1D model and WLF equation.

the project, we will work to establish a field calibration coefficient (F_{CC}) for HMAC to link the laboratory scale to the road scale.

5.1 Pavement design simulations

To illustrate the benefit of the HMAC ($\varepsilon_{6,10 °C-25 Hz} \geq 130$ µm/m; $|E^*_{10 °C-10 Hz}| \geq 14,000$ MPa) at the pavement structure scale, M-EPD methods must be used. However, the M-EPD softwares available have their own models and calibration coefficients that are not directly compatible with each other (Huang 2004, NCHRP 2004, SETRA-LCPC 1997). Numerical simulations were done using three softwares (Alizé, PavementME, and OPECC) to compare different pavement designs with and without HMAC as base layer. The objective was to correlate the inputs of each software to facilitate the transition from one software to another.

M-EPD methods have a mechanistic component and an empirical component. The former includes the calculation of the stresses and strains in the pavement structure, and it considers the thermomechanical performances of the mix and the accumulation of damage. The empirical component directly refers to F_{CC}. The F_{CC} links the lab fatigue performance (N_f) to the on-site performance (observed fatigue cracking). The F_{CC} ensures that the fatigue cracking (FC) percentages predicted by the M-EPD method are accurate to the on-site measurements. To determine the F_{CC} of the HMAC, a trial section will be built and monitored according to the following steps.

5.2 Initial pavement design

An initial pavement structure was designed according to M-EPD methods. To monitor FC on the trial section, around 300,000 to 2,000,000 equivalent single axle loads (ESAL) should pass on the trial section. To facilitate the measurement of the cracks, the pavement structure should not have a wearing course. Also, every type of damage that is not FC damage (e.g. freeze-thaw and rutting damages), should be minimized. The HMAC's modulus and fatigue resistance used in the design are determined from the laboratory mix.

5.3 Construction of the trial section

The construction of the trial section is a crucial step that requires the utmost care because the F_{CC}'s reliability depends on the quality of the construction. The trial section is instrumented, and different gauges are used to validate the stresses and strains and to monitor the damage evolution. Also, field tests, such as topographic surveys, cores, plate loading tests, falling weight deflectometer (FWD) and Benkelman beam (BB) tests, will be conducted during the construction. Most importantly, the HMAC will be sampled in the paver to characterize the rheological behavior of the on-site mix and identify its As-Built properties.

5.4 Monitoring of the trial section

An important part of the laboratory work done here on HMAC is not useful if it is not linked with field behavior. Because of this, it is very important to monitor the trial section. This part of the monitoring consists of monitoring the trial section's behavior as the level of damage increases. It will include the monitoring of the temperature, structural rigidity, strains, but, most importantly, the evolution of the fatigue cracks, the FC. As of today, the FC monitoring remains a meticulous visual inspection. Each crack must be identified, classified and measured. For each on-site measurement campaign, a fatigue cracking percentage (FC_P%) will be calculated.

5.5 Example of application to fix F_{CC}

Figure 3 shows an example of the As-Built design FC_P% predictions and the trial section FC_P% measurements leading to the determination of the F_{CC}. Note that this example is based on the results of the reference section of the BioRePavation project led by IFSTTAR-Nantes on their accelerated pavement testing (APT) facility (Blanc et al. 2019). The F_{CC} should properly link the laboratory and on-site behaviors, and it allows a more efficient use of the M-EPD method.

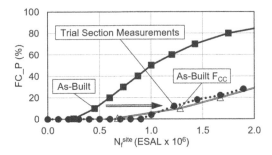

Figure 3. Application of the field calibration coefficient, F_{CC}, and the parameter of risk to obtain the As-Built F_{CC}.

6 CONCLUSIONS

The results presented in this paper are part of a collaborative research project done on the use of high modulus asphalt (HMAC) for cold region. More specifically, the aim of the work presented here was to characterize the fatigue law parameters and to evaluate the mixes sensitivity to variation in their productions. The fatigue study shows that the HMAC mixes have an excellent fatigue resistance, and similar Wöhler's parameters could be obtained from TC and 2PB tests. The sensitivity study reveals that the HMAC studied appears very promising from a manufacturing point of view. Laboratory investigation shows good to excellent thermomechanical properties of HMAC. Finally, it has been shown that the methodology proposed for determining the calibration coefficient is simple to apply and promising.

REFERENCES

Blanc, J., Hornych, P., Sotoodeh-Nia, Z., Williams, C., Porot, L., Pouget, S., …, Jimenez, A. 2019. Full-scale validation of bio-recycled asphalt mixtures for road pavements. Journal of Cleaner Production 227: 1068–1078.

Di Benedetto, H., de La Roche, C., Baaj, H., Pronk, A. & Lundström, R. 2004. Fatigue of bituminous mixtures, *Materials and Structures* 37(3): 202–216.

Huang, Y.H. 2004. *Pavement Analysis and Design*. Upper Saddle River: Pearson Education (US).

NCHRP. 2004. Guide for Mechanistic-Empirical Design of New and Rehabilitated Pavement Structures, Appendix II-1: Calibration of fatigue cracking models for flexible pavements. *Transportation Research Board, NRC*.

Perraton, D., Carter, A., Proteau, M., Lamothe, S. &, Pouteau, B. 2018. High Modulus Asphalt (EME) Mixes for Cold Climate: HPAC Ongoing Cooperative Research Project, Proceeding, *Canadian Technical Asphalt Association*, 59:1249–1268.

SETRA-LCPC. 1997. *French Design Manual for Pavement Structures*. [In French].

Tapsoba, N., Sauzéat, C. & Di Benedetto, H. 2013. Analysis of Fatigue Test for Bituminous Mixtures. *Journal of Materials in Civil Engineering* 25(6): 701–710.

Towards a mechanistic-empirical IRI prediction

H.S. Lee
Applied Research Associates, Champaign, IL, USA

ABSTRACT: International Roughness Index (IRI) is a widely accepted measure for pavement smoothness and is considered as one of the most important factors affecting the functional performance of pavement structures. As such, may researchers in the past have developed empirical IRI models to forecast future IRI as well as the remaining life of existing pavements. In this paper, a methodology for mechanistic-empirical IRI prediction is introduced. The methodology utilizes the pavement structure and the initial surface profile (rather than initial IRI) to predict the spatially varying dynamic load and rut depth values. The nonhomogeneous rut depth is then used to update the pavement surface profile from which the IRI is evaluated again. Following the description of the methodology, the paper provides a simple example comparing the mechanistic-empirical IRI results against those from an empirical model.

1 INTRODUCTION

Smooth pavements are not only recognized as good roadways by the driving public but also known to provide additional benefits such as increased fuel efficiency and improved safety (Robbins and Tran 2016). In addition, from a State Highway Agency's (SHA) point of view, a pavement initially constructed to be smooth not only remains smoother in the long term but also facilitates increased pavement life, reduced maintenance, and reduced life-cycle cost (Chatti & Zaabar, 2012; Massucco & Cagle 1999; Smith et al., 1997). Therefore, pavement smoothness measurements, such as those in terms of the International Roughness Index (IRI), are widely accepted as one of the most important factors affecting the structural and functional performance of pavement structures.

A vast number of efforts were made in the past to forecast the future IRI and to estimate the remaining life of in-service pavements. Most of these IRI prediction models were empirical in nature, with some of the frequently used equation forms shown in Table 1 (Qian et al. 2018).

The primary drawback of the above empirical IRI models is that pavement age and initial IRI (IRI_0) are the only independent variables. Although the coefficients of the above equations could be developed for various categories of pavements (e.g., based on thickness, foundation, region, traffic, etc.), they to not provide a direct link between IRI and other pavement distresses that may significantly affect the ride quality.

The IRI model adopted in the Mechanistic-Empirical (ME) Pavement Design Guide (i.e., AASHTOWare Pavement ME) was developed based on the principle that surface distresses have a significant effect on ride quality (AASHTO 2015). More specifically, the AASHTOWare IRI model for flexible pavements is a function of rut depth (RD), fatigue cracking (FC), transverse cracking (TC), and site factor (SF) which is another function of pavement age, precipitation, freezing index, and other material related variables:

$$IRI = IRI_0 + C_1(RD) + C_2(FC) + C_3(TC) + C_4(SF) \tag{1}$$

where C_i's are the calibration coefficients of the model.

Although the above AASHTOWare IRI model accounts for the surface distresses in IRI prediction, the nature of the above equation is still empirical, i.e., the IRI is predicted in the form of a regression equation.

In this paper, a methodology for mechanistic-empirical IRI prediction is introduced. The methodology is based on the spatially varying dynamic load simulated from vehicle-pavement interaction analysis and the rut depth (which also varies spatially) for updating the pavement surface profile. Then, a simple example is provided to illustrate the limitations of the empirical IRI models.

Table 1. Empirical IRI prediction forms.

Model Type	Functional Form
Exponential Function	$IRI = IRI_0 \cdot e^{(\beta \cdot Age)}$
Power Function	$IRI = IRI_0 \cdot Age^{\beta}$
Polynomial Function	$IRI = IRI_0 + \beta \cdot Age + \gamma \cdot Age^2$

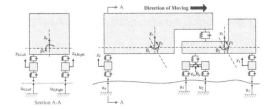

Figure 1. Three-Dimensional (3D) Truck-Trailer Model.

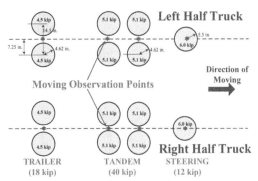

Figure 2. Configuration of Axle Weights and Moving Observation Points.

2 METHODOLOGY

2.1 Dynamic finite layer solution

In this study, a finite layer model known as ViscoWave (Lee 2013) was used for simulating the pavement responses under vehicle loads. ViscoWave was developed based on time-domain impulse loading & responses, and is capable of simulating the dynamic responses of elastic as well as viscoelastic layered structures. Although ViscoWave was originally developed for simulating the flexible pavement responses under impact loading (e.g., Falling Weight Deflectometer, FWD), the algorithm was recently enhanced for simulating the pavement response under moving loads (Lee & Steele 2017).

2.2 Moving frame analysis of pavement responses

ViscoWave allows for two different options for simulating the pavement responses under moving loads:

1. Fixed Point Analysis: This option is used for calculating the response at fixed observation points. In other words, the load is moving but the response is simulated at fixed points in the pavement that do not move (i.e., similar to an instrumentation embedded in the pavement).
2. Moving Frame Analysis: This option is used for calculating the response at observation points that move with the load. In other words, the response is simulated at fixed distances from the moving load (i.e, similar to the deflections measured from the Traffic Speed Deflection Devices, TSDD).

I n this study, the moving frame analysis was used for simulating the pavement responses under dynamic loading produced by the vehicle model described below.

2.3 Vehicle model and moving observation points

Figure 1 shows the vehicle model used for simulating the dynamic loading. As shown, the vehicle was a 3-dimentional truck & trailer system with a total of 14 degrees of freedom.

Figure 2 shows the static weights of each axle and tire of the above truck & trailer system. The figure also shows the locations of the moving observation points placed in each wheel path under each axle. A constant speed of 60 mph was used for simulation.

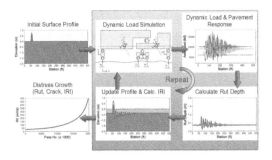

Figure 3. Schematic Procedure for Mechanistic-Empirical IRI Prediction.

2.4 Mechanistic-empirical model for IRI growth

As discussed previously, empirical IRI models are mostly simple functions of IRI_0, pavement age, and/or other pavement distresses. In other words, these empirical models do not involve the pavement profile from which the IRI is calculated. On the other hand, mechanistic-empirical IRI models generally involve (1) vehicle-pavement interaction analysis for simulation of dynamic loads, (2) calculating the pavement response and performance (e.g., rut depth) under the dynamic load that varies along the road, and (3) updating the pavement surface profile, and (4) re-evaluating the IRI based on the updated profile (Cebon 1999; Saleh et al. 2000). This procedure is schematically shown in Figure 3.

For this study, rutting within the Asphalt Concrete (AC) layer and within the subgrade layer was used for updating the pavement surface profile. The AC rutting (Rut_{AC}) was calculated using the AASHTOWare Pavement ME model (AASHTO 2015):

$$Rut_{AC} = \beta_1 k_z \varepsilon_{z,AC} 10^{k_1} n^{k_2 \beta_2} T^{k_3 \beta_3} \qquad (2)$$

where β_i's and k_i's are coefficients of the model and $\varepsilon_{z,AC}$ is the compressive strain at mid-depth of AC. The subgrade rutting (Rut_{SG}) was obtained from the Asphalt Institute (AI) model (Huang 1993):

$$Rut_{SG} = 1.365 \cdot 10^{-9} \varepsilon_{z,SG}{}^{-4.477} \qquad (3)$$

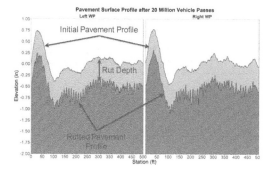

Figure 4. Initial Pavement Profile.

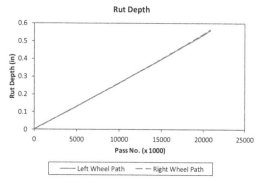

Figure 5. Average Rut Depth Predicted from Mechanistic-Empirical Model.

where $\varepsilon_{z,SG}$ is the compressive strain obtained at the top of the subgrade layer.

3 CASE STUDY EXAMPLE

The properties of a simple, 3-layer pavement structure used for this study are shown in Table 2. The AC layer was modeled as viscoelastic whose modulus depends on frequency (f). The temperature of the AC material was fixed at 70° F.

Figure 4 shows the initial pavement surface profile. The initial profile was obtained from the Long Term Pavement Performance (LTPP) section 04-0261 (in Arizona) after construction. The IRI values of the left and right wheel paths were 38.9 in/mi and 52.4 in/mi, respectively.

The above pavement structure (Table 2) and initial pavement profile (Figure 4) were used to simulate the IRI using the mechanistic-empirical model described above. The simulation was terminated after approximately 20 million passes of the truck & trailer system (Figure 1). To limit the required number of computations, the pavement profile was updated after every 50,000 passes of the vehicle.

Figure 4 also shows the rutted pavement profile after 20 million passes of the truck & trailer. The figure clearly shows that the rutted surface became rougher than the original pavement surface due to spatially varying rut depths.

Figure 5 shows the average rut depth in both the wheel paths that showed linear trends. This figure also shows that the average rut depth was approximately 0.5 in at the end of the mechanistic-empirical simulation.

The results of the mechanistic-empirical IRI prediction are shown in Figure 6. It is clearly demonstrated that the mechanistic-empirical IRI growth was relatively steady at the beginning of the simulation and until approximately 10 million passes of the truck & trailer system, after which the IRI started growing exponentially.

Also shown in Figure 6 are the empirical IRI values predicted from the AASHTOWare model [Equation (1)] that increase in a linear fashion and significantly

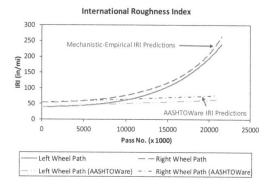

Figure 6. Predicted IRI.

underestimate the IRI from mechanistic-empirical simulation.

It should be noted that no environmental factors were introduced into the simulation. In other words, the SF value in Equation (1) is zero due to zero precipitation and zero freezing index. Although it is not shown in this paper for the sake of brevity, the fatigue damage calculated based on the horizontal strains at the bottom of the AC layer was less than 2.0 percent. As such, no fatigue cracking is expected at the surface of the pavement [i.e., $FC = 0$ in Equation (1)]. Furthermore, since the simulation was conducted at a constant temperature of 70°F, no transverse or thermal cracking is expected from the AASHTOWare model [i.e., $TC = 0$ in Equation (1)]. Consequently, Equation (1) simply becomes a linear function of average rut depth, as shown in Figure 6.

4 SUMMARY

Most of the existing IRI prediction models are empirical in nature. In other words, most of these models are simple regression functions that are based on initial IRI, pavement age, and or other surface distresses.

In this paper, a methodology for mechanistic-empirical IRI prediction has been introduced. The

Table 2. Pavement Layer Thickness and Modulus.

Layer / Parameter	Pavement Layer		
	AC	Base	Subgrade
Thickness (in.)	12.0	12.0	127.2
Modulus (ksi)	Frequency dependent dynamic modulus ($\|E*\|$) given as: $\log(\|E*\|) = 0.575 + \frac{2.918}{1+e^{-0.610-0.521\cdot\log(f)}}$ ($\|E*\|$ @ 10.0 Hz = 575.7 ksi)	25.5	8.4

methodology utilizes the pavement structure and the initial surface profile (rather than initial IRI) to predict the spatially varying dynamic load and the corresponding rut depth values. The nonhomogeneous rut depth is used to update the pavement surface profile from which the IRI is evaluated again.

A simple example has been provided to simulate the mechanistic-empirical IRI prediction. It was demonstrated that the mechanistic-empirical prediction may result in more realistic results under simple conditions (single temperature, constant speed, etc).

Nevertheless, it is the author's opinion that further research is needed before the mechanistic-empirical IRI prediction methodology can be implemented for routine, practical use: especially for the purpose of mechanistic-empirical pavement design. The primary reason behind this is that the initial pavement profile cannot be made available until the pavement is designed and constructed. It is also noted that mechanistic-empirical IRI prediction is time-consuming (compared to the empirical models) as the methodology involves repeated analyses of moving dynamic load, response, and performance (i.e., rut depth) simulations.

The mechanistic-empirical IRI prediction methodology introduced above was only based on spatially varying rut depth values. In other words, the effect of surface cracking on IRI was not addressed. Future research is needed to study the effect of crack width and depth on IRI before such a model can incorporate the effect of surface cracking. Furthermore, future research is also recommended to study the type of vehicle as well as its characteristics and parameters to used for mechanistic-empirical IRI prediction.

REFERENCES

AASHTO. 2015. Mechanistic-Empirical Pavement Design Guide, A Manual of Practice. American Association of State Highways and Transportation Officials. Washington, D.C.

Cebon, D. 1999. Handbook of Vehicle-Road Interaction. Swets & Zeitlinger Publishers. Netherlands.

Chatti, K. and Zaabar, I. 2012. Estimating the Effects of Pavement Condition on Vehicle Operating Costs. NCHRP Report 720. Transportation Research Board of the National Academies, Washington, D.C.

Flintsch, G., and McGhee, K. K. 2009. Quality Management of Pavement Condition Data Collection. NCHRP Synthesis No. 401, National Cooperative Highway Research Program (NCHRP), Transportation Research Board of the National Academies, Washington, D.C.

Huang, Y.H. 1993. Pavement Analysis and Design. Prentice Hall, NJ.

Lee, H. S. 2013. Development of a New Solution for Viscoelastic Wave Propagation of Pavement Structures and Its Use in Dynamic Backcalculation. Ph. D. Dissertation. Michigan State University.

Lee, H.S. and Steele, D. 2017. Dynamic Backcalculation of Asphalt Pavement Properties and Simulation of Pavement Response Under Moving Loads. Paper No. 17-01018. TRB 96th Annual Meeting Compendium of Papers, Transportation Research Board, Washington, D.C.

Lee, H.S. and Steele, D. 2018. Effect of Moving Dynamic Loads on Pavement Deflections and Backcalculated Modulus. Paper No. 18-01768. TRB 97th Annual Meeting Compendium of Papers, Transportation Research Board, Washington, D.C.

Massucco, J., and Cagle, J. 1999. Getting Smoother Pavement: An Arizona Success Story That's Adaptable Nationwide. Public Roads, Vol. 62, No. 5, FHWA, U.S. DOT.

Qian, J., Jin, C., Zhang, J., Ling, J, and Sun, C. 2018. International Roughness Index Prediction Model for Thin Hot Mix Asphalt Overlay Treatment of Flexible Pavements. Transportation Research Record, Vol. 2672, No. 40, pp. 7–13.

Robbins, M.M. and Tran, N. 2016. A Synthesis Report: Value of Pavement Smoothness and Ride Quality to Roadway Users and the Impact of Pavement Roughness on Vehicle Operating Costs. NCAT Report 16-03. National Center for Asphalt Technology, Auburn, AL.

Saleh, M.F., Mamlouk, M.S., and Owusu-Antwi, E.B. 2000. Mechanistic Roughness Model Based on Vehicle-Pavement Interaction. Transportation Research Record, Vol. 1699, pp. 114–120.

Smith, K. L., Smith, K. D., Evans, L. D., Hoerner, T. E., and Darter, M. I. 1997. Smoothness Specifications for Pavements. NCHRP Web Document 1. Transportation Research Board of the National Academies, Washington, D. C.

Advances in Materials and Pavement Performance Prediction II – Kumar et al. (eds)
© 2021 Taylor & Francis Group, London, ISBN 978-0-367-46169-0

Overview of pavement vehicle interaction research at the MIT concrete sustainability hub

J. Mack
CEMEX USA, Houston, Texas, USA

M. Akbarian, F.J. Ulm, Randolph Kirchain & Jeremy Gregory
Department of Civil and Environmental Engineering, Massachusetts Institute of Technology, Cambridge, MA, USA

A. Louhghalam
Department of Civil and Environmental Engineering, University of Massachusetts Dartmouth, Dartmouth, MA, USA

ABSTRACT: In 2009, the US cement and concrete industries established the Concrete Sustainability Hub at the Massachusetts Institute of Technology. A primary thrust of MIT's activities has been improving the Life Cycle Assessment practices to better quantify the environmental impacts over the life of a pavement. In their research, the MIT CSHub determined that the "use phase," and specifically Pavement Vehicle Interaction (PVI) has a very large impact on a Pavement's sustainability aspects. This paper will summarize the CSHub PVI research findings to date.

1 INTRODUCTION

In 2009, the United States cement and concrete industries established the Concrete Sustainability Hub (CSHub) at the Massachusetts Institute of Technology (MIT) to carry out a multi-year research program to evaluate and improve the environmental and economic impact of concrete in pavements and buildings. The goal of the effort is to develop more sustainable pavement infrastructure and buildings by (1) providing scientific basis for informed decisions; (2) demonstrating the benefits of a life-cycle view; and (3) transferring research into practice.

One focus area of MIT's pavement's research has been on improving the Life Cycle Assessment (LCA) practices to quantify the environmental impacts, energy consumption, material use, etc. throughout a pavement's life-time. While LCA's can be used to evaluate the environmental impact of a single product (e.g., a pavement) to reduce the impact of that product, they can also be used to compare two different options for a product (e.g. concrete and asphalt in pavement design) in the same way that life cycle cost analysis is used to compare costs. As such, it is important that the process be as comprehensive as possible to truly reflect the environmental impacts of a given pavement.

While the mechanics of performing a pavement LCA are not terribly difficult, it is extremely data intensive and it is essential that a standardized LCA

framework that includes the raw material production (extraction, processing, and manufacturing), construction, use, maintenance, and disposal or end of life phases be used. (Santero et al. 2011). This framework ensures that short term gains in the early stages do not come at the expense of long-term deficits at later stages.

In doing their research, and in reviewing results from others, MIT determined that the LCA results will always vary based on the context for the given scenario and are dependent on the pavement structural design (thickness and materials); the maintenance and rehabilitation treatments used; the anticipated traffic; and the environment that the pavement will be located.

Having said that, MIT did determine that while the overall impact varies based on context, the "use phase" almost always plays a major role (;APA 2004; MIT 2014), and is often times much larger than the other phases combined. MIT also found that a large portion of a use-phase environmental impacts come from the emissions by vehicles using the pavement due to excess rolling resistance between the pavement and the vehicle, which increases the fuel usage. This is known as pavement-vehicle interaction or PVI. The goal of the paper is highlight MIT's research findings on PVI for the last 9 years and show how it can be used and compared against the life cycle GHG emissions from the other pavement life cycle phases.

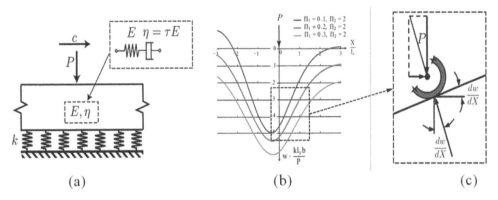

Figure 1. Overview of MIT Deflection-induced PVI Excess Fuel Consumption Model.

2 OVERVIEW OF PAVEMENT-VEHICLE INTERACTION

Previous research has shown that the three primary factors that contribute to PVI are (1) surface texture, (2) roughness or smoothness, and (3) deflection or dissipation of the pavement. In the US, roughness and deflection are considered the main contributors to pavement vehicle interaction. As there has been considerable study on roughness (Chatti 2012; Gillespie et al. 1980; Louhghalam et al. 2015; Sayers et al. 1986; Zaabar & Chatti 2010); MIT CSHub research focused on the deflection and the impact that the structural and material properties of the pavement, and how they change over time, effect fuel consumption and GHG emissions. This would then allow PVI Deflection to be used in project specific LCAs to compare the phase PVI impacts to the other phases in an LCA trade-off analysis.

3 PAVEMENT DEFLECTION MODELLING AND ITS IMPACTS ON EXCESS FUEL CONSUMPTION

Pavement deflection refers to the small dent that a vehicle creates in the pavement as it drives on a roadway. This deflection creates a slight, but perpetual uphill climb under the tire that results in a resisting force to the vehicle's motion. To maintain a constant speed, the vehicle must compensate for the added resistance by consuming excess fuel, whose magnitude depends on the steepness of the hill and is a function of the condition and structure of the pavement and the weight of the vehicle. Note that excess fuel consumption (EFC) is defined as the additional fuel consumption compared with a pavement that is perfectly smooth, rigid, and does not bend (i.e. non-dissipative).

It is important to note is that while the effect of PVI on an individual vehicle is small, its impact within a full pavement life cycle can be significant due to the large number of vehicles that travel over pavements.

This is especially true for high volume, heavy traffic roadways where PVI can easily surpass the energy consumption and emissions due to the materials, construction and maintenance phases of the roadway over its lifetime.

MIT's final PVI model shows that the dissipated energy in a viscoelastic pavement is directly related to the slope under the moving wheel (Akbarian 2012; Louhghalam et al. 2013; Louhghalam et al. 2014) and is conceptually a 2-step model like shown in Figure 1. The first part determines the slope of a pavement system based on its structure and material properties (the *stiffness* of the system) and the second part relates the rolling resistance (i.e. dissipated energy) needed to drive up this slope to EFC. To quantify the slope under a moving wheel, the pavement is modelled through an infinite viscoelastic beam over an elastic foundation (Figure 1(a)). A Maxwell model is used to describe the viscoelastic material's stress and strain relationship, where E is the Young's modulus, τ is the relaxation time of the viscoelastic top layer that accounts time of loading and temperature dependencies (i.e. cool vs hot, slow vs fast), and η is the material viscosity.

Figure 1(b) shows the pavement deflection basin, where the maximum deflection is behind the moving wheel due to the delayed viscoelastic response. The deflection basin depth (w) and wavelength (l_s) that define the grade slope are a function of the load and the pavement structure / material properties at the time of loading. Once the gradient slope is determined, the EFC of the vehicle is in direct relationship with the resistance, or excess energy, δE, that the vehicle must overcome to travel up the slope as shown in Figure 1(c).

Once MIT developed their theoretical model, they developed a small-scale, desk-top experiment that allowed them to calibrate and partially validate their model by isolating the interaction of the wheel and the pavement structure. The pavement system, a two-layered viscoelastic beam on an elastic subgrade, (Figure 2(a)) was experimentally represented through a two-layered silicone elastomer pavement shown in Figure 2(b). This experimental setup allowed for a range of top layer thicknesses, elastic moduli,

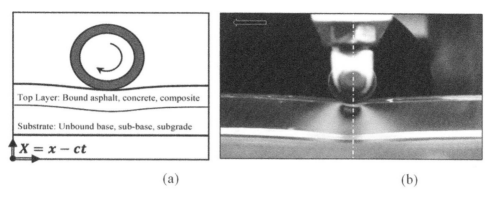

| (a) | (b) |

Figure 2. Link between the (a) Theoretical Model and (b) Desktop Experiment and photoelasticity showing the wheel moving left.

viscoelastic properties, and loading conditions to be tested. The use of polymers also allowed MIT to visually observe the pavements' response to the moving wheel and the resulting resisting horizontal force through technique called photoelasticity. The asymmetry of stresses inside the material shows that the wheel is always moving "up a hill."

In total, MIT ran nearly 200 experimental configurations, equivalent to 290 km (180 miles) of road testing, to investigate the impact of key PVI parameters on excess energy dissipation with varying loads, speeds, pavement modulus values, pavement thicknesses and viscoelasticities (relaxation time). Once the testing was completed, the scaling relationship between the theoretical deflection-induced PVI model were compared to the scaling of the dissipation forces from PVI desk-top model and the results were found to be consistent. This confirmed the contribution of pavement structure to EFC and that an increase in pavement stiffness minimizes the impact of deflection-induced PVI. It also verified that deflection-induced PVI impacts can be captured and can have a significant impact on life cycle energy use and emissions. (Mack et al. 2018)

4 APPLICATION OF PVI AT THE PROJECT LEVEL

If an agency wants to lower the PVI EFC/GHG emissions of a pavement system at the project level; they have two primary strategies that can be used:

1. Build and Maintain Smoother Pavements either by improving the maintenance activities that keep a pavement smooth over its lifetime, or by building pavements that stay smoother longer
2. Build stiffer, or stiffen the existing pavement so that it deflects less

In general, the relative importance "smoothness" and "stiffness" depends on the roughness conditions, traffic volumes and the specific pavement structures being evaluated. Improving the smoothness has a greater impact when the road is old, rough, and/or in need of repair, while stiffness is a fairly constant value and has a larger relative impact when the pavement is new or smooth, and when there is a large amount of trucks (light vehicles such as cars do not weigh enough to cause a large deflection). It is also important to note that PVI roughness impacts are primarily a function of how the agency maintains the road. If an agency keeps both pavement systems at the same smoothness level, the roughness based PVI impacts will be the same. Note pavement type does impact the rate of pavement deterioration, but the agency still decides when to maintain the pavement.

In contrast, stiffness impacts are dependent on pavement type and will be essentially constant for that given pavement design, material properties, and traffic volume. For concrete pavements, which are very stiff, PVI deflection is not a major factor in any traffic mix or pavement condition. However, for asphalt and composite structures, the PVI deflection impacts can be more than 2 orders of magnitude (100 times) higher when compared to concrete (Mack et al. 2018). It is important to note that while the magnitude of the PVI deflection impacts for asphalt and composite pavements will change as the visco-elastic properties of the asphalt material changes, the overall trend does not.

Often times, it is stated that if agencies just keep their pavements smooth, they will reduce EFC and have more sustainable pavements. While keeping the pavements smooth does help, the fact is that both smoothness and structure play a major role in lowering PVI impacts and for a given roadway condition (smoothness level) requirement, it will be the deflection PVI component that differentiates pavements use phase PVI GHG emissions.

As an example, Figure 3 shows a comparison of two equivalent asphalt and concrete pavements and the lifetime EFC for two different scenarios. The first scenario is a "typical smoothness scenario" and the second is an "enhanced smoothness scenario." On the left side of Figure 3 are the projected IRI;s for a local highway in Missouri using the AASHTO Pavement ME Design

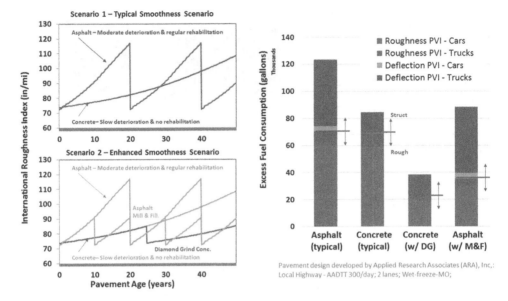

Figure 3. Contributions to EFC from roughness and deflection based PVI for Local highway in Missouri. Equivalent Asphalt and Concrete Pavement Designs, AADTT 300/day; 2 lanes; wet-freeze.

Procedure (MIT 2014) for both scenarios (typical is top and enhanced is bottom). In the top, typical scenario, the asphalt (red line) has moderate deterioration and is rehabilitated every 20 years to keep the road at a low level of roughness. The concrete pavement (purple line) has a slow rate of deterioration and provides a similar level of lifetime smoothness as the asphalt without rehabilitation. In the bottom graph, to maintain an even higher level of smoothness, the pavements are rehabilitated by applying a diamond grinding (blue line) to the concrete pavement at year 25, or a mill-and fill asphalt overlay (orange line) to the asphalt pavement every 10 years in order to lower the IRI.

On the right is the projected EFC for each pavement and scenario based on the PVI source of the EFC (roughness from cars, roughness from trucks, deflection from cars, and deflection from trucks). If one looks only at the roughness based PVI impacts on the typical scenario, one can see that they are basically the same for both pavements because the overall roughness (area under the curve) is similar.

However, once PVI structural EFC is included, there is a dramatic difference in the overall EFC for the concrete and asphalt structures due to concrete's increased stiffness. Likewise, on the enhanced smoothness scenario, one can see that while the roughness PVI EFC, and overall PVI are decreased substantially, the EFC from the deflection PVI impacts are unaffected by the IRI changes. In fact, for this case, the overall EFC emissions of the asphalt are higher than the concrete without diamond grinding solely due to the structural PVI impacts. The takeaway is that structure/deflection also plays a dominant role when the pavement is already smooth; and if one ignores the structural PVI

impacts, the conclusions about the sustainability of the different scenarios could be flawed.

5 SUMMARY

In 2009, the US cement and concrete industries established the Concrete Sustainability Hub at MIT to find breakthroughs that will lead to more sustainable and durable pavement infrastructure and buildings. With regards to pavements, MIT found that the "use phase," and specifically PVI plays a substantial role in the environmental impact of given pavement design, and that these impacts are often higher than the impacts associated with the pavement materials and construction.

With respect to this, MIT developed a mechanistic based PVI deflection model to determine excess fuel consumption of a pavement system based on its structure and material properties (the *stiffness* of the system), and then used a small-scale, desk-top experiment to partially calibrate and validate the PVI-Deflection (structure) model. MIT then applied the model at the project level and determined that while both smoothness and structure play a major role in lowering PVI impacts; for a given roadway condition (smoothness level) requirement, it will be the deflection PVI component that differentiates pavements use phase PVI GHG emissions.

REFERENCES

Akbarian M., M.-A. S.-J. (2012). Mechanistic Approach to Pavement-Vehicle Interaction and Its Impact on

Life-Cycle Assessment,. Transportation Research Record: Journal of the Transportation Research Board, No. 2306

Chatti, K. Z. (2012). Estimating the Effects of Pavement Condition on Vehicle Operating Costs, Project 1-45. National Cooperative Highway Research Program, Report 720.

EAPA (2004). Environmental Impacts and Fuel Efficiency of Road Pavements. Task Group Fuel Efficency.

Gillespie, T. D., Sayers, M., & and Segel, L. (1980). Calibration of Response-Type Road Roughness Measuring Systems.

Louhghalam, A., Akbarian, M., & Ulm, F.-J. (2013). Flügge's conjecture: dissipation-versus deflection-induced pavement–vehicle interactions. Journal of Engineering Mechanics 140.8(04014053).

Louhghalam, A., Akbarian, M., & Ulm, F.-J. (2014). Scaling Relationships of Dissipation-Induced Pavement–Vehicle Interactions. Transportation Research Record: Journal of the Transportation Research Board, No. 2457

Louhghalam, A., Akbarian, M., & Ulm, F.-J. (2015). Roughness-Induced Pavement–Vehicle Interactions: Key Parameters and Impact on Vehicle Fuel Consumption. Transportation Research Record: Journal of the Transportation Research Board 2525

Louhghalam, A., Tootkaboni, M., & Ulm, F.-J. (2015). Roughness-Induced Vehicle Energy Dissipation: Statistical Analysis and Scaling. Journal of Engineering Mechanics 141.11(04015046).

Mack, J. W., Akbarian, M., Ulm, F. J., & Louhghalam, A. (2018). Overview of Pavement-Vehicle Interaction Related Research at the MIT Concrete Sustainability Hub. Proceedings of the 13th International Symposium on Concrete Roads 2018. Berlin Germany

MIT. (2014). Scenario Analysis of Comparative Pavement Life Cycle Assessment Using a Probabilistic Approach and Supplementary Information for Comparative Pavement Life Cycle Assessment and Life Cycle Cost Analysis. MIT Concrete Sustainability Hub.

Santero, N., Masanet, E., & Horvath, A. (2011). Life-cycle assessment of pavements. Part I: Critical review. Resources, Conservation, and Recycling. 55(9–10)

Sayers, M. W., Gillespie, T. D., & Paterson, W. (1986). Guidelines for the Conduct and Calibration of Road Roughness Measurements. Washington, DC: The World Bank.

Zaabar, I., & Chatti, K. (2010). Calibration of HDM-4 Models for Estimating the Effect of Pavement Roughness on Fuel Consumption for U.S. Conditions. Transportation Research Record: Journal of the Transportation Research Board, No. 2155

Advances in Materials and Pavement Performance Prediction II – Kumar et al. (eds)
© 2021 Taylor & Francis Group, London, ISBN 978-0-367-46169-0

Parametric and sensitivity analysis of PVI-related surface characteristics models for fuel consumption

K. Mohanraj & D. Merritt
The Transtec Group Inc., Austin TX, USA

N. Sivaneswaran
Turner-Fairbank Highway Research Center, Federal Highway Administration (FHWA), McLean, VA, USA

H. Dylla
Federal Highway Administration (FHWA), Washington DC, USA

ABSTRACT: Pavement surface characteristics and the associated pavement vehicle interaction (PVI) affects vehicle operating costs (VOC) and excess fuel consumption (EFC). This paper summarizes an evaluation of six current PVI models that consider pavement roughness and/or texture through a parametric study of the model inputs. The intent of the study was to gain a better understanding of the sensitivity of the different models to the various PVI input parameters, and to further assess the impacts of pavement surface characteristics on EFC. In general, all models indicated that increase in roughness, texture, speed and vehicle weight results in increased fuel consumption. Exceptions are noted along with some unidentifiable contradictions that may be associated with the empirical nature of some of the models.

1 INTRODUCTION

Highway agencies are increasingly interested in utilizing pavement Life-Cycle Assessment (LCA) to analyze and quantify the environmental impacts of project and network level investment decisions. These efforts are intended to help steer decisionmaking processes towards strategies that are more sustainable. The VOC and the EFC associated with PVI after construction of a pavement, and over its service life, may be estimated using existing mechanistic and/or empirical models. Agencies may consider these costs while making investment decisions or in evaluating strategies during Life-Cycle Costs Analysis (LCCA) or the use-phase of pavement LCA.

Several studies have investigated factors influencing rolling resistance influencing such as pavement roughness, macrotexture, and structural responsiveness to develop models to predict VOC and EFC. The pavement surface characteristics considered as influential on rolling resistance are pavement roughness/unevenness and macrotexture.

This paper discusses an evaluation of six current PVI models that consider pavement roughness and/or texture through a parametric study of the model inputs. The primary intent of the study was to understand the sensitivity of the models to the various input parameters and to further assess the impacts of pavement surface characteristics on EFC (expressed in volume per distance). Based on the parametric study, the paper summarizes relevant suggestions related to the use of these models and the range of input parameters, both pavement-related (roughness and texture) and others (temperature, wind speed, vehicle speed and type, season), to be considered during implementation of these models to support investment decisionmaking.

2 BACKGROUND

PVI models were identified from a literature review and evaluated by documenting the model inputs, assumptions, gaps, and limitations considered during the development of the models. Six surface characteristic models were selected for further evaluation primarily based on availability and ease of use of the models for decisionmaking process by state highway agencies (SHAs).

The surface characteristics models selected are listed below:

NCHRP 1-45 model – This is the Highway Development and Management (HDM-4) model calibrated by Michigan State University (MSU) for U.S. conditions and is empirical in nature.

MIRIAM model – This is an empirical model developed in Europe as part of the Vejstandard Og Transport Omkostninger (VETO) program.

MIT Model – This is a mechanistic model developed by Massachusetts Institutes of Technology's (MIT) Concrete Sustainability Hub and funded by the Portland Cement Association.

NRC Phase III – These are a set of empirical models developed by National Research Council of Canada from field tests covering various seasons and pavement types using a passenger car and a heavy truck.

UIUC RSI Model – The University of Illinois at Urbana-Champaign (UIUC) developed this regression model using HDM-4 and Environmental Protection Agency's Motor Vehicle Emission Simulator (EPA MOVES) to evaluate roughness speed impacts (RSI).

New HERS model – The Federal Highway Administration's Highway Economic Requirement System (FHWA HERS) model was recently updated by the University of Nevada at Reno and Nevada Automotive Test Center using physics-based simulation and modelling.

3 STUDY FRAMEWORK

The input parameters identified as common to all surface characteristics PVI models were roughness and speed. Additionally, the NCHRP 1-45 model and the MIRIAM models consider texture as an input, and the NCHRP 1-45 and NRC Phase III models take in to account ambient/pavement temperature. Wind speed is an input for the NRC Phase III model only. The effects of these model-specific inputs (temperature, texture, wind speed) were evaluated separately before performing the parametric analysis on the parameters common to all models (roughness and speed).

Table 1 shows the range and interval of the input parameters that were varied in the study. The roughness input for all models is in the form of International Roughness Index (IRI) except for the MIT model that allows the use of Power Spectral Density (PSD) components of waviness number and unevenness index, or alternatively, IRI combined with an assumed value for waviness number. For the new HERS model, the roughness input is a range of IRI categorized as smooth to rough.

The NCHRP 1-45 model considers ambient temperature, whereas the NRC Phase III model allows ambient or pavement temperature without differentiating and allows them to be used interchangeably as an input.

The wind speed input in the NRC model considers an average wind speed relative to the direction of the vehicle with tail wind being positive and head wind being negative.

There are six seasonal models (winter, spring, summer day, summer night, fall, and all season) for trucks and two seasonal models (winter and summer) for passenger cars in the NRC models. Each of these models have an input to specify pavement type (flexible, rigid and composite) as well.

Table 1. Parametric study setup.

All Models						
Roughness, IRI (m/km)	1	2	3	4	5	
Speed (km/h)	30	50	70	90	110	120
NCHRP 1-45 and MIRIAM Models ONLY						
Texture, MPD (mm)	0.5	1	1.5	2	2.5	3
NCHRP 1-45 and NRC Phase III Models ONLY						
Temperature (°C)	−10	0	10	20	30	40
NRC Phase III Model ONLY						
Wind Speed (km/h)	−30	−20	−10	10	20	30

Table 2. Baseline values for input.

Input Parameter	Baseline/Default value
Roughness	1 m/km
Speed	70 km/h
Texture	1 mm
Temperature	25○C
Wind Speed	0 km/h

The effects of input parameters on fuel consumption were checked by varying the parameter per the increments shown in Table 1 and using baseline values shown in Table 2 for the remaining inputs. For NCHRP 1-45, MIRIAM, NRC Phase III, and the new HERS models, the inputs for roadway characteristics such as super elevation, curvature and grade were set to zero to simplify the analysis.

The vehicle weights used for each of the models were taken from supporting literature. Vehicle properties such as aerodynamic coefficient, projected frontal area and other relevant properties are incorporated in the models for limited vehicle types.

4 RESULTS

The results of the parametric analysis for each of the models are summarized below, and as an example, Figure 1 is shown for the trends observed in NCHRP 1-45 model only:

4.1 NCHRP 1-45 model

Temperature Effects: Fuel consumption decreases with increase in temperature at higher speed (100km/h). However, at lower speed (50 km/h) the rate of decrease is lower. Decrease at various speeds were approximately 0.1 to 0.2% per 1°C increase in temperature.

Texture Effects: Fuel consumption increases with increase in texture depth (mean profile depth, MPD).

For lighter vehicles (passenger cars and SUVs), this increase is approximately 0.5% per 1 mm increase in texture depth. For heavier vehicles, the increase is approximately 2% per 1 mm.

Speed Effects: Changes in fuel consumption with increase in speed varies and also depends on vehicle type (weight). For lighter vehicles and trucks, fuel consumption decreases with increase in speed up to 50 km/h and increases thereafter at an approximate rate of 1 to 2% per 1 km/h increase in speed. For heavy trucks fuel consumption increases with increase in speed at an approximate rate of 1.5% per 1 km/h increase in speed.

Roughness Effects: Fuel consumption increases at an approximate rate of 1 to 2% per 1 m/km increase in roughness.

4.2 MIRIAM model

Texture Effects: Fuel consumption increases with increase in texture depth for all vehicle types. For lighter vehicles such as passenger cars, the increase is approximately 2.5 to 3% per 1 mm increase in texture depth. For heavier trucks, the increase is approximately 5 to 8% per 1 mm.

Speed Effects: Fuel consumption increases approximately 0.5 to 1% per 1 km/h increase in speed, depending on vehicle type (weight).

Roughness Effects: Fuel consumption increases at an approximate rate of 0.2 to 3% per 1 m/km increase in roughness. The increase in fuel consumption due to roughness is a function of speed and vehicle type.

4.3 MIT model

Speed Effects: Fuel consumption increases with speed at a rate of 1% per 1 km/h increase in speed. And this impact of speed, with increase in roughness by 1m/km, the fuel consumption increases at an approximate rate of 1 to 2%.

Roughness Effects: The rate of increase in fuel consumption decreases at higher roughness levels. For any level of roughness, the percent rate of increase in fuel consumption depends on the roughness alone and is independent of vehicle type.

4.4 NRC phase III models

Wind Speed Effects: Fuel consumption increases with increase in headwind or decrease in tailwind speed. The increase is approximately 0.2 to 1.7% per 1 km/h increase in wind speed at 70 km/h speed, and the rate depends on both the vehicle speed and type.

Temperature Effects: Fuel consumption changes with increase in temperature and depends on vehicle speed and type. Fuel consumption increases at an approximate rate 0.5% per 1°C increase in temperature for passenger cars and decreases at an approximate rate 1.1% per 1°C increase in temperature for heavy

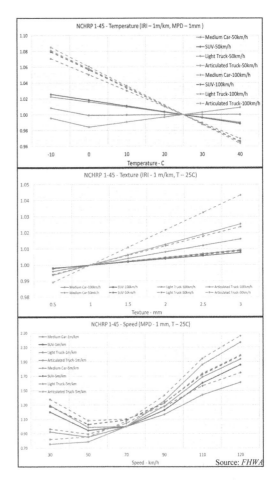

Figure 1. NCHRP 1-45 change in fuel consumption with roughness, texture, speed, temperature, and vehicle type.

trucks (except in winter, the trend is reversed and at an approximate rate 0.5% per 1 °C).

Speed Effects: Fuel consumption variation with speed depends on vehicle weight. Fuel consumption decreases at an approximate rate of 0.4% per 1 km/h increase in speed for light vehicles and trucks and increases at an approximate rate of 2.4% per 1 km/h increase in speed for heavy trucks.

Roughness Effects: Fuel consumption changes with increase in roughness and depends on the vehicle type. Fuel consumption decreases at an approximate rate of 0.9% per 1 m/km increase in roughness for passenger cars and increases at an approximate rate of 2.5% per 1 m/km increase in roughness for heavy trucks.

Seasonal Effects: The impact of season was analyzed on loaded trucks at a speed of 70 km/h at 25°C. The winter model was excluded from the analyses. For any pavement type, the minimum fuel consumption is from the summer day model and maximum in the fall models. The increase in fuel consumption from the summer day model to fall model is approximately 30 to 40% for loaded trucks, depending on pavement type.

Pavement Type Effects: Pavement type was analyzed for loaded trucks at a speed of 70 km/h at 25°C. In general, the minimum fuel consumption was observed for composite pavements, except for the summer day model for which rigid pavements exhibited the minimum fuel consumption. The maximum fuel consumption for flexible pavement is from the all-season, spring, and summer day models, whereas for rigid pavement, it is from the summer night and fall models.

4.5 UIUC RSI model

Speed Effects: Fuel consumption decreases up to a speed of 70 km/h and increases thereafter at an approximate rate of 0.1 to 0.4% per 1 km/h increase in speed.

Roughness Effects: An increase in roughness by 1 m/km increases fuel consumption from approximately 0.4 to 4.2% and increases with speed and depends largely on the vehicle type. The effects of roughness are greater on passenger cars compared to those on light to heavy trucks.

4.6 New HERS model

Speed Effects: Fuel consumption changes at an approximate rate of 0.1 to 1.2% per 1 km/h change in speed, depending on the vehicle type.

Roughness Effects: Fuel consumption increases by approximately 0.4 to 4.1% per increase in roughness category and increases with speed and depends largely on the vehicle type. The effects of roughness are greater on heavy trucks compared to those on lighter trucks or passenger cars.

5 CONCLUSIONS

The relevant conclusions to aid with the range to be considered for input parameters in the models, both pavement-related (roughness and texture) and other (temperature, wind speed, vehicle speed and type, season) are summarized below:

The NCHRP 1-45 and MIRIAM models both indicate that texture must be considered when evaluating fuel consumption at lower speed for heavier vehicles. The impact of texture reduces with increase in speed or when considering lighter vehicles such as passenger cars.

In general, all models indicate that with increase in roughness, speed, and vehicle weight, there is a greater impact on fuel consumption. However, there were some exceptions noted though the influencing factors for contradictions were not easily identifiable due to empirical nature of the models. Therefore, a wide range of roughness at various speeds and vehicle weights need to be considered in evaluating fuel consumption due to roughness.

When considering the impact of speed, although the various models indicate different trends, evaluating fuel consumption at a constant speed eliminates the need to account for the effect of change in speed. Therefore, it is suggested that the fuel consumption using each model be evaluated for more than one type of roadway such that a range of maximum speed limits are included.

Similar to speed, the impact of vehicle type was also observed to vary from model to model. It may be ideal to consider a wide range of vehicles when evaluating fuel consumption.

The NCHRP 1-45 model indicated that impact of temperature is more significant at higher speeds. The NRC model indicated opposite trends for cars and trucks, wherein the trucks trended similar to NCHRP 1-45 model and indicated the impact of temperature to be greater at higher speeds. Based on this, it may be concluded that the impact of temperature needs to be considered while using these models, particularly for both low and high temperatures at highway speeds (100 km/h and higher).

The NRC models suggests that the effect of wind speed must be accounted for lighter vehicles at higher speeds and heavier vehicles at lower speeds. Based on the results and the opposite trends, it may be necessary to consider wind speeds at all speeds for all vehicle types. The NRC models also indicate that season and pavement type can influence fuel consumption, however the empirical nature of the models make it difficult to identify the trends and suggest that a wide range of seasons and all pavement types must be considered in studying fuel consumption

REFERENCES

Chatti, K., & Zaabar, I. (2012). Estimating the Effects of Pavement Condition on Vehicle Operating Costs. NCHRP Report 720. National Academies Press.

Hajj, E. Y., Xu, H., Bailey, G., Sime, M., Chkaiban, R., Kazemi, S. F., and Sebaaly, P. E. (2017). "Enhanced Prediction of Vehicle Fuel Economy and other Operating Costs, Phase I: Modeling the Relationship between Vehicle Speed and Fuel." Technical Report, Federal Highway Administration, Washington, DC.

Hajj, E. Y., Sime, M., Chkaiban, R., Bailey, G., Xu, H., and Sebaaly, P. E. (2018). "Enhanced Prediction of Vehicle Fuel Economy and other Operating Costs, Phase II: Modeling the Relationship between Pavement Roughness, Speed, Roadway Characteristics and Vehicle Operating Costs," Technical Report, Federal Highway Administration, Washington, DC.

Karlsson, R., Carlson, A., & Dolk, E. (2012). Energy use generated by traffic and pavement maintenance: decision support for optimization of low rolling resistance maintenance treatments. Vti Notat, pp 43.

Louhghalam, A., Tootkaboni, M., & Ulm, F. J. (2015). Roughness-Induced Vehicle Energy Dissipation: Statistical Analysis and Scaling. Journal of Engineering Mechanics, 137(July), 826–833.

Louhghalam, A., Tootkaboni, M., Igusa, T., & Ulm, F. J. (2018). Closed-Form Solution of Road Roughness-Induced Vehicle Energy Dissipation. Journal of Applied Mechanics, 86(1), 011003.

Sandberg, U. (2011). Rolling resistance-Basic Information and State-of- the-Art on Measurement methods. Report from Models for rolling resistance in Road Infrastructure Asset Management systems (MIRIAM).

Sandberg, U., Bergiers, A., Ejsmont, J. A., Goubert, L., & Karlsson, R. (2011). Road surface influence on tyre/road rolling resistance. Report from Models for rolling resistance In Road Infrastructure Asset Management systems (MIRIAM).

Taylor, G. W., & Patten, J. D. (2006). Effects of Pavement Structure on Vehicle Fuel Consumption – Phase III.

Advances in Materials and Pavement Performance Prediction II – Kumar et al. (eds)
© 2021 Taylor & Francis Group, London, ISBN 978-0-367-46169-0

Evaluation of pavement service life reduction in overload corridors

Ali Morovatdar & Reza S. Ashtiani
The University of Texas at El Paso, Texas, USA

ABSTRACT: Texas has experienced a big boom in energy-related activities such as natural gas and crude oil production since 2008. These activities have created large volumes of Over-Weight (OW) truck operations that have adversely affected the longevity of the pavement infrastructures. This was the motivation to devise a framework for the mechanistic characterization of the Pavement Life Reduction (PLR) of the impacted pavements. To achieve this objective, initially, the authors collected the site-specific traffic information by deploying Portable Weight-in-Motion (P-WIM) devices to ten representative energy corridors in the Eagle Ford Shale region. Subsequently, the traffic information associated with the current and pre-energy development era, pavement layers properties from the field Non-Destructive Testing (NDT), as well as the climate information, were then incorporated in the Mechanistic-Empirical pavement design software to determine the PLR associated with the changes in traffic patterns. The analysis results indicated that the reduction of service life was more pronounced for Farm-to-Market (FM) roads, with less robust structural capacity compared to State and US highways.

1 INTRODUCTION

In recent years, Texas has experienced a substantial increase in the energy-related activities such as natural gas and crude oil productions. Energy development operations have provided favorable economic benefits to the state and the nation. However, such activities have significantly increased the volume and frequency of the Over-Weight (OW) truck traffic and Super Heavy Load (SHL) movements in the network, which adversely affected the longevity of transportation infrastructure systems such as pavements and bridges. Specifically, the overload corridors in the Permian Basin and the Eagle Ford Shale regions in West and South Texas have experienced an alarming increase in the OW truck operations since 2008, leading to substantial serviceability loss in pavements. Figure 1 illustrates comparisons of Equivalent Single Axle Load (ESAL) values over the last decade for the evaluated representative pavement sections in this study. The plot shows that FM468 roadway in Laredo District, and US281 and SH123 highways in Corpus Christi District have undergone a considerable increase in truck traffic (i.e., 2400%, 228%, and 790% increase, respectively) in the past decade. This alarming increase in the traffic volume can consume the service life of the highway infrastructures if the commensurate maintenance and rehabilitation (M&R) strategies are not adopted.

Correct prediction of loss of serviceability of the pavements can further optimize the M&R plans to mitigate the deterioration of ride quality in the impacted networks. This can potentially protect state assets by

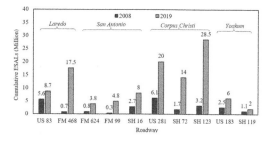

Figure 1. Cumulative 18-kip ESALs over 20-year design life for representative roadways in the Texas overload corridors.

reduction or elimination of M&R costs associated with the premature failure of the transportation facilities. Consequently, it is imperative to have a realistic assessment of the pavement life reduction (PLR) of pavements subjected to taxing loading conditions in overload corridors. An accurate account of the traffic characteristics is the precursor for reliable prediction of the service life of the pavements subjected to OW traffic loads. Current mechanistic-empirical (ME) design guides, such as the Texas Mechanistic-Empirical (TxME) flexible pavement design system, allow users to incorporate site-specific traffic data collected from Weigh-In-Motion (WIM) devices in the field. Using the site-specific traffic data, instead of default/national-level traffic input parameters can significantly improve the accuracy and reliability of the obtained results from the ME pavement analysis (Jasim et al. 2019).

Table 1. TxME pavement performance criteria limits.

Performance Criteria	Limit
Total Surface Rutting (inch)	0.5
Thermal Cracking (ft/mile)	2112
AC Fatigue Cracking Area (%)	50
Stabilized Base Fatigue Cracking (%)	50

Figure 2. Flowchart describing the procedure for backtracking the current traffic to pre-energy development traffic.

The primary objective of this research study was to develop a mechanistic approach for the characterization of the reduction of the service life of the pavements due to the drastic changes in truck traffic volumes in Texas overload corridors over the last decade. The proposed approach accounts for specific characteristics of vehicles operating in the network with consideration of environmental conditions and unique features of the transportation systems. Prominent features associated with the traffic patterns and material properties of pavement layers were directly determined from field data collection and nondestructive testing of ten representative pavement sections in energy corridors of the Eagle Ford Shale region.

2 APPROACH TO ASSES REDUCTION OF THE PAVEMENT SERVICE LIFE

TxME pavement design software was deployed for the ME analysis of the pavement sections in this study. Essentially, the TxME system requires different categories of inputs, i.e., pavement structure, traffic characteristics, and climate information. To characterize the design inputs, initially, the layer configurations and stiffness properties of the pavement layers were obtained from the post-processing of the Ground Penetrating Radar (GPR) and Falling Weight Deflectometer (FWD) data. The authors then deployed Portable WIM (P-WIM) devices to collect the site-specific traffic information in the evaluated representative sites.

The research team further developed a procedure for routine measurements of tire pressure, tire footprint, and tire spacing in the calibration process of P-WIM devices to have an accurate account of the loading conditions for further numerical simulations (Morovatdar et al. 2020). Afterward, the site-specific pavement layers properties and traffic loading information obtained from the field data collection efforts, and climatic data using the software pre-defined values were incorporated in the TxME pavement design software to obtain pavement responses, and the cumulative pavement distresses. It should be noted that the pavement distresses evaluated in this study include cumulative rutting at the surface, AC layer fatigue cracking, stabilized base layer fatigue cracking, and thermal cracking. Performance criteria limits set forth in the TxME design system are also shown in Table 1.

One of the significant steps in the followed approach is to incorporate accurate traffic information.

Essentially, in order to properly quantify the PLR due to increased overweight traffic loads, it is imperative to have an accurate account of current and pre-energy development traffic loading conditions. Hence, the authors obtained the projected ESAL values over a 20-year design life for both current and pre-energy development traffic loading conditions to further characterize the traffic in numerical simulations. Initially, the cumulative 18-kip ESAL values that correspond to the current traffic condition were calculated from Equation 1 as:

$$ESAL_{current} = \sum (EALF)_i n_i \tag{1}$$

where: $EALF_i$ =Modified EALF values; n_i =Number of passes of ith-axle load group.

Modified Equivalent Axle Load Factor (EALF) values were determined using the mechanistic quantification of the pavement damages procedure described in Morovatdar et al. (2019). As stated in the aforementioned study, the modified EALFs were substantially higher than traditional industry-standard axle load factors currently employed by the pavement design industry. The number of axle load repetitions were also derived from the site-specific axle load spectra databases. Then, the EALF values for each axle type were multiplied by the projected frequencies of load repetitions to calculate the projected ESAL values. Additionally, based on the analysis of traffic growth in the region, the majority of the representative pavement sections experienced drastic changes in truck operations between 2008 and 2012. Using available information in the PMIS, TxDOT's Traffic Count Database System (TCDS), and LTPP databases, the $ESAL_{Pre-EnergyDevelopment}$, representing the 2008 traffic volume, was found for each of the evaluated highways. Then, considering the traffic growth rates during the entire period, $ESAL_{Current}$ and $ESAL_{Pre-EnergyDevelopment}$ values were converted to the equivalent 18-kip axles corresponding to the reconstruction/rehabilitation year, to consider all the traffic load applications immediately after reconstruction, as shown in the flowchart in Figure 2.

Ultimately, the research team incorporated the converted ESAL values into the TxME and compared the corresponding service lives. The difference between these two results represents the pavement life reduction associated with the changes in traffic patterns for each pavement section, as shown in Figure 3.

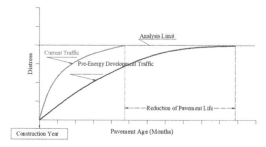

Figure 3. Schematic diagram of the Pavement-Life Reduction (PLR) analysis.

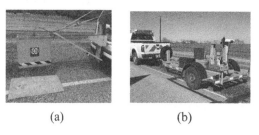

(a)   (b)

Figure 4. Nondestructive testing in Texas overload corridors (a) GPR testing, and (b) FWD testing.

2.1 *Nondestructive testing*

GPR and FWD tests were deployed to the field, as shown in Figure 4, for the determination of the layer configurations and layer moduli in representative pavement sections. This information was a direct input to the PLR analysis protocol. Comprehensive information on the layer thicknesses, layer profile, layer material, and back-calculated layer moduli attributed to all evaluated roadways is provided in Ashtiani et al. (2019).

2.2 *Deployment of the portable WIM devices to develop site-specific axle load spectra*

Portable WIM devices were deployed in selected FM, SH, and US roads within energy developing areas in Texas to collect the site-specific truck traffic information. Figure 5 shows the piezoelectric sensors inserted into specialized pocket tapes and installed on the pavement surface, as well as the data acquisition system, as principal components of the P-WIM units. Axle load spectra, vehicle class distributions, number of load repetitions, and axle configurations were the most relevant traffic information that was captured/derived by the P-WIM units.

3 FIELD DATA ANALYSIS AND RESULTS

Figure 6 shows an example of the analysis of pavement life reduction for the US281 highway. The plot shows the rutting performance for US281, considering the pre-energy development and current traffic

Figure 5. Deployment of the P-WIM devices in the field.

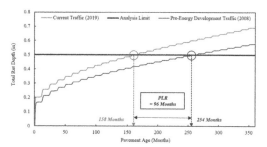

Figure 6. Reduction of pavement service life for US281 highway due to changes in traffic patterns.

patterns. Based on the internal distress algorithms in TxME, and pre-energy development traffic characteristics, it takes 254 months for US281 to develop 0.5 in. of rutting. However, if the 2019 traffic characteristics were incorporated in the TxME, it takes 158 months to develop 0.5 in. of cumulative rut depth. In other words, changes in the traffic patterns attributed to the energy developing activities have consumed 96 months (i.e., 254-158=96) of the service life of US281 highway.

Similar analyses were conducted for all the representative roadways in the network. Then, in order to better represent the analysis results, a new parameter, named "*rate of pavement life reduction*," was defined as shown in Equation 2:

$$Rate\ of\ pavement\ life\ reduction = (L_{PE} - L_C)/L_{PE} \times 100$$

$$(2)$$

where L_{PE}: expected pavement life corresponds to pre energy-development traffic characteristics; L_C: expected pavement life corresponds to current traffic characteristics.

Figure 7 illustrates the reduction rates of pavement service life associated with the studied overload corridors. As evidenced in the plot, FM roadways with less robust pavement profile were found to be more sensitive to the increasing traffic patterns, compared to SH and US highways. FM99, FM468, and ultimately FM624 roadways have been subjected to the greatest pavement life reductions as 66%, 51%, and 48%, respectively. The plot also revealed that SH123 and US281 highways with substantial recent changes in traffic patterns have also experienced remarkable

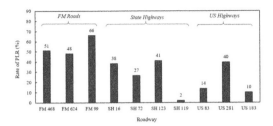

Figure 7. Reduction rate of pavement service life for the studied overload corridors in Texas.

reductions in pavement service life. This highlighted the influence of increased traffic frequencies on pavement life consumptions in heavily trafficked highways across the region.

Considering the detrimental influence of heavy axle weights in the serviceability of pavements, OW truck percentage was another contributing factor, besides traffic volume (i.e., ESAL repetitions), in the analysis of pavement life reduction. Based on the post-processing of the P-WIM traffic data, on average, 32% of the truck traffic in FM468 exceeds the Texas permissible axle weight load limits. SH123 and US281 highways also had alarming OW percentages of 36% and 48%, respectively. In this study, a direct relationship between the rate of PLR and the percentage of OW trucks passing in the network was found as described in Ashtiani et al. (2019).

Additionally, the analysis results indicated that the rutting was the primary source of distresses, causing premature failure of the pavement sections subjected to the movements of OW and superheavy load trucks. It is noted that only two sites (SH16, and SH119) with medium thicknesses of the asphalt layer, i.e., 3.5 in. (8.9 cm), experienced relatively high fatigue cracking, exceeding the failure criterion, among all the studied representative sites.

4 SUMMARY AND CONCLUSIONS

The authors developed a mechanistic algorithm for the assessment of the reduction of the service life of pavements due to changes in the traffic characteristics in Texas overload corridors over the last decade. A novel approach was proposed to retrace and backtrack the traffic distributions to the reconstruction year for each specific pavement section. Analysis of the reduction of the service life of ten representative pavement sections revealed the impact of the changes in the traffic demands in the past decade on the loss of serviceability of transportation facilities in Texas overload corridors. The reductions of service life were more pronounced for FM roads, with less robust structural capacity compared to SH and US highways.

Moreover, an informative index, i.e., rate of pavement life reduction, was defined to better characterize the serviceability loss of the pavements due to changes in traffic patterns. It was revealed that the service life of the studied pavement sections in FM, SH, and US highways have reduced as high as 66%, 41%, and 40%, respectively, due to the dramatic increase in the truck traffic volume.

Additionally, the post-processed results indicated that the primary source of distresses in the evaluated pavement sections in this study was associated with cumulative surface rutting due to overweight truck traffic. This in line with our expectations, as the primary culprit for premature failure of the overload corridors pertains to the passage of overweight trucks in the network.

REFERENCES

Ashtiani, R. S., Morovatdar, A., Licon, C., Tirado, C., Gonzales, J., & Rocha, S., 2019. Characterization and quantification of traffic load spectra in Texas overweight corridors and energy sector zones (No. FHWA/TX-19/0-6965-1).

Jasim, A. F., Wang, H., & Bennert, T., 2019. Evaluation of clustered traffic inputs for Mechanistic-Empirical Pavement Design: Case study in New Jersey. Transportation Research Record, 0361198119853557.

Morovatdar, A., Ashtiani, S. R., Licon, C., & Tirado, C., 2019. Development of a mechanistic approach to quantify pavement damage using axle load spectra from south Texas overload corridors. In Geo-Structural Aspects of Pavements, Railways, and Airfields Conference, (GAP 2019).

Morovatdar, A., Ashtiani, S. R., & Licon, C., 2020. Development of a Mechanistic Framework to Predict Pavement Service Life using Axle Load Spectra from Texas Overload Corridors. ASCE's International Conference on Transportation & Development.

Advances in Materials and Pavement Performance Prediction II – Kumar et al. (eds)
© 2021 Taylor & Francis Group, London, ISBN 978-0-367-46169-0

Experimental performance of buried concrete utility under flexible pavements subjected to heavy dynamic loads

H. Nabizadeh
Applied Research Associates, Inc. (ARA), IL, USA

S. Elfass, E.Y. Hajj & R.V. Siddharthan
Department of Civil and Environmental Engineering, University of Nevada, Reno, NV, USA

M. Nimeri
King County International Airport, WA, USA

M. Piratheepan
Department of Civil and Environmental Engineering, University of Nevada, Reno, NV, USA

ABSTRACT: Concrete culverts are underground conduits located under roadways. These culverts are required to withstand soil overburden as well as vehicular surface loads. Previous studies investigated the integrity of the culverts by applying surface loads directly on top of the soil which represents the worst-case scenario. While this may be a good design practice, for realistic buried utility assessment subjected to superheavy load (SHL) vehicles, the role of existing pavement layers should be addressed. To gain insight into the performance of buried culverts under SHL vehicle, two full-scale pavement experiments were designed and carried out at the University of Nevada, Reno. A concrete culvert was buried in the subgrade of a typical pavement structure constructed in $10 \times 10 \times 7$ ft box. Falling Weight Deflectometer (FWD) loads ranging from 9000 to 27000 lb were applied on top of the asphalt concrete layer. This paper describes the experiment and presents the captured culvert performance.

1 INTRODUCTION

Typical underground utilities that are often found near highway routes include sewer and drain lines, water mains, gas lines, electrical conduits, culverts, oil and coal slurry lines, and heat distribution lines. In general, underground utility structures are categorized as flexible or rigid. A flexible pipe should be able to withstand at least 2 percent deflection ratio (i.e., vertical deflection normalized with respect to the original size) without any significant structural distresses. The utility structures that do not meet this criterion are generally considered rigid. Steel, ductile or cast iron, plastic pipes are usually classified as flexible. Concrete and clay pipes are usually considered rigid (Moser & Folkman 2008).

The stresses induced on buried utilities from dead (i.e., overburden) and live (i.e., traffic) loads strongly depend on the stiffness properties of the utility and the surrounding soil. This phenomenon is commonly referred to as soil-structure interaction. In rigid utility structures, it is generally assumed that the vertical stresses are more critical, and horizontal stresses are often neglected (American Water Works Association 2008).

The Marston theory is routinely used to compute dead loads on rigid utilities (Marston & Anderson 1913). Based on this theory, the resultant load on an underground structure is computed as the weight of the material above the top of the conduit minus the shearing or friction forces along the sides of the trench. Several experimental and analytical attempts have been made to investigate the stress variation as a function of depth from surface traffic live loads. The classical Boussinesq solution and other solutions, such as spreading the load over an area as a linear function of depth (e.g., AASHTO Load and Resistance Factor Design (LRFD) Bridge Construction Specifications) are the most widely used calculation approaches (Petersen et al. 2010, AASHTO 2012). The AASHTO LRFD Bridge Construction Specifications spread live-load through homogenous subgrade soil irrespective of the characteristics of the buried structure. Applicability of classical solutions is often constrained to linear elastic, homogenous, and half-space soil conditions.

2 PROBLEM STATEMENT

A fair assessment of the induced stresses from dead and live loads is required to analyze the internal integrity of a buried utility. Although the state of practice uses methods that provide recommendations with respect to the load distribution, they are limited, especially when assessing the risk to buried utilities under a superheavy load (SHL) vehicle. The limitations associated with the state of practice are: *(a) Considering only the standard truck (mostly HS20) as live load and simulating it as a point load or as a rectangular tire footprint*: SHL-hauling units are much larger in size and weight compared to standard trucks and the spacing between tires and axles is not standard Consequently, the effects of closely spaced tires, nonuniform tire pressure distribution, and much heavier tire load cannot be addressed directly using the existing methods; *(b) Applying surface-tire loads directly at the surface of unpaved roads (i.e., on top of the subgrade)*: This case represents the worst-case scenario since asphalt concrete and base layers affect the stress distribution and can significantly reduce the stresses transferred to the utility. While this may be a good design practice, for a realistic buried utility assessment subjected to an SHL vehicle movement, the role of existing pavement layers should be addressed; *(c) Spreading the live load at a constant rate to the depth of subgrade cover*: This assumption is not valid when considering a multilayer pavement system with distinct stiffness material properties; *(d) Simulating and applying the live load as a static load*: The influence of speed on the viscoelastic behavior of an AC layer needs to be adequately accounted for the stress distribution estimation process under an SHL vehicle movement.

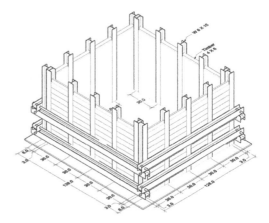

Figure 1. Three-dimensional (3D) schematic of PaveBox (dimensions in inches).

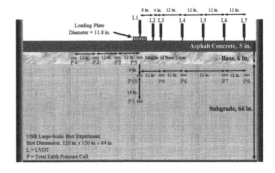

Figure 2. Pavement layer thicknesses and instrumentation plan in Experiment No. 3 (control experiment).

3 RESEARCH APPROACH

To gain insight into the performance of rigid buried utilities under an SHL vehicle, two separate full-scale pavement experiments were designed and carried out at the University of Nevada, Reno. A full-scale experiment with a concrete culvert buried in the subgrade of a typical pavement structure and a control experiment that had no buried utilities were conducted. A careful comparison between the two experiments identified the role of buried utilities in the stress distribution within a typical pavement structure and the induced pressure on the utilities. This paper is part of a Federal Highway Administration (FHWA) research study on "Analysis Procedures for Evaluating Superheavy Load Movement on Flexible Pavements" (Hajj et al. 2018).

4 FULL-SCALE EXPERIMENTS

The experimental program involved two separate full-scale pavement experiments. This experimental program utilized UNR's large-scale pavement/soil testing facility (PaveBox) which is a large-scale box with internal dimensions of 10×10×7 ft. Figure 1 shows a 3D schematic of the PaveBox. Detailed design, fabrication, and construction process of PaveBox are presented elsewhere (Nabizadeh Shahri 2017; Nimeri et al. 2018).

As shown in Figure 2, the control experiment (i.e., no buried utility), referred to as Experiment No. 3, consisted of 5 inch of asphalt concrete (AC), 6 inch of crushed aggregate base (CAB), and 66 inch of subgrade (SG). A similar pavement structure with underground utilities installed in the subgrade was constructed in Experiment No. 5. Figure 3 shows a schematic of the test setup in Experiment No. 5. It should be noted that effort was made to use the same materials and apply similar compaction practices in both experiments. A detailed discussion regarding the construction procedure, instrumentation, material properties can be found elsewhere (Nabizadeh Shahri 2017; Nimeri et al. 2018).

As illustrated in Figure 3, two types of buried utilities, a rigid (i.e., concrete culvert) and flexible (i.e., steel pipe), were installed in Experiment No. 5. However, this paper only presents the performance of the rigid concrete culvert, which is a 12 inch square cross-section area by 9-ft long by 1 inch thick concrete culvert.

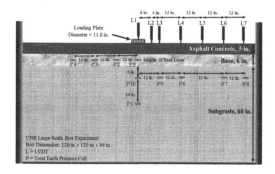

Figure 3. Pavement layer thicknesses and instrumentation plan in Experiment No. 5 (buried utility experiment).

In both experiments, Falling Weight Deflectometer (FWD) at various load levels ($\sim$ 9000, 12000, 16000, 21000, 25000 lb) were applied at the pavement surfaces (i.e., top of the AC layer). A hydraulic actuator was used to apply dynamic surface loads on an FWD loading plate. Experiment No. 5 was divided into three phases. In the first and the second phase, the FWD loads were applied at the surface directly above the centerline of the steel pipe (Location A) and the centerline of the concrete culvert (Location B), respectively. In the third phase, the same surface loads were applied between the two buried structures (Location C).

As shown in Figures 2 and 3, surface deflections using Linear Variable Differential Transformers (LVDTs) at the center of the loading plate and different locations away from the center of the plate were measured. In addition, several Total Earth Pressure Cells (TEPCs) with 4 in. diameter were installed in the base and subgrade layer to capture the load-induced vertical stresses during the load application. TEPCs were also installed on top of and underneath the buried utilities to capture the induced pressure on the utilities.

5 EXPERIMENTAL RESULTS AND OBSERVATIONS

To understand the impact of buried utilities on the stress distribution within the pavement structure, the measured vertical stresses at the location of TEPCs in Experiment No. 5 were compared to the corresponding measured stresses in Experiment No. 3. In other words, the stress distributions in these two pavement structures were compared by monitoring the TEPCs measurements that are located at similar positions relative to the applied surface load.

Figures 4 and 5 show the recorded vertical stresses measured by P10 and P1 in Experiment No. 3 and P10B and P1B in Experiment No. 5, respectively. All these TEPCs were located at the centerline of the FWD load. P10 and P10B were installed 6 inch from the subgrade surface, while P1 and P1B were placed 20 inch

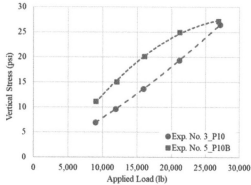

Figure 4. Measured vertical stresses by P10 in Experiment No. 3 and P10B in Experiment No. 5 (load applied at Location B).

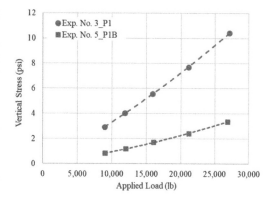

Figure 5. Measured vertical stresses by P1 in Experiment No. 3 and P1B in Experiment No. 5 (load applied at Location B).

from the subgrade surface. P10B and P1B in experiment No. 5 were installed on the top and the bottom of the concrete culvert respectively.

Figure 6 shows the vertical stresses measured by P9 in Experiment No. 3 and P9B, which is installed on the top of the concrete culvert (i.e., 6 inch from the subgrade surface). Both of these TEPCs were at 12 inch offset from the centerline of the load.

Figure 7 depicts the recorded vertical stresses measured by P6 in Experiment No. 3 and P10B in Experiment No. 5 while the load was applied at Location C. In other words, the concrete culvert was not located directly under the load and the TEPCs were at 24 inch offset from the centerline of the load.

As presented in Figures 4, 6 and 7, the load-induced vertical stresses in Experiment No. 5, measured by P10B, were higher than the measurements by P10 in experiment No. 3. However, substantially lower vertical stresses were measured by P1B compared to the measurements by P1, as shown in Figure 5. These observations are attributed to the soil-structure interaction. Concrete culvert, because of larger stiffness parameters and rigidity relative to the sounding

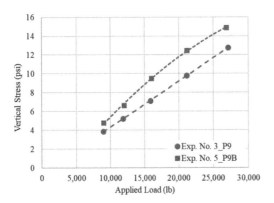

Figure 6. Measured vertical stresses by P9 in Experiment No. 3 and P9B in Experiment No. 5 (load applied at Location B).

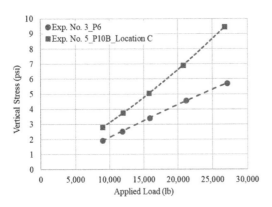

Figure 7. Measured vertical stresses by P6 in Experiment No. 3 and P10B in Experiment No. 5 (load applied at Location C).

soil, draws excess pressure through negative arching. In other words, negative arching means the concrete culvert attracts the dead- and live-induced stresses.

To investigate the internal integrity of the buried concrete culvert in Experiment No. 5, the structural adequacy analysis of the concrete culvert was conducted. The calculated maximum induced moment in the top slab, bottom slab, and side walls were 3832, 3865, and 3877 lb-inch. However, the flexural capacity of the members of the concrete culvert was 3011. The higher maximum induced moment than the flexural capacity indicates a possibility of failure. Such an observation is consistent with the nonlinear behavior of rigid culvert response which was observed as skewness in the measured vertical stresses on top of the culvert at the highest FWD load level of 27000 lb (see Figure 4). It should be mentioned that the concrete culvert was purposely designed in a way to experience distresses at the higher load levels of surface FWD loading.

6 SUMMARY AND CONCLUSION

While every utility has certain specific design considerations, two common steps are being followed in the existing design methods. In the first step, the load distribution on the buried utility structure due to the dead (i.e., soil overburden) and live (i.e., traffic) loads is determined. There exist common practices to accomplish the first step are available. Subsequently, in step 2, the buried utility structure is designed in accordance with the specification unique to its type. There are four main limitations associated with the state of practice to estimate the live load distribution on an underground utility when it is buried in a roadway subjected to an SHL vehicle movement.

To gain insight into the role of buried utilities in the stress distribution within a typical pavement structure and the induced pressure on the utilities, two full-scale experiments were designed and conducted. The control experiment represented a full pavement structure without any buried utilities subjected to surface FWD loads at different intensities. A similar pavement structure with two types of buried utilities installed in the subgrade was subjected to the same surface FWD loads applied directly above the centerlines of the buried utilities.

It was found that the vertical stresses experienced by the concrete culvert are substantially higher, about 50 percent, compared to those measured in the subgrade in the control experiment (i.e., no buried utility). This is attributed to the negative arching effect in which the concrete culvert attracts the induced stresses because of its larger stiffness relative to the sounding soil stiffness.

REFERENCES

American Association of State Highway and Transportation Officials. 2012. *AASHTO LRFD Bridge Construction Specifications, Sixth Edition*. Washington, D.C.: AASHTO.

American Water Works Association. 2008. *Concrete Pressure Pipe: Manual of Water Supply Practices, AWWA Manual M9, Third Edition*. Denver: AWWA.

Hajj, E.Y., Siddharthan, R.V., Nabizadeh, H., Elfass, S., Nimeri, M., Kazemi, S.F., Batioja-Alvarez, D., & Piratheepan, M. 2018. *Analysis Procedures for Evaluating Superheavy Load Movement on Flexible Pavements, Volume I: Final Report*. Report No. FHWA-HRT-18-049. Washington, D.C.: FHWA, U.S. Department of Transportation.

Marston, A. & Anderson, A.O. 1913. *The Theory of Loads on Pipes in Ditches, and Tests of Cement and Clay Drain Tile and Sewer Pipe, Bulletin No. 31*. Ames: Iowa State College of Agriculture and Mechanic Arts.

Moser, A.P. & Folkman, S. 2008. *Buried Pipe Design, Third Edition*. New York City: McGraw-Hill Companies, Inc.

Nabizadeh Shahri, H. 2017. *Development of a Comprehensive Analysis Approach for Evaluating Superheavy Load Movement on Flexible Pavements*. University of Nevada, Reno: ProQuest Dissertations Publishing.

Nimeri, M., Nabizadeh, H., Hajj, E.Y., Siddharthan, R.V., Elfass, S., & Piratheepan, M. 2018. *Analysis Procedures for Evaluating Superheavy Load Movement on Flexible Pavements, Volume II: Appendix A: Experimental Program*. Report No. FHWA-HRT-18-050. Washington, D.C.: FHWA, U.S. Department of Transportation.

Petersen, D L., Nelson, C.R., Gang, L., McGrath, T.J., & Kitane, Y. 2010. *Recommended Design Specifications for Live Load Distribution to Buried Structures, NCHRP Report No. 673*. Washington, D.C.: National Cooperative Highway Research Program.

Advances in Materials and Pavement Performance Prediction II – Kumar et al. (eds)
© 2021 Taylor & Francis Group, London, ISBN 978-0-367-46169-0

Measurement of structural rolling resistance at two temperatures

N.R. Nielsen & T. Hecksher
Roskilde University, Roskilde, Denmark

C.P. Nielsen
Greenwood Engineering A/S, Brøndby, Denmark

P.G. Hjorth
DTU Compute, Lyngby, Denmark

ABSTRACT: In this study, we investigate how an increase in road temperature influences the structural rolling resistance of a heavy vehicle. The structural rolling resistance (SRR) is defined as the dissipated energy due to pavement deflection under a moving load. It is measured using a newly proposed method, which is based on the relationship between SRR and the slope of the deflection basin underneath the load. Using the Traffic Speed Deflectometer technology, we measured SRR on the same road under two different road temperatures, 18°C and 35°C respectively. On average, an increase in SRR of 59% was observed, with some areas of the road having up to 400% increase. This indicates that under warm road conditions SRR might have a significant effect on the overall rolling resistance of a heavy vehicle.

1 INTRODUCTION

When a pavement is subject to a moving load, it will deform underneath it. If the pavement is viscoelastic, the time delay in the deflection makes the maximum deflection appear behind the load. This results in an asymmetric deflection basin, as illustrated on Figure 1a. Consequently, the load experiences an uphill deflection slop (Fig. 1b) and has to do work in order to maintain a constant speed (Flügge 1975). The excess energy consumption due to deflection of the pavement is dependent on the pavement structure, and we will refer to it as structural rolling resistance (SRR). SRR can be calculated directly from the asymmetric deflection basin (Balzarini et al. 2018; Chupin et al. 2013).

Estimates of SRR are often derived by simulating the pavement response to a moving load with constant speed. The pavement parameters used in these simulations are obtained from either back-calculated falling weight deflectometer measurements or laboratory measurements on the pavement materials (Akbarian et al. 2012; Balzarini et al. 2017; Pouget et al. 2012). Moreover, indirect measurements of SRR have been conducted (Zaabar & Chatti 2014). However, it has been proven difficult to develop accurate and robust methods for measuring SRR directly.

In Nielsen et al. (accepted) we presented a new method for direct measurements of SRR, using the Traffic Speed Deflectometer (TSD) technology. The

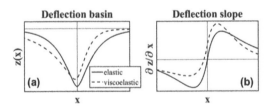

Figure 1. Pavement deflection (a) and associated deflection slope (b) of an elastic and viscoelastic pavement. Viscous properties make the deflection basin asymmetric, which results in a positive deflection slope underneath the load $(x = 0)$.

method is based on the relation between SRR and the slope of the deflection basin underneath a moving load. Using the TSD has the advantage that it mimics a full-size trailer and thus it measures the pavement deflection slope under realistic driving conditions. The method proved to be robust and measure SRR with high accuracy when repeated measurements where compared. Furthermore, it has the clear advantage that it does not require a model or prior knowledge about the pavement in order to calculate SRR.

The influence of temperature on SRR has been investigated in literature, by use of numerical simulations (Pouget et al. 2012; Shakiba et al. 2016). The magnitude of the found temperature effect differs between the studies is dependent on the applied pavement models. To the authors knowledge no direct measurements of the temperatures influence on SRR for heavy vehicles exist.

In this paper, we investigate the effect of road temperature on SRR. We expect that a higher road temperature will lead to a softer asphalt layer and an increased pavement response to the moving load. In addition, within the investigated temperature range we expect the viscoelastic damping of the asphalt to increase with increasing temperature (Pouget et al. 2012; Shakiba et al. 2016). Consequently, we expect that an increase in road temperature will lead to a higher SRR.

2 TRAFFIC SPEED DEFLECTOMETER DATA

2.1 *TSD principle*

The Traffic Speed Deflectometer (TSD) continuously measures the vertical velocity (v_d) of the pavement underneath the right rear-end trailer tires, while the truck is moving. This is done by means of Doppler lasers positioned between the tire set both in front and behind the axle. From this, the pavement deflection slope ($dz(x_n)/dx$) for each position (x_n) is obtained by dividing with the driving speed (v),

$$\frac{dz(x_n)}{dx} = \frac{v_d(x_n)}{v} \tag{1}$$

The principle is explained in more depth in Nielsen (2019).

Having measurements on both sides of the load enable us to determine the asymmetry in the deflection basin arising from viscoelastic properties in the pavement.

2.2 *Raw data*

For this study, two sets of measurements were made at a road section near Copenhagen, Denmark. The measurements were made on two days (15 months apart), where the pavement temperature was 18°C and 35°C respectively. Each set of measurements were repeated three times and a good reproducibility was seen with median standard deviations of 9% (18°C) and 5.5% (35°C). The TSD truck was at maximum axle load (10 tonnes) and driving speed was between 50-60 km/h, with the exact driving speed recorded continuously during the measurement rounds.

In Figure 2a, a plot of the measured deflection slope data in the beginning of the ~10 km measured road section is shown. The measured deflection slope for each sensor is an average over 10 m. An example of the deflection slopes at 2 km, as a function of distance from the load, is seen in Figure 2b. The center of the axle is at x=0. The deflection slope curve is characterized by a minimum located behind the load and a maximum in front of the load. The standard deviations are illustrated with error bars in Figure 2b. In some cases, the error bars are smaller than the markers and thus not visible.

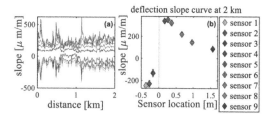

Figure 2. Example of raw data at 18°C. (a) Measured deflection slope for all sensors in the beginning of the measured road. (b) deflection slopes as a function of distance from the load. Standard deviations are shown with error bars.

3 SIMPLE METHOD FOR ESTIMATING THE STRUCTURAL ROLLING RESISTANCE

This method was presented for the first time in Nielsen et al. (accepted). In the simplest approach, we assume that the applied load is a point load, located at the center of the tire (x=0) with magnitude F_L. The dissipated energy in the pavement (P_{SRR}), can be calculated from the vertical pavement velocity underneath the load and the applied load,

$$P_{SRR} = F_L v_d(x=0) = F_L v \frac{dz}{dx}(x=0) \tag{2}$$

where the last expression comes from Equation 1. In the case of a perfectly elastic pavement the deflection slope under the load is zero, and therefore the dissipated energy is also zero, $P_{SRR} = 0$. On the other hand, if there is some damping in the pavement, the slope underneath the load will be larger than zero. In this case, energy is dissipated in the pavement i.e. P_{SRR} >0. Equation 2 requires knowledge of the deflection slope exactly underneath the load (x=0). However, due to the presence of the axle, it is not possible to measure in that point. Therefore, we have to estimate the slope at x=0 based on the surrounding data points. The simplest approach is to make a linear interpolation between the two sensor points closest to the load (sensor 3 and 4). Doing this we have that

$$P_{SRR} = F_L b v, \tag{3}$$

where b is the intersection at x = 0 for the linear interpolation. From this, we obtain the expression for the structural rolling resistance force, $F_{SRR} = P_{SRR}/v$. We can also derive the structural rolling resistance coefficient, defined as the ratio between the rolling resistance force and the applied load,

$$C_{SRR} = \frac{F_{SRR}}{F_L} = b. \tag{4}$$

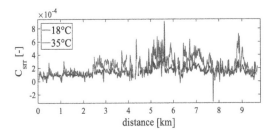

Figure 3. Calculated C_{SRR} at warm (35°C) and cold (18°C) road temperature. The median of the standard deviations are $0.13 \cdot 10^{-4}$(9%) for 18°C and $0.2 \cdot 10^{-4}$(5.5%) for 35°C. Note that the C_{SRR} values are negative in some points at 35°C, which is unphysical. This behavior is commented on in section 4.2.

4 RESULTS

4.1 *The influence of temperature on the structural rolling resistance coefficient*

Using two sets of deflection slope data measured at the road temperatures 18°C and 35°C, we study the influence of road temperature on SRR. The temperatures were measured at the surface using an infrared temperature sensor through all measurements. In Figure 3 C_{SRR} is plotted versus distance on the ~10 km measured road segment. In this plot, the mean C_{SRR} values after three repeated measurements are shown. There is a systematic increase in C_{SRR} as the road temperature increases. The median of the standard deviations in C_{SRR} are $0.13 \cdot 10^{-4}$(9%) for 18°C and $0.2 \cdot 10^{-4}$(5.5%) for 35°C. Thus, the calculated C_{SRR} values have a good precision.

In Figure 4, a histogram of the measured C_{SRR} values is shown. Here, we see that the mean C_{SRR} over the entire road increases from $1.4 \cdot 10^{-4}$ to $2.3 \cdot 10^{-4}$ when the temperature is increased. Furthermore, the distribution of C_{SRR} becomes broader with higher temperature. This means that the calculated C_{SRR} for 35°C varies more along the road. The total rolling resistance of a truck is typically on the order of 1% of the load. Based on this, the mean C_{SRR} found in this study are 1.4% (cold) and 2.3% (warm) of the typical total rolling resistance.

C_{SRR} varies considerably throughout the measured distance (from 0.01% of the load to 0.06%). This variation is completely reproducible within the three repeated measurement runs, and we see that spatial variations are similar for the two temperatures. A notable increase in C_{SRR} is seen around 2.5 km. There is no visible change in the asphalt in this area, and thus the change is due to a structural change in the underlying layers. The varying C_{SRR} values reflect the fact that the road measured on is not a homogeneous road, but a real road with varying pavement structure. Most likely the thickness of the asphalt layer and possibly also the type of asphalt differs along the road. As a result, SRR and its temperature dependence will also differ along the road. The method shows a good ability

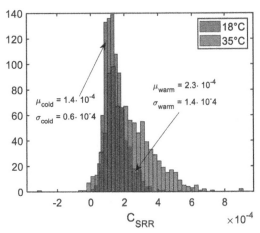

Figure 4. Histogram of measured structural rolling resistance coefficients for cold (18°C) and warm (35°C) road conditions. We see an increase in the mean (μ) C_{SRR} value, when the road temperature is increased. Furthermore, the distribution of measured values is broader under warm conditions (increased σ).

to reproducibly capture these changes in C_{SRR}, even in areas where C_{SRR} changes dramatically.

4.2 *Influence of temperature on the deflection slope curves*

By looking into some representative sets of deflection slopes, we can qualitatively investigate the change caused by increased road temperature. Figure 5a illustrates the most commonly encountered influence that the increased road temperature has on the deflection slope curves. Either the maximum deflection slope increases, the minimum deflection slope decreases or both effects occur at the same time (as seen in Figure 5a). An increase in maximum deflection slope value means that the deflection basin gets steeper in front of the load. This often leads to the deflection slope value underneath the load being increased (thus higher P_{SRR}). A decrease in the minimum deflection slope value means that the deflection basin becomes steeper behind the load. Furthermore, we often see that the minimum deflection slope is moved to the left (away from the load) in the warm data. This corresponds to the maximum deflection moving further behind the load, something which is associated with an increased effect of viscous damping.

In Figure 3, the calculated C_{SRR} is negative in a few places, which is unphysical. In Figure 5b, an example of such a deflection slope curve in a place with negative C_{SRR} is plotted. Here, a plot of the deflection slope values at 8.03 km is shown for both temperatures. The relative change in C_{SRR} from cold to warm data is −112.9%, as in the warm situation, we calculate a negative C_{SRR} when using linear interpolation. Note that the shape of the deflection slope curve changes dramatically around the minimum, when temperature

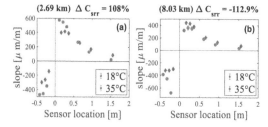

Figure 5. Representative sets of deflection slope curves. The relative change in C_{SRR} is listed for each plot as ΔC_{SRR}. Standard deviations are indicated with error bars (not visible when these are smaller than the markers). It should be noted that in between the two sets of measurements the sensor locations have been changed slightly. We only expect this to have a minor effect on the results of the analysis.

is increased. Furthermore, for the warm data we see that the magnitude of the deflection slope is higher behind the load than in front of the load. This behavior indicates that using linear interpolation to find the deflection slope underneath the load, in some cases is too simple to capture the actual slope assess.

5 SUMMARY AND CONCLUSION

In this study, we have presented measurements of structural rolling resistance of a $\sim$10 km road section, measured at two different road temperatures (18°C and 35°C). The method shows good reproducibility between the repeated measurements, with small standard deviations. Furthermore, it was also able to capture the spatial changes in C_{SRR}, which occur in data at 18°C and 35°C.

The found SRR values have a magnitude which is comparable with results found in empirical and numerical studies on the subject (Akbarian et al. 2012; Chupin et al. 2013; Pouget et al. 2012; Zaabar & Chatti 2014). On average, C_{SRR} increased with 59% over the measured distance when temperature increased, in some areas even up to 400% increase, showing that for warm weather conditions SRR have an effect on the overall energy consumption for heavy vehicles.

An increase in SRR with temperature was expected based on studies in the literature and our physical intuition. We observed a difference in the degree of which temperature influenced SRR, depending on which area of the measured road we looked at. This result in a broadening of the distribution of C_{SRR} for increasing road conditions. The general trend is, that the magnitude of the deflection slope maximum and minimum increases (separately or together) which shows that the deflection basin gets steeper and deeper. This is consistent with our expectation, that the asphalt layer becomes softer at higher temperatures. Furthermore, the maximum deflection moves further behind the load, indicating that the role of viscous damping in the pavement becomes greater.

Some unexpected behavior in the deflection slope curves was also observed. In some areas, the magnitude of the slope becomes bigger behind the load than in front of the load, when the temperature is increased. This odd behavior was fully reproducible within the three repeated measurements. We speculate that this behavior is due to a situation where the asphalt layer becomes much softer than usual. In this case, there will be a compression of the top layer in addition to the usual bending of the top layer. This leads to a non-intuitive behavior of the overall deflection basin, and thus an odd signal in the deformation slope. A better understanding of the temperature influence on the pavement response requires a model study, giving a more detailed insight into road temperature effects on the structural behavior.

REFERENCES

Akbarian, M., Moeini-Ardakani, S. S., Ulm, F. & Nazzal, M. 2012. Mechanistic Approach to Pavement-Vehicle Interaction and Its Impact on Life-Cycle Assessment. *Transport Res Rec.* No. 2306: 171–179.

Balzarini, D., Zaabar, I. & Chatti, K. 2018. Effect of Pavement structural response on rolling resistance and fuel economy using a mechanistic approach. *Advances in Material and Pavement Performance Predictions.* Vol. 10: 49–51.

Balzarini, D., Zaabar, I., & Chatti, K. 2017. Impact of Concrete Pavement Structural Response on Rolling Resistance and Vehicle Fuel Economy. *Transport Res Rec.* Vol. 2640(1): 84–94.

Chupin, O., Piau, J. & Chabot, A. 2013. Evaluation of the Structure-induced Rolling Resistance (SRR) for pavements including viscoelastic material layers. *Materials and Structures, Springer Verlag.* Vol. 46 (4): 683–696.

Flügger, W. (second ed.) 1975. *Viscoelasticity.* Springer-Verlag Berlin Heidelberg

Nielsen, N. R., Chatti, K., Nielsen, C. P., Zaabar, I., Hjorth, P. G. & Hecksher T. Method for Direct Measurement of Structural Rolling Resistance for Heavy Vehicles. *Accepted for publication in Transport Res Rec. (Feb. 28, 2020).*

Nielsen C. P. 2019. Visco-Elasic Back-Calculation of Traffic Speed Deflectometer Measurements. *Transport Res Rec.* Vol. 2673 (12); 439. 448

Pouget, S., Sauzéat, C., Benedetto, H.D. & Olard, F. 2012. Viscous Energy Dissipation in Asphalt Pavement Structures and Implication for Vehicle Fuel Consumption. *Journal of materials in civil engineering.* Vol. 24 (5): 568–576.

Shakiba, M., Ozer, H., Ziyadi, M. & Al-Qadi, I. 2016. Mechanics based model for predicting structure-induced rolling resistance (SRR) of the tire-pavement system. *Mech Time-depend Matter.* Vol. 20: 579–600

Zaabar, I. & Chatti, K. 2014. A field investigation of the effect of pavement type on fuel consumption. *T and Di Congress 2011.* Pp. 772–781.

Advances in Materials and Pavement Performance Prediction II – Kumar et al. (eds)
© 2021 Taylor & Francis Group, London, ISBN 978-0-367-46169-0

Mechanistic damage analysis of pavements: Concept and a case study

Mohammad Rahmani & Yong-Rak Kim
Zachry Department of Civil & Environmental Engineering, Texas A&M University, College Station, TX, USA

ABSTRACT: In this study, a fully mechanistic analysis framework is proposed to simulate and evaluate the damage evolution mostly due to cracking and resulting performance of pavements. To do so, the viscoelastic and fracture properties of designated pavement materials are obtained through experiments and a fully mechanistic damage analysis is carried out using a finite element method (FEM). Different pavement configurations and traffic loads are considered based on three main functional classes of roads suggested by FHWA i.e., arterial, collector and local. For each road type, three different material combinations for asphalt concrete (AC) and base layers are considered to study damage behavior of pavement. A concept of the approach is presented and a case study where three different material combinations for AC and base layers are considered is exemplified to study progressive damage behavior of pavements when mixture properties and layer configurations were altered.

1 INTRODUCTION

Understanding the failure mechanisms and any kinds of distresses leading the asphalt pavements to degrade in term of serviceability is necessary before taking steps toward finding a proper solution. Among the common distresses in bituminous pavements, cracking is one of the most important. The intrinsic heterogeneous nature of asphalt mixtures made their cracking behavior challenging to address. Several studies have been conducted by incorporating continuum and fracture mechanics to evaluate the cracking behavior of viscoelastic asphalt mixtures. (Aragão et al. 2017; Aragão et al. 2014; Im et al. 2013; Ozer et al. 2013; Rami et al. 2017; Rodrigues et al. 2019).To computationally model the fracture behavior of asphalt materials, Cohesive Zone Modelling (CZM) has recently captured researchers' interest. (Kim 2011; Kim et al. 2010; Kim et al. 2015; Rodrigues et al. 2019).In this study, a fully mechanistic analysis framework is applied to simulate and evaluate the damage evolution and performance of pavements with different asphalt concrete (AC) mixtures: reclaimed asphalt pavement (Rap) and virgin mixtures. To do so, a fully mechanistic damage analysis is carried out using a finite element method (FEM). Different pavement configurations and traffic loads are considered based on three main functional classes of roads suggested by FHWA i.e., arterial, collector and local. For each road type, three different material combinations for AC and base layers are considered to study damage behavior of pavements. The outcome of this study quantifies the damage extent in pavements with different alternatives of RAP usage. Moreover, the extended results can be used in predicting the pavement performance over time.

2 PAVEMENT CONFIGURATIONS AND FEM MODEL

Table 1 presents three different pavement configurations: arterial, collector and local road types. The FEM mesh of arterial is exemplified in Figure 1.

A critical zone is assumed under the tire loads where damage occurrence is more likely. Within this zone, in order to have a more accurate calculation of stress field

Table 1. Pavement configurations.

Road Type	Layer Thickness (mm)			
	Surface	Base	Subbase	Subgrade
Arterial	100	150	200	300
Collector	50	100	250	350
Local	50	50	250	400

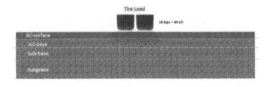

Figure 1. Mesh instance in finite element model.

over the elements and to get the damage status, finer mesh with automatic insertion of cohesive elements are employed.

3 GOVERNING EQUATIONS

The constitutive equation for isotropic linear thermo-viscoelastic materials is expressed as follows

$$\sigma_{ij}(x_m, t) = \int_0^\xi C_{ijkl}(x_m, \xi - t) \frac{\partial \xi_{kl}(x_m, \tau)}{\partial \tau} d\tau$$
$$- \int_0^\xi \beta_{ij}(x_m, \xi - t) \frac{\partial \theta(x_m, \tau)}{\partial \tau} d\tau \quad (1)$$

$$\beta_{ij}(x_m, \xi) = C_{ijkl}(x_m, \xi) \alpha_{kl}(x_m) \quad (2)$$

where σ_{ij} is the Cauchy stress tensor, x_m is the spatial coordinate, C_{ijkl} is the tensor of thermo-viscoelastic relaxation moduli, ε_{kl} is infinitesimal strain tensor, β_{ij} is the second-order tensor of relaxation moduli relating stress to temperature variations, θ is temperature, ξ is reduced time, τ is time integration variable, and α_{kl} is coefficient of thermal expansion.

The reduced time is defined as

$$\xi(t) = \frac{1}{a_T(\theta(\tau))} d\tau$$

$$\log_{10}(a_T) = \frac{-C_1(\theta - \theta^R)}{C_2 + (\theta - \theta^R)} \quad (3)$$

where t is time, a_T is temperature shift factor, θ^R is reference temperature and C_1 and C_2 are constants.

The viscoelastic relaxation modulus for thermo-viscoelastic asphalt mixture can be represented in a form of Prony series which can be characterized by conducting dynamic modulus test. It is expressed as:

$$E(t) = E_\infty + \sum_{i=1}^n E_i e^{-t/\rho i} \quad (4)$$

where E_i and E_∞ are spring constants in a generalized Maxwell model and ρ_i is relaxation time.

For simulating cracks, this study employed the nonlinear-viscoelastic cohesive zone model proposed by Allen and Searcy as shown in the following traction-displacement relation.

$$t_i(x_p, \xi) = \frac{1}{\lambda(x_p, \xi)} \frac{[u_i(x_p, \xi)]}{\delta_i^*(x_p)} [1 - \alpha_D(x_p, \xi)]$$

$$\left\{ \sigma_i^f(x_p, \xi - \xi') + \int_0^t E_{CZ}(x_p, \xi - \xi') \frac{\partial \lambda(x_p, \xi')}{\partial \xi'} \right\} \quad (5)$$

where t_i is the traction vector on cohesive element boundary, u_i is the cohesive zone opening displacement, δ_i^* is the material length parameter vector, α_D is

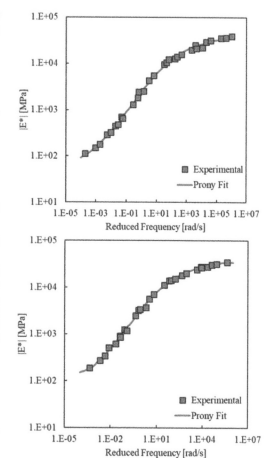

Figure 2. DM test result for AC mixture: top: without RAP, and bottom: with 30% RAP.

the internal damage parameter, σ_i^f is the stress level associated with damage initiation, E_{CZ} is the cohesive zone relaxation moduli and λ is the Euclidean norm of the cohesive zone displacement opening in normal and tangential directions.

4 MATERIAL PROPERTIES

In order to obtain the viscoelastic stiffness of AC surface and base mixtures, dynamic modulus (DM) tests were performed on two sets of virgin and 30% RAP asphalt mixtures. Figures 2 and 3 present the DM test results.

To obtain damage properties of AC surface mixtures, Indirect Tensile Test (IDT) were performed at the temperature of 25°C and finite element simulation was carried out to calibrate the cohesive zone model parameters. Figure 4 presents the IDT results and calibration process for RAP and virgin AC mixtures.

225

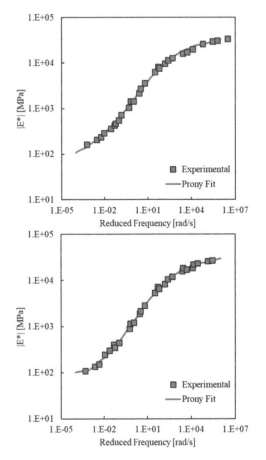

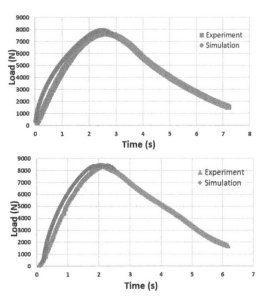

Figure 4. AC mixture IDT test result and calibration: top: without RAP, and bottom: with 30% RAP.

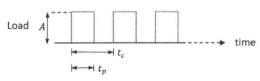

Figure 5. Sine pulse loading.

Figure 3. DM test result for base mixture: top: without RAP, and bottom: with 30% RAP.

5 LOADING AND BOUNDARY CONDITIONS

For arterial and collector roads, loadings are assumed 80 kN per axle and for local road, 6.7 kN single axle load was applied. To specify how the prescribed loads evolve in time, sine pulse time function was used. Mathematical description of the sine pulse time function is expressed as

$$f(t) = \begin{cases} A \sin\left[\frac{\pi}{t_p}(t - t_o)\right] & t_0 < t < t_0 + t_p \\ 0 & \text{otherwise} \end{cases}$$
$$t_0 = (cycle_i - 1) \times t_c \qquad (6)$$

where A is loading amplitude, t_p is pulse duration, t_c is cycle duration and $cycle_i$ is the ith loading cycle.

Figure 5 illustrates the definition of time function variables.

Regarding boundary conditions for the pavements, horizontal displacements at two sides of transverse cross section as well as vertical displacements at the bottom of subgrade layer were fixed in finite element model.

6 RESULTS AND DISCUSSION

For each road type (arterial, collector and local), three different cases of layer materials are studied to compare the damage evolution of surface layer over time.

Total 10,000 loading cycles were applied on the pavements and damage status of the surface layer (cohesive elements in critical zone) were obtained. Figure 6 shows the number of cohesive elements inserted in each case for arterial road as the number of loading cycles increase. Figure 7 illustrates the accumulated damage in cohesive elements over loading cycles.

As it is depicted in Figure 6, the number of cohesive elements inserted in the model increases as the loading history expands. As this is not necessarily a direct indicator for damage development in the pavement since there might be cohesive elements with zero percent damage, thus accumulated damage in cohesive elements were evaluated.

Figure 7 shows the extent of damage developed in cohesive elements during the loading period for different road types and different RAP usage alternatives.

As shown in Figure 7, the short-term (0-2,000 cycles) performance for all cases did not show significant case-specific damage evolution, while the

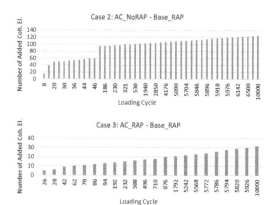

Figure 6. Arterial Road: number of inserted cohesive elements in FE model over loading cycles for three RAP usage alternatives.

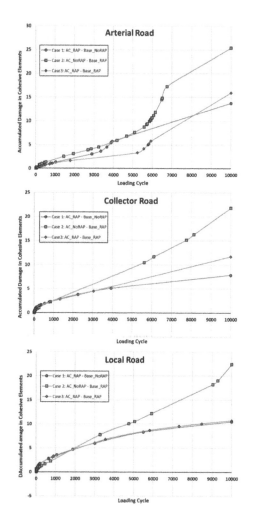

Figure 7. Accumulated damage in cohesive elements in different road types.

evolution of damage became quite different between cases when longer-term performance is considered. This case study, although limited, demonstrates how the mechanistic approach can be used to predict the progressive damage behavior of pavements when mixture properties and layer configurations were varied, which can be used for better selection and design of mixtures and pavement structures.

REFERENCES

Aragão, F. T. S. et al. 2014. Numerical–Experimental Approach to Characterize Fracture Properties of Asphalt Mixtures at Low Temperatures. *Transportation Research Record* 2447(1): 42–50.

Aragão, F. T. S. et al. 2017. Characterization of temperature- and rate-dependent fracture properties of fine aggregate bituminous mixtures using an integrated numerical-experimental approach. *Engineering Fracture Mechanics* 180: 195–212.

Im. S. et al. 2013. Rate-and temperature-dependent fracture characteristics of asphaltic paving mixtures. *Journal of Testing and Evaluation* 41(2): 257–268.

Ozer, H. et al. 2013. Performance characterization of asphalt mixtures at high asphalt binder replacement with recycled asphalt shingles. *Transportation Research Record* 2371(1): 105–112.

Rami, K. Z. et al. 2017. Modeling the 3D fracture-associated behavior of viscoelastic asphalt mixtures using 2D microstructures. *Engineering Fracture Mechanics* 182: 86–99.

Rodrigues. J. A. et al. 2019. Crack modeling of bituminous materials using extrinsic nonlinear viscoelastic cohesive zone (NVCZ) model. *Construction and Building Materials* 204: 520–529.

Kim, Y.-R. et al. 2010. Damage modeling of bituminous mixtures considering mixture microstructure, viscoelasticity, and cohesive zone fracture. *Canadian Journal of Civil Engineering* 37(8): 1125–1136.

Advances in Materials and Pavement Performance Prediction II – Kumar et al. (eds)
© 2021 Taylor & Francis Group, London, ISBN 978-0-367-46169-0

Mechanistic modeling of macro-texture's effect on rolling resistance

S. Rajaei
Intertek-PSI, Dallas, TX, USA

K. Chatti
Department of Civil and Environmental Engineering, Michigan State University, East Lansing, MI, USA

ABSTRACT: Rolling resistance of a vehicle plays an important role in its fuel consumption. There are different mechanisms involved in the rolling resistance, namely, vehicle dynamics, tire bending and deformation, and tire tread deformation. Pavement surface macro-texture is one of the factors that influences the rolling resistance mechanisms of tire bending and tire and tread deformations. The main aim of this study is to develop a mechanistic model that is able to capture the effect of macro-texture on the tire's rolling resistance. For this purpose, a 3D finite element model of a tire is developed. The profile macro-texture is extracted from measured profiles and characterized using the root mean square (RMS) parameter. The rolling resistance coefficient is then calculated for surfaces with different RMS values. A comparison of the results with previous experimental studies shows that the model is capable of capturing the effect of macro-texture accurately.

1 INTRODUCTION

The road surface profile consists of various scales of roughness and texture which can affect different aspects of tire-pavement interaction such as rolling resistance, friction, noise, etc. Rolling resistance plays an important role in vehicle fuel consumption while friction is essential for safety of the vehicle. There is a good understanding of the effect of pavement surface texture on friction. However, the effect of surface texture on rolling resistance has only been investigated in a few studies.

The effect of texture on tire rolling resistance can be captured by different approaches ranging from tire models and mechanistic approaches to experimental studies. A detailed FE tire model is able to calculate the tire deformation and the interaction of its components at different scales (Ghoreishy 2008). However, so far, the existing FE tire models barely consider the effect of pavement texture, due to the high computational costs. The other simplified mechanistic models also have some limitations. As an example, Wullens & Kropp (2004) proposed a 3D model of a simplified equivalent tire structure similar to the Pre-tensioned Kirchhoff's plate for capturing the effect of texture. However, their model does not consider the curvature of the tire. O'boy & Dowling (2009) developed a model for tire-road noise prediction considering the tire as a multi-layer viscoelastic cylinder belt which is able to account for the surface texture.

In another study, in addition to experimental measurements of the effect of texture on rolling resistance,

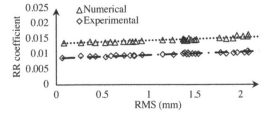

Figure 1. Boere experimental and numerical results at v=80km/h (Boere 2009).

Boere (2009) developed a numerical model by dividing the problem into two parts: (i) a FE tire model rolling on a smooth surface in a steady state condition and (ii) a mechanistic model of tire moving on a textured surface developed using modal analysis (based on Andersson & Kropp 2008). The model accounts for the texture by applying a nonlinear stiffness for the contact. The total energy dissipation was calculated as the summation of the energy dissipation of the two parts. The results of the model were compared to experimental results from measurements using a trailer on 30 test tracks. The measurements were reported in terms of rolling resistance coefficient (RRC), which is the ratio between the rolling resistance and axle forces. The root mean square (RMS) of the surface profile was used as the surface texture measure. Figure 1 shows the comparison of Boere's numerical and experimental results.

It can be seen that Boere's numerical model captures the effect of texture on rolling resistance very well.

However, there is a significant difference between the total value of rolling resistance of the numerical and experimental results.

The available experimental studies for the effect of texture on rolling resistance include the NCHRP 1-45 study (Chatti & Zaabar 2012) and the MIRIAM project (Hammarström et al 2012). In NCHRP 1-45, a 1 mm increase in mean profile depth (MPD) causes an increase of 4% in the rolling resistance of a car, while the MIRIAM project reported a significantly higher effect with 15.1% increase.

In this study, the main goal is to develop a single mechanistic tire model that is able to capture the effect of the surface macro-texture on the rolling resistance of the tire using finite element modeling. For this purpose, a finite element (FE) tire model is developed in ABAQUS software. The mesh size of the tire tread is chosen to be small enough to capture the effect of the texture accurately, while keeping the computational cost at minimum. The model is then verified by comparison with a series of controlled tests performed by Wei & Olatunbosun (2014). After performing a sensitivity analysis for finding the optimum mesh size for the contact, the tire is rolled over different surfaces and the rolling resistance force is calculated. The model results are then validated by comparing with Boere's experimental results.

2 SURFACE TEXTURE CHARACTERIZATION

A pavement surface profile can be divided into various scales of micro-texture (wavelength (λ)<0.5mm), macro-texture ($0.5 < \lambda < 50$mm and $0.1 <$ peak to peak amplitude (A)<20 mm), mega-texture ($50 < \lambda < 500$mm and $0.1 < A < 50$ mm), Roughness ($\lambda > 500$mm) (Wambold et al 1995).

In this study, the focus is on effect of macro-texture on rolling resistance of the tire. Two different surface profile data sets are available. Data from (i) chip seal surface profiles (with 43cm length) including macro-texture and mega-texture; and (ii) Data from the Danish Road Directorate (with 100m length), including all (four) scales.

To obtain the macro-texture of the profiles, the higher and lower wavelengths should be eliminated from the profiles so that the remainder of the profile only includes the macro-texture wavelengths. This operation is performed using filtering.

The surface profiles are defined in the FE model directly in the driving direction. However, for comparison of the results with Boere's experiments, the profiles should be characterized using the RMS parameter defined as

$$R_q^2 = \frac{1}{n}\sum_{i=1}^{n}(z_i)^2 \quad (1)$$

where z_i is the height of each profile point when the mean is set to zero, and n is the number of points.

The RMS values for the macro-texture in the two available profile databases are mostly below 2mm. Based on the obtained range, the final surfaces are selected with RMS values of 0.51, 1.1, 1.6 and 2.1 mm.

3 TIRE MODEL DEVELOPMENT AND VERIFICATION

Most of the published finite element models of tires consist of a simplified tire model including only the main parts of the tires such as ply, belts, beads, side-walls, tread, and rim (Hernandez 2015; Srirangam 2015; Wei & Olatunbosun 2016). For developing such models, the geometry of the different parts, their material properties, mesh type, the contact properties between the tire and the pavement surface, and the required boundary conditions for the problem should be determined. In this study, the modeling process for capturing the effect of macro-texture on rolling resistance is divided into two steps: (i) Verification of the model by comparison with experimental study by Wei & Olatunbosun (2014) where the tire size, material properties, and test setup are similar to their study; and (ii) comparison of the rolling resistance due to surface texture with Boere's experimental results where the size and the boundary conditions are matched with their experiments.

For performing dynamic analysis of the tire 5 different steps are followed. The first four steps, that are performed within ABAQUS standard, follow each other. The last step of transient analysis is performed in ABAQUS explicit by using step (iv) as the initial condition:

i. 2D model of half of the tire cross section. The model geometry, material properties, and cross section mesh is defined here. (see Figure 2 (a)).
ii. 3D half tire model by revolving the 2D cross section axi-symmetrically. The contact and the number of meshes in the driving direction are defined here (see Figure 2 (b)).
iii. 3D full tire model, by mirroring the 3D half tire model. The vertical load on the rim (axle load) reaches to the required value (see Figure 2 (c)).
iv. Steady state analysis of the tire is performed to increase the rolling velocity of the tire to the required velocity in free rolling condition.
v. Transient analysis of the tire

For tire model verification, the results of the model are compared to experimental results by Wei & Olatunbosun (2016). In their experiment, a tire is rolled over a step (10 * 25mm) at a velocity of 10km/h (see Figure 2 (d)). Figure 3 shows the results of this comparison. As it can be seen, the simulation is in a very good agreement with the experimental results.

Then, for capturing the effect of macro-texture on rolling resistance, the tire model is changed to match Boere's study. In Boere's experiments, the rolling resistance measurements are performed at a velocity of

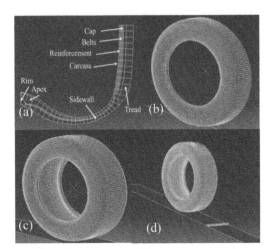

Figure 2. Tire model development, (a) tire cross section (b) 3D half tire, (c) 3D full tire, (d) tire rolling over a step in 10km/h velocity

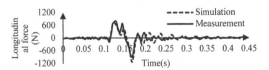

Figure 3. Tire model verification.

80km/h. Before rolling the tire over the textured profile, the transient model should reach steady state by rolling over a smooth surface. The length of the smooth surface (transition length) is dependent on the rolling velocity. For 80km/h this length is found to be 6m for the developed tire model. After reaching steady state condition, the tire is rolled over different texture surfaces and their corresponding rolling resistance coefficients are calculated and compared with experimental results.

Rolling resistance force is defined as the force applied to the rim required to compensate any force resisting the tire rolling at a certain velocity. So, in the tire FE model, the rolling resistance force is defined as the average work per unit length or average longitudinal force that should be applied to the centroid of the tire rim to roll it at a specific velocity.

$$RRF = \frac{\Sigma F \cdot d}{d} \qquad (2)$$

The total rolling resistance coefficient (RRC) is defined as the ratio between the rolling resistance forces in the tire and the applied normal force on the rim (axle load).

4 MESH SENSITIVITY ANALYSIS

As the wavelength (λ) of the profile and the distance between the two points on the profile decreases, the

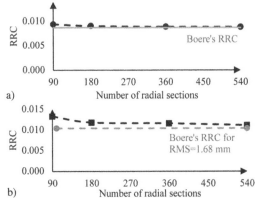

Figure 4. Mesh sensitivity analysis for (a) a smooth surface (b) texture profile with RMS = 1.68 mm.

mesh size in the driving direction and consequently the mesh size within the tire cross section should be reduced. For this reason, a mesh sensitivity analysis is performed to determine the minimum radial mesh size for capturing the energy dissipation of macro-texture profile efficiently. It should be noted that while decreasing the mesh can be beneficial, it also increases the computational time dramatically. So, a trade off is required between these two factors.

The mesh sensitivity analysis is performed on both smooth and textured surfaces with radial section numbers of 90 (element size of 2.2cm) to 540 (element size of 3.7mm) (see Figure 4). As it can be seen, RRC variation between 360 and 540 number of sections for smooth and macro-texture surface is small, and the results tend to a plateau (Figure 4 (a) and (b)). This is an indication that the effect of the wavelength less than 3.7mm on tire RRC is small. As a result, the number of sections of 540 should be sufficient for capturing the effect of macro-texture on RRC.

5 RESULTS

After performing the mesh sensitivity analysis, the tire model is rolled over textured surfaces with different RMS values, and the RRC values corresponding to each RMS are calculated (see Figure 5). The RMS value of zero corresponds to the tire rolling on the smooth surface.

As it can be seen, a linear relationship (with a slope of 0.0012) can be considered for the effect of macro-texture profile on FE tire RRC. The obtained results are very similar to Boere's study with the FE model giving slightly higher influence of the macro-texture on the tire RRC for higher RMS values. This difference can be related to the assumptions and simplifications considered in the model that may be different from the experimental set up. Another comparison is performed between the results of the FE model and two

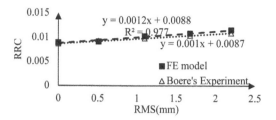

Figure 5. RRC and RMS relationship- Comparison of FE model and Boere Experimental studies.

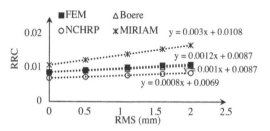

Figure 6. Comparison between different studies.

additional experimental studies: NCHRP-1-45 and MIRIAM projects (see Figure 6).

Regardless of the difference between the RRC values on the smooth surface (RMS=0), it is evident that MIRIAM project results, with a slope of 0.003, are showing a much higher effect of the surface texture on RRC in comparison to the other three studies, which have slopes between 0.0008 and 0.0012.

6 CONCLUSION

The effect of pavement surface profile macro-texture on tire rolling resistance is investigated in this study. To capture the effect of the texture on tire deformation, including tire bending and tread deformation, a full 3D finite element tire model is developed and verified. The higher and lower scales of surface profile (roughness, mega-, and micro-textures) are filtered from the profile. A mesh sensitivity analysis is performed to find the minimum mesh size in the driving direction that is capable of capturing the effect of macro-texture on rolling resistance.

For comparison of the results of the model with those from other experimental results, the RMS of the filtered profile height is used to characterize the

surface profile. The analysis shows that the slope of the linear relationship between the texture RMS and rolling resistance coefficient (RRC) of the tire is similar to those from Boere (2009) and NCHRP 1-45 project (within 20%). However, the slope from the MIRIAM model is three times as much, thus significantly overestimating the effect of texture in comparison to the other three models.

REFERENCES

Andersson P.B.U. & Kropp. W. 2008. "Time domain contact model for tyre/road interaction including nonlinear contact stiffness due to small-scale roughness. "Journal of Sound and Vibration, 318(1–2):296–312.

Boere, S. 2009. "Prediction of road texture influence on rolling resistance." Eindhoven University of Technology.

Chatti, K. & Zaabar, I. 2012. Estimating the effects of pavement condition on vehicle operating costs. NCHRP Report 720. Transportation Research Board.

Ghoreishi MH. 2008. A state of the art review of the finite element modelling of rolling tyres. Iranian Polymer Journal.17(8):571–97.

Hammarström, U., Eriksson, J., Karlsson, R., & Yahya, M. R. 2012. Rolling resistance model, fuel consumption model and the traffic energy saving potential from changed road surface conditions. Statens väg-och transportforskningsinstitut.

Hernandez, JA. 2015. Development of deformable tire-pavement interaction: contact stresses and rolling resistance prediction under various driving conditions. University of Illinois at Urbana-Champaign.

O'Boy DJ, Dowling AP. 2009. Tyre/road interaction noise— Numerical noise prediction of a patterned tyre on a rough road surface. Journal of Sound and Vibration. 5;323(1–2):270–91.

Srirangam, S. K. 2015.Numerical Simulation of Tire-Pavement Interaction.

Wei C, Olatunbosun OA. 2014. Transient dynamic behaviour of finite element tire traversing obstacles with different heights. Journal of Terramechanics. 1;56:1–6.

Wei C, Olatunbosun OA. 2016. The effects of tyre material and structure properties on relaxation length using finite element method. Materials & Design.15;102:14

Wambold JC, Antle CE, Henry JJ, Rado Z. 1995. PIARC (Permanent International Association of Road Congress) Report. International PIARC Experiment to Compare and Harmonize Texture and Skid Resistance Measurement, C-1 PIARC Technical Committee on Surface Characteristics, France.20.

Wullens F, Kropp W. 2004. A three-dimensional contact model for tyre/road interaction in rolling conditions. Acta Acustica united with Acustica.1;90(4):702–11.

Advances in Materials and Pavement Performance Prediction II – Kumar et al. (eds)
© 2021 Taylor & Francis Group, London, ISBN 978-0-367-46169-0

Pavement instrumentation with near surface LVDTs

A. Skar, J. Nielsen & E. Levenberg
Technical University of Denmark, Lyngby, Denmark

ABSTRACT: This paper presented an approach for mechanical pavement condition monitoring based on post-construction installation of near surface LVDTs. The concept calls for making shallow grooves or blind holes at the pavement surface, and then fixing horizontal LVDTs to measure any changes in groove length or hole diameter resulting from nearby vehicle passes. A field experiment was designed and executed to demonstrate the approach involving the deployment of two LVDTs in an existing asphalt road, and recording the effects of a passing truck. It is shown that the LVDT readings can be matched with a layered elastic pavement model, leading to the inference of in situ layer moduli. The investigation is considered a first step towards practical application of a real-world sensing platform for pavements.

1 INTRODUCTION

Monitoring the mechanical condition of pavement systems within the live transportation network is typically carried out periodically, e.g., by means of non-destructive testing equipment like the FWD. Given the economical and societal importance of pavements, and the ever-increasing reliance upon the service level they provide, the engineering community is faced with the technological challenge of increasing condition monitoring frequency while minimizing user disturbance. One solution approach is embedding pavement sensors that monitor responses triggered by passing vehicles, and subsequently utilizing their reading to estimate mechanical condition (Levenberg 2013). Embedded pavement sensing is not new (e.g., Freeman et al. 2001; Tabatabaee & Sebaaly 1990, with common off-the-shelf devices capable of measuring stresses, strains, and differential vertical displacements. Nonetheless, it is hard to envision wide-area usage of such technology (Levenberg et al. 2014); the main reason is the need for pre-construction installation whilst in most situations the infrastructure is already built. Another reason is the need for embedding wires to supply power and communicate measured data. Yet another shortcoming of embedded sensing is that the gear is not easily retrievable or accessible for repair and upgrades.

The current work attempts to alleviate some of the above listed limitations; the advocated idea is post-construction instrumentation of near surface sensing gear that is fixed roadside - where road studs are typically placed. Doing so means that a monitoring system can be attached to any exiting pavement system. Another potential advantage of deploying near surface roadside sensors is the technological feasibility of establishing wireless communication between

sensing nodes as well as utilizing solar energy to power the devices.

Common sensors that are suitable for post-construction installation are accelerometers (Levenberg 2012; 2015) and tiltmeters (Mohsen & Saleh 1998); geophones and gyroscopes are also deemed relevant though underexplored. Inspired by a French standard from the late 1970's called Ovalisation Test (Kobisch & Peyronne 1979), another type of sensing concept is herein proposed, based on Linear Variable Differential Transformer (LVDT) technology. The concept calls for making shallow grooves or blind holes at the pavement surface, and then fixing horizontal LVDTs to measure changes in groove length (or changes in diameter for the case of a hole) due to nearby loading (Nielsen 2019). Unlike accelerometers and geophones, LVDTs are capable of measuring the response from passing vehicles in the entire range of speeds, down to stationary conditions. Consequently, high quality response data can be collected to serve as basis for continuous inference of mechanical condition (as well as inference of traffic characteristics).

The objective of this paper is to: (i) explain the near surface LVDT instrumentation concept in a manner than can be replicated by others, and (ii) demonstrate the concept in a field experiment for evaluating mechanical condition based on the sensor readings.

2 INSTRUMENTATION CONCEPT

The concept of a near surface LVDT installed post-construction in an asphalt pavement is presented in Figure 1. This figure offers both side view and a top view sketches, as well as a picture taken from an actual installation. The specific LVDT utilized

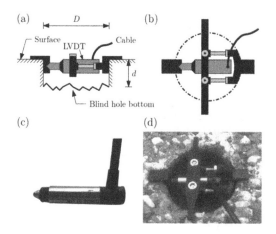

Figure 1. Installation of near surface LVDT: (a) side view ketch (b) plan view sketch, (c) picture of GT0500XRA LVDT, and (d) LVDT installed in an asphalt road.

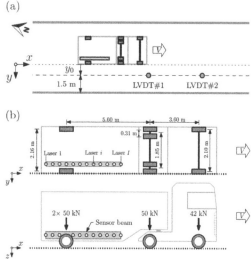

Figure 2. Experimental setup: (a) top view ($x - y$ plane) of the test section and (b) sketch of the Raptor truck.

was model GT0500XRA manufactured by RDP Electrosense, with body length of 40 mm and diameter of 8 mm (see Figure 1c). This sensor is characterized by a measurement range of ± 0.5 mm and measurement resolution better than 1 μm. The installation shown is inside a blind core-hole, with an internal diameter of $D = 54$ mm, i.e., slightly larger than the LVDT length, and depth of roughly $d = 20$ mm, i.e., slightly larger than the LVDT diameter. As can be seen, the aim is to monitor changes in D at the pavement surface.

A mounting device was designed and 3D-printed to horizontally fix the LVDT in the hole without touching the bottom. As can be deduced from Figure 1, the device was composed of three parts: a central part that holds the sensor body with 'wings' on each side resting freely on pavement surface, a rear adjustable part that can accommodate different hole diameters (i.e., designed for multiple uses) integrated with a right angle piece resting freely on the asphalt surface, and a front right angle piece that is resting on the edge of the hole and making contact with the measuring tip of the LVDT.

3 FIELD DEMONSTRATION

3.1 Instrumentation and measurements

In order to demonstrate the instrumentation concept, a field test was designed and carried out in a low-volume asphalt road. The paved width of the road was 8 m, and the structural thickness was 650 mm, composed of 120 mm asphalt concrete and 530 mm of combined unbound base and subbase. The subgrade was a relatively stiff sandy clay (overconsolidated) extending to a large depth and clear of underground utilities.

A flat section of the road was instrumented with two sensors named LVDT#1 and LVDT#2. Both were fixed to the asphalt surface according to the arrangement shown in Figure 1. Two interlinked USB signal

conditioner units (SD20 manufactured by Metrolog) were employed for powering the sensors and for data logging. These units offer 24-bit analog to digital conversion and acquisition rates of up to 880 Hz. Figure 2a offers a plan view of the test section indicating sensor positions. Both sensors were oriented to measure the horizontal differential displacement in the direction of travel (i.e. along the x-axis). As can be seen, the sensors were spaced 2 m apart in the travel direction, with an identical transverse offset from the road edge of 1.5 m (i.e., they were expected to provide identical readings).

Sensor readings were triggered by nearby passes of the Raptor – a heavy truck that hosts a beam of lasers (as well as other sensors) designed to perform pavement evaluation while driving. Figure 2a superposes a silhouette of the Raptor over the experimental arrangement; the travel speed is denoted by V (assumed constant) and the travel direction is assumed parallel to the x-axis. The symbol y_0 denotes the offset distance from the rear right tire edge to the sensor line. Due to the proposed installation configuration, the LVDT sensors should not be located within the truck's wheel path. Figure 2b offers side and top view sketches of the Raptor truck, showing tire configuration and indicating axle loads.

Data from a single truck pass with $V = 59.6$ km/h and $y_0 = 500$ mm (independently evaluated) is presented in Figure 3 utilizing a common timeline with arbitrary origin. It can be observed from the Figure that peaks appear in the sensor readings with timing corresponding to the truck's axle spacing, and magnitudes corresponding to the truck's axle weights. Considering the longitudinal spacing between the sensors, the time delay in the signal is directly related to the travel speed (Levenberg 2014); this information was used to obtain V. The signals are not identical, possibly due to differences in installation conditions in combination

233

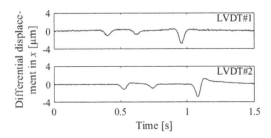

Figure 3. Measured traces from LVDT#1 and LVDT#2 in the field experiment.

with a truck that is not driving perfectly parallel to the sensor line.

3.2 Mechanical condition evaluation

An interpretation method is hereafter described, focused on evaluating the mechanical properties of the tested road based on the acquired LVDT measurements in Figure 3. In general terms, the evaluation is sought by best fitting measured traces with calculated traces obtained from a computational pavement model (see also Nielsen 2019). In this connection, the classic two-layered linear elastic model was chosen (Burmister 1945), with two unknowns to be inferred: E_1 - the Young's modulus of the top layer representing the structure (asphalt, base, and subbase layers), and E_2 - the Young's modulus of the half-space representing the sandy clay subgrade. To expedite the underlying computations, the numerical acceleration technique proposed in Andersen et al. (2018) was employed. Taken as known in the fitting procedure are the thickness of the top model layer (650 mm), the Poisson's ratios of the two layers $v_1 = 0.35$, and $v_2 = 0.40$, the travel speed of the truck $V = 59.6$ km/h, the truck's loading configuration (Figure 2b), as well as the offset $y_0 = 500$ mm (Figure 2a).

The fitting procedure was carried out in several steps: (i) for trial values of E_1 and E_2 the two-layered pavement model was engaged to calculate horizontal surface displacement traces at two surface points spaced apart by D resulting from the passage of a Raptor truck; (ii) the differential horizontal displacement between the two points was taken as representative of the LVDT readings; (iii) a so-called individual objective function was defined to quantify the mismatch between a given LVDT trace and the corresponding calculated trace; (iv) a global objective function was defined to combine the two individual objective functions into a single expression that represents the mismatch between measured and calculated traces; and (v) a search was performed to find an optimal moduli set that minimizes the global objective function. Ultimately, the optimal moduli values are deemed representative of the in situ mechanical condition of the pavement system in the vicinity of the sensor locations.

The individual objective function $\varphi_j = \varphi_j(E_1, E_2)$ for LVDT#j is defined as

$$\varphi_j = \frac{1}{N} \sum_{n=1}^{N} |f_{j,n} - g_{j,n}| \tag{1}$$

where $f_{j,n}$ is the measured signal of LVDT#j at time t_n, $g_{j,n}$ is the corresponding calculated (or model-simulated) signal for LVDT#j at time t_n, and N is the total number of data points utilized for the matching. Combining the matching information from both sensors, the global objective function $\Phi = \Phi(E_1, E_2)$ is defined as

$$\Phi = \sum_{j=1}^{2} \left(\frac{\varphi_j}{\varphi_j^0} - 1 \right) \tag{2}$$

where φ_j is the individual objective function defined in Equation 1, and φ_j^0 is the lowest value the objective function φ_j can attain when individually minimized with respect to E_1 and E_2. This formulation is essentially a min-max formulation (Osyczka 1978). The last analysis step is minimizing the global objective function with respect to E_1 and E_2 to arrive at an estimate for the in situ mechanical pavement properties E_1^* and E_2^*. Formally, this means

$$(E_1^*, E_2^*) = \underset{(E_1, E_2)}{\arg \min} \Phi \tag{3}$$

The required minimization tasks were carried out with a two-dimensional grid search method. The method was first applied for finding the minimum solution ϕ_j^0 of the individual objective functions, and second for finding the minimum of the global objective function. Figure 4 presents the gridded solution space for the global objective function as a contour plot in E_1 and E_2. A clear global minimum value can be seen (indicated by x), corresponding to $E_1^*=141$ MPa and $E_2^*=147$ MPa. These moduli values represent consensus among the two LVDTs; the associated measured and calculated traces are superposed in Figure 3.

Having obtained the mechanical properties, any critical pavement responses (e.g. critical horizontal strain or maximum deflection) can be deduced from a mechanical model allowing engineers to evaluate the prevailing mechanical pavement condition.

4 CONCLUSION

Based on the above-described investigation, the following conclusions are drawn: (i) post-construction near surface installation of LVDTs is feasible resulting in measurements in horizontal, differential displacement at pavement surface, and (ii) a multi-layered linear elastic half-space model can well reproduce LVDT readings, and (together with field measurements) be used in an inverse analysis for evaluation of the mechanical condition of a pavement structure.

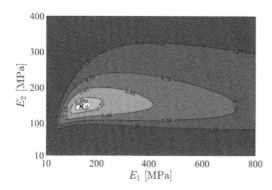

Figure 4. Contour plot of global objective function Φ [μm] as a function of E_1 and E_2, with optimum solution indicated by x.

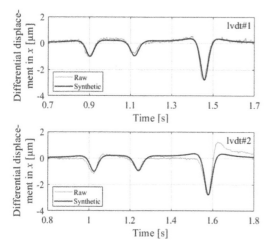

Figure 5. Field-measured and model-calculated LVDT traces corresponding optimum solution E_1^*=141 MPa and $E_2^* = 147$.

It should be noted that the suggested interpretation scheme required a priori knowledge of vehicle loading configuration as well as lateral offset from the sensing array. Hence, for application of a real-world sensing platform that offers continuous assessment of mechanical pavement conditions both aspects must be separately and independently measured. As a means of identifying vehicle loading, one idea is to utilize a weigh-in-motion system in combination with image recognition. Another idea is to carry out the condition assessment by periodically traversing the pavement network with a truck of known weight and dimensions. Offset estimation can be potentially achieved with proximity or laser distance sensors placed next to the LVDTs.

REFERENCES

Andersen, S., Levenberg, E., & Andersen, M.B. 2018. Efficient reevaluation of surface displacements in a layered elastic half-space. International Journal of Pavement Engineering, DOI: 10.1080/10298436.2018.1483502.
Burmister, D.M. 1945. The general theory of stresses and displacements in layered systems. Journal of applied physics, 16(2), 89–94 (Part I), 126–127(Part II), 296–302 (Part III).
Freeman, R.B., Tommy, C.H, McEwen, T., & Powell, R.B. 2001. Instrumentation at the national center for asphalt technology test track. Report No. ERDC TR-01-9, U.S. Army Corps of Engineers, Engineer Research and Development Center.
Kobisch, R., & Peyronne, C. 1979. L'ovalisation: Une nouvelle methode de mesure des deformations elastiques des chaussees. Bulletin de Liaison des Lab des Ponts et Chaussées, 102, 59–71.
Levenberg, E. 2012. Inferring pavement properties using an embedded accelerometer. International Journal of Transportation Science and Technology, 1(3), 229–246.
Levenberg, E. 2013. Inverse analysis of viscoelastic pavement properties using data from embedded instrumentation. International Journal for Numerical and Analytical Methods in Geomechanics, 37(9), 1016–1033.
Levenberg, E. 2014. Estimating vehicle speed with embedded inertial sensors. Transportation Research Part C, 46, 300–308.
Levenberg, E. 2015. Backcalculation with an implanted inertial sensor. Transportation Research Record, 2525, 3–12.
Levenberg, E., Shmuel, I., Orbach, M., and Mizrachi, B. 2014. Wireless Pavement Sensors for Wide-Area Instrumentation. Proceedings of the 3rd International Conference on Transportation Infrastructure (ICTI), Losa, M. and Papagiannakis, T. (eds.), CRC Press, Leiden, The Netherlands, pp. 307–319.
Mohsen, M.S., & Saleh, B. 1998. Evaluation of in-situ pavement moduli using high resolution tiltmeter. Journal of Transportation Engineering, 124(5), 443–447.
Nielsen, J. 2019. Inferring pavement layer properties from roadside sensors. MSc Thesis, Technical University of Denmark, Department of Civil Engineering, Kongens Lyngby, Denmark.
Osyczka, A. 1978. Approach to multicriterion optimization problems for engineering design. Computer Methods in Applied Mechanics and Engineering, 15(3), 309–333.
Tabatabaee, N. & Sebaaly, P. 1990. State-of-the-art pavement instrumentation. Transportation Research Record 1260, 246–255.

Advances in Materials and Pavement Performance Prediction II – Kumar et al. (eds)
© 2021 Taylor & Francis Group, London, ISBN 978-0-367-46169-0

A parameter back-calculation technique for pavements under moving loads

Zhaojie Sun, Cor Kasbergen, Karel N. van Dalen, Kumar Anupam & Sandra M.J.G. Erkens
Delft University of Technology, Delft, The Netherlands

Athanasios Skarpas
Khalifa University, Abu Dhabi, United Arab Emirates
Delft University of Technology, Delft, The Netherlands

ABSTRACT: Maintenance and rehabilitation strategies of pavements are usually made based on the results of performance evaluation. An efficient tool for pavement structural evaluation at network level is the traffic speed deflectometer (TSD) test. In order to deal with TSD measurements, this paper proposes a parameter back-calculation technique. Firstly, the sensitivity of the surface response for an elastic pavement structure with hysteretic damping to different structural parameters is investigated. Then, the ability of the parameter back-calculation technique is verified by conducting a case study. The results show that the proposed technique is able to back-calculate the structural parameters of pavements by analysing TSD measurements. The presented work contributes to the development of parameter back-calculation techniques for the TSD test.

1 INTRODUCTION

Maintenance and rehabilitation issues are very important in the whole lifespan of pavements. Accurate maintenance and rehabilitation strategies can restore the performance of pavements with minimum costs. In general, the formulation of the strategies are on the basis of functional and structural evaluation of the pavement performance. The pavement structural evaluation is usually achieved by conducting non-destructive tests (Al-Khoury et al. 2001a). A commonly used non-destructive test for pavement structural evaluation is the falling weight deflectometer (FWD) test (Marecos et al. 2017; Rabbi & Mishra 2019). However, the FWD test is quite time and resource consuming because of the stop-and-go measuring process. In addition, this measuring process also results in traffic disruption and safety issues (Liu et al. 2018). These limitations of the FWD test hinder its application at network level. In order to overcome these limitations, the traffic speed deflectometer (TSD) test has been developed recently. The TSD device can measure the pavement surface response at normal driving speeds, so it is a promising tool for pavement structural evaluation at network level (Shrestha et al. 2018). However, the TSD test is not widely used because of the lack of a proper parameter back-calculation technique. To solve this problem, the suitability of a spectral element method-based parameter back-calculation technique to deal with TSD measurements is investigated in this paper.

2 MODEL DESCRIPTION

Theoretically, the TSD test on pavements is modelled as a layered system subjected to a uniformly moving surface load. The load is considered to be a constant force which is evenly distributed over a pair of rectangular areas. The geometry of the loading areas is determined based on the footprint of the TSD wheel. The layered system consists of several layers and a half-space, which are well-defined elastic continua. For the layered system, each layer and the half-space are respectively simulated by a layer spectral element and a semi-infinite spectral element, which are developed by following the spectral element method-based procedure shown in Sun et al. (2019). It is assumed that the layered system has hysteretic damping, which is simulated by replacing the Young's modulus E with a complex Young's modulus $\tilde{E}(k_x,\omega)$ defined in the wavenumber-frequency domain related to a moving coordinate system that follows the load:

$$\tilde{E}(k_x,\omega) = E\left[1 + 2\mathrm{i}\xi\mathrm{sgn}(\omega + ck_x)\right] \qquad (1)$$

where i is the imaginary unit satisfying $\mathrm{i}^2 = -1$, ξ is the damping ratio, sgn(·) is the signum function, ω is the angular frequency, c is the driving speed of the TSD device, and k_x is the wavenumber in the driving direction.

The origin of the moving coordinate system is consistent with the centre of the loading area, and the

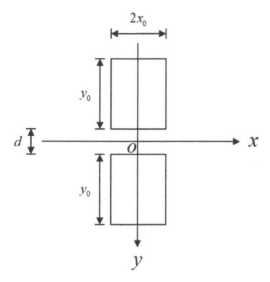

Figure 1. TSD loading configuration used in the simulations.

positive x-direction is the driving direction. In practice, the TSD device measures the vertical deflection slopes of fixed points in the moving coordinate system. The configuration of the loading area is shown in Figure 1, which can be described by a spatial distribution function $h_0(x,y)$ defined as follows:

$$h_0\left(x,y\right) = H\left(x_0 - |x|\right) H\left(\frac{y_0}{2} - \left|y + \frac{y_0+d}{2}\right|\right)$$

$$+ H\left(x_0 - |x|\right) H\left(\frac{y_0}{2} - \left|y - \frac{y_0+d}{2}\right|\right) \qquad (2)$$

in which $H(\cdot)$ is the Heaviside step function, $2x_0$ is the dimension of one rectangular area in x-direction, y_0 is the dimension of one rectangular area in y-direction, and d is the distance between two rectangular areas.

3 PARAMETER SENSITIVITY ANALYSIS

The accuracy of the back-calculated value for a certain parameter is related to the sensitivity of the response to this parameter. Hence, the parameter sensitivity analysis is necessary to select parameters which can be accurately back-calculated. To well represent the load applied by the TSD device, the following parameters are used:

- The driving speed c of the TSD device is 13.9 m/s (50 km/h);
- The magnitude of the loading pressure p_0 is 707 kPa;
- For the loading area, the parameter x_0 is 0.06316 m, y_0 is 0.27432 m, and d is 0.15 m.

It is assumed that the applied load has a rectangular influence area on the pavement surface with a length of 400 m in both x-direction and y-direction, and the

Table 1. Reference structural parameters of the pavement.

Layers	E MPa	ξ –	ν –	ρ kg/m^3	h m
Surface	2000	0.05	0.3	2200	0.1
Base	200	0.05	0.3	2000	0.3
Subgrade	50	0.05	0.3	1800	∞

Note: E is Young's modulus, ξ is damping ratio, ν is Poisson's ratio, ρ is density, and h is thickness.

centre of this area is the same as that of the loading area. The reference structural parameters of the considered pavement are shown in Table 1. The sensitivity of the slope curve of surface vertical displacement along the x-axis to different structural parameters (Young's modulus, damping ratio, and thickness) is investigated based on single factor analysis, and the result of the reference pavement structure is shown in solid lines. The variation of a certain parameter is set to be 50% of the reference value. In addition, the sensitivity to different parameters is divided into five levels: hardly sensitive, slightly sensitive, moderately sensitive, relatively sensitive, and highly sensitive. For brevity, the surface layer, base layer, and subgrade are represented by subscripts "1", "2", and "3", respectively.

3.1 Sensitivity to Young's modulus

The slope curves of surface vertical displacement for pavements with different Young's moduli of surface layer, base layer, and subgrade are shown in Figures 2(a), 2(b), and 2(c), respectively. The results show that the slope curve is relatively sensitive to the Young's modulus of the surface layer, and it is highly sensitive to the Young's moduli of the base layer and subgrade. In addition, the results also indicate that the slope of vertical displacement is zero (the vertical displacement is maximum) at a point behind the centre of loading area.

3.2 Sensitivity to damping ratio

The slope curves of surface vertical displacement for pavements with different damping ratios of surface layer, base layer, and subgrade are shown in Figures 3(a), 3(b), and 3(c), respectively. The results show that the slope curve is hardly sensitive to the damping ratio of the surface layer, and it is slightly sensitive to the damping ratios of the base layer and subgrade.

3.3 Sensitivity to thickness

The slope curves of surface vertical displacement for pavements with different thicknesses of surface layer and base layer are shown in Figures 4(a) and 4(b), respectively. The results show that the slope curve is highly sensitive to the thicknesses of the surface layer and base layer.

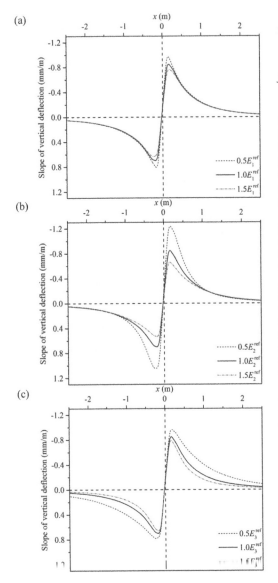

(a)

(b)

(c)

Figure 2. Sensitivity of the slope curve to the Young's modulus of: (a) surface layer, (b) base layer, and (c) subgrade.

4 PARAMETER BACK-CALCULATION

Generally, a parameter back-calculation technique is a combination of a forward calculation model with a minimisation algorithm. In this paper, the proposed spectral element method-based model (Sun et al. 2019) is combined with the Powell hybrid algorithm (Al-Khoury et al. 2001b) to back-calculate the structural parameters of pavements by analysing TSD measurements.

The objective function for the minimisation algorithm is:

$$f\left(\underline{p}\right) = \sqrt{\sum_{m=1}^{M}\left[\frac{s^{modelled}\left(x_m, y_m; \underline{p}\right)}{s^{measured}\left(x_m, y_m\right)} - 1\right]^2} \quad (3)$$

where $f(\underline{p})$ is the objective function, $\underline{p}$ is a vector that contains the parameters to be back-calculated, (x_m, y_m) are the coordinates of the m-th measuring point, M is the total number of measuring points, $s^{modelled}(x_m, y_m; \underline{p})$ is the modelled slope of vertical displacement at the m-th measuring point, and $s^{measured}(x_m, y_m)$ is the corresponding measured slope.

The most likely parameters which give good match between modelled and measured slopes can be obtained by minimising the value of the objective function. According to the results of the sensitivity analysis, the Young's moduli (E_1, E_2, and E_3) of all layers and the thicknesses (h_1 and h_2) of surface layer and base layer are chosen as unknown parameters to be back-calculated. The modelled slopes at five points on the surface of the reference pavement structure are set as synthetic measurements, which are used to back-calculate the unknown parameters by the proposed technique. The true values of these parameters are the values shown in Table 1.

In the back-calculation process, it is found that the initial guesses of the unknown parameters affect both the accuracy of the back-calculated results and the computational time. Hence, some auxiliary methods are recommended to be used to find a good set of initial guesses, such as referring to the design data of pavements. In this paper, the following initial guesses of the unknown parameters are used: $E_1 = 2500$ MPa, $E_2 = 150$ MPa, $E_3 = 40$ MPa, $h_1 = 0.07$ m, and $h_2 = 0.4$ m. The back-calculated parameters are: $E_1 = 2001.2$ MPa, $E_2 = 205.5$ MPa, $E_3 = 50.2$ MPa, $h_1 = 0.099$ m, and $h_2 = 0.30$ m. The relatively good agreement between the back-calculated and the true parameter values confirms the ability of the proposed technique to deal with TSD measurements.

5 CONCLUSIONS

A spectral element method-based parameter back-calculation technique for the traffic speed deflectometer (TSD) test is proposed in this paper. Firstly, the sensitivity of the slope curve of pavement surface vertical displacement to different structural parameters is investigated. Then, the ability of the proposed parameter back-calculation technique is verified via a case study. On the basis of the results and discussions above, the following conclusions can be drawn:

- The slope curve is relatively sensitive to the Young's modulus of the surface layer, and it is highly sensitive to the Young's moduli of the base layer and subgrade.

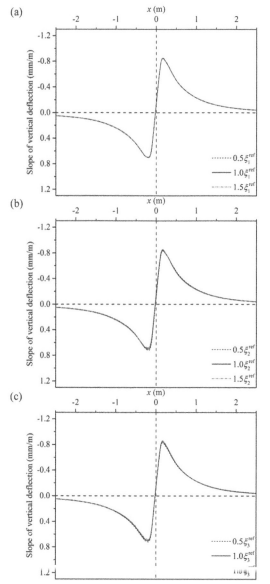

Figure 3. Sensitivity of the slope curve to the damping ratio of: (a) surface layer, (b) base layer, and (c) subgrade.

- The slope curve is hardly sensitive to the damping ratio of the surface layer, and it is slightly sensitive to the damping ratios of the base layer and subgrade.
- The slope curve is highly sensitive to the thicknesses of the surface layer and base layer.
- The combination of the spectral element method-based pavement model and the Powell hybrid algorithm may become a promising parameter back-calculation technique for the TSD test.

The work presented in this paper promotes the development of parameter back-calculation techniques based on TSD measurements.

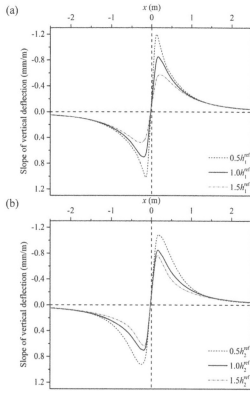

Figure 4. Sensitivity of the slope curve to the thickness of: (a) surface layer and (b) base layer.

ACKNOWLEDGEMENTS

This work is financially supported by the China Scholarship Council (No. 201608230114).

REFERENCES

Al-Khoury, R., Scarpas, A., Kasbergen, C. & Blaauwen-draad, J. 2001a. Spectral element technique for efficient parameter identification of layered media. I. Forward calculation. *International Journal of Solids and Structures* 38(9): 1605–1623.

Al-Khoury, R., Kasbergen, C., Scarpas, A. & Blaauwen-draad, J. 2001b. Spectral element technique for efficient parameter identification of layered media: Part II: Inverse calculation. *International Journal of Solids and Structures* 38(48–49): 8753–8772.

Liu, P., Wang, D., Otto, F., Hu, J. & Oeser, M. 2018. Application of semi-analytical finite element method to evaluate asphalt pavement bearing capacity. *International Journal of Pavement Engineering* 19(6): 479–488.

Marecos, V., Fontul, S., de Lurdes Antunes, M. & Solla, M. 2017. Evaluation of a highway pavement using non-destructive tests: Falling Weight Deflectometer and Ground Penetrating Radar. *Construction and Building Materials* 154: 1164–1172.

Rabbi, M.F. & Mishra, D. 2019. Using FWD deflection basin parameters for network-level assessment of flexible pavements. *International Journal of Pavement Engineering*: 1–15.

Shrestha, S., Katicha, S.W., Flintsch, G.W. & Thyagarajan, S. 2018. Application of traffic speed deflectometer for network-level pavement management. *Transportation research record* 2672(40): 348–359.

Sun, Z., Kasbergen, C., Skarpas, A., Anupam, K., van Dalen, K.N. & Erkens, S.M.J.G. 2019. Dynamic analysis of layered systems under a moving harmonic rectangular load based on the spectral element method. *International Journal of Solids and Structures* 180–181: 45–61.

Mixes

Moderators: Yong-Rak Kim (Texas A&M University),
Shane Underwood (North Carolina State University),
Zhen Leng (Hong Kong Poly. University)
and Eshan Dave (University of New Hampshire)

Advances in Materials and Pavement Performance Prediction II – Kumar et al. (eds)
© 2021 Taylor & Francis Group, London, ISBN 978-0-367-46169-0

Effect of recycling agents on the homogeneity of high RAP binders

A. Abdelaziz & A. Epps Martin
Zachry Department of Civil & Environmental Engineering, Texas A&M University, Texas, USA

E. Masad
Zachry Department of Civil & Environmental Engineering, Texas A&M University, Texas, USA
Mechanical Engineering Program, Texas A&M University at Qatar, Doha, Qatar

E. Arámbula-Mercado
Texas A&M Transportation Institute, Texas, USA

F. Kaseer
Florida Department of Transportation, State Materials Office, Gainesville, Florida, USA

ABSTRACT: Increased costs of virgin asphalt pavement materials have encouraged transportation agencies to allow higher quantities of reclaimed asphalt pavement (RAP) in their recycled asphalt mix design if performance can be maintained. Recycling agents or rejuvenators are organic materials used to reduce the stiffness and brittleness of RAP binders and make recycled mixtures less prone to cracking. Recycling agents can facilitate the use of higher quantities of RAP; however, there is limited understanding on their effect on the homogeneity of the final blend of virgin and recycled binders and recycling agent. This study utilized Atomic Force Microscopy (AFM) to assess the effect of recycling agents on the homogeneity of different blends of virgin and RAP binders. AFM results revealed that the addition of recycling agents improved the homogeneity of blends at unaged and short-term aging conditions; however, this effect diminished after long-term aging.

1 INTRODUCTION & OBJECTIVES

Increased costs of virgin asphalt pavement materials have encouraged transportation agencies to allow higher quantities of reclaimed asphalt pavement (RAP) in their designs if performance can be maintained. The use of RAP improves sustainability by conserving raw materials and reducing construction costs. However, using large amounts of RAP increases the stiffness and brittleness of asphalt mixtures and make them more prone to cracking. This effect can be mitigated with the use of softer virgin binders or the inclusion of recycling agents (Al-Qadi et al. 2007; Copeland 2011). Recycling agents (or rejuvenators) are organic materials that can partially restore the properties of RAP binder by rebalancing the chemical components in the recycled asphalt binder that gets disturbed during the aging process. The addition of recycling agents can enhance the workability and performance of asphalt mixtures; however, their dose must be carefully selected to avoid rutting issues (Arámbula-Mercado et al. 2018; Kaseer et al. 2019).

One of the main goals for the use of recycling agents is achieving homogeneity of the final blend of virgin binder, recycled binder, and recycling agent. Homogeneity depends on the level of blending or interaction between virgin and RAP binders, which is affected by aging condition, performance grade (PG) of the virgin binder, recycling agent type and dose and RAP content. Several studies have indicated that partial blending occurs between RAP and virgin materials (Gottumukkala et al. 2018; McDaniel et al. 2012; Nazzal et al. 2014; Shirodkar et al. 2011); however, the effect of recycling agents on the blending process and homogeneity of the final blend is still not well understood (Zaumanis et al. 2019).

This study evaluates the influence of recycling agents on the homogeneity of RAP and virgin binder blends through the performance of Atomic Force Microscopy (AFM) tests. AFM was used to characterize microstructural and micromechanical properties of high RAP binder blends at different recycling agent doses and aging conditions, and the results were used to assess homogeneity of the different binder blends.

2 MATERIALS

Four different binder blends were prepared as described in Table 1. These blends were obtained from a field project in Wisconsin associated with the National Cooperative Highway Research Program

Table 1. Material properties.

	Blend 1	Blend 2	Blend 3	Blend 4
Blend label	DOT Control, (27% RAP)	Recycled Control (36% RAP)	Rejuvenated (36% RAP) +1.2% recycling agent	Rejuvenated (36% RAP) +5.5% recycling agent
Binder content	5.6%	5.4%	5.4%	5.4%
RAP content	27%	36%	36%	36%
Recycling Agent Dose	0%	0%	1.2%	5.5%

Table 2. Rheological properties.

	Blend 1	Blend 2	Blend 3	Blend 4
G-R Parameter (kPa)				
RTFO	8.1	15.6	7.0	0.8
20 h PAV	89.3	117.6	79.5	12.6
40 h PAV	715.7	1,609.4	550.8	124.1
ΔT_c (°C)				
20 h PAV	−4.3	−5.3	−4.8	−3.1

(NCHRP) 9-58 Project (Epps Martin et al. 2019). All blends were produced using a PG 58-28 virgin binder. The DOT Control blend (blend 1) was designed with 27% RAP and no recycling agent to meet the requirements of the Wisconsin Department of Transportation (WisDOT). The Recycled Control blend (blend 2) was also prepared without a recycling agent, but with a higher RAP content (36%). The effect of a modified vegetable oil recycling agent was evaluated using two Rejuvenated Blends (3 and 4) at 1.2% and 5.5% doses, respectively, by weight of total binder. The recycling agent dose used in blend 3 was selected based on manufacturer recommendations to restore a PG 58-28 virgin binder to a substitute PG 52-34. The recycling agent dose selected for blend 4 was based on the method recommended by Arámbula-Mercado et al. (2018). In this method, recycling agents were added until the continuous high temperature performance grade (PGH) of the binder matched the target PGH of 58 based on Wisconsin climate and traffic conditions.

In addition to the unaged condition, the blends were conditioned by three different aging protocols: rolling thin film oven (RTFO), 20-hour Pressurized Aging Vessel (PAV) and 40-hour PAV. RTFO and PAV aging conditions were performed according to AASHTO T 240 and AASHTO R 28, respectively. Rheological properties of the different blends including the Glover-Rowe (G-R) parameter and ΔT_c are provided in Table 2. G-R is defined as $G^*(\cos\delta)^2/\sin\delta$, where G^* is the complex shear modulus and δ is the phase angle. ΔT_c represents the difference in continuous low temperature performance grades (PGL) for creep stiffness (S) and relaxation properties (mvalue) by AASHTO M 320. Stiffness and brittleness of asphalt binders increases as G-R increases and ΔT_c decreases. As expected, the Recycled Control blend (blend 2) exhibited higher stiffness and embrittlement properties compared to the Rejuvenated Blends (3 and 4).

3 LABORATORY TESTS

A Bruker Dimension Icon AFM was used to characterize the microscopic properties of the asphalt binder blends. AFM testing was performed in the PeakForce Quantitative Nanomechanical Mapping (PFQNM) mode. This mode allows for capturing both microstructural and micromechanical properties of the material. Microstructural properties were evaluated based on topography images, and micromechanical properties were evaluated based on Derjaguin-Muller-Toropov (DMT) and adhesion measurements. Samples were scanned using a TAP 150A tip with a spring constant of 5 N/m and a radius of 8 nm.

AFM samples were prepared using the heat cast method. In this method, a small amount (0.07 grams) of the binder blend was removed from a container and placed on a glass slide. The slide was put in an oven for 10 minutes at 160°C. Then, samples were removed from the oven and covered for at least 24 hours at room temperature prior to testing.

4 RESULTS & DISCUSSION

Three different phases were observed in the microstructure of the binder blends: bee structure, dispersed phase and matrix. Figure 1 (a) through (d) illustrates the effect of aging on the microstructure. The bee structure and dispersed phases increased in size with aging. No bee structures were observed at unaged or RTFO aging conditions; however, they appeared after PAV aging. The percent of area occupied by bee structures at the 20-hours and 40-hours PAV was found to be 1.0% and 2.4%, respectively. This finding agrees with Menapace et al. (2015) who also reported that dispersed and bee structure phases occupied larger areas after RTFO, and PAV aging compared to those in an unaged condition.

Some studies attributed the bee structures to the wax content (Blom et al. 2018), while others related it to the asphaltene content (Zhang et al. 2011).

Figure 2 demonstrates the change in asphalt binder morphology due to higher RAP content. Similar to aging, the addition of RAP also increased the amount of dispersed and bee structure phases within the matrix. In the unaged condition, blend 1 with 27%

244

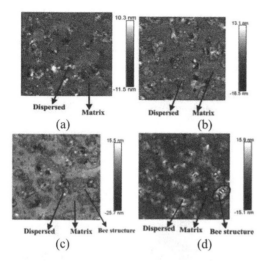

(a)   (b)

(c)   (d)

Figure 1. Surface morphology of blend 1 at (a) unaged, (b) RTFO, (c) 20-hours PAV, and (d) 40-hours PAV.

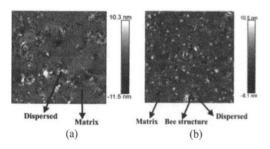

(a)   (b)

Figure 2. Surface morphology at unaged condition (a) blend 1 and (b) blend 2.

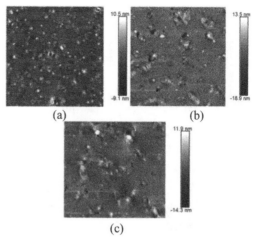

(a)   (b)

(c)

Figure 3. Surface morphology at unaged condition (a) blend 2 (b) blend 3, and (c) blend 4.

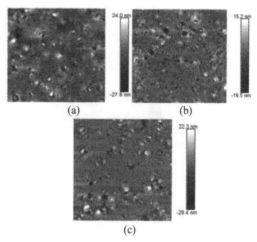

(a)   (b)

(c)

Figure 4. Surface morphology at 40-hours PAV (a) blend 2, (b) blend 3, and (c) blend 4.

RAP showed no bee structures; however, after increasing the RAP content to 36%, the percent of area occupied by bee structures became 3.2%.

Surface morphology maps of the Recycled Control (36% RAP) blend (blend 2) before and after the addition of recycling agent at the unaged condition are shown in Figure 3. The addition of recycling agent reduced the bee structures, particularly at the higher recycling agent dose. At 1.2% recycling agent dose, the percent of area occupied by bee structures was reduced from 3.2% to 0.9%. Increasing the recycling agent dose to 5.5%, further reduced this percentage to 0.5%. This indicates that the addition of recycling agents can result in the formation of a more homogeneous surface morphology. However, this effect was not maintained after long-term and extended aging. After 40-hours PAV aging (Figure 4), bee structures did not disappear after the addition of 1.2% or 5.5% recycling agent (blends 3 and 4). This could be a result of the additional aging or due to a decreased effectiveness of the recycling agent.

Histograms of DMT modulus and adhesion maps were created to characterize the effect of recycling agents on the homogeneity of the blends at different aging conditions. DMT modulus histograms of the Recycled Control (36% RAP) blend (blend 2) before and after the addition of 5.5% recycling agent (blend 4) at the unaged condition are shown in Figure 5. Before the addition of recycling agents, the distribution of DMT modulus was not uniform and the histograms showed two different phases. After the addition of recycling agents, these two phases were homogenized to one phase. This indicates that the addition of recycling agents enhanced the homogeneity of the blend.

Similar to the unaged condition, the addition of recycling agents after RTFO aging also led to a more uniform DMT modulus distribution over the scanned surface, as exhibited in Figure 6.

The effect of recycling agents on the homogeneity of the blend was reduced after long-term aging. As shown

245

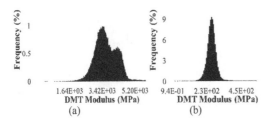

(a)
DMT Modulus (MPa)

(b)
DMT Modulus (MPa)

Figure 5. DMT modulus histograms at unaged condition of (a) blend 2, and (b) blend 4.

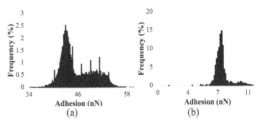

(a)
Adhesion (nN)

(b)
Adhesion (nN)

Figure 8. Adhesion histograms at RTFO condition of (a) blend 2, and (b) blend 4.

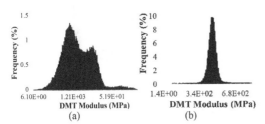

(a)
DMT Modulus (MPa)

(b)
DMT Modulus (MPa)

Figure 6. DMT modulus histograms at RTFO condition of (a) blend 2, and (b) blend 4.

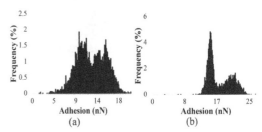

(a)
Adhesion (nN)

(b)
Adhesion (nN)

Figure 9. Adhesion histograms at 40-hours PAV condition of (a) blend 2, and (b) blend 4.

adhesion over scanned areas became more uniform with the addition of recycling agent. However, the effectiveness of the recycling agent was reduced after long-term aging. Future research is recommended to corroborate the findings of this study using different virgin and recycled binder sources and recycling agent types.

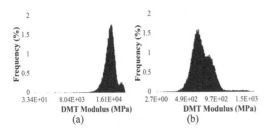

(a)
DMT Modulus (MPa)

(b)
DMT Modulus (MPa)

Figure 7. DMT modulus histograms at 40-hours PAV of (a) blend 2, and (b) blend 4.

in Figure 7, after 40-hours PAV aging, the two phases were not homogenized to one phase with the addition of recycling agent. This signifies that the effectiveness of the recycling agent might have been reduced with aging.

Histograms of adhesion measurements showed similar trends as DMT modulus histograms, as exhibited in Figures 8 and 9. The addition of recyling agents led to a more uniform adhesion distribution over the surface at less aged conditions. However, as aging increased, the effect of recycling agents diminished and the two phase phenomenon reappeared.

5 CONCLUSIONS

This study utilized AFM to investigate the effect of a modified vegetable oil recycling agent on the homogeneity of different blends of virgin and RAP asphalt binders. The addition of recycling agent improved the homogeneity of the blends by reducing the amount of dispersed and bee structure phases in the microstructure. Moreover, the distribution of DMT modulus and

REFERENCES

Al-Qadi, I.L., Elseifi, M. and Carpenter, S.H., 2007. *Reclaimed asphalt pavement—a literature review.*
Arámbula-Mercado, E., Kaseer, F., Martin, A.E., Yin, F. and Cucalon, L.G., 2018. Evaluation of recycling agent dosage selection and incorporation methods for asphalt mixtures with high RAP and RAS contents. *Construction and Building Materials* 158: 432–442.
Blom, J., Soenen, H., Katsiki, A., Van den Brande, N., Rahier, H. and van den Bergh, W., 2018. Investigation of the bulk and surface microstructure of bitumen by atomic force microscopy. Construction and Building Materials, 177, 158–169.
Copeland, A. 2011. Reclaimed asphalt pavement in asphalt mixtures: *State of the practice* (No. FHWA-HRT-11-021). United States. Federal Highway Administration. Office of Research, Development, and Technology.
Epps Martin, A., Kaseer, F., Arambula-Mercado, E., Bajaj, A., Cucalon, L.G., Yin, F., Chowdhury, A., Epps, J., Glover, C., Hajj, E.Y. and Morian, N., 2019. Evaluating the Effects of Recycling Agents on Asphalt Mixtures with High RAS and RAP Binder Ratios. *NCHRP Research Report*, (927).
Gottumukkala, B., Kusam, S.R., Tandon, V. and Muppireddy, A.R., 2018. Estimation of Blending of Rap Binder in a Recycled Asphalt Pavement Mix. *Journal of Materials in Civil Engineering* 30(8).
Kaseer, F., Martin, A.E. and Arámbula-Mercado, E., 2019. Use of recycling agents in asphalt mixtures with high

recycled materials contents in the United States: A literature review. *Construction and Building Materials* 21: 974–987.

McDaniel, R. S., Shah, A. and Huber, G. 2012. *Investigation of low-and high-temperature properties of plant-produced RAP mixtures* (No. FHWA-HRT-11-058). United States. Federal Highway Administration. Office of Pavement Technology.

Menapace, I., Masad, E., Bhasin, A. and Little, D., 2015. Micros tructural properties of warm mix asphalt before and after laboratory-simulated long-term ageing. *Road Materials* and Pavement Design 16(sup1): 2–20.

Nazzal, M.D., Mogawer, W., Kaya, S. and Bennert, T., 2014. Multiscale evaluation of the composite asphalt binder in high–reclaimed asphalt pavement mixtures. *Journal of Materials in Civil Engineering* 26(7).

Shirodkar, P., Mehta, Y., Nolan, A., Sonpal, K., Norton, A., Tomlinson, C., Dubois, E., Sullivan, P. and Sauber, R., 2011. A study to determine the degree of partial blending of reclaimed asphalt pavement (RAP) binder for high RAP hot mix asphalt. Construction and Building Materials 25(1): 150–155.

Zhang, H.L., Wang, H.C. and Yu, J.Y., 2011. Effect of aging on morphology of organo-montmorillonite modified bitmen by atomic force microscopy. *Journal of Microscopy*, 242(1), 37–45.

Advances in Materials and Pavement Performance Prediction II – Kumar et al. (eds)
© *2021 Taylor & Francis Group, London, ISBN 978-0-367-46169-0*

Rutting resistance and high-temperature PG of asphalt mixtures

Haleh Azari & Alaeddin Mohseni
Pavement Systems LLC, Bethesda, MD, USA

ABSTRACT: The Incremental Repeated Load Permanent Deformation (iRLPD) test is specified in AASHTO TP 116. The test is improvement over the Flow Number test in several aspects. The main advantage of the iRLPD test is that it determines resistance of asphalt mixtures to rutting at any temperature and stress level in half an hour. In this study, a modified method of iRLPD test, which is performed on volumetric samples in unconfined mode, is introduced. The test determines High-Temperature Performance Grade (HTPG) of the mixture as well as grade bump for traffic level. The modified method is included as Method A in TP 116. Several different materials, including mixtures with modified binders and RAP/RAS, are tested according to the Method A and the results are correlated to the known properties of the binder and mastic.

1 INTRODUCTION

The Incremental Repeated Load Permanent Deformation (iRLPD) test is specified in AASHTO TP 116 (1). The test is similar to the Flow Number (FN) test in terms of loading frequency (0.1 second of loading followed by 0.9 second of rest); however, there are several differences (2). One main difference is that the FN test is performed for 10,000 cycles (2 hrs. and 48 min.) at one stress level; while, the iRLPD test is performed for 2000 cycles (33 min.) at four stress levels (Figure 1a) and provides the mixture resistance to rutting at any temperature and stress level.

Following Method B of AASHTO TP 116, the m* values of 400 kPa, 600 kPa, and 800 kPa increments are used to create a damage curve for each mixture (the 200 kPa increment is for conditioning and is not used for the analysis). The damage curve, which represents resistance to rutting at any temperature and stress level, has the power function shown in Equation 1:

$$m* = a(T \times P)^b \tag{1}$$

Where:
m* (Minimum Strain Rate) = unit rutting damage,
T = temperature (°C),
P = Stress (MPa),
a and b= Coefficients of the power curve

The "a" coefficient is constant over the range of high temperatures and could be assumed to be 1. The "b" coefficient changes in a range of 2 to 3 for different mixtures. While m* changes with temperature (T) and stress (P), the "b" coefficient is a unique parameter of the mixture and independent from stress level and test temperature. For this reason, the "b" parameter has been successfully used to rank various mixtures

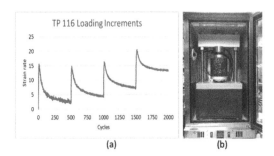

Figure 1. (a) Strain rate for four test increments of TP 116 at 200 kPa, 400 kPa, 600 kPa, and 800 kPa; (b) iRLPD test setup.

designed for different traffic levels and environments (3). In addition, the relationship between the parameter "b" of asphalt mixtures and Performance Grade (PG) of asphalt binder has been examined previously (4). Therefore, the "b" coefficient could be considered for determining the high temperature performance grade of asphalt mixture similar to those of asphalt binder.

Once the PG of the mixture is determined, it is then possible to establish the grade bump and the allowable traffic with consideration to the required PG of the environment (5).

In this study, the above findings will be used to determine mixture PG and corresponding allowable traffic included in TP 116. To further increase the practicality of TP 116, three features of the tests are modified and included as Method A. The modifications include: 1) use of volumetric samples, 2) performing the test unconfined, and 3) reducing the number of test increments. During the construction season, there is not sufficient resources to compact new samples for performance based Quality

Control/Quality Acceptance (QC/QA) process. Based on the modified rutting procedure, the volumetric samples, which are routinely compacted for bulk specific gravity measurements can be reused for performance testing. Figure 1b shows the iRLPD setup for testing of the volumetric samples unconfined.

The larger size of the volumetric samples with less air voids (150-mm diameter with 4% air voids) compared to the current TP 116 test specimens (100-mm in diameter with 7% air voids) allows performing the test unconfined.

With the knowledge of the power model for the rutting damage curve (Equation 1), it is now possible to determine the "b" coefficient of a mixture by performing only one test increment. This reduces the testing time by 75%, while allowing determination of mixture PG and allowable traffic.

Establishing the mixture PG and allowable traffic from performing the unconfined iRLPD test on volumetric samples would be a quick, reliable, and economical tool for performance based mix design, QC/QA, as well as pavement design.

2 OBJECTIVES AND SCOPE

The main objective of this study is to evaluate the iRLPD modified rutting procedure, currently Method A of AASHTO TP 116, using seven mixtures obtained from Missouri DOT (MoDOT) SPS10 LTPP sites and three mixtures from Maryland State Highway Administration (MDSHA). The following tasks will be accomplished:

1. Conducting the shortened version of the iRLPD rutting test on unconfined volumetric samples of the mixtures.
2. Determining the effective mixture high-temperature PG, grade bump, and environment PG with traffic for the mixtures.
3. Investigating the relationship between the mixture PG and known properties of the binder and mastic.

3 SIMPLIFIED IRLPD METHOD FOR RUTTING

It is desired to be able to perform a simple and quick test for determining effects of RAP/RAS on mixture's performance during the production control. The current iRLPD rutting procedure could be further simplified to serve for this purpose. The test is simplified in the following three areas:

3.1 Sample preparation

The first step is to simplify the sample preparation. Since compacting separate sets of samples for performance tests is cumbersome during production, use of volumetric samples could be explored for performance based QC/QA. As-compacted 150-mm diameter by 115-mm tall volumetric samples are routinely compacted for bulk specific gravity measurements and discarded afterward. Further use of the volumetric

samples for rutting evaluation would make possible performance based QC/QA without expensing significant amount of time and resources.

The additional advantages of using volumetric samples can be outlined here: 1) volumetric samples are extensively made by state DOTs and contractors and are readily available, 2) samples are better representatives of pavement material (about 75% more material), 3) volumetric samples provide improved aggregate size to specimen dimension ratio, 4) sample's diameter is better representational of truck tire contact area, and 5) height of volumetric samples is closer to pavement layer thickness.

The test setup in Figure 1b allows the volumetric samples to be conditioned at a temperature between 55°C to 60°C (determined using the equation provided in TP 116) to make them ready for the performance test.

3.2 Unconfined test

The second area which can be simplified for increased productivity is performing the test unconfined. Confined tests require extra care for providing and maintaining confinement pressure and special material such as latex membrane for conducting the test. The larger size and denser structure of the volumetric samples make possible performing the test unconfined.

3.3 Reducing testing time

The third factor for simplifying the test, is to perform the test with only one increment at 600 kPa (500 cycles) instead of four increments at four stress levels (2000 cycles). This reduces the testing time from 33 min. to less than 10 min. Using the power curve model (Equation 1), which was established based on multi-incremental test, the coefficient "b", as the parameter for rutting resistance of the mixture, can be determined from the m* value at 600 kPa and the test temperature, which has been determined based on Degree Day variable from LTPPBind V3.1 (6).

4 MIXTURE HIGH-TEMPERATURE PG

A valuable addition to TP 116 is determination of effective high-temperature mixture PG using iRLPD test parameter. As mentioned earlier, the "b" power coefficient of Equation 1 is unique for each mixture. Using the data from two published TRB papers (3,4), an equation was developed to determine mixture effective PG from the "b" coefficient.

4.1 Model to estimate mixture PG

The basis of the mixture PG model is the data derived from testing twelve dense graded mixtures prepared with the aggregates from NCHRP 9-29 study and 12 modified and unmodified binders from four suppliers as explained in the previous study (4). The model was verified using nine mixtures with known properties

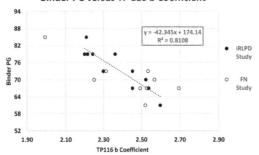

Binder PG versus TP 116 b Coefficient

$y = -42.345x + 174.14$
$R^2 = 0.8108$

• iRLPD Study

○ FN Study

Figure 2. Relationship between binder PG and mixture "b" coefficient.

from nine state DOTs (2,3). The resulted model for the effective mixture PG follows as:

$$PG = -42.345 \times b + 174.14 \qquad (2)$$

Figure 2 shows the relationship between the high-temperature PG of the binders and the "b" values of their corresponding mixtures for the 12 mixtures used in the development of the model and the nine mixtures for verifying the model. Note that the mixtures used for development and verification of Equation 2 were virgin mixtures without RAP/ RAS. Therefore, grade of mixtures could be correlated to the grade of binder. The properties of the mixtures were reported in previous studies (2,3,4). As indicated from the Figure, the "b" coefficient is well correlated with the binder PG with R^2 of 0.81.

4.2 PG grade bumping from LTPP bind

Determining allowable traffic level can now be accomplished through mixture PG, rather than MSR (m*), as in the original version of TP 116 (2015). The grade bumping table in LTPPBind V.3.1, which was developed based on Pavement Design Guide models for mixtures (5), is used to determine the environment PG with traffic. Table 1 provides the allowable traffic for the determined effective mix PG; alternatively, it provides the required PG for design traffic level. For example, if a mixture PG is determined to be 74 and the environment PG is 58, then the allowable traffic would be 10 to 30 MESAL. Alternatively, if environment PG is 58 and the design traffic is 10 to 30 MESAL, then effective PG of the mix should not be less than 73 and preferably not excessively more than 77.

5 MATERIALS TESTED IN THE STUDY

The materials tested for the evaluation of the modifications for TP 116, include seven mixtures from Missouri DOT (MoDOT) SPS10 collected from the LTPP sites and three mixtures from Maryland State Highway Administration (MDSHA) production. The selected mixtures are all dense graded and content a wide range of RAP/RAS amounts.

Table 1. Allowable traffic based on mixture PG and the environment PG in TP 116 (2020).

Traffic Level	Design ESAL (million)	Environment PG			
		52	58	64	70
Standard (S)	>1 to 3	52	58	64	70
Heavy (H)	>3 to 10	62	67	71	76
Very Heavy (V)	>10 to 30	69	73	77	81
Extreme (E)	>30	71	75	79	83

Table 2. Effective high-temperature mixture PG and calculated grade bump for MDSHA mixtures.

Property	MD #19	MD #22
Binder	64–22	64–22
Aggregate Nominal Size	12.5	19
Dust/Binder	1.36	1.51
RAP, %	32	18
RAS, %	– 5	
Binder, %	4.8	4.3
Effective Mix PG	65	85
Grade Bump	$65.1 - 64 = 1$	$85 - 64 = 21$

6 DISCUSSION OF RESULTS

Table 2 provides the effective PG of two of the MDSHA mixtures prepared for production control. As shown in the table, both mixtures #19 and #22 have PG 64-22 binder. However, mixture #22 has larger aggregate nominal size (19 mm vs. 12.5 mm), higher dust to binder ratio (1.51 vs. 1.36), and lower binder content (4.3% vs. 4.8%) than Mixture #19. In addition, Mixture #22 has 5% RAS, but lower percentage of RAP (18% vs. 32%). The effective grades of the two mixtures (65°C and 85°C) indicate that Mixture #19 is less rut resistant than Mixture #22. In addition, despite 32% RAP in Mixture #19, the bump in the high-temperature grade from RAP is only 1°C; while, for the Mixture #22, the bump in binder grade from combination of RAP and RAS is 21 C.

Table 3 provides the average effective mixture PG, the coefficient of variation (C.V. %) of mix PG, and the environment PG with traffic level of seven MoDOT SPS10 mixtures and three MDSHA production mixtures.

The M 320 PG or M 332 PG of the virgin binders and the percentages of RAP and RAS of the mixtures are also provided in the table. The PG with traffic is determined from comparison of the mix PG with the PG values in Table 1. As indicated from Table 3, PG with traffic either agree with the M 332 PG of the binders (Column 3) or exceed the grades.

7 VERIFICATION OF EFFECTIVE MIX PG WITH PG OF MASTIC

The PG of the 10 mixtures were compared with the PG determined for mastics separated from the mix-

Table 3. iRLPD mixture PG and PG with traffic for the MoDOT SPS10 and MDSHA QC mixtures.

Mix	Mix No	PG Mix M 320/ M 332	RAP %	RAS %	Avg Mix PG	Mix PG CV%	Mastic PG	PG + Traffic
MO DOT SPS10	1	64H22	25	–	74	3.2	75.7	64H
	2	64H22	25	–	75	2.6	72.0	64H
	4	64H22	–	–	75	1.8	72.6	64H
	5	64H22	34	–	90	4.0	95.5	64E
	6	58–28	–	19	73	2.7	68.5	64H
	7	58–28	19	16	74	1.0	74.0	64H
	8	58–28	35	–	73	6.9	68.5	64H
MD SHA QC	1	64–22	40	–	82	1.5	79.8	64E
	2	64–22	18	5	85	1.6	81.8	64E
	3	64–22	32	–	65	2.0	62.4	64S

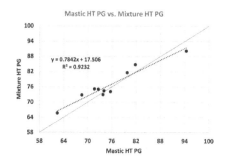

Figure 3. Relationship between high-temp. PG of mastic and mixture.

tures. The high-temperature test on mastic follows the same principal as the mixture iRLPD test (7); however performed on a DSR. The mastic PG values of the 10 mixtures are provided in Table 3 and the relationship between mastic and mixture PG is shown in Figure 3. As indicated from the Figure, PG of mixture is highly correlated with the PG of mastic ($R^2 = 0.93$) and the PG values are in average within 2°C. Figure 3 implies that the difference in mixture high-temperature PG is mainly due to the difference in the mastic property since the aggregate gradations were similar. The PG of mastic is affected by the type and amount of binder, RAP/RAS, rejuvenators, and filler; hence, it may be used to optimize the high-temperature mixture performance for a specific aggregate gradation.

8 SUMMARY AND CONCLUSIONS

To increase the usefulness of the rutting procedure in AASHTO TP 116 for production control, several additions/revisions were investigated, which are currently included in the test method. The modifications address several areas of the test method as follows:

1. Simplified the procedure by performing unconfined test on volumetric samples. The larger and denser volumetric samples compared to the current TP 116 test specimens allow performing the iRLPD test unconfined.
2. The test is further simplified by applying only one test increment at the stress level of 600 kPa instead of applying three increments. This reduces the testing time from 33 min. to less than 10 min.
3. The power model of the damage curve, which has been established from multi-incremental test in TP 116 can be used to determine the "b" parameter of the mixtures from applying only one test increment.
4. An equation for determining the effective Performance Grade (PG) of the mixtures using the mixture "b" parameter is provided.
5. A table for determining allowable traffic based on mixture PG has been developed. Alternatively, the required mixture PG for design traffic and environment can be determined from the table.

As results of the above revisions/additions, TP 116 can be reliably used for performance based mix design and QC/QA process without spending significant amount of time and resources. Using mixture PG, the change in grade of mixture from addition of RAP/ RAS and other additives can be easily quantified. The allowable traffic can be determined based on the mixture PG with consideration to the environment PG. This could serve as a robust tool for the design of asphalt pavement.

REFERENCES

Aashto TP 116: Standard Method of Test for Rutting and Fatigue Resistance of Asphalt Mixtures Using Incremental Repeated Load Permanent Deformation (iRLPD). American Association of State Highway Transportation Officials (AASHTO). Washington, D.C., 2020.

Azari, H. and Mohseni, M. Incremental Repeated Load Permanent Deformation Testing of Asphalt Mixtures. Report No. 12-4381, Transportation Research Board 91th Annual Meeting, 2012.

Azari, H., and M. Mohseni. Permanent Deformation Characterization of Asphalt Mixtures using Incremental Repeated Load Testing. In Asphalt Materials and Mixtures. Vol. 4. Transportation Research Board, Washington, DC, 2013.

Mohseni, A., and H. Azari. High-Temperature Characterization of Highly Modified Asphalt Binders and Mixtures. Transportation Research Record: Journal of the Transportation Research Board, Issue 2444, pp 38–51 2014.

Mohseni, A., Carpenter, S., and D'Angelo. Development of SUPERPAVE High-Temperature Performance Grade (PG) Based on Rutting Damage. Journal of the Association of Asphalt Paving Technologists (AAPT), 2005.

Mohseni, A., and H. Azari. *Effective Temperature for Permanent Deformation Testing of Asphalt Mixtures*. ISAP. International Society for Asphalt Pavements, Lino Lakes, MN, 2014

Mohseni, A., Azari, H. and Hesp, S., "Determining the Effect of Additives on Hot Mix Asphalt performance Through the Testing of recovered Composite Mastic", *7th Eurasphalt & Eurobitumin Congress*, Madrid, May 2020.

Advances in Materials and Pavement Performance Prediction II – Kumar et al. (eds)
© 2021 Taylor & Francis Group, London, ISBN 978-0-367-46169-0

A comparative study of SMA and dense graded HMA mixtures using a laboratory rutting test and accelerated pavement testing

Dario Batioja-Alvarez & Jeong Myung
Purdue University, USA

Yu Tian
Tongji University, P.R. China

Jusang Lee
Indiana Department of Transportation, USA

John E. Haddock
Purdue University, USA

ABSTRACT: The objective of the study was to evaluate stone matrix asphalt (SMA) and hot mix asphalt (HMA) rutting performance using accelerated pavement testing (APT) and the Hamburg wheel track test (HWTT) and to understand how their performance relates to rutting predictions obtained from the PavementME. To do this, rutting results from a recent APT study that analyzed the performance of both SMA and HMA in a typical Indiana full-depth pavement structure were compared to HWTT and PavementME rutting results. It was found that both the APT and HWTT provide clear distinctions between the SMA and HMA mixtures and both showed trends of relative rutting resistance. Results also indicate the HWTT and APT outcomes are not comparable in magnitude. This could be explained by the HWTT having a testing environment that is more aggressive than the APT. PavementME predictions were comparable to total pavement rutting measurements from the APT, but the software was not capable of identifying the superior rutting performance of SMA.

1 INTRODUCTION

1.1 Background

Stone matrix asphalt (SMA) mixtures have a gap-graded aggregate structure and rely on stone-on-stone contact to provide load-carrying capability and an asphalt binder rich mastic to ensure durability. SMA mixtures are premium products in which most of the voids between the aggregates are filled with the mastic, a mixture of fine aggregate, mineral filler and binder. Due to their increased performance compared to the conventional dense-graded hot mix asphalt (HMA), SMA mixtures have been extensively used by the Indiana Department of Transportation (INDOT) as pavement wearing surfaces for heavy traffic volume pavements.

In recent years INDOT has investigated the rutting behavior of full-depth flexible pavements using full-scale accelerated pavement tests (APT). In 2017 a study evaluated two types of wearing surface material (HMA and SMA) and implemented a novel mid-depth rut monitoring and automated laser profile system to monitor the evolution of the transverse rutting profiles

in each pavement structural layer during APT loading (Tian et al. 2017). The study found that SMA has greater rutting resistance than the regular dense graded HMA mixtures, when tested under the same conditions (Tian et al. 2017), (Nantung et al. 2018).

Due to its practicality, the Hamburg wheel-tracking test (HWTT) has gained wide use among state highway agencies for asphalt mixture laboratory rutting evaluation. HWTT was first used in Germany in the early 1970s and later introduced to the United States during the 1990s. HWTT can provide an indication of a mixture's rutting and moisture susceptibility. In multiple studies, researchers have used HWTT to identify the various influential factors that impact the resistance of asphalt mixtures, including SMA, to rutting and stripping (Mendez Larrain & Tarefder 2016; Sel & Ozhan 2014). Limited studies have compared and correlated results from both APT and HWTT conditions. For instance, Ling et.al compared the rutting potential of two typical dense graded HMA mixtures used for airport applications. In this same study, HWTT results were compared to rutting observed after APT loading (Ling et al. 2020).

In the study presented in this paper an important goal is to understand how laboratory and APT results compare to performance predictions from pavement analysis and design tools, such as AASHTOWare PavementME.

1.2 *Objectives*

The objective of this study is to evaluate a dense graded HMA and SMA rutting performance using APT and HWTT testing and understand how this laboratory performance relates to rutting predictions obtained from the PavementME, version 2.3.0.

2 TESTING APPROACH AND MATERIALS

2.1 *APT*

Full scale accelerated pavement testing was conducted at the INDOT full-scale APT facility. The APT test pit is 6.1 m wide by 6.1 m long and 1.8 m deep. For this study the test pit was divided into four independent test sections. Each test section was 1.5 m wide and 6.1 m long. A dual tire load of 4.1 tons. was applied using a one-half standard axle. Both tires were size 11R22.5 inflated to 0.69 Mpa. During the test, the pavement temperature at a depth of 3.8 cm (bottom of the asphalt layer) was 47° C. The INDOT APT building is an indoor facility that has the capability of conditioning the pavementsections at low or high temperatures as required by the testing program. In addition, each lane experienced 50,000 passes of the dual tire assembly.

The test sections were designed using the conventional INDOT full-depth flexible pavement structure with overall pavement thicknesses of either 32 cm. or 39 cm. The analysis presented in this paper only considers the 32 cm. structures. As shown in Figure 1, all test lanes had a 3.8-cm. wearing surface (SMA or dense graded HMA), 6.4-cm. intermediate asphalt layer, 6.4-cm., open-graded (OG) asphalt drainage layer, and 15.3-cm. of asphalt base placed in two separate 7.6-cm. lifts (lower and upper base layers) located below and above the OG drainage layer, respectively. The wearing course mixtures were a typical 9.5-mm Superpave-designed dense graded HMA asphalt mixture and a 9.5-mm SMA. The Superpave-designed mixtures design parameters are shown in Table 1. The intermediate course and base course were produced using the same job mix formula, varying only in asphalt binder performance grade (PG) (Nantung et al. 2018). It is noted that the only difference between both pavement section is the wearing course (either SMA or dense graded HMA). Despite this fact, it was worth investigating if these layers influenced the rutting development of the entire pavement structure.

A rutting monitoring system was implemented to investigate the rutting developed within each individual layer. A series of monitoring holes were drilled into the pavement, and different sets of monitoring holes drilled at different depths, one for each layer interface,

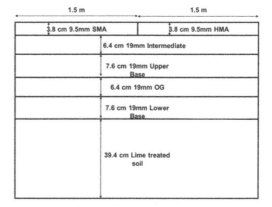

Figure 1. SMA and HMA APT sections.

Table 1. Mix design parameters for all APT sections.

Mix Course	Surface SMA	Surface HMA	Intermediate / Base	OG
PG	70-22	70-22	70-22/64-22	76-22
NMAS (mm)	9.5	9.5	19	19
N_{design}	100	100	100	20
P_b (%)	5.6	5.7	4.6	3.1
VMA @ N_{des} (%)	17.5	15.6	13.2	N/A
VFA @ N_{des} (%)	76.6	74.4	69.7	N/A

N/A: Not applicable

were necessary. By measuring the change in elevation at the bottom of the monitoring holes, the profile of each interface could be measured across the test lanes. The profiles were used to determine the rutting in each individual layer. An automated laser profile system was used to scan the pavement surface and the monitoring holes during pavement loading. Detailed information on the pavement rut monitoring system can be found in (Tian et al. 2017).

2.2 *Hamburg wheel track test*

Wearing course core specimens extracted from undamaged (outside the wheel path) areas of the APT test sections were evaluated using the HWTT at 50°C, in accordance with AASHTO T324-17. The standard specifies the cylindrical test specimens be inserted into plastic molds, to secure them into place during testing. The wearing course was cut from each full-depth APT core to evaluate the rutting characteristics of only the wearing courses (see Figure 2). Air voids content was measured for each of the wearing course specimens; results ranged from 6-8 percent. Since the wearing course core specimens were all thinner than the 62 mm standard requirement, spacer discs made of uncompressible material were used to level the specimens in the plastic molds. Testing continued until both HWTT specimens achieved a rut depth of 12.5 mm, or until 20,000 wheel passes were achieved, whichever came first. Test results are the number of wheel passes

Figure 2. APT cores from the HMA section.

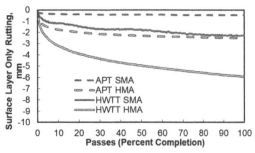

Figure 3. Wearing course rutting from APT and HWTT.

and final rut depths; both reported as the averaged values for each pair of tested specimens.

2.3 PavementME

APT traffic and climatic conditions were modeled in the PavementME software, version 2.3.0, to estimate APT section rutting. The climatic conditions were modeled by creating a climatic input file in which the pavement temperatures were kept constant at 47° C (APT testing temperature) for an entire year. The APT dual tires was modeled using the APT test speed and tire pressure. The corresponding dynamic modulus data of dense graded HMA and SMA materials and the global calibration factors were implemented for predictions in rutting performance. Pavement structures representing the two APT sections with their corresponding material properties and layer thicknesses were implemented in the analyses.

3 RESULTS AND DISCUSSION

3.1 APT rutting

Rut depths, as defined in the AASHTO R 48, *Standard Practice for Determining Rut Depth in Pavements*, were used in the study to compare and analyze the rutting performance of the APT test sections. The mid-depth rut monitoring system was used to separate individual wearing course rut depths from the total rutting. In Figure 3 the number of APT passes (50,000) has been normalized to show percent completion rather than number of wheel passes. As depicted in Figure 3, the APT rut depth measurements from the SMA wearing course section exhibit considerably less rutting than that does the HMA wearing course section. The SMA wearing course rutting was minimal, less than 1.0 mm.

3.2 HWTT rutting

In Figure 3, the number of HWTT passes (20,000) were also normalized to show the rut depth development in both wearing courses. Both the SMA and dense graded HMA final rut depths were below the 12.5 mm threshold, but differences can be observed in final rut depths. While the SMA has a final rut depth of 2.3

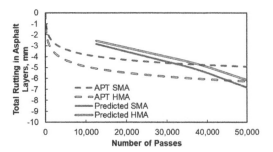

Figure 4. Total pavement rutting from PavementME and APT.

mm, the dense graded HMA experienced 6.0 mm. It is important to note that the observed trends are the same for the APT and HWTT results. The use of spacers to level the HWTT specimens did not influence the test results. However, a formal investigation to assess this factor is necessary to draw appropriate conclusions.

3.3 PavementME rutting predictions

The PavementME predicted total pavement rutting for the two APT sections as shown in Figure 4. These predictions represent only one year of service life, as it is believed this is about the service life evaluated in the APT testing. As seen in the figure, PavementME predicted more rutting for the SMA section than the dense graded HMA section, about a 1.0 mm difference. This finding suggests that PavementME was not able to replicate the trend indicating SMA's much better rutting performance. For comparison purposes, Figure 4 also presents Total pavement rutting (all layers) for both SMA and dense graded HMA sections, as obtained from the APT. The dense graded section has about 30% more total rutting that does the SMA section. Also, the final dense graded HMA and SMA rut depths are nearly identical.

3.4 Rutting results comparison

Figure 5 summarizes the rutting results. Both the APT and HWTT indicate the SMA provides better rutting resistance than does the dense graded HMA. While PavementME dense graded HMA pavement rutting

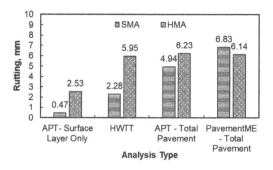

Figure 5. Total and wearing course rutting from SMA and HMA APT sections.

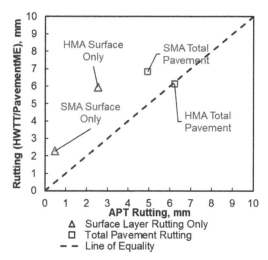

Figure 6. Total and wearing course rutting from SMA and HMA APT sections.

predictions are comparable to total pavement rutting in the APT test sections, PavementME SMA pavement rutting predictions were higher than those found for dense graded HMA, something that is atypical of in-service pavements.

HWTT rut depths substantially higher than APT rut depths may be due to several factors particular to testing conditions. The HWTT test temperature was 50°C, slightly higher than the APT test temperature of 47°C. Additionally, the HWTT used a water bath to hold test temperature. Therefore, HWTT results not only include rutting damage, but also possible moisture damage. There are also differences in pavement structures and boundary conditions. While the APT provides a full-scale pavement structure, the HWTT evaluates a 62 mm core. Finally, the APT and HWTT have very different wheel loading conditions. While the APT closely mimics real pavement loading conditions, the HWTT may be a much more aggressive loading. For this reason, rutting measurements from both APT and HWTT should not be expected to be similar or comparable in magnitude.

It is interesting to note that despite the differences in loading conditions and structures, both analyses show similar trends and indicate that SMA mixtures provide better anti-rutting characteristics than conventional HMA mixtures.

In Figure 6 final rut measurements and predictions are plotted and compared. It is observed that the HWTT overestimates SMA and dense graded HMA rutting, when compared to only wearing course rutting from APT measurements. Total dense graded HMA rutting predictions from PavementME are comparable to APT measurements. It is also observed that PavementME overestimates total SMA rutting when compared to actual APT measurements.

4 CONCLUSIONS

HMA and SMA rutting performance were evaluated using APT, HWTT, and PavementME, version 2.3.0. Given the results and analyses, the following conclusions can be drawn:

1. Both the APT and HWTT provide clear distinctions between SMA and the HMA wearing surfaces. Both test methods can be reliable in providing similar trends, or relative indications of rutting resistance characteristics.
2. Due to several factors inherent to each testing condition, results from the HWTT and APT are not comparable in magnitude. The HWTT testing environment is more aggressive than the APT and therefore tends to overestimate rutting.
3. PavementME did not predict differences between dense graded HMA and SMA rutting. A proper calibration of the PavementME rutting model is needed in order to distinguish the anti-rutting benefits of SMA.

REFERENCES

Ling, J., Wei, F., Chen, H., Zhao, H., Tian, Y., & Han, B. (2020). "Accelerated Pavement Testing for Rutting for Asphalt Overlay Under Hight Tire Pressure," *Journal of Transportation Engineering Part B: Pavements*, 146 (2).

Mendez Larrain, M. M., & Tarefder, R. A. (2016). "Effects of Asphalt Concrete Gradation, Air Voids, and Test Temperatures on Rutting Susceptibility by Using the Hamburg Wheel Tracking Device (HWTD)," *Geo-China*, 83–90.

Nantung, T., Lee, J., & Tian. (2018). *Efficient pavement thickness design for Indiana* (Joint Transportation Research Program Publication No. FHWA/IN/JTRP-2018/06). West Lafayette, IN: Purdue University.

Sel, I., Yildirim, Y., & Ozhan, H. B. (2014). "Effect of Test Temperature on Hamburg Wheel-Tracking Device Testing," *Journal of Materials in Civil Engineering*, 26 (8), 04014037.

Tian, Y., Lee, J., Nantung, T., & Haddock, J. E. (2017). "Development of a Mid-depth Profile Monitoring System for Accelerated Pavement Testing," *Construction and Building Materials*, 140, 1–9.

Advances in Materials and Pavement Performance Prediction II – Kumar et al. (eds)
© 2021 Taylor & Francis Group, London, ISBN 978-0-367-46169-0

Stress-strain-time response of emulsified cold recycled mixture in repeated haversine compression loading

Atanu Behera & J. Murali Krishnan
Indian Institute of Technology Madras, Chennai, India.

ABSTRACT: Emulsified cold recycled mixtures (ECRM) consists of reclaimed asphalt pavement materials, water, cement and bituminous emulsion. ECRM has a high air void of approximately 15%. Very little information is available on the stress-strain-confinement pressure-time response of ECRM and such data is necessary to prescribe an appropriate constitutive relationship. In this study, ECRM was subjected to repeated haversine compression loading at 0°C and 50°C, with and without confinement pressure over ten frequencies (25 Hz to 0.01 Hz). The influence of confinement pressure on the mechanical response of ECRM was observed to be significant at high temperature when compared to low temperature. The material exhibited diverse mechanical response, and it was seen that the phase angle increased as the frequency increased at high temperature without confinement pressure. The experimental data collected underlines the multiple retardation times of the material, and this will come in handy when developing appropriate constitutive models.

1 INTRODUCTION

Cold recycling is a promising technique for the rehabilitation of asphalt pavement. Cold recycling is carried out by pulverising the distressed asphalt pavement, known as Reclaimed asphalt pavement (RAP), and the RAP is mixed with fresh aggregate, water, active filler, and bituminous emulsion or foamed bitumen. The cold recycling using bituminous emulsion is termed as an emulsified cold recycled mixture (ECRM).

The mechanical response of ECRM needs to be evaluated so that the structural design of the pavement can be carried out. ECRM has higher air voids, RAP content, residual bitumen from the emulsion, water, and cement. Presence of RAP, which is active can impart viscoelastic response and presence of RAP which is merely a black rock, may make the material behave as granular material (pressure dependent). Addition of cement can lead to brittleness at low temperatures, improve resistance to permanent deformation at high temperature, and resistance to moisture damage (Issa et al. 2001; Kavussi & Modarres 2010). It is also understood that the existence of higher air voids can lead to viscoelastic response (Passman 1984; Nivedya et al. 2017) or pressure-dependent response (Zeiada et al. 2011). Presence of emulsion residue provides viscoelastic characteristics (Issa et al. 2001) to the ECRM.

The overall mechanical characterization of ECRM is challenging due to the reasons discussed above. It is believed that ECRM can behave like a granular material (pressure dependent) or like an asphalt material (loading rate dependent) depending on its constituents, condition of loading and temperature during testing. To establish a constitutive relation for ECRM, the mechanical response should be quantified, and this investigation presents the influence of confinement pressure, temperature, and frequency.

2 EXPERIMENTAL INVESTIGATION

2.1 Materials

The ECRM consists of RAP, active filler (cement), water and bituminous emulsion. The agglomerated RAP material was processed, and the particle size distribution was evaluated. A cationic slow-setting emulsion (SS-2), as prescribed by IS 8887-17 (2017) was prepared using a laboratory colloidal mill, for the current study. The emulsion was prepared with a VG10 (IS 73-13 2013) bituminous binder.

2.2 Mix design

The mix design of ECRM was carried out as per AASHTO PP-86 (2017). The trial mixture consisted of 100% RAP materials, 2% water content, 0.5% cement content, and the emulsion content was varied (2, 3, and 4%). Indirect tensile strength test with moisture susceptibility test (AASHTO T283 2014) was used to choose the optimum emulsion content. The samples were prepared using gyratory compactor with 100 mm diameter mould with 30 gyrations and cured at 60°C to obtain a constant mass (AASHTO PP-86 2017). Using the unconditioned and conditioned ITS values (shown

Table 1. ITS results of ECRM at 25°C.

Emulsion Content (%)	ITS (kPa)	
	Unconditioned Sample	Conditioned Sample
2	574.18	355.52
3	633.49	505.72
4	540.51	439.90

in Table 1), an emulsion content of 3% was determined as the optimum emulsion content.

2.3 Test protocol

The ECRM samples were prepared using a gyratory compactor with 600 kPa vertical pressure for 30 revolutions. The sample size was 100 mm in diameter and 150 mm in height. A repeated haversine compression loading was applied at two temperatures (0 and 50°C) and ten frequencies (25, 20, 10, 5, 2, 1, 0.5, 0.2, 0.1, and 0.01), with and without confinement pressure (200 and 0 kPa). The peak to peak strain at each cycle was maintained between 85 to 115 for all testing conditions. For each frequency, ten conditioning cycles and ten test cycles of loading were applied. Fifty data points were collected at each cycle of loading. The stress and strain response of ECRM was observed for the ten test cycles for each frequency and temperature.

3 RESULTS AND DISCUSSIONS

3.1 Influence of confinement pressure, frequency and temperature on phase angle of ECRM

Figure 1 shows the variation of phase angle as a function of frequency and confinement pressure at 0°C and 50°C. At 0°C, the phase angle increases with a decrease in frequency for both confined and unconfined condition. However, at 50°C, the phase angle increases and decreases for unconfined condition and decreases and increases for the confined condition with a decrease in frequency. Identical behaviour of phase angle variations have been reported by Deepa et al. (2019) for hot mix asphalt. The increasing and decreasing trend and decreasing trend of phase angle with the decrease in frequency were reported at higher temperatures (35, 45 and 55°C) for hot mix asphalt. As the material is subjected to a wide range of frequencies for a given temperature, a viscoelastic material with multiple retardation time can exhibit such behaviour (Deepa et al. 2019; Findley et al. 1989;.

It can be observed from Figure 1 that at 0°C, the influence of confinement pressure is negligible, but at 50 °C the influence is significant. The confinement pressure results in a reduction in the phase angle. In such cases, the analysis of strain data can lead to a

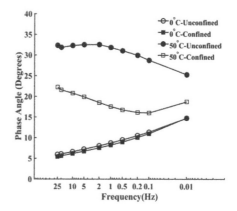

Figure 1. Plots of phase angle as a function of frequency at two different temperatures.

better understanding of the mechanical response of ECRM.

3.2 Influence of confinement pressure, temperature, and frequency on strain response of ECRM

The strain response of ECRM with and without confinement are compared in Figure 2 for 0°C and Figure 3 for 50°C at two frequencies (25, and 0.01 Hz). The total strain values were normalised with the initial strain value of each frequency. For the target strain value of 85 to 115 micro-strain for each cycle, at a given frequency, the stress value keeps on changing as a function of loading cycles.

Figure 2 shows that at 0°C, the confinement pressure has a negligible influence on the strain response at 25 Hz and 0.01 Hz frequency, however the frequency of loading influences the strain response. At 0.01 Hz frequency, the increase in total strain with loading time is more as compared to 25 Hz frequency.

The influence of confinement pressure and frequency of loading on the strain response of ECRM at 50°C can be observed from Figure 3. At 25 Hz frequency, the total strain keeps increasing with loading time irrespective of confinement pressure, but the total strain is more for the unconfined condition as expected. However, at 0.01 Hz frequency, the total strain increases for confined conditions and decreases for unconfined conditions. The decreasing trend of total strain with loading time is due to multiple retardation time of the material as discussed earlier (Deepa et al. 2019).

3.3 Influence of confinement pressure, temperature, and frequency on the stress response of ECRM

To obtain the target strain level of 85 to 115 microstrain in each cycle, the variation of the amplitude of stresses with frequencies with and without confinement are shown in Figure 4 for 0°C and 50°C.

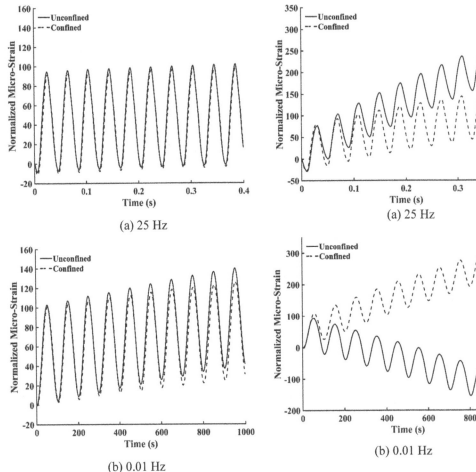

(a) 25 Hz

(a) 25 Hz

(b) 0.01 Hz

(b) 0.01 Hz

Figure 3.   Normalised strain variation at 50°C.

Figure 2.   Normalised strain variation at 0°C.

3.4   Influence of confinement pressure, temperature, and frequency on the hysteresis loop of ECRM

From Figure 4 (a), it can be seen that the influence of confinement pressure is negligible on the stress response of ECRM at 0°C as compared to 50°C. With the decrease in frequency, the stress amplitude reduces irrespective of the temperature and confinement pressure. The stress amplitude for 0°C at 25 Hz frequency is approximately 1000 kPa, and at 0.01 Hz it is 500 kPa in unconfined condition. However, the stress amplitude for 50°C (Figure 4 (b)) at 25 Hz frequency is 100 kPa, and at 0.01 Hz, it is 10 kPa. It can be seen that the stress applied at 0°C is much larger than the confinement pressure, and at 50°C, the stress amplitude is less as compared to the confinement pressure. Hence, one can conclude based on these experimental observations that the influence of confinement pressure is negligible at low temperature. However, the influence of confinement pressure is significant at high temperature.

One can glean substantial information in a graphical manner by plotting the centered stress and strain using the experimental data reported here. The shape of the hysteresis loop indicates the nature of the response of the material, and in case of viscoelastic response, the width of the ellipse and slope of the major axis is a measure of phase angle and dynamic modulus (Lakes 2009).

Figure 5 shows the *hysteresis loop* of ECRM at two different temperatures for 25 Hz and 0.01 Hz frequency for the tenth cycle. From Figure 5, one can observe the variation of stress, phase angle and modulus of the material. One can observe that the influence of confinement pressure is negligible at 0°C when compared to 50°C in terms of variation of the stress, phase angle, and dynamic modulus. At 0°C, the width of hysteresis loop for 25 Hz frequency is smaller compared to 0.01 Hz for both confinement condition (Figure 5 (a)). This indicates that the material has a smaller phase angle at 25 Hz as compared to

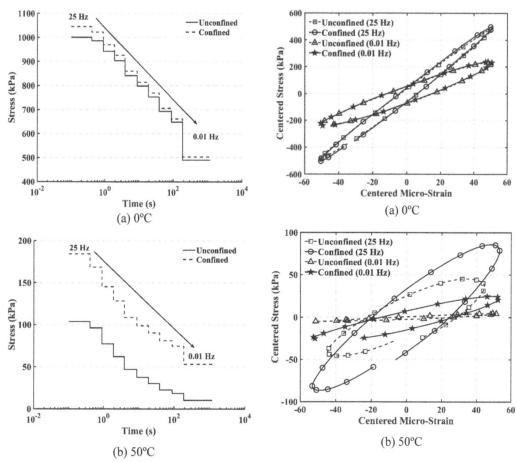

(a) 0°C

(b) 50°C

Figure 4. Variation of stress amplitude with frequencies.

(a) 0°C

(b) 50°C

Figure 5. The hysteresis loop of ECRM.

0.01 Hz frequency. At 50°C, the phase angle reduction from 25 Hz to 0.01 Hz for the unconfined condition is significant as compared to confined condition (Figure 5Figure 5 (b)). The variation of dynamic modulus from the hysteresis loop can also be observed. The dynamic modulus decreases with a decrease in frequency for both confinement condition. At 50°C, the confinement pressure leads to an increase in dynamic modulus of ECRM.

4 DISCUSSION

The stress-strain-time response of ECRM was reported in this paper. A repeated haversine compression loading was applied over ten frequencies at two temperatures, and the influence of confinement pressure was evaluated. The following observations are made:

1. For a targeted stain value, the stress applied at 0°C is much larger than the confinement pressure, and at 50°C, the stress amplitude is less as compared to the confinement pressure. Therefore, the influence of confinement pressure is negligible at 0°C, and at 50°C, the influence of the same was substantial.

2. The material exhibits response identical to that of a viscoelastic material with multiple retardation time, especially at 50°C. It is seen that the presence of confinement conditions strongly influences such responses.

3. Since it is not very clear whether the post-processing procedure stipulated for computing dynamic modulus for HMA can be applied ECRM, the use of hysteresis loop was demonstrated. It was seen that dynamic modulus as depicted by the slope of the hysteresis loop was found to be identical for both confinement condition at 0°C and increased at 50°C in the presence of confinement pressure. Also, the phase lag as depicted by the width of the hysteresis loop was found to be decreased with decrease in frequency (25 Hz to 0.01 Hz) and increase in temperature (0 to 50°C) for both confinement condition.

259

REFERENCES

AASHTO PP-86. 2017. "Standard Practice for Emulsified Asphalt Content of Cold Recycled Mixture Designs." *American Association of State Highway and Transportation Officials, Washington D.C., USA.*

AASHTO T283. 2014. "Standard Method of Test for Resistance of Compacted Hot Mix Asphalt (HMA) To Moisture-Induced Damage." *American Association of State Highway and Transportation Officials, Washington D.C., USA.*

Deepa, S., U. Saravanan, and J. Murali Krishnan. 2019. "On Measurement of Dynamic Modulus for Bituminous Mixtures." *International Journal of Pavement Engineering* 20(9):1073–89.

Findley, W. N., J. S. Lai, K. Onaran, and R. M. Christensen. 1989. *Creep and Relaxation of Nonlinear Viscoelastic Materials?: With an Introduction to Linear Viscoelasticity.* New York: Dover publications.

IS 73-13. 2013. "Paving Bitumen-Specification." *Bureau of Indian Standards,* New Delhi.

IS 8887-17. 2017. "Bitumen Emulsion for Roads (Cationic Type)." *Bureau of Indian Standards,* New Delhi.

Issa, Rita, Musharraf M. Zaman, Gerald A. Miller, and Lawrence J. Senkowski. 2001. "Characteristics of Cold Processed Asphalt Millings and Cement-Emulsion Mix." *Transportation Research Record* 1767:01–0249.

Kavussi, Amir and Amir Modarres. 2010. "A Model for Resilient Modulus Determination of Recycled Mixes with Bitumen Emulsion and Cement from ITS Testing Results." *Construction and Building Materials* 24(11):2252–59.

Lakes R. 2009. *Viscoelastic Materials.* Cambridge university press, New York.

Nivedya, M. K., A. Veeraragavan, Parag Ravindran, and J. Murali Krishnan. 2017. "Investigation on the Influence of Air Voids and Active Filler on the Mechanical Response of Bitumen Stabilized Material." *Journal of Materials in Civil Engineering* 30(3):4017293.

Passman, S. L. 1984. "Stress-Relaxation, Creep, Failure and Hysteresis in a Linear Elastic-Material with Voids." *Journal of Elasticity* 14(2):201–12.

Zeiada, Waleed Abd-Elaziz, Kamil E. Kaloush, Krishna Prapoorna Biligiri, Jordan Xavier Reed, and Jeffrey J. Stempihar. 2011. "Significance of Confined Dynamic Modulus Laboratory Testing for Asphalt Concrete: Conventional, Gap-Graded, and Open-Graded Mixtures." *Transportation Research Record: Journal of the Transportation Research Board* 2210:9–19.

Advances in Materials and Pavement Performance Prediction II – Kumar et al. (eds)
© 2021 Taylor & Francis Group, London, ISBN 978-0-367-46169-0

Multiscale study of the fracture properties of asphalt materials

L.M. Espinosa, S. Caro & J. Wills
Universidad de los Andes, Bogotá, Colombia

ABSTRACT: A proper mechanical characterization of asphalt mixtures is needed to improve the performance and durability of asphalt materials. This study aims at evaluating the influence of the air void content over the fracture properties of asphalt mixtures at two different length scales. To achieve this goal, Semi-Circular Bending (SCB) tests were conducted on a Hot Mix Asphalt (HMA) and its corresponding Fine Asphalt Matrix (FAM), under three different air void contents (i.e. 4, 7, and 10%). Peak load and fracture energy of FAM and HMA samples exhibited air voids dependency, while the area of its fracture zone did not. The fracture response of FAM samples could be used as an initial insight of the expected behavior of the full mixture, since the trend of the properties studied as a function of AV were quite similar in both materials.

1 INTRODUCTION

Hot Mix Asphalt (HMA) are complex and heterogeneous multiscale materials composed of coarse aggregates bonded by a softer phase of a Fine Aggregate Matrix (FAM). This softer phase consists of a combination of asphalt binder, fine particles (i.e. passing the sieve #16 or 1.18 mm) and air voids, and it is considered a key component of the mixtures since several critical damages, including fracture, initiates within this material (Kim et al. 2006).

Due to the importance of HMA, and the role that FAM plays in its behavior, several technics have been used to better understand the impact of the variation of its individual components on its response at different length scales (Aragão et al. 2010; Soares et al. 2003). Within these techniques, the Semi-Circular Bending (SCB) test has been widely used to experimentally quantify the fracture resistance of asphalt mixtures under the effect of different factors, such as: binder type, aggregate type, and air void content (AV), among others. Regarding AV, it has been found that the variation of this factor directly affects the resistance of the mixture under low and intermediate temperatures (Aliha et al. 2015; Kaseer et al. 2018; Li et al. 2008). This is due to the fact that lower AV contents create structures with fewer, smaller, and better-distributed voids, resulting in less stress concentration at large voids and, therefore, higher resistance to cracking (Harvey & Tsai 1996; Li et al. 2008).

Most of these studies have been conducted on full mixtures (i.e. HMA scale), which means that high amounts of time and material have been required to obtain reliable results. However, several researchers have suggested that the study of a mixture at a mortar scale (i.e. FAM scale) could give initial valuable information about the behavior of the whole mixture, while reducing testing time and costs (Karki et al. 2015; Kim et al. 2006; Underwood & Kim 2013). This study aims at assessing the sensitivity of FAM and HMA fracture responses to variations in the AV content, and at determining if the study of the fracture phenomena of an asphalt mixture at a mortar-scale could be used to understand or estimate the behavior of the full mixture.

2 MATERIALS AND SPECIMEN PREPARATION

This study was carried out using an HMA mixture with a nominal maximum aggregate size (NMAS) of 9.5 mm, and its corresponding asphalt mortar or FAM, which gradations are presented in Figure 1. Both mixtures were fabricated using an asphalt binder with penetration 60–70 (1/10 mm), at an optimum content of 5.3% for HMA and 10.6% for FAM.

Cylindrical specimens of 80 mm in height and 150 mm in diameter were compacted for each material

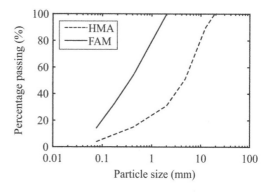

Figure 1. HMA and FAM gradation curves.

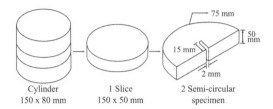

Figure 2. Semi-circular specimen fabrication.

Cylinder 150 x 80 mm   1 Slice 150 x 50 mm   2 Semi-circular specimen

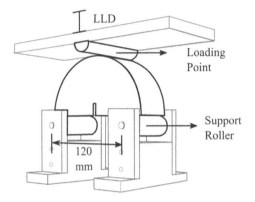

Figure 3. SCB test set-up.

using the Superpave gyratory compactor, at a target AV of 4±1%, 7±1% and 10±1%. Slices of 50 mm in height were cut from the middle of the cylinder, which were then cut symmetrically into two semi-circular specimens with a notch in its planar side of 155 mm length and 2 mm width, (Figure 2). In total, 42 SCB specimens were fabricated, 7 per scenario (i.e. FAM or HMA with each AV content)

3 EXPERIMENTAL PROCEDURE

The SCB test set up, presented in Figure 3, consists of two frictionless rollers supporting a semi-circular specimen that is subjected to flexion through a loading point at the center of its arched edge. The rollers have a space span of 120 mm, enabling a Mode I or pure flexion mode of failure.

The tests were conducted at a loading rate of 50 mm/min and a temperature of 24 ± 1°C. A load cell and a linear variable differential transducer (LVDT) were used to measure the applied load and displacement at the top center of the sample, respectively. The latter information was recorded and used to build a load-load line displacement (load-LLD) curve. Different fracture properties can be obtained from this curve, including the peak load (P_l) that is the maximum load recorded during the test which provides information about the strength of the material; or the fracture energy (G_f), which represents the amount of energy dissipated during the generation and propagation of a crack, and is calculated as the ratio between

the work of fracture (area under the load-LLD curve, W_f) and the fracture zone area (A).

4 MEASURMENT OF THE FRACTURE ZONE AREA

The fracture surface of composite materials, such as HMA, is conditioned by the unique characteristics of their microstructures (Roulin-Moloney 1986). It is then foreseeable that the fracture zone of HMA cannot be represented by a planar area, as commonly done, but by a three-dimensional (3D) one, accounting for the morphology of the mixture (Espinosa et al. 2019). To determine this 3D area, SCB cracked samples were digitalized using the Artec Space Spider® scanner, which uses blue light technology to create digital images with a resolution up to 0.1 mm and an accuracy of around 0.05 mm. These images were then exported and processed using the Autodesk Netfabb® software to isolate and compute the 3D area of the fracture zone (A).

FAM and HMA samples were scanned under the same resolution (in a scale of mm) to allow their comparison. This scale properly represents the morphology of the asphalt materials.

5 RESULTS AND DISCUSSION

To better understand the impact of AV on the properties of an HMA and its corresponding FAM, the area of the fracture zone (A),peak load (P_l), and fracture energy (G_f), of the 42 samples tested and scanned, are illustrated in Figures 4 to 6, in the form of mean values and coefficients of variation (COV).

Figure 4 illustrates the area of the fracture zone (A) of the two materials with different AV contents. The area of the fracture zone of HMA shows a positive correlation with the mixture's AV content,with values ranging from 37.3 cm² to 38.8 cm², for samples with AV with 4% and 10%, respectively. On the other hand, FAM exhibits a constant value of A,of approximately 34 cm², for AV values between 4% and 7%, and it tends to increase for samples with a 10% AV content, reaching a value of 36.06 cm².

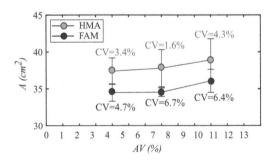

Figure 4. Area of the fracture zone of HMA and FAM as a function of the AV content.

To corroborate these observations, a one-way ANOVA, with a 95% of significance ($\alpha = 0.95$) was conducted for each consecutive pair of AV related results and for the two extreme AV values (i.e. 4%–7%, 7%–10%; 10%–4%). The *p-value* is the main output parameter of the test (i.e. values smaller than 1-α imply a significant difference among measurements). The ANOVA results of A, listed in Table 1, show there are only statistical differences in A as a function of AV in one specific case; which means that the A values of the

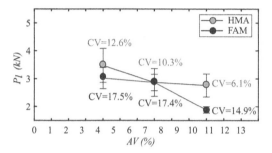

Figure 5. P_l of FAM and HMA mixtures as a function of the AV content.

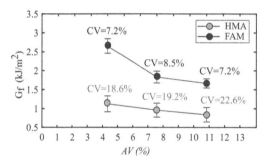

Figure 6. G_f of FAM and HMA mixtures as a function of the AV content.

full mixture under study, and its corresponding mortar, are not sensitive to the variation of the mixture's AV content.

However, A is highly sensitive to changes in the scale of the mixture, since relevant differences were found between the A values for FAM and HMA for a every AV value (up to 10%). Indeed, while the crack path in dense mixtures is conditioned by the coarse aggregates, in FAM mixtures it tends to propagate straighter under the loading point.

Figure 5 illustrates the peak load response of HMA and FAM under the variation of the mixture's AV. A negative correlation seems to exist between both variables, for both HMA and FAM. This result follows the findings of previous works that have stated that AV generates a reduction of the material's strength, since a mixture with more voids requires less driving force to be broken (Harvey & Tsai 1996; Kaseer et al. 2018).

After conducting an ANOVA test (Table 2), it was found that P_l in FAM is significantly sensitive to the variation of AV only when this value is higher than 7% or when its change is significant (i.e. between 4% and 10%). A similar behavior was found for the HMA, which only presents significant differences in P_l with relevant variations in AV. When comparing the P_l values of the HMA with those of its correspondent FAM, it was found P_l is statistical different between both materials only when they have an AV content of 10%. This means that the behavior of P_l depends on the mixture's AV content, but not on the mixture's scale.

Figure 6 presents the results for the fracture energy (G_f). It is observed that increments in AV generate reductions in the mixture's G_f values, for both FAM and HMA. However, the results of an ANOVA test on these values (Table 3) indicate that while G_f in FAM presents a significant reduction for consecutive increments of AV, for the full mixture these changes are only significant between extreme AV values (i.e. from 4% to 10%). Also, due to the high binder content of the FAM mixtures, FAM samples presented significantly

Table 1. ANOVA results for the significance of the A results for HMA and FAM.

	AV (%)	p-value	Significant		AV (%)	p-value	Significant		AV (%)	p-value	Significant
A of	4 vs 7	9×10^{-1}	No	A of	4 vs 7	7×10^{-1}	No	A of	4	4×10^{-3}	Yes
FAM	7 vs 10	3×10^{-2}	Yes	HMA	7 vs 10	5×10^{-1}	No	FAM vs	7	6×10^{-3}	Yes
	4 vs 10	5×10^{-2}	No		4 vs 10	3×10^{-1}	No	HMA	10	4×10^{-2}	Yes

*Sig = significant

Table 2. ANOVA results for the significance of the Pl results for HMA and FAM.

	AV (%)	p-value	Significant		AV (%)	p-value	Significant		AV (%)	p-value	Significant
Pl of	4 vs 7	4×10^{-1}	No	Pl of	4 vs 7	6×10^{-2}	No	Pl of	4	1×10^{-1}	No
FAM	7 vs 10	3×10^{-6}	Yes	HMA	7 vs 10	2×10^{-2}	No	FAM vs	7	1	No
	4 vs 10	7×10^{-6}	Yes		4 vs 10	7×10^{-1}	Yes	HMA	10	2×10^{-4}	Yes

*Sig = significant

Table 3. ANOVA results for the significance of the G_f results for HMA and FAM.

	AV (%)	p-value	Significant		AV (%)	p-value	Significant		AV (%)	p-value	Significant
FAM	4 vs 7	1×10^{-6}	Yes	HMA	4 vs 7	1×10^{-1}	No	FAM vs	4	7×10^{-9}	Yes
	7 vs 10	5×10^{-2}	Yes		7 vs 10	3×10^{-1}	No	HMA	7	6×10^{-7}	Yes
	4 vs 10	7×10^{-8}	Yes		4 vs 10	2×10^{-2}	Yes		10	5×10^{-7}	Yes

*Sig = significant

higher values of G_f than HMA mixtures (i.e., 49% to 58% larger). This means that the energy required to fracture an asphalt mixture is not only affected by the amount of AV but also by the scale on which is being studied, as it could be expected.

6 CONCLUSIONS

SCB tests were conducted in this study on asphalt mixtures at two different scales (HMA and FAM) with three different AV contents. The results of 7 replicates per case showed that the fracture zone (A) is not affected by AV, while it is affected by the mixture's scale. The peak load (P_l) and fracture energy (G_f) of the HMA were found to be significantly sensitive only to considerable changes in the mixture's AV content (i.e. from 4% to 10%). In the case of the FAM material, these parameters were found to be significant different between almost all consecutive changes in AV. Additionally, the FAM behavior was found to be equivalent to the full mixture in the case of P_l at AV values smaller than 10%. However, the general trend of all the fracture properties as a function of the AV was similar in both scales, indicating that the study of the fracture properties of mortars may provide an initial insight of the influence of certain volumetric variables on the expected fracture response of full mixtures.

It is noteworthy that this study only included one type of dense asphalt mixture. Therefore, it is recommended to conducted further studies on other types of mixtures or on mixtures with other materials (e.g. RAP, modified binder, etc.) in order to obtain more general conclusions.

REFERENCES

Aliha, M. R. M., Fazaeli, H., Aghajani, S., & Moghadas Nejad, F. (2015). Effect of temperature and air void on mixed mode fracture toughness of modified asphalt mixtures. *Construction and Building Materials*, 95, 545–555.

Aragão, F. T. S., Kim, Y. R., Asce, A. M., Lee, J., & Allen, D. H. (2010). Micromechanical Model for Heterogeneous Asphalt Concrete Mixtures Subjected to Fracture Failure. *Journal of Materials in Civil Engineering*, 23(1), 30–38.

Espinosa, L.M., Caro, S. and Wills, J. (2019). Influence of the morphology of the cracking zone on the fracture energy of HMA materials. Materials and Structures, Vol. 52: 35. https://doi.org/10.1617/s11527-019-1334-0

Harvey, J., & Tsai, B.-W. (1996). Effects of asphalt content and air void content on mix fatigue and stiffness. *Transportation Research Record: Journal of the Transportation Research Board*, 1543, 38–45.

Karki, P., Li, R., & Bhasin, A. (2015). Quantifying overall damage and healing behaviour of asphalt materials using continuum damage approach. *International Journal of Pavement Engineering*, 16(4), 350–362. https://doi.org/10.1080/10298436.2014.942993

Kaseer, F., Yin, F., Arámbula-mercado, E., Epps, A., Sias, J., & Salari, S. (2018). Development of an index to evaluate the cracking potential of asphalt mixtures using the semicircular bending test. *Construction and Building Materials*, 167, 286–298. https://doi.org/10.1016/j.conbuildmat.2018.02.014

Kim, Y. R., Lee, H. J., Little, D. N., Kim, Y. R., Gibson, N., King, G., …Fee, F. (2006). A simple testing method to evaluate fatigue fracture and damage performance of asphalt mixtures. In *Asphalt Paving Technology: Association of Asphalt Paving Technologists-Proceedings of the Technical Sessions 75* (pp. 755–788).

Li, X., Buttlar, W. G., Braham, A. F., Braham, A. F., & Marasteanu, M. O. (2008). Effect of factors affecting fracture energy of asphalt concrete at low temperature. *Road Materials and Pavement Design*, 9, 397–416. https://doi.org/10.1080/14680629.2008.9690176

Roulin-Moloney, A. (1986). *Fractography and failure mechanisms of polymers and composites*. London: Elsevier Applied Science.

Soares, J. B., De Freitas, F. A. C., & Allen, D. H. (2003). Considering Material Heterogeneity in Crack Modeling of Asphaltic Mixtures. *Transportation Research Record*, (1832), 113–120.

Underwood, B. S., & Kim, Y. R. (2013). Effect of volumetric factors on the mechanical behavior of asphalt fine aggregate matrix and the relationship to asphalt mixture properties. *Construction and Building Materials Journal*, 49, 672–681.

Advances in Materials and Pavement Performance Prediction II – Kumar et al. (eds)
© 2021 Taylor & Francis Group, London, ISBN 978-0-367-46169-0

Evaluating effects of volumetric properties on asphalt rutting using SSR test

A. Ghanbari, Y.D. Wang, B.S. Underwood & Y.R. Kim
North Carolina State University, Raleigh, North Carolina, USA

ABSTRACT: The stress sweep rutting (SSR) test is a test method to characterize the resistance of asphalt mixtures to rutting using the permanent deformation shift model. This test was developed at North Carolina State University. The SSR test measures the permanent strain characteristics of asphalt mixtures as a function of deviatoric stress, loading time, and temperature. The permanent deformation shift model is based on the permanent deformation behavior of an asphalt mixture in the primary and secondary regions. The results from two SSR tests at each of the high and low temperatures are used to develop the shift model. The shift model has been implemented into the pavement performance prediction program, FlexPAVE™, to predict the permanent deformation of asphalt layers under various deviatoric stress levels, loading times, and temperatures as a function of pavement depth and time. In this paper, the SSR and FlexPAVE™ model are used to show the effect of different volumetric properties on the rutting resistance of asphalt mixtures. A typical asphalt mixture in North Carolina was chosen as a case study, and the volumetric properties were changed systematically. The results state that the permanent deformation shift model can capture the effect of binder content, in-place density, and volumetric properties on the rutting behavior of asphalt mixture.

1 INTRODUCTION

Permanent deformation in roadways can lead to traffic accidents, especially in rainy or snowy weather conditions. Rain can cause dangerous driving hazards, such as hydroplaning, and large amounts of water spray can hinder visibility. Snow can cause similar problems because snow and ice can collect in the rutted wheel path. In order to predict the amount and rate of rutting, various models can be used to predict permanent deformation in the flexible pavements. These models can be classified into two categories: mechanistic viscoplastic models and power-law type models. The application of viscoelastic concepts demands high calibration and computing costs. Therefore, despite providing good predictions, mechanistic viscoplastic models have not been widely accepted by agencies (Choi et al. 2012). In contrast, power-law type models are relatively simple and easily implemented. The representative model is a strain ratio model implemented in the Pavement ME™ program.

1.1 Permanent deformation shift model

Permanent deformation modeling research at North Carolina State University (NCSU) has resulted in a viscoplastic model that is based on viscoelastic convolution integrals for explaining the behavior of asphalt mixture in compression under repeated loading. This model has been reduced further to a simplified form, the so-called *incremental model*, which is an advanced power-law type model that can represent both the primary and secondary regions in terms of the permanent strain growth behavior of asphalt mixture. (Kim & Kim 2017)

Based on the previous work at NCSU, a simple mechanistic permanent deformation model and accompanying test protocol are suggested for evaluating the rutting behavior of asphalt mixtures. The model can represent the effects of load time, stress, and temperature on the permanent deformation of asphalt mixture, and has been verified using complex loading histories and field-measured rut depths at various sites (Kim & Kim 2017).

In summary, the shift model employs shift factors which are comprised of the reference curve, the load time shift function, and the stress shift function. The reference curve is defined as the permanent strain growth curve obtained from a single triaxial repeated load permanent deformation (TRLPD) test under a reference loading condition. The incremental model then becomes the final reference curve form, which is considered to be the permanent strain mastercurve. Strains obtained from the reference curve are translated horizontally first for load time-shifting and again for stress shifting to construct the reference curve; accordingly, the total shift factor is a summation of two shift factors, reduce load time shift factor, and vertical stress shift factor.

Equation (1) expresses the shift model form (Ghanbari et al. 2020; Kim et al. 2018).

$$\varepsilon_{vp} = \frac{\varepsilon_0 N_{red}}{(N_l + N_{red})^\beta} \qquad (1)$$

$$a_{\xi_p} = p_1 \log(\xi_p) + p_2$$
$$a_{\sigma_v} = D * (\log(\sigma_v/P_a) - 0.877)$$

$$D = d_1 * T + d_2$$
$$N_{ref} = 10^{p_2} \cdot 10^{-0.877*D} \cdot N(\xi_p)^{p_1} \left(\frac{\sigma_v}{P_a}\right)^D$$

where
a_{ξ_p} = reduced load time shift factor,
a_{σ_v} = vertical stress shift factor,
D = vertical stress shift factor fitting coefficient,
T = test temperature, °C, and
$\varepsilon_0, N_I, \beta, p_1, p_2, d_1, d_2$ = linear regression coefficients,

1.2 SSR background

The SSR test evolved from a previous method developed at NCSU, the Triaxial Stress Sweep (TSS) test. The TSS test was advantageous over other existing models because it could simulate the effect of temperature, load time, and deviatoric stress. However, the TSS test protocol required eight specimens and two days of testing to calibrate a single mixture. The SSR test was proposed to overcome these limitations. This protocol reduces both the testing time and the number of specimens, without unnecessarily sacrificing model accuracy. First, the SSR test eliminated the need for a separate test to establish the reference curve. Instead, the first loading path at the high temperature serves as the reference curve. Also, in the SSR, experiments are conducted at two temperatures (T_H and T_L). The TSS test used three temperatures, but the third temperature was eliminated because it was the effect of temperature on the shift model could be estimated with tests at only the two conditions (Kim & Kim 2017).

The SSR test is performed on a 100 mm (4 in.) diameter by 150-mm (6-in.) tall specimen fabricated in accordance with AASHTO R 83. This standard is applicable to laboratory-prepared specimens of asphalt mixtures with a nominal maximum aggregate size that is less than or equal to 37.5 mm (1.5 in.).

At least two replicate specimens should be tested at T_H and T_L and the temperatures should be selected based on LTPPBind V. 3.1 software. The SSR low and high temperatures are based on the climatic condition and the temperature will be selected based on the location where the mixture will be used.

During the test, a constant confining pressure of 10 psi is applied. Vertical loading is applied for 600 cycles at three deviatoric stress levels for each temperatures. The testing order for the deviatoric stresses at T_L is 70, 100, and 130 psi, respectively. In a study by Kim and Kim, the deviatoric stress orders for T_H were reversed and 100, 70, and 130 psi were applied. Reversing the loading order allows the characterization of the

reference curve using the strain results from the 100 psi loading cycles. The load pulse is 0.4 s for each cycle. The rest periods are 1.6 s for T_L and 3.6 s for T_H. The permanent axial deformation that occurs at each load cycle is measured using actuator displacement, so this test protocol does not require on-specimen LVDTs. More information on the testing procedure as well as theories behind the test, can be found elsewhere (Kim et al. 2020; Kim & Kim 2017).

2 MATERIALS AND METHODOLOGY

In this study, a typical surface mixture in North Carolina was used. The RS9.5B mixture is a typical fine graded surface mixture with 9.5 mm as the Nominal Maximum Aggregate Size (NMAS). This mixture contains 30% RAP and the virgin binder is PG 58-28. In this study, the gradation, binder content, and in-place density were changed systematically and the SSR test was conducted on different conditions. The main goal of this study is to see the effect of volumetric properties on the SSR test results. In order to change the VMA, three gradations were designed. These gradations were labeled as G1, G2, and G3. Two binder contents and three in-place densities were considered at two of the designed gradations (G1 and G3). In summary, nine different conditions were tested. Table 1 shows the experimental design for this study. In this table, the design air void is changed in the range of 3% to 5% to see the effect of the design air void as well as the binder content. In order to study the effect of in-place density, the in-place air void was also changed from 2.9% to 7.3% within each gradation. More details on these mixtures can be found elsewhere (Wang et al. 2019).

2.1 Sample preparation for SSR test

All specimens have been fabricated following AASHTO R 83. The mixing and compaction temperatures were 152°C and 139°C respectively. The asphalt mixture specimens were compacted to the height of 180 mm and 150 mm diameter, and then cored and cut to 150 mm height and 100 mm diameter. The

Table 1. Experimental design.

Mix ID	Volumetric @ N_design			In-Place
	%AC	%AV	VMA	%AV
G1-33	6.0	3.0	15.3	2.9
G1-53	5.3	5.0	15.7	3.2
G1-55	5.3	5.0	15.7	4.7
G1-57	5.3	5.0	15.7	6.8
G2-44	5.8	4.7	16.3	4.2
G3-33	7.0	3.0	17.4	3.3
G3-54	6.1	5.0	17.2	3.9
G3-55	6.1	5.0	17.2	5.4
G3-57	6.1	5.0	17.2	7.3

cored and cut specimen were prepared to be tested in the Asphalt Mixture Performance Tester (AMPT) machine in accordance with AASHTO TP 134. The test temperatures were selected as 29°C and 51°C for low and high temperature respectively. These temperatures were selected based on Raleigh-Durham Airport section in LTPP Bind V3.1. Two replicates were tested at each temperature.

3 TEST RESULTS

Figure 1 and Figure 2 show the test results for different conditions listed in Table 1. For each mixture, two replicates were tested at both T_L and T_H. These figures show the average of two replicates for each mixture at each temperature. In these figures and in order to better present the results, different mixtures are sorted in the legend based on their permanent microstrain rankings.

The results show some meaningful trends within each gradation. For example, in Gradation 1 and at the two temperatures, permanent strain increases as the in-place air void increases. The same trend can be observed in Gradation 3. Also, a higher binder content shows a higher permanent strain in both gradations.

In the next step, the test results were used to predict the pavement performance using FlexPAVE™

software. One of the features of the permanent deformation shift model is to account for temperature, stress level, and loading time. The permanent deformation shift model has been implemented in the FlexPAVE™ program and the rut depth knowing the pavement structure, climatic location, and design traffic can be calculated using this program. In order to predict the rut depth for each mixture, the following structure was defined. In this structure, 4 inches (10 cm) of asphalt mixture on top of 8 inches (20 cm) of aggregate base and subgrade was subjected to 10 million ESAL of traffic for a design life of 20 years. In this study, the location was chosen as Raleigh, North Carolina and the rut depth was predicted in FlexPAVE™ software. Figure 3 shows the structure that was used for FlexPAVE™ simulation.

The FlexPAVE™ simulations were done on different pavements with fixed aggregate base and subgrade, but different surface layers with the mixtures listed in Table 1. Figure 4 shows the rut depth predicted by the

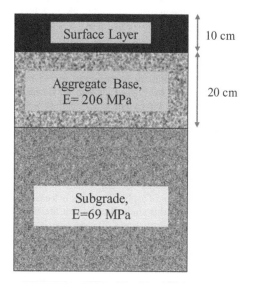

Figure 3.   Structure used for FlexPAVE™ simulation.

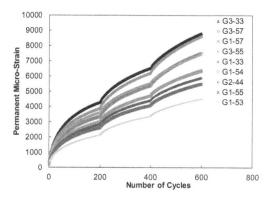

Figure 1.   SSR low temperature test results.

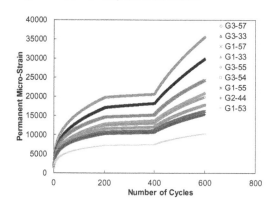

Figure 2.   SSR high temperature test results.

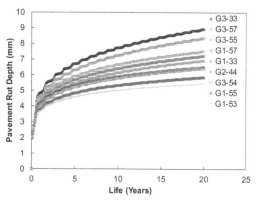

Figure 4.   Rut depth prediction using FlexPAVE™.

267

FlexPAVE™ program for the design life of 20 years. Based on these results, the G3-33 pavement has the highest rut depth amongst the tested mixtures. These results are somehow expected based on engineering intuition. In this study, Gradation 3 has the highest VMA amongst the three designed gradations, and this gradation is the finest gradation within three different gradations. G3-33 has the highest asphalt content (7.0%) and it has been accepted that a higher asphalt content will lead to a higher permanent deformation and less rutting resistance. The best performance in the tested mixtures belongs to G1-53. This mixture had the coarsest gradation, lowest asphalt content (5.3%), and high in-place density (96.8%). This is again compatible with the engineering expectation.

Other results are all along with engineering intuition as well. For example, in Gradation 3, and with decreasing the in-place density, the rut depth at the end of design life increases significantly. In this gradation, G3-54 has the least rut depth (6.4 mm) and G3-57 has the highest rut depth (8.4 mm). In other words, 87% increasing in the in-place air void (3.9% to 7.3%) can increase the pavement rut depth by 31% (6.4 mm to 8.4 mm).

Similar trend can be found in G1 mixtures. G1-53 has the best performance and G1-57 has the highest permanent deformation. In this case, increasing the in-place air void by 112% (3.2% to 6.8%) leads to 33% increment in the pavement rut depth (5.4 mm to 7.2 mm).

4 CONCLUSION

This study evaluated the effect of volumetric properties on rutting behavior of asphalt mixtures in a case study. The results from this study showed that the SSR test and permanent deformation shift model can be used as a method to predict the asphalt mixture rutting behavior. The results were along with engineering intuition. The findings in this study showed that the SSR test and permanent deformation shift model are sensitive to the change of asphalt content, gradation, and in-place density for this case study. The aforementioned parameters are very important quality characteristics during the construction. Therefore, this test can be employed as a tool to evaluate the rutting behavior of asphalt mixture during the construction. This research is still ongoing by testing more mixture types from different locations to generalize the aforementioned statement.

ACKNOWLEDGMENT

The authors would like to acknowledge the financial support from the Federal Highway Administration under the DTFH61-13-C-00025 and DTFH61-14-D-00008 Projects.

REFERENCES

Choi, Y. T., Subramanian, V., Guddati, M. N., & Kim, Y. R., 2012. Incremental model for prediction of permanent deformation of asphalt concrete in compression. *Transportation Research Record*, 2296(1), 24–35.

Ghanbari, A., Underwood, B. S., & Kim, Y. R., 2020. Development of Rutting Index Parameter Based on Stress Sweep Rutting Test and Permanent Deformation Shift Model. *International Journal of Pavement Engineering*, In publication.

Kim, Y. R., Castorena, C., Wang, Y., Ghanbari, A. and Jeong, J., 2018. *Comparing Performance of Full-depth Asphalt Pavements and Aggregate Base Pavements in NC* (No. FHWA/NC/2015-02).

Kim, Y. R., Guddati, M., Choi, M, Y., Kim, D., Norouzi, A., Wang, Y. D., Keshavarzi, B., Ashouri, M., Ghanbari, A., & Wargo, A., 2020. *Development of Asphalt Mixture Performance-Related Specifications*. Washington, D.C.: Federal Highway Administration (FHWA), In publication.

Kim, D., & Kim, Y. R., 2017. Development of Stress Sweep Rutting (SSR) test for permanent deformation characterization of asphalt mixture. *Construction and Building Materials*, 154, 373–383.

Wang, Y. D., Ghanbari, A., Underwood, B. S., & Kim, Y. R., 2019. Development of a Performance-Volumetric Relationship for Asphalt Mixtures. *Transportation Research Record*, 2673(6), pp.416–430.

Design of rubberized asphalt mixtures for noise and vibration damping layers

J. Huang, G. Cuciniello, P. Leandri & M. Losa
Department of Civil and Industrial Engineering, University of Pisa, Largo Lucio Lazzarino, Pisa, Italy

ABSTRACT: Traffic-induced vibrations challenge road authorities due to their harmful effects on humans and buildings. The use of rubberized binders seems to improve the damping properties of asphalt mixtures reducing the vibratory mechanism. However, a reliable design method lacks definition. This work aims at designing two mixtures for damping layer prepared with large volumes of rubberized binder. The mixtures were prepared by increasing the binder content and maintaining the same Dust Proportion of an Open Grade mixture used as reference. Tensile strength, moisture susceptibility, stiffness, and phase angle, and damping ratio were measured in laboratory. The latter was used to optimize damping properties. The mixtures show adequate tensile strength and, obviously, low moisture susceptibility. On the other hand, the increase in the binder content softens the mixtures. However, the reduction in stiffness is accompanied by an increase in damping ratio which indicates a higher energy dissipation under dynamic loadings.

1 INTRODUCTION

1.1 Background

Traffic-induced vibrations challenge road authorities due to their harmful effects on humans and buildings. The firsts are affected by low-frequency noise (below 1000 Hz) generated by the vibratory mechanism (Sandberg 1999), which, on the other side, undermines the stability and the durability of the seconds (Browne et al. 2012; Hunaidi et al. 2000; Ouis 2001;. Therefore, reducing vibrations is of primary interest.

In asphalt mixtures, the vibratory mechanism under dynamic loading is controlled by their damping properties (Lakes 2009). The more the material dissipates energy, the lower are the vibrations. Studies between the composition of rubberized asphalt mixtures (wet and dry processes) and their damping properties have been conducted for more than one decade (;e et al. 2012; Losa et al. 2012; Schubert et al. 2010; Wang et al. 2011;). Findings show that the use of large volumes of rubberized binders (Crumb Rubber Modified binders – CRMB) with crumb rubber from end of life tires (ELT) improves the damping properties of mixtures. In 2013, Biligiri et al. used the phase angle as an indicator of the noise-damping properties of asphalt mixtures in the field. The findings highlight that rubberized mixtures with high binder content, high porosity, and rubber inclusions, show improved damping and acoustic performances. In 2018, Huang and co-workers presented a theoretical model to predict the damping ratio of a road pavement specifically designed to mitigating traffic-induced vibrations; The pavement was composed of a damping layer at the interface between the binder and the base layer. Based on model results, to be effective in mitigating vibrations, they defined some target values of material damping composing the damping layer. Specifically, in the modeled pavement structure, to achieve a decrease of vibration accelerations at the soil surface away from the wheel track, they calculated at least a double value of the damping ratio of the damping material is needed compared to that of a conventional mix.

However, although the model results seem to be promising, a valid design method for optimizing the damping properties of asphalt mixtures lacks definition. The objective of this design method is to deliver specific mixtures capable of mitigating vibrations without compromising on site performance.

1.2 Scope of the work

This work focuses on the design of two mixtures for damping layers prepared with a high content of rubberized binder (Mix 1 and Mix 2). An Open Graded (OG) mixture was used as a reference.

Indirect tensile strength (ITS), moisture susceptibility, dynamic modulus ($|E^*|$), phase angle (δ), and the damping ratio (ζ) were measured in the laboratory. As discussed in §2.4, δ and ζ are representative of damping under certain conditions. Therefore, the scope of the mix design is to increase ζ while maintaining adequate levels of stability and stiffness, and a low moisture susceptibility. The work proposes to achieve this goal by increasing the rubberized binder content of the mixtures and by adjusting their volumetrics accordingly.

2 MATERIALS AND METHOD

2.1 Materials

The mixtures were prepared with basalt aggregates ($G_{sb} = 2.753$; Water absorption $= 1.39\%$), natural sand ($G_{sb} = 2.629$; Water absorption $= 0.86\%$) and mineral filler ($G_{sb} = 2.650$). The rubberized binder was supplied by a local manufacturer in Tuscany. The binder was produced according to the wet process by mixing a Pen 50–70 base bitumen with 20% of crumb rubber.

2.2 Design of mixtures

An Open-Graded (OG) mixture prepared with a rubberized binder was used as a reference mixture (Mix_{ref}). The gradation of Mix_{ref} is given in Figure 1.

OG mixtures are designed to achieve a volume of air voids (AV) between 20 and 25% after compaction. Furthermore, they provide a large volume of Voids in Mineral Aggregates (VMA) that can be filled with a high volume of binder (or mastic). Despite the high voids content, the strength is compensated by the aggregate interlock that derives from though and angular basalt aggregates. The mixtures were mixed at 180°C and compacted in the Superpave Gyratory Compactor (SGC) at the same temperature, as recommended by the binder supplier. The volumetrics of the control mixture (Mix_{ref}) are given in Table 1.

The gradations of Mix 1 and Mix 2 were prepared by adjusting the one of Mix_{ref} to accommodate large volumes of the binder. The VMA of Mix_{ref} (36.1% – Table 1) was saturated with a rubberized binder to maximize the Volume of Effective Binder (V_{be}). However, to maintain adequate stiffness and stability, the binder content was increased by maintaining the same Dust Proportion (D/P) of Mix_{ref}. Therefore, besides the binder, the amount of filler was increased. In this way, the stiffness of the mastic phase would not have been significantly reduced by the disproportionate amount of binder added to the mixtures. In the case of Mix 1, the amount of extra-filler and extra-binder were calculated by solving the system of Equations 1–2.

$$
\begin{cases}
\frac{Pf_i + \Delta P_f}{(Pbe_i + \Delta Pb_e)} \cong \left(\frac{D}{P}\right)_{Mix_{ref}} & (1) \\[2mm]
\frac{AV + (V_{be_i} + \Delta V_{be}) + (V_{f_i} + \Delta V_f)}{V_{agg} + V_b + AV} \cong VMA_{Mix_{ref}} & (2)
\end{cases}
$$

Where:
- f_if_i – %weight of filler (P#0.063 mm) Mix_{ref};
- $V_{be_i} V_{be_i}$ – volume of effective binder in Mix_{ref};
- $\Delta P_f \Delta P_f$ – %weight extra-filler;
- $\Delta Pb_e \Delta Pb_e$ – %weight of extra effective bitumen;
- $AV AV$ – % air voids in the compacted mix;
- $V_{aaa} V_{aaa}$ – %volume of aggregates in the mix;
- $V_b V_b$ – %volume of binder in the mix;

Equation (1) comprises the criteria of the stiffness of the mastic phase. Equation (2) indicates that in Mix 1, the VMA of the reference mixture was filled with rubberized mastic (rubberized binder + filler). Mix 2 was prepared by following the same criteria of Mix 1, but an extra amount of (+5%) of the binder was included adjusting the filler content accordingly. In this way, the gradations of Mix 1 and Mix 2 can be determined as shown in Figure 2.

The volumetrics of Mix 1 and Mix 2 are given in Table 2.

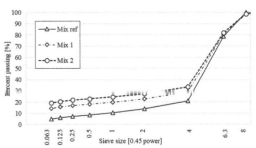

Figure 2. Gradations of Mixref, Mix 1 and Mix 2.

Table 2. Volumetrics of Mix 1 and Mix 2. (W – weight; V – Volume).

	s %	f %	b %	N_{des} 1	AV %	VMA %	VFA %	D/P 1
Mix 1								
(W)	77.2	9.9	12.9	50	2.8	29.8	90.3	0.97
(V)	63.2	7.1	26.9					
Mix 2								
(W)	68.3	15.0	16.7	50	2.9	35.9	92.2	0.97
(V)	54.5	9.6	33.0					

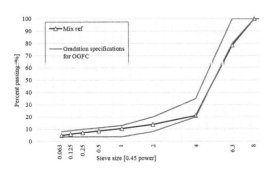

Figure 1. Mixref gradation.

Table 1. Volumetrics of control mix (OG). (W – Weight; V – Volume; s – Stone; f – filler; b – binder; VFA – Void filled with asphalt).

	s %	f %	b %	N_{des} 1	AV %	VMA %	VFA %	D/P 1
(W)	90.4	4.8	4.8	50	27.8	36.1	22.9	0.95
(V)	60.7	3.2	8.3					

2.3 Test methods

The indirect tensile strength (ITS) and the moisture susceptibility were measured according to the EN 12697-23. The dynamic modulus ($|E^*|$) and the phase angle (δ) were measured by following the AASHTO TP 79 standard. A strain level of 100 μs was used at unconfined conditions. Three temperatures (5, 20, 31°C), and eight loading frequencies (0.1, 0.5, 1.0, 2.0, 10, 20, 25 Hz) were used. The master curves of $|E^*|$ and δ were developed under the applicability of the Time-Temperature Superposition Principle (TTSP). The horizontal shift factors were calculated by the William-Landel-Ferry Equation (WLF) (Ferry 1980). The master curves of $|E^*|$ and δ were modeled by using the Modified Christensen-Anderson-Marasteanu Model (CAM – Model) (Zeng et al. 2001).

2.4 Damping ratio of the mixes (ζ)

In a single degree of freedom vibrating system, three parameters, namely, mass, viscous damping coefficient, and stiffness characterize the vibration of a structural element. This analogy applies also to the transmission of vibrations through the pavement material. Therefore, the damping response can be characterized by mass and stiffness of the material along with its inherent viscous damping characteristics. For simple harmonic excitation (e.g., tire rolling on a pavement surface) in a complex modulus material (such as asphalt mixtures), studies have shown that phase angle δ and damping ratio ζ can be related by Equation (3) which was used to derive the damping ratio master curve (Blake 2009; Biligiri 2013; Turner et al. 1991).

$$\zeta = \frac{1}{2} \tan \delta \qquad (3)$$

3 RESULTS AND DISCUSSION

3.1 Tensile strength and moisture susceptibility

The ITS and ITSR results are given in Table 3.

The mixtures are compliant with the specifications for the ITS$_{dry}$ and the ITSR. The first indicates adequate levels of stability under loading. In this case, the reduced ITS values of Mix 1 and Mix 2 compared to Mix$_{ref}$ depends on the higher volume of

Table 3. Results of ITS and ITSR tests (@25°C).

| Mix | ITS$_{dry}$ MPa | ITSR % | Specification limits | |
			ITS$_{dry}$	ITSR
Mix$_{ref}$	0.62 ($\pm$0.04)	82.3	$\geq$ 0.4 MPa	$\geq$ 80%
Mix 1	0.55 ($\pm$0.03)	87.3		
Mix 2	0.41 ($\pm$0.01)	87.8		

binder that reduces the grain-to grain contact of the aggregate skeleton. The ITSR values (Table 3) indicate a low moisture susceptibility of all the mixtures. This result is confirmed by other authors (Leandri et al. 2013; Sangiorgi et al. 2017). Rubberized binders tend to form a thick film of binder coating the aggregate, which prevents water from diffusing and causing debonding. The higher moisture resistance of Mix 1 and Mix 2 compared to Mix$_{ref}$ is given by the higher binder content.

3.2 $|E^*|$, δ and ζ master curves

The $|E^*|$, δ, and ζ master curves are shown in Figures 3–5.

Results were repeatable, with the coefficient of variation below 10% in all the cases. The horizontal shift

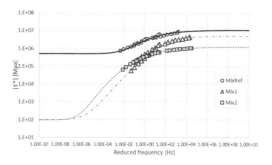

Figure 3. Dynamic modulus master curves.

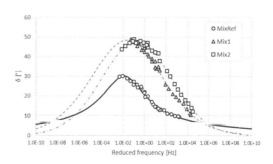

Figure 4. Phase angle master curves.

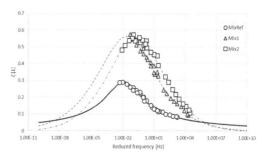

Figure 5. Damping ratio master curves.

factors were calculated at a reference temperature of 20°C on the modulus master curves and were applied to shift the δ isotherms. The CAM model provides an adequate accuracy ($R^2 = 97|E^*|$; $R^2 = 95\%$ δ) for all mixes. Mix$_{ref}$ shows the highest stiffness (Figure 3) and the lowest values of δ (Figure 4). Both aspects depend on the lower binder content compared with Mix 1 and Mix 2. Furthermore, results of stiffness are in agreement with the tensile strength values (Table 3). The difference between Mix$_{ref}$ and the two damping mixtures is marked in the low-frequencies region where the response is more controlled by the aggregate skeleton than the mastic phase. The low levels of stiffness of Mix 1 and Mix 2 allow thinking that they might more appropriately used in pavements as a thin interlayer rather than structural layers. However, both show similar levels of stiffness in the intermediate and low range of frequencies. Mix 1 becomes stiffer at lower temperatures (higher frequencies), and Mix 2 stiffer at higher temperatures (lower frequencies). The shape of the δ master curves is typical of mixtures that show a peak (Figure 4). As expected from Equation (3), the damping ratio master curve shows quite a similar trend as δ (Figure 5). Considering the scope of the work, hereinafter results are discussed in terms of the damping ratio. However, the proposed considerations apply to δ as well. At low frequencies, the aggregate skeleton has a stronger influence on the response, which is more elastic; for this reason, ζ decreases. At the very high frequencies, the mastic becomes stiffer and more elastic, lowering ζ again. In the intermediate region, the response is controlled by the solid skeleton and the mastic showing visible viscoelastic behavior (that is, time dependency). In this region, the peak is likely to represent the threshold between the effects of the two constituents. On the right side of the peak, the mastic phase controls more the response. On the left side, vice versa. It is worth to notice that Mix 2 shows the peak at higher frequencies than Mix 1 and Mix$_{ref}$. In this case, the higher amount of binder makes the mixture more temperature-susceptible and moves the peak towards higher frequencies (or lower temperatures). In other words, Mix 2 shows higher damping at lower temperatures. However, irrespective of frequency, Mix 1 and Mix 2 display consistently higher values of damping ratio than Mix$_{ref}$ showing a better capacity of mitigating the vibration mechanism; the value of ζ is about double compared to that of Mix$_{ref}$ in a wide interval of reduced frequencies, ranging from 10^{-2} to 10^{2} Hz.

4 CONCLUSIONS

The high levels of VMA in OG mixtures permit to accommodate large volumes of rubberized binders that allow improving damping while maintaining aggregate interlock. The volume of rubberized binder should be increased by maintaining the same D/P of the reference mixture to prevent the mastic phase from softening. Results of mixtures prepared by following these design criteria show that the mixtures for damping layers have adequate levels of tensile strength and low moisture susceptibility. The analysis of the viscoelastic response focuses on $|E^*|$ and ζ. The increase in the binder content softens Mix 1 and Mix 2, particularly in the range of high temperatures (low frequencies). On the other hand, the values of the damping ratio of the same mixtures are consistently higher than those of the reference mixture, highlighting a better ability to dissipate energy and to reduce vibrations under dynamic loadings. These results are not conclusive, and further investigations are needed. However, these findings provide a reasonably comprehensive guide on what are the parameters to consider while designing mixtures for damping layers.

REFERENCES

Blake, R. E. 2009. Basic vibration theory, Harris' shock and vibration handbook, McGraw-Hill Handbooks.

Biligiri, K. P. 2013. Effect of pavement materials' damping properties on tyre/road noise characteristics. *Construction and Building Materials*, 49, 223–232.

Browne, M., Allen, J., Nemoto, T., Patier, D., & Visser, J. 2012. Reducing social and environmental impacts of urban freight transport: A review of some major cities. *Procedia-Social and Behavioral Sciences*, 39, 19–33.

Ferry, J. D. 1980. Viscoelastic properties of polymers. New York: John Willey & Sons.

Huang, J., Losa, M., & Leandri, P. 2018. Determining the effect of damping layers in flexible pavements on traffic induced vibrations. *Advances in Materials and Pavement Perormance Prediction – Procedings of the International AM3P Conference*, ISBN 978-113831309-5. pp. 255–259.

Hunaidi, O., Guan, W., & Nicks, J. 2000. Building vibrations and dynamic pavement loads induced by transit buses. *Soil Dynamics and Earthquake Engineering*, 19(6), 435–453.

Lakes, R. 2009. Viscoelastic Materials, Cambridge University Press.

Leandri, P., Rocchio, P., & Losa, M. 2014. Identification of the more suitable warm mix additives for crumb rubber modified binders. *Sustainability, Eco-efficiency and Conservation in Transportation Infrastructure Asset Management*.

Li, Y. L., Ou, Y. J., Tan, Y. Q., & Lu, M. Y. 2012. Dynamic characteristics of rubber powder modified cement asphalt mortar. In *Advanced Engineering Forum* (Vol. 5, pp. 243–246). Trans Tech Publications.

Losa M., Leandri P., & Cerchiai M. 2012. Improvement of pavement sustainability by the use of crumb rubber modified asphalt concrete for wearing courses. *International Journal of Pavement Research and Technology*, 5(6), November ISSN 1977-1400, 395–404.

Ouis, D. 2001. Annoyance from road traffic noise: a review. *Journal of environmental psychology*, 21(1), 101–120.

Sandberg, U. 1999. Low noise road surfaces-a state-of-the-art review. *Journal of the Acoustical Society of Japan* (E), 20(1), 1–17.

Sangiorgi, C., Eskandarsefat, S., Tataranni, P., Simone, A., Vignali, V., Lantieri, C., & Dondi, G. 2017. A complete laboratory assessment of crumb rubber porous asphalt. *Construction and Building Materials*, 132, 500–507.

Schubert, S., Gsell, D., Steiger, R., & Feltrin, G. 2010. Influence of asphalt pavement on damping ratio and resonance frequencies of timber bridges. *Engineering Structures*, 32(10), 3122–3129.

Turner, J. D., & Pretlove, A. J. 1991. Acoustic for Engineers. Macmillan Education Ltd., UK.

Wang, Z. Y., Mei, G. X., & Yu, X. B. 2011. Dynamic shear modulus and damping ratio of waste granular rubber and cement soil mixtures. *Advanced Materials Research* (Vol. 243, pp. 2091–2094). Trans Tech Publications.

Zeng, M., Bahia, H. U., Zhai, H., Anderson, M. R., & Turner, P. 2001. Rheological modeling of modified asphalt binders and mixtures (with discussion). *Journal of the Association of Asphalt Paving Technologists*, 70.

Advances in Materials and Pavement Performance Prediction II – Kumar et al. (eds)
© 2021 Taylor & Francis Group, London, ISBN 978-0-367-46169-0

Laboratory permeability testing of chipseal road surfaces

S.P. Huszak & T.F.P. Henning
University of Auckland, Auckland, New Zealand

P. Herrington
WSP Ltd, Wellington, New Zealand

ABSTRACT: Thin bituminous chipseal surfaces are used on low volume roads throughout the world. Chipseals provide the required surface functionality for the traffic spectrum between unsealed roads and higher trafficked roads that requires asphalt surfacing. Increasing traffic loadings and a changing climate are increasing the demand on these seals to perform and protect the underlying pavement. It is becoming more accepted that thin bituminous surfaces are not waterproof, and the water resistance of various seals need to be measured. This paper presents the development of static and dynamic water permeability tests that measure the water resistance of thin bituminous road surfaces in a laboratory setting.

1 INTRODUCTION

1.1 Background

Thin bituminous road surfaces are used throughout the world, as the most efficient way of maintaining widespread low volume road assets in a serviceable state. Commonly known as Spray Seals or Chip Seals classify the thin surfacing options for low volume roads. In New Zealand, it is referred to as "chipseals" (NZ Transport Agency 2005a).

Sparsely populated countries such as New Zealand rely heavily on its road network for the safe and effective transportation of people and resources. Chipsealed road assets are vital to maintaining the country's economy as it carries an important range in the traffic spectrum that include a high portion of economic trips such as agricultural and industrial haulage.

Typically, chip seal surfaces are around 10 mm thick, comprising only a thin application of bituminous binder (typically 1.5 mm thick), overlain by a layer of uniformly sized aggregate "sealing chips". These surfaces are a fragile, complex system that are prone to surface defects such as chip loss and delamination.

The chipseal surface is a road pavement's primary defense against water ingress from the surface; which is a significant contributor of failures occurring before reaching design life. It is commonly assumed that chipseals are waterproof, however, there is evidence that this is not the case, even with newly constructed seals (Alabaster et al. 2015; Hussain et al. 2011; Towler & Ball 2001).

Defects in chipseals can exist in many different forms, manifesting for various reasons. One common situation that may result in defects is when water is present on the road surface, and passing vehicles generate a water pressure beneath their tyres that forces water down through the seal.

With climate change (Schellnhuber et al. 2012), an increase in the use of more water-sensitive marginal aggregates in road construction (Patrick 2009), and an increase in heavy traffic volumes (Ministry of Transport 2015) occurring, there is an increased need to understand factors contributing to leaking seals. The ability to measure the water permeability of chipseals will open the door to a better understanding of how they can be designed to adapt to the changing conditions.

1.2 Existing Methods

There are several papers that have attempted to develop or use permeability test methods (Ball et al. 1999; Brown & Cooley 1999; Cooley 1999; Cornwell 1983; King & King 2007; Towler & Ball 2001). Most commonly, static testing has been explored where a falling head test is used. In a falling head test some water of a certain height is applied to the chipseal surface, and the time it takes for the water to flow through the seal is measured. Typically a static test provides a constant of permeability (k), measured in metres per second (m/s).

The static test mentioned above is usually a field test, to be conducted on chipseals constructed on site. Field testing raises several challenges and limitations:

- A watertight seal must be formed between the base of the test apparatus and the chipseal surface. Chipseal surfaces tend to have a high amount of surface texture which makes this difficult to achieve.

- Traffic control is required to undertake the test; which either limits testing to situations where the road is not in service, or the test becomes expensive and impractical to measure any useful amount of data.
- Each test is only a snapshot of water permeability at that particular time considering those particular conditions. Bitumen contained in a chipseal flows over time, causing the characteristics of the seal to change with time.
- The history of the tested seal is unknown, and the seal may contain defects that are difficult to identify or quantify.

Furthermore, a major drawback of static testing is that the test conditions do not give an accurate representation of on-site water permeation. There are two conditions in which water can leak through a seal; a) when water is retained on the road surface and it passes slowly through under gravity; and, b) by being forced through under a passing vehicle's tyre. Static tests apply an unrealistically high pressure compared to water ponding on the road's surface, and an unrealistically low pressure for too long a time period to represent the water pressure generated beneath a passing vehicle tyre.

Bitumen's behavior, being a non-Newtonian fluid, depends on loading magnitude, time, and temperature. As each one of these is altered, the bitumen behaves differently. It is therefore important to be able to explore both the dynamic and static response of a seal, as the response of the bitumen is likely to differ.

Ball et al. (1999) developed a laboratory based dynamic test method to measure the water permeability of chipseals that had been extracted from the field under traffic-like rapid loads. A high dynamic pressure equal to truck tyre pressure was applied to the seals, but it was found that they failed and al-lowed significant water permeation well before this pressure was achieved. It was proven that chipseals are not waterproof under rapid loads, but useful quantification of this effect was not achieved.

2 DEVELOPMENT OF THE METHOD

Huszak, Henning & Herrington (2018) have developed a method to construct first coat chipseals in a laboratory setting. This method provides a good basis to allow the observation of a chipseal's permeability response to both static and dynamic type loadings. Permeability test methods were developed within the same laboratory, so various seals with different designs and construction techniques could be tested.

A test process was also developed that subjected the test specimens to periods of "curing" at a constant temperature, to allow the chipseals to "self-heal" between a series of static and dynamic tests.

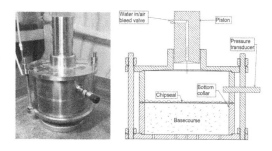

Figure 1. Permeability test apparatus.

2.1 *Equipment*

The test apparatus was designed around the test specimens that were produced from the construction method (Huszak et al. 2018). These consist of a compacted granular base course layer with a single or two-coat seal layer. The test specimens are 150 mm in diameter, 55 mm high, and contained within a circular steel mould, sitting on a perforated baseplate (to allow water to escape downwards). To measure water permeability, another circular steel collar was placed above the sample, and a top plate above that. A sealed space above the chipseal was created, which was then filled with water, and all air expelled. It is important that the system is completely sealed, so that water loss is only occurring by passing through the seal.

The top plate contains a plunger that is able to move upwards and downwards; it is through this plunger that pressure is applied in either a static or dynamic nature. A pressure transducer is attached to the upper collar, which allows the generated water pressure to be recorded.

It is important to undertake the entire test process at a controlled temperature, so testing and "curing" of the sample (as well as the seal construction) was all undertaken at the same temperature in various temperature-controlled units.

For the static tests, water pressure head was applied through a vertical tube, connected to an open valve in the plunger (through which air could be expelled). For the dynamic tests the valve was closed, and a hydraulic compression load frame was used to apply rapid load pulses with a break in between pulses.

2.2 *Static permeability testing*

For the measurement of the static permeability of the chipseals, the falling head permeability (Head 1982) test method was adopted. The test contained the following steps:

1. Place the top collar, top plate and plunger above the test specimen. Fill the void with water, expel all air through the plunger valve, and close the valve.
2. Fill the vertical tube with water and expel all air. Attach the vertical tube to the closed plunger valve.

3. Open the plunger valve, and record the time it takes for the water level in the vertical tube to drop from one point to another.
4. Close the valve, refill the tube and repeat step three above.

By repeating the test, the operator is able to record further information of the chipseals response to the static loading. It was observed that the rate of water permeation could either increase or decrease as the test was repeated.

With the test apparatus available at the university, the applied pressure head varied up to 1.5 m.

2.3 Dynamic permeability testing

No test method existed that allows the dynamic measurement of a chipseals permeability, so it was necessary to develop a new method. As previously mentioned, it is important that the loading magnitude, time, and temperature were all controlled in order to be confident the bitumen response was being controlled. The compression load frame was able to be programmed in such a way as to control the loading magnitude and time. The temperature was controlled by undertaking the test in a closed temperature-controlled unit.

With the same collar, top plate and plunger as for the static test, the following steps were followed:

1. Place the top collar, top plate and plunger above the test specimen. Fill the void with water, expel all air through the plunger valve, and close the valve.
2. Place the test specimen in the compression load frame, within the temperature-controlled unit. Ensure the pressure transducer is in connected.
3. With the valve still closed, undertake the dynamic test; which applies cycle loadings (further discussed below) to the top of the plunger, until the test must stop due to the plunger running out of travel.

The compression load frame was programmed to apply a series of loads as isolated pulses, with a 5 second gap in between pulses to ensure the applied pressure reduced to zero for that time.

Using this particular load frame, the time of loading was limited by the speed at which the hydraulic system was able to apply and remove the load. Therefore, whilst using the maximum loading rate, the magnitude applied was controlled by adjusting the loading time. It was found that selecting a load time of 0.02 seconds to apply the load, and the same to remove the load, a realistic water pressure was generated. The peak water pressure that was generated depended on how resistant the seal was to water permeation, and ranged from 40 to 460 kPa, with a median of 180 kPa. If a tyre contact length of 250 mm is assumed, a total load time of 0.04 seconds corresponds to a vehicle speed of 22.5 km/h.

Measurements of generated water pressures beneath a truck tyre were undertaken at the Canterbury Accelerated Pavement Testing Indoor Centre (Alabaster 2019). The results showed that the generated water pressures depended on tyre tread, vehicle speed, and the amount of water present on the surface. In the most extreme conditions, when aquaplaning occurred, the generated water pressure reached a level similar to the tyre pressure (approximately 700 kPa).

Therefore, the pressures generated during the dynamic test (typically 180 kPa at an equivalent vehicle speed of 22.5 km/h) are considered to give a better simulation of real life conditions than existing test methods.

3 RESULTS PROCESSING

3.1 Static test

Head's (1982) formula for calculating a constant of permeability (k), measured in metres per second (m/s) was adopted. It should be noted that k is a one dimensional parameter that is originally intended to apply to a porous medium of a given thickness (for example, a particular type of soil). To apply this formula to chipseals, being composite systems, it was necessary to determine the thickness of each seal.

The seal thickness is considered to be equal to the binder thickness contained within a seal; which varies between samples. Assuming the volume of voids in a single layer of chip is always 50% of the total volume, and the ratio of binder volume to volume of voids is equal to the ratio of binder thickness to seal height; the binder thickness contained in a seal can be assumed to be twice the binder application rate.

It was found that k values for each sample provided a good indicator of the waterproofness of each sample, and differences between variables were detectable by k value alone.

3.2 Dynamic test

The dynamic test output only two measurements; vertical displacement of the plunger, and the generated water pressure above the test specimen. As mentioned in section 2.1, vertical displacement of the plunger was assumed to represent water permeation through the seal. Therefore, displacements were converted to volume of water lost. The raw data was processed to determine several parameters which included the following:

– Median rate of water permeation during pulse loads, in ml/kPa/s;
– Median rate of water permeation between pulse loads (residual load only), in ml/kPa/s;
– k value during a pulse.

The rates of water permeation were converted to millilitres of water permeation that occurred per kilopascal of water pressure, per second of loading. With further information for a particular road (including traffic and weather data), the rate measurement could readily be converted to a meaningful prediction, such as litres of permeation per year.

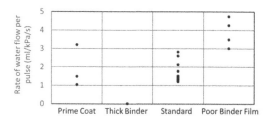

Figure 2. Rate of water permeation during dynamic pulse loads of various chipseal test specimens.

Figure 2 shows example results that validate the test method. Various variations to chipseal test specimens are observed to result in variations in water permeation rate as would be expected.

4 CONCLUSIONS

The developed permeability test methods allow useful parameters of chipseal water permeation to be measured; which provide a better understanding of how in-service chipseals under various conditions behave. Furthermore, predictions of water permeation are able to be made, and exploration into how design and construction techniques could optimise a chipseal's waterproofness is possible. It is, however, important to understand the limitations of the test methods, as outlined below:

- The dynamic tests only last as long as the travel of the plunger allows, causing the tests to vary in number of applied pulses;
- In the field, suction is created behind a passing vehicles tyre. This effect is not simulated during these tests;
- Between pulses, the pressure did not reduce to zero load;
- The compression load frame was unable to apply loads quickly enough to replicate traffic travelling faster than 22.5 km/h;
- As tests progress, water enters the basecourse, which may affect the behavior of the seal itself in the form of changing pore pressures within the basecourse.

REFERENCES

Alabaster, D. (2019, December 4). Personal Communication.

Alabaster, D., Patrick, J., Hussain, J., & Henning, T. F. P. (2015). Effects of water on chipseal and basecourse on high-volume roads March 2015. (Research Report No. 564). Wellington, New Zealand: NZ Transport Agency.

Ball, G. F. A., Logan, T. C., & Patrick, J. E. (1999). *Flushing processes in chipseals: Effect of water.* Transfund research report 156.

Brown, R. E., & Cooley, A. L. (1999). *Designing stone matrix asphalt mixtures for rut-resistant pavements: Part 1: Summary of research results, part 2: Mixture design method, construction guidelines, and quality control/quality assurance procedures.* (No. NCHRP report 425). Washington DC, USA: Transportation Research Board.

Cooley, A. L. (1999). Permeability of superpave mixtures: Evaluation of field permeameters. *National Center for Asphalt Technology, NCAT Report,* 99–91.

Cornwell, W. L. (1983). In-situ permeability of chipseal pavements. *Auckland Engineering Laboratory Report AEL, 83/78*

Head, K. H. (1982). Manual of soil laboratory testing, Vol 2, Pentech Press, ISBN 0-7273-1305-3

Hussain, J., Wilson, D. J., Henning, T. F. P., & Alabaster, D. (2011). What happens when it rains?: Performance of unbound flexible pavements in accelerated pavement testing. Road & Transport Research: A Journal of Australian and New Zealand Research and Practice, 20(4), 3.

Huszak, S. P., Henning, T. F. P., & Herrington, P. (2018). Developing a chipseal construction method for the laboratory. In Masad, E. (Ed.), Bhasin, A. (Ed.), Scarpas, T. (Ed.), Menapace, I. (Ed.), Kumar, A. (Ed.). (2018). *Advances in Materials and Pavement Prediction* (pp. 574–577). London: CRC Press, https://doi.org/10.1201/9780429457791

Kim, Y. R., & Adams, J. (2011). Development of a new chip seal mix design method. Final Report for HWY-2008-04. FHWA, North Carolina Department of Transportation.

King, G., & King, N. (2007). Spray applied polymer surface seals. *Cooperative Agreement no.DTFH61-01-0004 Final Report, Issued by the FHWA to the Foundation for Pavement Preservation.*

Ministry of Transport. (2015). Forecasts for the future – national freight demands study. Retrieved from http://www.transport.govt.nz/research/nationalfreightdemandsstudy/forecastsforthefuture-nationalfreightdemandsstudy/

New Zealand Standards. (1986). Methods for testing soils for civil engineering purposes. NZS 4402:1986,

NZ Transport Agency. (2005a). Chipsealing in New Zealand. Wellington, New Zealand:

NZ Transport Agency. (2005b). Specification for construction of unbound granular pavement layers TNZ B/02.

NZ Transport Agency. (2006). Specification for basecourse aggregate TNZ M/4.

Patrick, J. (2009). The waterproofness of first-coat chipseals. (Research Report No. 390). Wellington, New Zealand: NZ Transport Agency.

Schellnhuber, H. J., Hare, W., Serdeczny, O., Adams, S., Coumou, D., Frieler, K., Martin, M., Otto, I. M., Perrette, M., Robinson, A., Rocha, M., Schaeffer, M., Schewe, J., Wang, X., Warszawski, L. (2012). Turn down the heat–why a 4 C warmer world must be avoided. World Bank,

Towler, J. I., & Ball, G. F. A. (2001). Permeabilities of chipseals in New Zealand. Proceedings, 20th ARRB conference: Managing your transport assets (pp. 16). Melbourne, Australia: ARRB Group Limited.

Advances in Materials and Pavement Performance Prediction II – Kumar et al. (eds)
© 2021 Taylor & Francis Group, London, ISBN 978-0-367-46169-0

Effectiveness of polymer modified emulsion based rejuvenator

Siksha Swaroopa Kar, G. Bharath & Manoj Kumar Shukla
CSIR-Central Road Research Institute, New Delhi, India

ABSTRACT: The growth of economy of any country depends upon the development of transportation and there is a growing demand for air transportation in developing country. In runway construction, the flexible pavements typically have a functional life of 12 to 15 years between major maintenance treatments. Pavement preservation plays a significant role in maintaining the pavement infrastructure under severe budget constraints. Numerous methods are being employed for asphalt pavement preservation including rejuvenator, bitumen emulsion fog seals surface treatments (including slurry and micro surfacing technologies), and emerging bituminous thin overlay technologies. The present study investigated the field performance and mechanical properties of bituminous mixtures treated with polymer modified emulsion based rejuvenator.

1 INTRODUCTION

In runway construction, the flexible pavements typically have a functional life of 12 to 15 years between the major maintenance treatments; the actual interval depends largely on the environmental conditions, with bituminous surfaced pavements ageing faster in hot climates. Poor wet skid resistance is a common problem for aged concrete runways. Runways need to have adequate wet skid resistance in view of the very high speeds involved (Tabakovic et al. 2019). Hence, in between the cyclic resurfacings, repairs will be required to rectify defects observed during pavement inspections. It is anticipated that a surface rejuvenating treatment will prolong the pavement life.

There are numerous methods for flexible pavement preservation including surface application of rejuvenator, bitumen emulsion fog seals, other surface treatments (including slurry and micro surfacing technologies), and also emerging bituminous thin overlay technologies. To make the most of maintenance budgets, many agencies have resorted to the use of bituminous rejuvenators as an alternative to revive aging and brittle asphalt pavements. Use of bituminous rejuvenators to revive an aging pavement, is an economical method to extend the pavement life. Various studies conducted worldwide have proved that this type of asphalt pavement treatment has the potential to extend the life of an asphalt pavement for several years beyond the point where rehabilitation or major reconstruction would normally be required; thus significantly decreasing the pavements annual maintenance costs (Lin et al. 2012; Lin et al. 2014; Ghosh et al. 2018). According to White et al. (2019), surface enrichments or rejuvenations methodology involving light applications of cutback bitumen or dilute polymer modified

emulsion, is used to replace lost binder from the surface, fill fine cracks and protect the remaining exposed binder from direct oxidation and ultra-violet light.

Ghosh et al. (2018) evaluated low viscosity emulsion, modified emulsion and bio sealant based rejuvenator in terms of laboratory and field performance and concluded that modified emulsion and biosealant based rejuvenator performed better. Zadshir et al. (2018) examined the feasibility of using biomodifier and polymer modified rejuvenator application against UV oxidation and reported that polymer modifier rejuvenator application give better resistance towards aging of asphalt pavement. Studies show that aquaplaning or hydroplaning is a major contributor to runway overrun accidents due to loss of contact between the tyre and runway in presence of water or poor friction (ATSB 2009; ATSB 2001). CAA (2010) found that simple emulsion application for runway rejuvenation results a slippery surface leaving an excess amount of bitumen on the surface after the emulsion breaks without properly penetrating into the surface voids. In 2003 the US Army investigated the comparative field performance of various proprietary rejuvenators and seal coat materials over a period of more than 1 year and found that effectiveness of rejuvenation application depends on penetration of rejuvenator material and also on application methodology (Shoenberger 2003).

No study is found in the literature on the effectiveness of surface treatment in terms of surface texture. The study reported in this paper investigates the effectiveness of application of polymer modified emulsion (PME) based surface rejuvenation method in terms of surface texture. The goals of this paper are to measure surface textures, permeability and the friction coefficient of the pavement before and after rejuvenator application.

2 MATERIAL CHARACTERISATION AND APPLICATION

In this study, PME based rejuvenator was investigated. The selection was based on the availability of materials and construction project in which rejuvenator was used (Ambala Cant runway, India). Rejuvenator treatment was applied in June 2018 as follows:

- PME based rejuvenator (PMER) is a relatively low viscosity cationion emulsion constituting mineral dispersing agents such as bentonites clay particle. The physical properties of PMER is presented in Table 1.
- PMER was blended with mineral filler (lime) and water before spraying on field. First step was to prepare mineral powder solution in the range of 1:8 (Mineral Powder:Water) by weight. Then the equal amount of mineral powder solution is added to PMER and stirred continuously to make a homogeneous solution.
- The solution was then applied on surface at the rate of 1.0 ltr per square meter area for 400 m² through mechanized spraying unit. After 24 hr of curing, another coat of solution was applied on the surface.

To simulate this process in laboratory and to check the permeability and resistance to moisture damage, PMER solution was prepared and the Marshall mix samples (dense graded) were dipped into the solution for 1 min. The coated sample was kept for 24 hr in air for drying and curing. After that, the samples were tested. Laboratory coating of sample is shown in Figure 1.

Table 1. Physical properties of PMER.

S.No.	Properties of PME Rejuvenator	Test Methods	Test Results
i)	Residue on 600 micron IS sieve (% mass), max.	ASTM D6933	0.06
ii)	Viscosity by Saybolt Furol Viscometer, seconds, at 50°C	ASTM D7496	15.0
iii)	Storage stability after 120 h., %, max.	ASTM D 6930	3.4
iv)	Storage stability after 24 h., %, max.	ASTM D 6930	1.0
v)	Particle charge	ASTM D 244	+ve
vi)	Specific Gravity	ASTM D6937	0.98
vii)	pH	ASTM D 244	5.0
viii)	Test on residue:		
	(a) Residue by evaporation, %	ASTM D 5	44.5
	(b) Penetration, 25°C.	ASTM D 5	96.0
	(c) Penetration, 4°C.	–	64.0
	(d) Ductility, 27°C/cm.,	ASTM D7553	108
	(e) Solubility In Trichloroethylene, %	ASTM D 2872	99.0
	(f) RTFO, Mass change, %		2.0

Figure 1. (A) Coating of sample (B) View of coated and cured samples.

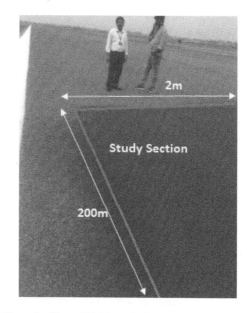

Figure 2. View of PMER applied section.

Skid resistance, sand patch and permeability tests were carried out before and after application of product at three different locations for six months. Mean value of three locations' results are presented in the paper. View of study area is shown in Figure 2.

3 LABORATORY PERFORMACE

3.1 Permeability

The Florida Test Method (FM 5-565) was followed to determine the laboratory permeability values. The laboratory permeability of the vacuum saturated test specimen was estimated using the Karol-Warner Asphalt permeameter using a falling head permeameter method. The coefficient of permeability, k, is determined using the following equation (1):

$$k = \frac{a \times L}{A \times t} \times \ln(h_1/h_2) \times t_c \qquad (1)$$

Where, k = coefficient of permeability, cm/s; a = inside cross-sectional area of the buret, cm²;

Table 2. Effect of PME coating on moisture resistance.

	M_R @ 35°C (MPa)			Tensile strength Ratio (After MIST*100/
Sample	Before MIST	After MIST	Difference	Before MIST), %
After Coating	3586	3462	124	96.5
Before Coating	3435	2033	526	59.2

Table 3. Effect of PME coating on aging of mix.

	M_R @ 25°C (MPa)			Aging Ratio (After aging MR/Before Ageing MR)
Sample	Before Aging	After Aging	Difference	
After Coating	7635	8399	764	1.1
Before Coating	6879	11006	4127	1.6

L = average thickness of the test specimen, cm; A = average cross-sectional area of the test specimen, cm^2; t = elapsed time between h1 and h2, s; h_1 = initial head across the test specimen, cm; h_2 = final head across the test specimen, cm; t_c = temperature correction for viscosity of water. A temperature of 20°C is used as the standard temperature.

Marshall samples prepared at 5% air voids, showed an average permeability of 42×10^{-5} cm/s, before coating and 7×10^{-5} cm/s after coating. It is evident that the permeability of Marshall sample after coating reduces by six times compared to uncoated sample. This is because the coating of PMER penetrates into the mix filling the voids and also PMER forms a layer reducing the ingress of water into the mix.

3.2 Resistance to moisture damage

To evaluate the moisture damage susceptibility of samples before and after coating, an accelerated test (MIST) was used in this study to simulate the action of traffic on wet pavement. The test is performed at elevated temperature of 60°C, at an applied pressure is 40 psi and test is conducted for 3500 cycles according to ASTM D 7870. The samples after conditioning in MIST for 3500 cycles are soaked in a water bath for 2 hour at 25°C and tested for resilient modulus as per ASTM D4123. the conditioned and unconditioned samples were subjected to repeat loading pulse width of 0.1sec, and pulse repetition period of 1 sec at an assumed Poisson's ratio of 0.35 to measure the resilient modulus value.

Table 2 shows the results of resilient modulus test conducted on specimens conditioned with the MIST. The results clearly show the improvement in stiffness when samples are coated with PMER. The relatively higher resilient modulus obtained for coated specimens is due to reduction of ingress of water into the mix due to coating. Hence, it seems that application of PMER over the pavement would be helpful in damage resistance of mix against repetitive action of moisture under traffic at high temperature.

3.3 Resistance to aging

To determine the long-term properties, the Marshall samples were long-term aged in an oven at 85°C for

Table 4. Field results.

Surface type	British Pendulum Number (BPN)		Mean Texture Depth (mm)	Permeability (cm/s)
	Dry Condition	Wet Condition		
Before Application (June, 2018)	80	65	1.3	0.902
After Application August 2018	85	50	1.2	0.0005
December 2018	84	58	1.15	0.0004

120 hours according to AASHTO R 30. Resilient modulus was determined on both unaged and the long-term aged specimens before and after coating of PMER to determine increase in stiffness of the long-term aged mixture due to aging. Result shows that, by applying the PMER, aging is reduced by 30%, showing better ageing resistance.

4 FIELD PERFORMANCE

The application of surface treatments generally reduces surface texture creating hydroplaning in runway. Hydroplaning is a major contributor to runway overrun accidents. It significantly reduces the runway friction coefficient (by up to 95 per cent compared to a dry runway) and braking action. There have been reports of aircraft skidding off runways where the surface texture and fiction has been adversely impacted by surface treatment (ATSB 2009; ATSB 2001). However, if the texture and friction are already marginal, then further reduction by the application of a surface treatment is unacceptable. Hence, it is required to perform friction test before and after a trial application, usually by a spot-tester such as the British pendulum and sand patch method.

The Cooper-Wessex Pendulum Skid Tester CRT-PENDULUM was used to evaluate the skid resistance in both dry and wet surface before and after applications. As per International Civil Aviation Organization (ICAO), acceptable value for skidding resistance is when British Pendulum (BP) number is between 35

Figure 3. View of testing (a) Skid resistance (b) Sand patch.

to 60 for wet surface condition. The skid test results are given in Table 4.

Volumetric approach of measuring pavement macrotexture was considered in the present study as per ASTM E965. In this study a known volume of glass spheres (50 ml) was spread evenly over the pavement surface to form a circle. The diameter of the circle was measured on four axes and the value was averaged. This value was used to calculate the mean texture depth (MTD) in mm as per following equation.

$$\text{MTD} = \frac{4000 \times V}{\pi D^2}$$

Where V is the exact volume of glass spheres in mL and D is the average diameter of the sand patch in mm. ICAO recommends that MTD should not be less than 1mm for construction of new pavement in runways. Average of MTD at three locations before and after application of PMER was determined and presented in Table 3. Results show that micro texture of runway after application is well within the permissible limit after six months of application.

The infiltration rate at field was calculated using the following equation

$$I = \frac{KM}{D^2 t} \tag{2}$$

Where, I = Infiltration rate, mm/h; M = Mass of infiltered water, Kg; D = inside diameter of the cylinder, mm; t = time required for water to flow between 0 and 80 mm line; K = Constant (4,583,666,000), mm^3.s/Kg.h

Permeability test results are also presented in Table 3 and it is found that with application of PMER, the runway attains almost zero porosity, resulting reduced ingress of water and reduced further damage to runway.

5 CONCLUSION

This paper has presented the results of a recent study on the field performance and mechanical properties of polymer modified emulsion based rejuvenator surface treatment process. Surface texture, skid resistance and permeability values of treated surface before and after application was observed for six months. Laboratory-treated samples were also prepared using an application process that mimicked the spraying of sealant in actual field conditions. The laboratory experimental work consisted of calculation of permeability and resistance towards moisture damage. From both lab and field testing, it was seen that permeability of surface is reduced after application of PMER and hence, damage to moisture resistance is also improved. It was observed that treatment of PMER has increased whole-of-life benefits of runways without affecting the friction and micro texture of runways.

REFERENCES

ATSB. 2009. Runway excursions Part 1: A worldwide review of commercial jet aircraft runway excursions. Aviation Research and Analysis Report, AR-2008-018(1). *Australian Transport Safety Bureau*, Canberra.

ATSB. 2001. Boeing 747-438 VH-OJH, Bangkok, Thailand, 23 September 1999. Aviation Safety Investigation Report 199904538. *Australian Transport Safety Bureau*, Canberra.

CAA. 2010 Interim Report Number 2 in respect of the Investigation into the cause(s) of an accident involving an Embraer 135-LR aircraft, ZS-SJW during landing at George Airport on 7 December 2009.

Ghosh, D., Turos, M., Johnson, E. and Marasteanu, M., 2018. Rheological characterization of asphalt binders treated with bio sealants for pavement preservation. *Canadian Journal of Civil Engineering*, 45(5), pp.407–412.

Lin, J., Guo, P., Wan, L. and Wu, S., 2012. Laboratory investigation of rejuvenator seal materials on performances of asphalt mixtures. *Construction and Building Materials*, 37, pp.41–45.

Lin, J., Hong, J., Huang, C., Liu, J. and Wu, S., 2014. Effectiveness of rejuvenator seal materials on performance of asphalt pavement. *Construction and Building Materials*, 55, pp.63–68.

Shoenberger, J. E. 2003. Rejuvenators, Rejuvenator/Sealers, and Seal Coats for Airfield Pavements. ERDC/GSL TR-03-1, U.S. *Army Engineer Research and Development Center, Geotechnical and Structures Laboratory*. Vicksburg, Mississippi.

Tabakovic, A., O'Prey, D., McKenna, D. and Woodward, D., 2019. Microwave self-healing technology as airfield porous asphalt friction course repair and maintenance system, *Case Studies in Construction Materials*. Under Press

White, G., McLachlan, F. and Wallace, S., 2019. Comparing Asphalt Preservation Products on a Grooved Runway. In *Bituminous Mixtures and Pavements VII: Proceedings of the 7th International*

Zadshir, M., Hosseinnezhad, S., Ortega, R., Chen, F., Hochstein, D., Xie, J., Yin, H., Parast, M. M. and Fini, E. H., 2018. Application of a biomodifier as fog sealants to delay ultraviolet aging of bituminous materials. *Journal of Materials in Civil Engineering*, 30(12), p.04018310

Advances in Materials and Pavement Performance Prediction II – Kumar et al. (eds)
© 2021 Taylor & Francis Group, London, ISBN 978-0-367-46169-0

Mix design approach of cold mix asphalt using response surface method

Siksha Swaroopa Kar, Satish Chandra & M.N. Nagabhushana
CSIR-Central Road Research Institute, New Delhi, India

ABSTRACT: Cold mix asphalt (CMA) has been increasingly recognized as an important alternative technology to hot mix asphalt (HMA) worldwide. The present study investigates the optimum proportions of fine content (FC) and emulsion content (EC) to gain suitable levels of both mechanical and volumetric properties. A factorial design with response surface method (RSM) was applied to optimize the mix design parameters. This work aimed to investigate the interaction effect between these parameters on the mechanical and volumetric properties. The stability and indirect tensile strength (ITS) tests were performed to obtain the mechanical response. The results indicate that the interaction of both FC and EC influences the bulk density. However, the EC tended to influence the ITS more significantly than FC. Further, the experimental results for the optimum mix design were in agreement with model predictions. It is found that optimization using RSM is an effective approach for mix design of CMA.

1 INTRODUCTION

The construction of flexible pavements using conventional hot mix asphalt (HMA) causes emissions and pollution to surrounding environment. During construction, a considerable amount of fumes are generated and emitted into the air. Cold Mix Asphalt (CMA) is an alternative technology for construction of flexible pavement, where the mixing and paving is done at ambient temperature with unheated mineral aggregate and bitumen emulsion. Many advantages could be achieved when CMA is used as an alternative to traditional hot mix asphalt such as pavement produces less environmental impact, is more cost effective, and requires less energy consumption (Al-Busaltan et al. 2012). Though the cold mix technology has several advantages it is lagging much behind in both research and applications both particularly in developing countries like India.

According to Brown and Needham (2000), mechanical properties of CMA are affected by a number of parameters, including void content, binder grade, additives like cement and curing time. Studies show that optimum emulsion content (OEC) depends on the aggregate surface area (Dash & Panda 2018; Swaroopa et al. 2015). Lump formation is one of the major issues associated with cold mix technology (Graziani et al. 2018). Olard and Perraton (2010) showed that the proportion of coarse and fine fractions in an aggregate blend has an effect on compactability, strength and stiffness of CMA. Raschia et al., (2019) concluded that compatibility and workability of emulsified mixes depend on the aggregate gradation. Fine content has an effect on optimum emulsion content and volumetric

properties of CMA (Graziani et al. 2018; Kuchiishi et al. 2019).

Unlike in hot mix design, universally accepted cold mix design procedure is not seen in the laboratory. The laboratory procedures to determine optimum emulsion content vary widely among researchers and agencies. As per Asphalt Institute Manual Series 14 (AI MS-14), initial emulsion content (IEC) is obtained from an empirical formula and then with IEC coating test is carried out to determine optimum pre-wetting water content (OPWC). Then the specimens are prepared with varying emulsion content, maintaining the OPWC value same in all batches and the optimum emulsion content (OEC) is determined based on stability values. Initial emulsion content has been determined in terms aggregate gradation.

Most of the studies reported in the literature on CMA have focused on using the method adopted by the Asphalt Institute. There is a potential to explore the use of a statistical tool to optimize the mixture design of CMA. In response to the above need, the present study has been undertaken in order to optimize fine content and emulsion content through design of experiment approach using response surface method (RSM).

2 MATERIAL AND METHOD

2.1 Materials

Bitumen emulsion and stone aggregate were used for production of the cold mixes and their physical properties along with specification limits are given in Table 1.

Table 1. Physical properties bitumen emulsion and aggregate.

S. No	Properties of designed bitumen emulsion	IS specification	Binder specification
1	Residue on 600 micron IS sieve, percent by mass, max	0.05	0.02
2	Viscosity by Saybolt Furol Viscometer, seconds	30–100	53
3	Coagulation of emulsion at low temperature	NIL	NIL
4	Storage stability after 24 h, %, max	1	0.3
5	Particle charge	+ve	+ve
6	Stability to mixing with cement (percentage coagulation)	Less than 2	1.2
7	Miscibility with water	NIL	NIL

S. No.	Test	Aggregate size (mm)	Test Results	Recommended value as per MoRTH
1	Specific gravity	20	2.8	–
		13.2	2.7	
		Stone dust	2.6	
2	Water absorption, %	20	1.0	2 max
		13.2	0.3	
3	Combined flakiness and elongation indices, %	20	21.0	30 max
		13.2	17.2	
4	Aggregate impact	20	19.2	27 max
		13.2	18.6	
5	Stripping, % coating		99	95 min

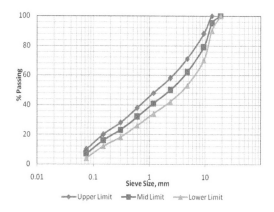

Figure 1. Aggregate gradation used in study.

Cationic slow setting (CSS-2) bitumen emulsion with 63.4% residual bitumen content was used as binder. CSS-2 was chosen as it was believed that the setting time involved in the emulsion would suit most appropriately the traffic behavior during execution in India (Das & Panda 2018). Bituminous concrete (BC), dense graded aggregate gradation same as used in hot mixes as per the available Indian specifications (MoRTH) have been used in this study (Figure 1). Both upper and lower limits along with mid limit have been considered in this study and were defined in terms of fine content (% fines passing through 2.36 mm sieve). CMA sample has been prepared as per MS 14 (Asphalt Institute) and before testing, curing of samples was done for 72 hr at 40°C.

2.2 Testing methodology

To assess the optimum emulsion content of the cold mix, volumetric properties and performance based tests were conducted in the laboratory. The Marshall stability of the mix is defined as the maximum load carried by the specimen at the specified standard test temperature. The resistance to plastic deformation of a compacted cylindrical specimen of cold bituminous mixture is measured when the specimen is loaded diametrically at a deformation rate of 50 mm per minute (ASTM D6927). The bulk density of the sample is determined through Corelok apparatus as per AASHTO TP 69. The indirect tensile test was conducted according to ASTM D 6931-12 standards. The test is used to evaluate the cohesive strength of asphalt mixes.

2.3 Design of experiment

RSM is defined as a mathematical and statistical technique used for designing experiments to establish relationships between multiple factors and to optimize the input parameters to predict the best responses (Montgomery 2008). In this study Design Expert© Software Version 10 (Stat-Ease Inc., Minneapolis, USA) was used for the design, mathematical modelling, statistical analysis, and optimization of the process parameters. Analysis of variance (ANOVA) was conducted in order to obtain the interaction among the different parameters and the influence of each individual parameter.

3 RESULT AND DISCUSSION

An experimental program was undertaken in order to consider the effect of aggregate gradation and emulsion content on performance of CMA. The input independent parameters with respective ranges were selected based on literatures and standard specifications. The range of levels of parameters is presented

Table 2. Details of input parameters and responses.

Input parameters	Code	Unit	Coded Parameters		
			−1	0	1
Fine Content (FC)	A	%	42	50	58
Emulsion Content (EC)	B	%	7	9	11

Responses	Code	Unit
Indirect Tensile strength	ITS	kg/cm²
Stability	Sability	kN
Bulk Density	BD	Gm/cc

Table 3. Matrix of experimental design by CCD.

Run No	A:FC (%)	B:EC (%)	BD (gm/cc)	Stability (kN)	ITS (kg/cm²)
1	42	11	2.30	10.93	2.87
2	50	7	2.21	12.60	4.01
3	42	7	2.20	13.20	3.50
4	50	11	2.25	11.50	2.79
5	58	7	2.01	9.52	3.05
6	42	11	2.30	10.93	2.87
7	58	11	2.18	11.53	2.78
8	50	9	2.27	15.56	3.76
9	42	11	2.30	10.93	2.87
10	58	9	2.26	12.45	3.58
11	58	9	2.26	12.45	3.58
12	42	9	2.31	12.81	3.51
13	58	9	2.26	12.45	3.58

in Table 2. Stability, Indirect tensile strength and Bulk density are considered as responses.

The total number of experiments carried out were 13 ($2^k + 2k + 5$), where k is the number of input parameters. eight different combinations were supplemented with 5 replicates of mean case to improve the precision of experiments and minimize human and any possible sources of error. The central composite design (CCD) matrix employed is presented in Table 3.

As per literature, polynomial model of the second order was adequate for evaluation of the impact of the aggregate and bitumen content on the hot mixtures properties such as stability, tensile strength, and resilient modulus (Bouraima & Qiu 2017; Galan et al. 2019). Hence, quadratic equation is applied in the present study too. In the present study, student t test and ANOVA were considered at significance level of 5% to determine significant effect of input parameter on responses (Bouraima & Qiu 2017; Zhang et al. 2010). Insignificant terms which have limited influence ($p > 0.1$), were ex cluded from the study to improve the models. The LOF (Lack of Fit) F-test was also used to evaluate the adequacy of the model. LOF

depicts the variation of data around the fitted model. It is worth noting that while LOF values were significant, reasonable agreement between predicted and adjusted R^2 were found for all responses such that it can be concluded that the suggested models for all responses can be used to navigate satisfactorily in the design space to find optimum mix design parameters. The final regression models, in terms of significant influencing factors, are expressed by the following second order polynomial equations, where A denotes fine content (FC), % and B denotes emulsion content (EC), %.

$$BD = 2.31 - 0.05 \times A + 0.05B - 0.1B^2 \quad (1)$$

$$Stability = 14.53 - 0.65A + 0.95AB - 1.41A^2 \quad (2)$$

$$ITS = 3.83 - 0.32B + 0.95AB - 0.476B^2 \quad (3)$$

Coefficient values of A and B show that both have impact on output parameters. In case of ITS, fine content has negligible effect compared to emulsion content. Emulsion content has significant effect on stability value compared to fine content. The contour plots generated for the BD, stability and ITS, are presented in Figure 6(a) to (c), respectively.

Results show that the maximum ITS is obtained at lower emulsion content compared to stability and BD. Also, optimum fine and emulsion content is required to obtain maximum mechanical properties of mix. With increase in emulsion content and fine content, both stability and ITS increases upto a certain limit and then decreases.

The response surface shows elliptical contours which is the pattern obtained when there is a perfect interaction between independent variables and responses. Accordingly, there is a region of maximum BD around 42 to 53% FC and 8 to 10.5% EC. In case of stability, the range of optimum EC decreases to 8 to 9.5% and FC to 45 to 50% for obtaining maximum stability. To obtain maximum ITS value, the range of optimum EC further decreases to 7.5 to 9.0%, whereas FC remains same as stability value.

An additional laboratory experiment was also performed to validate the optimum mix design proportions obtained by the RSM model and presented in Table 4.

The results in Table 4 show a limited variation of optimum mix design proportions. A residual analysis was carried out to test the goodness of fit and also to verify the suitability of the regression models. In Figure 3 the expected residuals against the observed ones assuming normal distribution with the same mean and variance are plotted for all responses. It is observed that the residues are normally distributed throughout the observed responses, confirming the validity of the fit.

Design Expert utilizes desirability function as optimisation function and the mathematical representation of the function is given in equation (4) (Yıldırım et al.

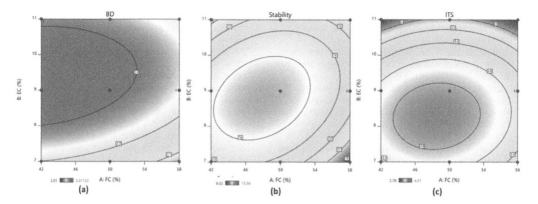

(a) (b) (c)

Figure 2. Contour plots (a) BD, gm/cc (b) Stability, kN and (c) ITS, kg/cm^2.

Table 4. Model prediction and laboratory response.

Parameters	Model Prediction	Laboratory response	Variability, %
FC (%)	52	52	–
EC (%)	8	8	–
BD (gm/cc)	2.24	2.25	0.44
Stability (kN)	13.74	13.25	3.69
ITS (kg/cm^2)	3.84	3.72	3.22

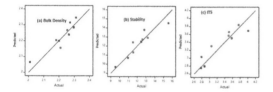

Figure 3. Normal predicted versus observed residuals for (a). BD, gm/cc (b) Stability, kN and (c) ITS, kg/cm^2.

2018). desirability value ranges between 0 (lowest) to 1(highest).

$$D = \left[\prod_{r=1}^{R} d_r \right]^{1/R} \tag{4}$$

Where D is the desirability value; R is the number of R-response variables and dr is the importance level of response variable.

The desirable mechanical responses were defined as being a maximum to achieve the highest performance. In the present study, maximum of each responses has been considered as required desirability factor. The largest desirability of 0.913 determines the optimum conditions with 48% FC and 8.5% of EC. The estimated response values are detected with BD of 2.31 g/cc, stability of 14.5 kN and ITS of 3.9 kg/m^2.

4 CONCLUSION

Use of cold mix technology is gaining popularity across the world. In this work, effect of aggregate gradation and emulsion content on volumetric and mechanical properties of cold mixes was evaluated through statistical approach. The RSM approach offers a more comprehensive view of the effect of the variation of each mix design parameter on the mechanical and volumetric responses of CMA than by other methods. It has the advantage that all parameters are investigated at one time. Response surface of second order can reasonably capture the effect of the fine content and emulsion content on the properties of cold bituminous mixtures. The mechanical and volumetric properties of cold bituminous mixtures are highly influenced by both fine content and emulsion content.

REFERENCES

Al-Busaltan, S. F., Al Nageim, H., Atherton, W., & Sharples, G. 2012. A comparative study for improving the mechanical properties of cold bituminous emulsion mixtures with cement and waste materials. *Construction and Building Materials*, 36, 743–748.

ASTM, A., 2015. D6927-15 Standard Test Method for Marshall Stability and Flow of Asphalt Mixtures. *ASTM International: West Conshohocken, PA, USA*.

ASTM, D., 6931, 2012. Indirect Tensile (IDT) Strength for bituminous mixtures. *ASTM International: West Conshohocken, PA, USA*.

AASHTO, T., 69-04 (2004). Standard Method of Test for Bulk Specific Gravity and Density of Compacted Asphalt Mixtures Using Automatic Vacuum Sealing Method, *American Association of State and Highway Transportation Officials, Washington, DC*.

Bouraima, M. B., & Qiu, Y., 2017. Investigation of influential factors on the tensile strength of cold recycled mixture with bitumen emulsion due to moisture conditioning. *Journal of Traffic and Transportation Engineering* (English edition), 4(2), 198–205.

Brown, S., & Needham, D., 2000. A study of cement modified bitumen emulsion mixtures. *Asphalt Paving Technology*, 69, 92–121.

Dash, S. S., & Panda, M., 2018. Influence of mix parameters on design of cold bituminous mix. *Construction and Building Materials*, 191, 376–385.

Galan, J. J., Silva, L. M., Pérez, I., & Pasandín, A. R., 2019. Mechanical behavior of hot-mix asphalt made with recycled concrete aggregates from construction and demolition waste: A design of experiments approach. *Sustainability*, 11(13), 3730.

Graziani, A., Virgili, A., & Cardone, F., 2018. Testing the bond strength between cold bitumen emulsion composites and aggregate substrate. *Materials and Structures*, 51(1), 14.

Kuchiishi, A. K., Vasconcelos, K., & Bariani Bernucci, L. L., 2019. Effect of mixture composition on the mechanical behaviour of cold recycled asphalt mixtures. *International Journal of Pavement Engineering*, 1–11.

Montgomery, D. C. 2008. Design and analysis of experiments. 7th ed. New York, USA: John Wiley & Sons, Inc.

MoRTH (Ministry of Road Transport and Highways), 2013. Specifications for road and bridge works. In *Indian Road Congress*. New Delhi, India: Author.

Olard, F., & Perraton, D. 2010. On the optimization of the aggregate packing characteristics for the design of high-performance asphalt concretes. *Road Materials and Pavement Design*, 11(Suppl. 1), 145–169.

Raschia, S., Badeli, S., Carter, A., Graziani, A., & Perraton, D. 2018. Recycled glass filler in cold recycled materials treated with bituminous emulsion, in *Transportation Research Board (TRB) 97th Annual Meeting*. 2018: Washington, DC, USA.

Swaroopa, S., Sravani, A., & Jain, P. K., 2015. Comparison of mechanistic characteristics of cold, mild warm and half warm mixes for bituminous road construction. *Indian Journal of Engineering and Material Science*, 22, 85–92.

Yıldırım, Z. B., Karacasu, M., & Okur, V., 2018. Optimisation of Marshall Design criteria with central composite design in asphalt concrete. *International Journal of Pavement Engineering*, 1–11.

Zhang, S. L., Zhang, Z. X., Xin, Z. X., Pal, K., & Kim, J. K. (2010). Prediction of mechanical properties of polypropylene/waste ground rubber tire powder treated by bitumen composites via uniform design and artificial neural networks. *Materials & Design*, 31(4), 1900–1905.

Advances in Materials and Pavement Performance Prediction II – Kumar et al. (eds)
© 2021 Taylor & Francis Group, London, ISBN 978-0-367-46169-0

Asphalt mixture compaction in gyratory compactor using bullet physics engine

Satyavati Komaragiri, Alex Gigliotti, Syeda Rahman & Amit Bhasin
Deptartment of Civil, Architectural and Environmental Engineering, The University of Texas at Austin, USA

ABSTRACT: Engineering and optimization of asphalt mixtures in laboratory is very important to achieve the desired performance of the pavements and to reduce the likelihood of expensive premature failures. Since there are several variables in play to produce an optimal mix, identifying an optimal design would require a systematic evaluation of innumerable combinations. However, such an effort is not feasible in a laboratory setting using conventional experimental methods. This study demonstrates the feasibility of using Bullet Physics engine to virtually compact the asphalt mixtures in a gyratory compactor. The model developed in this study can then be used to conduct broad parametric analyses, and together with computational tools, can serve as a screening tool for the optimization of asphalt mixtures.

1 INTRODUCTION

The performance of an asphalt mixture is determined by several variables (e.g. aggregate gradation, binder content, material type, etc.). Optimal asphalt mixture design involves preparing and evaluating various mixtures in the laboratory for different combinations of these variables. A laboratory test mixture is created by placing loose asphalt mixture into a cylindrical mold and compacting it using a gyratory compactor. However, the process of compacting test specimens in a laboratory setting for a large number of design variable combinations is time consuming and expensive. To overcome these shortcomings, this study explores the possibility of creating a computational model to virtually compact asphalt mixtures in a gyratory compactor with the intent of producing mixture geometry that can be used with computational tools for further analysis.

The Bullet Physics engine was used in this study to simulate mixture compaction. It is an open-source platform and has been used in a variety of different scenarios including robotics, game development, and visualization. It contains a robust rigid body simulator, which was utilized for the simulations in this study. Using Newton-Euler equations of motion, it accurately simulates translational and rotational dynamics of rigid bodies. Furthermore, it includes formulations and algorithms for friction, collision detection, and meshing (Coumans 2013). Some researchers (E. Izadi & Bezuijen 2015; Ehsan Izadi & Bezuijen 2018) have recently had success in simulating soil behavior and granular dynamics utilizing physics engines.

A few computational packing models have been developed in the past for asphalt mixtures (Chen et al. 2013; Liu & You 2009; Xu et al. 2010). However, these models use simplified materials as well as geometry, rendering them unworkable in a practical setting. The goal of this study is to create a computational model using realistic particle geometry for compaction in a gyratory compactor and examine the feasibility of such an approach. This computational model improves on previous models by using real aggregates and considering the influence of asphalt binder during the compaction process.

2 METHODOLOGY

Asphalt mixtures comprise of asphalt binder and aggregates. The asphalt mixture sample in this study was considered to be 50 mm in height and 150 mm in diameter, and, assumed to have 4% by volume of air voids and 13.87% by volume of binder. The following subsections describe the modeling of these materials as well as the simulation process.

2.1 Aggregates

The aggregates gradation used in this study is shown in Table 1. It is not feasible to simulate all the aggregate particles in the model due to the limitation in computational processing. Thus, only particles with size equal to or greater than 4.75 mm were simulated in this model. Aggregates smaller than 4.75 mm size were considered in the model as a coating around

Figure 1. Laser scanned aggregates.

Table 1. Aggregates gradation.

Sieve size (mm)	% Passing	Number of particles on this sieve
19	100	0
12.5	97.7	18
9.5	88.6	169
4.75	56.2	2302
2.36	40	Included in mastic
1.18	29.5	Included in mastic
0.6	21.3	Included in mastic
0.3	13.6	Included in mastic
0.15	7.9	Included in mastic
0.075	5	Included in mastic

the aggregates greater than 4.75 mm size. Therefore, it was assumed that 70% of the volume of the remaining aggregates was distributed on all aggregate particles greater than 4.75 mm as a film surrounding each aggregate particle. The thickness of the film on different aggregates was assumed to follow normal distribution with a spread of 0.1 standard deviation units away from the mean. Seven real aggregate particles (as shown in Figure 1) of different geometry were scanned using a laser and used in this model. These seven aggregates were scaled to obtain 19 mm, 12.5 mm, 9.5 mm, and 4.75 mm size aggregates. These aggregates are assumed to be rigid bodies in the simulation. Since the volume of each aggregate is already known, using the gradation, the number of particles for each sieve size is calculated and used in the simulation (Table 1).

2.2 Asphalt binder

In this model, the influence of asphalt binder, or more accurately the asphalt binder mastic comprising binder and the fine aggregates, between colliding aggregates has been considered by emulating the viscous and cohesive nature of asphalt binder. It was assumed that asphalt binder at compaction temperature will behave as a liquid. Due to its viscous and cohesive nature, two types of forces (as shown in Figure 2) were assumed to act between any two colliding aggregates: 1) Cohesive forces (normal to the contact), 2) Viscous forces (tangential to the contact).

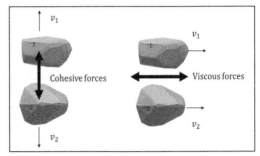

Figure 2. Asphalt binder as forces between aggregates, left: Cohesive forces, right: Viscous forces.

2.2.1 Cohesive forces

Cohesive forces act normal to the contact between two colliding aggregates and resist the separation between them. This relationship was developed using Buckingham π theorem and dimensional analysis. The parameters considered important for these cohesive forces model are: surface area of contact and asphalt binder surface tension.

The total cohesive force is proportional to the area of contact. Surface tension (of the mortar), being the fundamental property that keeps the contact surfaces together, can determine the resistance the binder itself provides when aggregates are pulled apart. Therefore, the cohesive force is assumed to be proportional to both these attributes and is defined as:

$$F_c = f(A, \gamma) \tag{1}$$

The goal is to find this force which is a function of the contact surface area A and the asphalt binder surface tension γ. The solution of the equation using the π theorem along with dimensional analysis is:

$$F_c = c\gamma\sqrt{A} \tag{2}$$

where, c is the proportionality constant. For the simulations in this study, the value of c is assumed to be 10^{-5} and the asphalt binder surface tension was assumed to be 2.5×10^{-5} N/mm. These values are only first estimates based on trials that result in a realistic compaction scenario and will be a subject of future studies.

Due to limitations of the Bullet Physics engine, two simplifications were made. 1) The exact contact surface area between colliding aggregates, A, is difficult to estimate. Therefore, as an approximation, half the surface area of the smaller aggregate in a pair of colliding aggregates was used as the contact surface area (this can be adjusted in future as a function of aggregate angularity). 2) Although, in reality this force would act locally on the surface of contact, in this study this force was assumed to act at the center of the colliding objects.

2.2.2 Viscous forces

These forces reflect the viscous nature of the asphalt binder which results in the damping of the linear and angular movements of the colliding aggregates. The asphalt mortar is assumed to be a Newtonian liquid at the compaction temperature. No-slip boundary condition at the binder aggregate interface and constant temperature throughout the compaction process were also assumed. For the viscous force damping model, two relationships were developed for: 1) linear velocity damping and 2) angular velocity damping.

2.2.2.1 Linear velocity damping

Newton's second law of motion along with Newton's law of viscosity were used to develop this relationship. The flow velocity of the asphalt mortar between two colliding aggregates here is assumed to be varying linearly (note that this relationship can easily be modified to reflect a non-linear behavior as needed). The following equations describe the Newton's second law of motion and Newton's law of viscosity, respectively:

$$-F_n = mdv/dt \qquad (3)$$

$$F_n = \eta Av/y \qquad (4)$$

where, F_n is the magnitude of viscous force and was obtained from the Newton's law of viscosity. In Equation 3 and 4, m represents mass of the aggregate, η represents asphalt binder viscosity, y represents the area of contact between the colliding aggregates and represents the effective distance between the colliding aggregates. Like the cohesive forces in the previous section, area of contact between two colliding aggregates for this relationship is also assumed to be half of the smaller aggregate's surface area.

All the variables in Equations 3 and 4, except the velocity, are assumed to be constants. Therefore, by solving these two equations, velocity of the colliding aggregates after damping is obtained as:

$$v = v_0 \exp\left(-\eta At/ym\right) \qquad (5)$$

where, v_0 is the initial velocity of the aggregate when the collision starts, v is the final velocity and t is the contact time.

2.2.2.2 Angular velocity damping

Analogous to Newton's second law of motion, angular motion is described by the following equation:

$$-F_n y = I \, d\omega/dt \qquad (6)$$

where, I is the moment of inertia of the aggregate, y is the effective distance between the two aggregates, and ω is the angular velocity. For computing I in this relationship, the shape of the aggregate is assumed to be an ellipsoid. Similar to linear damping, when Equations 4 and 6 are solved together the angular velocity of the colliding aggregate after damping is obtained as follows:

$$\omega = \omega_0 \exp\left(-\eta Ayt/I\right) \qquad (7)$$

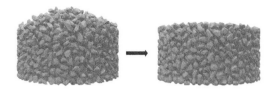

Figure 3. Left: aggregates after placing, Right: aggregates after compaction.

2.3 Simulations for gyratory compaction

Aggregates throughout the simulation uses rigid body mechanics and collision detection of Bullet Physics engine. The procedure of the simulation process (as shown in Figure 3) is described as follows:

1. A 3D modeling software was used to create the compactor and the mold.
2. The aggregates were randomly dropped in a random orientation into the mold to simulate the process where aggregates are discharged into the mold. A python script was created and used to execute the random dropping of the aggregates.
3. Once all the aggregates are placed. The viscous and cohesive forces between the colliding aggregates due to asphalt binder described in section 2.2 were activated along with compactor motion by using python scripts.
4. Position of the compactor was recorded along with the number of gyrations during the compaction process to obtain the densification curve.

3 RESULTS

Asphalt mixture compaction in a gyratory compactor has been widely accepted to be sensitive to the compactor gyratory angle. Typically, a higher gyratory angle compacts the asphalt mixture more densely due to the fact that higher gyratory angle produces more shear in mix thus compacting it more as compared to the lower gyratory angle. In this study, three different compactors were used to compact the same asphalt mixture. Each gyratory compactor had a different gyratory angle while the other compactor parameters, which are, vertical pressure on the compactor and gyration rate, were kept the same. The gyratory angles of these three compactors were 1.25°, 3° and 6°, respectively. The asphalt mixtures were compacted using the materials and method described in section 2 of this paper.

Height of the asphalt mixture sample was tracked in the simulation along with the number of gyrations of the compactor to obtain the densification curve. The curves obtained from the compaction under these three different scenarios were then compared to evaluate the sensitivity of the asphalt mixture compaction simulation with respect to the gyratory angle. These curves are shown in Figure 4.

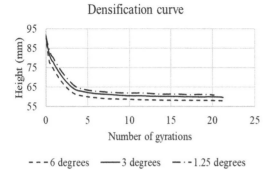

Densification curve

- - - 6 degrees ——— 3 degrees — · - 1.25 degrees

Figure 4. Densification curves obtained for three different gyratory angles of the compactor.

It can be seen from Figure 4 that as the gyratory angle of the compactor was increased the packing density of the aggregates increased. Therefore, it can be inferred that the compaction characteristics from the simulation agrees with the compaction characteristics of the asphalt mixtures compacted in the laboratory setting.

4 CONCLUSIONS AND FUTURE WORK

This study demonstrated the feasibility of using Bullet Physics engine to simulate the compaction process in a gyratory compactor. The same asphalt mixture was compacted using three compactors of different gyratory angles and the densification curves obtained from these three compactions were compared. Results show that the method used in this study has adequate sensitivity to differentiate the compaction from different gyratory angles of the compactor.

Additional ongoing work is focused on calibrating the cohesive and viscous forces relationships and validating this computational model against laboratory compacted asphalt mixtures. A variety of other improvements are also planned on the current work:

1. Including smaller aggregates in the simulation to improve upon the accuracy of the model.

2. Incorporation of high-performance computing algorithms to increase simulation efficiency.
3. Test on a variety of mixtures to evaluate robustness of this computational model.

ACKNOWLEDGEMENTS

Authors acknowledge the crucial contribution of Dr. Ramez Hajj. Authors also thank Mr. Wellington Lorran Gaia Ferreira and Ms. Ingrid Gabrielle Do Nascimento Camargo for their valuable insights and Ms. Vasundhara Komaragiri for her inputs and proofreading this article.

REFERENCES

Chen, J., Huang, B., & Shu, X. (2013). Air-void distribution analysis of asphalt mixture using discrete element method. *Journal of Materials in Civil Engineering*, 25(10), 1375–1385. https://doi.org/10.1061/(ASCE)MT.1943-5533.0000661

Coumans, E. (2013). *Bullet 2. 82 Physics SDK Manual Table of Contents*.

Izadi, E., & Bezuijen, A. (2015). Simulation of granular soil behaviour using the Bullet physics library. *Geomechanics from Micro to Macro - Proceedings of the TC105 ISS-MGE International Symposium on Geomechanics from Micro to Macro, IS-Cambridge 2014*, 2, 1565–1570. https://doi.org/10.1201/b17395-285

Izadi, E., & Bezuijen, A. (2018). Simulating direct shear tests with the Bullet physics library: A validation study. *PLoS ONE*, 13(4), 1–28. https://doi.org/10.1371/journal.pone.0195073

Liu, Y., & You, Z. (2009). Visualization and simulation of asphalt concrete with randomly generated three-dimensional models. *Journal of Computing in Civil Engineering*, 3801(March), 99–109. https://doi.org/10.1061/(ASCE)0887-3801(2009)23

Xu, R., Yang, X. H., Yin, A. Y., Yang, S. F., & Ye, Y. (2010). A three-dimensional aggregate generation and packing algorithm for modeling asphalt mixture with graded aggregates. *Journal of Mechanics*, 26(2), 165–171. https://doi.org/10.1017/S1727719100003026

Advances in Materials and Pavement Performance Prediction II – Kumar et al. (eds)
© 2021 Taylor & Francis Group, London, ISBN 978-0-367-46169-0

Simulation of performance of PA mixtures with different mastics

P. Liu, C. Wang, G. Lu, D. Wang & M. Oeser
Institute of Highway Engineering, RWTH Aachen University, Aachen, Germany

X. Yang
Department of Civil Engineering, Monash University, Clayton, Australia

S. Leischner
Institute of Urban and Pavement Engineering, TU Dresden, Dresden, Germany

ABSTRACT: In this study, the effect of mastics with four different mineral fillers (limestone, granodiorite, dolomite and rhyolite) on the mechanical performances of porous asphalt (PA) mixtures was investigated. X-ray computer tomography (X-ray CT) scanning and digital image processing (DIP) techniques were applied to detect and reconstruct the microstructure of PA specimens. Finite element (FE) simulations were conducted to simulate indirect tensile tests. The results show that the mastics with different fillers have significant influence on the mechanical performances of PA mixtures. The PA mixture with limestone exhibits the least load-bearing capacity while the specimen with granodiorite exhibits the highest values. Although the distribution and the magnitude of the creep strain in the different mastic are similar, the locations of the maximum creep strain are not the same. Further investigations should be carried out to facilitate the improvement of the pavement design process.

1 INTRODUCTION

To recover the natural hydrological cycle and lower the urban flood risk, void-rich pavement structures (permeable pavements), such as Porous Asphalt (PA) and Porous Concrete (PC) pavements, are implemented by directly allowing the rain water to seep through the pavement surface (Cooley et al. 2002; Kuang et al. 2011).

To minimize the degradation of pavement surfaces and increase the durability of the PA layers, there is a need to improve the conventional mastic (Zhang & Leng 2017). Multiple researchers have demonstrated that the geometrical, chemical and mechanical properties of fillers considerably influence the performance of the mastic as well as the final asphalt mixes (Antunes et al. 2015; Rieksts et al. 2018; Wang et al. 2019; Wang et al. 2011). The mechanical response of asphalt mixtures has been widely studied using experimental and numerical methods (Ban et al. 2017; Bhattacharjee & Mallick 2012; Chen et al. 2015; Liu et al. 2017a; Souza & Castro 2012; Wang & Al-Qadi 2010). The rapid development of computational capacity has allowed for numerical methods to gain popularity compared to intensive laboratory testing with high costs.

The objective of this research is to investigate the effect of different asphalt mastics on the mechanical performance of the PA mixtures. The conventional filler limestone and three other mineral fillers, granodiorite, dolomite and rhyolite, are used to prepare

asphalt mastic and the mastic rheological properties are measured. X-ray computed tomography (X-ray CT) scanning was carried out to obtain gray images of the PA. The microstructure of the asphalt specimens was reconstructed based on digital image processing (DIP) techniques and then used in finite element (FE) simulations. An indirect tensile test was simulated to investigate the evolution of creep strain, the stress states at the mortar-aggregate interface.

2 EXPERIMENTAL PROGRAM

2.1 *Preparation of the porous asphalt mixture*

A common used type of PA was selected in this study with maximum grain size of 8 mm (designated as PA 8) to derive the microstructure. The PA 8 specimens were composed of crushed diabase aggregate, mineral filler, and bonded by bitumen with a 50/70 penetration grade. In order to compare the influences of the different mastic on the mechanical performance of the PA 8, conventional mineral filler limestone and three other mineral fillers, granodiorite, dolomite and rhyolite, were used in this research. The PA 8 mixture is only used to derive the microstructure, and the microstructure is believed not to be influenced by the mastic with different fillers. As a result, only limestone was used to prepare the PA 8. The bitumen and the filler were blended in the laboratory to prepare asphalt mastics respectively with a binder-filler mass ratio of 1:1.6. The mixtures were prepared by means of Marshall Compaction (50 impacts per side).

2.2 X-ray CT scanning and DIP

The internal microstructure was detected through X-ray CT scanning. Scanning intervals were set to 0.1 mm. The resolution of the gray images was 1024 · 1024 pixel^2 with each pixel being 80 μm.

The gray values determined by the X-ray CT device range from 0 to 255 according to material density. Within the asphalt mixtures, aggregate resulted in the maximum gray value while air-voids returned the minimum gray values. The microstructure was extracted by means of DIP. The detailed information about the DIP to reconstruct the microstructure for the FE simulation can be found in the previous study (Liu et al. 2018).

3 DEVELOPMENT OF FINITE ELEMENT MODEL

The aggregate and asphalt mastic require appropriate material properties for the FE simulation in general-purpose FE software ABAQUS. Compared with the asphalt mastic, aggregates are normally assumed as linear elastic and independent on temperature. In this study, a Young's modulus of 55000 MPa and a Poisson's ratio of 0.25 were used (You et al. 2008). In this study, the asphalt mastic was considered to be linear viscoelastic by using generalized Maxwell model for the FE modelling in ABAQUS. The Prony series of asphalt mastic was used for the simulation at 20°C. They were derived from the test of strain and frequency sweep on the mastic. The detailed procedure to derive the viscoelastic parameters can be found in the previous researches (Blasl et al. 2019; Liu et al. 2017b).

For an explicit representation of aggregate in the two-dimensional (2D) model, the threshhold size to distinguish aggregate from filler was 0.5 mm, i.e., aggregates with the size smaller than 0.5 mm were ignored in the reconstruction process and are regarded as part of the mastic. After importing the geometry of air voids and aggregate grains into ABAQUS, the asphalt mastic is created by means of Boolean operations. The aggregate grains and asphalt mastic are then assembled to construct the microstructural model. Hard contact conditions are assumed for the interaction between aggregates and the aggregates were tied together with the asphalt mastic.

The asphalt mixture was discretized by linear triangle 3-node plane stress elements (CPS3). To simulate the indirect tensile test, the width of the loading and support strips was set to 12.7 mm. The loading was uniformly distributed and was conducted at a constant deformation rate of 50 mm/min for 2 s, which is believed not to cause damage in the specimens. The support strip was fixed. One exemplary model is shown in Figure 1. The reliability of the FE models has been validated in a previous investigation (Liu et al. 2017b).

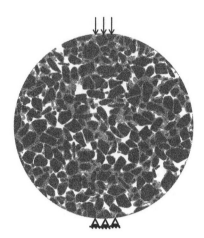

Figure 1. FE model in ABAQUS.

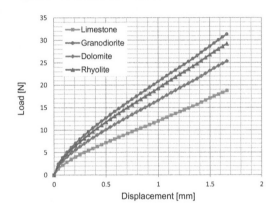

Figure 2. Comparison of load–displacement curves.

4 ANALYSIS AND DISCUSSION

4.1 Load- bearing capacity

In order to compare the load-bearing capacity of the PA with different asphalt mastics, the average values of the loads derived from the three 2D models are plotted in Figure 2. Given the fact that the load still gets increase until the simulation ends, the assumption can be made that no cracks emerge.

The initial stiffnesses (the slope of the curves) of the PA with granodiorite, dolomite and rhyolite are similar; also, the PA with limestone exhibits the least load-bearing capacity while the specimen with granodiorite exhibits the highest values.

4.2 Maximum principal stress

The failure theory of principal stress is commonly used, in which failure occurs when the maximum principal stress in a system reaches the value of the maximum strength corresponding to the elastic limit in simple tension (Abaqus 2014). The distribution of maximum principal stresses in the FE model is thus considered to analyse the effect of the different mastic with the aid of FE simulations, which is shown in Figure 3.

From these figures, it can be seen that the distributions of the maximum principal stresses in all specimens are similar, i.e., the maximum principal stresses mainly occur in the central region along the loading and support strips. Furthermore, the compressive stress is dominant in the specimens. The stress concentration is obvious at the interface between aggregates and mastic, especially around the aggregates with sharp corners, which is consistent with the previous study (Hu et al. 2019). In the testing mode of constant displacement rate at 20°C, the stiffness of the asphalt mastic with granodiorite has the highest value, while the limestone mastic has the lowest one. As a result, more larger and smaller maximum principal stresses occur in the mastic of the PA with granodiorite and limestone, respectively. While the stress states in the PA with dolomite and rhyolite are more similar.

4.3 Creep deformation

Rutting is a frequently observed phenomenon and is caused by creeping of the asphalt mastic; which is of high significance for pavement design and must be understood from a micromechanical perspective (Zhang et al. 2012, 2013).

Because the distribution and the magnitude of the creep strain in the four mastics are extremely similar, only an exemplar with the locations where the maximum creep strain occurs is shown in Figure 4. Although the distribution and the magnitude are similar, the locations of the maximum creep strain are not the same. The maximum creep strains tend to accumulate around narrow gaps between sharp aggregate edges.

The maximum creep strains in the four mastics with the limestone, granodiorite, dolomite and rhyolite at the computational time 2 s are 1.18, 1.13, 1.15 and 1.15, respectively. The trend is found that the limestone mastic shows the highest maximum creep strain while the granodiorite shows the lowest, and the maximum creep strains in dolomite and rhyolite mastic are in between and the values are close to each other.

5 CONCLUSIONS AND RECOMMENDATIONS

The influence of the asphalt mastic with different mineral fillers (limestone, granodiorite, dolomite and rhyolite) on the mechanical response of porous asphalt mixtures is studied at the microscale based on FE simulations. The FE simulations of indirect tensile tests under displacement-control mode are carried out to investigate the load-bearing capacity, stress state and creep deformation within the asphalt mixtures.

When testing with a constant displacement rate, the initial stiffnesses of the PA with granodiorite, dolomite and rhyolite are similar; also, the PA with limestone

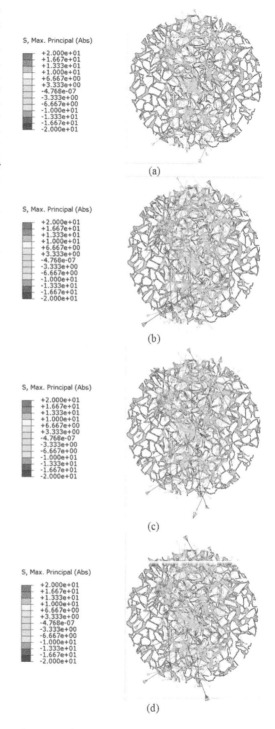

Figure 3. Distribution of maximum principal stress in mastic with different fillers at 2 s. (a) Limestone; (b) Granodiorite; (c) Dolomite; (d) Rhyolite.

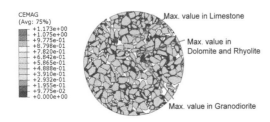

CEMAG
(Avg: 75%)
+1.173e+00
+1.075e+00
+9.775e-01
+8.798e-01
+7.820e-01
+6.842e-01
+5.865e-01
+4.888e-01
+3.910e-01
+2.932e-01
+1.955e-01
+9.775e-02
+0.000e+00

Max. value in Limestone

Max. value in
Dolomite and Rhyolite

Max. value in Granodiorite

Figure 4.   Distribution of creep strains in an exemplar.

exhibits the least load-bearing capacity while the specimen with granodiorite exhibits the highest values. The maximum principal stresses mainly occur in the central region along the loading and support strips. The stress concentration is obvious at the interface between aggregates and mastic, especially around the aggregates with sharp corners. Although the distribution and the magnitude of the creep strain in the different mastic are similar, the locations of the maximum creep strain are not the same. The maximum creep strains tend to accumulate around narrow gaps between sharp aggregate edges.

The asphalt mastic with different mineral fillers significantly influences the mechanical response of the asphalt mixture. The abovementioned conclusions contribute to the current knowledge, while further investigation needs to be carried out, i.e., a larger temperature range in the FE simulation will be adopted and more microstructural FE models with more advanced constitutive material laws will be simulated.

ACKNOWLEDGMENTS

The work underlying this project was carried out under the research grant numbers FOR 2089/2 (OE514/1-2, WE 1642/1-2 and LE 3649/1-2) and OE514/4-1, on behalf of the grant sponsor, the German Research Foundation (DFG). Besides, the authors also acknowledge the support of German Academic Exchange Service (DAAD) and Universities Australia (grant number 57446137).

REFERENCES

Abaqus Analysis User's Guide. 2014. Abaqus 6.14, Dassault Systèmes Simulia Corp.

Antunes, V. Freire, A. Quaresma, L. & Micaelo, R. 2015. Influence of the geometrical and physical properties of filler in the filler–bitumen interaction. *Construction and Building Materials* 76: 322–329.

Ban, H. Im, S. Kim, Y.R. & Jung J.S. 2017. Laboratory tests and finite element simulations to model thermally induced reflective cracking of composite pavements. *International Journal of Pavement Engineering* 1–11.

Bhattacharjee, S. & Mallick R.B. 2012. Effect of temperature on fatigue performance of hot mix asphalt tested under model mobile load simulator. *International Journal of Pavement Engineering* 13(2): 166–180.

Blasl, A. Khalili, M. Canon Falla, G. Oeser, M. Liu, P. & Wellner F. 2019. Rheological characterisation and modelling of bitumen containing reclaimed components. *International Journal of Pavement Engineering* 20(6): 638–648.

Chen, J.Q. Zhang, M. Wang, H. & Li L. 2015. Evaluation of thermal conductivity of asphalt mixture with heterogeneous microstructure. *Applied Thermal Engineering* 84: 368–374.

Cooley, L. Prowell, B.D. & Brown E.R. 2002. Issues pertaining to the permeability characteristics of coarse-graded Superpave mixes. *Asphalt Paving Technology* 71: 1–29.

Hu, J. Liu, P. Wang, D. Oeser, M. & Canon Falla G. 2019. Investigation on interface stripping damage at high-temperature using microstructural analysis. *International Journal of Pavement Engineering* 20(5): 544–556.

Kuang, X. Sansalone, J. Ying, G. & Ranieri V. 2011. Pore-structure models of hydraulic conductivity for permeable pavement. *Journal of Hydrology* 399(3-4): 148–157.

Liu, P. Xing, Q. Wang, D. & Oeser M. 2017a. Application of dynamic analysis in semi-analytical finite element method. *Materials* 10, 1010.

Liu, P. Hu, J. Wang, D. Oeser, M. Alber, S. Ressel, W. & Canon Falla G. 2017b. Modelling and evaluation of aggregate morphology on asphalt compression behaviour. *Construction and Building Materials* 133: 196–208.

Liu, P. Hu, J. Wang, H. Canon Falla, G. Wang, D. & Oeser M. 2018. Influence of temperature on the mechanical response of asphalt mixtures using microstructural analysis and finite-element simulations. *Journal of Materials in Civil Engineering* 30(12), 04018327.

Rieksts, K. Pettinari, M. & Haritonovs V. 2018. The influence of filler type and gradation on the rheological performance of mastics. *Road Materials and Pavement Design* 1–15.

Souza, F.V. & Castro L.S. 2012. Effect of temperature on the mechanical response of thermo-viscoelastic asphalt pavements. *Construction and Building Materials* 30: 574–582.

Wang, D. Liu, P. Oeser, M. Stanjek, H. & Kollmann J. 2019. Multi-scale study of the polishing behaviour of quartz and feldspar on road surfacing aggregate. *International Journal of Pavement Engineering* 20(1): 79–88.

Wang, H. & Al-Qadi I.L. 2010. Near-surface pavement failure under multiaxial stress state in thick asphalt pavement. *Transportation Research Record: Journal of the Transportation Research Board* 2154: 91–99.

Wang, H. Al-Qadi, I.L. Faheem, A.F. Bahia, H.U. Yang, S. & Reinke G.H. 2011. Effect of mineral filler characteristics on asphalt mastic and mixture rutting potential. *Transportation Research Record: Journal of the Transportation Research Board* 2208(1): 33–39. doi:10.3141/2208-05

You, Z. Adhikari, S. & Dai Q. 2008. Three-dimensional discrete element models for asphalt mixtures. *Journal of Engineering Mechanics* 134(12): 1053–1063.

Zhang, Y. Luo, R. & Lytton R.L. 2012. Characterizing permanent deformation and fracture of asphalt mixtures by using compressive dynamic modulus tests. *Journal of Materials in Civil Engineering* 24(7): 898–906.

Zhang, Y. Luo, R. & Lytton R.L. 2013. Mechanistic modeling of fracture in asphalt mixtures under compressive loading. *Journal of Materials in Civil Engineering* 25(9): 1189–1197.

Zhang, Y. & Leng Z. 2017. Quantification of bituminous mortar ageing and its application in ravelling evaluation of porous asphalt wearing courses. *Materials and Design* 119: 1–11.

Advances in Materials and Pavement Performance Prediction II – Kumar et al. (eds)
© 2021 Taylor & Francis Group, London, ISBN 978-0-367-46169-0

A step towards a multiscale model of frost damage in asphalt mixtures

L. Lövqvist, R. Balieu & N. Kringos
KTH Royal Institute of Technology, Stockholm, Sweden

ABSTRACT: Damage in the asphalt layers of pavements due to frost action can be a major problem during winter in cold and wet regions. To develop effective prevention measures it is important to understand the damage mechanism inside the material and how it interacts with other damage phenomena. To obtain this understanding, a multiscale model is currently developed in this research project. This paper presents the idea for the multiscale model framework, as well as utilizes a developed micromechanical model to investigate two parameters required for developing the damage envelope needed for the multiscale model: the two types of damage modes (adhesive and cohesive) and the increase of air void volume due to damage. From the results it was concluded that the relation between the two damage modes significantly, and sometimes counterintuitively, affects their combined effect, and that the relation between the air void content and the induced damage is nonlinear.

1 INTRODUCTION

Winter is an especially tough season for asphalt pavements in regions with cold and wet climates. The damage occurring during this period can showcase itself as cracks, potholes and uneven surfaces. These can be caused by many different factors, or a combination of them, such as low temperature, embrittlement of the bitumen and frost action in both the lower and upper layers of the pavement. The frost action in the asphalt layers of the pavement, also referred to as frost damage, can be particularly problematic in regions with much rain and snow and many freeze-thaw cycles (FTCs) each winter. In order to prevent or minimize this type of damage it is important to characterize the process and to identify which parameters that are dominant. However, due to the complexity of the material and the process, as well as the influence of other concurring damage processes, it is difficult to make this characterization experimentally.

Instead, modeling, and in particular microscale modeling, enables a detailed analysis of how the different material components react to the different loads and where the damage is most likely to occur. A limited number of such microscale models dealing with frost damage exist (e.g.Lövqvist et al. 2019; Varveri et al. 2014). However, these cannot make reliable damage predictions of the whole pavement which can also account for the effect of other damage processes occurring on a different scale. In order to do this, a multiscale model which makes predictions on the pavement scale based on the behavior on the microscale is a better choice. Currently no such model which specifically regards frost damage in asphalt exists to the best of our knowledge.

This paper presents the overall idea for the framework of a multiscale model of frost damage which is under development. Additionally, two necessary steps towards the multiscale model will be taken by using a developed micromechanical model. These case studies investigate the effect of two of the parameters required for the development of the multiscale approach: the type of damage and the air void content.

2 MULTISCALE MODEL FRAMEWORK

In order to formulate the multiscale approach, a relationship between the microscale and the macroscale needs to be established. A common technique of doing this is to use the homogenization principle, where the behavior on the microscale is averaged over the entire microstructure. In this project, the homogenization will be done through virtual testing of the microstructure. By testing the stiffness of the microstructure depending on different parameters, such as the number and duration of FTCs, the temperature, and moisture exposure, a damage envelope can be constructed. This damage envelope will control the asphalt behavior on the macroscale depending on the environmental parameters used as input.

The damage envelope will thus indicate how the material parameters depend on the damage, which in turn depends on the environmental parameters. For example, it is known that FTCs create damage inside the asphalt which means that the air void content is increased, allowing more moisture to infiltrate the asphalt and thus create even more damage in the following FTCs. This means that for this case the evolution of (i) the diffusion coefficient of moisture in the asphalt on the macroscale, and (ii) the FT damage curve from the microscale controlling the behavior of the macroscale, need to be determined depending on the number of FTCs. This determination is necessary since (i) a higher air void volume will increase the permeability of the asphalt and therefore will affect the diffusion coefficient, and (ii) an increased air void volume will increase the FT damage, as well as change its evolution.

By using the damage envelope it can be investigated how sensitive the overall behavior of the asphalt is with respect to the different environmental parameters.

3 CASE STUDIES

In order to develop the damage envelope, the relationships between the developed damage and the environmental parameters must be determined. This section presents two of the steps required for this determination. The first step is to determine the effect of each of the two damage modes (adhesive and cohesive damage) on the behavior of the whole microstructure and on each other depending on the number of FTCs. The second step is an investigation of the accelerating effect that the increasing damage, in the form of a higher air void content, has.

The simulations are computed using a developed micromechanical model (Lövqvist et al. 2019), which will be included in the multiscale model. The same microstructure is used in both simulations. The microstructure, shown in Figure 1, was obtained from an X-ray CT scan of a real asphalt microstructure and consists of 59.3 volume% aggregates, 29.4 volume% mastic (bitumen + aggregates smaller than 2.5 mm) and 11.3 volume% air voids.

The simulations for both cases assumes the air voids are fully saturated and include 10 FTCs consisting of 4 hours of freezing from 2°C to −2°C, 1 hour at a constant temperature of −2°C, 4 hours of thawing from −2°C to 2°C, and finally 1 hour at a constant temperature of 2°C.

3.1 *The effect of adhesive vs. cohesive damage*

Three separate simulation cases are computed to investigate the separate effects that the two damage modes (adhesive in the interface and cohesive inside the mastic) have on the structural behavior and on each other. The first simulation includes only adhesive damage,

the second only cohesive damage, and the third both adhesive –and cohesive damage. Additionally, in order to study the importance of the relation between the two damage modes, two relations are tested: one where the cohesive damage is larger and one where the adhesive damage is larger. This is performed by changing the damage initiation criteria for the cohesive damage. All other parameters are kept constant, thus maintaining the same behaviors for both cases. After each FTC in the three simulations, the microstructure is subjected to a pressure of 0.1 MPa, applied on the top of the microstructure. From this, the stiffness of the structure can be obtained and the effect of the separate damage modes can be evaluated. As presented in Figures 2 & 3, both the adhesive cohesive damages affect the structural stiffness of the microstructure for both relations of the damage modes.

For the case where the cohesive damage is dominant, it can be seen in Figure 2 that although both modes give a similar effect after the first FTC, where

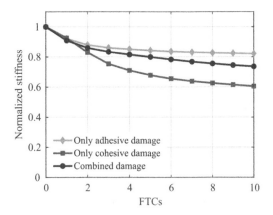

Figure 2. The stiffness of the microstructure as a function of FTCs for the different damage modes for the case when the cohesive damage is the largest.

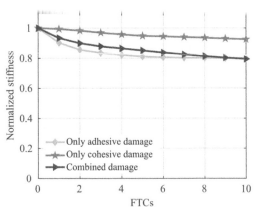

Figure 3. The stiffness of the microstructure as a function of FTCs for the different damage modes for the case when the adhesive damage is the largest.

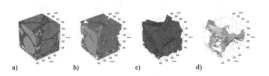

Figure 1. (a) The full microstructure, (b) Aggregates, (c) Mastic, and (d) Air voids.

the stiffness decreases about 8% compared to the original, during the following FTCs the effect of the cohesive damage continuously increases more than the effect of the adhesive damage. By the end of 10 FTCs, the stiffness decrease caused by only cohesive damage has exceeded more than 20% of that caused by solely the adhesive damage. This difference in the increased rate of the damage can be attributed to the viscoelastic behavior and viscous damage evolution of the mastic, which cause a continued deformation of the mastic during the entire period of loading. Due to this, the cohesive damage rate in the mastic will be higher than that of the adhesive damage in the interface, which would require an increase in load to continue to increase. In reality, such an increase could occur from a higher air void content due to damage, as discussed in section 3.2. In that case, the evolution of the adhesive damage could behave differently.

Figure 2 also reveals that, except for after the first FTC, the combined effect of the two damage modes on the stiffness of the microstructure is not larger than each of the separated effects. In fact, after 10 FTCs the stiffness of the microstructure, which included both damage modes, has only decreased around 9% more than the microstructure including only adhesive damage and thus, 13% less than the one with only cohesive damage. This indicates that instead of the two damage modes having an accelerated effect on each other when being present at the same time, they have somewhat of a counteracting effect on each other. A possible source for this counterintuitive effect could be that when the adhesive damage occurs (which in reality causes stripping, i.e., mastic/bitumen loosens from the stones), the mastic has more freedom to move. This freedom means that when the water in the air voids expand during freezing, the mastic will experience less stress and thus less damage. After the first FTC, the interface damage might not be so large that it would give a prominent effect on the behavior of the mastic or vice versa. This could indicate that the location of the damaged interface in relation to the damaged mastic is of importance.

As shown in Figure 3, the increase rate for the cohesive damage is still higher than the one of the adhesive damage for the case when the adhesive damage is the larger damage mode. Considering that only the damage initiation criteria is changed between the cases, this is expected. The behavior of the total damage however differs depending on which damage mode is the largest. For the case when the adhesive damage is larger, the combined effect of the two damage modes gives a decrease of stiffness of about 7% after the first FTC, which is in between the decreases given by only the cohesive damage (1%) and only the adhesive damage (10%). This again indicates that the damage modes have a counteracting effect on each other. However, contrary to the case when the cohesive damage was the largest, this counteracting effect is reduced during the FTCs when the adhesive damage is the largest. In fact, by the end of 10 FTCs, the stiffness decrease

caused by the combined effect exceeds the one caused by only the adhesive damage by almost 1%. This difference between the two cases of parameters shows that the combined effect of the two damage modes depends on the relation between them, i.e., the material properties. An equal adhesive and cohesive strength or a strong difference between them can thus imply different effects on the short and long term. The damage models for the two modes therefore need to be individually calibrated for each material that may be of interest in the multiscale model in order to develop the damage envelope.

For all of these results it is important to note that it is the behaviors and trends of the damage evolutions that are of importance in this paper and not the numbers since the two damage equations have yet to be calibrated for the same material. This is currently being worked on by the authors.

While the effect of the damage on the stiffness is of importance, the damage variables need to be calculated in order for it to be possible to incorporate the effect of the damage modes into the damage envelope under construction. The obtained damage variable evolutions are illustrated in Figure 4.

3.2 The accelerating effect of increasing damage

While the increase of damage can showcase itself as both the emergence of new air voids (i.e. microcracks) and the growth of existing ones, this paper only considers the second case and also assumes all air voids grow separately so that no merging of the air voids and improvement of the interconnectivity of the air void system occurs. The evaluation of this effect of an increasing air void volume is done by modifying the microstructure to consist of different air void volumes, which simulates the growth of existing air voids and emergence of new ones. The modified microstructures, showed as slices together with the original microstructure in Figure 5, have an air void volume of 8.6% and 13.6% respectively.

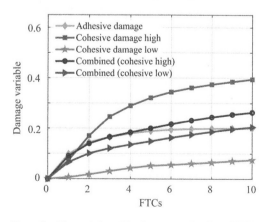

Figure 4. The evolution of the damage as a function of FTCs for the different damage modes for both cases.

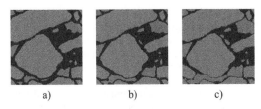

a)   b)   c)

Figure 5. Three versions of the same microstructure but with different air void contents. a) 8.6 volume%, b) 11.3 volume% (original), and c) 13.6 volume%.

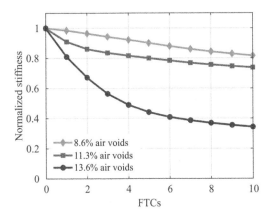

Figure 6. The stiffness as a function of FTCs for the microstructure with different air void contents.

As in the previous case, the microstructure is subjected to a pressure of 0.1 MPa, applied on the top of the microstructure after each FTC. It is shown in Figure 6 that the air void content has a significant effect on the normalized stiffness of the microstructures, as expected, as the microstructure with 13.6% air voids experiences a decrease in normalized stiffness which is 40% and 45% more than those of the original microstructure and the one with 8.6% air voids, respectively.

In order to include the effect of an increased volume of air voids due to damage in the damage envelope, the relation between the damage and the air void volume must be established. From these results it can be concluded that this relation is certainly a nonlinear one. The difference in the decrease of normalized stiffness between the microstructure the low air void content and the original microstructure is 8% with an air void content difference of 2.7%, while the difference

between the original microstructure and the one with a high air void content is 40% with a 2.3% difference in air void volume. Additionally, the point where the rate of change begins to decrease differs between the different versions of the microstructure, from 2 FTCs for the original to 4-5 FTCs for the one with 13.6% air voids. This also needs to be accounted for in the damage envelope.

4 CONCLUSIONS

This paper presented the framework for a multiscale model of winter damage in asphalt which is under development. In order to develop the damage envelope within this framework, the relation between the evolution of the damage and different environmental and material parameters needs to be determined. The effect of two of these parameters on the damage was in this paper investigated through simulated case studies: the effect of different damage modes and the effect of an increased air void volume that would be caused by an increase in damage.

It was found that the effect the adhesive and cohesive damage have on each other depends on which of the damage modes is dominant. Therefore, the two damage models need to be individually calibrated for each material which is of interest in the multiscale model so that the relation between the two modes and the combined effect of them can be included in the damage envelope. Additionally, the relation between the increasing air void content and the induced damage is of a nonlinear nature.

Future steps include calibrating both damage models, developing the damage envelope including the parameters investigated in this paper as well as other environmental and material parameters, and finalizing the multiscale model.

REFERENCES

Varveri, A., Avgerinopoulos, S., Kasbergen, C., Scarpas, A., & Collop, A. 2014. A constitutive model for simulation of water to ice phase change in asphalt mixtures. In Kim (ed.), *Proceedings of the International conference on asphalt pavements, ISAP 2014, Raleigh, North Carolina, 1–5 June 2014.* 531–539.

Lövqvist, L., Balieu, R., & Kringos, N. 2019. A micromechanical model of freeze-thaw damage in asphalt. *International Journal of Pavement Engineering.*

Advances in Materials and Pavement Performance Prediction II – Kumar et al. (eds)
© 2021 Taylor & Francis Group, London, ISBN 978-0-367-46169-0

Challenges in the generation of 2D Permeable Friction Courses (PFC) microstructures

L. Manrique-Sanchez, S. Caro, D.F. Tolosa & N. Estrada
Department of Civil and Environmental Engineering, Universidad de los Andes, Bogotá, Colombia

D. Castillo
Department of Civil Engineering, Aalto University, Finland

ABSTRACT: The use of computational mechanics tools has increased in pavement engineering due to their capacity to assess phenomena that are difficult to simulate in laboratory; and to their overall low cost in comparison to laboratory or field experiments. Thus, computational mechanics could be an efficient technique to understand the mechanical behavior of Permeable Friction Courses (PFC) mixtures and identify the causes of their main distresses. PFCs are gap-graded hot asphalt mixtures that are placed as thin layers over conventional pavements. Although PFC mixtures have several benefits, their main shortcoming is the difficulty to preserve their functionality and durability. This paper discusses the relevance of the geometry that is selected to represent the microstructure of PFCs mixtures in computational mechanics models and introduces a two-dimensional gravimetric method to numerically and randomly build those microstructures. In addition, this paper assesses different input parameters of the proposed gravimetric method and shows how those parameters impact different microstructural characteristics of the generated PFCs microstructures.

1 INTRODUCTION

Permeable Friction Courses (PFC) are hot mix asphalt materials placed as thin layers of 2 to 6 cm over conventional pavements. These mixtures present high Air Void (AV) contents (i.e. 18–25%) which brings several benefits in terms of safety under wet-rainy conditions and tire/pavement noise reduction (Kandhal 2002). However, PFCs present shortcomings in relation to their functionality and durability. In terms of functionality, 'clogging' reduces the effective AV content of the mixture, restraining water to drain within the PFC (Kandhal 2002). In terms of durability, the main distress is the loss of aggregates from the PFC layer surface, a phenomenon known as 'raveling'. These limitations make PFC mixtures less durable and more expensive than traditional dense Hot Mix Asphalt (HMA) materials due to their maintenance activities. (Cooley et al. 2009).

Several computational mechanics tools have been used to quantify the structural capacity of PFCs mixtures, and to understand the physical mechanisms associated with clogging and raveling. Some of these tools have allowed quantifying the influence of isolated parameters (e.g. load conditions, and material combinations) on the mechanical behavior and degradation of PFCs. Most of these works have been developed using two-dimensional (2D) Fine Element

(FE) models (;anrique-Sanchez et al. 2018; Manrique-Sanchez & Caro 2019; Mo et al. 2008), and Discrete Element (DE) theory (Alvarez et al. 2010). Although these works have demonstrated the potential of these models to characterize the mechanical behavior and deterioration of PFCs, they have also highlighted the challenges related to the development of high-quality PFC computational mechanics models.

The objective of this paper is to discuss the relevance of the geometry that describes the PFC microstructure in computational mechanics models, and to introduce a novel method to numerically obtain PFC microstructures through gravimetric techniques that combine a random generator of aggregate particles and DE modeling. In addition, this paper assesses the influence of some input parameters of the gravimetric method on the microstructural characteristics of the randomly generated PFC geometry.

2 PFC GEOMETRY

Considering that the strength of PFCs mixtures is developed within the stone-on-stone contact network (Alvarez et al. 2010), and that raveling is a mechanical phenomenon that occurs at the stone-on-stone or mortar-mortar contacts (Mo et al. 2007), it is important to rely on a numerical model that properly represents the aggregates morphological characteristics and contact network of PFCs mixtures.

The ideal model to study these mixtures would be a 3-Dimensional (3D) system. However, the computational cost of 3D models had forced researchers to make several simplifications (e.g. used of spheres as a simplified geometry of the aggregates) that could affect the accuracy of the results (Mo *et al.* 2008). Therefore, even with their several simplifications, the use of 2D geometrical models is a common alternative to model PFC mixtures, and it has proved to be efficient to evaluate specific aspects of PFC mixtures (e.g. Manrique-Sanchez *et al.* 2018; Manrique-Sanchez & Caro 2019). Next, a traditional and a new methodology to obtain realistic 2D PFC microstructures are described, including their advantages and shortcomings. It is noteworthy that these methods provide an initial geometrical arrangement of the coarse aggregates of the PFC but that each aggregate will need to be coated by mastic or asphalt mortar before conducting actual computational mechanics simulations.

2.1 PFC microstructures obtained from non-destructive techniques

This is the most common technique to obtain microstructure geometries from asphalt mixtures. The method is based on image analysis techniques, in which X-ray computed tomography (CT) is used to scan PFC samples. Then, several 2D images in different directions can be obtained to perform a 3D reconstruction of the sample. The 2D images obtained from the 3D reconstructed sample are usually analyzed in terms of their microstructural characteristics (number of aggregates, contacts, coordination number, etc.) and then used in computational mechanics models. Figure 1 illustrates a vertical cut of a PFC obtained using X-ray CT techniques, and the microstructural PFC geometry obtained from it.

2.2 Generation of PFC microstructures using gravimetric methods

This method was developed by the authors and consists in two main steps:

1. A set of random aggregates with an specific gradation and shape properties is created using the Asphalt Mixture's Random Generator (MG) developed by Castillo *et al.* (2015).
2. The set of random aggregates obtained from the previous step is subjected to gravity in order to create a contact network system. For this, the DE method software LMGC90, developed by the University of Montpellier (France), is used.

The MG method developed by Castillo *et al.* (2015) is able to randomly generate sets of aggregates that follow specific distribution sizes and morphological characteristics (i.e. shape and angularity). These characteristics can be modified and calibrated to comply with any design parameters. Thus, the MG randomly generates aggregates that are located in a predefined space (Figure 2a) and are used as an input parameter into LMGC90, where the aggregates are subjected to gravimetric methods. This DE software allows the user to set different parameters of the simulation, as the drop height of the aggregates, the friction of the particles and the specific gravity of the aggregates. The final microstructural characteristics are affected by the input parameters of the MG generator and LMGC90. Figure 2 exemplifies the initial set of aggregates created with the MG (Figure 2a) and a simulation of a PFC microstructure geometry obtained after subjecting the set of aggregates to gravity in LMGC90 (Figure 2b).

2.3 Comparison among the two PFC geometry generation methods

The PFC microstructures obtained through non-destructive methods are recognized for its easy and fast implementation, as well as for the possibility of obtaining multiple PFC microstructures from one scanned PFC sample. Nonetheless, this method creates differences between the original gradation and the one obtained in the 2D sample (Haft-Javaherian 2011) and, more importantly, the use of 2D PFC microstructures

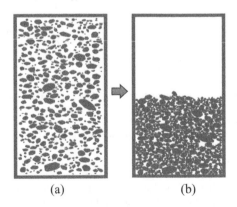

(a)   (b)

Figure 2. Generation of PFC microstructures using gravimetric methods: (a) MG set of aggregates and (b) PFC microstructure generated using the DE LMGC90 software.

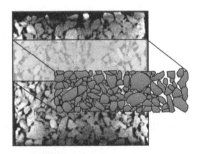

Figure 1. Microstructure geometry extracted from a PFC sample using X-Ray CT techniques.

obtained from non-destructive techniques underestimates the actual skeleton strength of the sample (i.e. the stone-on-stone contact network).

On the other hand, the main advantage of the generation of PFC microstructures using gravimetric methods is the inclusion of mechanical principles and the possibility of easily and randomly creating multiple replicates of the same mixture. However, its calibration is demanding, and it has a high computational cost.

3 SENSITIVITY ANALYSIS OF THE GRAVIMETRIC METHOD

Before subjecting a set of loose aggregates to gravimetric forces using the DE LMGC90 software, there are several input parameters that will affect the final microstructural characteristics of the PFCs.

Considering the work of Torres and Caro (2016), in which initial PFC microstructures were created in LMGC90 and some input parameters were evaluated, the authors chose the following parameters for the current sensitivity analysis:

1. Drop height of the aggregates.
2. Friction among the aggregates.
3. Restitution of the aggregates (i.e. ability to bounce once they touch other entity).

In all cases, the gradation used to generate the initial set of aggregates with the MG was that of a typical PFC-5 type of PFC specified by the Florida Department of Transportation (FDOT), commonly used in the United States.

Table 1 presents the values of the input parameters selected for this analysis. The analysis consisted in changing separately each one of the input parameters selected and assessing the results. The values in bold and italic are the control parameters. For example, when the drop height is changed, the friction and restitution values remain constant at 0.5 and 0.3.

To capture the heterogeneity of the PFC samples, ten different replicates were simulated per case. The PFC microstructures obtained through this method were analyzed in terms of two different parameters: 1) Air Void Content (AV), and 2) the Coordination Number (CN), defined as the average number of contacts per aggregate, which is considered a good indicator of network connectivity of the PFC microstructure.

Table 1. Input parameter values for the sensitivity analysis.

Drop height [mm]	Friction [–]	Restitution [–]
1	0.1	None
2	0.35	0.1
5	*0.5*	0.2
10	0.7	*0.3*

4 ANALYSIS OF RESULTS

A total of 120 simulations were evaluated for this analysis: 12 cases of analysis (i.e. drop height, friction and restitution) with 10 replicates per case. The results are presented in Figures 3 and 4.

An initial observation from these figures is that the variability among the 10 replicates per case is considered small, with a Coefficient of Variability (COV) for the AV and CN parameters lower than 5% in all the cases. This is partially because, even though the MG creates random sets of aggregates, they have the same morphological characteristics.

In terms of the effect of the input parameters on the final AV content of PFC microstructures, Figure 3a shows that AV decreased with an increase in the drop height of the aggregates. In fact, from 1 mm of height drop of the aggregates to 10 mm, the aggregates passed from having an AV of 21.7% to an AV of 13.2%, in average (a reduction of 39.2%). From Figure 3b, it is observed that the final AV of the PFC microstructures

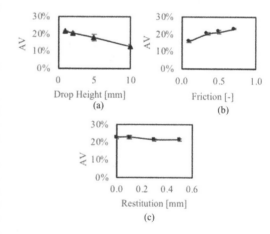

Figure 3. Effect of (a) drop height, (b) friction and (c) restitution on the final AV of the PFC microstructures.

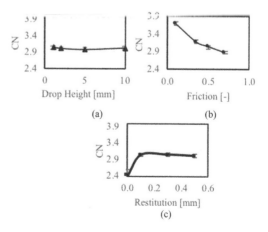

Figure 4. Effect of (a) drop height, (b) friction and (c) restitution on the CN of the PFC microstructures.

increased as the friction input parameter increased. With a friction coefficient of 0.1, the final AV content of the PFCs was, in average, 16.2%, and with a friction of 0.7, the final AV content of the PFCs was 23.6% (an increase of 31.4%). This result was expected, since higher values of friction increase the difficulty of the final arrangement of aggregates. In terms of the restitution input parameter, it was found that it does not affect the final AV of the PFCs (Figure 3b).

Regarding the effects of the input DE parameters on the CN of the PFC microstructures, the drop height of the aggregates did not impact the CN (Figure 4a). In fact, a change in the drop height from 1 to 10 mm, produced a difference in of only 0.4% in the CN. On the contrary, Figure 4b shows that the friction parameter strongly affects the CN of the PFCs: when the friction passed from 0.1 to 0.7, the CN of the PFCs decreased in 22.2%. This is related with the effect of this input parameter on the final AV of the PFCs, and the fact that higher AV contents usually lead to lower CN; in fact, there is a linear inverse correlation of 0.99 between the AV and CN of the PFCs with changes in the friction coefficient.

Figure 4c shows that when there is no restitution, the CN of the PFCs is 20.5% lower than with any other value. This is because when the aggregates are able to bounce, they can organize better than when they are not able to do so. In general, the there is no major differences (<1%) between the different restitution values and the resulting CN values.

In average, a PFC microstructure obtained from non-destructive techniques have a CN of 3.3. Using gravimetric methods with a friction of 0.35, the authors were able to obtain a CN of 3.2. However, PFC microstructures obtained from non-destructive techniques tend to underestimate the stone-on-stone contact network. For this reason, the input parameters generated through DE gravimetric methods should be mechanically calibrated. This means that the randomly generated PFC microstructure should have an equivalent mechanical behavior as that of a 3D PFC microstructure in order to be a proper representation of the mixture in 2D. The authors are currently working on this activity.

5 CONCLUSIONS

This paper discussed the generation of PFC microstructures through X-Ray CT and gravimetric methods, and the effect of three input parameters (i.e. drop height of the aggregates, friction and restitution) of the gravimetric method on the final AV content and CN of the PFC microstructures. The main conclusions are:

- The COV among the 10 replicates was lower than 5% for the AV and CN, which is mainly due to the same morphological characteristics and size distribution of the coarse aggregates.

- The drop height of the aggregates presented an inverse relation with the AV content of the aggregates, and this input parameter did not affect the final CN of the PFCs.
- The friction coefficient had a direct relation with AV, and an inverse relation with the CN of the PFCs. There was a correlation of −0.99 between the AV and the CN with changes in the friction.
- Restitution parameters higher than 0.0 did not impact the AV and CN of the PFCs. However, this parameter should be higher than 0.0 to obtain CN values near or higher than 3.0 and have a better network connectivity.

The random generation of PFCs microstructures trough gravimetric methods seems to be a good alternative to represent these mixtures in 2D. However, the resulting microstructures should be calibrated to guarantee that they have an equivalent mechanical behavior as that of 3D PFC mixtures.

REFERENCES

Alvarez, A. E. et al. (2010) 'Stone-on-Stone Contact of Permeable Friction Course Mixtures', Materials in Civil Engineer, 22(11), pp. 1129–1138.

Castillo, D. et al. (2015) 'Studying the effect of microstructural properties on the mechanical degradation of asphalt mixtures', Construction and Building Materials, 93, pp. 70–83.

Cooley, L. A. et al. (2009) 'Construction and Maintenance practices for Permeable Friction Courses', NCHR Report 640 – Project 09-41. The National Academies press: Washington, DC.

Haft-Javaherian, M. (2011) Virtual Generation of Asphaltic Mixtures. MSc. Thesis, University of Nebraska-Lincoln, Lincoln, Nebrasaka.

Kandhal, P. (2002) Design, Construction and Maintenance of Open-Graded Asphalt Friction Courses, Information Series No. 115, National Asphalt Pavement Association (NAPA). Lanham, MD.

Manrique-Sanchez, L. and Caro, S. (2019) 'Numerical assessment of the structural contribution of porous friction courses (PFC)', Construction and Building Materials, 225, pp. 754–764. doi: 10.1016/j.conbuildmat.2019.07.200.

Manrique-Sanchez, L., Caro, S. and Arámbula-Mercado, E. (2018) 'Numerical modelling of ravelling in Porous Friction Courses (PFC)', Road Materials and Pavement Design, 19(3), pp. 668–689.

Mo, L. T. et al. (2007) 'Investigation into stress states in porous asphalt concrete on the basis of FE-modelling', Finite Elements in Analysis and Design, 43, pp. 333–343.

Mo, L. T. et al. (2008) '2D and 3D meso-scale finite element models for ravelling analysis of porous asphalt concrete', Finite Elements in Analysis and Design, 44, pp. 186–196.

Torres, S. F. and Caro, S. (2016) Comparación entre metodologías para la generación de microestructuras de mezclas asfálticas porosas. Msc. Thesis, Universidad de los Andes, Bogotá, Colombia.

Advances in Materials and Pavement Performance Prediction II – Kumar et al. (eds)
© 2021 Taylor & Francis Group, London, ISBN 978-0-367-46169-0

Development of a damage growth rate-based fatigue criterion

R. Nemati
Department of Transportation, AECOM

E.V. Dave & J.E. Sias
Department of Civil and Environmental Engineering, University of New Hampshire, USA

ABSTRACT: The simplified viscoelastic continuum damage (S-VECD) theory to has gained wide-spread attention as a tool to characterize the fatigue behavior of asphalt concrete. One major challenge in using the S-VECD approach is determination of the crack localization point to compare mixtures' fatigue performance. In addition, the currently available index parameters from this approach are not highly correlated to the field performance as standalone parameters and they need to be paired with mechanistic pavement modeling to discriminate between performance of different asphalt mixtures. In this study, the three main components of the S-VECD approach: number of cycles to failure (N_f), accumulated decrease in pseudo stiffness ($\int_0^{N_f} (1 - C)dN$), and damage parameter (S) were used to develop a new damage growth rate-based fatigue criterion. An index parameter based on this approach was found to be highly correlated to the field fatigue performance for six mixtures.

1 INTRODUCTION AND BACKGROUND

One of the main outcomes of simplified viscoelastic continuum damage (S-VECD) theory and analysis is the damage characteristic curve (DCC), which describes a fundamental material relationship between material integrity, as pseudo stiffness (C), and accumulated damage (S) for a material. The DCC for asphalt mixtures has been shown to be temperature and mode of loading independent (Daniel & Kim 2002).

In order to develop the DCC curves, the direct tension cyclic fatigue test in accordance with AASHTO TP 107 should be conducted. However, the identification of a fatigue cracking performance index property that corresponds to the crack localization in the mixtures has been a major challenge in use of S-VECD approach within performance based mixture designs and specifications. With regards to the test results, different fatigue criteria have been proposed and used by researchers.

For example, to discriminate the mixtures' lab performance Sabouri and Kim proposed the rate of averaged dissipated pseudo strain energy (G^R) parameter (Sabouri & Kim 2014) in Equation 1.

$$G^R = \frac{\int_0^{N_f} W_C^R}{(N_f)^2} \tag{1}$$

Where: $\int_0^{N_f} W_C^R$ is the accumulated dissipated pseudo strain energy up to failure. Failure is defined as the

point when phase angle decreases, as per AASHTO specification.

The G^R parameter has a linear relationship with number of load cycles to failure (N_f) on a log-log scale and is independent of loading mode and temperature. The value of N_f at $G^R = 100$ is used as an index parameter to differentiate the mixture's performance. However, this parameter may not be reliably able to predict the mixture's field performance due to the variability in S-VECD test results and required extrapolation in log scale to match the field strain levels. This can result in significant errors and misprediction of the performance (Wang & Kim 2017).

To mitigate the shortcomings of the G^R parameter, Wang proposed the averaged reduction pseudo stiffness up to failure (D^R) criterion in Equation 2 (Wang 2017):

$$D^R = \frac{\int_0^{N_f} (1 - C)dN}{N_f} \tag{2}$$

Where: $\int_0^{N_f} (1 - C)dN$ is the accumulated decrease in pseudo stiffness up to failure. Lab test results indicated that the D^R parameter is a material property that is independent of testing conditions (Wang & Kim 2017). In general, mixtures with higher D^R value are considered to be more fatigue resistant. However, mixtures with significantly different DCCs could result in similar D^R values as the total damage (S_f) that different materials undergo before crack localization is not considered (Nemati et al. 2019). Therefore, D^R may

Table 1. Study mixture properties.

Mixture	Binder Grade	RAP (%)	AC (%)	Field Air Void (%)
Virgin-58-28	58-28	0	5.9	5.4
15%RAP-58-28	58-28	15	5.6	5.3
25%RAP-58-28	58-28	25	5.8	5.9
25%RAP-52-34	52-34	25	6.0	5.3
30%RAP-52-34	52-34	30	6.0	6.2
40%RAP-52-34	42-34	40	4.9	4.5

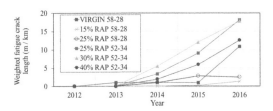

Figure 1. Weighted field fatigue cracking (Normalized with respect to section length).

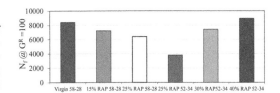

Figure 2. $N_f@G^R = 100$ Fatigue failure criteria.

not be able to discriminate the fatigue performance of mixtures as a standalone parameter. The apparent damage parameter (S_{app}) (Equation 3) has been recently proposed to overcome this deficiency by combining the effects of modulus and toughness in determining cracking susceptibility (Wang et al. 2018). This parameter is defined as the damage accumulation (S) when pseudo-stiffness (C) is equal to 1-D^R; a higher S_{app} value is more desirable.

$$S_{app} = \frac{1}{10000} \times \left(\frac{1}{C_{11}} \times D_R \right)^{\frac{1}{C_{12}}} \quad (3)$$

Where: C_{11} and C_{12} are the model coefficients (determined from DCC curve fitting)

Development of a reliable fatigue criterion to differentiate the mixtures' performance is critically important to design and place longer lasting mixtures. The objective of this study is to evaluate the discriminability of the aforementioned existing parameters and explore a new S-VECD based fatigue criterion that may be better correlated with the field performance of asphalt mixtures as a standalone parameter.

2 MATERIALS AND RESEARCH APPROACH

A set of six mixtures for which the field performance data is available was selected for evaluation. The 12.5 mm nominal maximum aggregate size mixtures were placed adjacently in the same cross sections on an interstate highway in 2011. The mixture design and properties are summarized in Table 1. The S-VECD analysis was performed on one-year old field cores tested in accordance with AASHTO TP 107 standard. In the table, RAP is the recycled asphalt pavement by weight of the mix and AC is the total asphalt content.

3 FIELD PERFORMANCE

Field performance of the sections has been monitored yearly since construction using an automated pavement data collection van by NHDOT (Daniel et al. 2019). The amount of fatigue cracking in each section is shown in Figure 1.

4 RESULTS OF THE DIRECT TENSION CYCLIC FATIGUE TEST

The mixtures are ranked and evaluated with respect to the three S-VECD failure criterion: (i) $N_f@G^R = 100$; (ii) D^R; and, (iii) S_{app}, in Figures 2–4, respectively. The ranking of mixtures from the three indices are quite different. The 40% RAP 52-34 is shown to have best fatigue performance with respect to N_f at $G^R = 100$, one of the lowest D^R values, and is ranked as the third best mixture using S_{app} but is the 4th ranked mixture with respect to field performance. Similar observations can be made for other mixtures, indicating that none of the indices have been able to rank the relative field performance of these mixtures. One main reason for this observation could be that with respect to continuum damage mechanics, it is the evolution and localization of micro-cracks that results in macro-cracks that form fatigue cracking. The magnitude and number of micro-cracks in the S-VECD analysis is quantified by the (S) value where neither the G^R nor D^R parameter explicitly take this into account. The S_{app} parameter incorporates the magnitude of the damage by considering the damage at an average integrity level of the mixture (C at $1 - D^R$). However, since the accumulation of damage as well as decrease in capacity is a nonlinear phenomenon, the use of average C and corresponding S value may not be an appropriate indicator to relate to field performance.

5 DEVELOPMENT OF THE DAMAGE-GROWTH RATE BASED FATIGUE CRITERION

For a given test specimen in the direct tension cyclic fatigue test, the decrease in pseudo stiffness (C) can be

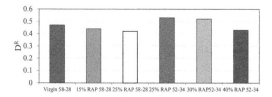

Figure 3. DR Fatigue failure criteria.

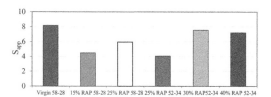

Figure 4. S_{app} Fatigue failure criteria.

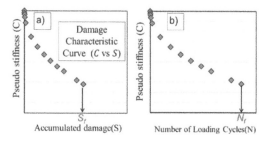

Figure 5. a) Pseudo stiffness versus damage accumulation (C vs S), b) Pseudo stiffness versus loading cycle (C vs N).

represented with respect to the accumulation of damage (C vs. S) in Figure 5(a), or with respect to loading cycles (C vs. N) in Figure 5(b). The area above the C vs. N curve indicates the accumulated decrease in material's capacity as discussed and used in the D^R criterion (Wang & Kim 2017).

In this study, the correlations between these three components of S-VECD theory: $\int_0^{N_f} (1 - C)dN$, N_f and S_f, were investigated to develop a new fatigue criterion. The Pearson's correlation coefficient matrix (Table 2) was used to determine the possible relationships between the three parameters. As the N_f decreases, the magnitude of the accumulated damage at failure increases. This correlation indicates that, for the majority of tested replicates for a mixture, a higher-level strain in the test will result in a higher amount of accumulated damage failure. In other words, at higher cyclic strains, the rate of development of micro-cracks and their localization to form a macro-crack is sufficiently high that the mixture is not able to utilize its full capacity to evenly disperse damage throughout the continuum to resist failure. This phenomenon is similar to a thermal shock occurrence for many other types of materials including asphalt mixtures where a sudden change in temperature results in premature cracks

Table 2. Pearson's correlation coefficients for the S-VECD based parameters.

S-VECD based parameters	N_f	S_f	$\int_0^{N_f} (1 - C)dN$
N_f	1.00	–	–
S_f	−0.53	1.00	–
$\int_0^{N_f} (1 - C)dN$	1.00	−0.53	1.00

in the material before the material is able to reorganize its microscopic or even molecular structure to accommodate the thermal strains. These observations indicate that not only is the magnitude of the damage at failure important, but the rate of increase in damage growth (governed by the strain levels at cyclic loading) along with reduction of pseudo-stiffness are important factors that should be taken into consideration in the development of a mixture fatigue performance index.

A new fatigue criterion based on damage growth rate is proposed in Equation 4.

$$C_{N_f}^S = \frac{\int_0^{N_f} (1 - C)dN}{S_f} \times m \qquad (4)$$

Where: $C_{N_f}^S$: Damage growth rate-based fatigue criterion,
m: Unit correction factor set to 10^3.

A higher S_f and lower $\int_0^{N_f} (1 - C)dN$ are more desirable for fatigue performance as they indicate that material is able to withstand higher amounts of damage with less disintegration, therefore a lower $C_{N_f}^S$ value is more desirable.

6 COMPARISON OF INDEX PARAMETERS

The $C_{N_f}^S$ versus number of cycles to failure (N_f) relationships are shown in Figure 6. The results indicate a linear relationship in arithmetic scale between $C_{N_f}^S$ and N_f as the best line of fit between the data points. This type of relationship minimizes extrapolation errors that may occur in a logarithmic scale. A smaller slope of the fitted trend line is more desirable. For the purposes of simplicity in applying the $C_{N_f}^S$ parameter to compare mixtures, the value of N_f @ $C_{N_f}^S$ =100 is suggested to be used in this study. The ranking order from all the available failure criterions with respect to field performance is presented in Table 3. According to the rankings the N_f @ $C_{N_f}^S$ =100 parameter has been able to reliably rank all 6 mixtures with respect to field observations, while other parameters such as N_f @ G^R =100 and S_{app} have only identified the worst mixture. The results from the comparisons indicate promise of the newly proposed fatigue criterion in discriminating the mixtures performance with respect to

305

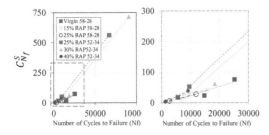

Figure 6. $C_{N_f}^S$ versus N_f plots.

Table 3. Mixture ranking using different fatigue criterion.

| | Ranking with respect to different parameters (1; best, 6; worst) | | | | |
Mixture	Field Rank (5 year after construction)	$N_f @ C_{N_f}^S = 100$	$N_f @ C_{N_f}^S = 100$	D^R	S_{app}
Virgin 58-28	3	3	3	3	1
15% RAP 58-28	1	1	4	4	5
25% RAP 58-28	2	2	5	6	4
25% RAP 52-34	6	6	6	1	6
30% RAP 52-34	5	5	3	2	2
40% RAP 52-34	4	4	1	5	3

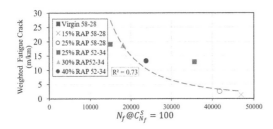

Figure 7. Weighted crack length vs $N_f @ C_{N_f}^S = 100$.

weighted normalized field crack length 5 years after construction.

The normalized fatigue crack lengths at the fifth year postconstruction were plotted versus the $N_f @ C_{N_f}^S = 100$ parameter to determine the statistical correlation in terms of the R^2 goodness of fit parameter for the data. As it is shown in Figure 7, a power function was fitted to the data which resulted in an $R^2 = 0.73$ indicating that the index parameter is capable of predicting the magnitude of cracking in the field in addition.

7 SUMMARY, CONCLUSION AND FUTURE EXTENSIONS

In this study, a new fatigue criterion called $C_{N_f}^S$ was developed and investigated. This criterion incorporates three components of the S-VECD theory ($\int_0^{N_f} (1 - C)dN$, N_f and S_f) to capture the mixture's disintegration with respect to damage growth rate. The $C_{N_f}^S$ is calculated for each tested replicate (minimum of two strain levels are required) and the results are plotted versus the number of cycles to failure (N_f). An index parameter of $N_f @ C_{N_f}^S = 100$ is selected to rank mixtures and is shown to discriminate the field performance of the test mixtures. Additional research is required to confirm that the proposed parameter is applicable to all types of mixtures and traffic levels and to determine the appropriate threshold value for use in performance engineered based mixture design approaches.

REFERENCES

Daniel, J. S., & Kim, Y. R. (2002). Development of a simplified fatigue test and analysis procedure using a viscoelastic continuum damage model. *Journal of the Association of Asphalt Paving Technologists*, 71, 619–650.

Daniel, J. S., Corrigan, M., Jacques, C., Nemati, R., Dave, E. V., & Congalton, A. (2019). Comparison of asphalt mixture specimen fabrication methods and binder tests for cracking evaluation of field mixtures. *Road Materials and Pavement Design*, 20(5), 1059–1075.

Nemati, R., Dave, E. V., Sias, J. E., Thibodeau, E. S., & Worsman, R. K. (2019). Evaluation of Laboratory Performance and Structural Contribution of Cold Recycled Versus Hot Mixed Intermediate and Base Course Asphalt Layers in New Hampshire. Transportation Research Record, 0361198119844761.

Sabouri, M., & Kim, Y. R. (2014). Development of a failure criterion for asphalt mixtures under different modes of fatigue loading. *Transportation Research Record*, 2447(1), 117–125.

Wang, Y. D. Keshavarzi, B., & Kim, Y. R. (2018). Fatigue performance analysis of pavements with RAP using viscoelastic continuum damage theory. *KSCE Journal of Civil Engineering*, 22(6), 2118–2125.

Wang, Y., & Kim, Y. R. (2017). Development of a pseudo strain energy-based fatigue failure criterion for asphalt mixtures. *International Journal of Pavement Engineering*, (2017), 1–11.

Advances in Materials and Pavement Performance Prediction II – Kumar et al. (eds)
© 2021 Taylor & Francis Group, London, ISBN 978-0-367-46169-0

Constitutive properties of materials from digital image correlation

Gabriel Nsengiyumva & Yong-Rak Kim
Texas A&M University, Texas, USA

ABSTRACT: This study proposes a method to characterize constitutive behavior and properties of materials by integrating non-linear optimization of local displacement data from DIC (digital image correlation) and finite element modeling (FEM). Toward that end, the Neader-Mead algorithm was used to minimize an objective function constructed from multiple discrete points from DIC and FEM simulation. The developed method was then used to determine constitutive properties of LE (linear elastic) and LVE (linear viscoelastic) materials. Polyetheretherketone (PEEK) and fine aggregate matrix (FAM) were selected for the LE and LVE investigations, respectively. The DIC-FEM method herein can converge to optimal solution in less than 60 iterations for both LE and LVE materials. The integrated DIC-FEM method can vastly improve testing efficiency by reducing the number of replicates and also provide a better understanding of localized behavior of materials such as cracking and damage, in particular of inelastic heterogeneous materials and composites such as asphaltic materials.

1 INTRODUCTION

Traditionally, mechanical properties of materials are derived from global results (i.e., load and displacement) that generalize material behavior over the entire specimen. As a result, several tests in different modes are required to reliably obtain material properties from global results. Fundamentally, global data are not unique given a material and testing conditions which require multiple tests to improve certainty of the results. For example, tension and shear testing are required to identify all elastic material constants using the relationship between the elastic and shear moduli (Boresi et al., 1985).

Local deformation data provide more information that can be used to complement the global data to more effectively characterize material properties. There are several methods of collecting the local data such as: strain gauges, LVDTs, photo-elasticity and the digital image correlation (DIC). While other methods typically measure deformation of a small location on a testing specimen, DIC can provide a continuous contour of deformation data over the entire specimen. As a result, DIC can be effective to examine constitutive behavior and deformation of materials in particular of heterogeneous materials.

2 DIC METHOD

The underlining principle of DIC involves taking successive images in time of a specimen under loading and deducing field deformations data by correlating (i.e., back-calculate) deformations. The correlation function is based on kinematics and assumes continuity between consecutive images.

A common subset DIC monitors a subset of pixels S in the reference image which becomes $\widetilde{S}$ in the current image. Inside the subset, mapping between a point S_P located at (x, y) inside the reference image and $\widetilde{S}_P$ located at $(\widetilde{x}, \widetilde{y})$ in the current image is achieved by equation (1):

$$\widetilde{x} = x + u(x, y)$$
$$\widetilde{y} = y + v(x, y) \tag{1}$$

where u and y are the displacement components which can be approximated by Taylor series expansion around a point (x_o, y_o) such as:

$$\widetilde{x} = x_o + u_o + u_x \Delta x + u_y \Delta y + \tfrac{1}{2} u_{xx} \Delta x^2$$
$$\qquad + \tfrac{1}{2} u_{yy} \Delta y^2 + u_{xy} \Delta x \Delta y$$
$$\widetilde{y} = y_o + v_o + v_y \Delta y + v_x \Delta x + \tfrac{1}{2} v_{yy} \Delta y^2 \tag{2}$$
$$\qquad + \tfrac{1}{2} v_{xx} \Delta x^2 + v_{xy} \Delta x \Delta y$$

where $\Delta x = (x - x_o)$ and $\Delta y = (y - y_o)$.

From equation (2), twelve parameters need to be identified $\{u_o, v_o, u_x, v_x, u_y, v_y, u_{xx}, v_{xx}, u_{yy}, v_{yy}, u_{xy}, v_{xy}\}$ which are components of the displacements at (x_o, y_o) the first order gradients and the second order gradients.

The gray-scale values of the reference and current images at mapped points are $g(S_p)$ and $\widetilde{g}(\widetilde{S}_p, \lambda)$, respectively. The correlation function C is:

$$C = \frac{\sum\limits_{S_p \in S} \left\{ g(S_p) - \widetilde{g}(\widetilde{S}_p, \lambda) \right\}^2}{\sum\limits_{S_p \in S} g^2(S_p)} \tag{3}$$

where: λ with ω being an additional parameter to account for offset in the gray-scale value

Equation (3) can be solved by the Newton-Raphson under consideration that at the solution the gradient $\nabla C(\lambda) = 0$. The problem then becomes an iterative loop of the form shown in equation (4)

$$\nabla \nabla C(\lambda_o)(\lambda - \lambda_o) = -\nabla C(\lambda_o) \qquad (4)$$

where λ_o is the initial guess, λ is the next approximate value and $\nabla \nabla C$ is the hessian matrix which can be approximated by assuming very small difference between $g(S_p)$ and $\widetilde{g}(\widetilde{S}_p, \lambda)$.

This subset DIC methods have been used to obtain measurement resolutions of 4.8 nm (Vendroux & Knauss, 1998) and 0.066 pixel (Lu & Cary, 2000). The high resolution of DIC is achieved through interpolation techniques such as bicubic spline function. It is noteworthy that convergence of DIC requires uniqueness of the subsets which can be achieved by applying a random, contrasting, isotropic and non-repetitive speckle pattern on the surface of the specimen.

To deduce material properties, displacement fields from DIC and FEM (finite element modeling) can be minimized using an optimization per equation (5):

$$\min_{\chi \in \mathrm{R}^N} \Psi_u(\chi) = \|\mathbf{u}_{DIC} - \mathbf{u}_{FEM}(\chi)\|_2 \qquad (5)$$

where, $\Psi_u(\chi)$ is the objective function, χ is a vector of size N material properties, $\mathbf{u}_{DIC}$ is experimental displacement results from DIC, $\mathbf{u}_{FEM}(\chi)$ is displacement from FEM at χ and, $\|\cdot\|_2$ is the Euclidian norm.

Although most engineering materials are time and temperature dependent, the DIC-FEM material characterization method have been mostly applied to elastic materials. There is a need of a DIC-FEM method applicable to time- and temperature-dependent viscoelastic materials such as asphaltic materials.

This study hypothesizes that kinematic deformation data from DIC can be integrated with a FEM framework to identify properties of simple elastic to more complex viscoelastic materials via an optimization technique.

The main objective of this study is to develop such FEM-DIC method based on kinematic data sets to obtain constitutive properties of both elastic and viscoelastic materials.

3 METHODOLOGY AND ALGORITHM DEVELOPMENT

3.1 Methodology

To achieve the goal stated above, the methodology shown in Figure 1 was adopted in this study. First, a laboratory test (e.g., three-point bending beam) is conducted and the DIC is used to measure surface deformation. Subsequently, the laboratory test is simulated with its finite element modeling.

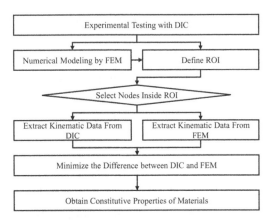

Figure 1. Research methodology used in this study.

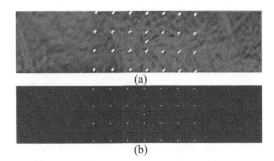

Figure 2. Exact points matching: (a) DIC and (b) FEM.

The objective function is then calculated using equation (5) by comparing both vertical and horizontal deformations resulting from testing and FE simulation.

Although conventional methods such as the Newton-Raphson can be used to solve equation (5), N-M (Nelder-Mead) solver is preferred due to its robustness and derivative-free nature (Lagarias et al., 1998). Due to the unconstrained nature of N-M, the penalty method which assigned a large number (e.g., 10^{20}) to the function value is used to enforce optimization constrains. The final objective function to be optimized is shown in equation (6).

$$\min_{\chi \in \mathrm{R}^N} \Psi_u(\chi) = \begin{cases} \|\mathbf{u}_{DIC} - \mathbf{u}_{FEM}(\chi)\|_2 & \text{for } \chi \in \Omega_\chi \\ 10^{20} & \text{for } \chi \notin \Omega_\chi \end{cases} \qquad (6)$$

Equation (6) is solved using N-M algorithm implemented in fminsearch function of Matlab™ software. Matlab™ calculates the objective function for a discrete matching location inside ROI (region of interest) as exemplified in Figure 2. Measurements from DIC are matched with FE model simulation results by assuming target material properties. The matching process is repeated by altering material properties until both agrees well. For the FE model simulation, a commercial FE package ABAQUS® was used in this study.

308

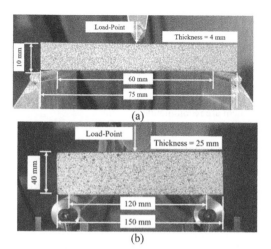

(a)

(b)

Figure 3. Testing set-ups for: (a) PEEK and (b) FAM.

4 MATERIALS AND TESTING SET-UP

Polyetheretherketone (PEEK) which is thermoplastic semi-crystalline polymer (El-Qoubaa & Othman, 2015) was selected as an elastic material to prove the concept and method.

For linear viscoelastic case, FAM (fine aggregate matrix) which is a viscoelastic component of AC (asphalt concrete) was adopted. FAM contains aggregates smaller than 1.19 mm and asphalt cement (Im et al., 2014).

Testing set-ups used for linear elastic PEEK and linear viscoelastic FAM are shown in Figure 3(a) and 3(b), respectively. A preliminary investigation was conducted to determine proper geometries that minimize load-point effect for both materials. In particular, it was found that a thickness less than 25 mm can result in localized high-compression area around the loading point in FAM.

5 RESULTS AND DISCUSSION

5.1 Linear elastic case

A loading rate of 10 mm/min was applied on the PEEK specimen for a duration of 20 seconds. A FE model simulation was also conducted after mesh convergence study which converged at 0.5mm mesh size using 3-node linear elements (CPS3). Boundary conditions were identical to the testing.

For isotropic and linearly elastic material in isothermal conditions, the constitutive equation is shown in equation (7):

$$\begin{cases} \sigma_{ij} = \lambda \varepsilon_{kk} + 2\mu \varepsilon_{ij} \\ \lambda = \dfrac{E\nu}{(1+\nu)(1-2\nu)}, \text{ and } \mu = \dfrac{E}{2(1+\nu)} \end{cases} \quad (7)$$

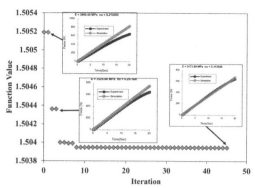

Figure 4. DIC-FEM results for PEEK: function evolution with corresponding global results (load vs. time).

where; λ and μ are Lamé constants, E and ν are the Young's modulus and Poisson's ratio. As known, only two constitutive properties (i.e., E and ν) are needed to characterize LE materials.

The results of the DIC-FEM for PEEK is shown in Figure 4. The evolution of function value for each iteration is shown along with corresponding global-level response (force vs. loading time). The stopping criteria were the relative error of the function and the variable to be less than 10^4 (i.e., error <=1e-4).

Figure 4 shows that the function converged relatively fast in the beginning before saturating towards the solution of $E = 3{,}173.69$ MPa and $\nu = 0.3135$ (Jaekel et al., 2011; Kurtz, 2019). The predicted PEEK material properties (E and ν) are satisfactorily similar to those found in litterature, which confims the capability of the method

5.2 Linear viscoelastic case

Creep testing was conducted for a duration of 3,000 seconds to determine LVE properties of FAM. Preliminary testing was conducted at increasing creep loads (e.g., 2, 4, 8, 16,...) in N to determine the level of creep load for LVE behavior in which the proportionality between load and middle point deflection was preserved (Wang et al., 1997). Consequently, 2.5N was determined to be the level of load that does not create nonlinear viscoelastic deformation of the FAM.

The uniaxial relaxation modulus $E(t)$ can be expressed in Prony series as such:

$$E(t) = E_\infty + \sum_{i=1}^{n} E_i e^{-t/\rho_i} \quad (8)$$

where E_∞ and E_i are spring constants in the generalized Maxwell model, t is loading time ρ_i is the relaxation time and n is the number of Maxwell units.

Poisson's ratio of FAM was assumed constant with a value of 0.35 (Im et al., 2013) due to its minimal effect to the beam deflection. The constant Poisson's ratio is considered a reasonable assumption since the FAM

309

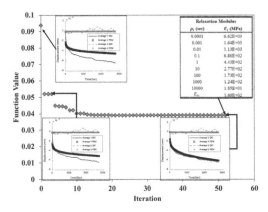

Figure 5. DIC-FEM results for LVE showing function evolution with corresponding overall average displacement results.

test conditions will not reach (near-) glassy domain (Sharpe, 2008).

Figure 5 shows the DIC-FEM results of FAM. As shown, the objective function rapidly converges iin the beginning to reach the optima at 57 iterations. When the target LVE properties were found through the nonlinear optimization, global results (deflection vs. time) at different locations in the beam show a good agreement between DIC measurements and FE simulation.

6 CONCLUSIONS

This study developed a DIC-FEM method to characterize constitutive material properties using nonlinear optimization. The method was then applied to linear elastic (PEEK) and linear viscoelastic (FAM) materials to prove the concept and approach. Based on the results, the following can be concluded:

- DIC results can be successfully integrated with FEM kinematic data.
- The developed method was robust by converging to optimal solution in less than 60 iterations for both LE and LVE cases.

- Obtained material properties from DIC-FEM successfully predicted the global-level responses.
- The method can be extended to more complicated cases such as heterogeneous materials and materials with fracture, which are in progress by the authors.
- Future studies about applying the developed DIC-FEM method outside controlled laboratory environment are recommended.

REFERENCES

Boresi, A. P., Schmidt, R. J. & Sidebottom, O. M. 1985. *Advanced mechanics of materials*, Wiley New York et al.
El-qoubaa, Z. & Othman, R. 2015. Characterization and modeling of the strain rate sensitivity of polyetheretherketone's compressive yield stress. *Materials & Design (1980–2015)*, 66, 336–345.
Im, S., Ban, H. & Kim, Y.-R. 2014. Characterization of mode-I and mode-II fracture properties of fine aggregate matrix using a semicircular specimen geometry. *Construction and Building Materials*, 52, 413–421.
Im, S., Kim, Y.-R. & Ban, H. 2013. Rate-and temperature-dependent fracture characteristics of asphaltic paving mixtures. *Journal of Testing and Evaluation*, 41, 257–268.
Jaekel, D. J., Macdonald, D. W. & Kurtz, S. M. 2011. Characterization of PEEK biomaterials using the small punch test. *Journal of the mechanical behavior of biomedical materials*, 4, 1275–1282.
Kurtz, S. M. 2019. *PEEK biomaterials handbook*, William Andrew.
Lagarias, J. C., Reeds, J. A., Wright, M. H. & Wright, P. E. 1998. Convergence properties of the Nelder–Mead simplex method in low dimensions. *SIAM Journal on optimization*, 9, 112–147.
Lu, H. & Cary, P. 2000. Deformation measurements by digital image correlation: implementation of a second-order displacement gradient. *Experimental mechanics*, 40, 393–400.
Sharpe, W. N. 2008. *Springer handbook of experimental solid mechanics*, Springer Science & Business Media.
Vendroux, G. & Knauss, W. 1998. Submicron deformation field measurements: Part 2. Improved digital image correlation. *Experimental Mechanics*, 38, 86–92.
Wang, C., Yang, T. & Lam, K. 1997. Viscoelastic Timoshenko beam solutions from Euler-Bernoulli solutions. *Journal of engineering mechanics*, 123, 746–748.

Advances in Materials and Pavement Performance Prediction II – Kumar et al. (eds)
© 2021 Taylor & Francis Group, London, ISBN 978-0-367-46169-0

Impacts of unblended reclaimed asphalt on volumetric mixture design

S.E. Pape & C.A. Castorena
Department of Civil, Construction, and Environmental Engineering

ABSTRACT: Increasing reclaimed asphalt pavement (RAP) contents in asphalt mixtures has economic bene-fits, however many assumptions which simplified the usage of lower RAP contents break down at higher RAP contents. One of the most fundamental is the assumption of complete and uniform blending between RAP and virgin materials. Energy dispersive spectroscopy (EDS) has been trialed in several previous studies to assess the blending between RAP and virgin binders and is used in this work to assess blending variability across a sample surface. The implications of incomplete blending on volumetric mixture design parameters is discussed as volumetric properties are used for mixture acceptance.

1 INTRODUCTION

There has been significant interest in the incorpora-tion of high amounts of reclaimed asphalt pavement (RAP) in new asphalt mixtures due spikes in asphalt prices in 2006 and 2008, combined with difficulty sourcing quality aggregates in many areas (Copeland 2011). There are many challenges to increasing RAP contents in asphalt mixtures, one of the most complex being uncertainty in the blending between recycled and virgin binders. Uncertainty in blending is not a new issue; it was noted as far back as 1980 (Epps et al. 1980). Most RAP mix design methods assume com-plete binder blending, including the Superpave method (Epps et al. 1980; AASHTO R 35). This assumption simplifies the design procedure and does not appear to pose a problem in mixtures with low RAP contents. McDaniel and Anderson (2001) concluded that RAP does not behave as a black rock; however, they were also not convinced that complete binder blending ever occurs. It is likely that not all the binder in the recy-cled material is contributing to the mixture, and thus the resulting mixture may have insufficient asphalt if a volumetric mix design procedure is used with an assumption of complete blending (Stroup-Gardiner 2016).

Evaluating the blending of binders is a challenge in recycled mixtures as the two binders are indis-tinguishable under most conditions. Typical paving asphalt binders are black and opaque, so the virgin and recycled are not visually distinguishable. Different methods to infer blending have been attempted, how-ever none have been widely accepted. It is inferred that this is due to the assumptions and limitations inherent to the various approaches that have been tried.

Lee et al. (1983) were the first to attempt to measure blending within recycled mixtures using a scanning electron microscope (SEM) with an energy dispersive X-ray spectroscopy (EDS) detector. While an SEM alone can only generate images of relative composi-tion, EDS allows for the detection of specific elements. EDS can be used to generate a quantitative estimate of the elemental mass composition on an area of a sample surface or approximate maps of an element over the sample surface. Lee et al. (1983) added trace amounts of titanium to the virgin binder to distinguish it from the recycled binder within compacted asphalt mixtures using EDS-SEM. Their intent was only to validate a different technique though, and with the limitations of SEMs at that time, it appears the method was not investigated further.

With increasing usage of recycled materials, and major advancements in SEM equipment and data col-lection, several researchers have tried tracer-based EDS-SEM methods to investigate RAP materials in recent years (Bressi et al. 2015; Castorena et al. 2016; Jiang et al. 2018; Rinaldini et al. 2014). Rinaldini et al. (2014) developed an EDS image of recycled asphalt concrete as part of a multiscale investigation of blend-ing between RAP and virgin materials. Only a single EDS image was presented, which indicated good blending as the titanium dioxide was well-distributed between the virgin and RAP aggregates.

Bressi et al. (2015) used EDS to investigate RAP particle clustering and attempted to differentiate clus-ters that existed before mixing from those that were formed during mixing. Their results pointed to incom-plete blending, however the mixtures used to quantify clustering were very unrealistic with only coarse virgin aggregates and only fine RAP.

Castorena et al. (2016) built on previous work, investigating RAP blending in a SEM with a strong EDS detector that could develop EDS maps more effi-ciently than in the past studies. Castorena et al. (2016)

attempted to quantify the visual observations made from the EDS maps, using local elemental spectra of specific areas of interest on the map. These local spectra were within the mastic phase of the asphalt but contained varying degrees of small aggregates. To minimize the effect of the aggregate content, the measured titanium concentration was normalized against the carbon concentration to develop a Ti:C ratio. Due to a lack of calibration specimens, it was not possible to reliably estimate a degree of blending from these local ratios, except in locations where the values were approximately zero, indicating no blending.

Jiang et al. (2018) also conducted EDS-SEM investigations of RAP blending, using a similar procedure to Castorena et al. (2016). Findings of the EDS investigations were consistent with expectations, showing partial blending for unaged, unmodified mixtures, and increasing interaction between the RAP and virgin binders with prolonged conditioning at elevated temperature, rejuvenation, or both. Jiang et al. (2018) normalized the titanium concentration to the sulfur concentration to quantitatively infer blending levels. Sulfur was used for normalization instead of carbon because carbon can be present in aggregates. They proposed an equation to estimate the degree of blending based on differences in the initial Ti:S ratio of the virgin binder and the final Ti:S ratio of the blended mixture, taking into account the total asphalt content, RAP content, RAP binder content, and sulfur content of both the RAP and virgin binders.

These past studies have all sought to use EDS to investigate RAP materials. While multiple institutions have replicated the concept, and the experimental protocol has been greatly improved through continued work, the interpretation of the EDS-SEM micrographs and elemental spectra is not yet settled. Most have sought to measure the degree of blending between the two binders within the mixture, however measuring a degree of blending for an entire specimen from a small number of microscopy images appears to be more complex than initially anticipated, and a single value may not adequately represent the entire picture of the material.

2 OBJECTIVES

- Investigate the concept of the degree of blending.
- Identify the parameters in volumetric mixture design that are impacted by assumptions of complete blending.

3 MATERIALS AND METHODS

3.1 Materials

A 9.5 mm nominal maximum aggregate size surface mixture with 45 percent RAP aggregates was developed using local materials, to approximate a typical gradation. Asphalt gyratory specimens were prepared from this mixture following AASHTO T 312, with the titanium dioxide tracer added to the virgin binder. The mixture was not short-term aged to provide insight on the minimum case of blending.

The tracer was added to the virgin asphalt binder using a high shear mixer at a dosage of 10 percent. High shear is required as the titanium dioxide microparticles (average size 0.2 microns) will agglomerate when added to liquid and high shear is required to break the agglomerations to create a stable blend. The mass of virgin binder added to the mixture was adjusted to maintain the same volume while accounting for the density of the tracer. The composition of both the RAP and virgin binders was evaluated using EDS-SEM, and the sulfur content was 4.5 percent for both.

3.2 Methods

To prepare asphalt mixture specimens for EDS analysis, small prismatic specimens (roughly 25 mm × 25 mm × 12 mm) were sawn from the gyratory specimens using water-cooled saws. For EDS analysis, specimen surfaces must be smooth to minimize the effects of surface topography. Thus, the samples were then ground with increasing grit silicon carbide papers on a water-cooled polishing system and rinsed and sonicated between grinding steps to remove any debris and embedded grinding media. Following polishing, specimens were cleaned and dried.

For conducting EDS-SEM, a Hitachi S3200N VPSEM outfitted with an Oxford X-Max silicon drift detector was used to avoid the need for sample coating while acquiring EDS data efficiently. The blending ratio proposed by Jiang et al. (2018) was evaluated as a measure of blending. The equation to determine the blending ratio (γ) is as follows:

$$\gamma = \frac{\left(\dfrac{i_{Ti:S}}{f_{Ti:S}} - 1\right) \times \left(OAC - \sum \alpha_i \beta_i\right) \times \dfrac{S_v}{S_a}}{\sum \alpha_i \beta_i} \times 100\%$$

(1)

where $i_{Ti:S}$ = initial Ti:S ratio in virgin binder; $f_{Ti:S}$ = final Ti:S ratio in blended binder; OAC = optimum asphalt content of the asphalt mixture; α_i = proportion of RAP stockpile i in the mixture; β_i = asphalt content of RAP stockpile i; S_v = sulfur content of the virgin binder; and S_a = sulfur content of the RAP binder.

4 RESULTS

The first step in reliably using the SEM with asphalt is identifying areas of interest. Prepared samples are roughly 625 square mm; however, the imaging area at 100× is roughly 1 square mm, so imaging the entire sample surface is unreasonable. However, in industrial applications, titanium dioxide is used as a whitening

Table 1. Calculated degree of blending for each site.

Site	Titanium (%)	Sulfur (%)	Ti:S	Blending Ratio
1	0.5	0.6	0.8	130%
2	0.4	0.6	0.7	220%
3	0.5	0.5	1.0	60%
4	0.6	0.6	1.0	60%
5	0.7	0.7	1.0	60%

and opacity agent and when added to asphalt it turns develops a brown hue. This allows visual observation of areas of poorly blended RAP, which are solidly black as they have not blended with the brown virgin binder. This can be seen in Figure 1, where a heavy dark shadow surrounds the triangular aggregate pointed to by the flag labelled "2", despite the sample being polished smooth. This aggregate is surrounded by a thick layer of what appears to be unblended RAP mastic, not only a single aggregate.

These visual observations cannot stand alone, as aggregates show wide color variations which complicate optical assessment of the asphalt mixture. Instead, the visual observations require support from EDS-SEM to verify the material composition and provide qualitative estimates of the binder interaction. The points of the flags in Figure 1 identify the approximate location of the sites where EDS measurements were taken. The tabular EDS results are shown in Table 1. The blending ratio was calculated using Equation (1). It compares the Ti:S value to the ideal perfect blend calculated from the composition of each binder and the relative composition of each binder in the mixture. Blending ratio values of 100 percent indicate perfect blending, while values deviating from 100 percent indicate an imbalance of the RAP and virgin binders. Values below 100 percent indicate virgin binder with incomplete contribution of RAP, while values above 100 percent indicate predominantly RAP with only limited interaction of virgin binder.

Electron images from the SEM and elemental maps for titanium and sulfur for Sites 1-3 are shown in Figure 2. Sites 1 and 3 show little difference between the titanium and sulfur, indicating a relatively uniform blend throughout the area. However, while Site 2 shows matching shadows surrounding aggregates in the upper left corner of both the titanium and sulfur images, in the lower right corner the titanium signal diminishes to the background levels indicating a coating of unblended RAP material adhered to the aggregate in the lower right corner.

The Superpave volumetric mixture design procedure requires aggregate properties to calculate approval criteria, including the dust proportion (DP), voids in mineral admixture (VMA), and voids filled with asphalt (VFA). Calculation of these properties requires the RAP aggregate gradation and RAP aggregate bulk specific gravity, which are both difficult to

measure. Copeland (2011) identified that recovered aggregate from binder content measurements could not approximate the true bulk specific gravity sufficiently, and recommended instead treating the RAP as mixture and measuring the theoretical maximum specific gravity, calculating the effective specific gravity of the aggregates, then assuming the binder content and absorption to estimate the bulk specific gravity. To measure gradation, recovered aggregates from ignition or solvent extraction are used, which assumes that any clusters of RAP aggregates will be broken down completely during mixing. This is especially complicated when the RAP aggregates are recovered by ignition oven. Kvasnak (2010) reported the aggregates may experience degradation in the ignition oven, which would possibly lead to overestimating the dust content. The effective binder content is impacted by the assumption of complete blending, and any uncertainty in the RAP binder content measurements, in addition to the issues noted with the aggregate bulk specific gravity. These errors from dust content measurement and effective binder content are combined to calculate the dust proportion.

Balanced mix design methods, where volumetric parameters are balanced with performance testing, may minimize some of the concerns. However, these methods still include volumetric parameters which are complicated by incomplete blending of binders, and fundamental questions of blending are still not answered.

5 CONCLUSIONS AND RESEARCH NEEDS

- RAP exhibits a mixture of both blending and behaving as black rock.
- EDS-SEM can quantify this behavior, however additional work is needed to identify how to tie microscopic evaluations to mixture quality and performance.
- Identification of black rock clusters of RAP pose questions about how RAP is incorporated in volumetric mixture design procedures, as unblended clusters will impact both mixture acceptance properties and the overall binder content and quality available in the mixture.
- Investigation into the role of RAP clusters in mixture performance is needed, to identify the practical implications of the EDS-SEM results. Fatigue performance and failure surfaces are of most importance as fatigue is the primary form of distress associated with increased RAP usage.

ACKNOWLEDGEMENTS

The authors would like to thank the North Carolina Department of Transportation for financial support of this work, and Trimat Materials Testing Inc. for assistance in sourcing materials.

REFERENCES

American Association of State Highway and Transportation Officials (AASHTO). 2015a. Preparing and Determining the Density of Asphalt Mixture Specimens by Means of the Superpave Gyratory Compactor. AASHTO T 312, Washington D.C.

American Association of State Highway and Transportation Officials (AASHTO). 2015b. Superpave Volumetric Design for Asphalt Mixtures. AASHTO R 35, Washington D.C.

Castorena, C., Pape, S., and Mooney, C. 2016. Blending Measurements in Mixtures with Reclaimed Asphalt: Use of Scanning Electron Microscopy with X-Ray Analysis. *Transportation Research Record*, No. 2574, pp. 5763.

Copeland, A. 2011. Reclaimed Asphalt Pavement in Asphalt Mixtures: State of the Practice. *Publication FHWA-HRT-11-021*. McLean, VA: FHWA.

Epps, J.A., Little, D.N., Holmgreen, R.J., and Terrel, R.L. 1980. Guidelines for Recycling Pavement Materials, NCHRP Report No. 224. *Transportation Research Board*, Washington, D.C.

Jiang, Y., Gu, X., Zhou, Z., Ni, F., and Dong, Q. 2018. Laboratory Observation and Evaluation of Asphalt Blends of Reclaimed Asphalt Pavement Binder with Virgin Binder using SEM/EDS. *Transportation Research Record,* 2018.

Kvasnak, A. 2010. What to Consider when Designing a High RAP Content Mix. *Hot Mix Asphalt Technology*, National Asphalt Pavement Association, Vol. 18, pp. 18–19, 21.

McDaniel, R., and Anderson, R.M. 2001. NCHRP Report 452: Recommended Use of Reclaimed Asphalt Pavement in The Superpave Mix Design Method: Technician's Manual. *Transportation Research Board*, Washington, D.C.

National Center for Asphalt Technology (NCAT). 2014. *NCAT Researchers Explore Multiple Uses of Rejuvenators.* Asphalt Technology News, Vol. 26, No. 1.

Rinaldini, E., Schuetz, P., Partl, M.N., Tebaldi, G., and Poulikakos, L.D. 2014. Investigating the Blending of Reclaimed Asphalt with Virgin Materials Using Rheology, Electron Microscopy and Computer Tomography. *Composites Part B*, Vol. 67, pp. 579587.

Stroup-Gardiner, M. 2016. Use of Reclaimed Asphalt Pavement and Recycled Asphalt Shingles in Asphalt Mixtures, NCHRP Synthesis 495, *Transportation Research Board,* Washington, D.C.

Advances in Materials and Pavement Performance Prediction II – Kumar et al. (eds)
© 2021 Taylor & Francis Group, London, ISBN 978-0-367-46169-0

Mix design and linear viscoelastic characterization of Fine Aggregate Matrix (FAM)

C. Pratelli, P. Leandri & M. Losa
University of Pisa, Pisa, Italy

M.P. Wistuba
Braunschweig Pavement Engineering Centre-ISBS Technische Universität Braunschweig, Braunschweig, Germany

ABSTRACT: The testing of Fine Aggregate Matrix (FAM) is gaining wide attention in the asphalt pavement research community since this phase is considered the most representative one of the overall asphalt mixture response. Moreover, it is characterized by a notable testing efficiency. In this study, an experimental campaign has been carried out on FAM specimens to investigate their rheological behavior within the linear viscoelastic range over a wide range of testing temperatures and frequencies. The experimental data, in terms of complex shear modulus and phase angle, were modelled with the 2S2P1D model. Results highlighted that the 2S2P1D model simulates in an excellent way the linear viscoelastic properties of FAM, both made with unmodified and modified binders. For each tested FAM the Time-Temperature Superposition principle was validated as well.

1 INTRODUCTION

Fine aggregates, filler, binder and entrained voids compose the FAM phase (Kim 2003), which represents the intermediate scale between mastic and asphalt mixture. Many researchers have highlighted the critical role of FAM in the overall performance evaluation of the corresponding asphalt mixtures (Gudipudi & Underwood 2015; Kim et al. 2002; Nabizadeh 2015; Riccardi et al. 2018), and consequently in the road pavement design process. The simplicity, efficiency and lower costs/times required for FAM testing make it a very attractive specification-type approach. However, despite the growing interest on FAM testing, there are still some concerns about the proper FAM mix design.

The objectives of this study are to identify a proper method for FAM mix design definition and to analyze the linear viscoelastic properties of FAM, made both with unmodified and modified asphalt binder, modelling them with the analogical 2S2P1D model.

2 TESTED MATERIALS

One of the main objectives of this research was to reproduce the FAM phase as it exists between the coarse aggregates of a corresponding asphalt mixture, by the identification of a proper design method.

Four different FAM mixes were produced, corresponding to four different asphalt mixtures. The mix design of the asphalt mixtures was the same

Table 1. Naming scheme.

Asphalt mixture	Corresponding FAM	Type of binder
MIX_111	FAM_111	PEN 50-70
MIX_211	FAM_211	PMB 25-55-55 A
MIX_321	FAM_321	PEN 70-100
MIX_421	FAM_421	PMB 45-80-65 A

and consisted of 6.0 w.% of binder content and the remaining 94.0 w.% of aggregates, with a distribution typical of a main course, according to the German standards (FGSV). The full graded asphalt mixtures were blended from five stockpiles of Gabbro, in different proportions, and limestone filler and four asphalt binders: two unmodified (PEN 50-70 and PEN 70-100) and two SBS polymer modified asphalts (PMB 25-55-55 A and PMB 45-80-65 A). The naming scheme is reported in Table 1. The Nominal Maximum Aggregate Size (NMAS) of the mixtures was 11.2 mm.

In this study the FAM mix design was related to the aggregate gradation and composition of the full graded asphalt mixture. The mix design of the corresponding FAM (binder content and aggregate gradation) was based on the meso-gravimetric results obtained by Underwood and Kim (Underwood & Kim 2013). They suggested that the delineation between coarse aggregates and fine aggregates (Fine Aggregate Initial Break-FAIB) depends on the NMAS of the corresponding asphalt mixture and it could be defined based on the aggregate packing principles proposed by

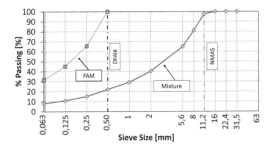

Figure 1. Aggregate gradation curve of the asphalt mixture and the corresponding FAM.

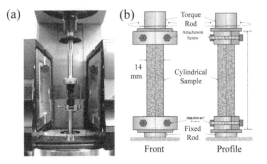

Figure 2. Picture (a) and scheme (b) of setup for testing FAM specimens using a torsion bar fixture.

Table 2. Summary of asphalt mixture and corresponding FAM mix designs.

Sieve size mm	Asphalt mixture gradation %	FAM gradation %
16.0	100	100
11.2	97.50	100
8.0	81.23	100
5.6	65.00	100
2.0	40.30	100
1.0	29.19	100
0.5	22.13	100
0.25	15.07	65.55
0.125	10.89	45.12
0.063	8.04	31.23
Binder content of the asphalt mixture		5.9 %
Air void content of asphalt mixture		3.5±1.0 %
Binder content of FAM		16.22 %
Mastic volumetric concentration		40.36 %
Thickness of mastic film		23.68 μm

Vavrik (Vavrik et al. 2002). According to this study, for an asphalt mixture with a NMAS of 11.2 mm, the corresponding FAM maximum aggregate size should be 0.50 mm (DFAIB). Then the aggregate gradation and the asphalt content of FAM were calculated considering the volume of filler in the mastic and the volume of mastic within the asphalt mixture (Underwood & Kim 2013). The binder content of FAM mixes in this study was calculated as 16.22% for a corresponding binder content of 5.9% in the full graded asphalt mixture. The reliability of this analytical method was empirically verified by means of solvent extraction of the sieved loose asphalt mixture. Figure 1 shows the gradation curves of both asphalt mixture and FAM. While Table 2 summarized their mix designs.

Samples compaction also plays a fundamental role on the representativeness of the FAM phase and on the engineering properties of the FAM specimens (Izadi et al. 2011). Previous researches highlighted that gyratory compacted FAM samples have a better microstructural resemblance to the FAM phase within the asphalt mixture (Izadi et al. 2011; Underwood & Kim 2013). In this research, FAM samples of 150

mm in diameter and 130 mm in height were compacted (Gudipudi & Underwood 2015; Gudipudi & Underwood 2017). Small cylindrical specimens with a diameter of 14 mm and a height of about 50 mm were drilled from the gyratory compacted samples.

Previous researches suggested, based on microstructural investigation, a target air voids content range for FAM between 50 and 75% of the total air voids of the asphalt mixture. The air voids content of FAM specimens was measured before testing and the specimens with air voids content differing from the target values were discarded and not used for the following tests.

3 EXPERIMENTAL PROCEDURE AND MODELLING

3.1 Tests and testing procedure

The cylindrical FAM specimens were tested using a torsion bar fixture in a Dynamic Shear Rheometer (DSR). Figure 2 shows a picture and a scheme of the test setup.

FAM specimens were clamped between the movable and the stationary fixture and then enclosed in a thermal chamber. When performing the test, special attention was given to ensure that the specimen was correctly aligned and clamped in the DSR.

Temperature-frequency (T-f) sweep tests were performed to determine the rheological properties of FAM in the linear viscoelastic range (LVE). These tests were performed by applying oscillatory torsional deformation in strain-controlled mode. Testing frequencies ranged from 0.1 to 10 Hz, while temperatures ranged between −10 and 40°C, with a step of 10°C. The imposed strain amplitudes used to perform T-f tests were selected based on the results of strain sweep tests. Strain sweep tests were performed to determining the limits of the linear viscoelastic (LVE) range, by increasing the strain amplitude logarithmically at four different frequency-temperature combinations. Indeed, the range of testing conditions was too wide to define a single optimal strain value. In the range of temperature between −10

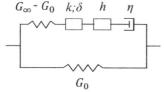

$$G_\infty - G_0 \quad k;\delta \quad h \quad \eta$$

$$G_0$$

Figure 3. 2S2P1D analogical model.

Table 3. WLF model parameters for FAM.

FAM	c_1 [–]	c_2 [°C]
111	13.2	124.38
211	14.5	122.88
321	17.4	141.04
421	24.3	198.04

and 10°C the stiffer condition (lower temperature-higher frequency), −10°C-10 Hz and the compliant one (higher temperature and lower frequency), 10°C-0.1 Hz, were investigated. Similarly, in the temperature range between 10 and 40°C the strain sweep tests were performed at 10°C-10 Hz and 40°C-0.1 Hz. The limits of the LVE range were defined according to the standards for asphalt binders tests (EN 14770, 2012), which claim that the LVE range is limited by a drop of 5% of the initial value of the complex shear modulus over the increasing level of strains. In order to prevent any non-linear behavior, the T-f sweep tests were performed at the imposed strain amplitude of $0.0001\% = 1.00~\mu\varepsilon$ in the temperature range between −10 and 10°C, and of $0.001\% = 10.00~\mu\varepsilon$ between 10 and 40°C.

T-f sweep tests performed on FAM specimens showed a good repeatability, with an average coefficient of variation of 3.26%. The number of test replicates was n=5.

3.2 2S2P1D model

Experimental data were fitted with 2S2P1D model (Di Benedetto et al. 2004; Olard & Di Benedetto 2003). This analogical linear viscoelastic model, composed of two springs, two parabolic creep elements and one linear dashpot (Fig. 3), is able to describe the linear viscoelastic rheological behavior of binders and asphalt mixtures.

According to the 2S2P1D model, at a given combination of temperature T and frequency ω, the complex shear modulus can be expressed as:

$$G^*(\omega, T) = G_\infty + \frac{G_0 - G_\infty}{1 + \delta(i\omega\tau(T))^{-k} + (i\omega\tau(T))^{-h} + (i\omega\tau(T))^{-1}} \tag{1}$$

Where G_0 is the static asymptotic shear modulus when $\omega \to 0$; G_∞ is the glassy asymptotic shear modulus when $\omega \to \infty$; k and h are dimensionless exponents such as $0 < k < h < 1$, while δ and β are dimensionless constants, in particular β, is linked to the Newtonian viscosity of the material η and $\tau(T)$ is the characteristic time depending on temperature.

If the Time-Temperature superposition principle (TTSP) is validated, master curves of bituminous materials can be obtained at a given reference temperature T_{ref}, by shifting the experimentally obtained

isotherm curves according to the temperature shift factor a_T. Based on the TTSP, the characteristic time τ can be defined as shown in Equation 2 below:

$$\tau(T) = a_T \cdot \tau(T_{ref}) \tag{2}$$

Where a_T = temperature shift factor at temperature T for the reference temperature T_{ref}. It can be modelled, as function of the temperature, using the Williams, Landel and Ferry (WLF) equation (Williams et al. 1955), Equation 3:

$$\log a_T(T) = \frac{C_1(T - T_{ref})}{C_2 + (T - T_{ref})} \tag{3}$$

Where C_1 and C_2 are empirical constants depending on the material.

4 RESULTS AND ANALYSIS

The rheological experimental data of complex shear moduli $|G^*|$ and phase angles φ obtained for all tested FAM specimens are presented in Figures 4–6.

The Time-Temperature superposition principle (TTSP) resulted verified for all the FAM mixes, as their curves in the Black space (phase angle vs complex modulus) were unique (Fig. 6b). The WLF constants for each FAM mix, used to fit the temperature shift factor values a_T are listed in Table 3. The complex shear modulus and the phase angle master curves at the reference temperature of 20°C (obtained using the a_T values calculated according to the WLF equation) of each FAM mix, are respectively shown in Figures 4a and 4b.

The 2S2P1D model successfully fits the experimental data. The 2S2P1D model parameters are summarized in Table 4. In Table 4 are also reported the correlation coefficient (R^2) for each FAM, to evaluate the model capability in predicting the experimental data. Values of R^2 close to 1, confirm that the 2S2P1D model fits well experimental data of complex shear moduli and phase angles.

5 CONCLUSIONS

The aim of this study was to investigate the rheological properties of FAM in the LVE range. Four FAM

Table 4. 2S2P1D model parameters for FAM.

| FAM | At any temperature | | | | | | At | |
	G_0 MPa	G_∞ MPa	k –	h –	β –	$\log^\tau$ s	R^2	T_{ref} R^2
111	0.46	7418	4.60	0.271	0.61	950	-3.77	0.999
211	1.36	7893	4.81	0.262	0.60	506	-3.64	1.000
321	0.21	7033	5.44	0.282	0.70	1410	-3.99	1.000
421	0.84	7729	5.92	0.237	0.57	4141	-4.05	0.997

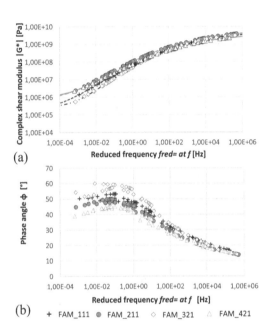

(a)

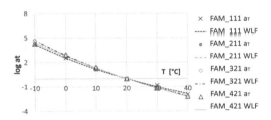

(b)

Figure 4. (a) Complex shear modulus and (b) phase angle master curves of FAM at the reference temperature of 20°C.

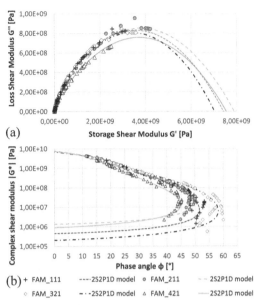

Figure 5. Temperature shift factors for each FAM mix and WLF modelling.

mixes were produced and tested using a torsion bar fixture in a DSR. T-f sweep tests were performed on five specimens for each mix, over a wide range of temperatures and frequencies. The experimental data shown a good repeatability. From the experimental results the following conclusions can be drawn. The Time-Temperature Superposition Principle is validated for every tested FAM and the analogical 2S2P1D model successfully fits the experimental data within the LVE

Figure 6. Experimental results and 2S2P1D model fitting for FAM in (a) Cole-Cole plane and (b) Black space.

range. High values of the correlation coefficient confirm the model capability in simulating the rheological behavior of FAM, which is able to describe the rheological behavior of FAM, as it does for the other asphaltic phases.

REFERENCES

EN 14770. 2012. Bitumen and bituminous binders - Determination of complex shear modulus and phase angle - Dynamic Shear Rheometer (DSR). European Standard.

Gudipudi, P. & Underwood, B.S. 2015. Testing and Modeling of Fine Aggregate Matrix and Its Relationship to Asphalt Concrete Mix. *Transportation Research Record: Journal of the Transportation Research Board*, 2507, 120–127.

Gudipudi, P. & Underwood, B.S. 2017. Development of Modulus and Fatigue Test Protocol for Fine Aggregate Matrix for Axial Direction of Loading. *Journal of Testing and Evaluation*, 45(2), 497–508.

Izadi, A., Motamed, A. & Bhasin, A. 2011. Designing Fine Aggregate Mixtures to Evaluate Fatigue Crack-Growth in Asphalt Mixtures. *Report 161022-1 Center for Transportation Research University of Texas*, 7(1), 54.

Kim, Y.R. 2003. Mechanistic Fatigue Characterization and Damage Modeling of Asphalt Mixtures. *PhD Thesis at Texas A&M Univ.*

Kim, Y.R., Little, D.N. & Lytton, R.L. 2002. Use of Dynamic Mechanical Analysis (DMA) to evaluate the fatigue and healing potential of asphalt binders in sand asphalt mixtures. *J. Assoc. Asphalt Paving Technology*, 71, 176–206.

Nabizadeh, H. 2015. Viscoelastic, Fatigue Damage, and Permanent Deformation Characterization of High RAP Bituminous Mixtures Using Fine Aggregate Matrix (FAM). *Thesis at University of Nebraska at Lincoln.*

Olard, F. & Di Benedetto, H. 2003. General "2S2P1D" Model and Relation between the Linear Viscoelastic Behaviours of Bituminous Binders and Mixes. *Road Materials and Pavement Design*, 4(2), 185–224.

Riccardi, C., Cannone Falchetto, A. & Wistuba, M.P. 2018. Experimental Investigation of Rutting in the Different Phases of Asphalt Mixtures. In *RILEM 252-CMB 2018*.

Underwood, B.S. & Kim, Y.R. 2013. Effect of Volumetric Factors on The Mechanical Behavior of Asphalt Fine Aggregate Matrix and the Relationship to Asphalt Mixture Properties. *Construction and Building Materials*, 49, 672–681.

Vavrik, W.R., Huber, G., Pine, W.J., Carpenter, S.H. & Bailey, R. 2002. Bailey Method for Gradation Selection in Hot-Mix Asphalt Mixture Design. *Transportation Research CIRCULAR, E-C044*.

Williams, M.L., Landel, R.F. & Ferry, J.D. 1955. The Temperature Dependence of Relaxation Mechanisms in Amorphous Polymers and Other Glass-Forming Liquids. *Journal of the American Chemical Society*, 77(14).

Advances in Materials and Pavement Performance Prediction II – Kumar et al. (eds)
© *2021 Taylor & Francis Group, London, ISBN 978-0-367-46169-0*

Energy dissipation approach to quantify the fatigue damage of glass fibre grid inlaid asphalt mixture

Arbin Raj, B.S. Abhijith & J. Murali Krishnan
Department of Civil Engineering, IIT Madras, Chennai, India

ABSTRACT: The glass fibre grid in asphalt mixtures can act as a stiffener as well as enhance the fatigue life. Four-point beam (4PB) bending test is usually carried out to evaluate the fatigue life of such composite mixtures. While there are several post-processing methods used to define the fatigue failure criterion from the 4PB test data, the total energy dissipation approach is predominantly used. In this paper, the fatigue dissipation for the computation of fatigue life is used instead of total energy dissipation. In this investigation, 4PB tests were carried out on asphalt mixtures integrated with and without the grid at 600 and 800 $\mu\varepsilon$. The fatigue life computed using fatigue dissipation shows that the grid is more effective at 800 $\mu\varepsilon$, and the fatigue life of the grid inlaid sample is 5.7 times higher than the samples without the grid.

1 INTRODUCTION

During the rehabilitation of asphalt concrete pavement, the glass fibre grid is laid between the existing layer and the newly laid layer to improve the fatigue life. While few laboratory investigations predict the fatigue life of such composite mixtures, most of them use an ad-hoc criterion such as a 50% reduction in modulus (Arsenie et al. 2017; Kim et al. 2010). The fatigue life computed in this manner may not give a proper understanding of the role of the grid integrated within the asphalt mixture. Therefore, it is important to re-look into the post-processing method used to quantify fatigue damage of such composite mixtures.

There are several approaches to quantify the fatigue damage of asphalt concrete in which the dissipated energy approach is predominantly used. When a viscoelastic material is subjected to repeated load cycles, it mainly exhibits two modes of dissipation, namely viscous dissipation and dissipation due to fatigue damage. To quantify the beneficial effect due to grid, it is necessary to separate the fatigue dissipation from the total dissipation. The fatigue dissipation can then be used to compare the fatigue damage between the control sample and the grid inlaid sample.

Among several approaches to quantify fatigue dissipation, a dissipated pseudo strain energy method (DPSE) is widely used (Kim & Little 1989). In this approach, the physical strain is replaced by a pseudo strain to transform the viscoelastic response of the material to an elastic response. Thus, as the fatigue test progresses, the damage in the material is captured using the area under the stress-pseudo strain curve. However, the limitation of the DPSE approach is

shown by Varma et al. (2019) in which the total dissipation is observed to be less than the sum of viscous dissipation and dissipation due to fatigue damage, which is an untenable scenario.

In the current investigation, the fatigue dissipation for the control and grid inlaid asphalt mixture is computed using the approach proposed by Varma et al. (2019). Unlike the DPSE approach, the phase angle corresponding to a purely viscoelastic response calculated using this approach (Varma et al. 2019) changes with fatigue damage in the material. The cumulative fatigue dissipation is also found to be increasing at the end of the fatigue test as expected.

In this approach, for the computation of fatigue dissipation, a constitutive assumption is made for the viscoelastic behaviour of damaged asphalt mixtures.

In a strain-controlled fatigue test, suppose the applied strain ε at any instant of time t on the beam is given by

$$\varepsilon(t) = \varepsilon_o \sin(\omega t), \tag{1}$$

where 'ε_o' is the peak strain; and ω is the angular frequency, the corresponding stress response σ of the damaged material can be expressed as:

$$\sigma_n(t) = \sigma_{n_o} \sin(\omega t + \delta_n), \tag{2}$$

where σ_{n_o} is the peak stress amplitude at loading cycle n and, δ_n is the phase angle at the nth cycle.

Now, if this stress $\sigma_n(t)$ is applied to an undamaged material, the strain response of the material is as follows,

$$\varepsilon_{ve}(t) = \frac{\sigma_n}{|E_{ve}^*|} \sin(\omega t + \delta_n - \delta_{v6}), \tag{3}$$

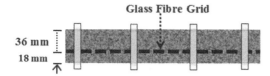

Glass Fibre Grid

36 mm

18 mm

Figure 1. Schematic representation of grid inlaid sample.

where ε_{ve} is the viscoelastic strain at the nth cycle, $|E_{ve}^*|$ and δ_{ve} are the dynamic modulus and phase angle corresponding to linear viscoelastic material.

Thus, using the above expressions, the viscous dissipation can be calculated as,

$$W_{viscous} = \oint \sigma_n(t)\dot{\varepsilon}_{ve_n}(t)dt, \qquad (4)$$

The viscous dissipation computed in this manner changes with the damage in the material, unlike the DPSE approach which assumes a constant viscous dissipation throughout the fatigue test.

The total dissipation can be calculated using the following equation,

$$W_{total} = \oint \sigma_n(t)\dot{\varepsilon}(t)dt, \qquad (5)$$

By knowing the total dissipation (W_{total}) and the viscous dissipation ($W_{viscous}$), the fatigue dissipation can be determined using the equation given below:

$$W_{fatious} = W_{total} - W_{viscous}. \qquad (6)$$

2 MATERIALS

For this study, unmodified binder VG30, which conforms to Indian Specification IS: 73 (2013), is used for the preparation of asphalt mixture specimens. The gradation of the asphalt mixture was chosen based on the Indian specifications (MoRTH 2013). Bituminous concrete grade-2 with a nominal maximum aggregate size of 13.2 mm, with median grading and a binder content of 5 % was used for the fabrication of test specimens. Shear box compactor (ASTM D7981 2015) is used to produce asphalt concrete beam samples of size 450 mm × 150 mm × 170 mm and the compacted beam sample is sliced to the required beam sample of size 380 mm × 63 mm × 54 mm for 4PB test. The detailed grid inlaid asphalt mixture sample preparation procedure using shear box compactor is explained by Raj et al. (2020). Samples with 4 ± 0.5 % air voids were selected for testing.

In the case of the grid inlaid sample, the glass fibre grid is integrated at two-third depth from the top of the beam sample (36mm from the top) as shown in Figure 1.

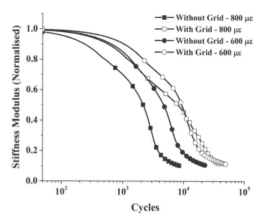

Figure 2. Stiffness modulus (normalised) variation for control and grid inlaid sample across different strain level.

3 EXPERIMENTAL INVESTIGATION

Strain-controlled four-point beam bending test was conducted using 4PB jig manufactured by M/s IPC Global, Australia. The test was conducted on asphalt concrete beam samples integrated with grid and control samples (without grid) at 20°C in accordance with AASHTO T321 (2007). The test was carried out at 600 and 800 $\mu\varepsilon$ with a sinusoidal load waveform at 10 Hz frequency. The beam samples were subjected to one million cycles or till 20 % of the initial stiffness whichever reached earlier. At every 0.001 seconds, the load-displacement data was recorded and using the recorded data, stiffness modulus was estimated. Figure 2 shows the variation of stiffness modulus normalized with respect to the initial modulus with loading cycles at two different strain levels for control and grid integrated samples. From the evolution of stiffness modulus (normalized) with the number of loading cycles, it is observed that the reduction in the stiffness modulus value is lower for the grid inlaid sample than the control sample. Further, the fatigue life computed from the 50% stiffness criterion (AASHTO T321 2007) indicates that the application of the grid significantly enhances the fatigue life of the asphalt mixture. For 600 $\mu\varepsilon$, the grid inlaid sample shows twice fatigue life compared to the control sample. For 800 $\mu\varepsilon$, the grid inlaid sample show seven times higher fatigue life compared to the control sample (Table 1). Thus, as the strain level increases, the role of the integrated grid in asphalt mixtures is found to be evident. To get additional insight into the role of the grid in asphalt mixtures in resisting fatigue damage, it is essential to look at the evolution of energy dissipated due to damage with the number of loading cycles.

4 TEST RESULTS AND DISCUSSION

Based on the alternative approach proposed by Varma et al. 2019, the viscous dissipation is computed using

Equation 4. From the total energy dissipation (Equation 5), Equation 6 is used to calculate fatigue dissipation. The evolution of cumulative fatigue, cumulative viscous, and cumulative total energy dissipation with the number of loading cycles is shown in Figure 3.

As reported by Varma et al. (2019), the cumulative fatigue dissipation curve follows the expected trend. The cumulative fatigue dissipation increases towards the end of the fatigue test for control and grid inlaid samples tested at both 600 and 800 με. In addition to that, for control and grid samples at both strain levels, the cumulative viscous dissipation reaches a constant value and dissipation due to fatigue damage increases.

Further, using the evolution of fatigue dissipation with the number of loading cycles, a new criterion to determine fatigue life is proposed, which is explained in section 4.1.

4.1 Fatigue dissipated energy ratio (FDER)

In the current study, the fatigue life was computed using three fatigue failure criteria; the energy ratio, AASHTO T321 (2007), and fatigue dissipated energy ratio (FDER). Energy ratio at any given cycle is defined as the ratio of the energy dissipated in the initial cycle to the energy dissipated at the current cycle (Rowe & Bouldin 2000). The energy ratio is computed using the following equation,

$$Energy\ Ratio = \frac{n \times W_0}{W_n} \qquad (7)$$

where n represents the cycle number, W_0 is the dissipated energy in the initial cycle, and W_n is the dissipated energy in the nth cycle.

Further, a new failure criterion is proposed in which the fatigue dissipation computed using Equation 6 is used instead of total energy dissipation in Equation 7. The following expression proposed to compute fatigue dissipated energy ratio (FDER) for asphalt mixtures;

$$FDER = \frac{n \times W_{f(ref)}}{W_{f(n)}} \qquad (8)$$

where $W_{f(ref)}$ is the fatigue dissipation in the reference cycle, and $W_{f(n)}$ is the fatigue dissipation in the nth cycle.

The variation of energy ratio and FDER with the number of loading cycles at 800 με for the glass fibre grid inlaid sample is shown in Figure 4. Two straight lines are fitted for the initial and final portions of the curve. The number of cycles corresponding to the intersection of two straight lines is taken to be the point of failure by fatigue. It can be observed that the fatigue life obtained from the FDER approach is higher than that of the energy ratio approach.

Table 1 provides the details regarding the fatigue life computed using AASHTO T321 (2007), Energy ratio (Rowe & Bouldin 2000) and FDER approach.

The fatigue life determined from the FDER approach is found to be nearly 1.4 to 2 times more than

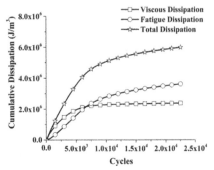

(a) Control sample at 600 με

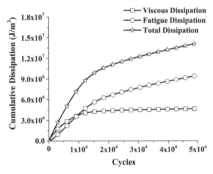

(b) Grid inlaid sample at 600 με

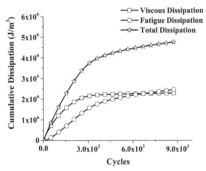

(c) Control sample at 800 με

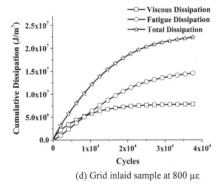

(d) Grid inlaid sample at 800 με

Figure 3. Cumulative dissipation for BC-VG30.

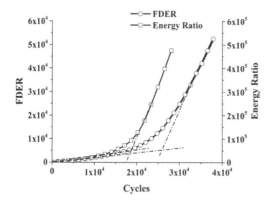

Figure 4. FDER and energy ratio curve for grid inlaid sample at 800 με.

Table 1. Comparison of fatigue life based on different approaches.

		Fatigue life	
Sample Type	Approach	800 με	600 με
Control	AASHTO T321 (2007)	2160	4680
	Energy Ratio	2070	4570
	FDER	4410	10170
With Grid	AASHTO T321	17200	9280
	Energy Ratio	17800	8950
	FDER	25700	17360

the energy ratio and AASHTO method. This increase in fatigue life based on the FDER approach can be due to the elimination of viscous dissipation from the total dissipated energy. Further, it is observed that at 800 με, the fatigue life of the grid inlaid sample is 5.7 times higher than the control sample. In the case of 600 με, the fatigue life of the grid inlaid sample is 1.7 times more than the control sample. From the above findings, it is clear that, as the strain level increases, the improvement in fatigue life between the grid and control sample also increases.

5 SUMMARY AND CONCLUSION

In this investigation, the dissipated energy approach was used to compare the fatigue performance of the control sample and glass fibre grid inlaid sample. For this purpose, strain-controlled four-point bending fatigue tests were carried at two strain levels, 600 and 800 με, respectively. The mixtures were subjected to sinusoidal loading waveform at 10 Hz frequency.

The energy dissipated due to fatigue damage was delineated from viscous dissipation using the approach proposed by Varma et al. (2019). The fatigue dissipation computed in this manner was used to predict the fatigue life of asphalt mixtures. A new fatigue failure criterion, fatigue dissipated energy ratio (FDER) was proposed by modifying the existing energy ratio approach. From the analysis of test results, it was observed that the fatigue life obtained based on the energy ratio and AASHTO T321 approach underestimates the actual fatigue life of the material. Further, the glass fibre grid is found to play a critical role in enhancing fatigue life of asphalt mixtures, specifically at higher strain levels.

REFERENCES

AASHTO T321 2007. Standard method of test for determining the fatigue life of compacted asphalt mixtures subjected repeated flexural bending. *American Association of State Highway and Transportation Officials*, Washington D.C., USA.
Arsenie, I. M., Cyrille, C., Jean-Louis, D., & Pierre, H. 2017. Laboratory characterization of the fatigue behaviour of a glass fibre geid reinforced asphalt concrete using 4PB tests. *Road Materials and Pavement Design*. 18 (1). 168–180.
ASTM D7981 2015. Standard practice for compaction of prismatic asphalt specimens by means of the shear box compactor. *American Society for Testing and Materials*, West Conshohocken, Pennsylvania, USA.
IS 73 2013. Paving bitumen- specifications. Fourth revision, *Bureau of Indian Standards*, New Delhi.
Kim, Y. R., & Little, D. N. 1989. Evaluation of healing in asphalt concrete by means of the theory of nonlinear viscoelasticity. *Transportation Research Record*. 1228. 198–210.
Kim, H., Partl, M. N., Pimenta, R., & Sivotha, H. 2010. Experimental investigation of grid reinforced asphalt composites using four-point bending tests. *Journal of Composite Materials*. 44 (5). 575–592.
MoRTH 2013. Specification for roads and bridge work. Fifth revision, *Indian Road Congress*, New Delhi, India.
Raj, A., Varma, R., & Krishnan, J. M. (2020). Influence of glass fibre grid and its placement on the fatigue damage of asphalt mixture. *Journal of Construction and Building Materials* (under review).
Rowe, G. M., & Bouldin, M. G. 2000. Improved techniques to evaluate the fatigue resistance of asphalt mixtures. *In Proceeding of 2nd Eurasphalt and Eurobitume Congress*. Breukelen, Netherlands. 754–763.
Varma, R., Atul Narayan, S. P., & Murali Krishnan, J. 2019. Quantification of viscous and fatigue dissipation of asphalt concrete in four-point bending tests. *Journal of Materials in Civil Engineering*. 31 (12). 1–12.

Advances in Materials and Pavement Performance Prediction II – Kumar et al. (eds)
© 2021 Taylor & Francis Group, London, ISBN 978-0-367-46169-0

Low density polyethylene for asphalt binder modification

K. Lakshmi Roja
Department of Mechanical Engineering, Texas A&M University at Qatar, Doha, Qatar

Amara Rehman
Department of Chemical Engineering, Texas A&M University at Qatar, Doha, Qatar

Eyad Masad
Department of Mechanical Engineering, Texas A&M University at Qatar, Doha, Qatar

Ahmed Abdala
Department of Chemical Engineering, Texas A&M University at Qatar, Doha, Qatar

ABSTRACT: Nowadays, the use of polyethylene (PE) for bitumen modification has become widespread due to the major advantage of environmental sustainability. For the application of PE modified binder, it is necessary to first study the compatibility of polyethylene and bitumen. In this study, low density polyethylene (LDPE) was blended with the binder in five different dosages (1, 2, 3, 4 and 5%) and PE wax (3% and 2% LDPE+1% Wax) was added to the binder in two dosages (3% and 2% LDPE+1%wax). The dispersion of polyethylene in binder was observed using an optical microscopy and images were captured at different time intervals of 30 and 150 minutes. It was noticed that the equivalent diameter of PE particles increased with time. When PE wax introduced to the blend, the total polymer content was maintained as 3%, because the polymer started forming large size agglomerates above 3% dosage. The addition of PE wax improved the polymer dispersity in blend. To understand the phase separation, viscosity-temperature profile of these binders were studied using temperature sweep experiment, in which the viscosity values were measured during temperature decrement from 160 to 100°C and increment from to 100–160°C. The higher difference in viscosity values represent the instability of binder and the binder with higher LDPE content was found to be more instable.

1 INTRODUCTION

Using polyethylene to modify binder is an eco-friendly way of utilizing virgin/waste plastics. The addition of PE not only reduces the required quantity of bitumen, but also it improves the mechanical properties of binders [1, 2, 3]. In 1980's, researchers studied the performance of polyethylene-modified asphalt and reported the addition of PE improved the permanent deformation of asphalt mixes [4, 5]. Later, other researchers [6, 7] conducted studies on PE modified binders to understand the interaction between polyethylene and bitumen. However, the compatibility and properties of polymer-modified binder depends on many factors such as type of polymer, bitumen source, molecular weight of polymer, polymer size and blending conditions. The phase separation depends on the capability of polymer swelling with maltene molecules and asphaltene content [8]. Some of the polymers like polyethylene have limited miscibility/solubility in bitumen, which results in separation of the polymer and bitumen phases.

The aim of this study is to analyze the phase separation of LDPE modified binders by improving the compatibility between LDPE and bitumen with the addition of PE wax. For this purpose, the following tasks were carried out:

1. Measuring the dispersion of polyethylene in bitumen by capturing the microscopic images at fixed time intervals
2. Constructing the viscosity-temperature profiles of LDPE blended binders.

2 MATERIALS AND BLENDING METHODOLOGY

In this study, a base binder of pen 60–70 grade was used. Low density polyethylene (LDPE) and polyethylene wax (PE wax) (density of LDPE and PE wax are 0.923 and 0.880 g/cc, respectively) were blended with base binder in 7 different proportions. The binders used in this study are listed in Table 1.

Table 1. Binders used in this study.

Polymer content	Binder designation
Base binder (0% polymer)	pen 60–70
1% LDPE content	1% LDPE + pen 60–70
2% LDPE content	2% LDPE + pen 60–70
3% LDPE content	3% LDPE + pen 60–70
4% LDPE content	4% LDPE + pen 60–70
5% LDPE content	5% LDPE + pen 60–70
3% PE wax	3% PE Wax + pen 60–70
2% LDPE & 1% PE wax	2%LDPE + 1%PE wax + pen 60–70

All the binders were studied in an unaged condition. The LDPE pellets and PE wax samples were blended with asphalt binder at 180°C for 1.5 hours at 2000 rpm speed.

3 MICROSCOPY

For microscopy experiments, the samples were prepared using heat and cast approach. The bitumen was heated to 180°C and small quantity was poured on the glass slide and kept it in the oven at 135°C for 5 minutes. The microscopic images of all the binders were captured with Zeiss AxiVert 40 MAT fitted with ERc5s camera. The size of each picture was 1083 ìm length and 809 ìm height (2560×1920 pixels). For each material, the photos were taken at minimum five spots with a time interval of 30 and 150 minutes. Examples of such images are shown in Figure 1 for LDPE and PE wax blends after 30 minutes of blending. As the dosage of LDPE increased in the blend, the PE particles started agglomerating and formed large particles within 30 minutes of blending (Figure 1(f)). For each blend, all the five images were analyzed with Avizo software and the equivalent diameter of LDPE particles were determined through image thresholding and reported in Figure 2.

As can be seen in Figure 2 that the equivalent diameter of polymer increased with the time duration and the increased dosage of LDPE. For example, after 30 minutes, the large polymer particle size of 1% LDPE blend was in the range of 51–100 ìm and 5% LDPE has the range of 151–200 ìm. This polymer size increased after 150 minutes and the shift in the range of polymer size from 30 to 150 minutes was 51–100 ìm to 101–150 ìm for LDPE 1% and 151–200 ìm to 201–250 ìm for LDPE 5% blend. This increase in size of polymer represents the phase separation of LDPE and bitumen materials [7].

To reduce the phase separation and make the polymer and bitumen more compatible, PE wax was introduced to the blend. With the addition of above 3% LDPE dosage, the polymer agglomeration increased rapidly with time and hence, we limited the total polymer content (including PE wax) to 3%. In a first attempt, 3% PE wax was blended with the binder and a white layer on the surface of binder was observed

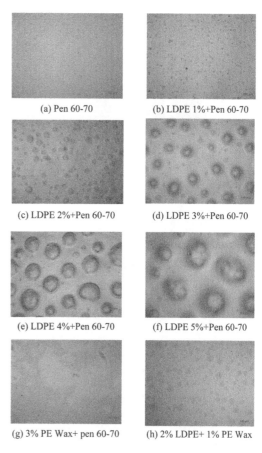

(a) Pen 60-70   (b) LDPE 1%+Pen 60-70

(c) LDPE 2%+Pen 60-70   (d) LDPE 3%+Pen 60-70

(e) LDPE 4%+Pen 60-70   (f) LDPE 5%+Pen 60-70

(g) 3% PE Wax+ pen 60-70   (h) 2% LDPE+ 1% PE Wax

Figure 1. Microscopy images of binders after 30 minutes of LDPE blending.

(Figure 1(g)). The addition of 1% wax to 2% LDPE improved the dispersion of the polymer in the binder (Figure 1(h)). To analyze the dispersion quantitatively, the equivalent polymer particles diameter of 2% LDPE + 1% PE wax blend (Figure 2(f)) were compared with 3% LDPE blend (Figure 2(c)). For the 3% LDPE blend, the maximum diameter of the polymer particles were in the range of 151–200 ìm after 150 minutes and it was reduced to 101–150 ìm after adding 1% PE wax to the blend. Similar behavior was reported by Ho et al., (2006) [6]. This shows that the polymer dispersion improved slightly with the addition of PE wax.

4 TEMPERATURE SWEEP TEST

To measure viscosity, Brookfield rotational viscometer was used with Thermosel temperature controller and SC 4-27 spindle. For measuring viscosity, 10.5 grams of bitumen sample was poured in rotational viscometer containers. To understand the phase

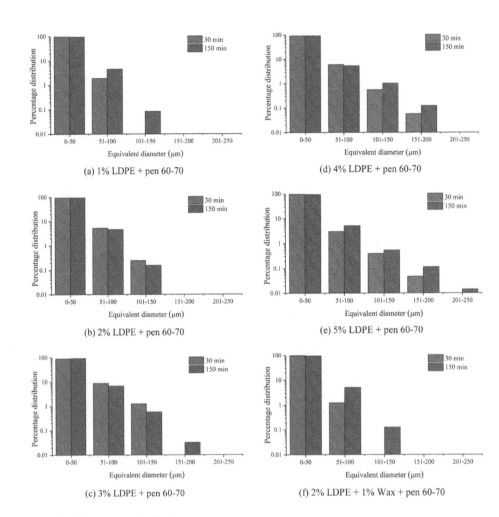

Figure 2.   Polyethylene particle distribution.

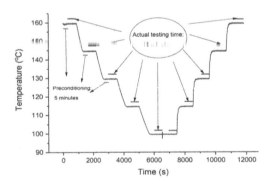

Figure 3.   Temperature sweep protocol.

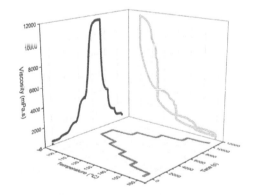

Figure 4.   Time, temperature, viscosity of 3% LDPE + pen 60–70.

separation of PE blended binders, the viscosity of the binders were measured during temperature decrement and increment and the difference in viscosity were used to represent the phase separation of the binders.

In this test method, the binder samples were sheared at a constant speed of 20 rpm at different temperatures of 160, 145, 130, 115, 100, 100, 115, 130, 145 and 160°C and the samples were sheared for 15 minutes. Out of which, the first 5 minutes were considered as pre-conditioning/ pre-shearing and the last 10 minutes

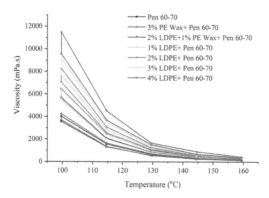

Figure 5. Viscosity of PE blended binders.

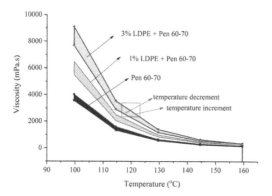

Figure 6. Area calculation of different binders.

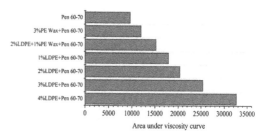

Figure 7. Area under viscosity curves.

improved compatibility of the LDPE blend with the addition of PE Wax.

5 CONCLUSIONS

In this study, the phase separation of LDPE and bitumen were analyzed using microscopic and rheological measurements.

In all the blends, the diameter of LDPE particles increased with time and LDPE dosage. However, such increment was more rapid for higher dosages of LDPE (above 3%) and the polymer agglomerates in the binder in a short time (less than 30 mins) leading to the phase separation of these two components.

Moreover, the addition of PE wax to the LDPE blend improved the dispersion of polymer as clearly observed from the diameter of polymer particles and the area of viscosity hysteresis curves. However, to further support this conclusion, it is necessary to check the dynamic rheological properties and the performance of the binders. Currently, there is ongoing work to develop binders modified using LDPE in combination with other chemicals to reduce elimination, enhance stability, and improve the binder rheological properties.

ACKNOWLEDGEMENT

This work was supported by the Qatar National Research Fund (QNRF): NPRP11S-1128-170041. All the statements are those of the authors. The authors acknowledge the support of Qatar Petrochemical Company in various aspects of providing materials and technical information about LDPE materials.

were the actual testing time as shown in Figure 3. The continuous data capture of time, temperature, viscosity of 3% LDPE + pen 60–70 binder are shown in Figure 4. In the test data, the average of the viscosity at 6, 7, and 8th minute was taken as the viscosity at the corresponding temperature. This test was not conducted for 5% LDPE + pen 60–70 binder. Because, the torque value exceeded 90% at 100°C with the same spindle speed of 20 rpm.

Figure 5 shows the average viscosity values of all LDPE/PE wax/bitumen blends. For each blend, two viscosity curves can be seen (Figure 6). The lower viscosity curve was obtained during temperature decrement (160 to 100°C) and the upper curve was captured during temperature increment (100 to 160°C). The area between these two curves as shown in Figure 6 was measured and reported in Figure 7.

As the LDPE dosage increases, the area between the viscosity curves increases. The area of 4% LDPE binder was 3.4 times higher than the Pen 60–70 binder. Moreover, the wax blends (3% PE Wax and 2% LDPE + 1% PE Wax) has 1.2- 1.6 times higher area than Pen 60–70 binder. These hysteresis values represent the instability of the blends with higher percentages of LDPE and in addition, it confirm the

REFERENCES

Dalhat, M. A., & Al-Abdul Wahhab, H. I. (2017). Performance of recycled plastic waste modified asphalt binder in Saudi Arabia. International Journal of Pavement Engineering, 18(4), 349–357.

Ghuzlan, K. A., Al-Khateeb, G. G., & Qasem, Y. (2013). Rheological properties of polyethylene-modified asphalt binder. In the 3rd Annual International Conference on Civil Engineering.

Hussein, I. A., Iqbal, M. H., & Al-Abdul-Wahhab, H. I. (2005). Influence of M w of LDPE and vinyl acetate content of EVA on the rheology of polymer modified asphalt. *Rheologica acta*, 45(1), 92–104.

D. N. Little, (1991) Performance assessment of binder-rich polyethylene-modified asphalt concrete mixtures (novophalt), Transportation Research Record (1317).

J. W. Button & S. G. Phillips, (1993) Effect of asphalt additives on pavement performance.

S. Ho, R. Church, K. Klassen, B. Law, D. MacLeod & L. Zanzotto, (2006) Study of recycled polyethylene materials as asphalt modifiers, Canadian journal of civil engineering 33 (8) 968–981.

M. Liang, X. Xin, W. Fan, H. Wang, H. Jiang, J. Zhang & Z. Yao, (2019) Phase behavior and hot storage characteristics of asphalt modified with various polyethylene: Experimental and numerical characterizations, Construction and Building Materials 203 608–620.

Upadhyay, S., Mallikarjunan, V., Subbaraj, V. K., & Varughese, S. (2008). Swelling and diffusion characteristics of polar and nonpolar polymers in asphalt. *Journal of applied polymer science*, 109(1), 135–143.

Advances in Materials and Pavement Performance Prediction II – Kumar et al. (eds)
© 2021 Taylor & Francis Group, London, ISBN 978-0-367-46169-0

Design of poroelastic wearing course with the use of direct shear test

D. Rys, P. Jaskula, M. Stienss & C. Szydlowski
Faculty of Civil and Environmental Engineering, Gdansk University of Technology, Gdansk, Poland

ABSTRACT: Poroelastic Road Surfaces (PERS) are characterized by porous structure with at least 20% of air void content and stiffness almost 10 times lower than typical asphalt course. Such properties enable noise reduction up to 12 dB in comparison to SMA 11 mixture. However, the main disadvantage of previously used poroelastic mixtures, based on resin type binders, was their low durability, which resulted in raveling and delamination from the lower layer. This paper presents initial results obtained for new type of PERS mixture, based on highly modified bitumen as a binder instead of resin type binder. The direct shear test was applied to estimate resistance of the mixture to raveling as well as to evaluate interlayer bond quality. Observations of first short test sections with different compositions of new PERS mixtures yielded promising results.

1 INTRODUCTION

1.1 Background

Poroelastic mixtures for road pavements contain about 20% of crumb rubber and should allow to obtain open (porous) structure of the constructed layer, which is almost exclusively wearing course. Such pavements are referred to as Poroelastic Road Surfaces (PERS). PERS technology originates from Swedish research conducted in the 1970s. From 1994 research efforts concerning poroelastic pavements were also conducted in Japan, where few generations of PERS were developed between 1994 and 2009 (Sandberg *et al.* 2010). First trials resulted in reduction of pavement noise by 5 dB, while further studies enabled a decrease even by 12 dB compared to reference SMA wearing course (Świeczko-Żurek *et al.* 2018). The biggest drawback noted during previous studies was very low durability – the pavement lasted only for a few weeks before deterioration. The sources of insufficient durability of poroelastic mixtures observed in previous studies (Bendtsen 2015) were reveling and debonding from lower layer. Despite excellent properties in noise reduction, the insufficient durability still makes PERS useless. Another unsolved problem is finding proper test method, which would allow to design and assess the quality of PERS mixture and layer efficiently.

1.2 Objective and scope

Presented results were obtained throughout realization of research project called SEPOR, which aims to improve durability of Poroelastic Road Surfaces (PERS). This paper describes investigation phase of poroelastic mixture composition and different types of interlayer bonding techniques. Direct shear test was applied both to estimate resistance to raveling and interlayer bond quality.

2 MATERIALS

2.1 Poroelastic mixture composition

During optimization of poroelastic mixtures 13 different mixtures of aggregate and crumb rubber added in dry process with two different highly modified bitumen were evaluated. More details concerning properties and optimization process are described in previous study (Jaskula *et al.* 2019). In this paper the range of results is limited to poroelastic mixture labelled as PSMA (poroelastic SMA). PSMA consisted of mineral and rubber aggregate, limestone filler and highly modified asphalt binder 45/80-80 with at least 7% content of SBS polymer. The proportions of mineral aggregate and crumb rubber are given in Table 1. The four contents of bitumen are marked in Table 1 by B1, B2, B3 and B4.

Three mixtures marked as PSMA5 W4 were selected after laboratory testing phase (see 3 p.) to be produced in full scale. While the mineral and crumb rubber composition remained the same for each mixture, the amount of bitumen was slightly different for each composition. PERS mixtures were produced with the use of ordinary asphalt batch plant. Crumb rubber was added to the pugmill by means of additional conveyor which is normally used for adding reclaimed asphalt pavement. Laying and compacting of poroelastic mixture did not require any modifications in the equipment (Figure 1).

Figure 1. Paving of poroelastic mixture PSMA 5 with highly modified bitumen 45/80-80.

Figure 2. Surfaces of lower layer: A) AC16 B) SMA11 C) SMA11 grooved (after milling) D) SMA11 with geogrid reinforcement.

Table 1. Composition of the poroelastic mixture PSMA 5 tested at laboratory stage and produced for field trials.

Aggregate		Content (% mass of aggregate)			
Type	Sieve [mm]	PSMA5 W3	PSMA5 W4[1]	PSMA8 W3	PSMA8 W4
Mineral	5/8	0	**0**	0	0
aggregate	2/5	60	**72**	72	78
(Gneiss)	0/2	10	**6**	6	13
Filler	<0.063	15	**7**	9	9
(Limestone)					
Crumb	4/7	0	**0**	4	4
rubber	1/4	10	**10**	10	8
	0.5/2	5	**5**	3	3

Bitumen		Content (% mass of mixture)			
45/80-80	B1	10.0	**9.0[1]**	10.0	9.0
(HiMA)	B2	12.0	**11.0[1]**	12.0	11.0
	B3	14.0	**13.0[1]**	14.0	13.0
	B4	–	15.0	16.0	15.0

1) Combinations produced in full scale.

2.2 Laboratory tests of interlayer bonding

Poroelastic mixture was laid on previously prepared slabs made of two typical asphalt mixtures: denser and stiffer for AC16 and more open and less stiff for SMA 11. Two different mixtures for lower layer were used in order to vary the surface texture that would be in contact with poroelastic mixture. Moreover, the effect of grooved texture obtained by milling of the lower layer, and effect of geogrid reinforcement were also considered. Figure 2 presents four various surfaces of lower layer.

The interface layer was applied as a tack coat over the lower layer. Cationic bituminous emulsions with following bitumens were applied: three SBR-modified bitumens 35/50, 50/70 and 70/100 as well as one

Table 2. Combinations of bonding techniques used in the studies.

Type of the lower layer	Type of bitumen used in bitumen emulsion for tack coat	Content of residual bitumen [%]
AC16	35/50+SBR	0.1, 0.2, 0.3
	50/70+SBR	0.1, 0.2, 0.3
	70/100	0.1, 0.2, 0.3
	70/100 + SBR	**0.2[1]**
SMA11	35/50+SBR	0.1, 0.2, 0.3
	50/70+SBR	0.1, 0.2, 0.3
	70/100	0.1, 0.2, 0.3
	70/100 + SBR	0.1, **0.2[1]**, 0.3
SMA11 after milling	70/100 + SBR	0.15, **0.3[1]**
SMA11 after milling and with geogrid	70/100 + SBR	0.15, **0.3[1]**

1) Combinations which were chosen to be used in full scale.

neat bitumen 70/100. The amount of residual bitumen equaled from 0.1 to 0.3 kg/m². Combinations of interlayer bonding techniques are summarized in Table 2.

3 LABORATORY TESTS

3.1 Specimen preparation

Cylindrical specimens for direct shear tests were prepared as follows:

1) By compaction with the use of Marshall compactor (specimens for optimization of mixture composition, with a diameter of 100 mm).
2) By drilling out from two layer slabs compacted in laboratory roller compactor (specimens for laboratory interlayer bonding evaluation, with a diameter size of 150 mm). After compaction of

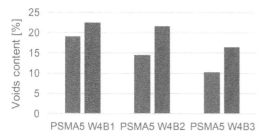

Figure 3. Air void content of the poroelastic mixture PSMA 5 compacted in laboratory and in the field.

the first, lower slab, its surface was covered with bitumen emulsion. After required time needed to obtain emulsion breakdown it was covered with loose poroelastic mixture and the entire set was again subjected to compaction in laboratory roller compactor. Lower slabs of selected specimens were grooved with the use of full-scale milling machine.

3) By drilling out cores from full scale field sections (specimens for field interlayer bonding evaluation, with a diameter size of 150 mm).

3.2 Volumetric properties

Figure 3 shows comparison of air void content in three mixtures of PSMA 5 compacted in laboratory and full scale conditions. In general, compaction obtained in the field was lower than in the laboratory conditions. Such behavior can be caused by elastic deformation of crumb rubber aggregate and relief of hot mixture compression between passes of roller, while laboratory roller compactor applies constant pressure with less ability to relief of compression of rubber aggregate.

3.3 Selected mechanical properties of PERS

Some tests performed during mixture optimization delivered insufficient results in terms of suspected performance. The Cantabro test, which is commonly used to simulate resistance of porous asphalt mixtures to abrasion and raveling, resulted in very low mass loss, bellow 2% for PSMA 5. Wheel tracking test at 60° C according to EN 12697-22 method B caused extremely fast distress of specimens and the proportional rut depth reached approximately 160 mm after 10000 wheel passes while the result of reference SMA 5 mixture was only 3,4%.. The poroelastic mixture PSMA 5 exhibited much lower stiffness modulus (around 200 MPa) in comparison to 1400 MPa obtained for the reference SMA 5 (IT-CY test at 25° C, according to EN 12697-26). These results implicate that the same performance tests that are used for asphalt mixtures may not be valid for poroelastic mixtures.

3.4 Direct shear test for PERS

3.4.1 Justification of choice of direct shear test for evaluation of PERS

The loss of aggregate from the pavement surface, which is commonly called as raveling, is mostly unrelated to the pavement structural design, as it primarily depends on surface-contact mechanics and quality of the mixture aggregate skeleton (Huurman et al. 2009)(Manrique-Sanchez et al. 2018). In the case of typical porous asphalt, the process of raveling can be attributed to the excessive amount of weak rock material in the aggregate. Obviously, replacement of the part of mineral aggregate with crumb rubber in poroelastic mixture has an adverse effect on resistance of the mixture to raveling.

The source of raveling arises from shear stresses caused by vehicle loads and to low internal mixture cohesion. The internal mixture cohesion impact on mixture shear strength too. It can be expected that increase in the internal (inlayer) shear strength will contribute to increase in the resistant to raveling. Direct shear test is also a well known method for evaluation of interlayer bonding quality.

3.4.2 Test procedure

Direct shear tests (Leutner 1979) were performed at 20° C, according to EN 12697-48 with constant rate of deformation 50.8 mm/min. For the purpose of this paper, maximum shear strength τ_{max} was considered both as a measure of internal cohesion of mixture and inlayer bonding quality. The difference was the plane of applied shear stress: in the middle of specimen height in case of testing internal cohesion or in joint between two layers in case of testing interlayer bonding quality. The values presented further represents average values calculated for at least two results obtained from test.

3.4.3 Results of inlayer shear strength of PERS

The average inlayer shear strength vesrus air void content for various mixture combinations are presented in Figure 4. The shear strength of reference mixture SMA 5 was at the level of 1.81 MPa. By comparing this value with result obtained for PSMA it can be concluded that tested poroelastic mixture has about 2.5 times lower inlayer shear strength than typical asphalt mixtures. It should also noted that air void content above 15% has an adverse effect on inlayer shear strength. The significant variability in the air voids results from bitumen content and its effect on compaction of poroelastic mixture.

Figure 5 presents comparison of inlayer shear strength obtained for specimens compacted in laboratory and full scale conditions. Inlayer shear strength of mixture compacted in the field is significantly lower which can be caused by higher air void content (compare to Figure 3 and 4).

After several months of service raveling was observed only on section with PSMA 5 W4B1, with 9,0% of binder content. It confirms that mixtures with

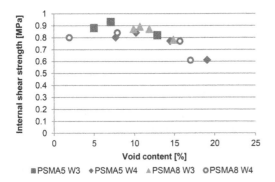

Figure 4. Impact of air void content of poroelastic mixture on inlayer shear strength.

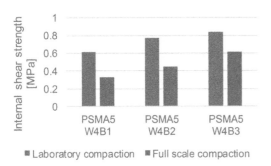

Figure 5. Inlayer shear strength of the poroelastic mixture PSMA 5 compacted in laboratory and full scale conditions.

lower inlayer shear strength are more vulnerable to raveling. Further field observations will allow to verify what is the acceptable level of inlayer shear strength.

3.5 Results of interlayer bonding

The results of interlayer bonding strength are presented in Figure 6 and Figure 7, for specimens prepared in laboratory and full scale conditions respectively.

Results (Figure 7) indicate that obtaining similar interlayer bonding quality in the field as in laboratory conditions can be problematic, despite properly prepared lower layer and application of tack coat. However, after several months of service of trial sections any distresses caused by delamination of PSMA layer have not been observed.

4 SUMMARY

Highly polymer-modified asphalt binders are promising in terms of obtaining reliable and durable poroelastic mixtures.

While Cantabro test and wheel tracking test did not provide reasonable results, direct shear strength test

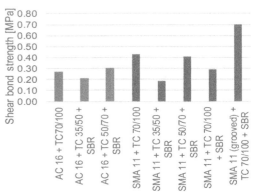

Figure 6. Interlayer bonding shear strength of specimens prepared in laboratory (TC – tack coat).

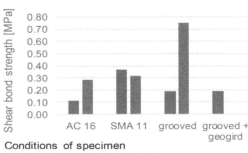

Figure 7. Interlayer bonding shear strength of specimens prepared in full scale and in laboratory.

can be used both to assess internal cohesion, resistance to raveling of poroelastic mixtures and interlayer bonding quality.

The problem of raveling occurred on one out of three full scale test sections. Poroelastic mixture used in this section had the lowest bitumen content (9% by mass) and lowest shear strength (0.33 MPa) simultaneously.

The problem of debonding of poroelastic layer was not reported on full scale sections regardless of type of the layer beneath.

REFERENCES

Bendtsen, H., 2015. Performance og PERS. *Final PER-SUADE seminar*.
Huurman, M., Mo, L.T., & Woldekidan, M.F., 2009. Unraveling Porous Asphalt Concrete, Towards a Mechanistic Material Design Tool. *Road Materials and Pavement Design*, 10 (sup1), 233–262.
Jaskula, P., Szydlowski, C., Stienss, M., Rys, D., & Jaczewski, M., 2019. Durable poroelastic wearing course SEPOR with highly modified bitumen. *In: AIIT International Congress on Transport Infrastructure and Systems in a changing World, TIS Roma 2019*.

Leutner, R., 1979. Untersuchung des Schichtenverbundes Oberbau. *Bitumen*, 3, 84–91.

Manrique-Sanchez, L., Caro, S., & Arámbula-Mercado, E., 2018. Numerical modelling of raveling in porous friction courses (PFC). *Road Materials and Pavement Design*, 19 (3), 668–689.

Sandberg, U., Goubert, L., Biligiri, K.P., & Kalman, B., 2010. *State-of-the-Art regarding poroelastic road surface, WP 8 Information Management.*

Świeczko-⁻urek, B., Goubert, L., Ejsmont, J.A., & Ronowski, G., 2018. Poroelastic Road Surfaces State of the Art. *In: Proceedings of the Rubberized Asphalt Asphalt Rubber 2018 Conference.* 625–643.

Advances in Materials and Pavement Performance Prediction II – Kumar et al. (eds)
© 2021 Taylor & Francis Group, London, ISBN 978-0-367-46169-0

Modeling changes in asphalt mixture properties with RAP content

N.F. Saleh, D. Mocelin, C. Castorena & Y.R. Kim
North Carolina State University, Raleigh, USA

ABSTRACT: This paper investigates whether an existing framework that predicts the changes of asphalt mixture properties as a result of changes in asphalt binder modulus caused by oxidative aging can be expanded to predict changes caused by the inclusion of recycled asphalt pavements (RAP). This study stipulates that mastercurves of mixtures of similar gradation and component material sources but with different RAP contents can coincide if shifted horizontally along the log frequency axis, such that the shift factor can be related to the change in binder modulus. Changes in mixture fatigue properties are also shown to be a result of changes in asphalt binder properties. Under the right conditions, this study shows that the properties of a mixture containing a certain RAP content and of a certain age level can be predicted from the short-term aged properties of a mixture containing no RAP, or vice-versa.

1 INTRODUCTION

Asphalt binder undergoes long-term oxidative aging during its service life. Oxidative aging, consequently, increases the asphalt mixture's stiffness and decreases its cracking resistance, which directly impacts the asphalt pavement performance. At the end of their service life, aged asphalt pavements are reclaimed and repurposed for use in constructing new asphalt pavements. The effect of either (oxidative aging or the addition of reclaimed asphalt pavement (RAP)) on the performance of asphalt mixtures is of interest to the pavement community.

The NCHRP project 09-54 has proposed a framework by which changes in the asphalt mixture linear viscoelastic and fatigue properties can be predicted as a result of changes in the asphalt binder properties caused by oxidative aging (Kim et al. 2019, Saleh et al. 2019).

The framework relies on the time-aging superposition concept, which for mixture modulus, implies that mastercurves of different aging levels coincide when shifted horizontally on the log reduced frequency axis. The horizontal shift factor is calculated as a function of the change in binder modulus from the reference condition, considered in their research as the short-term aging (STA) condition. The change in fatigue properties derived from the cyclic fatigue test, i.e. the damage characteristic curve and the energy-based failure criterion D^R, can also be predicted using a similar approach. The damage characteristic curve, which is a function of the pseudo stiffness denoted by C (integrity of the specimen) in terms of the damage parameter S (amount of fatigue damage in the specimen) can be shifted by rescaling S such that dC/dS decreases with increasing age level. The rescaling of S and the decrease of D^R with aging can both be related to the change in binder modulus.

In predicting the change in mixture properties, the change in binder modulus can be considered a state-variable such that source of change is irrelevant. A lower temperature, a higher frequency or aging level can cause a similar change in binder modulus and thus induce a single change in mixture property.

This paper investigates whether the change of mixture linear viscoelastic and fatigue properties can be predicted as a result of changes of asphalt binder modulus due to the addition of RAP. In other words, the source of change in binder modulus can possibly be expanded to include the RAP content as well as the change in temperature, frequency, and age level.

2 METHODOLOGY

2.1 Materials

Three laboratory-mixed, laboratory-compacted mixtures were evaluated in this study. The mixtures were designed in the laboratory in accordance with North Carolina Department of Transportation (NCDOT) specifications. The mixtures IDs and their properties (% RAP, RBR, NMAS, % AC, and virgin binder PG) are shown in Table 1. The % RAP shown in Table 1 is defined as the % RAP by weight of total aggregates. The same source of aggregates and RAP were used for all mixtures. The RS9.5B 30% and RS9.5B 50% mixtures were designed with a softer binder grade than RS9.5B 0% following NCDOT specifications. All three mixtures were designed to match a single gradation.

Table 1. Identification and properties of selected mixtures.

Mixture ID	% RAP	RBR	NMAS	% AC	Binder Grade
RS9.5B 0%	0%	0%	9.5 mm	6.6%	PG 64-22
RS9.5B 30%	30%	∼23%	9.5 mm	5.8%	PG 58-28
RS9.5B 50%	50%	∼40%	9.5 mm	5.2%	PG 58-28

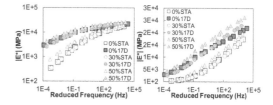

Figure 1. Mastercurves in log-log scale (left) and semi-log scale (right) for three mixtures at two aging levels.

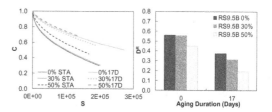

Figure 2. Damage characteristic curves (left) and D^R values (right) for three mixtures at two aging levels.

2.2 Sample preparation and test methods

2.2.1 Mixture aging

The mixtures were subjected to STA at 135° C for 4 hours prior to long-term aging. The NCHRP 09-54 long-term aging method was followed (Elwardany et al. 2017, Yousefi Rad et al. 2017). The STA loose mixtures were thinly spread into pans ($18''L \times 13''W \times 1''H$) and aged in a forced air draft oven at 95°C for 17 days. After long-term aging, the materials were taken out of the oven and mixed together in order to obtain a uniform mixture, and then the mixture was left to cool to room temperature before being reheated again to the compaction temperature and compacted.

2.2.2 Micro-extraction and recovery

To obtain the binder modulus, a sample of each mixture at each aging level was obtained and the binder was extracted and recovered before testing using a dynamic shear rheometer (DSR). Extraction was done using a centrifuge, and recovery using a rotary evaporator with a solvent mixture of toluene and ethanol (85:15) following the methodology proposed by Farrar et al. (2015) at WRI.

2.2.3 Binder testing

The binder was tested to obtain the aging index property proposed by NCHRP 09-54, which is $\log |G*|$ at 64°C, 10 rad/s (Kim et al. 2018). Anton Paar MCR 302 rheometer with parallel plate geometry and 1% strain was used without noteworthy deviation from the linear viscoelastic range.

2.2.4 Mixture testing

Small specimens were fabricated following AASHTO PP 99 and testing using an Asphalt Mixture Performance Tester (AMPT). Dynamic modulus testing was conducted following AASHTO TP 132 at 4°C, 20°C, and 40°C at 0.1 Hz, 0.5 Hz, 1 Hz, 5 Hz, 10 Hz, and 25 Hz. Cyclic fatigue testing was conducted following AASHTO TP 133 at 10 Hz and 18°C. Two replicates for dynamic modulus testing and three replicates for cyclic fatigue were obtained.

3 RESULTS

Figure 1 presents the dynamic modulus mastercurves in log-log and semi-log scales. The results demonstrate an increase in modulus value with aging, as expected.

The RS9.5B 0% and 30% mixtures exhibit similar modulus at both STA at 17 days of aging, whereas RS9.5B 50% mastercurve is slightly higher at both aging levels.

The damage characteristic curves and D^R are shown in Figure 2. It shows an upward shift of the damage characteristic curves as the aging level increases for all mixtures. Whereas aging increases the material' stiffness such that, for a given S value, the aged material exhibits greater stiffness or higher C values, the aged material actually becomes more prone to damage (i.e., the damage evolution is faster). This phenomenon can be reflected by higher C values at failure, indicating that the material becomes less tolerant to damage compared to the STA condition. Again, RS9.5B 0% and 30% have similar damage characteristic curves at STA whereas RS9.5B 50% curve is located higher. At 17 days of aging, all three mixtures exhibit similar damage characteristic curves. An indicator of the material's diminishing toughness with age is the D^R value, which exhibits a decreasing trend with aging. A higher D^R value generally indicates better fatigue resistance compared to a lower D^R value. RS9.5B 50% mixture has a lower D^R at both age levels as compared to the other mixtures.

Finally, the binder testing yielded a $\log |G*|$ value at 64°C, 10 rad/s of extracted and recovered binders from STA mixture of 0.940, 1.167, and 1.462 for RS9.5B 0%, 30%, and 50% respectively at STA. The $\log |G*|$ values at long-term aging (LTA) for the three mixtures are 2.751, 2.578, and 2.684 respectively.

4 DISCUSSION AND ANALYSIS

According to the NCHRP project 09-54, the time-aging shift factor obtained from Equation (1), which

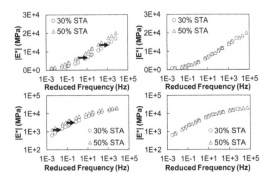

Figure 3. Mastercurves at STA in both semi-log and log-log scales before shifting (left) and after shifting (right).

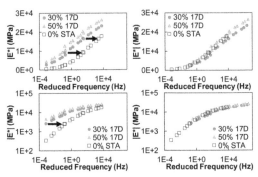

Figure 4. Mastercurves of RS9.5B 30% and 50% at LTA in both semi-log and log-log scales before shifting (left) and after shifting to collapse with mastercurve of RS9.5B 0% at STA (right).

is a function of the change of binder modulus with respect to a reference condition, and the time-temperature shift factor from Equation (2) are used to calculate a reduced frequency as shown in Equation (3), which if plugged into Equation (4) yields the mixture modulus at the combination of age level, temperature, and frequency of interest.

$$\log(a_A) = c \times (\log |G^*|_{LTA,Tref,10rad/s}$$
$$- \log |G^*|_{STA,Tref,10rad/s}) \qquad (1)$$

$$a_T = 10^{\alpha_1 \left(T - T_{ref}\right)^2 + \alpha_2 \left(T - T_{ref}\right)} \qquad (2)$$

$$f_r = f \times a_A \times a_T \qquad (3)$$

$$\log(|E^*|) = \delta + \frac{\alpha}{1 + e^{\beta + \gamma \log(f_r)}} \qquad (4)$$

where a_A = time-aging shift factor; G^* = binder dynamic shear modulus; a_T = time-temperature shift factor, c = slope of the line passing through zero of mixture a_T and corresponding difference of log $|G^*|$. T = temperature; T_{ref} = reference temperature; f_r = reduced frequency; f = frequency; $|E^*|$ = dynamic modulus; $\alpha_1, \alpha_2, \delta, \alpha, \beta, \gamma$ = fitting parameters.

The goal of this investigation is to check whether a shift factor based on the change in binder modulus can be utilized to predict the change in the mixture modulus with a certain RAP content. For a first trial, RS9.5B 30% STA and RS9.5B 50% STA are considered. Both mixtures have the same aggregate and RAP source, gradation, and virgin binder grade. The difference in log $|G^*|$ between RS9.5B 30% and 50% is 0.294. As shown in Figure 3, a good collapse between the mastercurves of the two mixtures can be achieved with an optimized c value of 3.1, indicating that the modulus of a mixture containing 50% RAP can be predicted by shifting the mastercurve of a mixture with 30% RAP, or vice versa.

Since RS9.5B 30% and RS9.5B 0% have similar modulus values from Figure 1, it can be deduced that

RS9.5B 50% mastercurve can also be made to collapse with RS9.5B 0% mastercurve, although the latter has a different virgin binder grade. However, in this shifting, the optimized c value is approximately 1 since the difference in log $|G^*|$ between RS9.5B 50% and 0% is greater than the difference in log $|G^*|$ between RS9.5B 50% and RS9.5B 30%. This implies that RS9.5B 30% cannot be properly predicted from RS9.5B 0% because their log $|G^*|$ values are different whereas their mastercurves are similar. This indicates that the effect of adding low to moderate RAP quantities is directly evident through the increase in binder modulus, but might not be similarly manifested in the mixture modulus, especially with the use of a softer virgin binder grade. Hence, this approach is useful in predicting the change in mixture modulus caused by the addition of RAP when the same virgin binder grade is used, or when a softer virgin binder is used and high RAP content is added. However, the same c value cannot be generalized.

Figure 4 shows that a good collapse can be achieved between the 17D mastercurves of both RS9.5B 30% and 50% and the STA mastercurve of RS9.5B 0% using a c value of 1.55, which is the value reported by NCHRP project 09-54 for RS9.5B 0%. This indicates that the mastercurve of a mixture at STA can be used to predict the mastercurves of similar mixtures with different RAP contents at STA and LTA with softer virgin binder grade by knowing only the difference in binder modulus.

The NCHRP project 09-54 proposed the use of Equations (5), (6), and (7) to predict the change of the damage characteristic curves with aging. The difference in binder modulus can be used to calculate a scaling factor for S as shown in Equations (5) and (6), which if plugged into Equation (7), yields C at the aging duration of interest.

$$\log(a_A) = 0.2025 \times (\log |G^*|_{LTA,Tref,10rad/s}$$
$$- \log |G^*|_{STA,Tref,10rad/s}) \qquad (5)$$

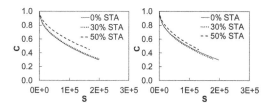

Figure 5. Damage characteristic curves at STA before shifting (left) and after shifting (right).

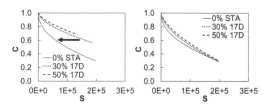

Figure 6. Damage characteristic curves of RS9.5B 30% and 50% at 17D before shifting (left) and after shifting (right).

$$S_r = S/a_A \qquad (6)$$

$$C = 1 - C_{11}(S_r)^{C_{12}} \qquad (7)$$

where a_A = scaling factor, S = damage parameter, C = pseudo stiffness, C_{11}, C_{12} = fitting parameters.

To calculate the scaling factor from the difference in binder modulus between RS9.5B 50% and RS9.5B 30%, it can be shown in Figure 5 that relatively good collapse can be achieved using the given regression parameter in Equation (5). This indicates that damage characteristic curves of mixtures containing 30% and 50% RAP with a softer virgin binder can be predicted knowing the 0% RAP mixture damage characteristic curve and the binder modulus.

Figure 6 shows that a good collapse can be achieved between the 17D damage characteristic curves of both RS9.5B 30% and 50% and the STA curve of RS9.5B 0%. This indicates that a curve of a mixture at STA can be used to predict the curves of similar mixtures containing different RAP contents at STA and LTA using only the difference in binder modulus.

The following equation reported by NCHRP project 09-54 was used to predict the change in D^R.

$$\frac{D^R_{LTA} - D^R_{STA}}{D^R_{STA}} = -0.3023 \times \qquad (8)$$

$$(\log |G^*|_{LTA,Tref,10rad/s} - \log |G^*|_{STA,Tref,10rad/s})$$

The D^R at STA of RS9.5B 30% and 50% can be predicted from the D^R of RS9.5B 0% at STA using the difference in binder modulus. Similarly, the D^R of RS9.5B 30% and 50% at 17D can be predicted from the D^R of RS9.5B 0% at STA. The results can be shown in Figure 7.

It is clear that D^R is not predicted with good accuracy, especially for RS9.5B 50%. The results of

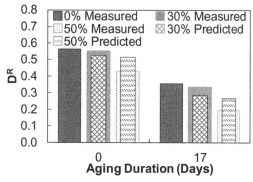

Figure 7. Measured and predicted D^R for STA and LTA.

NCHRP project 09-54 found that the change in D^R is not well correlated with the change in binder modulus. So, it is not surprising that a change in binder modulus to increasing RAP content will also not yield good predictions. Other functional forms relating D^R to the difference in binder modulus might need to be investigated in the future.

5 CONCLUSIONS

The conclusions of this study are as follows:

– The changes in mixture modulus can be predicted as a result of changes in binder modulus caused by the presence of RAP or due to oxidative aging if similar sources of materials and similar gradation are followed.
– If a softer virgin binder is used, prediction of changes in mixture modulus due to low or moderate RAP quantities might not be accurate, since the changes in the binder might not reflect the effect of RAP on the mixture modulus.
– The change in damage characteristic curves can be predicted as a result of changes in binder modulus due to the presence of RAP or due to oxidative aging.
– The prediction accuracy of the change of D^R as a result of the change in binder modulus was found to be low. Other means of predicting D^R should be investigated in the future.
– More mixtures and RAP contents should be investigated to support the findings of this study.

REFERENCES

Elwardany, M. D., Yousefi Rad, F., Castorena, C., & Kim, Y. R. 2017. Evaluation of asphalt mixture laboratory long-term aging methods for performance testing and prediction. *Road Materials and Pavement Design* 18(1): 28–61.

Farrar, M. J., Grimes, R. W., Wiseman, S., & Planche, J. P. 2015. Asphalt pavement-micro-sampling and micro-extraction methods, Fundamental Properties of Asphalts and Modified Asphalt III, Quarterly Technical Report.

Federal Highway Administration (FHWA), Contract No. DTFH61-07-D-00005.

Kim, Y. R., Castorena, C., Elwardany, M. D., Yousefi Rad, F., Underwood, B.S., Gundha, A., Gudipudi, P., Farrar, M., & Glaser, R. 2018. NCHRP Report 871: Long-term Aging of Asphalt Mixtures for Performance Testing and Prediction. *Transportation Research Board, Washington D.C.*

Saleh, N. F., Mocelin, D., Yousefi Rad, F., Castorena, C., Underwood, B. S., & Kim, Y. R. 2019. A predictive framework for modeling changes in asphalt mixture moduli with oxidative aging. *Transportation Research Record, Journal of the Transportation Research Board.* In Press.

Yousefi Rad, F., Elwardany, M. D., Castorena, C., & Kim, Y. R. 2017. Investigation of proper long-term laboratory aging temperature for performance testing of asphalt concrete. *Construction and Building Materials* 147: 616–629.

Advances in Materials and Pavement Performance Prediction II – Kumar et al. (eds)
© 2021 Taylor & Francis Group, London, ISBN 978-0-367-46169-0

Determining optimum doses of palm oil rejuvenators for recycled blends

D.B. Sánchez & S. Caro
Universidad de los Andes, Bogotá, Colombia

A.E. Alvarez
Universidad Industrial de Santander (UIS), Colombia

ABSTRACT: This study aims at exploring different methodologies to select the amount of two palm-oil based rejuvenators that allows maximizing the restored properties of recycled blends; that is, the blend of virgin binder, Reclaimed Asphalt Pavement (RAP) binder and rejuvenator. Three dosage selection methods based on Penetration, Softening Point and High Temperature Superpave Performance Grade (PGH) of the recycled blends in their unaged and aged states were evaluated. In addition, an approach based on the thermodynamic properties of the recycled blends was also explored. The dosage selection was determined as a function of the target virgin binder properties in unaged and aged conditions. The optimum palm-oil rejuvenator dose recommended corresponds to the average dose that restores penetration in unaged conditions and the PGH of the recycled blend.

1 INTRODUCTION

The scarcity and increased cost of virgin aggregates and binders used in the construction of flexible pavements, along with more rigorous environmental regulations, have motivated to increase the amount of reclaimed asphalt pavement (RAP).

Recycled mixtures containing high quantities of RAP (i.e. more than 20% by weight of the total asphalt mixture) are stiffer and more brittle as compared to their virgin counterparts, generating mixtures with improved rutting resistance but reduced fatigue and thermal cracking resistance (Copeland 2011; Epps et al. 2018). The most popular approach to mitigate this includes the use of softer (high penetration grade) binders. This is a useful alternative when the amount of RAP in the recycled mixtures is low (i.e. less than 15% by total weight of the asphalt mixture). However, when higher proportions of RAP are incorporated, or the bitumen in the RAP material is heavily aged, the effect of counteracting the increase in stiffness becomes less evident (West et al. 2013) and impractical (i.e. requirement of a very soft binder, which is pricey and not easily available). In these cases, the use of rejuvenators becomes a better alternative.

Rejuvenators are typically oils, extracts or greases aimed to restore the mechanical performance of the RAP material for another life-service period as part of a new road or rehabilitation of an existing one. Their use in recycled mixtures is commonly associated with enhanced workability, along with environmental and economic benefits (Kaseer et al 2019). However, the mixture design must be adjusted to determine the optimal rejuvenator dosage to restore the recycled

mixtures properties based on the available properties of the RAP and virgin binders.

Various rejuvenator dosage selection methods have been explored in the literature. The most common is based on restoring the physical properties (i.e. penetration, softening point and/or viscosity) of the RAP binder with various amounts of rejuvenators (e.g. Koudelka et al. 2018). Other researchers have used the Superpave performance grade (PG) system to evaluate the changes in the stiffness of the recycled binder due to the addition of rejuvenators (Epps et al. 2018). In this method, typically a minimum and a maximum dose are obtained to ensure low temperature cracking and sufficient rutting resistance. However, defining the proper dosage selection method is a difficult task, particularly because each methodology might provide a different optimum dose for a particular rejuvenator and RAP content, and some methodologies (i.e. those based on the physical properties) do not usually include the various ageing states to which the asphalt binder is exposed during its service life.

This study aims at selecting the amount of two palm-oil based rejuvenators doses that restore as many properties as possible of the recycled blends. The rejuvenators' dosage selection methods are based on Penetration, Softening Point, High Temperature Superpave Performance Grade (PGH) and Surface Free Energy (SFE) of the recycled blends. These methods assessed the properties of the recycled blends in different aging conditions.

This work is particularly pertinent in Colombia, since the country has over 500,000 hectares of palm oil crops, with an annual production of around 1.6 million tons, which ranks the country as the fourth largest producer of palm oil in the world and the first in America (Fedepalma 2019). As one of the world's richest countries in this natural resource, it is paramount to provide an alternative use to palm oil and its by-products. It is expected that these bio-materials could be used as rejuvenators to increase the amount of RAP material added into Hot Mix Asphalt (HMA), as a mean to promote more sustainable construction practices in the country.

2 MATERIALS

A virgin binder with a penetration of 62 (1/10 mm) and a softening point of 48°C was used as the control binder. This binder was also used to artificially produce the 'RAP binder' used in this study by subjecting it to the Rolling Thin Film Oven (RTFO) test, according to ASTM D2872, followed by the Pressure Aging Vessel (PAV) test for 20h, according to ASTM D6521.

Three rejuvenators were used: 1) crude palm oil, 2) hard stearin, which constitutes one of the solid by-products of the palm oil after a fractioning and refinery process, and a 3) vegetable-based rejuvenator labelled as "control rejuvenator", commercially available and widely used in the United States.

These materials were used to prepare recycled asphalt blends comprised of 70% virgin binder, 30% RAP binder and various combinations of the rejuvenator. The decision to rejuvenate recycled blends and not directly the aged RAP binder was based on previous studies (Epps et al. 2018), who found that the presence of virgin binder in the effective binder blend had an impact on the required amount of rejuvenator.

Since the asphalt blend, or effective binder, in an asphalt mixture depends mainly on the RAP content, the RAP binder contribution and on the mixing temperature, previous calculations based on typical volumetric proportions in an asphalt mixture, estimate that the selected percentages of RAP and virgin binders are equivalent to a typical asphalt mixture containing between 40 to 70% RAP material.

3 EXPERIMENTAL METHODS

Penetration at 25°C (ASTM D5) and Softening Point (ASTM 36) tests were performed on the virgin binder and on the recycled blends containing different doses of rejuvenators, in their unaged, short (RTFO) and long term (PAV) aged conditions.

In addition, Dynamic Shear Rheometer (DSR) tests were performed on the virgin binder and the recycled blends containing two doses of each rejuvenator in their unaged and short-aged states according to AASHTO M 320 to obtain their PGH. Only the PGH was determined, as the behavior of the materials at high temperatures is the concern in a tropical country like Colombia that does not reach temperatures below 0°C.

The SFE components of the virgin binder and the asphalt blends with different doses of rejuvenators in unaged conditions were measured by means of the Wilhelmy Plate method (Hefer et al. 2006). The SFE is defined as the energy required to create one unit of area of the material in vacuum conditions, and it is strongly related to the adhesion quality and durability of asphalt-aggregate systems (Bhasin et al. 2007). Based on the Good-Van Oss-Chaudhury theory, the total SFE (Γ) can be decomposed in three components: a) a monopolar basic component, Γ^-, b) a monopolar acid component, Γ^+, and c) a non-polar component, Γ^{LW}, as follows:

4 REJUVENATOR DOSAGE SELECTION METHODS

Based on experiences reported in the literature, recycled blends were prepared with a low recycling agent dose of 3% and a high recycling agent dose of 10% by weight of the virgin binder (i.e. part of the virgin binder was replaced by a certain amount of rejuvenator). In addition, an asphalt blend with no rejuvenator was used as reference. The following sections explain the different methods explored to select the optimum rejuvenator dosage.

4.1 Method 1: restoring the penetration of the virgin binder in unaged and aged states

In this method, the target or minimum reference value for rejuvenation of the recycled blends in their unaged, RTFO and PAV aged conditions, are determined as a function of the virgin binder penetration in its unaged or short and long-aged states, accordingly (Figure 1). The error bars in Figure 1, represent the standard deviation of the three measurments conducted for each recycled blend in each condition.

The results show that the ageing condition has a significant effect on the dosages for restoring penetration of the recycled blends. For instance, Figure 1a and b show that all rejuvenators reduce the penetration of the recycled blends to the target level of the virgin binder in its unaged, and RTFO aged state respectively, but not that of the PAV state (Figure 1c). This might indicate that the rejuvenated blends are less susceptible to ageing than the RAP binder, meaning that they are less stiff, and therefore their fatigue and thermal cracking resistance could be improved in the long-term, when used as part of a recycled mixture. Figure 1 also shows that the rejuvenators dose needed to restore the properties of the virgin binder in unaged state are higher (3.7 − 4.4%) than those required to target the virgin binder properties in RTFO state (0.9 − 3.0%). At this stage, the rejuvenators' doses below 3% might not have a pronounced effect on the recycled mixture's resistance; therefore the rejuvenators' dose results obtained to the target level of the virgin binder in its unaged state might

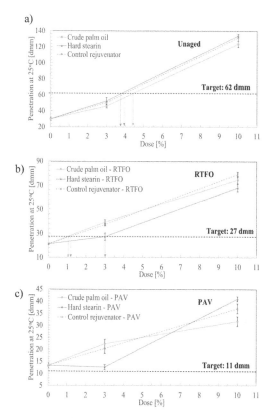

Figure 1. Penetration of the recycled blends in a) unaged, b) short-aged, and c) long-aged states.

provide better results to restore the physical properties of the virgin binder.

4.2 Method 2: restoring the softening point of the virgin binder in unaged and aged states

Similarly to the previous method, the target or reference minimum value for rejuvenation of the recycled blends was determined as a function of the virgin binder softening point in its unaged or short and long-aged states, as presented in Figure 2. This figure shows that all rejuvenators can target the softening point of the virgin binder in its original, but not that corresponding to its RTFO and PAV aged states. The dosage of each rejuvenator required to target the virgin binder properties in its unaged state are higher than those obtained targeting the penetration of the virgin binder in both unaged and short-aged conditions.

4.3 Method 3: restoring the High Performance Grade (PGH) of the virgin binder

For this method, the DSR results were used to determine the PGH of the recycled blends by the rutting parameter $|G*|/sin\delta$, according to the Superpave binder specifications (AASHTO M320), which

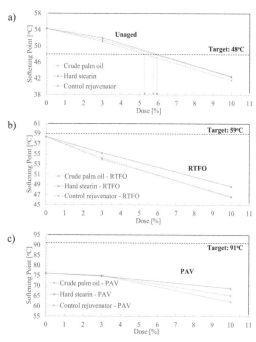

Figure 2. Softening point of the recycled blends in a) unaged, b) short-aged, and c) long-aged states.

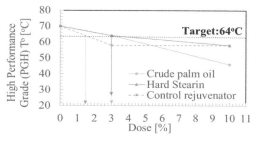

Figure 3. PGH of the recycled blends.

includes different requirements of the materials at unaged and short-aged states, (Figure 3).

Results from this figure show that the Crude palm oil and the Hard stearin require the same dose of rejuvenator (3% by weight of the virgin binder) to target the virgin binder PGH, while the Control rejuvenator requires a lower dose (1.5% by weight of the virgin binder) to deliver the same virgin binder PGH. In all cases, the rejuvenators dosage obtained from this method are lower than those obtained from the previous two methodologies.

4.4 Method 4: restoring the thermodynamic properties of the virgin binder

Five recycled blends were prepared containing 0, 3, 7, 10 and 12% of rejuvenators, by weight of virgin binder. Figure 4 shows that no clear trend in the total SFE was observed in any of the three rejuvenators to attain the

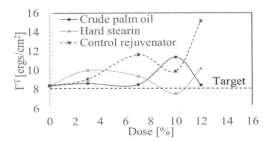

Figure 4. Total SFE of the recycled blends at different rejuvenator's dosages.

Table 1. Selected palm oil based rejuvenators doses.

Rejuvenator	Methodology		
	Penetration at 25°C Unaged	High T° PG Grade (PGH)	Average
Crude palm oil	4.4%	3.0%	3.7%
Hard Stearin	3.7%	3.0%	3.4%
Control rejuvenator	3.9%	1.5%	2.7%

same thermodynamic properties as the virgin binder at a particular dose. Similar trends were observed for the individual components of SFE (Eq. 1), which are not showed here for brevity. Therefore, this method is not suitable to select the dosage of rejuvenator required to deliver the same adhesion quality of the virgin binder.

5 OPTIMUM DOSE SELECTION

The proposed palm-oil based rejuvenators' dose required to maximize the amount of restored properties of the recycled blends is the average dose that restores the penetration of the virgin binder in unaged state and the PGH (Table 1). Because penetration provides information about a physical property of the material and it is still used in several countries to classify asphalt binders, and the PGH includes the two fundamental viscoelastic properties of the material (i.e. $|G * |, \delta$) in unaged and short-aged states, such dose would allow to restore the physical and rheological properties of the recycled blends. In addition, although the amount of work is increased by testing both penetration and PGH, it should be noted that satisfying penetration (in unaged state), does not neccesarily mean that the PGH is restored. The optimum doses obtained from the penetration in unaged condition for all rejuvenators are higher (3.7–4.4%) than those obtained through the PGH method (1.5–3.0%), thus using only penetration results will end up in recycled blends with lower PGH values than the virgin binder.

6 CONCLUSIONS

In terms of the first method the rejuvenators were able to restore penetration of the unaged and RTFO aged virgin binder, with lower rejuvenator doses at RTFO. Similarly, in the second method, the rejuvenators showed to restore the softening point properties in the unaged state, yielding larger doses than in the first method. However, in the short and long- aged states, the virgin binder properties were not restored. These results are positive because the rejuvenated blends have lower stiffness than their counterparts, which increases their fatigue and thermal cracking resistance in the long-term.

The third method –restore the PGH of the virgin binder–provided the lowest rejuvenator dosages. However, this methodology includes fundamental viscoelastic properties ($|G * |, \delta$) of the recycled blends in unaged and short-aged states, which in turn captures a more comprehensive behavior of the materials. The method based on SFE properties, failed to provide a selected dose of rejuvenators to target the virgin binder properties.

Based on these results, to maximize the restored properties of the recycled blends, the palm-oil based rejuvenator's dose is recommended as the average dose that restores the penetration in unaged conditions and the PGH. Notice that this selection does not guarantee maintaining mechanical properties in the long-term aged state of the samples. Although fatigue is not a main concern in the binder-rejuvenator blends due to their low stiffness, evaluation of this response is recommended in future studies.

REFERENCES

Bhasin et al. (2007). Surface free energy to identify moisture sensitivity of materials for asphalt mixes. Transportation Research Record: Journal of the Transportation Research Board, 2001, 37–45.
Copeland, A (2011). Reclaimed asphalt pavement in asphalt mixtures. State of the practice (FHWA-HRT-11-021
Epps et al. (2018). The effects of recycling agents on asphalt mixtures with high RAS and RAP binder ratios. NCHRP 9-58. Transportation Research Board. College Station, Texas.
Fedepalma (2019). El Palmicultor. August Edition. No. 570.
Hefer et al. (2006). Bitumen surface energy characterization using contact angle approach. Journal of Materials in Civil Engineering. 18 (6): 759–767.
Kaseer et al. (2019). Use of recycling agents in asphalt mixtures with high recycled materials contents in the United States: A literature review. Construction and Building Materials. 211, 974–987.
Koudelka et al. (2018). The use of rejuvenators as an effective way to restore aged binder properties. Proceedings of 7th Transport Research Arena TRA. Vienna, Austria.
West et al. (2013). Improved mix design, evaluation, and materials management practices for hot mix asphalt with high Reclaimed Asphalt Pavement content. NCHRP Report.

Advances in Materials and Pavement Performance Prediction II – Kumar et al. (eds)
© 2021 Taylor & Francis Group, London, ISBN 978-0-367-46169-0

Effect of biochar on the basic characteristics of asphalt mixtures

X. Sanchez, T. Somers, H. Dhasmana & O. Owolabi
University of New Brunswick, Fredericton, NB, Canada

ABSTRACT: This study explores the potential of using biochar, which is a renewable, more accessible, and sustainable asphalt additive. Fine biochar passing a 75μ sieve was used in different percentages to prepare asphalt mixes in the laboratory. 4, 8, and 12% by weight of biochar was added to a mix designed for a surface course. The theoretical maximum specific gravity of the asphalt mixes decreased with an increase in biochar percentage and reached an inflection point between 8 and 12%. An improvement in workability in samples with biochar was found during compaction. To study the moisture susceptibility of asphalt mixes made with biochar, Modified Lottman test was also conducted. No signs of stripping were observed in any specimen. Results revealed an improvement in the tensile strength ratio with the inclusion of biochar.

1 INTRODUCTION

Biomass-derived char or Biochar is the product of chemical or thermal transformation of the original feedstock. It is produced at a small or large scale all over the world for different purposes. According to a World Bank sponsored study conducted by the International Biochar Initiative (2014) and Cornell University, almost 43 countries led biochar projects that aimed at increasing soil health, energy efficiency and conservation, climate stability, and successful reuse of biomass residuals. Countries in Africa, South-east Asia, South Asia, and South America used biochar to improve agricultural sustainability and productivity, water filtration, energy supply etc. Biochar has been used for soil remediation in the rainforests of Brazil for thousands of years (Glaser & Birk 2012). Other applications of biochar include waste management, power generation, and as an adsorbent of heavy metals. The thermochemical techniques used to convert the organic material into bio-oil and biochar can be classified into three main categories: pyrolysis, gasification, and hydrothermal carbonization.

Pyrolysis of biomass for the creation of biofuels has become increasingly attractive as a sustainable technique (Macquarrie, Clark, & Fitzpatrick 2012). During pyrolysis, the feedstock is heated between 300 and 1000° C in the absence of oxygen. Different pyrolysis techniques commonly used are fast, slow, and flash pyrolysis (Motasemi & Afzal 2013). A conventional heating method, in which the biomass is heated from the surface to the center of the biomass, or a microwave radiation heating method that heats the biomass from the center outward, can be followed. Although microwave pyrolysis has many advantages as a biochar production technique, it has not been able

to scale well into the production of larger amounts of material as readily as more conventional techniques (Motasemi & Afzal 2013). This study produced biochar at the Bioenergy and Bioproducts Research Lab (BBRL) at the University of New Brunswick that was recently able to perform the first kilogram-scale conversion of biomass to biochar using microwave pyrolysis technique (Salema, Afzal, & Bennanmoun 2017).

Several studies have used biochar as a mineral filler or asphalt binder modifier and concluded that biochar can have a positive impact on the durability and fatigue resistance of asphaltic pavements (Walters, Begum, Fini, & Abu-Lebdeh 2015) (Zhao, Huang, Ye, Shu, & Jia 2014). A 2014 study by Sheng Zhao noted that the use of carbon black or charcoal, both carbon-rich materials, to improve mix workability as an asphalt modifier goes back to at least the mid-1960s in North America (Zhao, Huang, Ye, Shu, & Jia 2014). Their study focused on a rheological analysis of asphalt binder with biochar produced from different pyrolysis techniques blended into asphalt using a high-shear blender. On subjecting the modified asphalt binder to long-term aging, it was found that biochar might have the ability to partially offset the oxidation process during aging. Biochar also stiffened the asphalt concrete mixtures and showed much higher resistance to rutting. It also performed better than activated charcoal. One of the major findings from this study was that the method of pyrolysis did not seem to have a significant effect on the asphalt concrete mix performance. However, the size of biochar particles (below 75 μ) had the most impact. Unlike other recycled industrial by-products such as furnace slag or fly ash which have the potential for heavy metals leaching into the subgrade over time, biochar can act as a carbon sink and remove toxic elements from the environment (Transportation Association of Canada 2013).

2 MATERIALS AND TEST METHODS

2.1 Biochar properties

Biochar was produced from wood pellets, made from spruce sawdust, using microwave pyrolysis technique. It was characterized for its basic fuel properties, and bulk density. Elemental composition analysis showed that the biochar had 69.6–85.7% carbon, 2.4–3.3% hydrogen, and 10.4–26.3% oxygen. It was also noticed that the reduction of volatile matter during pyrolysis resulted in higher heating values and smaller bulk density values for the biochar in comparison to raw wood pellets. Surface morphological analysis of the biochar showed possible adsorption applications of the material (Nhuchhen et al. 2018).

2.2 Mineral aggregate and asphalt binder

PG 58-28 asphalt binder, which is the typical asphalt grade used in the province of New Brunswick (NB), was provided by a local supplier. Respective mixing and compaction temperatures of 148°C and 135°C were used as recommended by the manufacturer. Limestone aggregate was collected from a quarry in Fredericton, NB. Mix design for a Superpave SP 12.5 mix used in the surface layer, provided by the asphalt plant, was followed. The mix design had an aggregate size distribution with 63.3% aggregate passing the 4.75 mm sieve. The optimum asphalt content was 6.1% for all the mix samples. Biochar was incorporated as an additive in dosages of 4, 8, and 12% by weight of the asphalt binder. It was heated and mixed with the aggregate and asphalt binder to make 100 mm diameter briquettes, which were tested for their ability to withstand moisture damage.

3 RESULTS

3.1 Effect on mix workability

The compaction effort to obtain an air void content of 7% ± 0.5% was determined for the control mix. The same compaction effort was applied to the 4% biochar mixes; however, the average air void content dropped to 4%. For the 8% and 12% biochar mixes, the compaction effort was estimated to be approximately 40% less than that applied for the control mixes. It is proposed that biochar could be improving the heat transfer during mixing process and lubricating the mix to achieve the same level of compaction with less effort. Warm mix asphalt additives are frequently used in New Brunswick as mix compaction aid. It is important to note that previous findings concluded that the asphalt binder becomes stiffer and less workable as the

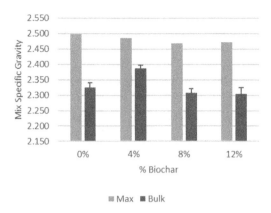

Figure 1. Maximum and bulk specific gravity vs. biochar percentage.

bio-char content increase above 10% by weight (Zhao, Huang, Ye, Shu, & Jia 2014).

3.2 Effect on mix specific gravity

This study followed ASTM D2041 standard to determine the Theoretical Maximum Specific Gravity and ASTM D2726 standard to determine the bulk specific gravity for each mix blend.

Figure 1 shows the impact of adding biochar on specific gravity of the mix. The effect of difference in compaction effort between control and 4% biochar mix samples is evident. Unfortunately, due to a shortage of materials, testing on this mix could not be repeated and it was decided to proceed with the existing results. It is also observed that addition of biochar reduced mix density. This could be due to light weight of the biochar particles.

3.3 Effect on mix volumetric properties

Considering similar mix design for all mixes, changes in volumetric properties were not expected. Voids in Mineral Aggregates (VMA) value for 4% biochar briquettes is not uniform due to low air void content of the mix. For other mixes, it is observed that average VMA values lied between 18 and 19% and the Voids Filled with Asphalt (VFA) values varied from 60 to 65% (Figure 2). The bulk specific gravity of combined aggregates was 2.662 and specific gravity of the binder was 1.02. According to sieve analysis, the percentage of aggregates passing 0.075 mm sieve was 5.4%. Based on this information, other volumetric properties were estimated (Table 1).

3.4 Effect on Indirect Tensile Strength

The asphalt mix samples were tested for moisture damage using ASTM D4867 standard, also referred to as the modified Lottman test. The tensile strength ratio was estimated as the ratio of between the strength of samples subjected to vacuum saturation and a

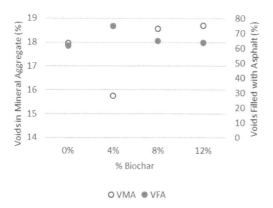

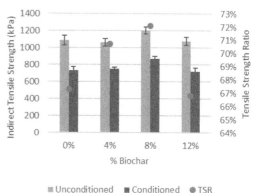

Figure 2. Voids in mineral aggregates and voids filled with asphalt vs. biochar percentage.

Figure 3. Indirect tensile strength values for unconditioned and conditioned mix samples vs. biochar percentage.

Table 1. Air voids, effective and absorbed Asphalt Cement (AC) for different biochar percentages.

Biochar	Air Voids %	Effective AC %	Absorbed AC%
0%	6.9	4.9	1.33
4%	3.9	5.1	1.12
8%	6.5	5.3	0.82
12%	6.8	5.3	0.87

Table 2. TSR values for different biochar percentages.

Biochar	TSR Value (%)
0%	67.3
4%	70.7
8%	72.1
12%	66.8

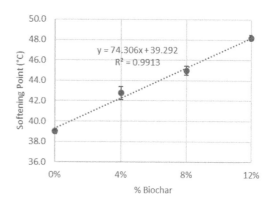

Figure 4. Softening point values vs. biochar percentage.

standard. An increase in the softening point of asphalt binder was observed with an increase in biochar content. The effect of biochar on the stiffness of the asphalt binder is going to be studied in the future using dynamic shear rheometer.

freeze/thaw cycle and a control batch to determine the ability of mix samples to resist moisture damage. This test is an industry standard and part of the battery of tests used by the province to approve or reject contractor mix designs.

TSR values obtained for different mix samples are summarized in Table 2. The TSR value in every case is below 80% however, there was no visible stripping of asphalt binder from aggregate, which can happen during the moisture conditioning of briquettes.

It is observed from Figure 3 that addition of biochar has a significant effect on the Indirect Tensile Strength of conditioned samples. Modification of asphalt mixes with 8% biochar makes the mix less susceptible to moisture-related damage.

3.5 Effect on Asphalt Binder Consistency

The Ring and Ball test was performed to determine the consistency of asphalt binder using ASTM D36

4 CONCLUSIONS

The effect of addition of biochar as a modifier in asphalt mixes was studied. For this purpose, three different biochar percentage values were selected. Standard tests were conducted to determine the workability, volumetric properties, and moisture susceptibility of the asphalt mix specimens. Following conclusions were made from this study:

– It was noticed that increased biochar modification resulted in reduced compaction effort. The improved compactability of the mix could help in achieving lower air void contents during construction.
– Due to the nature of biochar particles, specific gravity values decreased, while VMA increased for the modified mixes in comparison to the control mixes.

- The tensile strength ratio increased with the addition of the biochar content, but it would show an inflection point between 8% and 12%.
- Overall, it was noticed that 12% biochar modified asphalt mix samples demonstrated values almost similar to those of control mix samples.

Future studies could compare the effect of biochar with that of conventional antistripping additives. It is also recommended to investigate the capacity of biochar in reducing the optimum asphalt cement content.

REFERENCES

Glaser, B., & Birk, J. J. (2012). State of the scientific knowledge on properties and genesis of anthropogenic dark earths in central amazonia (terra preta de índio). *Geochimica Et Cosmochimica Acta*, 12.

International Biochar Initiative (2014). Biochar in Emerging and Developing Countries, Retrieved from https://biochar-international.org/biochar-in-developing-countries/. Last accessed on Feb 07, 2020.

Macquarrie, D., Clark, J. H., & Fitzpatrick, E. (2012). The microwave pyrolysis of biomass. *Biofuels, Bioprod. Biorefining*, 11.

Motasemi, F., & Afzal, M. T. (2013). A review on the microwave-assisted pyrolysis technique. *Renewable and Sustainable Energy Reviews*, 13.

Nhuchhen, D. R., Afzal, M. T., Dreise, T., & Salema. A. A. (2018). Characteristics of biochar and bio-oil produced from wood pellets pyrolysis using a bench scale fixed bed, microwave reactor. Biomass and Bioenergy, 119, 293–303. DOI: 10.1016/j.biombioe.2018.09.035.

Salema, A. A., Afzal, M. T., & Bennanmoun, L. (2017). Pyrolysis of corn stalk biomass briquettes in a scaled-up microwave. *Bioresource Technology*, 9.

Transportation Association of Canada. (2013). *Pavement asset design and management guide*. Ottawa: Transportation Association of Canada.

Walters, R., Begum, S., Fini, E., & Abu-Lebdeh, T. (2015). Investigating bio-char as flow modifier and water treatment agent for sustainable pavement design. *American Journal of Engineering and Applied Sciences*, 138-146.

Zhao, S., Huang, B., Ye, X., Shu, X., & Jia, X. (2014). Utilizing bio-char as a bio-modifier for asphalt cement: A sustainable application of bio-fuel by-product. *Fuel*, 52–62.

Advances in Materials and Pavement Performance Prediction II – Kumar et al. (eds)
© 2021 Taylor & Francis Group, London, ISBN 978-0-367-46169-0

A dissipated energy approach for flow number determination

Hamzeh Saqer[1], Munir D. Nazzal[2], Mohammad Al-Khasawneh[3], Ala Abbas[4] & Sang Soo Kim[5]
[1] *Former Research Assistant*
[2] *Professor, University of Cincinnati, Cincinnati, Ohio*
[3] *Research Assistant, University of Cincinnati, Cincinnati, Ohio*
[4] *Professor, University of Akron, Akron, Ohio*
[5] *Associate Professor, Ohio University, Athens, Ohio*

ABSTRACT: The dissipated energy concept was used to investigate the permanent deformation behavior of asphalt mixes. It was found that the dissipated energy can be used to identify the stages of asphalt mixture permanent deformation accumulation. In addition, the dissipated energy can provide better information about the degree of internal damage occurring during the secondary stage of permanent deformation accumulation that leads to the tertiary flow. A new proposed method of determining the flow number was developed based on the dissipated energy. One main advantage of proposed dissipated energy method is that it reflects the changes of the strength and the deformational state of the samples with the number of load cycles. The developed method was able to accurately determine the flow number and reduce variability in the obtained flow number values.

1 BACKGROUND

Repeated application of traffic loading induces damage in asphalt pavement. One of the major distresses in asphalt roads that is primarily related to traffic loads and stresses is rutting. Rutting can be defined as the accumulation of permanent deformations in asphalt pavement along the wheel path (Matthews & Monismith 1992). Depending on the magnitude and frequency of loading, and the properties of asphalt mixture, rutting starts to develop as a result of plastic flow in the surface asphalt layer. For every loading cycle, a certain amount of energy is being input into the asphalt surface layer, and a part of the energy causes deformation. A part of the deformation can be released as elastic rebound, and the rest is dissipated as permanent deformation, fatigue damage, or heat Widyatmoko et al. (1999). Therefore, studying the dissipated energy is essential for a better understanding of pavement rutting behavior.

The dissipated energy concept has been widely used by many researchers to investigate the fatigue cracking resistance of asphalt mixtures (Van Dijk 1975; Rowe 1993; Ghuzlan & Carpenter 2006; and Carpenter & Shen 2006). However, its application in permanent deformation of asphalt pavement is very limited.

The dissipated energy concept implies that when the material is subjected to loading, a certain amount of energy is dissipated by causing internal damage to the material. The area under the stress-strain diagram represents the amount of energy being input into that material. For a purely elastic material, when the load

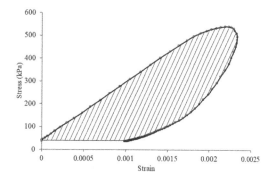

Figure 1. Hysteresis loop for one loading cycle.

is removed, the unloading curve on the stress-strain diagram is identical to the loading curve, therefore the energy is stored elastically, and the total dissipated energy is zero. However, asphalt is a viscoelastic material, and its response to loading has both elastic and plastic components. Accordingly, the loading and unloading curves for HMA do not overlap, and this is referred to as a hysteresis loop. An illustration of the hysteresis loop for a one loading cycle during repeated axial load test is shown in Figure 1. The area inside the hysteresis loop on the stress-strain curve represents the dissipated energy for one loading cycle.

The main advantage of this approach is considering both stresses and strains to characterize pavement behavior, which reflects strength and deformation states of asphalt mixtures under loading, indicating

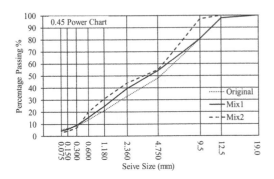

Figure 2. Aggregate gradations on 0.45-power chart.

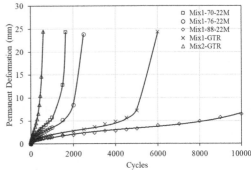

Figure 3. Permanent deformation curves from samples included in the study.

the degree of internal damage occurring under cycling loading. Thus, it can provide a more accurate prediction of long-term rutting behavior, while the strain rate methods might not as they are based on the observed rutting behavior during the test only. It is noted that Tao et al. (2010) have successfully used the dissipated energy concept to characterize the rutting behavior of base course materials under different stress levels.

2 LABORATORY TESTING PROGRAM

2.1 Materials

In this study, a laboratory testing program was conducted to evaluate the rutting performance of different asphalt mixtures. Four plant-produced asphalt mixtures were evaluated, where the aggregate gradation was similar for all mixtures. The Bailey method was used to modify the aggregate gradation for an optimal rutting resistance. Three styrene-butadiene-styrene (SBS) polymer modified binders and one ground tire rubber (GTR) modified binder were used. The polymer-modified asphalt binders meet the performance grades PG 70-22M, PG 76-22M, and PG 88-22M, while the GTR-modified binder was graded as PG 70-22M. Using different binder grades allowed to investigate the contribution of asphalt binder towards rutting resistance. A control mix from a dense-graded asphalt mixture typically used for surface course in Ohio was also included in this study to provide a baseline for comparison. Aggregate gradations used are shown in Figure 2.

2.2 Flow number test

For each mix, four cylindrical specimens measuring 4 in. (100 mm) in diameter and 6 in. (150 mm) in height were fabricated using the gyratory compactor with a target air void content of 7.0±0.5%. Testing was conducted based on AASHTO TP79-13 standard test method for flow number determination.

Testing was performed in a temperature-controlled environmental chamber at a high temperature of 130°F (54°C). Each specimen was conditioned for three hours prior to running the test. The test consists of applying 10,000 load cycles, while each cycle consists of about 0.1 second loading/unloading time and 0.9 second of rest time to generate creep. The repeated axial load was set at 1100 lbs., equivalent to 600 kPa. A contact stress of 30 kPa was maintained during the test to keep the specimen in contact with the loading platens. The resulting axial deformation for each cycle was recorded every 2 milliseconds using an accurate LVDT system. The test was terminated when the total axial deformation of the specimen reached about 1 inch.

3 RESULTS AND DISCUSSION

3.1 Flow number determination based on permanent strain rate

Figure 3 presents the permanent deformation curves for mixes tested in the study. The effect of the binder type had a significant on the permanent deformation behavior of asphalt mixes. As the binder polymer content increased, the resistance to permeant deformation improved.

According to AASHTO TP79-13, the flow number (FN) is defined as the number of loading cycles to reach the tertiary flow (failure). Francken's model, a mathematical fitting model that was recommended by Biligiri et al. (2007) was used to fit the test outputs. The flow number was then determined as the cycle number at which the slope of the permanent deformation curve changes from negative to positive marking the begging of failure as illustrated in Figure 4. Francken's model is provided in Equation (1), where ε_p is the permanent deformation, N is the cycle number, and A, B, C, and D are model fitting parameters.

$$\varepsilon_p = AN^B + C(e^{DN} - 1) \tag{1}$$

The average flow number for each mix determined using Francken's model is shown in Figure 7.

Based on these results, Mix1-88-22M had the best performance in terms of rutting resistance expressed in

348

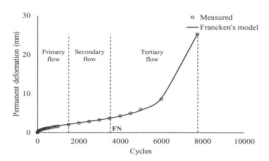

Figure 4. Permanent deformation versus number of cycles using Francken's model.

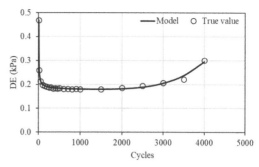

Figure 5. Proposed model for flow number determination.

terms of the average flow number. However, it shows the highest variability of the flow number value among all mixes. Mix2-GTR had the lowest flow number, indicating poor resistance to rutting as compared to the other mixes. It can be observed that the binder grade used had a significant contribution towards permanent deformation resistance of the asphalt mixture. Also, for the same binder grade, the GTR-modified mix performed better than the polymer-modified mix.

3.2 Application of dissipated energy concept

The dissipated energy for each load cycle was determined as the area of the open loop in the stress-strain diagram as indicated in Figure 1. An excel macro program was developed to estimate the dissipated energy using the trapezoidal rule represented in Equation (2), where DE_i is the dissipated energy, σ_a is the maximum axial stress and ε_a is the permanent axial strain for the ith cycle. Note that the second term of the integral was eliminated due to only axial stress being applied in the test (confining pressure, $\sigma_3 = 0$). A plot of dissipated energy per cycle versus number of load cycles was then generated for each specimen tested as shown in Figure 5.

$$DE_i = \int \sigma_i d\varepsilon_i = \int (\sigma_1 d\varepsilon_a + 2\sigma_3 d\varepsilon_h)$$

$$DE_i = \frac{1}{2} \sum_{i=1}^{n-1} (\sigma_{a,i}) + \sigma_{a,i+1})\varepsilon_{a,i+1} - \varepsilon_{a,i} \qquad (2)$$

A new method for determining the flow number is presented herein. The proposed method is based on the dissipated energy change determined from repeated axial load test. The flow number was determined as the number of cycles where the rate of change in dissipated energy changes from a relatively constant value and starts increasing, corresponding to the start of failure as shown in Figure 5. A mathematical model was developed to fit the data as provided in Equation (3), where DE is the estimated dissipated energy per cycle (kPa), N is the cycle number, and A, B, C, and D are model fitting parameters. The model was considered

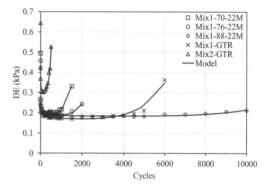

Figure 6. Dissipated energy curves from samples included in the study.

a good fit for the data of all mixes and provided a high coefficient of determination (R^2) of 99.0±0.5%.

$$DE = \frac{A}{N^B} + (CN)^5 + D \qquad (3)$$

The dissipated energy curves are presented in Figure 6. It can be observed that most of the energy dissipates within the first 100-200 load cycles, which represents the consolidation (compaction) stage. In this stage, the energy being input into the specimen is accommodated by particles relocation and sliding and thus, a reduction in the air voids. The dissipated energy had the largest value for the first few cycles and then started to decrease reaching a steady state with a relatively constant value. The length of this stage and the number of load cycles required by each specimen to reach this stage depends on the stability of the mixture. In this stage, the dissipated energy has the lowest values throughout the test, and this value depends on the stability of the mix as well. The dissipated energy in this stage is accommodated mostly by very small strains, which gradually accumulate at a relatively constant rate. By the end of this stage, the dissipated energy per cycle starts to increase drastically, resulting in large plastic deformations and shear distortion of the specimen until failure.

The average flow numbers obtained using the proposed method are also shown in Figure 7. Compared

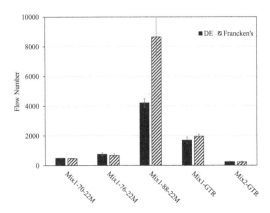

Figure 7. Flow number results from strain rate and proposed method.

to the results from strain rate analysis, the dissipated energy method leads to relatively more conservative flow number values. However, it generally reduced the variability of the results. Another advantage is a more accurate detection of the flow number for specimens that the strain rate analysis was not able to capture, such as in the case of Mix1-88-22M mixture.

Generally, the application of dissipated energy method was able to identify the three zones in asphalt pavement behavior under repetitive loading: primary, secondary, and tertiary stages. The results of the proposed method are in good agreement with the results of strain rate analysis. The variability of the results was generally reduced using the dissipated energy approach.

4 CONCLUSIONS

The flow number is an effective parameter is assessing the permanent deformation resistance of asphalt mixtures. The proposed method was able to detect the flow number based on the rate of change of dissipated energy. The flow number was determined as the point where the first derivative is equal to zero. The main advantage of using the dissipated energy approach is that it reflects both strength and deformational states of the specimen as the number of load cycles increases. It is worth noting that the proposed method was effective in determining the flow number for specimens that Francken's model was not able to detect. The flow number obtained from the dissipated energy curve is generally a bit more conservative as compared to

using Francken's model flow number determination. Overall, the proposed method provided a similar ranking of mixtures with Mix1-88-22M mix as the best performer.

The dissipated energy approach provides insightful information about the degree of internal damage occurring in the asphalt specimen as the test proceeds. It can be used as helpful tool to accurately determine the flow number under repeated loading. The dissipated energy approach also provides an accurate prediction of the long-term permanent deformation behavior of asphalt mixes beyond that observed in the test. It provides efficient ranking of mixtures in terms of their permanent deformation resistance. Overall, it can be concluded that to limit rutting, the dissipated energy should be minimized.

REFERENCES

Biligiri, K., Kaloush, K., Mamlouk, M., & Witczak, M. (2007). Rational modeling of tertiary flow for asphalt mixtures. Transportation Research Record: Journal of the Transportation Research Board, (2001), 63–72.
Carpenter, S.H., & Shen, S. (2006). Dissipated energy approach to study hot-mix asphalt healing in fatigue. Transportation Research Record, 1970(1), 178–185.
Ghuzlan, K.A., & Carpenter, S.H. (2006). Fatigue damage analysis in asphalt concrete mixtures using the dissipated energy approach. Canadian Journal of Civil Engineering, 33(7), 890–901.
Matthews, J.M., & Monismith, C.L. (1992). The effect of aggregate gradation on the creep response of asphalt mixtures and pavement rutting estimates. In Effects of Aggregates and Mineral Fillers on Asphalt Mixture Performance. ASTM International.
Rowe, G.M. (1993). Performance of asphalt mixtures in the trapezoidal fatigue test. Asphalt Paving Technology, 62, 344–344.
Tao, M., Mohammad, L.N., Nazzal, M.D., Zhang, Z., & Wu, Z. (2010). Application of shakedown theory in characterizing traditional and recycled pavement base materials. Journal of Transportation Engineering, 136(3), 214–222.
Van Dijk, W. (1975). Practical fatigue characterization of bituminous mixes. Journal of the Association of Asphalt Paving Technologists, 44, 38–72.
Widyatmoko, I., Ellis, C., & Read, J.M. (1999). Energy dissipation and the deformation resistance of bituminous mixtures. Materials and Structures, 32(3), 218–223.

Advances in Materials and Pavement Performance Prediction II – Kumar et al. (eds)
© 2021 Taylor & Francis Group, London, ISBN 978-0-367-46169-0

Effect of load eccentricity on uniaxial fatigue test results for asphalt concrete mixtures using FE modeling

A. Seitllari & M.E. Kutay
Michigan State University, East Lansing, MI, USA

ABSTRACT: Fatigue phenomenon in asphaltic layers is caused by repeated traffic loading applications and predominantly happens at intermediate temperatures. To better understand and assess the resistance of asphalt mixtures to fatigue cracking, running uniaxial fatigue tests became quite common. However, the most challenging issue with the uniaxial testing is premature end-failures, which is primarily due to load eccentricity effects. While studies provide very valuable technical insights towards reduction of load eccentricity, an investigation of the effects of the eccentricity in fatigue life of asphalt mixtures has not been conducted. The objective of this study was to evaluate the effects of load eccentricity on stress and strain distribution obtained through 3D finite element approach. In general, it was observed that the maximum axial stresses and corresponding strains at the sample ends increases sharply with the increase in load eccentricity, but follow a steady state after load eccentricity of 50%. On the other hand, the corresponding strains at the center of the sample decrease with the increasing load eccentricity. These central strains are the LVDT-measured strains typically used for evaluation of fatigue performance of asphalt mixtures. In the presence of load eccentricity, their use may indicate false fatigue performance of the asphalt mixture. Overall, this research study highlighted the impact of load eccentricity of stress-strain distribution in cylindrical samples used for evaluation of fatigue performance of asphalt mixtures. With the use of efficient finite element modelling approach, there is a potential in development of correction factors for eliminating the effects of load eccentricity for loaded cylindrical samples.

1 INTRODUCTION

Involves the cylindrical geometry of asphalt samples subjected to three-point tension-compression Fatigue phenomenon in asphaltic layers is caused by loading. Timoshenko beam theory and viscoelastic continuum damage theory were coupled together to analyze and simulate the fatigue performance of the various mixture at different testing conditions. This method has proven to be promising, however, it is still considered to be under development (Seitllari & Kutay, 2019). repeated traffic loading applications and predominantly happens at intermediate temperatures.

Excessive tensile strains at the bottom and top of the asphalt layers lead to microcracks, which eventually grow, coalesce and lead to serious structural deterioration (Seitllari, Boz, Habbouche, & Diefenderfer, 2020). Generally, two types of fatigue cracking can occur, depending upon the place the cracks initiate. While bottom-up fatigue cracking is mostly observed in relatively thin asphalt layers because of the flexural bending, top-down fatigue cracks can be seen in both thick and thin asphalt layers on the wheel path evolving between the tire edge and the asphalt layer. To better understand and assess the resistance of asphalt mixtures to fatigue cracking, numerous laboratory tests

have been developed to simulate the traffic load applications in the field. It should be noted that laboratory tests do not fully reflect field conditions. Numerous variables such as specimen preparation, loading mode, stress-strain distribution etc., vitally influence the test results and usually, a shift factor is used to relate laboratory performance to field performance. Recently, Seitllari and Kutay developed a new method called the three-point bending cylinder (3PBC) test, to characterize fatigue cracking performance of asphalt mixtures. The method

Traditionally, the four-point bending fatigue test mode has been the most common test method to characterize the fatigue resistance of asphalt mixtures (Huurman & Pronk, 2012). United States practice for four-point bending beam fatigue (4PBB) test follows AASHTO T 321 *"Standard Method of Test for Determining the Fatigue Life of Compacted Asphalt Mixtures Subjected to Repeated Flexural Bending"*. Nonetheless, these tests are lengthy, cumbersome and expensive. As an alternative, uniaxial fatigue tests (Kutay, Gibson, Youtcheff, & Dongré, 2009; Seitllari, Lanotte, & Kutay, 2019; Zeiada, Kaloush, Underwood, & Mamlouk, 2016) have been gaining wide acceptance for fatigue evaluation of asphalt pavements because of their advantages. Uniform state of stress-strain throughout the sample and ease of application of constitutive models (i.e., Viscoelastic Continuum Damage Theory) to uniaxial testing geometry has been a great advantage.

However, end-failures and end-platens the height of 180 mm and a diameter of 150 mm. Cores were extracted from the center of the gyratory samples using a diamond-coring stand, while the ends of the cores were trimmed by using a masonry saw to obtain smooth and parallel end surfaces. The target air void content for the |E*| samples was 7.0 ± 0.5 % and all samples outside this range were discarded.

The |E*| tests were conducted in general accordance with the AASHTO R 84 "*Developing Dynamic Modulus Master Curves for Asphalt Mixtures Using the Asphalt Mixture Performance Tester (AMPT)*". Samples were subjected to an axial haversine compressive stress at three temperatures (4, 20 and 40°C) and three frequencies (10, 1, and 0.1 Hz) per each temperature. The dynamic modulus master curve was determined through the application of the time-temperature superposition principle (TTS). A single dynamic modulus master curve was detachment are the most challenging issues with uniaxial testing when the sample is subjected to loading. Instead, for the data to be valid, the location of the failure plan should be within the linear variable differential transformer (LVDT) length. The occurrence of a failure plan outside this range leads to inaccurate conclusions regarding the resistance of the samples to fatigue. Failure to properly perform parallel end cuts and/or gluing are the major reasons for premature failure of the test, which leads to excessive sample preparation time and consumption of material. From the practical perspective, obtaining perfectly trimmed ends of the sample followed by proper gluing operation of end platens using a gluing jig can be cumbersome and prone to operator error (Kutay, Gibson, & Youtcheff, 2008).

The presence of load eccentricity on a loaded specimen has serious effects on the stress-strain relationship. LVDT measurements can easily recognize the presence of load eccentricity. High variability of captured specimen response among the attached LVDTs indicates the presence of this phenomenon. Nevertheless, it is a quite complicated task to quantify its effects through experimental investigation. This partially also due to the non-homogenous nature of asphalt mixtures. finally obtained at a reference temperature (T = 21°C) by shifting horizontally the |E*| values recorded during the test. The amount of shift is different at each temperature and defined by the so-called shift factor coefficient a(T). A second-order polynomial equation was used to develop the relationship between shift factors and the corresponding temperature. During the shifting process, the shift factors were varied until a good sigmoid fit to the | E*| data at all the temperatures were obtained. The following sigmoidal function was used:

Load eccentricity is the major parameter governing the validity of the uniaxial fatigue data. While studies provide very valuable technical insights towards the reduction of load eccentricity, an investigation of the effects of the eccentricity in fatigue life of asphalt mixtures have not been conducted yet. The objective of this study is to evaluate the effects of load eccentricity on stress and strain distribution obtained through the 3D finite element approach.

$$\log(|E^*|) = c_1 + \frac{c_2}{1 + e^{(-c_3 - c_4 \log(f_R))}} \tag{1}$$

where c_1, c_2, c_3, c_4 are the sigmoid coefficient; and f_R is the reduced frequency ($f_R = f \cdot a(T)$).

3D finite element analyses were performed to investigate the effects of load eccentricity in uniaxial fatigue test results. For this purpose, the uniaxial test setup and tension-compression cyclic loading mode with zero-mean were modeled in the ABAQUS environment. The dimensions of the developed model reflect the specimen size used for the uniaxial tension-compression test. The boundary conditions of the model reflect the laboratory testing conditions. Since the test is run in tension-compression cyclic loading with zero-mean at high frequency (i.e., 5 Hz) and intermediate temperature

(i.e., 20°C) elastic properties were assigned to the elements of the developed model. The simulation's

2 MODEL DEVELOPMENT

A dense-graded 5E3 (9.5 mm NMAS top course designed to maximum traffic of 3 million ESALs) surface mixture was utilized for the completion of this study. The mixture consisted of a neat binder with performance grade of PG 64-28. The laboratory mix design was performed in accordance with the Superpave mix design specifications.

Initially, the linear viscoelastic properties of the mixtures were measured. The performance test specimens were prepared in general accordance with the AASHTO R 83 "*Preparation of Cylindrical Performance Test Specimens Using the Superpave Gyratory Compactor (SGC)*". Dynamic modulus (|E*|) samples were obtained from Superpave Gyratory Compactor (SGC) specimens compacted at strain level was approximately 150×10^{-6} m/m. The corresponding |E*| at this temperature/frequency combination were extracted from the |E*| master curve of the asphalt samples measured using the AMPT and it was used as an input for material properties in ABAQUS. In order to evaluate the effects of load eccentricity on stress and strain distribution of uniaxial tests, numerous load eccentricities were simulated ranging from 0% to 90%. Note, 10% eccentricity means 10% of specimen radius of the cylinder (i.e., radius = 34 mm, eccentricity = 3.4 mm).

Sample were also obtained. It is worth noting that in ABAQUS, axial stress and strain values can be obtained from the field output except for the Von Mises equivalent strain (herein Von Mises strain). In order to obtain the Von Mises strain, individual strains at the nodes were used to calculate the equivalent Von Mises strain value (Smith, 2009).

The results revealed that the corresponding peripheral axial strains at the edge of the platen- sample contact area (εedge) are higher than the strain within the LVDT length (εcenter). As the load eccentricity increases, this difference is anticipated to become more pronounced.

3 RESULTS AND DISCUSSION

The maximum uniaxial stress and strain values were evaluated for different load eccentricities for the cylindrical asphalt sample subjected to cyclic loading. In general, at 0% load eccentricity, at the amplitude of the loading cycle, a stress.

The measured axial strain values for the proposed load eccentricities are illustrated in Figure 2a. With the increase in eccentricity, an increase in the magnitude of the εedge is observed, which will eventually cause the test sample to experience premature end failure. This magnitude reaches a steady state when the induced load eccentricity exceeds 50 %. Further increases in eccentricity have minor effects on εedge after this point. On the other hand, the εcenter decreases with the increasing load eccentricity. The εcenter corresponds to the LVDT measured strain and is typically used for the evaluation of fatigue performance of asphalt mixtures. It can be observed that in the presence of load eccentricity, the use of εcenter for further analysis may be misleading, resulting in false fatigue performance of the asphalt mixture. concentration at the ends of the specimen was observed (see Figure 1). In theory, the maximum axial stress and strain are assumed to occur at the mid-height of the cylindrical sample when subjected to loading.

The ratio of the axial strains at the edge and the axial strains at the center of the loaded object (i.e., ε edge/ εcenter) were also investigated and results are plotted in Figure 2b. Based on these results, the axial strain ratio experiences a linearly increasing rate with the changing load eccentricity.

However, the results obtained from the simulations in ABAQUS do not agree with this assumption. One potential reason for this behavior includes the difference in materials stiffness between the glued end platens and the asphalt mixture. In other words, the material of the end platens has higher stiffness compared to the stiffness value of the asphalt sample. Such a difference in material stiffness creates stress concentration regions when a sample is subjected to loading as shown in Figure 1.

Varying load eccentricities were simulated and axial stress and strain results were analyzed for each load eccentricity scenario. Initially, the maximum axial stress and the corresponding strain were located and

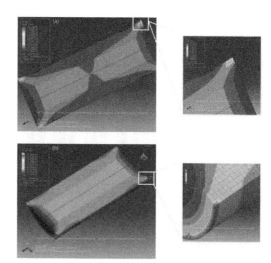

Figure 1. General (a) stress and (b) strain profile analysis for 0% load eccentricity.

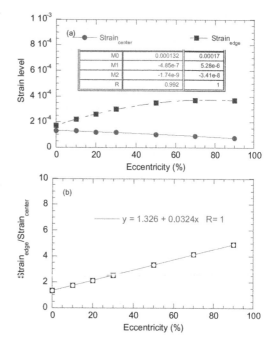

Figure 2. General (a) axial strain at the edge and center, (b) ε edge/ εcenter ratio for different load eccentricities.

extracted for each simulation. Likewise, axial stress and strain at the center of the asphalt

4 CONCLUSIONS

The objective of this study was to evaluate the effect of load eccentricity on uniaxial fatigue samples via the 3D finite element modeling approach.

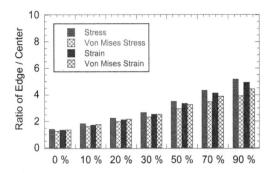

Figure 3. The ratio of edge and center stress-strain values.

The experimental program involved the preparation of cylindrical characterization of asphalt mixture. The following conclusions were derived from this study:

- The maximum axial stresses and strains at the platen-sample contact surfaces increase sharply with the increase in load eccentricity but follow a steady state after load eccentricity of 50 %.
- The opposite trend was observed for axial stresses and strains at the center of the cylindrical sample.
- The ε_{center} decreases with the increasing load eccentricity. The ε_{center} corresponds to the LVDT measured strain and is typically used for evaluation of fatigue performance of asphalt mixtures. This is clear evidence that the LVDT measured strain in the presence of load eccentricity may be misleading on the estimation of fatigue performance of asphalt mixtures.

Overall, this research study highlighted the impact of load eccentricity on the stress-strain distribution of cylindrical samples used for evaluation of fatigue performance of asphalt mixtures. In the future, the obtained effects of load eccentricity on fatigue performance of asphalt mixtures should be extended and compared to laboratory data, which can result in developing a correction factor for eliminating the effects of load eccentricity on loaded cylindrical samples.

Figure 3 illustrates axial and Von Mises stress and strain ratios for the ABAQUS simulations with varying load eccentricity values. It is important to note the difference between the axial stress and Von Mises stress. While axial stress is both tensile and compressive stress and in parallel to the direction of load application, Von Mises stress is a 3D stress counting for tensile, bending and twisting effects. The latest may be especially helpful in the presence of load eccentricity. In general, with the increase in load eccentricity,

the ratio between the edge stresses and center stress becomes more pronounced, leading to a shorter fatigue resistance of the mixture. Likewise, the corresponding axial strain and Von Mises strain values demonstrate the same trend. The ultimate stress and strain values of 50 % eccentrically loaded sample are more than three times higher than that of the same sample under concentric loading. With the increase in load eccentricity, stresses and strains at the end sides of the sample increase significantly, leading to misreading of the fatigue data and premature end failure of the fatigue test.

REFERENCES

Huurman, M., & Pronk, A.C. (2012). A detailed FEM simulation of a 4-point bending test device. In J. Harvey & J. Pais (Eds.), *Four-Point Bending* (pp. 3–12). DAVIS: CRC Press.

Kutay, M.E., Gibson, N., & Youtcheff, J. (2008). Conventional and Viscoelastic Continuum Damage (VECD) - Based Fatigue Analysis of Polymer Modified Asphalt Pavements. *Journal of the Association of Asphalt Paving Technologists*, 77, 1–32.

Kutay, M.E., Gibson, N., Youtcheff, J., & Dongré, R. (2009). Use of Small Samples to Predict Fatigue Lives of Field Cores Newly Developed Formulation Based on Viscoelastic Continuum Damage Theory. *Transportation Research Record: Journal of the Transportation Research Board* 2127, 90–97. https://doi.org/10.3141/2127-11

Seitllari, A., Boz, I., Habbouche, J., & Diefenderfer, S.D. (2020). Assessment of cracking performance indices of asphalt mixtures at intermediate temperatures. *International Journal of Pavement Engineering*, 1–10. https://doi.org/10.1080/10298436.2020.173083

Seitllari, A., & Kutay, M.E. (2019). Development of 3-Point Bending Beam Fatigue Test System and Implementation of Viscoelastic Continuum Damage (VECD) Theory. *Journal of the Association of Asphalt Paving Technologists* 88, pre-print.

Seitllari, A., Lanotte, M.A., & Kutay, M.E. (2019). Comparison of uniaxial tension-compression fatigue test results with SCB test performance indicators developed for performance-based mix design procedure. (A. F. Nikolaides & E. Manthos, Eds.), *Bituminous Mixtures and Pavements VII: Proceedings of the 7th International Conference 'Bituminous Mixtures and Pavements'(7ICONFBMP)*. Thessaloniki, Greece: CRC Press.

Smith, M. (2009). *ABAQUS/Standard User's Manual, Version 6.9*. Providence, Rhode Island.

Zeiada, W.A., Kaloush, K.E., Underwood, B.S., & Mamlouk, M. (2016). Development of a Test Protocol to Measure Uniaxial Fatigue Damage and Healing. *Transportation Research Record: Journal of the Transportation Research Board* 2576, 10–18. https://doi.org/10.3141/2576-02

Advances in Materials and Pavement Performance Prediction II – Kumar et al. (eds)
© 2021 Taylor & Francis Group, London, ISBN 978-0-367-46169-0

Determining fracture properties of reclaimed asphalt pavement-based cement mortar using semicircular bending test

X. Shi & Z.C. Grasley
Center for Infrastructure Renewal, Texas A&M University, College Station, Texas, USA

ABSTRACT: While critical stress intensity factor (K_{Ic}^s) and critical crack tip opening displacement ($CTOD_c$) are conventionally tested using single edge notched beam (SEN(B)) specimens, the test method suffers some drawbacks. A new test to characterize the fracture properties of the two-parameter fracture model (TPFM) using semi-circular bending (SCB) specimens has recently been developed by the authors. In comparison with the SEN(B) specimens, the SCB specimens can be readily made in lab from the commonly used cylindrical specimens. The geometry is also more suitable for field applications due to the ease of extracting cylinder specimens from concrete structures. In this study, the fracture properties of the TPFM for a reclaimed asphalt pavement (RAP) based cement mortar and a plain mortar at different curing ages were characterized using the SCB fracture test. The test results show that the RAP-Mortar (cement mortar containing 100% RAP) has lower K_{Ic}^s than the CON-Mortar (cement mortar containing 100% natural sand). However, the reduction of K_{Ic}^s is less significant than that of the compressive strength or splitting tensile strength. On the other hand, the RAP-Mortar was found to have higher $CTOD_c$ than the CON-Mortar. The increased $CTOD_c$ indicates that cement mortar containing RAP could potentially have a higher capacity to sustain cracking.

1 INTRODUCTION

The two-parameter fracture model (TPFM) proposed by Jenq and Shah (1985) has been widely used to predict fracture behavior of cementitious materials. While the major model inputs, critical stress intensity factor (K_{Ic}^s) and critical crack tip opening displacement ($CTOD_c$), are conventionally characterized using single edge notched beam specimens based on an RILEM recommendation (Shah 1990), the test method suffers some drawbacks: first, the fabrication of beam specimens is material- and labor-intensive. Second, it is usually difficult to extract beam samples from the field, so this test method is relatively impractical to use to evaluate field concrete.

A new test to characterize the fracture properties of the TPFM using semi-circular bending (SCB) specimens has recently been developed by the authors (Shi et al. 2019). In comparison with the SEN(B) specimens, the SCB specimens can be readily made in lab from the commonly used cylindrical specimens. The geometry is also more suitable for field applications due to the ease of extracting cylinder specimens from concrete structures. Furthermore, fracture tests for different fracture modes can be designed by changing the support location and crack angle of the SCB specimen (Mirsayar et al. 2017).

The objective of this study is to determine the fracture properties of the TPFM for a reclaimed asphalt pavement (RAP) based cement mortar and a plain mortar using the SCB fracture test. RAP is a demolition waste from asphalt concrete pavement.

While RAP has been conventionally reused to produce hot mix asphalt, the limitation and challenge of this practice have motivated the investigation of using RAP as an aggregate replacement in concrete (Shi et al. 2017; Debbarma et al. 2019).

2 MATERIALS AND MIX DESIGN

The RAP based mortar (RAP-Mortar) and the plain mortar (CON-Mortar) both used a 0.40 water to cementitious material mass ratio. The fine aggregate (natural sand or fine RAP) to cementitious material volume ratio is 2.8. For the RAP-Mortar, a locally collected fine RAP was used to replace 100% of the natural sand on a volumetric basis. A commercially available Type I/II cement was used, and a class F fly ash served as a supplementary cementing material to replace 20% of the cement (by weight). The mix proportions are shown in Table 1. During the mixing, the RAP-mortar was found to have better workability than CON-mortar possibly due to the lubricating effect of asphalt.

3 EXPERIMENTAL PROGRAM

3.1 Mechanical property test

Before the SCB fracture test, the mechanical properties that are conventionally determined in lab were tested according to the related ASTM standards for the CON-Mortar and RAP-Mortar. The tested mechanical properties are shown in Table 2.

Table 1. Mix proportions for the studied cement mortars.

Materials	CON-Mortar	RAP-Mortar
Cement (kg/m^3)	467	467
Fly Ash (kg/m^3)	117	117
Natural sand (kg/m^3)	1458	0
Fine RAP (kg/m^3)	0	1168
Water (kg/m^3)	234	234

Table 2. Mechanical property tests.

Test	Standard	Specimen size
Modulus of elasticity	ASTM C469	75×150 mm cylinders
Compressive strength	ASTM C39	75×150 mm cylinders
Splitting tensile strength	ASTM C496	75×150 mm cylinders

3.2 SCB fracture test

To make the SCB specimens, 150×300 mm cylindrical specimens were sliced into eight 37.5-mm disks, and sixteen SCB specimens were made by cutting each disk specimen into two SCB specimens. A 3 mm wide and 38 mm long notch was made in the middle of the SCB specimens with a small blade. Two knife edges were then epoxy glued to each side of the notch bottom. The epoxy was cured for 24 hours to ensure it gained sufficient strength.

The SCB fracture test was conducted using a stiff, high resolution MTS machine. A clip-on displacement gage was mounted to the pre-installed knife edges to record the crack mouth opening displacement (CMOD) with a data recording rate of 10 Hz during the test. The bending span to specimen radius ratio was 1.6. The testing procedures followed similar procedures specified in the RILEM method (Shah 1990):

Step 1: Load the specimen monotonically up to the maximum load at the constant crosshead displacement rate of 0.05 mm/min.

Step 2: Release the load when the load passes the peak value but is still within 95% of the peak at the same displacement rate.

Step 3: When the specimen is unloaded to approximately 10% of the peak load, reload the specimen till the specimen only carries 10%.

The SCB fracture test is shown in Figure 1.

4 RESULTS

4.1 Mechanical properties

The means of modulus of elasticity, compressive strength, and splitting tensile strength for the studied cement mortars are shown in Table 3. The RAP

Figure 1. The SCB fracture test.

Table 3. Mechanical properties for the studied cement mortars.

Mix type	Days	CON-Mortar	RAP-Mortar
Modulus of elasticity (GPa)	3	26.6 (2%)	8.5 (3%) ↓ 68%
	7	29.6 (4%)	9.8 (1%) ↓ 67%
	28	34.1 (2%)	10.8 (0.4%) ↓ 68%
Compressive strength (MPa)	3	27.4 (5%)	13.0 (5%) ↓ 52%
	7	36.4 (5%)	17.0 (1%) ↓ 53%
	28	47.9 (1%)	20.2 (2%) ↓ 58%
Splitting tensile strength (MPa)	3	4.2 (4%)	2.0 (10%) ↓ 51%
	7	5.8 (0.5%)	2.4 (5%) ↓ 58%
	28	6.4 (9%)	3.0 (6%) ↓ 52%

based cement mortar has lower modulus of elasticity, compressive strength, and splitting tensile strength than the plain mortar at all three ages. The reductions in the modulus of elasticity by the RAP addition are reported as 68%, 67%, and 68% at 3, 7, and 28 days, respectively. This is because asphalt is a significantly softer material compared with the cement paste or natural sand. The percent reductions for the compressive strength are 52%, 53%, and 58% for 3, 7, and 28 days, respectively. The percent reductions for the splitting tensile strength are 51%, 58%, and 52% for 3, 7, and 28 days, respectively. The strength reductions of the RAP-Mortar are attributed to asphalt cohesive failure (Mukhopadhyay and Shi 2019; Shi et al. 2020).

Note: only mean is shown. The coefficient of variance is indicated in the parenthesis. ↓x% = the property of the RAP-Mortar decreases by x% in comparison with that of the CON-Mortar.

4.2 Fracture properties

Typical load-CMOD curves for CON-Mortar and RAP-Mortar is shown in Figure 2. The RAP-Mortar specimens invariably have a lower peak load compared with the CON-Mortar.

Critical stress intensity factor

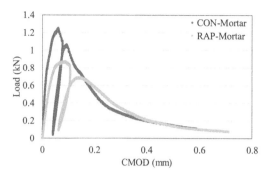

Figure 2. Typical load-CMOD curves for CON-mortar and RAP-mortar.

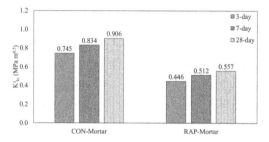

Figure 3. Critical stress intensity factor results.

The critical stress intensity factor, , is one of two major parameters in the TPFM; it is considered a "real" material property of cementitious materials that has no size dependency (Shah et al. 1995). To determine the from the SCB fracture test, a critical effective elastic crack length is first calculated by assuming that the compliance determined from the loading and unloading portions of the load vs CMOD curve are equal. Theis then calculated using the critical effective elastic crack based on linear elastic fracture mechanics (LEFM). The procedures to calculate the can be found in the appendix of the authors' paper (Shi et al. 2019).

Figure 3 compares thebetween the CON-Mortar and RAP-Mortar for different ages. The longer the curing ages, the higher thefor both the cement mortars. The RAP-Mortar specimens showed lowervalues compared the CON-Mortar specimens, irrespective of curing age. The reductions inby the RAP addition are 40%, 39% and 39% for 3-day, 7-day, and 28-day specimens, respectively.

Critical crack tip opening displacement

The critical crack tip opening displacement,, is the other key input for the TPFM. According to the TPFM, at the critical fracture of material, not only the stress intensity factor criterion needs to be satisfied, the CTOD criterion shall also be met.

The can be readily determined according to Shi et al. (2019). The results for the studied cement mortars at different ages are shown in Figure 4. In general, aging has a positive effect on the development of . The addition of the fine RAP was found to noticeably increase theof the cement mortar by 44%, 54%,

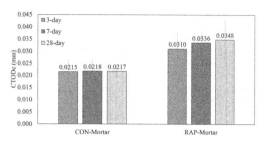

Figure 4. Critical crack tip opening displacement results.

and 60% at the 3, 7, and 28 days, respectively. The increasedsuggests that the RAP based cement mortar could allow a higher amount of separation at the crack tip before a catastrophic failure. This indicates that the RAP based cement mortar has a higher capacity to sustain cracking than the plain cement mortar.

5 CONCLUSIONS

The fracture properties of the TPFM for the RAP based cement mortar and its control mortar were characterized using the newly developed SCB fracture test. The major findings of this study are:

The developed SCB fracture is sensitive enough to pick up the fracture property changes caused by the RAP addition in cement mortar.

The RAP-Mortar has lowerthan the CON-Mortar. However, the reduction for theis less significant than that for the compressive strength or splitting tensile strength.

The RAP-Mortar has higherthan the CON-Mortar. The increasedindicates that the RAP based cement mortar has a higher capacity to sustain cracking.

REFERENCES

Debbarma, S., Ransinchung, G. & Singh, S. 2019. Feasibility of roller compacted concrete pavement containing different fractions of reclaimed asphalt pavement. Construction and Building Materials, 199, 508–525.

Jenq, Y. & Shah, S.P. 1985. Two parameter fracture model for concrete. Journal of engineering mechanics, 111, 1227–1241.

Mirsayar, M., Shi, X. & Zollinger, D. 2017. Evaluation of interfacial bond strength between Portland cement concrete and asphalt concrete layers using bi-material SCB test specimen. Engineering Solid Mechanics, 5, 293–306.

Mukhopadhyay, A. & Shi, X. 2019. Microstructural Characterization of Portland Cement Concrete Containing Reclaimed Asphalt Pavement Aggregates Using Conventional and Advanced Petrographic Techniques. ASTM International Selected Technical Papers.

Shah, S.P. 1990. Determination of fracture parameters (K Ic s and CTOD c) of plain concrete using three-point bend tests. Materials and Structures, 23, 457–460.

Shah, S.P., Swartz, S.E. & Ouyang, C. 1995. Fracture mechanics of concrete: applications of fracture mechanics to concrete, rock and other quasi-brittle materials, John Wiley & Sons.

Shi, X., Grasley, Z., Hogancamp, J., Brescia-Norambuena, L., Mukhopadhyay, A. & Zollinger, D. 2020. Microstructural, Mechanical, and Shrinkage Characteristics of Cement Mortar Containing Fine Reclaimed Asphalt Pavement. Journal of Materials in Civil Engineering. 32(4), 04020050.

Shi, X., Mirsayar, M., Mukhopadhyay, A.K. & Zollinger, D.G. 2019. Characterization of Two-Parameter Fracture Properties of Portland Cement Concrete Containing Reclaimed Asphalt Pavement Aggregates by Semicircular Bending Specimens. Cement & Concrete Composites, 95, 56–69.

Shi, X., Mukhopadhyay, A. & Liu, K.-W. 2017. Mix design formulation and evaluation of portland cement concrete paving mixtures containing reclaimed asphalt pavement. Construction and Building Materials, 152, 756–768.

Advances in Materials and Pavement Performance Prediction II – Kumar et al. (eds)
© 2021 Taylor & Francis Group, London, ISBN 978-0-367-46169-0

Workability quantification of bituminous mixtures using an improved workability meter

S. Sudhakar
Department of Engineering Design, Indian Institute of Technology Madras, India

V.T. Thushara & J. Murali Krishnan
Department of Civil Engineering, Indian Institute of Technology Madras, India

Shankar C. Subramanian
Department of Engineering Design, Indian Institute of Technology Madras, India

ABSTRACT: Quantification of the workability of bituminous mixtures has always been challenging due to the associated complexity during mixing, laying and compaction. The transitory response of the rheological behavior of the binder, the particle size distribution and the quantum of work applied during mixing and compaction are a few of these factors. A few attempts exist in the literature to quantify the workability of bituminous mixtures, and such attempts have computed the torque required indirectly at a constant angular speed and constant temperature. Also, the relationship between workability during mixing and during compaction requires detailed investigation. This study presents initial results of workability equipment fabricated in-house at IIT Madras, India. Direct torque measurements recorded for different bituminous mixtures at different temperature conditions are reported and their sensitivity is discussed.

1 INTRODUCTION

The constructability of bituminous layers is highly dependent on their workability and compactibility. Workability is the ease with which the bituminous mixture can be mixed and placed in the pavement. The workability of the bituminous mixtures is influenced by the type, amount, and viscosity of the bitumen, as well as the characteristics of aggregate gradation. Compactability during the construction phase is related to the stability and resistance of the mixture to densification. To a large extent, it depends on the ability of the aggregate particles to slide past each other (Koneru et al. 2008).

In order to identify a suitable temperature window for mixing and placing of mixtures, one needs to understand the mechanics during mixing and compaction. In the mixing phase, the aggregates are coated by the bituminous film, which in turn provides lubrication for the aggregates to get rearranged in the mixture. The mixing efficiency is related to how well the binder can coat all the aggregate particles. During the compaction phase, the temperature of the mixture reduces, and viscosity increases, which in turn offers resistance to particle packing. Based on the binder and mixture rheology, different mixtures will respond differently. Therefore, the temperature window in which mixing and compaction occur is highly influential for the proper construction of bituminous pavement.

Various newer bituminous mixture types with different combinations of bituminous binders and aggregate gradations are used in pavement construction. This includes mixtures with elastomer, plastomer, and crumb rubber modified binders; Warm Mix Asphalt mixtures (WMA), emulsion-based cold mix asphalt, reclaimed asphalt pavement mixtures (RAP) and Stone Mastic Asphalt (SMA). For conventional bituminous mixtures, the mixing and compaction temperature regimes are reasonably well understood (due to a significant empirical data), whereas it is not so for the newer bituminous mixtures. Currently, all the bituminous mixtures are compacted identically more or less in the same temperature range (excepting WMA wherein the temperature window is wider). In order to optimize the mixing and compaction operation, models based on mixture rheology needs to be developed, which require data at different scales such as binder viscosity, binder coating thickness as well as the mixture workability.

Laboratory-scale workability measurements, based on torque, will provide data to compare the rheological properties among the bituminous mixtures. Also, they help to identify the temperature window of the various mixtures so that the effective packing of aggregates for a given compactive effort can be achieved. However, only very few attempts are reported in the literature related to the workability measurement of bituminous mixtures.

Marvillet and Bougalt (1979) made the first attempt to investigate the workability of bituminous mixtures by developing a workability meter, which could measure the resistance of the mixture using a potentiometer and converted into torque readings. Gudimettla et al. (2004) evaluated the workability of mixes by pushing a paddle through Hot Mix Asphalt (HMA) and measured the torque required to maintain a given rate of revolution. They defined the workability in terms of current demand, measured using an ammeter to run the electric motor, i.e., more the current, tougher the mix, and vice versa. Tao and Mallick (2013) developed a workability meter having 18 kg capacity in which a torque wrench was used to determine the workability of RAP mixtures. Poeran and Sluer (2016) designed an improved workability meter with a movable mixing drum and an electric motor driven paddle. The workability of HMA was measured through an inline torque sensor.

In most of the workability meters reported in the literature, the workability was measured indirectly based on spring stiffness, ammeter readings, and torque wrench only. In all these attempts, the temperature of the mixing chamber/ drum did not have any provision of heater arrangement to control or to maintain the desired temperature of the mixing chamber. Also, in most cases, the fresh aggregate and the binder were mixed elsewhere, and the mixture was placed on the workability equipment to carry out further workability measurements. The effort required to coat the aggregate particles to a suitable binder thickness is excluded in all these studies.

The present study aims to measure the workability of different bituminous mixtures in terms of direct torque measurements using a newly developed in-house temperature controlled workability device, which can measure the torque required to subject the mixture to various angular speeds of the paddle, at the desired mixture temperatures. The measurements made through an inline torque sensor during mixing would give more accurate and repeatable results by eliminating uncertainties in the torque measurements. The direct torque measurement would provide a relative indication of the mixture's workability as well as its rheological characteristics at various temperatures and various shear rates.

2 EQUIPMENT

A newly developed workability meter capable of mixing 35 kilograms of material in a temperature-

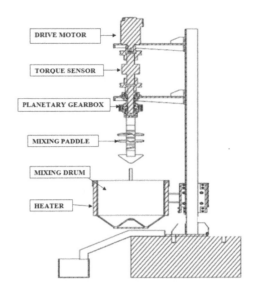

Figure 1.   Schematic diagram of workability meter.

controlled mixing drum is developed in this study. The mixing drum consists of two modular mixing paddle assemblies, which facilitates the measurement of torque simultaneously during the mixing. The paddle assembly used for the mixing process has both rotary and planetary motion for efficient mixing and hence the uniform coating of bitumen over the aggregates, without creating any fixed shear plane, during mixing is ensured. An inline rotary torque sensor, with a capacity of 500 Nm and a data acquisition system that acquires data at an interval of 0.01 s, is used for capturing torque data. A unique hydraulic drive system is used to operate the paddles at any given speed (0 to 50 RPM) consistently irrespective of the change of load on the paddles, and also to move the mixing drum up and down to introduce / withdraw the bituminous mixture over the paddles. The schematic diagram of the equipment is given in Figure1.

3   EXPERIMENTAL INVESTIGATIONS

Torque measurements are carried out for different bituminous mixtures. For each sample, twenty-seven kilograms of batched aggregate is mixed with a 5% binder by weight of the total mixture. The graded aggregates are fed into the mixing drum in 4 layers, and after each layer is fed, one-fourth of the total binder required is poured to it. The effort required to move the paddles at an angular speed of 25 rpm, through the mixture is captured for torque measurements. The collected raw torque data have fluctuations due to the heterogeneous nature of the mixture. A Gaussian smoothening algorithm is adopted to smoothen the captured torque data. Different bituminous mixtures with modified and unmodified binders at various

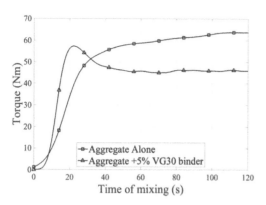

Figure 2. Lubrication.

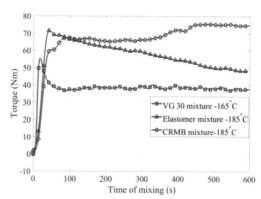

Figure 3. Workability for different binders.

mixing temperatures are quantified for workability as given in the following subsections.

3.1 The lubricating effect of Binders

Bitumen is added at a certain viscosity for lubricating aggregates so that they can glide past each other smoothly during mixing and compaction phases. The reduction in effort to work with the bituminous mixtures upon the addition of bitumen will be a measure of interest for pavement engineers.

Two sets of torque data are measured in order to quantify the lubricating effect of bituminous binders on the bituminous mixtures. In the first trial, the effort required to move the mixing paddles through only the aggregates graded to the bituminous concrete mid-gradation as per Indian specifications (MoRTH 2013) is collected. In the second trial, 5% of bitumen is added to the aggregates at the mixing temperature, and the torque is measured during the mixing process. The data is presented in Figure 2. The data captures the reduction in the effort required to move the mixing paddles through the sample with the addition of bituminous binder.

3.2 Comparison of various binders

Bituminous mixtures with one unmodified binder designated as VG30 and two modified binders (elastomer and crumb rubber) are prepared with Bituminous concrete mid-gradation (MoRTH 2013) and the torque during mixing is captured. Aggregates are preheated to 170°C, 190°C, and 190°C, for preparing VG30, crumb rubber (CRMB), and elastomer mixtures, respectively. Bitumen is heated to 160°C, 180°C, and 180°C, corresponding to the mixing temperatures determined for VG 30, CRMB, and elastomer binders in order to achieve similar consistency. After mixing, the resulting bituminous mixtures attained 165, 185 and 185°C, respectively. To capture the rheological behavior at steady shear, the mixing process is prolonged for 10 minutes.

Figure 3 shows the comparison of torque data of mixtures with various types of binders. It could be interpreted that the VG30 mixture exhibits a time-independent response after initial mixing with the binder, whereas the elastomer mixture becomes more workable, and the CRMB mixture shows a slight reduction in workability with prolonged mixing.

3.3 Comparison of mixing temperature

The temperature at which bituminous mixtures are prepared has a prominent role in its workability, as the binder viscosity varies with temperature. In order to quantify the same, torque data is collected at various mixing temperatures. The VG30 mixture data is collected for a temperature range of 100°C-170°C and the CRMB mixture in the range of 130°C-180°C at every 10°C rise in temperature. Two data collection schemes are adopted. In the first data collection scheme, the bituminous mixtures are prepared afresh at each measurement temperature considered. This data corresponds to torque measured over time for a given angular speed when the mixing was carried out at different temperatures. In the second data collection scheme, the fresh bituminous mixture is prepared at 100°C for VG 30 mixture and 130°C for the CRMB mixture. These temperatures correspond to the lowest measurement temperatures considered in this particular study. Now, these prepared mixtures are allowed to get heated in the mixing drum without shearing. Once the temperature rise by 10°C, the shearing is restarted, and the torque data is captured. A similar procedure is adopted for collecting data at each further 10°C rise in temperature.

Figures 4a and 4b represent the data of VG 30 mixture for both the measurement schemes.

The data from scheme 1 (Fig. 4a) indicate that torque required is significantly high during the mixing phase of the material as observed from the torque magnitude in the initial 100 s of mixing. The torque magnitudes become steady after about 100 s, for all the temperatures. It is interpreted that, once the binder

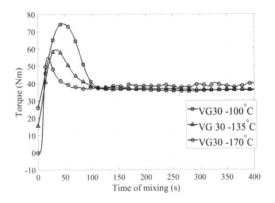

Figure 4. a. VG30 mixture at varying temperature – Scheme 1.

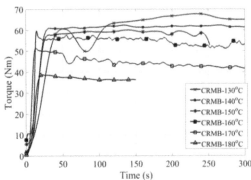

Figure 5. Workability data of CRMB mixture at various temperatures – Scheme 2.

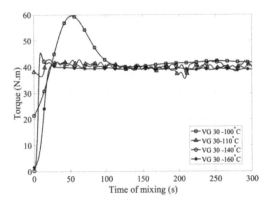

Figure 4. b. VG 30 mixture at varying temperature – Scheme 2.

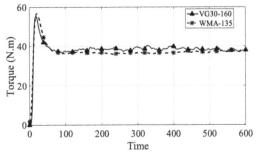

Figure 6. Workability data of VG30 and WMA mixtures.

is coated with the aggregate and when a mastic phase is formed in the mixture, the effort to mix the bituminous mixtures remains almost the same irrespective of the temperature for the VG 30 mixture.

Figure 4b gives the torque data of VG30 mixture when fresh aggregates are mixed with binder at 100°C, and the data is collected as per scheme 2. The measurements give insight that the torque required is the same for unmodified VG 30 mixture irrespective of the temperature and measurement scheme, soon after the mixing phase of the bitumen with aggregates. Also, during the mixing phase, the torque requirement is substantially high when compared to the effort to agitate the mix further.

Torque data is collected for CRMB mixture as per scheme two, and is plotted in Figure 5. The torque measurements capture the relative easiness to work with the CRMB mixture at various temperatures. Compared to the VG30 mixture, the CRMB mixture torque requirement is not the same for all the measurement temperatures after mixing. It can be interpreted that the binder viscosity is influencing even after the mastic phase is formed for the CRMB mixture, and the selection of the compaction temperature window is much sensitive in this case.

3.4 Comparison of workability of Warm Mix Asphalt with unmodified VG30 mixture

Warm Mix Asphalt (WMA) is a special form of the bituminous mixture in which either viscosity modifying additives such as SASOBIT or interfacial friction-reducing additives such as EVOTHERM are added to bituminous mixtures so that the mixing and compaction temperature window becomes wider compared to the regular Hot Mix Asphalt. WMA mixtures are recommended to be mixed and placed at around 30°C less than HMA. Figure 6 shows the torque data of the VG30 mixture at 160°C and WMA mixture (prepared with 0.4% Evotherm by weight of binder) at 135°C. The data shows that the ease of working with a WMA mixture at 135°C is the same as to work with a VG30 mixture at 160°C.

4 CONCLUSIONS

The workability of different bituminous mixtures is quantified using direct torque measurements in a new workability equipment developed in the pavement engineering laboratory at IIT Madras. The torque measurements of bituminous mixtures with different binders at varying mixing temperatures and mixing schemes are captured and analyzed. The study quantifies the laboratory-scale degree of easiness to work with various bituminous mixtures, which give valuable insight into the mixture behavior. These measurements will provide data to develop rheological models that can be used to identify the optimum mixing and compaction window of the bituminous mixture in hand.

REFERENCES

Gudimettla, J.M., Cooley Jr, L.A. & Brown, E.R., 2004. Workability of Hot-Mix Asphalt. *Transportation Research Record*, 1891(1): 229–237.

Koneru, S., Masad, E., & Rajagopal, K.R. 2008. A Thermomechanical Framework for Modeling the Compaction of Asphalt Mixes. *Mechanics of Materials*, 40: 846–864.

Marvillet, J., & P. Bougault. 1979. Workability of Bituminous Mixes-Development of a Workability Meter (With Discussion). In *Association of Asphalt Paving Technologists Proceedings*, 48:91–110.

MoRTH 2013. Specification for Road & Bridge Works, Fifth Revision. New Delhi, India: *IRC publications*, Ministry of Road Transport & Highways (2013).

Poeran, N. & Sluer, B. 2016. Workability of Asphalt Mixtures. E&E Congress, 6th Eurasphalt & Eurobitume Congress, Prague, Czech Republic. dx.doi.org/10.14311/EE.2016.057

Tao, M., & Mallick, R.B. (2009). Effects of Warm-Mix Asphalt Additives on Workability and Mechanical Properties of Reclaimed Asphalt Pavement Material. *Transportation Research Record*, 2126(1): 151–160.

Rejuvenation of Reclaimed Asphalt Pavement (RAP) binder appealing to design of experiments

M. Thirumalavenkatesh, V.T. Thushara & J. Murali Krishnan
Department of Civil Engineering, Indian Institute of Technology Madras, India

ABSTRACT: Reclaimed asphalt binder blending involves a lot of mixture variables and experimentation. The objective of this study is to adopt statistical experimental design framework to design a ternary mixture consisting of Reclaimed Asphalt Pavement (RAP) binder, vacuum tower bottom (VTB) and rejuvenator to meet target steady shear viscosity, Performance Grade (PG) pass/fail temperature and softening point of a typical Viscosity Grade 30(VG30) binder. I-Optimal design algorithm is used to design the experimental matrix. Response prediction models are developed for steady shear viscosity, high temperature performance grade (PG) (pass/fail) and softening point based on the measurements carried out on the prepared blends. From the predicted response models, the possible RAP binder blend space which yield target responses is identified.

1 INTRODUCTION

Reclaimed Asphalt Pavement (RAP) is re-processed road material containing aggregate and asphalt binder. Superpave recommends (McDaniel & Anderson 2001) that one can use up to 10 to 20% RAP without any change in the fresh binder grade. For RAP percentage greater than 20%, it recommends to characterize the extracted and recovered RAP binder, along with the usage of empirical blending charts (Swiertz et al. 2011) to determine the performance grade of fresh binder to be used in the mix design.

In hot mix asphalt recycling process aged binder is expected to get combined with a fresh binder, leading to grade change of resultant binder.

A lot of mixture variables and experimentation is required to fix the blend proportions, which yield target properties. The interactions among the components of the blended RAP binder might be of a higher order. To formulate the blending proportions and to understand the interactions, a few studies (Chavez-Valencia et al. 2005; Hamzah et al. 2013) have used a statistical-based experimental design.

The objective of the study is to find the optimum proportions of the RAP binder, vacuum tower bottom (VTB) which is a residue obtained from refinery units, and rejuvenator such that prepared blends will have target properties. The target is to meet steady shear viscosity, PG pass/fail temperature and softening point within the experimental design space. The principles of design of experiments are used to understand the effects of mixture variables on the properties of the blended RAP binder. The steady shear viscosity,

high-temperature PG (pass/fail), and softening point are measured as the responses.

2 EXPERIMENTAL INVESTIGATION

2.1 Materials

In this study, extracted and recovered RAP binder, vacuum tower bottom (VTB), and rejuvenator are used to prepare blends with RAP binder. The RAP is collected from a bituminous concrete (BC) pavement in Trichy-Dindugal National Highway, Tamilnadu, India. To account for variability in RAP binder extraction and recovery, samples are taken from 7 different places of the stockpile. The RAP binders are extracted and recovered as per ASTM D2172-17, 2017, and ASTM D5404-17, 2017, respectively. VTB is the straight Vacuum Tower Bottom readily available from a vacuum distillation of crude oil and its effectiveness in reducing high-temperature PG and viscosity of asphalt binders is reported in the literature (Amini and Imani-nasab 2018). One commercially available rejuvenator, which consists of fatty acid derivatives, is used as the third component for blending.

2.2 Experimental design

Based on the preliminary experiments, review of the literature (Al-Qadi et al. 2007), as well as to maximize RAP usage and minimize the dosage of rejuvenator, the following constraints as given in Equation (1) are adopted for the three components. The consistency of constraints is ensured as per Cornell (2002).

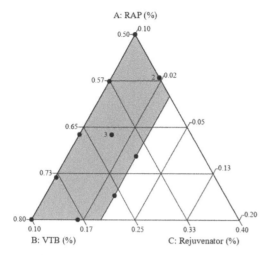

A: RAP (%)

B: VTB (%)   C: Rejuvenator (%)

Figure 1. Mixture design space diagram.

$$A + B + C = 1$$
$$0.5 \leq A \leq 0.8$$
$$0.1 \leq B \leq 0.4 \tag{1}$$
$$0 \leq C \leq 0.1$$

where, A, B, and C are the weight proportions of RAP binder, VTB, and rejuvenator, respectively. Constrained ternary mixture design using Design Expert 12 software (Montgomery & Runger 2010) is used to identify the blend proportions. A flexible design structure based on the I-Optimal technique, which identifies design points for blending so as to minimize prediction variance, is adopted to accommodate the constrained design space (Design Expert 12 2019; Goos et al. 2016). Thirteen blends are designed and are presented in the reduced ternary space diagram in Figure 1. Among the design blend points, some are repetitions so that the variability within the measurements is also captured.

2.3 Tests conducted

The RAP binder and VTB are preheated to pouring consistency, and subsequently, the blends are prepared by hand mixing for 5 minutes at 140°C. For all the blends, three responses were measured, and they are steady shear viscosity (η_{60}) at 60 degrees with a shear rate of 1 reciprocal second, high-temperature PG (T_{PG}) and softening point (T_{SP}) and shown in Table 1.

3 DEVELOPMENT OF MODELS

A sequential model-fitting approach is adopted to determine the suitable response prediction model for each of the three responses. The rationale is that model terms which are significant by certain significance level will only be incorporated in the model. Analysis

Table 1. Measured responses of the blends.

Blend No.	Weight Proportion A	B	C	(Poise)	T_{PG} (°C)	T_{SP} (°C)
1	0.8	0.17	0.03	45909	87.4	66.5
2	0.58	0.32	0.10	2092	63.4	44
3	0.57	0.40	0.03	15705	76.8	57.5
4	0.67	0.28	0.05	14248	77.6	57.5
5	0.70	0.30	0.00	61731	92	71
6	0.8	0.10	0.10	7709	73.6	53
7	0.67	0.28	0.05	14130	77	57
8	0.57	0.4	0.03	14361	76.7	59
9	0.66	0.24	0.10	3157	68.2	48
10	0.73	0.17	0.10	3453	68.4	48.5
11	0.67	0.28	0.05	14200	78.1	57.5
12	0.50	0.40	0.10	1315	52.3	43
13	0.76	0.24	0.00	157110	96.8	73.5

of variance is conducted to quantify the mean square of the model and the mean square of the residuals (Montgomery & Runger 2010). The standard models are modified by adding or/and removing terms until the lack of fit is insignificant. Along with this, R^2, adjusted R^2, predicted R^2, and the prediction residual error sum of squares (PRESS) are considered to check the efficacy of the models. The statistical significance of each of the model coefficients is estimated at a 5% significance level. After analyzing different models, the following three equations are selected and represented in Equations (2–4).

$$\sqrt{\eta_{60}} = 1082A - 430B + 224559C - 1779AB$$
$$- 258731AC - 251507BC + 142543A^2BC$$
$$+ 167922AB^2C - 1102512ABC^2 \tag{2}$$

$$T_{PG} = 165A - 675B - 1925C + 1292AB$$
$$+ 3460AC + 1773BC - 1095AB(A - B)$$
$$- 2433AC(A - C) \tag{3}$$

$$T_{SP} = 81A + 45B - 171C \tag{4}$$

The response prediction models are given in Equation 2, and Equation 3 could capture the higher-order interactions along with individual effects through their significant model terms. The and T_{PG} values vary over a wide range for the measured blends, with a slight change in the proportion of the components. For instance, it is in the range of 1315 to 157110 poise, and T_{PG} is varying from 54 to 74°C across the design space. Also, from the coefficients, it can be interpreted that the rejuvenator (variable C) is having a predominant effect in the response compared to the other components. In the case of softening point, the effects are mainly due to individual components and are captured in Equation 4. The fit statistics of the selected response prediction models are presented in Table 2.

Table 2. Model fit statistics.

Response	$\sqrt{\eta_{60}}$	T_{PG}	T_{SP}
Model Form	Special quatric	Reduced cubic	Linear
R²	0.9999	0.9974	0.9904
Adjusted R²	0.9996	0.9936	0.9884
PRESS	4238.09	264.25	21.05
Predicted R²	0.9648	0.842	0.9808

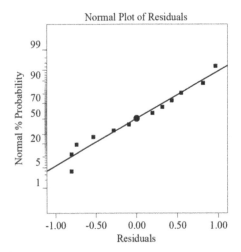

Figure 2. PG Pass/Fail temperature.

The Predicted R² of all the three models are in reasonable agreement with Adjusted R² since the difference is less than 0.2 (Design Expert 12, 2019). This indicates the good predictive capability of the three selected models within the experimental regime. The normal probability plot of PG pass/fail temperature response model residuals are shown in Figure 2.

From Figure 2, it can be interpreted that the residuals are normally distributed as their values are having the least deviation from the normal probability line. This is another measure to check the efficacy of the selected response model, as it indicates that the residuals are only due to the randomness involved in the measurements. Similarly, normal probability plots of residuals of the other two responses are also verified for their randomness. The response prediction contour plots using the selected models for the T_{PG}, and T_{SP} are given in Figures 3–5, respectively.

Figure 3 shows the influence of various mixture components on the of the blended RAP binder. From the results (Figure 3), it is seen that the contribution of the rejuvenator and its interactions are significant compared to the other components. The role of rejuvenator in viscosity reduction can be inferred from this trend. As expected, higher-order interactions are present among components, and curvature of the contour plot captures these interactions.

The curvature in the contour plots of T_{PG} as given in Figure 4 indicates that the effect of individual

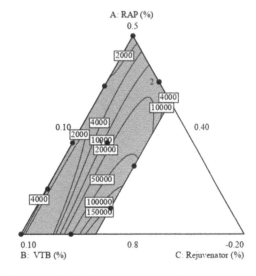

Figure 3. Predicted steady shear viscosity.

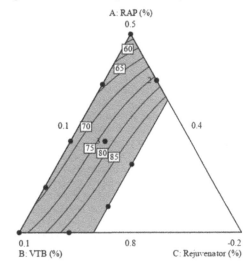

Figure 4. Predicted pass/fail temperature.

components and their interactions on the pass/fail temperature is less as that compared to the, as in Figure 3.

The softening point prediction contour for the design space is shown in Figure 5. The linear contour lines indicate that the interaction among the components is insignificant, with 95% confidence on the softening point of the blended RAP binders.

4 IDENTIFICATION OF DESIRABILITY REGION

The predicted response models are used to identify possible ranges by optimization using target ranges of responses so that the resulting blend will have

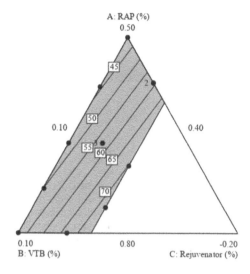

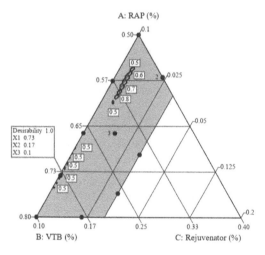

Figure 5. Contour plot for predicted softening point.

Figure 6. Desirability plot.

5 CONCLUSIONS

It can be concluded that the rejuvenator used is effective in the reduction of viscosity as well as PG grade and the most influential component in the property variation of the blended RAP binder. It can also be concluded that component interactions having a significant effect in the measured steady shear viscosity across the design space, whereas the component interactions are insignificant in the case of softening point.

The present study is focused to quantify the effects of ingredient proportions on the prepared blend response using a statistical framework, and hence the quantitative trends of the component interactions are only reported. The study can be extended to study the chemical/physical/mechanical interactions of the components.

REFERENCES

Al-Qadi, I.L., Elseifi, M., & Carpenter, S.H. 2007. Reclaimed asphalt pavement—a literature review. ICT-07-001, Technical Report, FHWA, Illinois Center for Transportation, Rantoul, IL.

Amini, A. & Imaninasab, R. 2018. Investigating the effectiveness of Vacuum Tower Bottoms for Asphalt Rubber Binder based on performance properties and statistical analysis, *Journal of Cleaner Production*, 171, 1101–1110.

Chavez-Valencia, L.E., Manzano-Ramírez, A., Luna-Barcenas, G., & Alonso-Guzman, E. 2005. Modeling of the performance of asphalt pavement using response surface methodology. *Building and Environment*, 40(8): 1140–1149.

Cornell, J.A. 2002. Experiments with mixtures: designs, models, and the analysis of mixture data, Third Edition, New York, John Wiley & Sons.

Design-Expert 12. 2019 Minneapolis, MN, USA. Stat-Ease, Inc.

Hamzah, M.O., Golchin, B., & Tye, C.T. 2013. Determination of the optimum binder content of warm mix asphalt incorporating Rediset using response surface method. *Construction and Building Materials*, 47: 1328–1336.

McDaniel, R., & Anderson, R.M. 2001. NCHRP Report 452: Recommended Use of reclaimed asphalt pavement in the Superpave mix design method: *technician's manual. TRB, National Research Council*, Washington, DC.

Montgomery, D.C., & Runger, G.C. 2010. Applied statistics and probability for engineers. Fifth Edition, USA, John Wiley & Sons.

Peter Goos, Bradley Jones & Utami Syafitri (2016) I-Optimal Design of Mixture Experiments, Journal of the American Statistical Association, 111:514, 899–911.

Swiertz, D., Mahmoud, E., & Bahia, H.U. 2011. Estimating the effect of recycled asphalt pavements and asphalt shingles on fresh binder, low-temperature properties without extraction and recovery. *Transportation Research Record: Journal of the Transportation Research Board*, 2208, 48–55.

properties similar to the fresh binder. The optimization technique is capable of identifying blend space corresponding to any combination of targets within the measured response ranges. For illustration, a typical VG30 binder similar to AC 30 as per ASTM: D3381(2018) is targeted. Based on laboratory data, this binder has a steady shear viscosity of 4300 poise at 60°C, and a shear rate of 1 s^{-1}, high-temperature performance grade of 64°C and softening point exceeding 47°C. For these target inputs, ten solutions are obtained which satisfy all the given criteria (desirability=1). The blend region with desirability more than 0.5, for the solution with a maximum of RAP binder is shown in Figure 6. The desirability value of 1 in this figure corresponds to a blend proportion with 73% RAP binder, 17% VTB and 10% rejuvenator.

Advances in Materials and Pavement Performance Prediction II – Kumar et al. (eds)
© 2021 Taylor & Francis Group, London, ISBN 978-0-367-46169-0

Improved testing and analysis of uniaxial tension fatigue test

J. Uzan
Professor Emeritus, Technion, Israel Institute of Technology, Haifa, Israel

ABSTRACT: The paper presents two major improvements to the testing and analysis of the uniaxial tension fatigue test. A displacement transducer is proposed to control the strain on the specimen and avoid the problems caused by the crosshead displacement control. The fatigue test, with more than one million data points, is analyzed using the numerical integration proposed by Uzan (2020). The analysis of one fatigue test suggest that the pseudo stiffness, calculated from the peaks of the stress and pseudo strain, may not represent the ma-terial state at the end of loading cycle.

1 INTRODUCTION

Modelling the fatigue performance of asphalt con-crete (AC) using continuum damage (CD) theo-ries is an attractive approach that can (a) explain dif-ferences in behavior when tested in different modes of loading, controlled stress or strai; (b) reduce the testing amount and (c) hopefully improve the predic-tion of cracking in pavements. One such model, the viscoelastic continuum damage (VECD) model, was developed, expanded and implemented by Richard Y. Kim (NCSU), Dallas Little and Robert Lytton (Texas A&M) and their students (Kim & Little 1990; Kim et al 199; Lee & Kim 199; Daniel & Kim 2002; Underwood et al. 2010; Luo et al. 2013, 2016 among many others). It was adopted by other re-searchers (a) To develop simplifications for analyz-ing the beam fatigue test (Christensen & Bonaquist 200; Pronk et al. 2017) and (b) To investigate and derive the endurance limit (Witczak et al. 2013).

The beam fatigue test was recommended by NCHRP (National Cooperative Highway Research Program report) mainly because: (a) the uniaxial test is more time consuming than the beam fatigue test and requires an enormous attention to details in terms of alignment and controlling the load, and (b) The beam fatigue test is more established with a larger database available in the literature than the uniaxial test. How-ever, the deformation and stress conditions are non-uniform in the beam test, confin-ing the analysis of the test to phenomenological and approximate models (Christensen & Bonaquist 2009, Pronk et al 2017).

The loading conditions at the bottom of an AC layer are slightly different in the longitudinal (direc-tion of travel) and transversal directions. In the lon-gitudinal direction, the bottom of the AC layer goes into com-pression as the load approaches the section of interest, in tension when the load crosses the sec-tion and

again in compression as the load moves away from the section. The level of the tension is about 3-6 times larger than the level in compression. In the lateral direction, the compression is not pre-sent and only tension develops at the bottom of the layer. Another observation differentiates between the longitudinal and lateral directions: In the longi-tudinal direction, the strain before and after the load application must be zero and return to zero. In the transversal direction, the material can move, expand and contract and the strain can accumulate. The level of accumulation depends on the distance from the shoulders, material stiffness and other conditions. Under normal traffic speed, the loading time is less than 0.1 second which is the usual loading time in the lab. Finally, the shape of the strain history re-sembles the haversine curve

The VECD model is an application of the corre-spondence principles developed by Schapery (1984) which are aimed at transforming a time dependent vis-coelastic system to a time independent elastic sys-tem. In the one-dimensional case, the pseudo strain, repre-senting the strain in the transformed body, is computed using the hereditary integral for thermo-rheologically simple materials as:

$$\varepsilon^R = (1/E_R) \int_i^0 E(\psi - \psi')\{\partial \varepsilon(\psi')/\partial \psi'\}d\psi' \quad (1)$$

where ε^R = pseudo strain, $\varepsilon(\psi')$ = actual strai history, E_R = arbitrary constant with dimension of modulus, usuall take as 1, $E(\psi)$ = relaxation modulus, ψ = reduced time. The ratio is the elastic modulus in the transformed body and is referred as the pseudo stiff-ness. It is a measure of the damage that develops in the material. The pseudo strain is like the effective stress in the continuum damage mechanics (Lemaitre 1996) and C is like the continuity function. All the above is formulated for linear viscoelasticity and elasticity, with or without damage. However, AC materials are reported to ex-hibit nonlinear behavior (Levenberg & Uzan 201; Luo et al. 2016). Some nonlinearity can be included via the reduced time. Si et al (2002) pro-posed to in-clude nonlinearity via the constant E_D. As it is very difficult to differentiate intrinsic nonlinear-ity from damage plus healing, Levenberg (2006) and Uzan & Levenberg (2007) lumped the nonlinearity into the damage and healing model. In Equation (1), it is as-sumed that damage does not alter the form of the re-laxation modulus. However, it has been reported that, during fatigue testing, the phase angle increases and thus, introduces a time lag between the pseudo strain and the stress and difficulties in the computa-tion of the internal variable S which is associated with damage and given by integrating the following:

$$dS = [\,-\,1/2(\varepsilon R)2dC/d\psi^{\alpha/(1+\alpha)}\,d\psi \qquad (2)$$

where: $\alpha =$ material property. In view all the above, the paper presents an improvement to the uniaxial tension fatigue testing and analysis.

2 TESTING PROGRA

2.1 The test includes:

(1) Mounting a displacement measuring device on the specimen to control the strain. The device is composed of two transducers with one output aver-aging the displacements of the transducers. The device shown in Figure 1 was manufactured by Epsilon Technology Corp USA. Similar devices can be found in MTS and Instron catalogs.
(2) Using a trapezoidal specimen to allow to test field specimen (Uzan 2020).
(3) Conducting the test in the following se-quences (in addition to the complete dynam-ic modulus test):

 (a) Conduct a set of sinusoidal loadings to obtain dynamic modulus at the test tem-perature at different frequencies. The test is performed using the displacement trans-ducer to control the strain to +/–30 mi-crostrains.
 (b) Conduct an additional set of two sinus-oidal loadings at two frequencies.
 (c) Conduct the fatigue test for the number of repetitions to produce 50 percent dam-age, using a tension haversine controlled by the displacement transducer, with or without rest periods.
 (d) Repeat step b.
 (e) Wait few hours (or overnight) to allow for complete healing.
 (f) Repeat step b.
 (g) Continue the fatigue test to complete failure as needed.

3 TESTING ANALYSIS

(1) All dynamic test results are processed to get dynamic moduli at different frequencies and the relaxation modulus.

Figure 1. Test setup.

(2) The fatigue test results are processed to compute the **whole** pseudo strain history (Eq. 1) using a computational scheme like the one presented by Uzan (2020).
(3) The pseudo stiffness Cmax at each cycle is com-puted as the ratio of the maximum measured stress to the maximum computed pseudo strain. Other indicators of the dam-age are also computed.

4 RESULT

The results of one fatigue test, conducted at 21 C, are presented and analyzed. The loading included a 0.1 second haversine of about 200 microstrains fol-lowed by a 0.2 second rest period (zero microstrain)

Figure 2 presents the results of the dynamic mod-ulus tests. It is seen that the moduli after the test is smaller than the ones before the test. The ratio is 0.6. The phase angle increased by 2 degrees. It is also seen that the moduli recover to a ratio of 0.7 after overnight rest

Figures 3a and 3b present the results of the first and last (40,000) cycles. It is seen that: (a) In the first cycle, the measured stress and the computed pseudo strain coincide for up to 0.5 MPa, suggesting that the damage begins to develop after this stress leve; (b) In the last cycle, the measured and the pseudo strain do not coincide, indicating that the ma-terial is already damaged; (c) In the last cycle, the measured stress is shifted with respect to the pseudo strain, indicating that the phase angle had increased during the tes; (d) Both the measured stress and the pseudo strain exhibit a compression state when the applied strain goes to zero. This is due to the viscoe-lastic nature of the mate-rial and to permanent strain that may develop at 21 C.

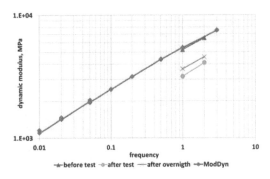

Figure 2.   Results of dynamic modulus tests.

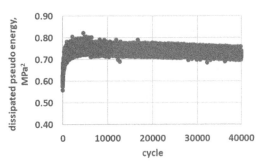

Figure 5.   Dissipated pseudo energy history.

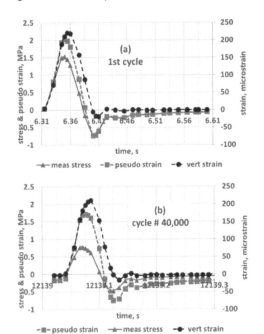

Figure 3.   Pseudo strain, stress and strain histories.

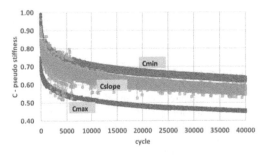

Figure 4.   Damage versus number of cycles.

The hydraulic system is forcing the recovery of the viscoelastic strain which theoretically recovers fully after very long tim; (e) During the 0.2 second rest period, the strain is zero and the compressive stress relaxes.

Figure 4 presents the pseudo stiffness versus number of load cycles, where the Cmax curve repre-sents the ratio of the measured stress and pseudo strain peaks. Cmax represents the maximum damage that the material experiences every cycle and it reaches the value of 0.45, much lower than the ratio of 0.6 from the dynamic modulus testing. However, this ratio does not correspond to the material state at the end of the 0.1 second loading or of the cycle. Figure 4 presents two additional indicators of the material state: (a) Cmin at the end of the 0.1 second as the ratio of the minima of the measured stress and pseudo strain and (b) Cslope at the end of the cycle, as the ratio of the initial moduli at the beginning of the following cycle. These indicators reach values of 0.59 and 0.64 for Cmin and Cslope respectively, close to the ratio of the dynamic moduli, before and after testing. This result seems to support the allega-tion that the viscoelastic strain may be related to damage and healing (Levenberg 200; Uzan & Le-venberg 2007) and indicates that Cmax may not represent the material state at the end of the 0.1 second.

The evaluation of the damage internal variable S, using Equation (2) is straight forward for a continuous loading test. This is not the case for a cyclic loading where phase angle change and healing (during the unloading step) are not considered in the equation. Underwood et al. (2010) proposed a simplified ap-proach to compute S.

The transformation in Equation (1) did not pro-duced a linear relation between the pseudo strain and stress. The hysteresis is due to nonlinear behavior and damage. According to Luo et al. (2013), the non-linearity occurs in the strain range between 40 and 60 microstrains. Therefore, and in the present case of applying about 200 microstrains, most of the hys-teresis is due to damage. Figure 5 presents the area of the hysteresis loop, the dissipated pseudo strain energy during the test. It is seen that: (a) The dissi-pated energy increases with increasing number of cycles, in the range where the pseudo stiffness de-creases sharply (primary behavior, up to the 5000th load cycle) and (b) It remains relatively constant (or decreases slightly) in the range where the pseudo stiffness decreases monotonously (secondary behavior, after 5000 load cycles).

5 CONCLUSION

Two improvements are proposed for the testing and analysis of the uniaxial tension fatigue test:

(1) In order to avoid the problems caused by the crosshead loading control, it is proposed to use a displacement transducer, mounted on the specimen, for the strain control. It is possible to simulate the strain shape measured in the pavement.

(2) The pseudo strain in Equation (1) can be computed using the numerical integration proposed by Uzan (2020). All data points in the data acquisition, usually more than one million, are used to compute the pseudo strain history.

The detailed analysis of one fatigue test suggest that the pseudo stiffness, calculated from the peaks of the stress and pseudo strain, may not represent the material state at the end of loading. Other pseudo stiffness indicators, Cmin and Cslope seem to better characterize the material state at the end of the loading.

As the strain-controlled fatigue test is usually terminated before complete failure, it is recommended to perfor fingerprin test (suc as th dynamimic modulus tests) before and after the fatigue test, to validate the results of the analysis.

REFERENCES

Christensen, D.W., & Bonaquist, R.F. 2009. Analysis of HMA Fatigue Data Using the Concepts of Reduced Load-ing Cycles and Endurance Limit, *Journal of the Association of Asphalt Paving Technologists*. 78: 377–416.

Daniel J.S. & Kim Y.R. 2002. Development of a Simplified Fa-tigue Test and Analysis Procedure Using a Viscoelastic Continuum Damage Model, *Journal of the Association of As-phalt Paving Technologists*, 71 619–650.

Kim, Y.R., Lee, Y.C. & Lee. H.J. 1995. Correspon-dence Principle for Characterization of Asphalt Concrete, *ASCE Journal of Materials in Civil Engineering.* 7(1) 59–68.

Kim, Y.R. & Little, D.N. 1990. One-Dimensional Consti-tutive Modeling of Asphalt Concrete, *ASCE Journal of Engineering Mechanics*: 116(4): 751–772.

Lee, H.J. & Kim Y.R. 1998. A Viscoelastic Continuum Damage Model of Asphalt Concrete with Healing, *ASCE Journal of Engineering Mechanics*, 124(11) 1224–1232.

Lemaitre, J. 1996. *A Course on Damage Mechanics*, 2nd edition, Berlin: Springer-Verlag.

Levenberg, E. 2006. Constitutive modeling of asphalt-aggregate mixes with damage and healing. *PhD dissertation. Technion, Israel Institute of Technology: Haifa*

Levenberg, E., & Uzan, J. 2012. Exposing the nonlinear vis-coelastic behavior of asphalt-aggregate mixes. *Mech. Time-Depend. Mater.* 16(2), 129–143.

Luo, R., Liu, H. & Zhang, Y. 2016. Characterization of linear viscoelastic, nonlinear viscoelastic and damage stages of asphalt mixtures. *Construction and Building Materials*, 125, 72–80.

Luo, X., Luo, R., & Lytton, R.L. 2013. Characterization of fatigue damage in asphalt mixtures using pseudos-train ener-gy. *Journal of Materials in Civil Engineering, American Society of Civil Engineers (ASCE)*, 25 (2) 208–218.

Pronk, A.C., Gajewski, M. & Bankowski, W. 2017. Pro-cessing the four-point bending test results for visco-elasticity and fatigue models. *International Journal of Pavement Engineering*. 20(10): 1226–1230.

Schapery, R.A. 1984. Correspondence Principles and a Gener-alized J-integral for Large Deformation and Frac-ture Analy-sis of Viscoelastic Media, *International Jour-nal of Fracture*, 25(3) 195–223.

Si, Z. Little, D.N. & Lytton, R.L. 2002. Characterization of Microdamage and Healing of Asphalt Concrete Mixtures, *J. Mater. Civ. Eng.*, 14(6): 461–470.

Underwood, B.S., Kim, Y.R. & Guddati, M.N. 2010 Im-proved calculation method of damage parameter in visco-lastic continuum damage model, *International Journal of Pavement Engineering*, 11:6 459–476.

Uzan, J. 2020. A new approach to characterizing asphalt con-crete in the linear/nonlinear small strain domain. *Mech. Time-Depend. Mater.*, To be published.

Uzan, J. & Levenberg, 2007. Advanced Testing and Char-acterization of Asphalt Concrete Materials in Tension, *ASCE International Journal of Geomechanics*, 7(2) 158–165.

Witczak, M.W., Mamlouk, M., Souliman, M. & Zeiada, W. 2013. Laboratory Validation of an Endurance Limit for As-phalt Pavements. *NCHRP Report 762, National Academy of Sciences, Transportation Research Board*. p.26.

Advances in Materials and Pavement Performance Prediction II – Kumar et al. (eds)
© 2021 Taylor & Francis Group, London, ISBN 978-0-367-46169-0

Assessing permanent deformation of reclaimed asphalt blended binders using non-linear viscoelasticity theory

Bhaskar Vajipeyajula
Department of Mechanical Engineering, Texas A&M University, USA

K. Lakshmi Roja & Eyad Masad
Department of Mechanical Engineering, Texas A&M University at Qatar, Doha, Qatar

Kumbakonam R. Rajagopal
Department of Mechanical Engineering, Texas A&M University, USA

ABSTRACT: In this study, the resistance of various RAP blended binders to permanent deformation was analyzed with consideration to the nonlinear response of asphalt binders. Multiple creep and recovery (MSCR) and repeated creep and recovery with multiple stress levels (RCRMS) were performed on the RAP blended binders at 64°C. For the binder used in this study, it was found that the linear viscoelasticity framework is sufficient to capture the MSCR results. However, the use of a nonlinear viscoelastic model was necessary to predict the response of the binders when subjected to the RCRMS protocol. The results highlight the advantages of using a nonlinear viscoelasticity framework instead of relying only on the non-recoverable creep compliance (J_{nr}) to evaluate the rutting resistance of asphalt binders.

1 INTRODUCTION

Rutting is one of the most common distresses in asphalt pavements. It is caused primarily by densification in the initial stages of the pavement life followed by degradation in the shear strength of the mixture. Several test protocols were developed to analyze the resistance of asphalt binders to permanent deformation or rutting. Traditionally, the loss modulus, which is a measure of the dissipated viscoelastic energy in each cycle, (G*/sinδ) was used as a measure of susceptibility to permanent deformation. This parameter was mostly suitable for quantify rutting in unmodified binders but did not work well to rank modified binders [1–3].

The multiple stress creep and recovery test (MSCR) was developed to address the shortcomings of the Superpave high temperature binder parameter (G*/sinδ) [4]. In MSCR, the binder is subjected to stress levels of 0.1 and 3.2 kPa with 1 second creep time followed by 9 seconds recovery time for 10 cycles at each stress level. The average non-recoverable strain and J_{nr} are calculated from the strain response of the binders. However, several researches [5–7] have reported that the binder response is linear even at the higher loading conditions of 3.2 kPa and also the recovery time of 9 seconds is not enough for the binder to recover completely.

The main objective of this paper is to analyze the effects of RAP on the virgin binder using linear and nonlinear viscoelastic models and creep and recovery experiments. The following tasks were carried out to achieve this objective:

- Perform MSCR and RCRMS tests on the various RAP blend binders (containing various amounts of RAP).
- Analyze the test data using traditional protocol (calculation of J_{nr}) as well as linear and nonlinear viscoelastic models.
- Determine the need for using a nonlinear viscoelastic framework to analyze permanent deformation of asphalt binders.

2 MATERIAL

A virgin binder of penetration grade Pen 60/70 (PG 64-22) was used in this study. RAP binders were extracted from mixtures that were milled from Qatar roads. The grade of the RAP binder was PG 94-0. Asphalt mixtures were produced with different RAP proportions (0, 15, 25 and 35%). Then, binders extracted and recovered (AASHTO T 164, 2014) from these mixtures were used in this study.

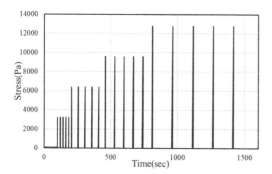

Figure 1. RCRMS test protocol.

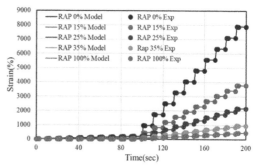

Figure 2. Burgers model prediction of MSCR data.

Table 1. Burgers model parameters.

RAP (%)	G_m	μ_m	G_k	μ_k	R^2
0%	1.50E+4	5.03E+2	2.67E+2	2.82E+3	0.98
15%	3.55E+8	6.89E+2	2.31E+4	1.40E+4	0.98
25%	1.05E+5	1.42E+3	1.20E+4	1.28E+4	0.99
35%	1.05E+5	4.43E+3	1.20E+4	1.28E+4	0.99
100%	1.09E+5	7.92E+3	1.18E+4	1.60E+4	0.97

3 EXPERIMENTAL PROCEDURE

All measurements were performed at 64°C using a dynamic shear rheometer. Each experiment was conducted in a parallel plate geometry with a diameter of 25 mm and height of 1 mm. To ensure repeatability, each experiment was carried out twice and the error between the two trials to their mean was found to be less than 3%.

3.1 Multiple stress creep and recovery (MSCR)

As mentioned earlier in this paper, the RAP blended binders were subjected to a loading of 0.1 and 3.2 kPa with a creep and recovery time of 1 second and 9 seconds respectively, and at each stress levels 10 cycles were carried out AASHTO T 350 (2014) [8].

3.2 Repeated creep and recovery with multiple stress levels (RCRMS)

Several researchers [5–7] reported limitations of the MSCR in capturing the nonlinear response of binders and that the recovery time of 9 seconds is not long enough for the material to completely recover. Due to these limitations, several modifications of the traditional MSCR protocol were suggested.

In the RCRMS protocol, the binders were subjected to stress levels of 0.1, 3.2, 6.4, 9.6, 12.8 kPa with a creep loading time of 1 second and with a recovery time of 9, 20, 50, 70, 150 seconds, respectively, as shown in Figure 1.The first two stress levels were selected to be the same as the MSCR standard protocol, while the next three stress levels were selected in such a way that one can do linear scaling and superposition checks.

4 MODELING

4.1 Modeling of MSCR

The Burgers linear viscoelastic model was used to analyze the strain response of the RAP blend binders subjected to the MSCR protocol. The modeling was

carried out in the creep loading (using eq 1) and recovery period (using eq 2).

$$\varepsilon_c = \frac{\sigma}{G_m} + \frac{\sigma_t}{\mu m} + \frac{\sigma}{G_k}\left(1 - e^{\frac{-tG_k}{\mu_k}}\right), \tag{1}$$

$$\varepsilon_r = \varepsilon_c + \frac{\sigma}{G_k}\left(1 - e^{\frac{-tG_k}{\mu_k}}\right) + \frac{\sigma t_c}{\mu_m} \\ - \frac{\sigma}{G_k}\left(1 - e^{\frac{-(t-t_c)G_k}{\mu_k}}\right), \tag{2}$$

where t is time since the beginning of the cycle, t_c is creep time, G is the shear modulus and μ is the viscosity, ε is strain and σ is stress. The subscript k indicates Kelvin-Voigt element, M indicates Maxwell elements, c indicates creep and r indicates recovery. The model was able to capture the material response very well as shown Figure 2. The obtained parameters are tabulated in Table 1.

4.2 Modeling RCRMS

The parameters of the burgers model obtained from the analysis of the MSCR were used to predict the RAP blended binders response when subjected to RCRMS. The model underpredicted the material response as shown in Figure 3.

4.2.1 Nonlinearity check
A nonlinearity check was performed to determine if the material behavior is nonlinear. The material response is considered nonlinear if the normalized strain (strain/torque) is dependent on the applied

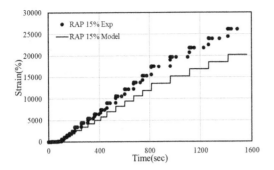

Figure 3. Burgers model prediction of RCRMS data for RAP 15%.

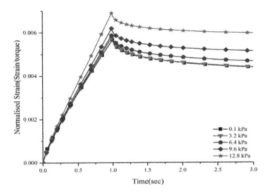

Figure 4. Normalized strain of RAP blend binders for RAP 100%.

torque. The normalized strain response of the first cycle at each strain levels are shown in Figure 4.

The normalized strain response was found to be dependent on the torque being applied; therefore, the material response is considered to nonlinear at higher stress levels (higher than 6.4 kPa).

4.2.2 Nonlinear modeling

To model the response of the RAP blend binders when subjected to RCRMS, a nonlinear model derived by Malek, Rajagopal [9] was used. The model is given by the equations below

$$T = -pI + 2\mu_3 D + G_1 B^d_{\kappa p(t)_1} + G_2 B^d_{\kappa p(t)_2} \quad (3)$$

$$\breve{B}_{\kappa p(t)_i} = -\frac{G_i}{\mu_i} B_{\kappa p(t)_i} B^d_{\kappa p(t)_i} \quad (4)$$

where T Cauchy stress tensor, μ_3 is the viscosity and G_1 and G_2 are the shear moduli, $\left(\frac{G_1}{\mu_1}\right)^{-1}$ and $\left(\frac{G_2}{\mu_2}\right)^{-1}$ are the two relaxation times. A^d is the deviatoric part of the tensor and $\breve{A}$ is the Oldroyd derivative of the tensor. p is the Lagrange multiplier owing to the assumption of incompressibility of the asphalt binders. A detailed description of the nonlinear model and how the modeling is carried out can be found in Malek, Rajagopal [9] and Vajipeyajula, Masad [10].

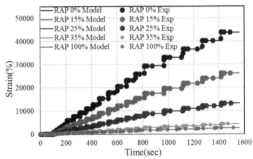

Figure 5. Nonlinear model prediction of RCRMS data.

Table 2. Nonlinear viscoelastic model parameters.

RAP (%)	G_1	μ_1	G_2	μ_2	μ_3	R^2
0%	577	201	350	100	295	0.98
15%	1023	756	700	320	380	0.99
25%	1374	774	1500	700	800	0.98
35%	3516	2400	2400	1900	2310	0.98
100%	5558	6797	2609	2308	2602	0.97

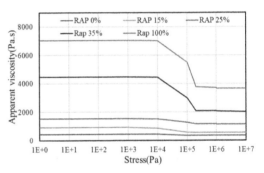

Figure 6. Apparent viscosity of RAP blend binders.

The above nonlinear model was curve fitted with the strain response using solvers and curve fitting tools in Matlab. The model was able to capture the material response as shown in Figure 5.

5 APPARENT VISCOSITY

Several researches have attempted to recommend a single parameter to rank binders based on their resistance to permanent deformation. Atul Narayan, Little [11] suggested the use of apparent viscosity of a nonlinear viscoelastic material as the parameter to quantify rutting. The apparent viscosity is defined as ratio of stress applied during creep to the steady state strain rate as given in the equation below

$$\mu(\tau) = \frac{\tau}{\dot{\varepsilon}} \quad (5)$$

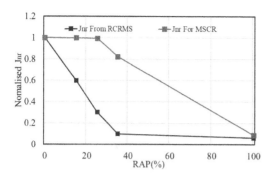

Figure 7. Normalized Jnr Values (at 3.2 kPa).

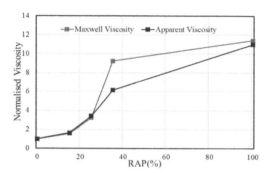

Figure 8. Normalized viscosity values.

where μ is the apparent viscosity, τ is the shear stress and is the steady state strain rate. The apparent viscosity is calculated for creep stress ranging from 1 to 10^7 Pa as shown in Figure 6

6 RESULTS AND DISCUSSION

The J_{nr} calculated from the MSCR and RCRMS test protocols as well as the Maxwell and apparent viscosity parameters are normalized with respect to the RAP 0% values and are shown in Figures 7 and 8. In the standard post processing method, the J_{nr} is calculated based on the three strain values (strain at the beginning of loading, strain at the end of loading and at the end of recovery). In the standard MSCR protocol, the recovery time is constant irrespective of the stress levels. In such cases, binder does not completely recover leading to higher residual strain and erroneous J_{nr} values in the MSCR test protocol. In the case of RCRMS, the recovery time is varied based on the stress levels, allowing the binder to recover completely. This protocol lowers the residual strain, which reduces the J_{nr} values. Figure 7 shows the J_{nr} values obtained from both testing protocols.

Nonlinear and linear parameters like Apparent viscosity and Maxwell viscosity were used to assess resistance to permanent deformation. These parameters are obtained from curve fitting the entire material strain response to creep and recovery loading. The

trends of the two viscosities are close to each other as shown in Figure 8 but are different than the trend of the J_{nr} values. Therefore, it is suggested to use the apparent viscosity parameter to assess permanent deformation as it accounts for both the linear and nonlinear response of the material.

7 CONCLUSION

Loading conditions in the standard MSCR test protocol are not ideal to analyze rutting in RAP blended binders because applied stress levels are not sufficient to capture their nonlinear viscoelastic behavior. In addition, the recovery time in MSCR is constant irrespective of the load magnitude. This time may not be sufficient for the binder to recovery completely at high loading conditions. The standard protocol of calculating the J_{nr} value has shown disagreement in the values obtained from the MSCR and RCRMS test. In the case of the MSCR test, the J_{nr} value was constant up to 25% RAP after which its value decreased with the addition of more RAP content. In the case of the RCRMS test, J_{nr} value gradually decreased with the addition of RAP until 35% RAP.

Unlike the J_{nr}, the viscosity parameter obtained from the rheological analysis has shown agreement between the values obtained from the MSCR and RCRMS tests. The response of the RCRMS test was captured quite well using the nonlinear viscoelastic model presented in this paper. The apparent viscosity obtained from the rheological analysis increased gradually with the addition of RAP up to 15% RAP after which there was a rapid increase in the viscosity for 25% RAP and beyond.

REFERENCES

Anderson, D.A. & T.W. Kennedy, *Development of SHRP binder specification (with discussion)*. Journal of the Association of Asphalt Paving Technologists, 1993. **62**.

Bahia, H.U., et al., *Characterization of modified asphalt binders in superpave mix design*. 2001.

D'ANGELO, J. & R. Dongr. *Superpave binder specifications and their performance relationship to modified binders*. in *Proceedings of the Forty-Seventh Annual Conference of the Canadian Technical Asphalt Association (CTAA): Calgary, Alberta*. 2002.

D'Angelo, J.A., *The relationship of the MSCR test to rutting*. Road Materials and Pavement Design, 2009. **10**(sup1): p. 61–80.

Delgadillo, R., *Nonlinearity of asphalt binders and the relationship with asphalt mixture permanent deformation*. 2008.

Loizos, A., et al., *A new performance related test method for rutting prediction: MSCRT*, in *Advanced Testing and Characterization of Bituminous Materials, Two Volume Set*. 2009, CRC Press. p. 1003–1012.

Dreessen, S., J. Planche & V. Gardel, *A new performance related test method for rutting prediction: MSCRT*. Advanced testing and characterization of bituminous materials, 2009. **1**: p. 971–980.

AASHTO, T., *350-14, Standard Method of Test for Multiple Stress Creep Recovery (MSCR) Test of Asphalt Binder Using a Dynamic Shear Rheometer (DSR)*. American Association of State and Highway Transportation Officials, 2014.

Malek, J., K.R. Rajagopal & K. Tuma, *A thermodynamically compatible model for describing the response of asphalt binders*. International Journal of Pavement Engineering, 2015. **16**(1): p. 297–314.

Vajipeyajula, B., et al., *A two-constituent nonlinear viscoelastic model for asphalt mixtures*. Road Materials and Pavement Design, 2019: p. 1–15.

Atul Narayan, S., D.N. Little & K.R. Rajagopal, *Analysis of rutting prediction criteria using a nonlinear viscoelastic model*. Journal of Materials in Civil Engineering, 2014. **27**(3): p. 04014137.

Advances in Materials and Pavement Performance Prediction II – Kumar et al. (eds)
© 2021 Taylor & Francis Group, London, ISBN 978-0-367-46169-0

Evaluation of volumetric effects on asphalt fatigue performance with VECD theory

Y.D. Wang, A. Ghanbari, B.S. Underwood & Y.R. Kim
North Carolina State University, Raleigh, North Carolina, USA

ABSTRACT: Volumetric properties have been widely used in the routine-based activities in the asphalt industry. The paper utilized the Simplified ViscoElastic Continuum Damage (S-VECD) model to evaluate the effects of the mixture volumetric property changes on their fatigue performance in a case study. A systematic experimental design was conducted, and performance tests were done on one asphalt mixture with nine different volumetric conditions. The index parameter, S_{app}, was applied to evaluate the mixture fatigue resistance on the material level while structural level analyses were performed using FlexPAVE™program. The test results indicate that the volumetric properties have great impacts on mixture fatigue resistance, and using the index parameter and structural performance simulations are both effective approaches to evaluate asphalt mixture fatigue performance.

1 INTRODUCTION

Fatigue cracking is one of the most common distresses on pavements. Pavement engineers have been using volumetric properties to ensure acceptable mixture fatigue performance. In the current Superpave design procedure, parameters such as voids in mineral aggregates (VMA) and voids filled with asphalt (VFA) are utilized to ensure sufficient air and effective binder in mixtures. During construction, most of the quality assurance specifications use the volumetric parameters as acceptance quality characteristics (AQCs) to adjust the pay factor. Applying volumetric parameters has practical meanings in such daily activities, and in the near future, those parameters will still be used during asphalt production (Kim et al. 2019; West et al. 2018;). Accordingly, the effectiveness of using volumetric properties to control mixture performance should be evaluated rigorously. The relationship between the volumetric parameters and mixture performance in pavements should be systematically evaluated. As new performance-related tests being considered in the future mix design and quality assurance specification, using models and testing methods with more mechanistic background would be beneficial to industry.

In this paper, an experiment including one mixture with nine different volumetric conditions was systematically designed. The fundamental mechanistic model, the S-VECD model, was used to characterize the mixture fatigue properties at different volumetric conditions. The index parameter, S_{app}, which was developed based on the S-VECD model, was applied to evaluate the mixture fatigue resistance on the material

level. Meanwhile, on the structural level, pavement performance was predicted using the FlexPAVE™program. The volumetric effects on asphalt mixtures were evaluated using such performance-related mechanical tools.

2 BACKGROUND

2.1 Simplified ViscoElastic Continuum Damage (S-VECD) model

The S-VECD model is a fundamental mechanistic model describing the changes in material constitutive relationship as fatigue damage accumulates. It identifies the relationship between the material integrity, C, and the damage, S. The damage is calculated based on the accumulated dissipated pseudostrain energy; the relationship between C and S is a fundamental material property which is independent of the mode of loading and loading temperatures (Underwood et al. 2012). A failure criterion, D^R, was developed, which enables the model to predict the macro-crack occurrence (Wang & Kim 2017).

2.2 Index parameter, Sapp, and FlexPAVE™program

The FlexPAVE™program simulates the performance of pavement structures using a three-dimensional finite element method and viscoelastic analyses with moving loads. The S-VECD model is implemented in the program to predict the fatigue evolution under massive passes of traffic loads. As the realistic climate conditions and pavement structures are modelled, the

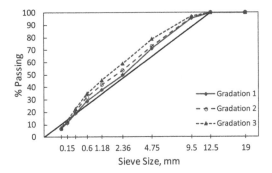

Figure 1. The design gradations in the experiment.

Table 1. The volumetric properties in the experimental design.

| | | Volumetric @ N_{design} | | | | Compaction Level |
	% AC	AV*	VMA	VFA	Vbe**	% Gmm
G1-33	6.0	3.0	15.3	80.4	12.3	97
G1-53	5.3	5.0	15.7	68.1	10.7	97
G1-55	5.3	5.0	15.7	68.1	10.7	95
G1-57	5.3	5.0	15.7	68.1	10.7	93
G2-54	5.8	4.7	16.3	71.2	11.6	96
G3-33	7.0	3.0	17.4	82.8	14.4	97
G3-54	6.1	5.0	17.2	70.9	12.2	96
G3-55	6.1	5.0	17.2	70.9	12.2	95
G3-57	6.1	5.0	17.2	70.9	12.2	93

*AV = % air void.
**Vbe = % volume of effect binder.

program has successes in predicting the amount of fatigue cracking on pavement sections (Wang et al. 2018). The prediction results have been calibrated against the field observations. On the other hand, the index parameter, S_{app}, has been developed to evaluate the fatigue resistance addressing the practical needs when the pavement structure is not clear, for instance, during pavement design. The values of S_{app} varies from 0 to 40, and higher values indicate better fatigue resistance. [FHWA 2019; Wang et al. 2019]

Table 2. The Sapp and % Cracking at different volumetric conditions.

	S_{app}	% Cracking
G1-33	22.2	12.9
G1-53	20.8	11.3
G1-55	12.6	27.0
G1-57	15.0	29.9
G2-54	21.5	13.1
G3-33	33.6	4.2
G3-54	26.1	6.7
G3-55	22.5	11.9
G3-57	22.8	12.1

3 MATERIALS AND METHODOLGY

A fine-graded surface mixture with 9.5 mm Normal Maximum Aggregate Size (NMAS) was used in this study. The mixture contained PG 58-28 virgin binder and 30% reclaimed asphalt pavement (RAP). In order to correlate the mixture performance changes with volumetric properties, an experiment with nine volumetric conditions was systematically designed. Three gradations were compacted to introduce a variation in VMA, as shown in Figure 1. At Gradation 1 and Gradation 3, two binder content levels were applied so that at each of the two gradations, the air void content at the compaction level of N_{design} was controlled to be 3% or 5%. As the performance specimens were fabricated, at some of the design conditions, the specimens were compacted to 3%, 5%, or 7% air void (corresponding to 97%, 95%, and 93% G_{mm}) to evaluate the effects of compaction levels on mixture performance. Table 1 presents the volumetric status of the nine design conditions. The first part the Mix ID expresses the gradation of the mixture; the first digit after the hyphen indicates the %AV at N_{design} and the last digit indicates the % AV in the performance test specimen. Among the nine conditions, the Condition G2-44 followed the original Superpave mix design in the job mix formula (Wang et al. 2019).

The dynamic modulus tests following the AASHTO TP 132 specification were performed to obtain the linear viscoelastic material properties. Cyclic fatigue tests with the respect to AASHTO TP 133 were conducted to characterize the material fatigue properties with the Asphalt Mixture Performance Tester (AMPT). Cylindrical specimens with 38 mm in diameter and 110 mm in height were cored and cut from gyratory compacted samples. The specimens were subjected to direct tension loads at 10 Hz. During the test, the machine actuator displacement was controlled to be constant. The on-specimen strain amplitude were determined before testing based on the modulus of the mixture; the range of the on-specimen strain amplitude was usually between 250 to 500 micro-strain. According the specification, the test temperature was determined based on the performance grade of the binder in the mix.

4 RESULTS AND DISCUSSION

The S-VECD model coefficients were calibrated using the test results at each condition. The testing data were processed using the excel-based program, Flex-MAT™. The values of the material-level index parameter, S_{app}, at different volumetric conditions are listed

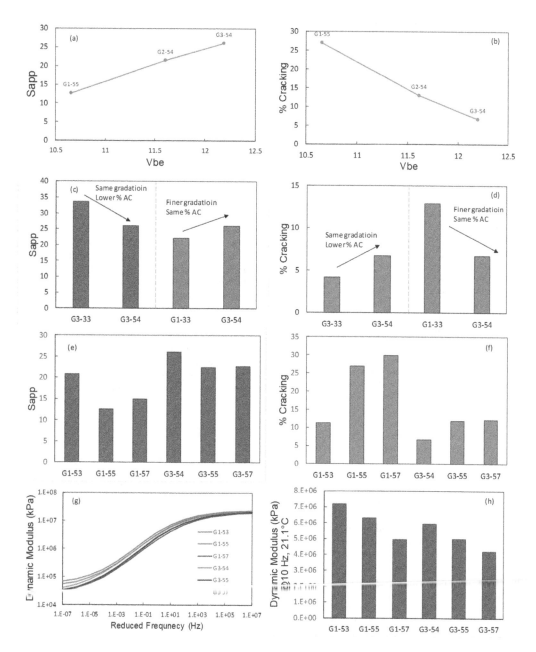

Figure 2. S_{app} and % Cracking at different volumetric conditions: (a) changes in S_{app} as Vbe increases; (b) trend in the predicted %Cracking as Vbe changes; (c) S_{app} values with different gradations and binder contents; (d) % Cracking with different gradations and binder contents; (e) S_{app} variations with compaction levels; (f) % Cracking variations with compaction levels; (g) dynamic modulus matercurves at different volumetric conditions; and (h) dynamic modulus values at 10 Hz and 21.1°C for mixtures at different volumetric conditions.

in Table 2. Furthermore, structural analyses were performed using the FlexPAVE™program. The mixture was placed in a 10 cm asphalt layer on top of 20 cm aggregate base with modulus of 206 MPa and subgrade with modulus of 69 MPa. The predicted % Cracking are presented in Table 2.

Figure 2 presents the variations of S_{app} and % Cracking as the material volumetric property changes. Figure 2 (a) and (b) show the trends in S_{app} and % Cracking as the volume of effective binder (Vbe) increases. It seems that higher Vbe yields greater fatigue resistance and less cracking on pavements.

However, one should realize that fatigue resistance is a result of multiple factors. Figure 2 (c) and (d) present the fatigue performance of Condition G1-33 and G3-54. Those two conditions have the same level of Vbe or binder content but different gradation. In Figure 2, the condition with finer gradation exhibits higher fatigue resistance according to the index parameter, S_{app}, and yields significantly less % Cracking on pavements. Therefore, using the parameter, Vbe, only is not sufficient to control the fatigue performance of an asphalt mixture.

The effects of compaction levels have been evaluated using S_{app} and predicted cracking levels as well. Figure 2 (e) and (f) present the variations in S_{app} and % Cracking as the compaction level changes. There is a general reduction trend in S_{app} and % Cracking as the compaction level decreases or the air void content increases. The amount of reduction in fatigue resistance from 5% to 7% air void in this mixture, however, is much lower than from 3% to 5%. This trend can be observed in both S_{app} and % Cracking in the structural simulation. Figure 2 (g) and (h) present the dynamic modulus mastercurves and values of the mixtures. It is clearly indicates that higher modulus does not necessarily yield better fatigue performance. However, in the AASHTO Pavement ME Design Software, as users do not usually change the material-specific fatigue parameters, kf_1, kf_2, and kf_3, stiffer materials would be shown to have lower predicted cracking. Besides, comparing Mixture G1-55 and G3-55 which both passed the volumetric criteria in Superpave Mix Design, S_{app} and % Cracking would suggest the latter one be an better design candidate.

As for the two evaluation methods used in this study: the material index parameter and the structural performance simulation, it can be observed that the evaluation results from the two methods have a good agreement. The structure simulation shows slightly higher agreements with engineers' expectation when used to evaluate the effects of compaction levels on fatigue performance. However, in some applications of index parameters, for example, in the balanced or performance-engineered mix design where the compaction level in the field is not primarily considered, the index parameter can be an effective tool to determine the optimum gradation and binder content.

5 CONCLUSION AND FUTURE WORK

This study evaluated the effects of volumetric properties on fatigue performance of asphalt mixtures in a case study. The mixture was performance-tested at different volumetric conditions with a systematic experimental design. The index parameter, S_{app}, and the FlexPAVE™program were utilized to evaluate the fatigue performance on the material-level

and the structural level, respectively. The following conclusions have been highlighted from the study:

1. The index parameter, S_{app}, and the FlexPAVE™program are both effective tools in evaluating mixture fatigue performance. Using structural performance simulation could be more suitable to evaluate the effects of compaction level. These mechanical modeling methods can be applied in mixture design in the future.
2. The volumetric parameter, Vbe, has great effects on fatigue performance; however, the fatigue performance of a mixture is a function of multiple factors, such as the aggregate gradation, binder content, and its compaction level.
3. The testing results indicate that in general, finer gradation, higher binder content, and higher compaction level can yield higher fatigue resistance.
4. Mixtures with higher modulus do not necessarily have better fatigue performance in pavements. The test results highlight a flaw in the current ME design software if users only dynamic modulus as material inputs without updating fatigue properties.
5. In future work, more than one mixtures will be used to have a full evaluation of volumetric parameter effects on asphalt fatigue performance.
6. An uncertainty study will be conducted in the future to statistically compare the fatigue resistance at different volumetric conditions.

ACKNOWLEDGMENT

The authors would like to acknowledge the financial support from the Federal Highway Administration under the DTFH61-13-C-00025 and DTFH61-14-D-00008 Projects.

REFERENCES

Federal Highway Administration (FHWA), 2019. *Cyclic Fatigue Index Parameter (S_{app}) for Asphalt Performance Engineered Mixture Design.* Tech Brief. FHWA-HIF-19-091.

Kim, Y. R., Guddati, M., Choi, M, Y., Kim, D., Norouzi, A., Wang, Y. D., Keshavarzi, B., Ashouri, Ghanbari, M., A., & Wargo, A., 2019. *Development of Asphalt Mixture Performance-Related Specifications.* Washington, D.C.: Federal Highway Administration (FHWA).

Underwood, B. S., Baek, C., & Kim, Y. R., 2012. Simplified Viscoelastic Continuum Damage Model as Platform for Asphalt Concrete Fatigue Analysis. *Transportation Research Record,* 2296, 35-45.

Wang, Y. D., Ghanbari, A., Underwood, B. S., & Kim, Y. R., 2019. Development of a Performance-volumetric Relationship for Asphalt Mixtures. *Transportation Research Record.*

Wang, Y. D., Keshavarzi, B., & Kim, Y. R., 2018. Fatigue Performance Predictions of Asphalt Pavements Using FlexPAVE^{TM} with the S-VECD Model and DR Failure Criterion. *Transportation Research Record.*

Wang, Y. D. & Kim, Y. R., 2017. Development of a Pseudo Strain Energy-Based Fatigue Failure Criterion for Asphalt Mixtures. *International Journal of Pavement Engineering*. DOI: 10.1080/10298436.2017.1394100.

Wang, Y. D., Underwood, B. S., & Kim, Y. R., 2019. Development of Fatigue Index Parameter for Asphalt Mixes Using Viscoelastic Continuum Damage Theory. *International Journal of Pavement Engineering*. Under Review.

West, R., Rodezno, C., Leiva, F., & Yin, F., 2018. *Final Report to the NCHRP, Project NCHRP 20-07/Task 405: Development of a Framework for Balanced Mix Design*. National Center for asphalt Technology at Auburn University, Auburn, AL.

Advances in Materials and Pavement Performance Prediction II – Kumar et al. (eds)
© 2021 Taylor & Francis Group, London, ISBN 978-0-367-46169-0

Aging and moisture induced adhesion reduction of high viscosity modified bitumen in porous asphalt mixture

M. Zhang, K. Zhong & M. Sun
Research Institute of Highway Ministry of Transport, Beijing, China
Key Laboratory of Transport Industry of Road Structure and Material, Beijing, China

Y. Zhang
School of Highway, Chang'an University, Xi'an, China

R. Jing
Delft University of Technology, Delft, The Netherlands

ABSTRACT: Aging and moisture induced adhesion reduction of high viscosity modified bitumen in porous asphalt mixture was studied. Specifically, PAV and water bath were adopted to simulate the aging and moisture damage process of porous asphalt mixture in the field respectively. Meanwhile, photoelectric colorimetry were utilized to measure the adhesion rates between high viscosity modified bitumen and basalt aggregates after aging and moisture conditioning. Moreover, pull-out test were conducted to test the adhesion strength of aggregate-bitumen-aggregate (ABA) system after aging and moisture conditioning. Results show that after same water bath duration, the adhesion rates of high viscosity modified bitumen and the adhesion strength of ABA system decreased 16.0% and 58.3% respectively in the first 20h PAV aging. The first 5 days' water bath had obvious effects on adhesion reduction of high viscosity modified bitumen (reduced by 47.7% for average). Moreover, the adhesion strength of ABA system would reduce obviously in the first 10d water bath (reduced by 48.4% for average). Compared with aging induced damage, moisture induced damage had more significant effects on the adhesion rates of high viscosity modified bitumen. Meanwhile, both aging and moisture induced damage can obviously lead to the adhesion strength reduction of ABA system.

1 INTRODUCTION

Raveling is one of typical damages in porous asphalt pavement (Rodríguez-Hernández et al. 2015), which is induced by the damage of bitumen-coarse aggregate interface (Mo et al. 2011). Meanwhile, because the air void ratio of porous asphalt pavement is usually more than 20%, bitumen-coarse aggregate interface are easy to be damaged by aging and moisture under environment conditioning (Mo et al. 2009).

In terms of aging induced adhesion reduction of bitumen binder, rolling thin-film oven and pressure aging vessel test (PAV) were taken to simulate the short-term and long-term aging of bitumen binder respectively (Wasiuddin et al. 2007; Jing et al. 2018; Riaz et al. 2013). As for moisture induced adhesion reduction of bitumen binder, water bath, freeze-thaw test and moisture induced sensitivity tester were conducted to simulate the constant temperature soaking, unconstant temperature soaking and hydrodynamic pressure damage in porous asphalt pavement respectively (Grenfell et al. 2015; Zhang et al. 2019; Zhang & Qian 2017; Tarefder et al. 2014). To measure the strength of adhesion, the bitumen bond strength test and contact angle measurements between asphalt binder and the aggregate surface by means of gomiometry were used (Aguiar-Moya et al. 2015).

However, the stability of bitumen-coarse aggregate interface coupling aging and moisture was rarely studied in recent papers (Bi et al. 2019;Hu et al. 2020). In this paper, aging and moisture induced adhesion reduction of high viscosity modified bitumen in porous asphalt mixture was studied. Specifically, PAV and water bath were adopted to simulate the aging and moisture damage process of porous asphalt mixture in the field respectively. Meanwhile, photoelectric colorimetry were utilized to measure the adhesion rates between high viscosity modified bitumen and basalt aggregates after aging and moisture conditioning. Moreover, pull-out test were conducted to test the adhesion strength of aggregate-bitumen-aggregate (ABA) system after aging and moisture conditioning.

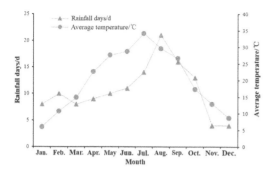

Figure 1. The Average Temperature and Rainfall Days Of NanJing, China In 2017.

Figure 2. Spectrophotometer.

2 RAW MATERIALS

High viscosity modified bitumen and basalt aggregate (BA) were usually utilized in porous pavement in China. Specifically, adding high viscosity asphalt modifier into pure bitumen can obviously improve its viscosity, so that to improve the high temperature stability and fatigue resistance of asphalt mixture.

In this research, high viscosity modified bitumen produced by Ruida Rubber Co., Ltd and basalt aggregate produced by Longteng Stone Industry Co., Ltd. were adopted for studying the aging and moisture induced adhesion reduction of high viscosity modified bitumen in porous asphalt mixture.

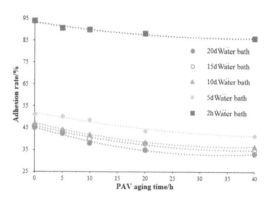

Figure 3. Adhesive rate V.S. PAV aging time.

3 AGING AND MOISTURE CONDITIONING

Pressure aging and water bath was usually adopted in laboratory to simulate the long-term aging process and moisture induced damage of asphalt pavement in the field respectively (Wu et al. 2017). Specifically, the aging level of bitumen after 20 hours' PAV at 100° and 2.1MPa was similar to that after 10 years' aging in the field. Moreover, the service life of porous asphalt pavement should be longer than 20 years according to the JTG D50-2017 (Specifications for Design of Highway Asphalt Pavement). So as for the aging condition in this paper, the high viscosity modified bitumen were aged by PAV for 0h, 5h, 10h, 20h, 40h respectively, in order to simulate 0~20 years' field aging.

In terms of moisture condition, the average temperature and rainfall days of Nanjing, China in 2017 were investigated, which is shown in Figure 1. According to Figure 1, August had the maximum rainfall days in 2017 for Nanjing, and there were 21 rainy days in August, 2017. Meanwhile, the average temperature of Nanjing in August, 2017 was 29.55°. So in this paper, water bath test was adopted to simulate the moisture condition of Nanjing in August, 2017. Specifically, the temperature of water bath was 30°. Meanwhile, the water bath duration of pull-out test were 0 days, 5 days, 10 days, 15 days, 20 days. Moreover, the water

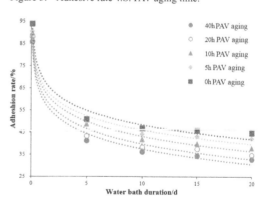

Figure 4. Adhesive rate V.S. Water bath duration.

bath duration for adhesion rate test is 2 hours, 5 days, 10 days, 15 days, 20 days

4 TESTING METHOD FOR THE ADHESION RATES OF BITUMEN-AGGREGATE SYSTEM

In this paper, photoelectric colorimetry and spectrophotometer (Hao et al. 2003; Zhang et al. 2018) were used to obtain the adhesion rates between high

Figure 5. Aggregate-Bitumen-Aggregate System.

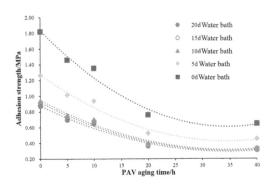

Figure 6. Adhesion strength V.S. PAV aging time.

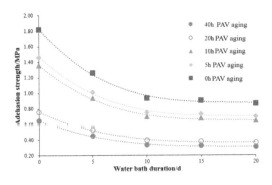

Figure 7. Adhesion strength V.S. Water bath duration.

viscosity modified bitumen and basalt aggregates, which is shown in Figure 2. Specifically, the high viscosity modified bitumen were aged in PAV for 0h, 5h, 10h, 20h, 40h respectively, and then conditioned in water bath for 2 hours, 5 days, 10 days, 15 days, 20 days respectively. The adhesion rates between high viscosity modified bitumen and basalt aggregates after aging and moisture conditioning are shown in Figures 3 and 4.

5 TESTING METHOD FOR THE ADHESION STRENGTH OF AGGREGATE-BITUMEN-AGGREGATE SYSTEM

Pull-out test were usually adopted to measure the adhesion strength of ABA system (Mo et al. 2008). In terms of the fabrication for ABA system, dynamic shear rheometer (DSR) was utilized to control the thickness of bitumen film between two aggregate cylinders, which is shown in Figure 5. Specifically, according to the existed research (Mo et al. 2010), the thickness of bitumen film between aggregates in asphalt mixture is $5\mu m\sim20\mu m$, and in this paper, $15\mu m$ was adopted as the thickness of bitumen film between two aggregate cylinders.

Before fabricating ABA system, the high viscosity modified bitumen were aged in PAV for 0h, 5h, 10h, 20h, 40h respectively. Meanwhile, after fabricated ABA system, specimens were conditioned by water bath for 0 days, 5 days, 10 days, 15 days, 20 days respectively. Finally, DSR was adopted as the pull-out tester for measuring the adhesion strength of ABA system after aging and moisture conditioning, and the results are shown in Figures 6 and 7.

6 RESULTS AND DISSCUSION

According to Figure 3 and Figure 6, after same water bath duration, both the adhesion rates of high viscosity modified bitumen and the adhesion strength of ABA system decreased with aging conditioning, while there was a negative correlation between the decreasing speed and the aging time. Specifically, after same water bath duration, the adhesion rates of high viscosity modified bitumen and the adhesion strength of ABA system decreased 16.0% and 58.3% respectively in the first 20h PAV aging. Meanwhile, the decrease of adhesion rates for high viscosity modified bitumen and adhesion strength of ABA system after aging might be induced by the failure of high viscosity modifier in high viscosity modified asphalt. Moreover, aging had negative effects on the adhesion stability of bitumen-aggregate interface, which can lead to the reveling damage of porous asphalt pavement.

According to Figure 4 and Figure 7, after same PAV aging condition, water bath had negative effects on both the adhesion rates of high viscosity modified bitumen and the adhesion strength of ABA system. Meanwhile, there was a negative correlation between the decreasing speed and the water bath duration. Specifically, the first 5 days' water bath had obvious effects on adhesion reduction of high viscosity modified bitumen (reduced by 47.7% for average). Moreover, the adhesion strength of ABA system would reduce obviously in the first 10d water bath (reduced by 48.4% for average). Furthermore, moisture induced damage had negative effects on the adhesion stability of bitumen-aggregate interface, which can lead to the reveling damage of porous asphalt pavement.

According to Figure 3 and Figure 4, compared with aging induced damage, moisture induced damage had more significant effects on the adhesion rates of high viscosity modified bitumen. Meanwhile, according to Figure 6 and Figure 7, both aging and moisture induced damage can obviously lead to the adhesion strength reduction of ABA system. Moreover, aging in porous asphalt pavement in the field is a long-term damage process, while the moisture induced damage in porous asphalt pavement in the field propagates quickly. So the adhesion stability reduction of bitumen-aggregate interface is mainly induced by moisture. In the area with heavy rainfall, water in porous asphalt pavement should be drained off in time, so that to prevent the raveling damage of porous asphalt pavement.

7 CONCLUSIONS

(1) After same water bath duration, the adhesion rates of high viscosity modified bitumen and the adhesion strength of ABA system decreased 16.0% and 58.3% respectively in the first 20h PAV aging.
(2) The first 5 days' water bath had obvious effects on adhesion reduction of high viscosity modified bitumen (reduced by 47.7% for average). Moreover, the adhesion strength of ABA system would reduce obviously in the first 10d water bath (reduced by 48.4% for average).
(3) Compared with aging induced damage, moisture induced damage had more significant effects on the adhesion rates of high viscosity modified bitumen. Meanwhile, both aging and moisture induced damage can obviously lead to the adhesion strength reduction of ABA system.
(4) The adhesion stability reduction of bitumen-aggregate interface is mainly induced by moisture. In the area with heavy rainfall, water in porous asphalt pavement should be drained off in time, so that to prevent the raveling damage of porous asphalt pavement.

ACKNOWLEDGMENT

The research work described herein was funded by the Fundamental Research Funds for the Central Research Institute (Grant No. 2019-0121). This financial support is gratefully acknowledged.

REFERENCES

Aguiar-Moya, J.P., Salazar-Delgado, J., Baldi-Sevilla, A., et al., 2015. Effect of aging on adhesion properties of asphalt mixtures with the use of bitumen bond strength and surface energy measurement tests. *Transportation Research Record: Journal of the Transportation Research Board* 2505: 57–65.

Bi, J., Zheng, N., Dong, S., et al. 2019. Effects of Asphalt and Aggregates on Water Stability Performance of Asphalt Mixture Using Large Samples Analysis. *Cailiao Daobao/Materials Reports* 33(12):4098–4101.

Grenfell, J., Apeagyei, A., Airey, G., 2015. Moisture damage assessment using surface energy, bitumen stripping and the SATS moisture conditioning procedure. *International Journal of Pavement Engineering* 16(5): 411–431.

Hao, P.W., Li, Y., Liu, J.Q. 2003. Evaluation of water susceptibility of asphalt aggregate mixture. *Journal of Wuhan University of Technology* (03): 13–16.

Hu, M., Sun, G., Su, D., et al. 2020. Effect of ther mal aging on high viscosity modified asphalt binder: Rheological property, chemical composition and phase morphology. *Construction and Building Materials* 241.

Jing, R., Varveri, A., Liu, X., et al., 2018. Chemo-mechanics of Ageing on Bituminous Materials. *97th Transportation Research Board Annual Meeting*.

Mo, L., Huurman, M., Woldekidan, M.F., et al., 2010. Investigation into material optimization and development for improved raveling resistant porous asphalt concrete. *Materials and Design* 31(7): 3194–3206.

Mo, L., Huurman, M., Wu, S., et al., 2008. 2D and 3D meso-scale finite element models for ravelling analysis of porous asphalt concrete. *Finite elements in analysis and design* 44(4): 186–196.

Mo, L., Huurman, M., Wu, S., et al., 2009. Ravelling investigation of porous asphalt concrete based on fatigue characteristics of bitumen-stone adhesion and mortar. *Materials & Design* 30(1): 170–179.

Mo, L., Huurman, M., Wu, S., et al., 2011. Bitumen-stone adhesive zone damage model for the meso-mechanical mixture design of ravelling resistant porous asphalt concrete. *International Journal of Fatigue* 33(11):1490–1503.

Moraes, R., Bahia, H.U., 2013. Effects of curing and oxidative aging on raveling in emulsion chip seals. *Transportation Research Record: Journal of the Transportation Research Board* 2361: 69–79.

Moraes, R., Velasquez, R., Bahia, H.U., 2011. Measuring the effect of moisture on asphalt-aggregate bond with the bitumen bond strength test. Transportation Research Record: *Journal of the Transportation Research Board* 2209: 70–81.

Riaz, K., Hafeez, I., Khitab, A., et al., 2013. Comparison of Neat and Modified Asphalt Binders Using Rheological Parameters under Virgin, RTFO and PAV Aged condition. *Life Science Journal* 10(3).

Rodriguez-Hernandez, J., Andrés-Valeri, V.C., Calzada-Pérez, M.A., et al., 2015. Study of the raveling resistance of porous asphalt pavements used in sustainable drainage systems affected by hydrocarbon spills. *Sustainability* 7(12): 16226–16236.

Wu, S., Zhao, Z., Xiao, Y., et al., 2017. Evaluation of mechanical properties and aging index of 10-year field aged asphalt materials. *Construction and Building Materials* 155: 1158–1167.

Zhang, M., Qian, Z., 2017. Effects of freeze-thaw cycles on fracture behavior of epoxy asphalt concrete. *Journal of Southeast University (English Edition)* 33(1):96–100.

Zhang, M., Qian, Z., Huang, Q., 2019. Test and Evaluation for Effects of Freeze-thaw Cycles on Fracture Performance of Epoxy Asphalt Concrete Composite Structure. *Journal of Testing and Evaluation* 47(1): 556–572.

Zhang, M., Qian, Z., Liu, Y., et al., 2018. Research on the Adhesion Property of Asphalt-Ceramisite System. *Advances in Materials and Pavement Prediction* 209.

Advances in Materials and Pavement Performance Prediction II – Kumar et al. (eds)
© 2021 Taylor & Francis Group, London, ISBN 978-0-367-46169-0

Micromechanical modelling of porous asphalt mixes at high temperatures

K. Anupam & H. Zhang
Faculty of Civil Engineering & Geosciences, Delft University of Technology, Delft, The Netherlands

T. Scarpas
Faculty of Civil Engineering & Geosciences, Delft University of Technology, Delft, The Netherlands
Khalifa University of Science and Technology, Abu Dhabi, UAE

C. Kasbergen, S. Erkens, H. Wang & P. Apostolidis
Faculty of Civil Engineering & Geosciences, Delft University of Technology, Delft, The Netherlands

ABSTRACT: Micromechanical modelling has been widely used to predict the properties of asphalt mixes. In comparison to numerical micromechanical models, analytical micromechanical models have the benefits of consuming much less time and facilities. The most commonly used analytical micromechanical models are the Eshelby-based micromechanical models. However, without the consideration of particles' interactions, these models fail to accurately predict the properties of asphalt mixes, especially at high temperatures. In porous asphalt (PA) mixes, due to the formation of an interconnected aggregate network, the particles' interactions under a compressive loading condition at high temperatures mainly refer to the packing effects. In order to describe the behavior of packed aggregates in PA mixes, Walton's model which predicts the effective moduli of a pack of spherical particles is possible to be a suitable way. However, to the best of authors' knowledge, this model has not been utilized for asphalt mixes. Therefore, this paper aims to investigate the application of Walton's model for predicting the properties of PA mixes at high temperatures.

1 INTRODUCTION

Micromechanical modelling has been widely used to predict the properties of asphalt mixes (Wang et al., 2020). By using micromechanical models, the effective properties of the mix can be predicted on the basis of the properties of its individual phases. In comparison to the traditional empirical models, micromechanical modelling does not require any calibration factors, and more importantly, it provides insight into the physical mechanisms behind the behavior of the mix.

Numerical and analytical techniques are often used to solve complex formulation of micromechanical models. Although numerical models based on the finite element technique and/or discrete element technique have been widely used, generally, these models require powerful computational facilities that may not be readily available.

In comparison to the numerical models, the analytical models provide an attractive alternative (Underwood and Kim, 2013). The most commonly used analytical micromechanical models are the Eshelby-based micromechanical models, such as the Mori-Tanaka model, the Self-consistent model, the generalized Self-consistent model, etc. These models were developed

on the basis of the Eshelby solution to an inhomogeneity problem where an inclusion is embedded into an infinite matrix.

The application of Eshelby-based micromechanical models to predict the properties of asphalt mixes has been widely evaluated. It has been generally concluded that the predictions match well with the experimental results at lower temperatures. However, at higher temperatures, since these models do not consider the particles' interactions, the predictions are significantly different from the experimental results (Aigner et al., 2009, Shu and Huang, 2009, Zhang et al., 2018a).

In order to improve the predictions at high temperatures, the interactions between different particles need to be taken into account. According to the authors, this necessitates the need for understanding the mechanisms behind the particles' interactions in a mix.

In this study, the mechanism behind the particles' interactions of porous asphalt (PA) mixes under compressive loading conditions at high temperatures was focused. In comparison to dense asphalt (DA) mixes, PA mixes have much higher contents of coarse aggregates and air voids. A large number of coarse aggregates form an interconnected aggregate network which highly influences the mechanical behavior

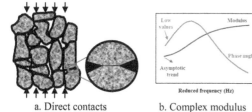

a. Direct contacts   b. Complex modulus

Figure 1.   Properties of PA mixes at high temperatures.

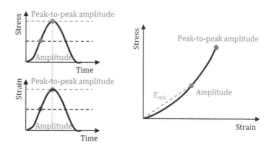

Figure 2.   Relationship between stress and strain.

of PA mixes (Zhang et al., 2019, Alvarez et al., 2010).

As highlighted in the previous work of the authors (Zhang et al., 2018b), for PA mixes the particles' interactions under a compressive loading condition at higher temperatures mainly refer to the packing effects between different particles, see Figure 1a. Due to the compressive load, the binder layer deforms, and the aggregates start to contact each other directly. Since the direct contacts form, the mix behaves more like an elastic material, which can be seen from the measured asymptotic trend of the mix's modulus and the low values of its phase angle, see Figure 1b.

The Hertzian contact theory (Fischer-Cripps, 1999) has been widely used to describe the behavior of two contacting particles. However, this theory just provides the solution for the normal direction. Later on, Walton (Walton, 1978) derived the solution for the contact behavior of two particles in the oblique compression. On the basis of this solution, Walton (Walton, 1987) further proposed an approach to calculate the moduli of a random packing of spherical particles with uniform sizes.

Although Walton's model has been widely used to predict the properties of granular materials, to the best of the authors' knowledge, it has not been used to predict the properties of asphalt mixes. Therefore, the aim of this paper is to use this model to predict the moduli of PA mixes at high temperatures.

2   BACKGROUND

2.1 *Proposed method for the application of Walton's model on PA mixes*

Walton's model describes the effective moduli of a random packing of spherical particles (perfectly rough or perfectly smooth) as a function of a) the initial loading condition, b) the mechanical and c) the volumetric properties of the particles. This model can be used to calculate both the tangent and the secant effective stiffness of the packing.

In this study, the cyclic compressive forces applied on the mixes followed such a manner that it increased from 0 to the peak-to-peak amplitude and further deceased from this peak value to 0. The stress/strain-time curve and the stress-strain curve are shown in

Figure 2a and 2b, respectively. Two points are worthy of being noted here:

- there is no significant phase lag between stress and strain because as mentioned above, the mix behaves like an elastic material at high temperatures;
- the stress and the strain are in a nonlinear relationship because the behavior of direct contacts which dominate the mix's behavior is nonlinear.

As shown in Figure 2b, the dynamic Young's modulus of the mix E_{mix} was calculated using the amplitude of stress divided by the amplitude of strain (AASHTO, 2015). It is obvious that the value of E_{mix} is similar to the secant effective stiffness when the stress applied on the packing system in the model is equal to the amplitude of the stress applied on the mix.

Furthermore, since the forces applied on the mix were in a uniaxial direction, the uniaxial compression case in the model is more suitable to predict the value of E_{mix}.

2.2 *Secant effective stiffness from walton's model*

When uniaxial compressive stress $\langle \sigma_3 \rangle$ is applied on a pack of spherical particles, the secant effective stiffness tensor $\mathbf{C}_{eff}^{S}$ can be represented as a matrix:

$$\mathbf{C}_{eff}^{S} = \begin{bmatrix} C_{11} & C_{12} & C_{13} & 0 & 0 & 0 \\ C_{12} & C_{11} & C_{13} & & & \\ C_{13} & C_{13} & C_{33} & & & \\ & & & C_{44} & & \\ & & & & C_{44} & \\ & & & & & (C_{11}-C_{12})/2 \end{bmatrix} \quad (1)$$

For perfectly rough particles, the values of all the elements are given by Equation (2)–(4), while for perfectly smooth particles, these values are given by Equation (5)—(6). The corresponding Young's modulus in the direction where the load is applied is given by Equation (7)

$$C_{11} = 3(\alpha + 2\beta), C_{12} = \alpha - 2\beta, C_{13} = 2C_{12}$$
$$C_{33} = 8(\alpha + \beta), C_{44} = \alpha + 7\beta \quad (2)$$

$$\alpha = \frac{1}{48\pi^2} \left[\frac{6\pi^2 \phi^2 n^2 (2B + C)\langle \sigma_3 \rangle}{(3B + C)B^2} \right]^{1/3} \quad (3)$$

387

$$\beta = \frac{1}{48\pi^2} \left[\frac{6\pi^2 \phi^2 n^2 B \langle \sigma_3 \rangle}{(3B + C)(2B + C)^2} \right]^{1/3} \qquad (4)$$

$$C_{11} = 3\alpha, C_{12} = C_{44} = \alpha, C_{13} = 2\alpha, C_{33} = 8\alpha, \qquad (5)$$

$$\alpha = \frac{1}{48\pi^2} \left[\frac{6\pi^2 \phi^2 n^2 \langle \sigma_3 \rangle}{B^2} \right]^{1/3} \qquad (6)$$

$$E_{\mathrm{mix}} = C_{33} - \frac{2C_{13}C_{13}}{C_{11} + C_{12}} \qquad (7)$$

with

$$B = \frac{1 - \nu^2}{\pi E}, \quad C = \frac{\nu(1 - \nu)}{\pi E} \qquad (8)$$

where E and ν are the Young's modulus and the Poisson's ratio of the particles; n is the contact number per particle; and ϕ is the volume fraction of the particles.

3 MATERIALS AND TESTS

3.1 Material properties and specimens preparation

The gradation of aggregates for making PA mix specimens is shown in Table 1 (CROW, 2015). The content of asphalt binder (Pen 70-100) was 4.3% by the total weight of the mix, and its density was assumed to be 1032 kg/m³. The air voids content of the specimens was controlled as 18%.

The specimens were prepared using the AASHTO standard method (AASHTO, 2015). The PA mix materials were compacted using a gyratory compactor to be specimens of D150 mm × H170 mm. These specimens were further cored and cut to testing specimens of D100 mm × H150 mm.

3.2 Amplitude sweep tests

In the amplitude sweep test, the PA specimens were loaded at a fixed low frequency of 0.1 Hz and a fixed high temperature of 37°C. Five levels of compressive forces F_c, 0.02kN, 0.04kN, 0.08kN, 0.12kN and 0.16kN, were applied. At each level, five cycles were applied, and there was no relaxation time between consecutive levels, see Figure 3.

Table 1. Properties of PA mix specimens.

Size (mm)	16	11.2	8	5.6	0.5	0.063	Filler
% Passing	98	77	44	22	14	4	0
Density (kg/m³)	2686	2686	2678	2670	2658	2658	2638

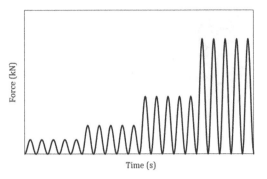

Figure 3. Applied force in the amplitude sweep tests.

Table 2. Calculated results of $\langle \sigma_3 \rangle$.

F_c (kN)	0.02	0.04	0.08	0.12	0.16
$\langle \sigma_3 \rangle$ (MPa)	0.0025	0.0051	0.0102	0.0153	0.0204

4 RESULTS AND DISCUSSIONS

4.1 Specified input parameters

The coarse aggregates with a size larger than 0.5mm were considered to form a skeleton framework in the mix (R.N. Khedoe, 2007). The volume fraction of the skeleton φ was calculated as 62%. The values of E and ν were assumed to be 53 GPa and 0.27, respectively. The value of n was assumed to be 8.5 according to the work of other researchers (Chang et al., 1999). The values of $\langle \sigma_3 \rangle$ (MPa) at different levels of F_c (kN) were computed by using Equation (9), and the obtained results are shown in Table 2.

$$\langle \sigma_3 \rangle = \frac{F_c}{1000 \times \pi \times (0.05)^2} \qquad (9)$$

4.2 Experimental results of effective stiffness

The measured values of E_{mix} under different levels of stress are shown in Figure 4. It can be seen that the values of E_{mix} increase with the increase of the compressive stress. This indicates that the behavior of the PA mix at high temperatures is not independent of the amplitude of the load, which is different from the behavior of a linear viscoelastic material. On the contrary, the mix behaves more like a granular material the properties of which are load-dependent.

4.3 Predicted results of effective stiffness

By substituting the specified input parameters into Equation (2)-(8), the predicted values of E_{mix} for both rough spheres and smooth spheres were obtained, see Figure 5.

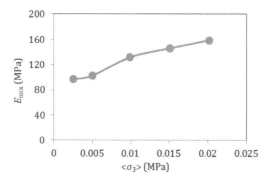

Figure 4. Experimental results of effective modulus.

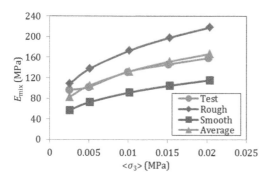

Figure 5. Comparison of predictions to experimental results.

It can be seen that the experimental results are located between the predicted values for perfectly rough particles and those for perfectly smooth particles. This observation can be explained by the following reasons:

- the crushed aggregates for preparing PA mixes are angular and have a lot of rough textures. Hence, they cannot be considered as perfectly smooth particles;
- on the other hand, there is asphalt binder covering the surface of the particles, which makes the particles less angular and decreases the number of textures. Therefore, these particles cannot be considered as totally rough either.

In order to obtain better predictions, the average values of the predicted results of E_{mix} for rough particles and those for perfectly smooth particles were calculated, see Figure 5. It can be concluded that the averaged values match well with the experimental results.

5 CONCLUSIONS

This study investigated the application of Walton's model on predicting the properties of PA mixes at high temperatures. Based on the obtained results, the following conclusions can be drawn:

- It was highlighted that under a compressive loading condition at high temperatures, the particles' interactions in PA mixes mainly refer to the packing effects among particles. Due to the compressive load, the mortar layer deforms, and the direct contacts between particles may form.
- The dynamic Young's modulus of the PA mix increases with the increase of the compressive stress. This indicates that under a compressive loading condition at high temperatures, a PA mix behaves more like a granular material instead of a linear viscoelastic material.
- The experimental values of the dynamic Young's modulus of the mix are located between the predicted values for perfectly rough particles and those for perfectly smooth particles. This can be explained by the fact that the aggregate particles for preparing PA mixes are not perfectly smooth due to lots of textures on their surface. Meanwhile, the asphalt binder covering on the surface of the particles reduces the roughness of these particles, and thus, they cannot be considered as perfectly rough as well.
- The averaged values of the predictions for perfectly rough and those for perfectly smooth are in good agreement with the experimental results.

REFERENCES

AASHTO 2015. Standard Method of Test for Determining Dynamic Modulus of Hot Mix Asphalt (HMA). Washington, D.C.
Aigner, E., Lackner, R. & Pichler, C. 2009. Multiscale Prediction of Viscoelastic Properties of Asphalt Concrete. *Journal of Materials in Civil Engineering*, 21, 771–780.
Alvarez, A. E., Mahmoud, E., Martin, A. E., Masad, E. & Estakhri, C. 2010. Stone-on-Stone Contact of Permeable Friction Course Mixtures. *Journal of Materials in Civil Engineering*, 22, 1129–1138.
Chang, C. S., Shi, Q. S. & ZHU, H. 1999. Microstructural Modeling for Elastic Moduli of Bonded Granules. *Journal of Engineering Mechanics*, 125, 648–653.
Crow 2015. Standaard RAW Bepalingen. The Netherlands.
Fischer-cripps, A. C. 1999. The Hertzian contact surface. *Journal of Materials Science*, 34, 129–137.
R.N. Khedoe, J. M. 2007. Lifetime Optimisation Tool: Sample preparation and laboratory testing for the LOTresearch program Delft: TU Delft.
Shu, X. & Huang, B. 2009. Predicting Dynamic Modulus of Asphalt Mixtures with Differential Method. *Road Materials and Pavement Design*, 10, 337–359.
Underwood, B. S. & KIM, Y. R. 2013. Microstructural Association Model for Upscaling Prediction of Asphalt Concrete Dynamic Modulus. *Journal of Materials in Civil Engineering*, 25, 1153–1161.
Walton, K. 1978. The oblique compression of two elastic spheres. *Journal of the Mechanics and Physics of Solids*, 26, 139–150.
WALTON, K. 1987. The effective elastic moduli of a random packing of spheres. *Journal of the Mechanics and Physics of Solids*, 35, 213–226.

Wang, H., Liu, X., Zhang, H., Apostolidis, P., Erkens, S. & Skarpas, A. 2020. Micromechanical modelling of complex shear modulus of crumb rubber modified bitumen. *Materials & Design*, 188, 108467.

Zhang, H., Anupam, K., Scarpas, A. & Kasbergen, C. 2018a. Comparison of Different Micromechanical Models for Predicting the Effective Properties of Open Graded Mixes. *Transportation Research Record*, 2672, 404–415.

Zhang, H., Anupam, K., Scarpas, A. & Kasbergen, C. 2018b. Issues in the Prediction of the Mechanical Properties of Open Graded Mixes. *Transportation Research Record*, 2672, 32–40.

Zhang, H., Anupam, K., Scarpas, A., Kasbergen, C. & Erkens, S. 2019. Effect of stone-on-stone contact on porous asphalt mixes: micromechanical analysis. *International Journal of Pavement Engineering*, 1–12.

Binders
Moderators: Michael Greenfield (University of Rhode Island),
Kamilla Vasconcelos (University of Sao Paolo), Katerina Varveri
(TU Delft) and Silvia Caro (University de Los Andes)

Advances in Materials and Pavement Performance Prediction II – Kumar et al. (eds)
© 2021 Taylor & Francis Group, London, ISBN 978-0-367-46169-0

Assessment of different long-term aging effect on FAM mixtures

A. Ningappa & S.N. Suresha
Department of Civil Engineering, National Institute of Technology Karnataka, Surathkal, India

ABSTRACT: Aging is considered as one of the major factors which increase stiffness and brittleness to asphaltic mixture. This study aimed at evaluating the effect of different aging protocol on viscoelastic and fatigue cracking of Fine Aggregate Matrix (FAM) mixtures. To evaluate this, six different long-term aging levels were considered. Linear Visco-Elastic (LVE) limit of each FAM mixtures was initially determined by conducting strain sweep test. Viscoelastic properties ($|G^*|$ and δ) and master curve shape parameters of FAM mixtures were further determined from temperature and frequency sweep test. Fatigue cracking of FAM mixtures was evaluated using G-R parameter. Irrespective of the aging level applied to the FAM specimen, LVE limit was found almost constant for all FAM mixtures. Viscoelastic properties for FAM specimen aged for 24 hrs at 135°C, and 12 days at 95°C aged FAM mixtures showed similar results from the master curve plots. Despite of the similar viscoelastic properties, the FAM mixtures with 12 days at 95°C and 24 hrs at 135°C were not shown similar crack potential.

1 INTRODUCTION

Fatigue cracking is considered as one of the major distress types to the flexible pavement. Fatigue failure occurs due to repeated application of load and it becomes especially critical for asphaltic pavements laid in intermediate to low temperature regions. Along with climatic temperature, the aging characteristics of asphaltic mixture also play an important role in controlling such cracks. An asphaltic mixture with a higher degree of susceptibility to aging is expected to have lower design life and vice versa. Therefore, it is important to understand the impact of aging at asphaltic mixture level. Further, a number of studies has been reported recently for the corresponding phenomena at asphaltic mixture level (Elwardany et al. 2017; Rahbar-Rastega et al. 2019). It is a usual practice to carry out aging in the laboratory using different temperatures and aging periods to simulate the field aging of asphaltic mixture. To simulate the Short Term Aging (STA), it is recommended to condition the asphaltic mixture at 135°C for 4 hrs before compaction. Similarly, conditioning at 85°C for 5 days on compacted specimens has been recommended for simulating Long Term Aging (LTA) of the asphaltic mixture (AASHTO R30). Additionally, NCHRP 09-54 recommended laboratory aging of the loose mixture at 95°C and 135°C for different aging durations and has been reported by various researchers to study the aging effect on viscoelastic and fatigue cracking properties of asphalt concrete mixtures (Chen et al. 2018; Chen & Solaimanian 2019; Rad 2018).

It is important to note that evaluating different viscoelastic properties, especially fatigue performance of

asphaltic mixture in the laboratory demands larger amount of materials, expensive equipment and an appreciable amount of time for specimen preparation, aging simulation, and performance testing. This has led the researchers in this area to look into developing other methods which can address the limitations in evaluating the corresponding viscoelastic and fatigue cracking properties at asphaltic mixture level to reduce the amount of material and required time (Freire et al. 2017; Masad et al. 2008; Suresha & Ningappa 2018). The first attempt in this direction was made by Kim et al. (2003) through examining viscoelastic properties of Fine Aggregate Matrix (FAM) and this approach was reported to be cheaper, relatively simple, repeatable and less time consuming. Based on the extensive review of literature and to the best of the knowledge to the authors, limited research work till date has been reported that aimed at understanding the effect of LTA on FAM mixture (Zhu et al. 2017). It is important to note that the aging process can be accelerated with the help of increased conditioning temperature in order to reduce the conditioning time and to achieve the same degree of aging (Chen et al. 2018). Though the increase in conditioning temperature can decrease the conditioning time, it is important to quantify the decrease in conditioning time with a corresponding increase in conditioning temperature considering its influence on long term performance parameter. This motivated the authors to investigate the impact of different aging and conditioning time for different FAM to simulate the aging (long term aging) of asphaltic mixture and the corresponding effect on different viscoelastic properties and long term performance.

Table 1. Aggregate gradation for HMA and FAM mix design.

Sieve size, mm	HMA mix design		FAM mix design	
	% Passing	% Retain	% Passing	% Retain
26.5	100	0	–	–
19.0	95	5	–	–
13.2	69	26	–	–
9.50	62	7	–	–
4.75	45	17	–	–
2.36	36	9	100	0
1.18	27	9	75	25
0.60	21	6	58	17
0.30	15	6	42	16
0.15	9	6	25	17
0.075	5	4	14	11
Pan	0	5	0	14
Total		100		100

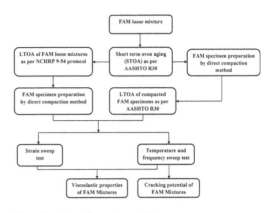

Figure 1. Overall experimental plan.

Therefore, the main objective of this study is to evaluate the various viscoelastic and fatigue cracking properties of FAM mixtures aged using varied level of conditioning, including AASHTO R30 and newly recommended NCHRP 09-54 protocols for the same.

2 MATERIALS AND FAM SPECIMEN PREPARATION

Viscosity grade asphalt binder (VG-30) provided by M/s Mangalore Refinery and Petrochemicals Ltd. was used as base binder. Basic properties of base binder are satisfied the various requirements set by IS: 73-2013 for (VG-30) and ASTM D3381 for (AC-30).

To prepare FAM specimens throughout this study, single source granite aggregate was used. The fine aggregates smaller than 2.36 mm sieve size was found to have specific gravity of 2.67. Similarly, water absorption value was found to be 0.54%, satisfying the criteria ($\leq$ 2%) set by the Ministry of Road Transportation and Highways (MoRTH 2013).

The FAM aggregate gradation was designed based on the dense graded asphalt mixture with a nominal maximum aggregate size of 19.0 mm. The FAM consists of the fine portion of the full asphalt mixture with aggregates passing sieve 2.36 mm. The optimum binder content for the FAM mixtures was determined with the help of surface area method (Li 2018a, Ng et al. 2018). The optimum binder content for FAM mixture was found to be 7.3%. The target air void for various FAM specimens was 4% ($\pm$1%). It is to be noted that the proportioning of fine aggregate present in the FAM mixtures were kept the same as in the full HMA mixture aggregate gradation. The HMA mix design and FAM gradation are shown in Table 1.

3 LABORATORY EXPERIMENTAL PLAN

This study aimed at evaluating the effect of different aging (loose mix aging and compacted specimen aging) levels on viscoelastic and fatigue properties of FAM mixtures. Short term aging on FAM loose mixture was carried out at 135°C for 4 hrs before compaction as per AASHTO R30 recommendation. To simulate long term aging, AASHTO R30's current protocol recommends to carry out aging for 5 days at 85°C on compacted FAM specimens. Additionally, unlike AASHTO R30 recommendation, NCHRP 9-54 recommends long term aging on loose FAM for different aging duration and temperature (6 hrs at 135°C, 12 hrs at 135°C, 24 hrs at 135°C, 5 days at 95°C and 12 days at 95°C). A, A1, A2, A3, B1, B2 refers to loose mixture aging, whereas, C' refers to compacted specimen aging.

Therefore, along with AASHTO R30's recommended protocol, long term aging of loose FAM was also carried out as per NCHRP 9-54 recommendation. Flowchart for the overall experimental plan is presented in Figure 1.

4 TEST METHODS AND RESULT ANALYSIS

Strain sweep test was conducted at a single temperature and frequency level of 25°C and 10 Hz respectively by changing the strain levels from 0.0001% - 0.1%. Strain corresponding to a 10% drop in the $|G^*|$ was considered as maximum strain level under LVE range. Temperature and frequency sweep test was conducted to determine the viscoelastic properties of each FAM mixtures. Based on the LVE test results, constant strain level well within LVE range was selected. The temperature was varied from 15°C to 65°C at the interval of 10°C, whereas, frequency level was varied from 0.1 Hz to 25 Hz. The master curve for complex shear modulus $|G^*|$ and phase angle δ was subsequently drawn. Williams-Landel-Ferry (WLF) equation was used for finding out the reduced frequency at the reference temperature of 25°C (Equation 3) (Underwood & Kim 2011). Logarithm sigmoidal model was further used for drawing the master curve for $|G^*|$ at the reference temperature (Sánchez et al. 2019; Yusoff et al. 2013). Equation 1 shows the mathematical form of the

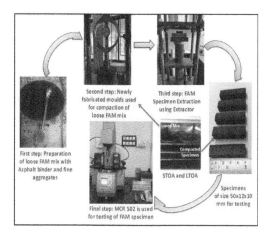

Figure 2.   Rectangular FAM specimen preparation.

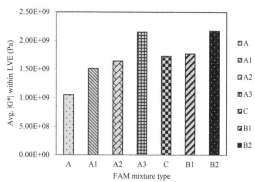

Figure 3.   Average |G*| within LVE region at different aging levels.

logarithmic sigmoidal model of |G*|. In this study, the Lorentzian peak equation was used to model the δ master curve accurately. Equation 2 shows the Lorentzian peak model equation for drawing the master curve for δ (Nemati & Dave 2018).

$$\log |G^*| = \alpha + \left(\frac{\beta}{1 + e^{\gamma + \kappa(\log \omega_r)}} \right) \qquad (1)$$

where, $\alpha, \beta, \gamma, \kappa$ are the sigmoidal fitting coefficients which describe the shape of the |G*| master curve and ωr is the reduced frequency.

$$\text{Phase angle}(\delta\delta) = \left(\frac{a \times b^2}{(\log \omega_r - c)^2 + b^2)} \right) \qquad (2)$$

The fit coefficients a, b, and c are termed as: 'a' indicates the peak value, 'b' controls the transition length, and 'c' is connected to the peak point horizontal position, δ is phase angle and ωr is the reduced frequency.

$$\text{Log } a_T = \left(\frac{C_1(T + T_R)}{C_2 + T - T_R} \right) \qquad (3)$$

where, T refers testing temperature (°C), TR is the reference temperature (25°C) and C_1, C_2 are the fitting coefficients. All tests were carried out on STA and LTA FAM specimens to characterise the viscoelastic and fatigue cracking potential of FAM mixtures. Figure 2 shows the various stages adopted during FAM specimen preparation.

4.1   Viscoelastic properties

The results of the strain sweep test on various FAM specimens are presented in Figure 3. Further, average |G*| within LVE region at different aging levels as a stiffness indicator is compared and shown in Figure 3. Strain corresponds to 10% drop in initial

|G*| was considered as LVE range. It is clear from the plot that, as the aging level increased, the stiffness value also increased. Moreover, it is also to be noted that the |G*| plot for specimen aged for 12 days at 95°C showed almost similar result with the corresponding plot with specimen aged at 135°C for 24 hrs (A3) within the LVE range. Irrespective of the aging level applied to the specimen, it is to be noted that the LVE range remained almost constant (≈0.006%) where LVE range was found to be insensitive to the induced aging level to the FAM specimens (Li 2017).

Based on the outcome obtained from strain sweep test as discussed earlier, temperature and frequency sweep test was carried out by applying an amplitude strain level within the LVE range. The master curve for |G*| for seven different FAM mixtures with STA and LTA (A, A1, A2, A3, C, B1 and B2) was drawn at a reference temperature of 25°C with the help of time-temperature superposition principle and logarithmic sigmoidal model. Figure 4 shows the master curve plot for different FAM mixtures. Increase in |G*| value with increase in frequency level can be observed as expected. Each plot can be seen as approaching towards a single value at higher frequency level which can be attributed to attainment of glassy state of the mixture and this in line with findings of FAM mixtures reported by different researchers (Li et al. 2018b; Sánchez et al. 2019). On the other hand, the apparent effect of aging can be clearly seen in relatively lower frequency zone. The lowest value of |G*| can be observed for STA specimen which subsequently increased with the increase in different aging level. Further, as in the case of amplitude sweep test (conducted at a frequency level of 10 Hz), where |G*| variation for specimen B2 (conditioned at 95°C for 12 days) and specimen A3 (conditioned at 135°C for 24 hrs) in LVE range was found to be similar.

The master curve of δ for each STA and LTA specimen was also drawn as shown in Figure 5. Unlike the variation of |G*|, a distinct change in peak δ with a change in reduced frequency can be seen. It is clear from the plot that as the aging level increased to the

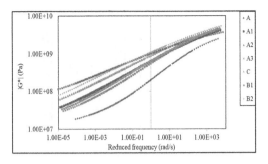

Figure 4. Complex shear modulus |G*| of STOA and LTOA FAM mixtures at reference temperature 25°C.

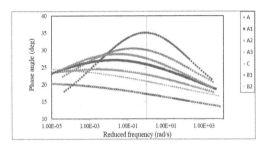

Figure 5. Phase angle of STA and LTA FAM mixtures at a reference temperature of 25°C.

Table 2. Goodness-of-fit results of |G*| and δ from master curve analysis.

Mix Type	|G*|, R^2	Acceptance criteria, (Yusoff et al. 2013)	δ, R^2	Acceptance criteria, (Yusoff et al. 2013)
A	0.997	Excellent (≥0.90)	0.870	Good (0.70-0.89)
A1	0.990	Excellent (≥0.90)	0.710	Good (0.70-0.89)
A2	0.827	Good (0.70-0.89)	0.710	Good (0.70-0.89)
A3	0.900	Excellent (≥0.90)	0.930	Excellent (≥0.90)
C	0.975	Excellent (≥0.90)	0.730	Good (0.70-0.89)
B1	0.974	Excellent (≥0.90)	0.700	Good (0.70-0.89)
B2	0.990	Excellent (≥0.90)	0.840	Good (0.70-0.89)

FAM mixture, the corresponding value at a particular frequency level decreased, indicating decreases in relaxation property of the FAM specimen. Such a response can be attributed to the viscoelastic nature of the FAM mixture which further indicates the increase in susceptibility of FAM towards cracking with an increase in aging level.

It is also interesting to note that as the aging level of FAM specimen increases, the flatness of the corresponding plot for δ value increases. For example, among A, A1, A2, and A3 specimens, the flatness of δ plot over reduced frequency is highest for A3. Finally, the various values obtained from the modeling were statistically compared with the corresponding experimental values and the goodness of fit was evaluated

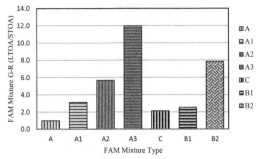

Figure 6. G-R ratio of FAM mixture G-R$_{LTOA}$ / Mixture G-R$_{STOA}$ (15°C and 0.005 rad/sec).

for each FAM specimen. The statistical analysis results are provided in Table 2 and variations was found to be within the acceptable range.

4.2 Cracking potential of FAM mixtures

Glover-Rowe parameter is used in this study to assess the cracking potential of FAM mixtures aged at different aging levels. The bases of this parameter have developed by Glover et al. (2005). Further, the mixture based G-R parameter was first introduced by Mensching et al. (2015) to assess the low temperature cracking resistance of asphalt mixtures. This same parameter for FAM mixtures was measured in this analysis at the combination of temperature-frequency of 15°C-0.005 rad/sec. The ratio of the FAM mixture G-R parameter in LTOA condition to the STOA condition is shown in Figure 6. The FAM mixture G-R parameter is expected to increase with changes in aging levels from short to medium and then to high aging levels, which indicates an increase in sensitivity to cracking. The lowest ratio of G-R parameter observed for STA specimen which subsequently increased with the increase in different aging level. For example, G-R parameter ratio of A3 loose mixture aged at 24 hrs at 135°C is almost 12 times more than the STA FAM mixture. Moreover, specimen C (conditioned at 85°C for 5 days) was found to have almost similar response as that of specimen B1 (conditioned at 95°C for 5 days) where the corresponding G-R parameter ratio for B1 can be seen as slightly higher than C. Such a response can be attributed to the stiffening effect to the FAM specimen with an increase in aging temperature and/or corresponding aging duration. This also indicates that FAM mixtures with higher G-R parameter ratio are expected to be more susceptible to cracking.

5 SUMMARY OF FINDINGS

The findings of this study are summarized below:
The LVE limit for STA and LTA aged FAM mixtures was found to be almost constant. This indicates that the

LVE range was found to be insensitive to the induced aging level of the FAM specimens.

The |G*| value of A3 specimen from master curve was found to be higher than B2 at lower frequency zone. Whereas, |G*| value of B2 was observed more than A3 at higher frequency levels. This indicates that specimen A3 may perform better than B2 specimen at higher temperature condition and specimen B2 may perform better at low to intermediate temperature conditions.

FAM mixtures aged at 95°C for 12 days and 135°C for 24 hrs showed similar response for |G*| variation in lower and intermediate frequency zone, indicating the equivalencies of their respective aging protocol.

Increase in aging level decreased the degree of dependency of δ value on loading frequency. Among A, A1, A2, and A3 specimens, the flatness of δ plot over reduced frequency was observed to be highest for A3. Such a response may be attributed to the increased effect of elastic aggregate structure in overall material response.

Based on the ratio calculated by using G-R parameter, FAM mixtures aged at higher aging levels have more sensitivity to cracking resistance.

ACKNOWLEDGEMENT

Authors would like to acknowledge the financial support extended by the Department of Science and Technology, Government of India under the scheme 'Fund for Improvement of Science & Technology infrastructure' (No.SR/FST/ETI-356/2013) for the creation of required research facilities at the Advanced Asphalt Characterisation and Rheology Laboratory, Department of Civil Engineering, National Institute of Technology Karnataka, Surathkal, India.

REFERENCES

Chen, C., Yin, F., Turner, P., West, R.C., & Tran, N. 2018. Selecting a Laboratory Loose Mix Aging Protocol for the NCAT Top-Down Cracking Experiment. *Transportation Research Record* 2672(28): 359–371.

Chen, X. & Solaimanian, M. 2019. Effect of Long-Term Aging on Fracture Properties of Virgin and Recycled Asphalt Concrete *Advances in Civil Engineering Materials* 8(1): 527–543.

Elwardany, M.D., Rad, F.Y., Castorena, C., & Kim, Y.R. 2017. Evaluation of asphalt mixture laboratory long-term aging methods for performance testing and prediction. *Road Materials and Pavement Design* 18(sup1): 28–61.

Freire, R.A., F. A. L. Babadopulos, L., T. F. Castelo Branco, V., & Bhasin, A. 2017. Aggregate Maximum Nominal Sizes' Influence on Fatigue Damage Performance Using Different Scales. *Journal of Materials in Civil Engineering*, 29 (8), 04017067.

Glover, C. J., Davison, R. R., Domke, C. H., Ruan, Y., Juristyarini, P., Knorr, D. B., & Jung, S. H. 2005. Development of a new method for assessing asphalt binder durability with field validation. *Texas Dept Transport*, 1872, 1–334.

Kim, Y.-R., Little, D., & Song, I. 2003. Effect of Mineral Fillers on Fatigue Resistance and Fundamental Material Characteristics: Mechanistic Evaluation. *Transportation Research Record: Journal of the Transportation Research Board*, 1832 (03), 1–8.

Li, Q., Chen, X., Li, G., & Zhang, S. 2018a. Fatigue resistance investigation of warm-mix recycled asphalt binder, mastic, and fine aggregate matrix. *Fatigue & Fracture of Engineering Materials & Structures* 41(2): 400–411.

Li, Q., Li, G., Ma, X., & Zhang, S. 2018b. Linear viscoelastic properties of warm-mix recycled asphalt binder, mastic, and fine aggregate matrix under different aging levels. *Construction and Building Materials*, 192, 99–109.

Masad, E., Castelo Branco, V. T. F., Little, D. N., & Lytton, R. 2008. A unified method for the analysis of controlled-strain and controlled-stress fatigue testing. *International Journal of Pavement Engineering* 9(4): 233–246.

Mensching, D. J., Rowe, G. M., Daniel, J. S., & Bennert, T. 2015. Exploring low-temperature performance in Black Space. Road Materials and Pavement Design, 16(sup2), 230–253.

Nemati, R. & Dave, E. V, 2018. Nominal property based predictive models for asphalt mixture complex modulus (dynamic modulus and phase angle), *Construction and Building Materials*, 158: 308–319.

Ng, A.K.Y., Vale, A.C., Gigante, A.C., & Faxina, A.L. 2018. Determination of the Binder Content of Fine Aggregate Matrices Prepared with Modified Binders. *Journal of Materials in Civil Engineering, ASCE*, 30 (4): 1–12.

Rad, F.Y. 2018. *Evaluation of the Effect of Oxidative Aging on Asphalt Mixtures for Pavement Performance Prediction.* Doctoral Dissertation, North Carolina State University, Raleigh, North Carolina, USA.

Rahbar-Rastega, R., Zhang, R., Sias, J.E., & Dave, E. V. 2019. Evaluation of laboratory ageing procedures on cracking performance of asphalt mixtures. *Road Materials and Pavement Design*, 20(sup2): S647–S662.

Suresha, S. N., & Ningappa, A. 2018. Recent trends and laboratory performance studies on FAM mixtures: A state-of-the-art review. *Construction and Building Materials* 174: 496–506.

Sánchez, D. B., Airey, G., Caro, S., & Grenfell, J. 2019. Effect of foaming technique and mixing temperature on the rheological characteristics of fine RAP-foamed bitumen mixtures. *Road Materials and Pavement Design* 1–17.

Underwood, B. S., & Kim, Y. R. 2011. Experimental investigation into the multiscale behaviour of asphalt concrete. *International Journal of Pavement Engineering* 12(4): 357–370.

Yusoff, N. I. M., Jakarni, F. M., Nguyen, V. H., Hainin, M. R., & Airey, G. D. 2013. Modelling the rheological properties of bituminous binders using mathematical equations. *Construction and Building Materials* 40: 174–188.

Zhu, J., Alavi, M. Z., Harvey, J., Sun, L., & He, Y. 2017. Evaluating fatigue performance of fine aggregate matrix of asphalt mix containing recycled asphalt shingles. *Construction and Building Materials* 139: 203–211.

Kinetics of epoxy-asphalt oxidation

Panos Apostolidis, Xueyan Liu & Sandra Erkens
Section of Pavement Engineering, Faculty of Civil Engineering and Geosciences, Delft University of Technology, Delft, The Netherlands

Athanasios Scarpas
Section of Pavement Engineering, Faculty of Civil Engineering and Geosciences, Delft University of Technology, Delft, The Netherlands
Department of Civil Infrastructure and Environmental Engineering, Khalifa University of Science and Technology, Abu Dhabi, United Arab Emirates

ABSTRACT: In-depth understanding of the temperature effect on oxidative aging in epoxy-asphalt blends is needed to enable accurate predictions on material response through their service life. Details of the significance of developing prediction models and tools on oxidative aging of pavement materials are presented in a companion paper (Apostolidis et al., *Oxidation Simulation of Thin Bitumen Film. AM3P*). In this research, the chemical compositional changes of epoxy modified asphalt binders, with and without filler, were analysed after oven-conditioning by means of Fourier transform infrared (FTIR) spectroscopy. With the carbonyl and sulfoxide compounds as aging indices, the sensitivity of chemical compositional changes of bituminous and epoxy-based systems due to the applied temperatures was observed.

1 INTRODUCTION

Nowadays, the asphalt industry has been focused on exploring new binders and modification technologies in an effort to offer a panoply to pavements against the climate change and the continuously increasing traffic volumes. Bitumen is product of petroleum refining process comprising in majority of non-polar, high molecular weight hydrocarbons, and currently it is the most dominant binding material used in pavements. Nevertheless, most asphalt pavements exhibit shortcomings in terms of durability and together with the top priority of road administrations to minimize the regular maintenance and reconstruction operations, super-durable pavement materials are highly demanding.

In this context, long-life pavements have started to attract the interest of road agencies and policy makers all over the world. Pavement structures of enhanced longevity would be expected to reduce the major repair needs justifying their high initial costs. New or relatively new binding systems specially designed to produce durable and long-lasting pavement materials have been proposed. One quite promising technology is the epoxy-based polymers which are able to lower the fracture (Widyatmoko et al. 2006; Youtcheff et al. 2006) and aging sensitivity of asphalt materials (Apostolidis et al. 2019b & 2020; Herrington & Alabaster 2008), and they have been utilized successfully for bridge and road pavements (ITF 2017; Lu & Bors 2015).

In comparison with the thermoplastic block co-polymers, epoxy-based polymers are thermo-hardening materials which cannot be re-melted once they are fully cured (Apostolidis et al. 2018, 2019a). Previous studies gave special emphasis on assessing the incorporating chemistry of epoxy-based polymers used in bitumen and the way this affects the evolution of physico-mechanical properties of asphalt paving materials under certain energy conditions (Apostolidis et al. 2018, 2019a). The impact of epoxy-based polymers in bitumen with and without filler on its aging performance was evaluated as well (Apostolidis et al. 2019b, 2020). However, fundamental understanding of oxidation mechanism of epoxy polymers is still needed because of the different nature of epoxy-based polymers comparing the conventional binders.

Especially, quantitative information on the rate of chemical compositional, physical and mechanical changes of epoxy-asphalt materials as function of time is important to predict precisely the evolution of material properties through their service life. Until now, predictions on the long-term performance of epoxy asphalt materials were with single-point data sets on the time scale at 130°C in the lab, much higher than the actual pavement temperatures (Apostolidis et al. 2019b, 2020). Useful lab data with pragmatic implications are missing to simulate the exact aging mechanism of such polymeric materials. Thus, kinetic data are important to assist on more reliable predictions of the long-term pavement performance, and this was the main scope of this research; to provide the kinetics parameters of oxidizing epoxy-modified asphalt materials.

2 MATERIALS AND METHODS

A 70–100 pengrade bitumen was selected for this research. A non-reactive filler passed through the 0.075-mm sieve were used to formulate the studied mastics. The epoxy-asphalt (EA) binder, supplied by ChemCo Systems, is formulated from two liquid parts; (i) the Part A (epoxy resin formed from epichlorhydrin and bisphenol-A) and (ii) Part B (fatty acid hardening agent in 70 pen bitumen). According to the supplier, Part A and B were oven-heated separately for 1 hour, at 85°C and 110°C, respectively. Afterward, Part A and B (weight ratio of 20:80) were mixed together for approximately 10 to 20 seconds, and the EA binder was produced.

Bitumen is always in conjunction with mineral particles (i.e., filler and aggregates) of different sizes in asphalt pavements. Filler particles together with bitumen form the mastic which is the binding system between the aggregates and may act as catalyst or inhibitor through the bitumen oxidation process, a possibility which needs further investigation. Thus, as in (Apostolidis et al. 2020), mastic samples were prepared as well by mixing fillers with the newly formulated binders with 56:44 weight ratio. Four mastic samples were studied of different weight ratio of 0:100 (EBF0), 20:80 (EBF20), 50:50 (EBF50) and 100:0 (EBF100) of EA binder and bitumen. To assess the influence of filler on oxidation kinetics of studied materials, the neat bitumen and EA binder were aged as well. All samples were placed in a refrigerator at −10°C to prevent any reaction.

Samples were subjected to oven-conditioning (0.1-MPa) over 0, 2, 5, 8, 24, 120, 240 and 480 hrs lengths of time at 80, 90 and 100°C. After each time period, the chemical compositional changes due to aging were measured as function of time through Fourier Transform Infrared (FT-IR) spectroscopy. IR spectra with wavenumber from 4,000 to 600 cm^{-1} were recorded and collected for all the samples as in (Apostolidis et al. 2019b, 2020). A minimum of three sub-samples were investigated for each sample and 20 scans per sub-sample were performed with a fixed instrument resolution of 4 cm^{-1}. Carbonyl (CO) and sulfoxide (SO) compounds are the typical aging indices which are defined as the integrated peak area from 1753–1660 cm^{-1} and 1047–995 cm^{-1}, respectively. Herein, their values were calculated by using the area method representing the extent of age of neat and epoxy-based systems. The calculation is performed by dividing the area under a specific location of the spectrum by the sum of other specific areas.

3 RESULTS AND DISCUSSION

Conditioning performed in an oven under 0.1 MPa of air pressure at three different temperatures; 80, 90 and

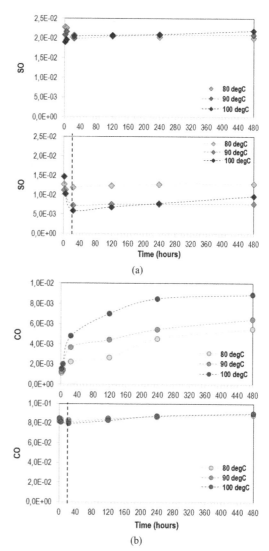

Figure 1. Change of (a) SO and (b) CO compounds of bitumen (top) and EA binder (bottom) over oven-conditioning at 80, 90 and 100 °C.

100°C. As mentioned earlier, the ultimate scope was to determine the oxidation reaction kinetic parameters of epoxy-modified asphalt binders and mastics. The evolution of SO and CO in bitumen and EA binders over oven-conditioning at different temperatures is demonstrated in Figure 1(a) and (b), respectively. The total amount of these compounds in the studied binders did not change dramatically at 80°C, however remarkable effect of temperature was noticed with increasing the applied temperature to 100°C. This trend is shown also when filler was added in these binders, see Figure 2.

A decreasing trend of CO and SO compounds of EA systems at the early phase of oven-conditioning (first 24 hrs) is shown in Figure 2, possibly due to the epoxy polymerization. The SO compounds

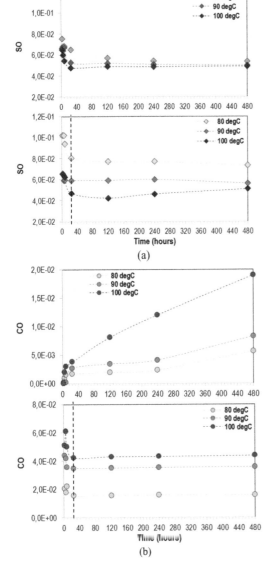

Figure 2. Change of (a) SO and (b) CO compounds of bituminous (top) and EA mastic (bottom) over oven-conditioning at 80, 90 and 100 °C.

the bituminous materials showing that CO of epoxy-based polymers were more temperature sensitive than of bitumen. Therefore, the activation energy of CO formation in EA systems is expected to be lower than of bituminous ones. Incremental values of CO compounds have been used for the determination of oxidation kinetic parameters, which are provided in the following sub-section.

3.1 Oxidation Kinetic Parameters

The chemical reaction of oxidation process in studied materials is expressed as

$$\frac{\partial x}{\partial y} = k(1-x)^n \tag{1}$$

Thus, the first-order rate expression of reaction kinetics is as

$$\frac{\partial x}{\partial t} = k(1-x) \tag{2}$$

and by calculating its integral

$$\ln(1-x) = \ln(1-x_0) - kt \tag{3}$$

where x is the reacted carbon or carbonyl compounds at different times, k is the reaction rate coefficient, n is the reaction order number, t is the oxidation time, x_0 is the initial carbonyl content.

Because conditioning temperature affects the chemical reactions and the transformation rate of carbon to carbonyl compounds, these effects can be incorporated into the reaction rate coefficients by using the Arrhenius approach as

$$\ln k = \frac{E_a}{RT} + \ln A \tag{4}$$

where E_a is the activation energy, R is the universal gas constant, T is the absolute temperature, k is the reaction rate coefficient, A is the reaction factor.

Based on the data generated by the incremental values of CO in both EA and bituminous materials, Figure 3 plots the relationship between $-\ln(k)$ and $1/T$. Figure 4 demonstrates the relationship of E_a and A values, which were derived from Figure 3, with epoxy proportion in bitumen obtained from Equation 4. From Equation 4 and 3, the final form of the oxidation kinetic equation of binders becomes as

$$\ln(1-x) = \ln(1-x_0) - Ae^{\left(\frac{E_a}{RT}\right)}t \tag{5}$$

All studied materials indicate similar curve-fitting results (Figure 3) and the values of Arrhenius coefficients (i.e., activation energy and reaction rate) were calculated and provided in Table 1. In Figure 3, the reaction rates of all materials illustrate a reasonably fit to the Arrhenius temperature dependency. Further, the temperature sensitivity of materials is represented

of bitumen and bituminous mastic remain almost unchanged through the aging at 80 and 90°C. However, the sulfur reacting species of studied mastics have shown an increasing trend at 100°C, a phenomenon which coincides with the increase attribute of SO in epoxy-modified materials at 130°C (Apostolidis et al. 2019b, 2020).

For quantitative reason, the incremental values of CO compounds, which are not demonstrated herein, used to determine the kinetic parameters of epoxy asphalt aging. The extent of CO incremental values of EA systems increased more rapidly than of

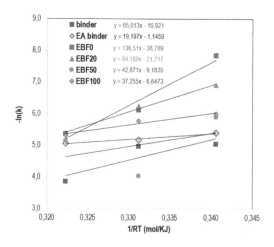

Figure 3.   -ln(k) versus $1/RT$ through oven-conditioning for the studied (a) binders and (b) mastics.

Table 1.   Kinetic parameters of studied materials.

Percentage of EA	A (1/hrs)	E_a (KJ/mol)
without filler		
0	2,23E+07	65,013
100	3,15E+00	19,197
with filler		
0	7,01E+16	136,510
20	2,70E+09	84,189
50	9,74E+03	42,871
100	7,71E+02	37,255

by the values of activation energy. As indicated in previous studies (Branthaver et al. 1993; Petersen et al. 1993; Petersen & Harnsberger 1998), the activation energy of CO development in bitumen is generally higher than of SO. So, the CO formation is less temperature sensitive than of SO. Hence, the low values of activation energy of epoxy modified mastics correspond to more temperature sensitive systems comparing the unmodified bituminous materials. Increase of activation energy of bitumen and epoxy-modified

binders with filler was noticed in Table 1. Thus, decrease of overall oxidation sensitivity is indicated with the addition of filler at any given temperature. Overall, the increase of epoxy proportion in bitumen, with and without filler, leads to decrease of activation energy and reaction rate of CO formation, and thus to temperature sensitive and oxygen resistant systems, respectively.

REFERENCES

Apostolidis, P., et al. 2018. Chemo-rheological study of hardening of epoxy modified bituminous binders with the finite element method. *Transportation Research Record 2672*, pp. 190–199.

Apostolidis, P., et al. 2019a. Kinetic viscoelasticity of crosslinking epoxy asphalt. *Transportation Research Record 2673*, pp. 551–560.

Apostolidis, P., et al. 2019b. Evaluation of epoxy modification in bitumen. *Construction and Building Materials 208*, pp. 361–368.

Apostolidis, P., et al. 2020. Evaluation of epoxy modification in asphalt mastic. *Construction and Building Materials*.

Branthaver, J.F., et al. 1993. *Binder characterization and evaluation, Volume 2: Chemistry.* Report SHRP-A-368.

Herrington, P., & D. Alabaster. 2008. Epoxy modified open-graded porous asphalt. *Road Materials and Pavement Design 9(3)*, pp. 481–498.

Long-life surfacings for roads: Field test results. 2017. International Transport Forum, OECD, Paris, France.

Lu, Q., & J. Bors. 2015. Alternate uses of epoxy asphalt on bridge decks and roadways. *Construction and Building Materials 78*, pp. 18–25.

Petersen, J.C., et al. 1993. Effects of physicochemical factors on asphalt oxidation kinetics. *Transportation Research Record 1391*, pp. 1–10.

Petersen, J.C., & P.M. Harnsberger. 1998. Asphalt aging: Dual oxidation mechanism and its interrelationships with asphalt composition and oxidative age hardening. *Transportation Research Record 1638,* pp. 47–55.

Widyatmoko, I., et al. 2006. Curing characteristics and the performance of epoxy asphalts. *Proceedings of the Canadian Technical Asphalt Association 51.*

Youtcheff, J., et al. 2006. The evaluation of epoxy asphalt and epoxy asphalt mixtures. *Proceedings of the Canadian Technical Asphalt Association 51.*

Advances in Materials and Pavement Performance Prediction II – Kumar et al. (eds)
© 2021 Taylor & Francis Group, London, ISBN 978-0-367-46169-0

Oxidation simulation of thin bitumen film

P. Apostolidis[1], H. Wang[1], H. Zhang[1], X. Liu[1], S. Erkens[1] & A. Scarpas[1,2]

[1] Section of Pavement Engineering, Faculty of Civil Engineering and Geosciences, Delft University of Technology, Delft, The Netherlands

[2] Department of Civil Infrastructure and Environmental Engineering, Khalifa University of Science and Technology, Abu Dhabi, United Arab Emirates

ABSTRACT: Oxidative aging is a complex phenomenon in bitumen and its fundamental understanding is needed to optimize paving materials with long-lasting characteristics. This research reports on a diffuse-reaction model for predicting the oxidation of bituminous binders over time and under different conditions. As known, the oxidation of bitumen is affected by the material chemistry, film thickness and temperature. Thus, these factors were considered in this research to simulate the oxidation of a thin bitumen film. Carbon compounds were assumed as the oxidation index of a model bitumen and analyses were performed enabling prediction of chemical compositional changes. In the future, the current model can be used to simulate the actual oxidative aging in (un)modified binders, such as epoxy modified asphalt, presented in a companion paper (Apostolidis et al., *Kinetics of Epoxy-Asphalt Oxidation*. AM3P).

1 INTRODUCTION

According to Greek mythology, the Cyclops traded one of their eyes with the god Hades in return to be able to foresee the future. However, Hades tricked them by the only vision to see their last day. In comparison with Cyclops, the exact end-of-life of asphalt pavements is unknown, however this did not prohibit the efforts of asphalt scientists to develop a solid research foundation on the aging of bitumen to predict the life of pavement structures (Apostolidis et al. 2017).

As everything in nature, aging in bitumen is a simply inevitable process. The chemical composition of bitumen, which depends on their crude oil source and distillation processing, is strongly influenced from oxygen over time through a chemical process named oxidative aging. Oxidative aging leads to chemo-mechanical changes that influence negatively the material performance. In chemistry terms, bitumen can be separated based on differences in solubility and polarity into four primary chemical classes: saturates, aromatics, resins, and asphaltenes. During the oxidation of bitumen, a decrease in aromatic content and subsequently an increase in resins are caused, together with a higher asphaltene content (Petersen & Harnsberger 1998). In mechanics terms, aging is accompanied by stiffening and embrittlement of binders, which contributes to the deterioration of asphalt pavements. Therefore, modelling the oxidation process of bitumen will enable to predict chemical compositional changes and subsequently the degradation of asphaltic materials over time.

In this perspective, the scope of this research is to provide insight information on the oxidation mechanism of a thin bitumen film by performing numerical analyses enabling prediction of changes in material chemistry. In the future, this information could be used as input set of data in a model to couple the oxidative age hardening and ultimately the material failure, in a way to control or prohibit aging, or design asphaltic materials with long-lasting characteristics.

2 DIFFUSE-REACTION MODEL

Typically, bitumen oxidation is distinguished in two phases, a non-linear fast-rate reaction and a linear constant-rate reaction phase (see Figure 1) (Liu et al. 1996; Van Oort 1956). Previous works focused on constant-rate kinetics of aging bitumen as well due to the fact that the duration of the fast rate period was speculated to be relatively short and with small impact on long-term asphalt performance (Liu et al. 1996; Domke et al. 1999). For the purposes of this research, a one-step reaction diffuse-reaction oxidation model has been developed and presented herein considering the carbon species as oxidizing species in bitumen.

Particularly, oxidation of bitumen is principally the conversion of carbon groups to carbonyl [C=O] groups. These are the only significant reactions that cause increase in modulus and at longer times, carbonyl formation dominates the stiffening of binder. The oxidized compounds are highly polar species

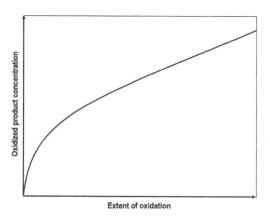

Figure 1. Schematic illustration of typical bitumen oxidation in-service conditions.

which change the mechanical properties, such as increase of modulus and decrease of phase angle by interacting with other nearby polar groups. The unreacted sited can be exposed to more oxygen to produce higher conversion and thus more inter-molecular interaction. In this research, it was assumed that the concentration-independent chemical reaction of producing carbon oxidized species is defined as

$$[C] + [O_2] \overset{k_C}{\rightarrow} [C = O] \tag{1}$$

The general expression of such irreversible reaction including the time rate of change of concentration [C] of carbon compound is as

$$\frac{d[c]}{dt} = \sum_{i=1}^{M} \omega_i k_i \prod_{\omega_{i}>0} [C^{\omega_i}] \tag{2}$$

Decomposition of single species produces compounds at a rate proportional to the amount of the decomposing carbon species can be similarly motivated. Most reactions consist of many simple sub-steps where only pairs of reactants need to be considered. For simplification, a one-step irreversible reaction of reactive species by oxygen (i.e., [O_2]) to carbon species (i.e., [C]) is considered as a perfect stoichiometric system.

Since the oxidation reactions are affected by the aging temperature, these effects can be incorporated into the reaction rate coefficients by following the Arrhenius approach as

$$k_C(T) = K_{0C}(T) \exp\left(-\frac{E_{aC}}{RT}\right) \tag{3}$$

where K_{0C} denotes the pre-exponential frequency factor (m^3/mol) of the carbon oxidation reaction, E_{aC} is the activation energy (J/mol) and R the gas constant (i.e., 8.314 J/(mol·K)).

The time-dependent mass balance per species, otherwise species transport through diffusion mechanisms, solves the mass conservation equation for one or more chemical species i

$$\frac{\partial[C]}{\partial t} + \nabla \cdot (-D\nabla[C]) = R_i \tag{4}$$

where c_i is the concentration of product carbon species (mol/m^3), D_i is the diffusion coefficient (m^2/s), which is modelled herein as in Arrhenius form for the O$_2$ in asphalt, and R_i is a reaction rate for the species (mol/(m^3·s)). The first term on the left side of Equation 4 corresponds to the consumption of the species. The second term accounts for the diffusive transport, accounting for the interaction between the perfectly mixed dilute species and the solvent.

The governing equation of the transient heat conduction within the bitumen is described by

$$\rho C_p \nabla T - \nabla \cdot (k\nabla T) = Q \tag{5}$$

where ρ is the mass density of asphalt binder, k denotes the thermal conductivity, c_p is the heat capacity, T is the temperature, Q represents the endothermic heat source because of aging (kW/m^3). It was assumed that the convection and radiation heat do not have important impact on the energy balance of the system. The heat absorbed due to oxidation was assumed negligible ($Q=0$ kW/m^3).

3 FINITE ELEMENT SIMULATIONS AND DISCUSSION

The oxidation rate of bitumen is affected by the material chemistry, film thickness and temperature (Herrington 2012). In this study, the parameters of diffusion coefficient (m^2/s) and the initial concentration of oxygen (mol/m^3) and reactive components in bitumen (mol/m^3) were obtained from (Jing 2019). These parameters determined by performing aging studies on a 200μm film thickness of bitumen and thus a model binder of the same film thickness was simulated herein at 75, 100 and 125°C as well (see Table 1). The reaction rate and the activation energy of the studied model was obtained from the available literature (1e10 1/d and 80 kJ/mol) (Jin et al. 2011). Carbon oxidation was speculated the oxidation as one-step reaction process. As in the reality, oxygen concentration was modelled unlimited by diffusion following the general behaviour shown in Figure 1.

The finite element mesh and predicted profile of oxidized carbon species is shown in Figure 2(a) and (b), respectively. The development of oxidized species gradients of the model binder film is illustrated in Figure 3[(top)] indicating that the oxidation rate over the depth of sample is affected by the applied temperature during oxidation. According to the model prediction of oxidized species depth profiles, the higher the applied

Table 1. Model parameters for the studied materials (Jing 2019).

Aging T (°C)	Initial $[O_2]$ (mol/m^3)	Initial $[C]$ (mol/m^3)	D_C (m^2/s)
75	1.10E-1	38	3.31E-13
100	1.31E-1	38	5.13E-13
125	1.49E-1	38	9.25E-13

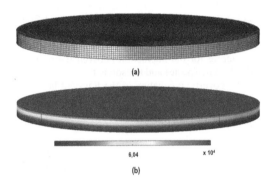

(a)

(b)

Figure 2. Finite element geometry: (a) mesh (number of elements: 33700 hexahedron and 14800 quad), and (b) predicted concentration of oxidized carbon species in thin bitumen film after 3000 hrs at 125°C.

temperature, the faster the oxidation reaction over the film thickness (see Figure 3$^{(top)}$). In this framework, the modulus gradients are expected to follow the sample pattern as of oxidized species in the model binder, allowing predictions of reasonable precision on the chemo-mechanical response of aged systems. Finally, the oxidation binder was more pronounced at high temperatures, and it was considerably greater at 125 °C than that of oxidation after 3000 hrs at 75 and 100 °C (see Figure 3$^{(bottom)}$).

The oxidation model presented here was based on a foundation of decades of research in the aging-related chemistry of bitumen. The concurrent diffuse-reaction oxidation model developed herein provides an improvement in the implementation of the current state-of-the-art on robust predicting schemes of oxidation chemistry of bitumen in a way to be extended in the future to simulate coupling phenomena through the oxidative age hardening in asphaltic materials. Oxidation studies performed enabling interpretation of the effect of environmental conditions and binder film thickness on oxidation rate and oxidized species concentration profiles through aging. The used multi-physics method for the purposes of this study offers a robust modelling tool to simulate the oxidation process in bitumen considering the diffusion phenomena together with the oxidation reactions.

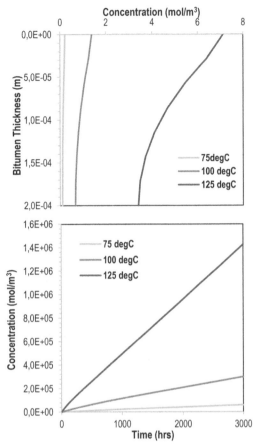

Figure 3. Predicted concentration of oxidized species along the thickness after 3000 hrs (top) and in bitumen (bottom).

REFERENCES

Apostolidis, P., et al. 2017. Synthesis of asphalt binder aging and the state of the art of antiaging technologies. *Transportation Research Record 2633*, pp. 147–153.

Domke, C.H., et al. 1999. Effect of oxidation pressure on asphalt hardening susceptibility. *Transportation Research Record 1661*, pp. 114–121.

Herrington, P.R. 2012. Diffusion and reaction of oxygen in bitumen films. *Fuel 94*, pp. 86–92.

Jin, X., et al. 2011. Fast-rate – constant-rate oxidation kinetics model for asphalt binders. *Ind. Eng. Che. Res. 50*, pp. 13373–79.

Jing, R. 2019. *Ageing of bituminous materials: Experimental and numerical characterization*. PhD Thesis, TU Delft.

Liu, M., et al. 1996. The kinetics of carbonyl formation in asphalt. *AIChE Journal 42*, pp. 1069–76.

Petersen, J.C., & P.M. Harnsberger. 1998. Asphalt aging: Dual oxidation mechanism and its interrelationships with asphalt composition and oxidative age hardening. *Transportation Research Record 1638*, pp. 47–55.

Van Oort, W.P. 1956. Durability of asphalt. *Industrial and Engineering Chemistry 48*, pp. 1196–01.

Advances in Materials and Pavement Performance Prediction II – Kumar et al. (eds)
© 2021 Taylor & Francis Group, London, ISBN 978-0-367-46169-0

Limitations of using sulfoxide index as a metric to quantify asphalt aging

S. Arafat & N.M. Wasiuddin
Department of Civil Engineering, Louisiana Tech University, Ruston, Louisiana, USA

D. Salomon
Pavement Preservation Systems, LLC, Garden City, Idaho, USA

ABSTRACT: Fourier Transform Infrared (FT-IR) spectroscopy is a convenient and quick way to investigate chemical changes for asphalt aging. The aging process changes the concentration of both Carbonyl and Sulfoxide in an asphalt binder. In this study, the limitation of considering the sulfoxide as an aging indicator for asphalt binders is investigated. Laboratory binder and mix aging showed that, increase in Sulfoxide index is dependent on laboratory aging method. Inconsistent increase of sulfoxide index made it unreliable to be considered as a metric for quantification of aging. Furthermore, Sulfoxide index for aged mixtures showed unusually higher values contributed by the presence of fines in the mixtures. This is also another disadvantage to establish this index as an indicator to quantify mixture aging. Although Sulfoxide index changes with the extent of aging, this study affirms that it should not be considered as a reliable indicator to detect the aging of an asphalt binder or a mixture.

1 INTRODUCTION

Aging of asphalt binder or mix can be identified in the laboratory utilizing different mechanical, rheological and chemical tests by measuring the change in certain properties. Fourier Transform Infrared (FT-TR) spectroscopy topped all other tests considering the ease of sample preparation and faster processing time. When an infrared beam (wavenumber between 600 cm^{-1} and 4000 cm^{-1}) passes through a material, a portion of the beam is absorbed by it. According to Beer's law, this absorbance intensity is proportional to the concentration of the molecule. From this relationship, the concentration of a particular functional group can be determined (Derrick et al. 2000).

Previous studies with FT-IR reported the increase in both the carbonyl (C=O) and sulfoxide (S=O) concentration due to aging (Huang and Grimes 2010; Liang et al. 2019). Researchers also showed that total absorbance in C=O and S=O of FT-IR spectra had a correlation with rheological parameters for different aging conditions (Ge et al. 2019; Jing et al. 2018; Qin et al. 2014). Hofko et al. conducted an extensive study on FT-IR to visualize its capability in different analysis methods for quantifying C=O and S=O aging indices (Hofko et al. 2018). Liang et al. showed that C=O absorbance area in FT-IR spectra had a linear relationship with aging duration at PAV condition (Liang et al. 2019). It is reported that, production rate of sulfoxide

in early stage of aging at higher temperature is faster but the rate slows down in later stages (Petersen and Glaser 2011). Although, both the carbonyl and sulfoxide increase with aging process, they do not follow the similar increasing trend. This study is focused on investigating the effect of aging on sulfoxide index. There is hardly any study available where the aging of mix is determined using sulfoxide index calculated from spectral analysis of FT-IR. Yut et al. determined the mix aging utilizing the carbonyl index (Yut et al. 2015). In this study applicability of sulfoxide index in mix aging is also investigated.

Laboratory binder and mix aging were performed in this study and sulfoxide indices were calculated for all the aged binders and mixes. This study will help to choose a convenient index for further analysis of FT-IR spectra to quantify aging.

1.1 *Objective*

Objective of this study is to investigate if the sulfoxide index is a suitable parameter to quantify the aging of the asphalt binder and mixture.

2 MATERIALS AND EXPERIMENTAL PLAN

2.1 *Materials*

In this study, eight unmodified/ neat binders of four different performance grades and from four different

Table 1. Laboratory aging plan for the study.

Materials	Aging Method	Duration (hours)	Binder Types
Binder	Unaged		Additional: PG58-28(LA), PG64-22(LA), PG67-22(MS)
	RTFO	1,2,4,8,6,24	PG52-34(LA), PG58-28 (TX),
	PAV	20,40,60,80	PG64-22(MS, NC), PG67-22(TX)
Mix	Unaged		PG52-34 (LA),
	STOA	2,4,12,24	PG58-28 (TX)
	LTOA	24, 72, 120	PG64-22 (NC)

sources were used to investigate if the aging indices are dependent of binder grades or sources. The binders are named by the grade followed by the location where it came from. The names of the binder are provided in the Table 1. Hot mix asphalt was produced in the lab using $1/2$ inch NMS aggregate following the SUPERPAVE aggregate gradation requirement. The design asphalt content was found to be 4.6%.

2.2 Experimental plan

ATR-FTIR spectra of binder and extracted mix were recorded in the laboratory. Different binders were subjected to standard and extended RTFO aging and PAV aging and afterward the absorbance spectra were collected for analysis. Similar spectra were obtained for laboratory aged mix also. Table 1 describes the different types of aging methods applied for binder and mix aging in this study.

3 EXPERIMENTAL METHODS

3.1 Laboratory aging of binder and mixture

Binders were aged using RTFO and PAV whereas, the mixture was aged in forced draft oven. RTFO aging was performed at 163^oC and PAV aging was performed at 100^oC under 2.1 MPa pressure. Duration for RTFO and PAV aging are provided in the Table 1.

Loose mixes were subjected to short-term oven aging (STOA) as well as long-term oven aging (LTOA). Temperature for STOA and LTOA were selected as 135^oC and 85^oC respectively and duration of mixture aging is given in Table 1.

Binder from mixture was extracted using dichloromethane (DCM). Detailed field extraction process is provided elsewhere (Arafat et al. 2020). For the extraction process 100 grams of loose mix and 100 milliliters of DCM was required for each mix sample. Whole extraction process can be performed in the lab or

field within 15 minutes by one person having minimal laboratory support.

3.2 Method of FT-IR data analysis

Agilent 4300 handheld FT-IR was used for the spectroscopy. A single reflection diamond ATR (Attenuated Total Reflectance) sample interface was used for data collection. Each spectrum was collected in the spectral region of 600 cm^{-1} to 4000 cm^{-1} at 4 cm^{-1} resolution by the default Microlab Mobile software equipped with the instrument and reported as an average of 24 spectra.

Quantitative analysis was performed by measuring the peak heights of C=O and S=O functional group at 1695 cm^{-1} and 1030 cm^{-1} respectively. The corresponding peak height was then divided by the height of asymmetric stretching vibration of CH_2 at 2920 cm^{-1} which was considered unsusceptible to aging (Yut et al. 2015). This quantification process for C=O and S=O functional groups were symbolized as carbonyl index (I_{CO}) and sulfoxide index (I_{SO}). For measuring peak height, baseline was drawn by connecting the valleys on both sides of the peak. It was observed that the location of the valleys remained unchanged irrespective of the binder type or intensity of aging. Baseline for C=O functional group was extended from the wavenumber 1684 to 1718 cm^{-1}. For S=O and asymmetric stretching vibration of CH_2, the limits of the base lines were 978 to 1080 cm^{-1} and 2753 to 2995 cm^{-1} respectively.

4 RESULTS AND DISCUSSION

4.1 Inconsistent increase in Sulfoxide Index with duration of aging

Concentration of both carbonyl and sulfoxide increased due to aging. Increase in carbonyl index for different hours of PAV aged binder are shown in the box plot (Fig. 1). Box plot corresponding to unaged and aged binder contain I_{CO} for eight and five different binders respectively. A gradual increase in I_{CO} for PAV aged binder is noticeable from the Figure 1a. Similar pattern is observed in RTFO aged binder too which is not shown here. In case of sulfoxide index, the increasing trend ceases after a certain level of aging and then starts decreasing as observed from Figure 1b. Extended duration of accelerated aging may cause some sulfoxide to decompose to sulfones and lower the I_{SO} (Huang and Grimes 2010). For this reason, a known value of I_{SO} cannot be directly correlated to the extent of aging of a binder. With the increase in aging duration, unlike carbonyl index, the variability of sulfoxide index value due to different binder grade increases. As, the increasing trend of sulfoxide index is not consistent and varies a lot because of the binder grade, this index might not be suitable for quantifying aging of unknown binder.

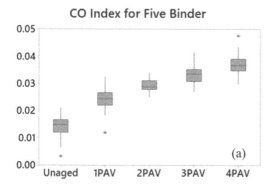

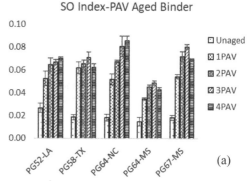

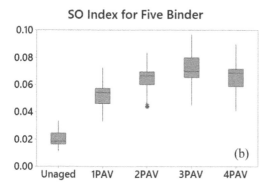

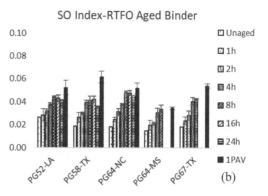

Figure 1. (a) Carbonyl (CO) and (b) Sulfoxide (SO) index for PAV aged binder.

Figure 3. Sulfoxide index for (a) RTFO aged binder and (b) Short-term oven aged mixture.

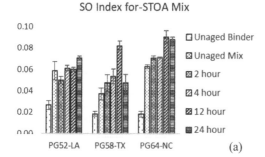

4.2 Unusually high Sulfoxide index of unaged mixture

In this study, I_{SO} of short-term and long-term aged mixture was determined using ATR-FTIR. It was observed that I_{SO} of unaged mix was much higher than the unaged binder (Fig. 2a) and this value varies considerably from mix to mix. Moreover, no noticeable trend is observed in STOA mixture. Similar observation was made in LTOA mixture too. It can be noted that small amount of binder was extracted in the field to determine the I_{SO} of the mix. There might be a chance that the I_{SO} is influenced by the mixing process.

Five different extractions were performed using the same binder and the I_{SO} was calculated. It is observed from the Figure 2b that the solvent does not have any effect on the I_{SO} as both the neat binder and binder residue from the solvent have the similar sulfoxide index. There exists a large increase in I_{SO} when binder is extracted from the mixture. Usually the dissolved binder from the mixture is filtered through an 80-micron nylon filter. If it was filtered using a 25-micron filter the I_{SO} decreases. Binder extracted from mix made without #200 and fine particles reduces the I_{SO} significantly. Still it was much higher than that of the unaged binder. Some fine particles present in the mix

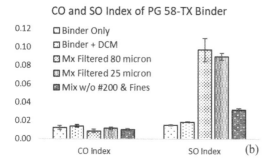

Figure 2. (a) Sulfoxide index of short-term oven aged mix, (b) Effect of fines on carbonyl and sulfoxide index.

can be present in the extracted binder which influences the I_{SO}. Interference of the molecular vibration of silicon oxide (at wavenumber 1000 cm^{-1}) from the fines and sulfoxide (at wavenumber 1030 cm^{-1}) creates higher peak value in the spectra and consequently results to higher sulfoxide index. Determination of carbonyl index is not influenced by the fines present in the mixture (Fig. 2b). That is why, sulfoxide index cannot be considered as a reliable metric to quantify the aging of the mixture.

4.3 Different behavior in RTFO and PAV aging

Sulfoxide index for the PAV and RTFO aged of five binders are provided in Figure 3a and 3b. It is observed that I_{SO} for PAV aged binders are much higher than the RTFO aged binder. Even sulfoxide index of 1PAV aged binder is much higher that the 24-hour of RTFO aged binder. Another study performed by the authors verified that, 8-hour of RTFO aging created equivalent effect to 1PAV aging in terms of carbonyl index. But similar types of relationship cannot be established for I_{SO} index. Moreover, I_{SO} depends on the aging method. Aging at higher pressure and longer duration (in PAV) creates more sulfoxide than the aging at higher temperature (in RTFO). Since the concentration of sulfoxide is dependent on the aging process, this index may be unreliable to quantify aging of a mixture in the field or aging of RAP.

5 CONCLUSION

Monitoring the concentration of sulfoxide with the extent of asphalt binder or mixture aging will provide inconsistent changes. For some binders, at some extent of aging the concentration of Sulfoxide stars to drop. The rate of change in Sulfoxide index largely depends on the method of aging. Even after 24-hours of RTFO aging the Sulfoxide index cannot reach the similar index value if it is aged in PAV. Moreover, in case of mixture aging, the index values are influenced by the fines present in the mixture. Therefore, Sulfoxide index should not be considered a reliable indicator to quantify the aging rate of an asphalt binder or mixture.

ACKNOWLEDGEMENT

This study was funded by the Louisiana Transportation Research Center (LTRC) under the project "Field Implementation of Handheld FTIR Spectrometer for Polymer Content Determination and for Quality Control of RAP Mixtures".

REFERENCES

Arafat, S., Noor, L., Wasiuddin, N., and Salomon. D., 2020. Implementation of Handheld FT-IR Spectrometer to Determine the RAP Content for Quality Control of Plant Mix. In 99th Annual Meeting of the Transp. Res. Board, Washington, DC.

Derrick, M. R., D. Stulik, and J. M. Landry. 2000. *Infrared spectroscopy in conservation science*. Getty Publications.

Ge, D., Chen, S., You, Z, Yang, X, Yao, H, Ye, M, and Yap, Y. K., 2019. Correlation of DSR Results and FTIR's Carbonyl and Sulfoxide Indexes: Effect of Aging Temperature on Asphalt Rheology. *Journal of Materials in Civil Engineering*, 31(7), p04019115.

Hofko, B., L. Porot, A. F. Cannone, L. Poulikakos, L. Huber, X. Lu, and H. Grothe. 2018. FTIR spectral analysis of bituminous binders: reproducibility and impact of ageing temperature. *Materials and Structures*, 51(2), p45.

Huang, S. C., and Grimes, W., 2010. Influence of aging temperature on rheological and chemical properties of asphalt binders. *Transportation Research Record*, 2179(1), p39–48.

Jing, R., A. Varveri, X. Liu, T. Scarpas, and S. Erkens. 2018. Chemo-mechanics of Ageing on Bituminous Materials. In 97th Annual Meeting of the Transp. Res. Board, Washington, DC.

Liang, Y., R. Wu, J. T. Harvey, D. Jones, and M. Z. Alavi. 2019. Investigation into the Oxidative Aging of Asphalt Binders. *Transportation Research Record*, 2673(6), p368–378.

Petersen, J. C. and R. Glaser. 2011. Asphalt oxidation mechanisms and the role of oxidation products on age hardening revisited. *Road Materials and Pavement Design*, 12(4), p795–819.

Qin, Q., J. F. Schabron, R. B. Boysen, and M. J. Farrar. 2014. Field aging effect on chemistry and rheology of asphalt binders and rheological predictions for field aging. *Fuel*, 121, p86–94.

Yut, I., Bernier, A., and Zofka, A., 2015. Field applications of portable infrared spectroscopy to asphalt products. *Introduction to Unmanned Aircraft Systems*, 127.

Advances in Materials and Pavement Performance Prediction II – Kumar et al. (eds)
© 2021 Taylor & Francis Group, London, ISBN 978-0-367-46169-0

Improved test method for low-temperature PG of binders

Haleh Azari & Alaeddin Mohseni
Pavement Systems LLC, Bethesda, MD, USA

ABSTRACT: A new low-temperature test on DSR, called iCCL, is introduced in this paper. It was shown that iCCL provides the same Low-Temperature Performance Grade (LTPG) as BBR for 583 binders of 20 different grades. The results of round robin testing in six different laboratories was used to determine precision estimates of iCCL, which showed that iCCL is at least two times more precise than BBR. iCCL was found to be significantly faster, safer, and more practical than BBR since it only takes 30 minutes to conduct the test and it does not need any chemicals, air pressure, or liquid bath. iCCL is also able to provide a good estimate of LTPG using original binders, which makes it useful for asphalt plants and terminals. This study shows that iCCL significantly improves the current asphalt binder testing by providing more precise, safer, more practical, and cost effective method.

1 INTRODUCTION

The low-temperature characterization of asphalt binder was developed during Superpave using Bending Bean Rheometer (BBR). The method is currently included in AASHTO T 313 and M 320 (1,2) and is widely used by laboratories today to verify the binder Performance Grade (PG). The continuous PG of asphalt binder is determined using AASHTO R 29 (3) which involves performing BBR tests at two different temperatures.

The BBR test requires extensive laboratory equipment and technician training and is rather labor intensive, time consuming, and thus costly to perform. For this reason, state agencies could only afford to perform a single test for verification of the binder grade but does not have enough resources to determine continuous PG. Due to many manual steps for the conduct of the test, the results of BBR test is quite variable. For this reason, results from different laboratories may not agree well. There are also safety concerns since hazardous chemicals and solvents used by BBR require special handling. For these reasons, asphalt industry has long been looking for an improved test for LTPG determination.

2 ICCL LOW-TEMPERATURE TEST

The low-temperature test, referred to as Incremental Creep at Low Temperature (iCCL), is performed on a DSR with 8-mm parallel plate geometry and 0.5-mm. A constant creep of 60 seconds is applied to the specimen similar to BBR. Test can be performed at several increments of multiple subzero temperatures or at multiple stresses in a fixed subzero temperature (e.g., –5°C). The duration of the test is about 30 mins. The

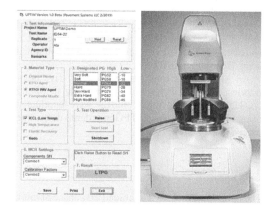

Figure 1. MCR 72 UPTiM and iCCL software.

continuous low-temperature PG similar to AASHTO R 29 is determined by comparing the DSR creep results to the BBR data. More details on the iCCL test can be found elsewhere (4).

Figure 1 shows the test setup (device and software). Sample preparations includes scooping binder from the cup at room temperature and weighting 32±1 mg. The sample is placed on the lower plate and test is started. Within half an hour, iCCL provides LTPG of binder equivalent to BBR. All aspects of the test including mounting, trimming, and post-processing are automated.

3 OBJECTIVES AND SCOPE

The main objective of this study is to investigate the applicability, reliability, and precision of iCCL for determining Low-Temperature Performance Grade (LTPG) of asphalt binders.

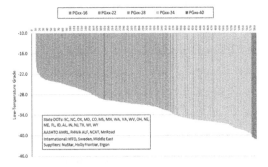

Figure 2. iCCL low-temperature PG determined for 583 binders of various grades.

Another objective is to perform round robin studies to determine precision estimates of iCCL.

This study also examines the possibility of using original and RTFO aged materials to approximate the LTPG equivalent to AASHTO R 29.

The scope of the study is to test 583 PG graded binders from 40 agencies using iCCL and compare the PG to the manufacturer specified PG.

Another scope is testing sixty RTFO/PAV aged asphalt binders from three state DOTs using iCCL and compare the results to the continuous PG obtained using BBR and AASHTO R 29.

An Alpha round robin study involving five laboratories and one binder and a beta study involving six laboratories and eight binders were conducted using binders with known LTPG in order to determine precision estimates of iCCL.

4 MATERIALS

The development of iCCL included a database of more than a thousand original, RTFO, and PAV aged binders

of various PGs collected from around 40 agencies including 23 state DOTs, five suppliers, and a few international agencies. Four sets of asphalt binders were used in this study. 1) Sixty PAV aged binders that were tested for continuous LTPG using BBR and were obtained from three State DOTs (40 from OHDOT, 10 from MnDOT and 10 from DelDOT). The PG for these binders were 52-34, 58-28, 58-34, 58H-34, 64-22, 64H-22, 64-28, 70-22, 76-22, and 64E-22. These binders were tested using iCCL and results were compared with BBR. 2) Eight binders from the first pool were selected for determining precision estimates. 3) A pool of 523 PAV aged binders with designated HTPG ranging from 46 to 88 and LTPG from −40 to −10 were used to verify that iCCL provides the same LTPG as BBR. 4) A pool of 248 binders from the third pool for determining LTPG from of original and RTFO aged binders.

5 RESULTS OF ICCL TESTING

Figure 2 shows iCCL LTPG for 583 binders with four nominal LTPGs (−22 to −40), shown with different shades. Figure 2 shows that some binders may be at higher grade than designated since State DOTs normally conduct one BBR testing to verify the grade but do not determine the actual grade.

As data in Figure 2 shows, iCCL can clearly distinguish different LTPG and provides the same grade as provided by BBR.

6 ICCL ALPHA AND BETA TESTING

The alpha testing of iCCL was conducted in 2016 between five operators using a single binder as shown in Table 1. The binder was a PG64-22 from South Carolina and the round robin testing resulted in an average C.V. of 1.5% for LTPG. The beta testing of iCCL test method started in February 2019 by supplying iCCL

Table 1. Coefficient of variation (C.V.) for the round Robin testing.

Round Robin	Company/Agency	No. of Tests	No. of binders	C.V., %
First Study Alpha	Anton Paar, Stuttgart	30	1	1.1
	Testing (2016) Anton Paar, Graz	15	1	1.97
	Anton Paar, USA	9	1	0.8
	Pavement Systems DSR 1	9	1	0.54
	Pavement Systems DSR 2	9	1	0.62
Second Study Beta	Delaware DOT, Dover	126	45	0.82
Testing (2019)	Minnesota DOT	99	33	1.21
	Ingevity, Tulsa, OK	57	23	1.15
	Rutgers University	16	8	0.69
	Indiana DOT	10	2	2.52
	Mississippi DOT	42	16	1.20
	Pavement Systems DSR 1	50	12	1.03
	Pavement Systems DSR 2	112	20	1.41

testing device to seven laboratories participated in the study. A total of 60 binders from Delaware, Ohio, and Minnesota were targeted for the round robin testing. Table 1 includes statistics such as number of tests, number of binders and average C.V. for each laboratory. The average C.V. between all materials tested at seven laboratories was 1.59% which was a little more than the result of alpha testing.

7 COMPARING ICCL PG WITH BBR PG

The low-temperature PG of iCCL was compared with the BBR continuous LT PG in AASHTO R 29 in order to evaluate the equivalency of the results. Figure 3 shows LTPG for BBR versus iCCL for the 60 binders. The BBR tests were performed at the state DOT laboratory and the iCCL values are an average value for all tests conducted in all seven laboratories according to Table 1 (beta testing). Figure 3 shows that the average iCCL PG was the same as BBR and that there is not a significance bias in the data. The average difference between BBR PG and iCCL PG for all tests was 2.6%. An internal study in one of the laboratories showed that the percent difference between the central lab BBR results and the contractors' BBR results was 6.6%. In this regards, the percent difference of iCCL between labs is much smaller than those for BBR and iCCL can determine the binder quality with higher precision.

Most of the variability shown in Figure 3 is due to the BBR variability. AASHTO R 29 does not provide precision of continuous PG and the d2s reported for BBR m-value in AASHTO T 313 is 6.8%, therefore, the d2s of continuous PG, which is calculated based on ASTM D4460, would be 9.6%.

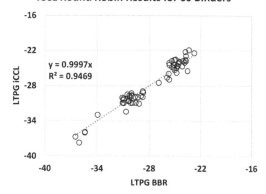

Figure 3. Continuous LTPG using BBR versus iCCL for 60 binders from three state DOTs.

Table 2 shows the average of two replicates of iCCL PG for each of the binders tested in six participant laboratories. The beta testing showed C.V. of 1.6%, which was consistent with the alpha testing (1.5%).

ASTM C670 procedure (5) for bias was conducted on the data set in Table 2 and the bias of the test was found to lie within 95% confidence limits.

Subsequently, the precision statement of the iCCL test was developed following ASTM C670. Table 3 provides the single-operator and multilaboratory standard deviations of continuous Low-temperature PG for iCCL. As shown in the table, both within and between-laboratory variability of the iCCL test is very small. The maximum difference between two replicate continuous grade values is 0.65°C and the maximum difference between two laboratory continuous grade values is 1.33°C.

8 ICCL PRECISION ESTIMATES

One of the objectives of the beta testing was to conduct an interlaboratory study to determine precision estimates of the test. Eight binders of various widely used grades (PG58-28, PG64-22, PG64-28, PG76-22, and PG46E-22) from two state DOTs (DelDOT and OHDOT) were used for the study.

9 DETERMINING LOW-TEMPERATURE PG USING STRAIGHT RUN BINDERS

The iCCL method of computing PG using a database of BBR results enables estimation of PG using original and RTFO aged binders. Asphalt aging has significant effect on the low-temperature properties of binders; however, these effects can be modeled to some extent.

Table 2. Results of iCCL beta testing for eight binders tested in six different laboratories.

State	PG	Del DOT	PS 1	PS 2	Ingevity	MnDOT	Rutgers
OHDOT	58-28	−29.3	−30.3	−30	−29.6	−29.7	−28.9
OHDOT	64-22	−23.4	−24.1	−24	−23.2	−23.9	−23.1
OHDOT	64-28	−30.5	−29.9	−30	−29.9	−30.6	−29.9
OHDOT	76-22	−22.5	−23	−23	−22.8	−23.4	−22.5
DelDOT	58-28	−30.6	−30.5	−30.7	−31.2	−31.0	−31.7
DelDOT	64-28	−29.6	−29	−28.6	−29	−29.8	−29.8
DelDOT	64-22	−23.7	−23.8	−24.2	−23.8	−24.1	−24.6
DelDOT	64E-22	−24.4	−25	−24.6	−24.1	−	−24.9

Table 3. Preliminary precision estimate for iCCL using PAV material.

Preliminary Precision Statement for Binder Continuous Low-Temperature Performance Grade (PG)		
Precision	Standard Deviation, °C (1s limit)A	Acceptable Range of Two Test Results, °C (d2s limit)A
Single-Operator Precision:		
Continuous Low-Temperature Grade (PG)	0.23	0.65
Multi-Laboratory Precision:		
Continuous Low-Temperature Grade (PG)	0.47	1.33

The precision estimates above are based on analysis of paired test results collected from six laboratories testing eight PAV aged binders: two sets of PG 58-28, two sets of PG 64-22, two sets of PG 64-28, one set of PG 76-22, and one set of PG 64E-22.

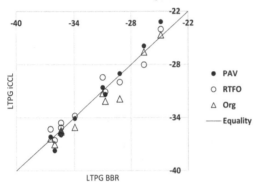

Figure 4. BBR LTPG versus iCCL for original, RTFO and RTFO/PAV using MnDOT data.

Figure 4 shows BBR LTPG versus iCCL PG using original, RTFO and RTFO/PAV aged materials from 10 Minnesota DOT (MnDOT) binders ranging from -40 to -22. The BBR and iCCL tests were conducted at MnDOT. Figure 4 shows there is a good correlation between LTPG using BBR with iCCL when using PAV, RTFO, or original binders. The average percent difference between BBR and iCCL LTPG was 2.9% for PAV, 2.6% for RTFO, and 3.3% for original binders.

Figure 5 shows the iCCL LTPG using PAV versus using original and RTFO binder for 248 binders. The dashed line is one way 98% confidence level. As can be seen from this figure, iCCL can estimate continuous LTPG of BBR from testing original binder within 2°C. The figure also shows similar trend for using RTFO aged binders.

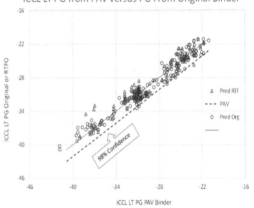

Figure 5. LTPG determined using PAV binder versus LTPG using original or RTFO binder.

using a portable device that does not require laboratory facility such as compressed air or any chemicals.

There is a significant improvement in testing time since as many as 12 tests may be conducted in one day. Each test takes about 5 minutes of technician time or one hour per day, which can result in significant reduction in the cost of testing.

The results of this study show that iCCL is significantly more precise than BBR. The main reason for improved precision of iCCL is that DSR has significantly better temperature and load control.

Unlike BBR, iCCL is able to provide an estimate of LTPG using original (unaged) and RTFO materials. This makes the test useful for quality control at asphalt plants. Table 4 lists advantages of iCCL over BBR.

10 ADVANTAGES OF ICCL OVER BBR

There are several advantages of performing iCCL test. The iCCL test is significantly faster, safer, and more practical than BBR since it is conducted in half an hour

11 SUMMARY AND CONCLUSIONS

The main objective of this study was to investigate the applicability of iCCL for quick determination of

Table 4. Comparing test features of BBR and iCCL.

Test Features	BBR	iCCL
Sample Prep. and Testing/ Sample	1.5 hrs.	40 min.
Continuous PG	Two Tests	One Test
Technician Time/ Sample	1 hr.	5 min.
Test Variability (D2S)	7%	2.1%
Calibration Check	Every Day	3 Months
Use of Hazardous Liquids	Coolant	None
Testing Original Binder	No	Yes
Air Pressure	Yes	None
Sample Storage Limit	Two hrs.	None
Molding/Demolding	Required	None
Sample Size	> 10 grams	30 mg
Small Extracted Material	No	Yes

low-temperature grade of binders. Another objective was to determine the precision of iCCL. Following is a summary of the findings:

• The iCCL test determined the same low-temperature PG as the BBR test for 583 binders of 20 different grades
• Results of round robin testing involving 60 binders and seven laboratories were presented
• Precision estimates determined for iCCL using eight binders tested at six laboratories showed that iCCL is at least two times more precise than BBR

• The results of this study shows that iCCL significantly improves the current asphalt binder testing technology by providing a safer, more precise, more practical, less time consuming, and cost effective method to determine the currently established Superpave binder low-temperature PG.

REFERENCES

AASHTO M 320: Standard Specification for Performance-Graded Asphalt Binder. American Association of State Highway and Transportation Officials (AASHTO), Washington D.C., 2017.
AASHTO T 313: Standard Method of Test for Determining Flexural Creep Stiffness of Asphalt Binder Using the Bending Beam Rheometer (BBR). AASHTO, Washington, D.C., 2012.
AASHTO R 29: Standard Practice for Grading or Verifying the Performance Grade (PG) of an Asphalt Binder. AASHTO, Washington D.C., 2015.
ASTM C670: Standard Practice for Preparing Precision and Bias Statements for Test Methods for Construction Materials. ASTM, West Conshohocken, PA, 2015
Azari, H. and Mohseni, A., "A Quick Asphalt Binder Low-Temperature Grade Determined by iCCL Test Using DSR," ASCE International Airfield and Highway Pavement Preceding, In Press 2020.

Advances in Materials and Pavement Performance Prediction II – Kumar et al. (eds)
© 2021 Taylor & Francis Group, London, ISBN 978-0-367-46169-0

Discrete fracture modelling of rubber-modified binder

Xunhao Ding
Southeast University, Nanjing, Jiangsu, China

Punyaslok Rath & William G. Buttlar
University of Missouri, Columbia, Missouri, USA

ABSTRACT: A strength-softening model combined with Burger's model was applied to investigate the fracture behavior of rubber-modified binder in this study. Based on the commercial software Particle Flow Code in 3-Dimensions (PFC3D), the interaction between rubber and binder was visually illustrated. Results from a virtual bending beam test and a newly-designed binder cracking test were integrated to arrive at the required parameters for viscoelastic and fracture constitutive models in the DEM simulation. Simulation results revealed a close agreement with experimental data, which verify the modelling methods used herein. It was found that the presence of rubber particles in asphalt leads to a crack-pinning toughening behavior that enhances the ability of the asphalt specimen to inhibit cracking at low temperatures.

1 INTRODUCTION

Ground Tire Rubber (GTR) has been widely used to modify asphalt mixtures for decades due to its environmental sustainability and performance benefits (Rath et al. 2019). The modified mixtures can be manufactured using either a wet or dry process (Sienkiewicz et al 2017). In either case, the physical reaction between asphalt binder and rubber leads to the swelling of rubber particles, which in turn results in a significant change of physical properties in asphalt mastic and mixture systems. Additionally, it is known that rubber particles do not completely dissolve in the asphalt binder and retain some granular, particulate structure in the mixture, especially in the dry process. Due to the complex interaction between rubber and binder, very limited studies have explicitly studied the physical cracking mechanism present in rubberized asphalt mastic and mixture systems (Lee et al. 2008).

Using existing experimental methods, only macro properties of GTR-modified binders have been evaluated in prior research. As compared to experimental campaigns, numerical methods allow the investigation and estimation of intrinsic material properties, mode I fracture constitutive models at low temperatures. A powerful numerical modeling scheme, the discrete-element method (DEM) developed by Cundall (Cundall & Strack 1980) has been utilized by various researchers to characterize micro-scale fracture behavior in asphalt mixtures (Kim et al. 2008; Buttlar & You 2001;.

The objective of this study is to investigate the fracture mechanism of GTR-modified binder at low temperature based on DEM and a new asphalt mastic fracture test.

Table 1. Selected properties of Illinois PG64-22 binder.

Rotational Viscometer	Viscosity (Pa.s) 135°C	Viscosity (Pa.s) 165°C
	0.084	0.023
BBR at −12°C	Stiffness (MPa) 162	m-value 0.365

2 METHODOLOGY AND EXPERIMENTAL

2.1 Materials

A PG64-22 binder, obtained from Illinois, was selected as a base binder and its properties were tested according to ASTM D6648 (2016) and ASTM D4402 (2000), as shown in Table 1. Dry ground rubber tire particles, 500-600 in diameter, or finer (#30 mesh) were obtained, and modified binder was made by mixing the rubber particles with base asphalt under high shear for 30 minutes at 176°C. Three GTR-modified binders with rubber content of 8%, 10% and 12% by weight of binder were prepared in the laboratory.

2.2 Experimental tests

For this study, two experimental tests were conducted-the standard BBR test at −12°C and a new binder fracture test, developed based on fracture mechanics principles, closely following the DC(T) test. The DC(T) mixture test was designed following the ASTM E399 (2013), often used for fracture characterization of metals. A compact tension (CT) geometry was adopted for mastic testing, based on previous literature (Hakimzadeh et al. 2017). As shown in Figure 1(a), an

Figure 1. Binder CT test design: (a) test mold; (b) binder specimens; (c) loading construction.

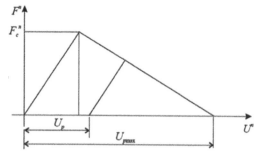

Figure 2. Strength-softening model in DEM.

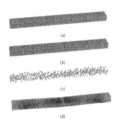

Figure 3. BBR modelling: (a) unmodified binder; (b) rubber-modified binder; (c) random generation of rubber particles (10% content); (d) contact force distribution in BBR simulations (red line is tension force while black line is compression force).

aluminum mold was used to cast asphalt binder and mastic specimens of dimensions 145*139.2*40mm. Two loading holes and a notch creating insert were placed in the mold before pouring the binder. After the GTR-modified binders were poured into the mold, the specimens were kept at room temperature for 90 minutes, then placed in a cooling chamber at −15°C for 30 minutes before demolding, as shown in Figure 1(b). The binder CT specimen was fractured at −22°C at a loading rate of 0.2 mm/min. The load-CMOD curves were recorded to evaluate the low-temperature cracking resistance of GTR-modified binders. More details on the test were recently reported (Rath et al. 2020).

2.3 Constitutive laws in DEM

The prediction accuracy in DEM is associated with the sphere packing arrangements, sphere size, constitutive laws etc. (Ding et al 2020). In this study, face-centered packing arrangements were used. To maintain consistency in BBR and binder CT modelling, all the spheres were set with a radius of 0.0005m. Burger's model and strength-softening model were applied to characterize the viscoelastic and fracture behavior of GTR-modified binder respectively. The Burger's model was assigned to all spheres in BBR test and majority (except for ligaments) in binder CT test. The strength-softening model was used for modeling interaction within the ligament area (post-peak curve in binder CT test) instead of the conventional bonding model. Nine parameters were needed in Burger's model including friction coefficient, stiffness, and viscosity in normal/shear directions as shown in Table 2.

The basic form of strength-softening model in DEM is illustrated in Figure 2. As shown, U_p represents the currently accumulated plastic displacement. When the contact force exceeds the strength, the material strength is linearly degraded, as shown in Eq. 1.

$$F_{max} = \left(1 - \frac{2a}{\pi}\right) \cdot F_c^n + \frac{2a}{\pi} \cdot F_c^s \qquad (1)$$

where, F_{max} is the contact strength; a is the angle between the directions of the contact force and the line segment connecting the centers of the spheres; F_c^n and F_c^s are two strength parameters as a function of the current orientation of the contact force.

3 RESULTS AND DISCUSSION

3.1 Viscoelastic predictions

Four GTR-modified BBR specimens were simulated in PFC-3D, including 0, 8, 10, 12% of GTR modification (by wt. of binder). The Burger model parameters were calibrated as follows: the experimental and simulation test of unmodified binder were conducted to obtain the viscoelastic parameters for the base asphalt binder, and based on the results, randomly-placed rubber particles were generated as shown in Figure 3(b) & (c). Using a simple trial-and-error approach, final model input parameters were inversely determined as shown in Table 2.

Figure 4 presents verification of the Burger's model used in the BBR tests. As shown in Figure 4, based on the experimental data, the rubber plays a positive role in softening the stiff binder at low temperature. With more rubber added, the final deflection increases. The inversely-obtained bulk viscoelastic parameters were then applied to the binder CT test simulation.

3.2 Fracture modelling verification

Binder CT simulations were developed in PFC3D, as shown in Figure 5. As running 3D model simulations is time-consuming, a relatively thin slice (1 mm) was used in preliminary 3D models to accelerate the fracture simulations (Figure 5). The strength-softening model was assigned in the ligament area and the contacts between rubber and binder were differentiated. The calibrated parameters of strength-softening model were summarized in Table 3.

Table 2. Parameters of Burger's model in DEM.

Type	Asphalt + asphalt	Asphalt + rubber
Kkn	8e4	5e4
Ckn	3.7e5	3e5
Kmn	4e6	9e5
Cmn	1.5e7	8e6
Kks	8e4	5e4
Cks	3.7e5	3e5
Kms	4e6	9e5
Cms	1.5e7	8e6
Fric	0.5	0.5

Kkn- normal stiffness for Kelvin section (Ks); *Ckn*- normal viscosity for Ks; *Kmn*- normal stiffness for Maxwell section (Ms); *Cmn*- normal viscosity for Ms; *Kks*- shear stiffness for Ks; *Cks*- shear viscosity for Ks; *Kms*- shear stiffness for Ms; *Cms*- shear viscosity for Ms; *Fric*- friction coefficient.

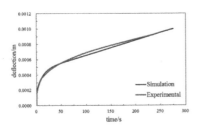

(a)

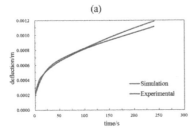

(b)

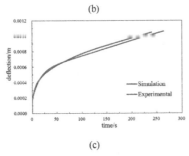

(c)

Figure 4. Verification of burger's model: (a) 64-22+8%GTR (by wt. of binder); (b) 64-22+10%GTR; (c) 64-22+12%GTR.

Figure 6 shows the fracture process zone in the binder CT specimen. The red, black, and yellow lines represent the tension force, compression force and micro-cracks, respectively. The experimental and simulation results are summarized in Figure 7. As shown, with the combined use of Burger's model and strength-softening model, the global response of the binder CT

Figure 5. Binder CT fracture test development.

Table 3. Parameters of Strength-softening model in DEM.

Type	Asphalt + asphalt	Asphalt + rubber
Sof_knc	1e5	3e5
Sof_knt	8e4	1e5
Sof_kns	1e5	2e5
Sof_ftmax	1.5	3
Sof_fsmax	1	1.5
Sof_fric	0.5	0.5
Sof_uplim	8e-4	9e-4

Sof_knc-normal stiffness in compression; *Sof_knt*-normal stiffness in tension; *Sof_kns*-shear stiffness; *Sof_ftmax*-tensile strength; *Sof_fsmax*-shear strength; *Sof_fric*-friction coefficient; *Sof_uplim*-accumulated plastic displacement for which the bond strength softens to zero.

Figure 6. Fracture process zone modelling.

test could be reasonably simulated, with a slight over-prediction of the post-peak softening response for this GTR-modified mastic with a ductile-type failure.

3.3 Interactions between rubber and binder

Images of crack propagation captured using a digital camera alongside DEM simulations results are illustrated in Figures 8 (a) & (b), respectively. The results revealed that the rubber particles played a positive role in improving the crack resistance of the asphalt binder. As shown in Figure 8b, the propagating crack encounters an embedded rubber particle (orange particle) of a higher stiffness and fracture energy than the base binder matrix (blue particles) and require extra force and energy for local separation to occur. The crack pinning mechanism shown herein via experiments and 3D DEM modeling provides additional insight into

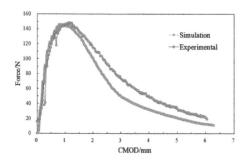

Figure 7. Verification of the combined strength-softening model and Burger's model in fracture modelling (10% rubber by wt. of binder).

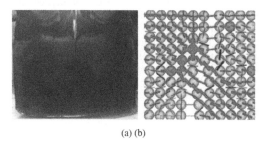

(a) (b)

Figure 8. Experimental and modelling of crack propagation: (a) Images captured by camera; (b) crack pinning toughening mechanism revealed in DEM.

this previously reported phenomenon (Smith & Hesp 2000).

4 CONCLUSIONS

This study provides additional insight towards the crack pinning toughening mechanism that appears to exist in GTR-modified binder systems, based on new experimental techniques presented alongside 3D DEM simulations. The main conclusions are summarized as follows.

(a) The combined use of Burger's model and a strength-softening model in PFC-3D was successfully calibrated to novel GTR-binder CT fracture experiments, after first calibrating GTR-binder stiffness parameters from BBR tests;

(b) Rubber-modification of asphalt can increase the energy needed for crack propagation at low temperatures, apparently due to crack pinning and increased fracture surface tortuosity. The rubber-modified binder shows a ductile-type failure in a fracture test in comparison to the brittle-type failure observed in the unmodified PG64-22 binder.

It is hoped that the results presented herein will provide motivation for asphalt mixture designers to experiment with durable, sustainable, and economical designs of modern paving mixtures with recycled ground tire rubber.

ACKNOWLEDGMENT

The study is financially supported by the Graduate Innovation Project of Jiangsu Province (KYCX18_0147) and the Chinese Scholarship Council (No.20190 6090216).

REFERENCES

American Society for Testing and Materials International, 2000, Standard Test Method for Viscosity Determination of Asphalt at Elevated Temperatures Using a Rotational Viscometer, West Conshohocken, PA, ASTM D4402.

American Society for Testing and Materials International, 2013. Standard Test Method for Linear-Elastic Plane-Strain Fracture Toughness K_{Ic} of Metallic Materials, West Conshohocken, PA, ASTM E399.

American Society for Testing and Materials International, 2016. Standard Test Method for Determining the Flexural Creep Stiffness of Asphalt Binder Using the Bending Beam Rheometer (BBR), West Conshohocken, PA, ASTM D6648.

Buttlar, W.G., and Z. You, "Discrete Element Modeling of Asphalt Concrete: A Mirco-Fabric Approach," Journal of the TRB, No. 1757, National Research Council, National Academy Press, Washington, D.C., pp. 111–118, 2001.

Cundall, P. A., & Strack. O. D. L. 1980. Discussion: a discrete numerical model for granular assemblies. Geotechnique, 30(3), 331–336.

Ding, X., Ma, T., Gu, L., Zhang, D., & Huang, X. 2020. Discrete Element Methods for Characterizing the Elastic Behavior of the Granular Particles. Journal of Testing and Evaluation, 48(3).

Hakimzadeh, S., Behnia, B., Buttlar, W. G., & Reis, H. 2017. Implementation of nondestructive testing and mechanical performance approaches to assess low temperature fracture properties of asphalt binders. International Journal of Pavement Research and Technology, 10(3), 219–227.

Kim H., Wagoner M.P., and Buttlar, W. G. 2008. Simulation of Fracture Behavior in Asphalt Concrete Using a Heterogeneous Cohesive Zone Discrete Element Model. Journal of Materials in Civil Engineering, v 20, n 8, p 552–563.

Lee, S. J., Kim, H., Akisetty, C. K., & Amirkhanian, S. N. 2008. Laboratory characterization of recycled crumb-rubber-modified asphalt mixture after extended aging. Canadian Journal of Civil Engineering, 35(11), 1308–1317.

Rath, P., Love, J.E., Buttlar, W.G. and Reis, H., 2019. Performance analysis of asphalt mixtures modified with ground tire rubber modifiers and recycled materials. Sustainability, 11(6), p.1792.

Rath, P., Chen, S., Buttlar, W., 2020. Investigation of Cracking Mechanisms in Rubber-Modified Asphalt through Fracture Testing of Mastic Specimens. Road Materials and Pavement Design (Submitted).

Sienkiewicz, M., Borzêdowska-Labuda, K., Wojtkiewicz, A., & Janik, H. 2017. Development of methods improving storage stability of bitumen modified with ground tire rubber: A review. Fuel Processing Technology, 159, 272–279.

Smith, B., and Hesp, S., 2000. Crack Pinning in Asphalt Mastic and Concrete: Regular Fatigue Studies. Transportation Research Record 1728, p 75–81.

Advances in Materials and Pavement Performance Prediction II – Kumar et al. (eds)
© 2021 Taylor & Francis Group, London, ISBN 978-0-367-46169-0

Investigating the factors which influence the asphalt-aggregate bond

M.N. Fakhreddine & G. Chehab
American University of Beirut, Beirut, Lebanon

ABSTRACT: Moisture damage is a problem that has plagued asphalt concrete pavements ever since their introduction in the late nineteenth century. Water infiltrates into the pavement structure and weakens its structural integrity by degrading the asphalt-aggregate adhesive bond, or by weakening the cohesive bond between the molecules of the asphalt binder itself. Despite being a topic of research since the 1930s, and with multiple methods currently available to test the moisture susceptibility of asphalt mixtures, it remains a critical issue till this day as results from the existing methods do not correlate properly with the field performance. The objective of this research is to introduce a new method for measuring the moisture susceptibility of asphalt mixtures. The tests presented within this paper will focus on studying the effect of the strain rate and asphalt film thickness, which are usually taken as constants for any mix design when conducting moisture susceptibility testing, on the strength and mode of failure of the samples. Testing was done on dry samples and this study will be followed up by another which includes moisture conditioning.

1   INTRODUCTION

Moisture damage is one of the main problems that threaten to deteriorate an asphalt concrete pavement long before its anticipated maintenance and reconstruction dates. Water infiltrates into the structure through the surface, capillary rise of subsurface water, or diffusion of water vapor, and leads to stripping by weakening the adhesive and cohesive bonds of the materials. Adhesive failure occurs at the level of the asphalt-aggregate interface and leads to their separation from one another; Cohesive failure occurs within the asphalt binder due to the loss of stiffness (Diab et al. 2014). Moisture damage leads to the dislodging of aggregates from the pavement and accelerates the formation of other distresses such as rutting, cracking, and raveling (Kakar et al. 2015). The issue of moisture damage persists although it has been researched for decades, and several methods are currently being practiced to determine the moisture susceptibility of asphalt mixtures. Therefore, there is a need to develop a new method which is simple and robust.

2   LITERATURE REVIEW

One of the most common tests to evaluate the moisture susceptibility of asphalt mixtures is the AASHTO T283. Three samples are soaked in a water bath for 24 hours at 60 degrees Celsius before conditioning at 25 degrees Celsius for 2 hours and then testing in indirect tension at a rate of 2 inches per minute. The results are compared with those of 3 other samples

which have not been subject to moisture conditioning. This test, however, does not isolate the stripping phenomenon at the level of the asphalt-aggregate interface since it uses compacted asphalt samples (Al Basiouni Al Masri et al. 2019). It also a fixed conditioning time and load rate which does not accurately replicate field conditions. Kringos et al. (2009) conducted finite element simulations of a dry indirect tension test and concluded that the specimen experiences at all times continuously changing strain rates. Other tests such as the Hamburg Wheel Tracking Device (AASHTO T324), Environmental Conditioning System (ECS) (AASHTO TP34), Asphalt Pavement Analyzer (APA), and Moisture Induced Sensitivity Test (MIST) also exist but are rarely used due to the lack of standardization and complexity of the procedures and sample preparation (Taib et al. 2019).

Researchers have proposed other methods to evaluate moisture sensitivity by using the Dynamic Shear Rheometer (DSR) and different variations of pull-off tests. Cho & Bahia (2010) proposed the wet to dry yield shear stress (W/D YSS) ratio as a parameter to quantify moisture damage by using the Dynamic Shear Rheometer (DSR). They concluded that while the W/D YSS ratio is sensitive enough to measure moisture effects, other factors limit the use of the DSR. Zhang et al. (2016) also used the DSR to control the film thickness of the samples by adding specially manufactured parts for it. The samples were then tested using a Universal Testing Machine (UTM) by also using custom fixtures to hold the samples, which makes the testing setup both very expensive and complicated to use. Rahim et al. (2019) also modified a 10 KN Universal

Testing Machine (UTM) and the authors reported that further improvement for the gap assembly is needed for repeatability and practicability reasons. One of the most common tests to evaluate moisture damage is the Bitumen Bond Strength (BBS) test which uses the Pneumatic Adhesion Tensile Testing Instrument (PATTI). The BBS test consists of pulling a metal stub attached to an aggregate substrate by a film of binder and recording the maximum pulling force. The test was reported by Moraes et al. (2011) to be successful in measuring the effects of moisture conditioning timing and asphalt modification on the bond strength of asphalt-aggregate systems. The BBS test was used to investigate several issues such as the effect of the bitumen stiffness on the adhesive strength, (Moraes et al. 2012) and the adhesive and cohesive properties of asphalt-aggregate systems using different combinations of aggregates, binders, and mineral fillers (Canestrari et al. 2010; Chaturabong et al. (2018). Despite the popularity of the BBS test, it does have some drawbacks such as using a high film thickness of 0.8 mm, not all elements of the system being at the same temperature at the time of the sample preparation, and the limitation of the test in terms of output where only the strength can be obtained while other data such as strain cannot (Cala et al. 2019).

Despite the significant research to assess the moisture susceptibility of asphalt mixes, little has been done to investigate the sensitivity of test-related factors on the result of the test.

3 OBJECTIVES

The objective of this study is to investigate the factors that affect the asphalt aggregate bond such as the asphalt film thickness and loading rate by running sensitivity analyses. The factors will be tested based on the direct tension test using a simple testing and sample preparation procedure. The results of this sensitivity study on the key affecting factors are used to come up with optimal testing and conditioning procedures that promote adhesive failure.

4 MATERIALS

For the purpose of the sensitivity study, two types of asphalt binder are used:

1) an unmodified PG64-16 binder
2) a polymer modified PG76-28 binder (PMB)

Limestone is used as an aggregate. Furthermore, limestone filler (less than 75 μm particle size) is mixed with both binders to produce asphalt mastics.

5 SAMPLE PREPARATION AND TEST PROCEDURE

To produce the asphalt-aggregate samples used in the tensile test, limestone cylinders are first extracted by coring large limestone blocks obtained from a known quarry. The cylinders are then sliced into 5mm-thick aggregate discs, each of which is epoxied onto an aluminum endplate. The exposed aggregate surface is cleaned using a clean cloth dipped in a small amount of acetone to remove any residues or oils that might affect the adhesion with the asphalt binder. Meanwhile, asphalt binder is heated until it liquefies, poured in a round silicon mold, and left to cool down at room temperature. In order to control the asphalt film thickness, the zero-gap is defined by fixing two aggregate discs into the tensile machine and lowering the actuator until a 1 N contact load is recorded.

The aggregates are then removed and placed on top of a hot plate to reach the binder application temperature which should be similar to the mixing temperature for an asphalt mix with the same binder. After the temperature of the aggregates stabilizes, they are quickly fixed in the tensile machine and the binder disc is directly attached to the hot aggregate surface. The actuator is then lowered until the specified film thickness is reached. Finally, the excess binder is trimmed, and the sample is left to cool down to room temperature before testing or removing from the machine. The temperature of the sample is monitored using a thermocouple inserted at the middle of a dummy specimen; it was found that 30 minutes are enough to bring the entire sample back to room temperature. It should be noted that the time between heating the aggregates and lowering the actuator should not be more than 1 minute to avoid excessive loss of temperature of the aggregate surface which can cause inadequate adhesion with the binder.

5.1 *Problems faced with sample preparation and testing*

- For samples that were tested after moisture conditioning, it was observed that some gained strength. The issue was traced back to the resting time after the samples were fabricated. The asphalt-aggregate bond continued to gain strength even after cooling to room temperature and seemed to stabilize after 24 hours. As such, it is recommended that all samples

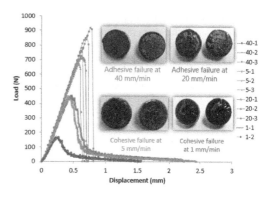

Figure 1. Load vs. displacement for different strain rates.

419

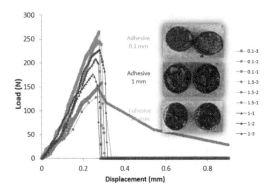

Figure 2. Load vs. Displacement for Unmodified Binder.

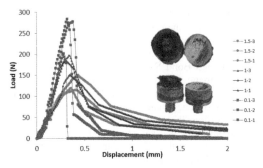

Figure 3. Load vs. displacement for PMB.

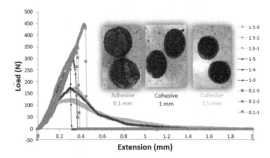

Figure 4. Load vs. displacement for mastic.

be left to rest for at least 24 hours before dry testing or moisture conditioning.

- Due to the viscoelastic nature of asphalt, the binder film decreased in thickness under the weight of the top plate and aggregate over the resting period. The decreased film thickness affected the strength of the bond and thus all samples were modified after fabrication by adding fixtures to their sides so that no load is applied on the asphalt film
- Some samples exhibited low strength after testing due to the low temperature at which the binder was attached to the aggregate. It is necessary to ensure that the aggregates are sufficiently hot to allow for proper adhesion of the binder to the surface of the aggregates.
- Upon observation of the aggregate surface after testing, it was noticed that some samples were failing in adhesion particularly from one edge of the disk. This was because the surfaces of the 2 aggregates were not completely parallel and thus the sample had a tapered film thickness. The surfaces of the aggregates must be parallel to ensure a uniform film thickness.

6 RESULTS

6.1 Effect of rate of loading

23The behavior of asphalt, being a viscoelastic material, depends on the strain rate being applied to it. It is known that as the strain rate increases, the stiffness of the binder also increases, and vice versa; and given that a pavement in the field experiences different strain rates depending on the speed of passing trucks and heavy vehicles, then it makes sense to test for different strain rates when checking for the moisture susceptibility of an asphalt mix. To verify that the strain rate influences the strength of the asphalt-aggregate bond and its mode of failure, 3 replicates were tested for each of the chosen strain rates and the results are as shown in Figure 1.

As the strain rate increases, the strength of the asphalt-aggregate bond also increases. The mode of failure, which was determined by visual observation

of the sample, also tends to be adhesive for higher strain rates and cohesive for low strain rates.

6.2 Effect of asphalt film thickness

To check the sensitivity of the test for the asphalt film thickness, thicknesses of 0.1 mm, 1 mm, and 1.5 mm were tested for samples using unmodified binder , polymer modified binder, and mastic using the unmodified binder and limestone filler for a ratio of 1 to 1 by mass and at 20 mm/min.

The results show that as the asphalt film thickness decreases, the strength of the asphalt aggregate bond increases and the mode of failure tends to be adhesive, and vice versa for higher film thicknesses. In the case of the polymer modified binder, the same mode of failure which was completely adhesive was observed for all film thicknesses, the authors suspect that the wetting temperature is the main reason behind this observation.

7 CONCLUSION

This study confirmed that the asphalt film thickness and loading rate are important factors that must be included as parameters when measuring the moisture susceptibility of asphalt mixes. An increase in the loading rate increases strength and promotes adhesive failure, and an increase in asphalt film thickness decreases strength and promotes cohesive failure. A similar conclusion was made by Huang & Lv (2016) by who studied the sensitivity of these factors for the BBS test. This study will be followed up by another which includes moisture conditioning and a more rigorous testing plan.

REFERENCES

Al Basiouni Al Masri, Z., Alarab, A., Chehab, G., & Tehrani-Bagha, A. (2019). Assessing Moisture Damage of Asphalt-Aggregate Systems Using Principles of Thermodynamics: Effects of Recycled Materials and Binder Aging. *Journal of Materials in Civil Engineering*, 31(9), 4019190.

Cala, A., Caro, S., Lleras, M., & Rojas-Agramonte, Y. (2019). Impact of the chemical composition of aggregates on the adhesion quality and durability of asphalt-aggregate systems. *Construction and Building Materials*, 216, 661–672.

Canestrari, F., Cardone, F., Graziani, A., Santagata, F. A., & Bahia, H. U. (2010). Adhesive and cohesive properties of asphalt-aggregate systems subjected to moisture damage. *Road Materials and Pavement Design*, 11(sup1), 11–32.

Chaturabong, P., & Bahia, H.U. (2018). Effect of moisture on the cohesion of asphalt mastics and bonding with surface of aggregates. *Road Materials and Pavement Design*, 19(3), 741–753.

Cho, D.-W., & Bahia, H.U. (2010). New parameter to evaluate moisture damage of asphalt-aggregate bond in using dynamic shear rheometer. *Journal of Materials in Civil Engineering*, 22(3), 267–276.

Diab, A., You, Z., Hossain, Z., & Zaman, M. (2014). Moisture Susceptibility Evaluation of Nanosize Hydrated Lime-Modified Asphalt–Aggregate Systems Based on Surface Free Energy Concept. *Transportation Research Record*, 2446(1), 52–59.

Huang, W., & Lv, Q. (2016). Investigation of Critical Factors Determining the Accuracy of Binder Bond Strength Test to Evaluate Adhesion Properties of Asphalt Binders. *Journal of Testing and Evaluation*, 45(4), 1270–1279.

Kakar, M.R., Hamzah, M.O., & Valentin, J. (2015). A review on moisture damages of hot and warm mix asphalt and related investigations. *Journal of Cleaner Production*, 99, 39–58.

Kringos, N., Azari, H., & Scarpas, A. (2009). Identification of parameters related to moisture conditioning that cause variability in modified Lottman test. *Transportation Research Record*, 2127(1), 1–11.

Moraes, R., Velasquez, R., & Bahia, H. (2012). *The Effect of Bitumen Stiffness on the Adhesive Strength Measured by the Bitumen Bond Strength Test*.

Moraes, R., Velasquez, R., & Bahia, H. U. (2011). Measuring the effect of moisture on asphalt–aggregate bond with the bitumen bond strength test. *Transportation Research Record*, 2209(1), 70–81.

Rahim, A., Thom, N., & Airey, G. (2019). Development of compression pull-off test (CPOT) to assess bond strength of bitumen. *Construction and Building Materials*, 207, 412–421.

Taib, A., Jakarni, F. M., Rosli, M. F., Yusoff, N. I. M., & Aziz, M. A. (2019). Comparative study of moisture damage performance test. *IOP Conference Series: Materials Science and Engineering*, 512(1), 12008. IOP Publishing.

Zhang, J., Apeagyei, A. K., Grenfell, J., & Airey, G. D. (2016). Experimental Study of Moisture Sensitivity of Aggregate-Bitumen Bonding Strength Using a New Pull-Off Test. *8th RILEM International Symposium on Testing and Characterization of Sustainable and Innovative Bituminous Materials*, 719–733. Springer.

Advances in Materials and Pavement Performance Prediction II – Kumar et al. (eds)
© 2021 Taylor & Francis Group, London, ISBN 978-0-367-46169-0

Aging differences within RAP binder layers

W.L.G. Ferreira
Universidade Federal Rural do Semi-Árido (UFERSA), Campus Caraúbas - Caraúbas, Brasil

V.T.F. Castelo Branco
Universidade Federal do Ceará (UFC), Campus do PICI - Bloco Fortaleza, Brasil

K. Vasconcelos
Universidade de São Paulo (USP), Escola Politécnica, São Paulo, Brasil

A. Bhasin & A. Sreeram
The University of Texas at Austin, Austin, USA

ABSTRACT: The degree of binder activation of reclaimed asphalt pavement (RAP) is a critical aspect the design of the recycled asphalt mix-tures. Assuming RAP binder activation of less than 100% insinuates that part of the RAP binder will be inactive in the mixing process. It is im-portant to comprehend if there is any difference between the active and the inactive binder as the recycling agent to be used could be chosen considering only the active binder characteristics. The objective of this study is to evaluate aging differences within different RAP binder layers. Stages of binder extraction were performed on two different RAP materials in order to simulate different binder activation levels. After each ex-traction, the binder recovered was evaluated in terms of its rheological properties and the saturate-aromatic-resin-asphaltene (SARA) fractions. The results imply that the outer layer (external RAP binder layer) is more oxidized as compared to the inner layer (binder close to the aggregate). This indicates that the more oxidized RAP binder will be available to be blended.

Keywords: RAP, recycled asphalt mixture, aging, SARA.

1 INTRODUCTION

The partial activation of the binders in reclaimed asphalt pavement (RAP) is the most acceptable scenar-io for recycled asphalt mixtures (RAM) (Abed et al. 201, Al-Qadi et al., 200; Huang et al., 200; Kasser et al. 2019; Shirodkar et al. 2011;. In this con-text, part of the aged binder will be inactive, and its effect on the mixtur's long-term performance in unclear. Some diffusion between the blended binder (active binder + recycling agent) and the inactive binder might occur which could affect the RAM's performance. Although assuming 100% of RAP binder availability is not realistic, this assumption is common in RAMs design. In a recent survey from national cooperative highway research project (NCHRP), Synthesis 495 (2016), 77% of the state highways in the United States consider 100% RAP availability.

In general, for RAM with up to 20% of RAP, the virgin asphalt binder used is identical to the control asphalt mixture, i.e., without RAP. For small amounts of RAP, the effect of the RAP binder on the blended binder is reduced. However, for high amounts of RAP (above 20–25%), in the United States, it is common to use blending charts where virgin binde's performance grade (PG) is chosen based on the RAP quantity, as presented in AASH-TO M323-17. In Europe, the same concept is used; however, other parameters are used in order to choose the recycling agent, such as penetration and softening point (EN 13108-8:2005).

Usually, the RAP binder is characterized after the complete extraction and recovery procedures, after which no difference between active binder and inactive binder is considered. However, even in virgin asphalt mixtures, it is possible to find aging differences within asphalt layers covering the aggregates (Gaspar et al. 2017). Cui et al. (2014) and Sirin, Paul and Kassem (2018) state that the volatilization is one important mechanism that occurs during the hot mix preparation of asphalt mixtures. When as-phalt binder comes into contact with aggregates at high temperatures (150°C, for example), aromatic fractions tend to evaporate, and asphaltene fractions generally increase (Sirin et al. 2018). This phenomenon depends on the contact surface ar-ea and also the aggregate poros-ity and temperature. Lee et al. (1990) stated that the

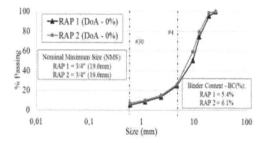

Figure 1. RAPs size distribution.

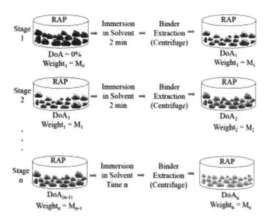

Figure 2. Illustration of the stages of binder extraction procedure.

polar fraction of as-phalt binder tends to be absorbed by the aggregates. Cui et al. (2014) claimed that the volatilization could cause significant increase in the dynamic shear modulus and decrease in the phase angle for asphalt binders

In this context, it is essential to characterize any differences between active binder and inactive bind-er. This information could be used to define the re-cycling agent that matches with the active binder. Besides, the characterization of the inactive binder could help to better understand the impact of it on the RAM's long-term performance. Normally, the recycling agent is usually chosen considering that all the available RAP binder is homogeneous, and a fulblending will hap-pen between RAP binder and recy-cling agents. These two assumptions could result in RAM with uncertain long-term performance. Thus, the main objective of this study is to evaluate aging differences among RAP binder layers

2 MATERIALS AND METHOD

The RAP materials used in this study were collected from different stockpiles at locations in Texas, USA. Firstly, around 80kg of each RAP was collected in the field and then the material was homogenized. Sec-ondly, the material was reduced to small quanti-ties to run the tests. Two RAPs were analyzed. Figure 1 present the aggregate size distribution of both.

2.1 Method

In order to reach the objectives previously presented, the degree of binder activity (DoA), which indicates the percentage of active RAP binder that a mix de-signer can consider for a selected RAP, was obtained using stages of binder extraction to activate the RAP binder. The procedure consists of keeping the RAP sample (around 1000g) immersed in solvent (toluene in this study) for 2 minutes, and then the centrifuga-tion method (ASTM D2172-18, method A) was used to separate the solution (asphalt binder + toluene) from aggregates. The first step of this process is to calculate the total binder content (BC). The BC was obtained fol-lowing the method ASTM D2172-18 where stages of binder extraction should be done un-til only the solvent

is obtained, which indicates that there is no more trace of asphalt binder. The total binder content represents 100% of DoA, which re-fers to the stage where all the RAP binder covering aggregates have been extracted. Figure 2 illustrates the stages of extraction performed in this study. The DoA was calculated after each stage by using the differences of aggregate's masses before and after binder extraction.

Four stages of extractions were performed, and the number of stages depends on the RAP characteristics mainly in terms of its gradation and binder ag-ing. The solvent immersion time was 2 minutes for stages 1, 2 and 3. Two minutes is the minimum time necessary to prepare the centrifuge equipment before starting the centrifugation. In stage 4, the immersion time for some RAPs was around 6 hours in order to extract as much binder as possible considering that after 3 stages of extraction almost 90% of the binder was activated for those RAPs. The DoA ranges from 0%, which repre-sents the RAP before any binder ex-traction, to 100% after two or more binder extractions (like illustrated in Figure 2).

The asphalt binders extracted after each stage were recovered using a rotary evaporator (RE), fol-lowing the method presented in AASTHO TP2-01. In general, 500ml of binder-toluene solution was poured inside the round flask. The round flask is ful-ly submerged in an oil bath and rotated. The oil bath temperature starts at 80°C and is raised over time up to 165°C. During the process (around 3h), a vacuum pump is turned on to reach a vacuum of at least 65cmHg in order to avoid binder oxidation. The tol-uene starts to evaporate and is collected in another round flask.

The asphalt binder recovered after each stage was analyzed in two ways. First, a frequency sweep test was conducted with frequencies ranging from 0.1Hz to 15Hz and temperatures sweep ranging from -6°C to 40°C using the DSR. Due to the small amount of extracted binder among the stages, the 4mm diameter parallel plate geometry and a 2 mm thick speci-men was used. For high temperature (above 40°C in this

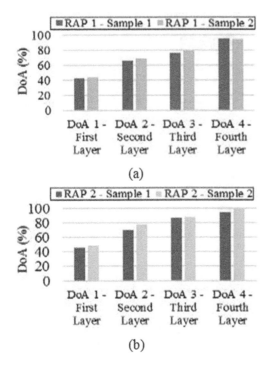

(a)

(b)

Figure 3. DoA obtained after each stage of binder extraction for RAP 1 (a) and for RAP 2 (b).

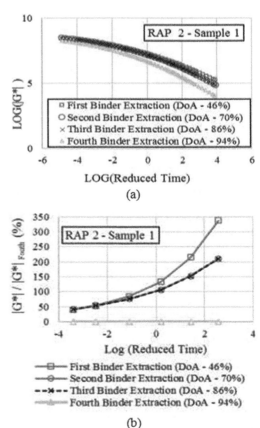

(a)

(b)

Figure 4. Master curve for RAP 2 (sample 1) (a) and per-centage increase on |G*| value in related to the fourth binder extraction (b) for RAP 2 (sample 1).

particular case), the low viscosity and small amount of material might affect the test. The master curves were constructed at a reference temperature of 20°C using the time-temperature superposition principle and also using a sigmoidal function. Second, the *saturate-aromatic-resin-asphaltene* (SARA) fraction test was performed. The SARA organic components can be divided into *asphaltenes* and *maltenes*. Maltenes can be subdivided into *satu-rates, aromatics and resins* (SAR). Due to the small amount of binder obtained after each stage of extraction, SARA fraction test proposed by Sakib & Bhasin (2018) was performed. For this test, only 400mg ± 20mg of asphalt binder is required. In gen-eral, first the binder solution (*n*-heptane + binder) is filtered using a syringe filter (25-30mm, PTFE, 0.2-0.22μm pore size) in order to separate the asphaltene component. Then, 15ml of maltene solution (mal-tene +*n*-heptane) is filtered using a solid phase ex-traction (SPE) cartridge (5g of silica gel, 20 ml capacity) combining different solvents with specific concentrations. Further details of this procedure can be found in Sakib & Bhasin (2018).

3 RESULT

Figure 3 presents the DoA after each stage of binder extraction for RAPs 1 and 2, considering 2 samples for each RAP. It shows that after the first extraction,

around 40% of the binder is activated for both RAPs. The results between samples present good replicabil-ity, except for DoA2 (Second Layer, RAP 2) where this difference is almost 7% (the biggest one).

Figure 4a presents the results of the rheological characterization for RAP 2 (samples 1). In order to quantify the differences among |G*| values, as the results presented in Figure 4a are in log scale, Figure 4b presents, in percentage, how much the binder |G*| on the first, second and third extraction differ from the fourth layer, on average, for different reduced time. As seen in Figures 4a and 4b, the outer layer (first binder extraction) is stiffer than the inner laye (closer to the aggregate, fourth binder extraction). The first binder layer is up to 350% stiffer than the fourth layer, especially for results at higher temperatures (30°C and 40°C).

Overall, the results for RAP 1 (samples 1 and 2) were similar to RAP 2. However, the first binder layer is up to 150% stiffer than the fourth layer for RAP 1 (samples 1 and 2). Besides the rheological data, the SARA fractions results are presented in Figure 5 for RAP 1 (sample 1) and RAP 2 (sample 1).

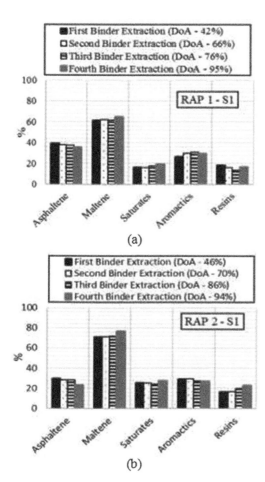

(a)

(b)

Figure 5. SARA fraction results for RAP 1 sample 1 (a), and RAP 2 sample 1 (b).

The SARA fraction results are in agreement with the rheological results. The asphaltene percentage tends to decrease after each stage of binder extrac-tion. The first binder layer is more oxidized than the fourth layer. The difference between asphaltene values for the first and the fourth layers is approximately 4% for both RAPs (1 and 2). The binders from the second and the third extraction show similarities (the binder from third extraction presents slightly (1%) less asphaltene than the second extraction on average). The results between two samples for the same RAP binder are usually identical when using this method (within 1% range of each other). However, more samples are under analysis in order to evaluat-ed this statistically. Among RAPs, RAP 1 presents higher values of asphaltene (around 40%) as com-pared with RAP 2 (around 30%) indicating that RAP 1 is more oxidized than RAP 2. The authors do not have information about the fresh binder, but consid-ering the method used, it is gener-ally expected that an unaged binder shows asphaltene content between 15–20%. It is important to mention

that, due to the volatilization phenomenon, the exter-nal layer of binder seems to be more oxidized because this layer is more exposed to oxygen and moisture in the field. Besides, there is no particular trend for the saturates, aromatics and resins fractions so far. More tests need to be done in that regard.

4 CONCLUSION

The results indicated that the asphalt binder closer to the aggregate (possible inactive binder) is less vis-cous than the external asphalt binder (possible active binder). It implies that, considering the partial RAP binder activation scenario, the RAP binder that is more oxidized will be available to be blended in a mixing scenario. More RAP materials are currently under analysis as part of this study in order to pro-vide concrete conclusions. In this way, taking into consideration the partial RAP binder activation, quan-tification of aging differences within RAP may help to appropriately design RAMs

REFERENCES

Abed, A., Thom, N., & Lo Presti, D. 2018. Design Considera-tions of High RAP-Content Asphalt Produced at Reduced Temperatures. *Materials and Structures*, 51(4), DOI:10.1617/s11527-018-1220-1.

Al-Qadi, I. L., S. H. Carpenter, G. Roberts, H. Ozer, & Q. Au-rangzeb, M. Elseifi, J. Trepanier. 2009. Determi-nation of Usable Residual Asphalt Binder in RAP. *In: Report FHWA-ICT-09-31*, Illinois Center for Transporta-tion, Rantoul, IL.

Cui, P.C., Wu, S., Xiao, Y., & Zhang, H. 2014. Study on the Deteriorations of Bituminous Binder Resulted from Vola-tile Organic Compounds Emissions, *Con-struction and Building Materials*, vol. 68, pp. 644–649. DOI:10.1016/j.conbuildmat.2014.06.067.

Gaspar, M. S., Vasconcelos, K. L., Lopes, M. M., Lo Presti, D., Bernucci, L. B., & Augusto Júnior, F. 2017. Procedi-mento de Extração em Etapas para Avaliação de Interação entre Ligantes na Reciclagem a Quente e Morna, *In XIX Con-gresso Ibero-latinoamericano Del Asfalto (CILA)*, Medel-lín, Colômbia.

Huang, B., Li, G., Vukosavljevic, D., Shu, X., & Egan, B. 2005. Laboratory Investigation of Mixing Hot-Mix Asphalt with Reclaimed Asphalt Pavement. *Transporta-tion Research Record: Journal of the Transportation Research Board*, 1929, 37–45. DOI:10.3141/1929-05.

Kaseer, Arámbula-mercado, & Martin, A. E. 2019. A Method to Quantify Reclaimed Asphalt Pavement Binder Avail-ability (Effective RAP Binder) in Recycled Asphalt Mixes. *Transportation Research Record: Journal of the Trans-portation*. DOI: doi.org/10.1177/0361198118821366

Lee, D.Y. et al. 1990. Absorption of Asphalt into Porous Aggregates. *In: Report SHRPA/UIR-90-009, Strategic High-way Research Program*. Washington, D.C, EUA

National Cooperative Highway Research Program – NCHRP, 2016. Synthesis 495: Use of Reclaimed Asphalt Pavement and Recycled Asphalt Shingles in Asphalt Mixtures. Trans-portation Research Board. DOI: doi.org/10.17226/23641.

Sakib, N., & Bhasin, A. 2018. Measuring Polarity-Based Distri-butions (SARA) of Bitumen Using Simplified Chromato-graphic Techniques. *International Journal of Pavement Engineering*. DOI:10.1080/10298436.2018.1428972.

Sirin, O., Paul, D.K., & Kassen, E. 2018. State of the Art Study on Aging of Asphalt Mixtures and Use of Antioxidant Additives. *Advances in Civil Engineering*, Vol. 2018, 18 p., DOI: https://doi.org/10.1155/2018/3428961

Shirodkar, P., Mehta, Y., Nolan, A., Sonpal, K., Norton, A., Tomlinson, C., & Sauber, R. 2011. A study to Determine the Degree of Partial Blending of Reclaimed Asphalt Pavement (RAP) Binder for High RAP Hot Mix Asphalt. *Construc-tion and Building Materials*, 25(1), 150–155. DOI: 10.1016/j.conbuildmat.2010.06.045.

Fourier transform rheology of asphalt binders

Saqib Gulzar & Shane Underwood
Department of Civil, Construction, and Environmental Engineering, North Carolina State University, Raleigh, NC, USA

ABSTRACT: In general, asphalt binders are characterized in a *low strain* regime where the response is linear viscoelastic (LVE). The shear stress response is independent of the strain amplitude within the LVE limit and the frequency of the response is same as the applied frequency. However, when the asphalt binder is subjected to strains beyond the LVE limit, the response becomes nonlinear. In this regime, the response is stress or strain level dependent and may be comprised of higher order frequencies that contribute to the overall nonlinear response. In this study, a crumb rubber modified asphalt binder is tested at large strains of 30%, 40% and 50% at an applied frequency of 0.5 Hz and testing temperature of 30°C. The total stress response is analyzed using Fourier transform rheology. It is found that there is an appearance of higher harmonics in the total stress response. The contribution of higher harmonic is characterized as a function of strain amplitude and it appears to follow a sigmoidal fit of strain amplitude.

1 INTRODUCTION

Traditionally. asphalt binder characterization has been based on linear viscoelasticity (LVE), mostly comprising of oscillatory shear rheological measurements under small strains. However, when used in service, the binder is subjected to larger strains as part of the asphalt concrete. Thus, it is important to study the behaviour of asphalt binders under large strains beyond the LVE limit. With the use of modified binders such as rubber, polymer nanocomposites, etc, the material behavior has become more complex and there is a need to characterize the binder in the nonlinear region (Gulzar & Underwood 2019a). Most of the current rheological measurements in the pavement community use the peak stress-strain data; however, in the authors' opinion, it is essential to consider the complete waveform of the viscoelastic response to characterize linear and nonlinear response of asphalt binders under large strains.

Under large strains, the shear stress response is no longer sinusoidal and has higher harmonic contributions (Shan et al. 2018). Fourier transform (FT) rheology is a tool to decipher the complete waveform of complex nonlinear material response. It enables to delineate the contributions from higher harmonics and compute relative nonlinearities. Gulzar & Underwood (2019b) used FT rheology to evaluate the contributions of higher harmonics in the binder response and found significant nonlinearity due to third harmonic. In another work, Gulzar and Underwood (2020) evaluated the complete stress waveforms of the nonlinear response of modified asphalt binders

using large amplitude oscillatory shear protocol under different conditions and found that nonlinearity is a function of temperature, applied frequency, and input strain amplitude.

2 MATERIALS

In this work, a terminally blended crumb-rubber modified asphalt binder has been used to study the nonlinear response under large strains. This binder was used in two previous Transportation Pooled Fund (TPF) research projects, TPF-5(019): *Full-Scale Accelerated Performance Testing for Superpave and Structural Validation* and SPR-2(174): *Accelerated Pavement Testing of Crumb Rubber Modified Asphalt Pavements*. The crumb rubber terminal blend (CR-TB) modified asphalt binder was produced by a hybrid technique with a combination of recycled tire crumb rubber (5.5%) and new SBS rubber (1.8 %) blended at an asphalt terminal (Ginson et al. 2012). A dynamic shear rheometer (Anton Paar^{TM}, EC-502 Twist) was used for obtaining the complete stress waveforms under the strain levels of 30%, 40%, and 50% at an applied frequency of 0.5 Hz and a testing temperature of 30°C. The PG grade of the binder was 76–28.

3 THEORETICAL BACKGROUND

3.1 *Linear vs nonlinear material response*

In general, oscillatory shear has been widely used to characterize the mechanical response of asphalt binders. When a sinusoidal strain is applied as an input, the stress is measured as a response of the asphalt binder. Since the applied strain is sinusoidal with a constant frequency, say ω_o, the measured stress becomes a periodic function with the same applied frequency ω_o and a constant amplitude after some time when the material reaches a steady state. Since the input strain is applied periodically, the measured stress output is a function of time. Depending upon the applied strain level, the output stress response may or may not be sinusoidal. If it is sinusoidal like the input strain, the material is linear, else it is nonlinear.

3.2 Fourier transform rheology

As explained in Sec. 3.1, the measured stress response is a periodic function with a period of $P\ (=2\pi/\omega_o)$. The Fourier series expansion of a periodic function $s(t)$ is written as:

$$s(t) = \frac{a_o}{2} + \sum_{n=1}^{\infty} a_n \cos\left(\frac{2n\pi t}{P}\right)$$
$$+ \sum_{n=1}^{\infty} b_n \sin\left(\frac{2n\pi t}{P}\right) \tag{1}$$

where

$$a_n = \frac{2}{P} \int_{c+P}^{P} s(t) \cos\left(\frac{2n\pi t}{P}\right) dt;$$

$$b_n = \frac{2}{P} \int_{c+P}^{P} s(t) \sin\left(\frac{2n\pi t}{P}\right) dt; n = 0, 1, 2, \ldots \tag{2}$$

Using the Fourier expansion (1) for stress response under large strains, we have

$$\sigma(t) = \frac{\sigma'_o}{2} + \sum_{n=1}^{\infty} \sigma'_n \cos\left(n\omega_o t\right)$$
$$+ \sum_{n=1}^{\infty} \sigma''_n \sin\left(n\omega_o t\right) \tag{3}$$

or

$$\sigma(t) = \frac{\sigma'_o}{2} + \sum_{n=1}^{\infty} \sigma_n \sin\left(n\omega_o t + \delta_n\right) \tag{4}$$

where

$$\sigma_n = \sqrt{\left(\sigma'_n\right)^2 + \left(\sigma''_n\right)^2}; \tan\left(\delta_n\right) = \frac{\sigma'_n}{\sigma''_n} \tag{5}$$

As can be seen from (4), the stress response under large strains is a function of time and has inherent contributions of multiple amplitudes and phases as a

function of applied frequency, ω_o. In order to implement FT-rheology, a discrete, complex and half-sided Fourier transform is applied to the time- dependent response function of the shear stress (similar to (4)) to obtain the contributions of higher frequencies. The Fourier transform for any time function, $s(t)$ is given as:

$$\hat{s}(\omega) = F\left(s(t)\right) = \int_{-\infty}^{+\infty} s(t) e^{-i\omega t} dt \tag{6}$$

Applying (6) to (3), we have

$$\hat{\sigma}(\omega) = \pi\sigma'_o \delta(\omega) + \pi \sum_{n=1}^{\infty} \sigma'_n[\delta(\omega - n\omega_o)$$
$$+\delta(\omega + n\omega_o)]\ldots - \pi i \sum_{n=1}^{\infty} \sigma''_n[\delta(\omega - n\omega_o)$$
$$-\delta(\omega + n\omega_o)] \tag{7}$$

where Dirac-delta function, $\delta(\omega - \omega_o)$ is defined as

$$\delta\left(\omega - \omega_o\right) = \frac{1}{2\pi} \int_{-\infty}^{+\infty} e^{-i(\omega - \omega_o)t} dt = \left\{F\left(e^{i\omega_o t}\right)\right\} \tag{8}$$

We define intensity in the frequency domain as

$\hat{I}\left(\omega\right) = \sqrt{\hat{\sigma}\left(\omega\right)\overline{\hat{\sigma}}\left(\omega\right)}$ where $\overline{\hat{\sigma}}\left(\omega\right)$ is the complex conjugate of $\hat{\sigma}\left(\omega\right)$, then we have in (7),

$$\frac{\hat{I}\left(\omega\right)}{\pi} = \left|\sigma'_o\right|\delta(\omega) + \sum_{n=1}^{\infty} \sigma_n \delta(\omega - n\omega_o)$$
$$+ \sum_{n=1}^{\infty} \sigma_n \delta(\omega + n\omega_o) \tag{9}$$

For $\omega > 0$, equation (9) indicates that the plot of $\hat{I}\left(\omega\right)$ is a series of spikes having an intensity of $\hat{I}_n = \pi\sigma_n$ at $\omega = n\omega_o$, called as the Fourier intensity of the nth harmonic

It should be noted that the input strain is applied at only one frequency, ω_o to ensure that any output contribution at frequencies other than ω_o is associated with the nonlinearity in the stress response. Further, it is assumed that the stress response is shear symmetric and the sign of shear stress changes as the sign of shearing changes, thus, the shear stress must be an odd function with respect to shearing deformation.(Hyun et al., 2011) Hence, only odd harmonics in the Fourier spectra are associated to the material nonlinearity under large strains. Further, in the Fourier expansion of the stress output, the stress amplitudes (σ_n) and phases (δ_n), are, in general, functions of the two input parameters such as input frequency (ω_o) and strain amplitude (γ_o). Thus, equation (4) can be re-written as

$$\sigma(t) = \sum_{n=1,odd}^{\infty} \sigma_n(\gamma_o, \omega_o) \sin\left[n\omega_o t + \delta_n(\gamma_o, \omega_o)\right] \tag{10}$$

428

Of all the odd harmonics present in the shear stress response, the first few harmonics such as the third, fifth, seventh, etc. are of prime importance. The shear stress response can also be represented by a power series expansion to separate strain dependence from frequency dependence. The Fourier intensity of the n^{th}harmonic, $\hat{I}_n(\gamma_o, \omega)$, is related to the strain amplitude over the entire frequency spectrum as follows,

$$\hat{I}_k(\gamma_o, \omega) \propto \gamma_o^k \tag{11}$$

In order to compute the relative intensity of the third-harmonic from FT-rheology, equation (10) can be expanded as

$$\sigma(t) = \sigma_1 \sin(\omega_o t + \delta_1) + \sigma_3 \sin(3\omega_o t + \delta_3) + \ldots$$

$$= \sigma_1 \cos \delta_1 \sin \omega_o t + \sigma_1 \sin \delta_1 \cos \omega_o t$$

$$+ \sigma_3 \cos \delta_3 \sin 3\omega_o t + \sigma_3 \sin \delta_3 \cos 3\omega_o t + \ldots \tag{12}$$

Thus, the relative intensity of the third harmonic is given as

$$I_{3/1} = \frac{I_3}{I_1} = \frac{\sigma_3}{\sigma_1} = \frac{\sqrt{(\sigma_3 \cos \delta_3)^2 + (\sigma_3 \sin \delta_3)^2}}{\sqrt{(\sigma_1 \cos \delta_1)^2 + (\sigma_1 \sin \delta_1)^2}} \tag{13}$$

Equation (13) is further solved in terms of power representation of stress response and finally, we obtain

$$\frac{I_3}{I_1} = Q(\omega_o, \gamma_o) \times \gamma_o^2 \tag{14}$$

where Q is a nonlinear coefficient which provides insights into the material behaviour as it develops and makes transitions from linear to nonlinear region. The Q-coefficient characterizes FT-rheology and is a function of both input frequency (ω_o) and strain amplitude (γ_o). It is expected that this nonlinear coefficient will approach an asymptotic limiting constant value when the strain approaches zero and can be used to quantify the intrinsic nonlinearity of the material, called as zero strain nonlinearity and is given as

$$Q_o(\omega) = \lim_{\gamma_o \to 0} Q(\omega, \gamma_o) \tag{15}$$

As evident from Equation (14), the relative nonlinearity, $I_{3/1}$, emerges quadratically as a function of strain amplitude. However, as the strain amplitude increases beyond a critical value (which is a function of the material), there are deviations from the quadratic proportionality. It has been empirically shown (Wilhelm, 2002) to obey a sigmoidal relationship with strain amplitude as follows,

$$\frac{I_3}{I_1}(\gamma_o) = A\left(1 - \frac{1}{1 + B\gamma_o^C}\right) \tag{16}$$

where A, B, and C are function fit parameters such that A represents maximum intensity at high strain amplitudes, B represents an approximate point of inflexion and C reflects the scaling exponent.

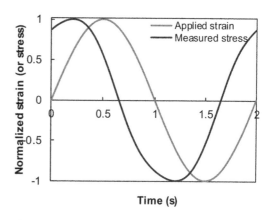

Figure 1. One cycle of applied strain and measured stress response at 30% strain level.

4 RESULTS AND DISCUSSION

4.1 Fourier transform analysis

Figure 1 shows one cycle complete cycle of normalized applied strain and measured stress response at a frequency of 0.5 Hz and 30% strain level. By carefully examining the two curves, it can be seen that the strain input is sinusoidal while the stress response is a periodic function of time but is not sinusoidal, hence there is the presence of nonlinearity. In other words, the stress response has contributions from odd higher harmonics. In order to compute the intensity of higher harmonics, Fourier transform is done for 10 cycles of the stress response as explained in Section 3.2. Figure 2 shows the Fourier spectra of the stress response at the strain levels of 30%, 40%, and 50%. It can be clearly seen that there is a presence of odd higher harmonics at each strain level, hence evidence for nonlinearity.

Further, it can be seen that the highest contribution is from 3rd harmonic, followed by 5th and 7th harmonic respectively for all strain levels. Figure 3 shows the comparative relative intensities of 3rd, 5th, and 7th harmonics at all the tested strain levels. It can be seen that the intensity of the higher harmonics increases with the strain level irrespective of the order of the harmonic considered. So, 50% strain level has the highest contributions for all odd harmonics followed by 40% and 30% strain levels respectively.

4.2 Strain dependence of nonlinearity

The relative nonlinearity ($I_{n/1}$) can in general be defined for each odd harmonic contributing to the overall nonlinear response of the material, similar to the third harmonic relative nonlinearity as shown in equations (13) and (14) as evident from equation (11). The variation of the relative intensity found in the current study is plotted as a function of strain amplitude for each significant odd harmonic, and the sigmoidal model as shown in equation (11) is fit to the measured

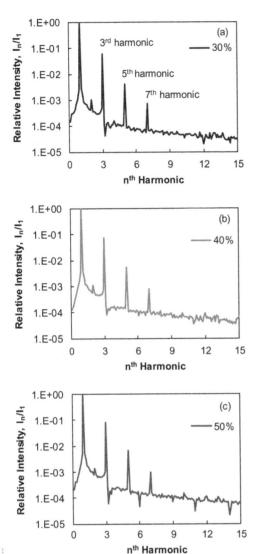

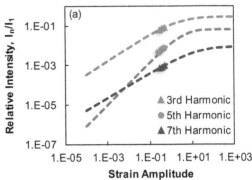

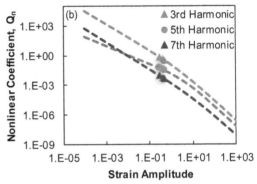

Figure 2. Fourier spectra of the stress response at the strain level of (a) 30%, (b) 40%, and (c) 50%.

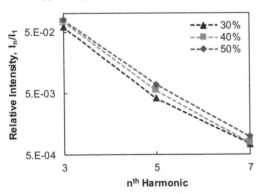

Figure 3. Relative intensity of 3rd, 5th, and 7th harmonics at different strain levels of 30%, 40%, and 50%.

Figure 4. Strain dependence of nonlinearity for observable and significant odd harmonics.

data as shown in Figure 4(a). It can be seen that relative nonlinearity increases with increase in the strain amplitude. Further, the nonlinear coefficient defined for third harmonic as in equation (14) can also be defined in a similar for higher harmonics using Equation (11). Although, the nonlinear coefficient, Q for the third harmonic has been widely used and reported, Figure 4(b) shows the nonlinear coefficient (Q_n) for all the observed odd harmonics. Since only limited data has been presented, the shapes of the curves for different conditions are qualitative; however, some trends are evident. There is very little variation in Q for the tested conditions for each harmonic, however based on the modeling, these appear to decrease with increasing strain amplitude. Further, it can be seen that nonlinear coefficient associated with third harmonic is highest for all the strain range. There is an interchange of 5th and 7th nonlinearity coefficient at very low strain amplitudes. It is possible that the trend may be apparent due to the scarcity of the data used. However, this variation is due to the difference in slopes between the observed measurements and can be attributed to experimental sensitivity.

5 CONCLUSION

In this study, the crumb rubber asphalt binder has been tested for large strains and the measured stress

430

response is investigated using Fourier transform rheology. An exploratory analysis of the Fourier spectra shows the presence of higher odd harmonics and provides evidence for the presence of nonlinearity in the binder response. The LVE measures such as dynamic shear modulus and phase angle lose meaning in such a case, hence there is a need to characterize the asphalt binder response beyond LVE. Further, contribution of higher harmonics was studied, and it was found that the third harmonic plays a major role and follows a sigmoidal relationship with strain amplitude. Since the asphalt binder was tested for only one frequency of 0.5 Hz, the frequency dependence of zero-strain nonlinearity has not been evaluated.

REFERENCES

Gibson, N., Qi, X., Shenoy, A., Al-Khateeb, G., Kutay, M.E., Andriescu, A., Stuart, K., Youtcheff, J. and Harman, T. 2012. *Performance testing for superpave and structural validation.* (No. FHWA-HRT-11-045). United States. Federal Highway Administration.

Gulzar, S, Underwood, S. 2019a. Use of polymer nanocomposites in asphalt binder modification.In *Advanced Functional Textiles and Polymers* (Shahid-ul-Islam, B.S. Butola, ed.), John Wiley-Scrivener USA, pp. 405–431.

Gulzar, S, Underwood, S. 2019b. Nonlinear rheological behavior of asphalt binders.In *91st Annual Meeting of The Society of Rheology*, Raleigh, USA.

Gulzar, S, Underwood, S. 2020. Nonlinear viscoelastic response of crumb rubber modified asphalt binder under large strains. *Transportation Research Record*, p.0361198120907097.

Hyun, K., Wilhelm, M., Klein, C.O., Cho, K.S., Nam, J.G., Ahn, K.H., Lee, S.J., Ewoldt, R.H. and McKinley, G.H. 2011. A review of nonlinear oscillatory shear tests: Analysis and application of large amplitude oscillatory shear (LAOS). *Progress in PolymerScience*: 36(12), pp. 1697–1753.

Shan, L., He, H., Wagner, N.J. and Li, Z., 2018. Nonlinear rheological behavior of bitumen under LAOS stress. *Journal of Rheology*, 62(4), pp. 975–989.

Wilhelm, M. 2002. Fourier-transform rheology. *Macromolecular materials and engineering*, 2002. 287(2), pp. 83–105.

Advances in Materials and Pavement Performance Prediction II – Kumar et al. (eds)
© 2021 Taylor & Francis Group, London, ISBN 978-0-367-46169-0

Large amplitude oscillatory shear of modified asphalt binder

Saqib Gulzar & Shane Underwood
Department of Civil, Construction, and Environmental Engineering, North Carolina State University, Raleigh, NC, USA

ABSTRACT: The current rheological characterization of asphalt binders mostly uses oscillatory shear based peak stress-strain data. The linear viscoelastic (LVE) limit is also estimated using such data. In order to characterize the complete response of the asphalt binder, it is essential to consider the entire waveform recorded from oscillatory shear testing. In this paper, the response of a modified binder is studied based on the complete waveform recorded during oscillatory shear testing. The waveform was recorded for strain amplitude of 30%, 40% and 50% at the testing temperature of 30°C. Viscoelastic moduli based on the geometry of Lissajous-Bowditch plots were used to study the behaviour of the binder and it was found that the asphalt binder exhibits nonlinearity under such strains. There is a need to study and characterize the nonlinear response of the asphalt binders and incorporate it into the current specifications.

1 INTRODUCTION

In order to characterize the rheological response of asphalt binders, oscillatory shear has been used widely since the introduction of performance grade specifications. The response of asphalt binder is considered to be viscoelastic and most of the conventional binders fall under linear viscoelasticity region. With the introduction of modifiers, additives, surfactants, etc. into the asphalt binders to improve their performance characteristics, the rheological response has become complex. Further, with increasing traffic and changing climate, the asphalt binders as apart of asphalt concrete in flexible pavements experience large strains beyond their LVE limit. In either case, the response of binders is expected to be nonlinear and there is a need to characterize the nonlinear response of asphalt binders (Gulzar & Underwood, 2019a,b,c).

In a dynamic oscillatory shear rheological testing, a sinusoidal strain signal, $\gamma(t)$ is applied to a sample at a frequency, ω_o, and strain amplitude, γ_o, as shown in Equation (1), the resulting stress response, $\sigma(t)$, is also sinusoidal with frequency same as applied frequency but with a phase difference, δ_o, for small amplitudes within the LVE limit (SAOS rheology) and is represented as shown in equation (2),

$$\gamma(t) = \gamma_0 \sin(\omega_0 t) \tag{1}$$

$$\sigma(t) = \sigma_0 \sin(\omega_0 t + \delta_0) \tag{2}$$

Equation (2) can be re-written as

$$\sigma(t) = \gamma_0 [G'_1 \sin(\omega_0 t) + G''_1(\omega_0 t)] \tag{3}$$

where G'_1 is the storage modulus or the in-phase component of stress response while $G?_1$ is the loss modulus or the out-of-phase component of stress response. These viscoelastic measures are based on the fundamental harmonic (1st harmonic), are a function of ω_o only, and are independent of γ_o. However, in LAOS, the strain amplitude is beyond the LVE limit where strain dependence of stress response manifests. These viscoelastic measures lose meaning beyond the LVE limit and the stress response becomes nonlinear. Beyond this limit, the response contains the contribution from higher harmonics, which cannot be ignored. It has been shown that only odd harmonics appear in the LAOS response (Hyun et al., 2011) while the appearance of even harmonics (if any) is due to a number of reasons such fluid inertia, misalignment, or imperfect excitation. The nonlinear stress response can be represented as

$$\sigma(t) = \sum_{n=0}^{\infty} \sigma_{2n+1}(\gamma_0, \omega_0) \sin[(2n+1)\omega_0 t + \delta_{2n+1}(\gamma_0, \omega_0)] \tag{4}$$

One way to find the contributions of higher harmonics in the total stress response is to analyze the response in the frequency domain using Fourier transform (FT) rheology (Gulzar & Underwood, 2020). However, the high sensitivity and arbitrariness of FT rheology does not result in a clear physical interpretation and many a times fails to capture the rich nonlinearities present in the raw waveform of the response (Ewoldt et al., 2008). On the other hand, some new nonlinear moduli have been proposed to gain deeper insights into the nonlinearities associated with the response (Ewoldt et al., 2008). In this study, these new measures have been used to decipher the nonlinearities associated with asphalt binder rheology beyond the LVE limit.

2 MATERIALS

In this work, a terminally blended crumb-rubber modified asphalt binder has been used to study the nonlinear response under large strains. This binder was used in two previous Transportation Pooled Fund (TPF) research projects, TPF-5(019): *Full-Scale Accelerated Performance Testing for Superpave and Structural Validation* and SPR-2(174): *Accelerated Pavement Testing of Crumb Rubber Modified Asphalt Pavements*. The crumb rubber terminal blend (CR-TB) modified asphalt binder was produced by a hybrid technique with a combination of recycled tire crumb rubber (5.5%) and new SBS rubber (1.8%) blended at an asphalt terminal (Ginson et al., 2012). A dynamic shear rheometer (Anton Paar^{TM}, EC-502 Twist) was used for obtaining the complete stress waveforms under the strain levels of 30%, 40%, and 50% at an applied frequency of 0.5Hz and a testing temperature of 30°C. The PG grade of the binder is 76–28.

3 THEORETICAL BACKGROUND

In order to examine the elastic and viscous characteristics of any material, it is common to use an oscillatory shear strain as shown in equation (1) as an input signal. It simultaneously imposes an orthogonal strain rate on the sample as shown in equation

$$\gamma(t) = \gamma_0 \omega_0 \cos(\omega_0 t) \tag{5}$$

As explained in section (1), at small strain amplitudes, the material is characterized by storage and loss moduli. However, under large strains, the stress response contains higher harmonic contributions as shown in equation (4). In terms of Fourier series representation, the equation (4) can be alternative written in terms of expressed either in elastic scaling as shown in equation (6) or in viscous scaling as shown in (7)

$$\sigma(t) = \gamma_0 \sum_{n_{odd}} [G'_n(\omega_0, \gamma_0) \sin(n\omega_0 t)$$
$$+ G''_n(\omega_0, \gamma_0) \cos(n\omega_0 t)] \tag{6}$$

$$\sigma(t) = \gamma_0 \sum_{n_{odd}} [\eta''_n(\omega_0, \gamma_0) \sin(n\omega_0 t)$$
$$+ \eta'_n(\omega_0, \gamma_0) \cos(n\omega_0 t)] \tag{7}$$

The first-harmonic coefficients G'_1 and $G?_1$ that are widely used in linear viscoelasticity fail to capture the nonlinearities beyond LVE limit and additional measures are needed which should reduce to these unique material characteristics in the LVE limit but should diverge in the nonlinear region to enable further physical insights into the response. Among numerous measures proposed in the literature, the ones used in this study were proposed by Ewoldt et al. (2008) and are discussed briefly in the subsequent sections.

3.1 Elastic modulus

In the graphical analysis, the nonlinear behavior of asphalt binder can be identified by studying the shape of the Lissajous-Bowditch curves wherein a distorted ellipse is an indication of the presence of nonlinearity in the viscoelastic stress response. Ewoldt et al. (2008) proposed local nonlinear viscoelastic moduli based on material response in an oscillatory cycle under both small and large strains or strain rates. These are given as below:

1. Dynamic modulus measured at minimum strain:

This modulus is based on the graphical analysis of L-B plots and is measured at zero strain level where the instantaneous strain-rate is also maximized). It is a tangent modulus and relates to the Fourier coefficients of equation (6) as shown in equation (8).

$$G'_M = \frac{d\sigma}{d\gamma}\Big|_{\gamma=0} = \sum_{n_{odd}} nG'_n \tag{8}$$

2. Dynamic modulus measured at maximum strain:

This modulus is measured at maximum strain level where the instantaneous strain-rate is minimized to zero. It is a secant modulus and is also referred to as large strain modulus. It relates to the Fourier coefficients of equation (6) as shown in equation.

$$G'_L = \frac{\sigma}{\gamma}\Big|_{\gamma=\pm\gamma_0} = \sum_{n_{odd}} G'_n(-1)^{\frac{(n-1)}{2}} \tag{9}$$

These nonlinear measures G'_M and G'_L converge to linear viscoelastic moduli under small amplitudes in LVE regime. Thus, dynamic moduli in linear region equate to storage modulus, i.e., $G'_M = G'_L = G'_1$.

3.2 Dynamic viscosity

Similar to elastic moduli, the nonlinear equivalent of loss modulus should be a representative of viscous dissipation but as a function of shear rates in an oscillatory shear deformation. Thus, the new nonlinear measures represent instantaneous viscosity at smallest and largest shear rates as defined below:

1. Dynamic viscosity measured at minimum strain-rate:

This viscosity is measured at minimum instantaneous strain-rate. At such a point, the instantaneous strain is maximum, and viscosity is measured as a tangent. It relates to the Fourier coefficients of equation (6) as shown in equation (10).

$$\eta^p_M rime = \frac{d\sigma}{d\gamma}\Big|_{\gamma=0} = \frac{1}{\omega} \sum_{n_{odd}} nG'_n(-1)^{\frac{(n-1)}{2}} \tag{10}$$

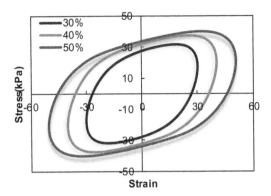

Figure 1. Elastic Lissajous-Bowditch plots for the strain levels of 30%, 40%, and 50%.

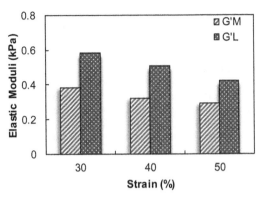

Figure 2. Minimum- and large-strain moduli for the strain levels of 30%, 40%, and 50%.

2. Dynamic viscosity measured at maximum strain-rate:

This viscosity is measured at maximum strain-rate where also the instantaneous strain is minimized. This dynamic viscosity is measured as a secant and relates to the Fourier coefficients of equation (6) as shown in equation (11).

$$\eta'_L = \frac{\sigma}{\gamma} \mid = \frac{1}{\omega} \sum_{n_{odd}} G'_n \qquad (11)$$

Similar to elastic moduli, the nonlinear measures of dynamic viscosities also become equal in the LVE regime and relate to loss modulus, i.e., $\eta'_M = \eta'_L = \eta'_1 = G''_1 / \omega_0$.

3.3 Nonlinearity Indices

The local nonlinear viscoelastic measures discussed in sections (3.1) and (3.2) can be used to quantify intra-cycle elastic and viscous nonlinearity using strain-stiffening ratio, S, and shear-thickening ratio, T, respectively, as given under:

$$S = \frac{G'_L - G'_M}{G'_L} \qquad (12)$$

$$T = \frac{\eta'_L - \eta'_M}{\eta'_L} \qquad (13)$$

These dimensionless ratios help to quantify the distortions in the Lissajous-Bowditch plots from a linear viscoelastic ellipse. The sign of these indices indicates the nature of nonlinearity present in the response.

The elastic nonlinearity is strain-softening and strain-stiffening for $S < 0$ and $S > 0$ respectively. Similarly, the viscous nonlinearity is shear-thinning and shear-thickening for $T < 0$ and $T > 0$ respectively. More detailed description is presented elsewhere in literature (Hyun et al., 2011). In addition to studying intra-cycle nonlinearity, the origin and evolution of nonlinearity with increasing applied strain amplitudes can be studied (Ewoldt et al., 2008).

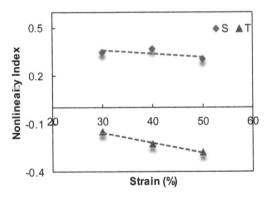

Figure 3. Viscous Lissajous-Bowditch plots for the strain levels of 30%, 40%, and 50%.

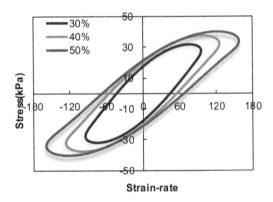

Figure 4. Strain-stiffening (S) and shear-thickening (T) ratios as a function of applied strain.

4 RESULTS AND DISCUSSION

4.1 Elastic Moduli

The elastic Lissajous-Bowditch plots for the tested binder at all the three strain levels of 30%, 40%, and 50% are shown in Figure 1. It can be seen that the L-B plots are distorted elliptical in shape, hence the LVE measures won't be sufficient to capture the complete response. The nonlinear elastic moduli presented in section (3.1) calculated as per equations (8) and (9) are used to capture the distortions and hence nonlinearity. Figure 2 shows the minimum and large strain moduli for all the strain levels and it can be seen that two different significantly showing the importance of having two different moduli to capture the nonlinearity based on extremes. Further, there is trend with increasing strain amplitude for both the moduli indicating the degree of distortion increases with the strain level.

4.2 Dynamic viscosity

The viscous Lissajous-Bowditch plots for the tested binder for all the three strain levels of 30%, 40%, and 50% are shown in Figure 3. Here again, it can be seen that nonlinearity manifests discernibly and the distortions manifests significantly as strain level increases. In order to capture the viscous characteristics, the dynamic viscosities at minimum strain-rate and maximum strain rate are computed as per equations (10) and (11), and are plotted in Figure 5. It can be seen that the both the viscosities differ significantly at all the strain levels. Further, it can be observed that both these nonlinear measures of dynamic viscosity decrease with increasing strain amplitude. This observation complements the similar observation made in case of elastic moduli.

4.3 Nonlinearity Indices

In order to characterize the intra-cycle nonlinearity, strain-stiffening and shear-thickening ratios are used at a particular strain level as per the equations (12) and (13). Further, to capture the evolution of nonlinearity with increasing strain amplitude, both these ratios are plotted as a function of applied strain amplitude as shown in Figure 4. It can be seen that nonlinearity index, S, is positive for all strain levels indicating that the elastic nonlinearity is strain-stiffening. Further, the second nonlinearity index, T, is negative for all the strain levels indicating that the viscous nonlinearity is shear-thinning for all the strain levels. Further, a both S and T have a general decreasing trend with increasing strain amplitude, however, at 40% strain level, the strain-stiffening ratio is slightly higher than that of 30% strain level.

5 CONCLUSION

In this study, the crumb rubber asphalt binder has been tested using the large amplitude oscillatory shear testing protocol. The nonlinear measures were used to quantify the nonlinearity present in the response beyond the LVE limit. The crumb-rubber asphalt binder showed strain-stiffening and shear-thinning behaviour for all the applied strain levels. The distortions in the Lissajous-Bowditch plots indicate that the binder response is nonlinear which cant be captured by traditional linear viscoelastic measures. There is a need to consider the nonlinear response of the binders in performance grade specifications. Further, constitute models to capture this nonlinear behavior should be developed.

REFERENCES

Ewoldt, R.H., Hosoi, A.E. and McKinley, G.H. 2008. New measures for characterizing nonlinear viscoelasticity in large amplitude oscillatory shear. *Journal of Rheology*, 52(6), pp.1427–1458.

Gibson, N., Qi, X., Shenoy, A., Al-Khateeb, G., Kutay, M.E., Andriescu, A., Stuart, K., Youtcheff, J. and Harman, T. 2012. *Performance testing for superpave and structural validation.* (No. FHWA-HRT-11-045). United States. Federal Highway Administration.

Gulzar, S, Underwood, S. 2019a. Use of polymer nanocomposites in asphalt binder modification. In *Advanced Functional Textiles and Polymers* (Shahid-ul-Islam, B.S. Butola, ed.), John Wiley-Scrivener USA, pp. 405–431.

Gulzar, S, Underwood, S. 2019b. Nonlinear rheological behavior of asphalt binders. In *91st Annual Meeting of The Society of Rheology*, Raleigh, USA.

Gulzar, S, Underwood, S. 2020. Nonlinear viscoelastic response of crumb rubber modified asphalt binder under large strains. *Transportation Research Record*, p.0361198120907097.

Gulzar, S, Underwood, S. 2020. Fourier transform rheology of asphalt binder. Advances in Materials and Pavement Performance Prediction, May 2020, USA.

Hyun, K., Wilhelm, M., Klein, C.O., Cho, K.S., Nam, J.G., Ahn, K.H., Lee, S.J., Ewoldt, R.H. and McKinley, G.H. 2011. A review of nonlinear oscillatory shear tests: Analysis and application of large amplitude oscillatory shear (LAOS). *Progress in Polymer Science*: 36(12), pp. 1697–1753.

Advances in Materials and Pavement Performance Prediction II – Kumar et al. (eds)
© 2021 Taylor & Francis Group, London, ISBN 978-0-367-46169-0

Stress decomposition of nonlinear response of modified asphalt binder under large strains

Saqib Gulzar & Shane Underwood
Department of Civil, Construction, and Environmental Engineering, North Carolina State University, Raleigh, NC, USA

ABSTRACT: The current characterization of asphalt binders is based on linear viscoelasticity wherein the testing is performed under small strains. It ensures that the binder stress response is independent of the strain amplitude and is sinusoidal in nature with frequency same as input frequency. However, the asphalt binders may be subjected to large stains as part of asphalt mixture in the pavement under field conditions and the response may well be nonlinear. The linear viscoelastic measures lose their meaning in the nonlinear range and it is essential to decipher the nonlinear binder response under large stains. In this study, a crumb rubber modified asphalt binder is tested at large strains of 30%, 40% and 50% at an applied frequency of 0.5 Hz and testing temperature of 30°C. A stress decomposition technique based on geometric interpretation of nonlinear response is utilized to obtain the elastic and viscous contributions in the total response. Further, Chebyshev polynomials of the first kind are used to obtain meaningful viscoelastic measures in the nonlinear regime. It is observed that third order elastic and viscous contribution appear in the nonlinear response, and the binder shows strain stiffening and shear thinning behavior under the tested conditions of large strains.

1 INTRODUCTION

In United States, agencies have been using Superpave characterization metrics for specification purposes. The Superpave methods use linear viscoelastic measures for fatigue, rutting, and thermal performance criteria. It recommends that the asphalt binders be tested within the linear viscoelastic range (LVER). If the material is subjected to strains or stresses beyond LVER, then the response of that material is classified as nonlinear. The nonlinearity, as per the definition of AASHTO, points to the strain dependence of dynamic shear modulus (more than 5% of its initial value), but it only relies only on peak measurements. It is important to note that the asphalt binder is subject to strain levels beyond the LVER as part of asphalt mixture in the field and hence it becomes essential to study its response beyond LVER. Further, the use of modified binders such as rubber, polymer nanocomposites, etc, the material behavior becomes more complex and the need to characterize the binder in the nonlinear region becomes more important (Gulzar & Underwood 2019a). Most of the current rheological measurements in the pavement community use the peak stress-strain data; however, in the nonlinear region, the LVE measures lose their meaning and it is essential to consider the complete waveform of the viscoelastic response to characterize linear and nonlinear response of asphalt binders under large strains.

The shear stress response under large strains may no longer be sinusoidal and can contain higher harmonic contributions. One way to quantitatively evaluate these higher order contributions is to use Fourier transform (FT) rheology. This approach enables one to delineate the contributions from higher harmonics and compute relative nonlinearities. Gulzar & Underwood (2019b) used FT rheology to evaluate the contributions of higher harmonics in the binder response and found significant nonlinearity due to third harmonic. In another work, Gulzar & Underwood (2020) evaluated the complete stress waveforms of the nonlinear response of modified asphalt binders using large amplitude oscillatory shear protocol under different conditions and found that nonlinearity is a function of temperature, applied frequency, and input strain amplitude. Another way of looking at nonlinear response is plotting Lissajous-Bowditch loops and compute the minimum and large strain moduli or dynamic viscosities (Ewoldt et al. 2008). Tacking these moduli can help in understanding both inter and intra cycle nonlinearities. A third method, which is discussed in this study, is known as stress decomposition and has been purported to help in the physical interpretation of the nonlinear response. This approach aims to explain the higher harmonics obtained from Fourier Transform analysis and geometrically plotted using Lissajous-Bowditch plots. It tries to aid in the physical interpretation of the shape traced by the elastic and viscous Lissajous-Bowditch plots. In this paper, the stress decomposition method proposed by Cho et al. (2005), which uses symmetry arguments to decompose the stress into two contributions known as elastic and viscous stress contributions, is briefly presented. It is then applied to decipher the nonlinear response of a modified asphalt binder at large strains and the results are discussed from a pavement engineering perspective.

2 MATERIALS

In this work, a terminally blended crumb-rubber modified asphalt binder has been used to study the nonlinear response under large strains. This binder was used in two previous Transportation Pooled Fund (TPF) research projects, TPF-5(019): *Full-Scale Accelerated Performance Testing for Superpave and Structural Validation* and SPR-2(174): *Accelerated Pavement Testing of Crumb Rubber Modified Asphalt Pavements*. The crumb rubber terminal blend (CR-TB) modified asphalt binder was produced by a hybrid technique with a combination of recycled tire crumb rubber (5.5%) and new SBS rubber (1.8 %) blended at an asphalt terminal (Ginson et al. 2012). A dynamic shear rheometer (Anton Paar^{TM}, EC-502 Twist) was used for obtaining the complete stress waveforms under the strain levels of 30%, 40%, and 50% at an applied frequency of 0.5 Hz and a testing temperature of 30°C. The PG grade of the binder was 76-28.

3 THEORETICAL BACKGROUND

3.1 Large amplitude oscillatory shear rheology

The most common rheological methods used to characterize the asphalt binders as part of Superpave characterization protocols are done at small strains, known as small amplitude oscillatory shear rheology (SAOS). Under SAOS, a sinusoidal input is applied at fixed frequency and the output is also a sinusoidal response at the same input frequency with a different amplitude and some phase lag. Although a time dependent waveform is applied to the material, only peak responses are measured and used further in the analysis. The linear viscoelastic measures such as dynamic shear modulus and phase angle are defined using these peak response measures as shown in Figure 1. It can be seen that the shear modulus can easily be decoupled into elastic and viscous components, known as storage and loss modulus respectively. The Superpave parameters for rutting and fatigue performance reflect this decoupled understanding of asphalt binder response under SAOS rheology. One of the drawbacks of this methodology is that only peak-to-peak responses are considered for characterizing the entire binder response, hence only a part of the response is considered. Secondly, the testing is done under small strains to ensure linearity, but the binder may well undergo large strains in the field and exhibit nonlinear response.

In continuation to our previous studies (Gulzar & Underwood 2020), the authors propose a possible advancement in asphalt binder viscoelastic characterization under large strains, known as large

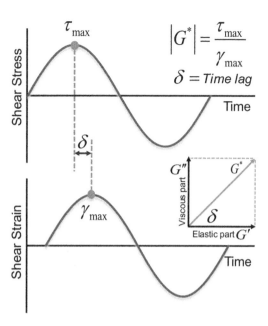

Figure 1. LVE measures under SAOS rheology used in asphalt binders.

amplitude oscillatory shear rheology (LAOS). Under this protocol, the asphalt binder is subject to strain or stress levels beyond LVE and the complete waveform response is measured. In order to have a meaningful interpretation of the nonlinear response, different analysis methods are employed. Hyun et al. (2011) has compiled the state-of-the-art of LAOS rheology and further details of the testing protocols and analysis methods used can be found there. In this study, the use of stress decomposition method to understand nonlinear binder response under large strains is discussed.

3.2 Stress decomposition method

The stress decomposition method for nonlinear response draws its inspiration from the viscoelastic representation of linear response. In the LVE region, for an input strain, as shown in Equation (1), the output stress response can be represented as shown in Equation (2),

$$\gamma(\omega, t) = \gamma_o \sin(\omega t) \tag{1}$$

$$\sigma(\omega, t) = \sigma_o \sin(\omega t + \delta) \tag{2}$$

or

$$\sigma(\omega, t) = \sigma_o \sin(\omega t)cos(\delta) + \sigma_o \cos(\omega t) \sin(\delta) \tag{3}$$

Using (1) and its time derivative (strain-rate) in (3), we get

$$\sigma(\omega, t) = \frac{\sigma_o}{\gamma_o} \cos(\delta)\gamma(\omega, t) + \frac{\sigma_o}{\gamma_o} \sin(\delta)\frac{\gamma(\omega, t)}{\omega} \tag{4}$$

Equation (4) can finally be written in terms of storage and loss modulus (see Figure 1) as

$$\sigma(\omega, t) = G'(\omega)\gamma(\omega, t) + G''(\omega)\frac{\gamma(\omega, t)}{\omega} \quad (5)$$

where G' and G'' are the storage and loss modulus respectively. The first term of Equation (5) represents the elastic part while the second term represents the viscous part. Using the notation, $x = \gamma$ and $y = \gamma/\omega$, Equation (5) can be re-written as

$$\sigma(x, y) = \sigma'(x, y_o) + \sigma''(y, y_o) \quad (6)$$

where $\sigma' - G'x(t)$ referred as elastic stress an $\sigma'' = G''y(t)$ referred as viscous stress. Equation (6) allows a clear decomposition of stress in the linear range to elastic and viscous components and provides storage and loss moduli as meaning measures to physically interpret the binder response.

However, in the nonlinear range, for the similar input strain as shown in Equation (1), the output stress response has contributions from higher harmonics and depending upon the intensity of each, the stress signal will be deviate from usual sinusoidal behavior, and can be generally represented as shown in Equation (7)

$$\sigma(t) = \sum_{n=1, odd}^{\infty} \sigma_n(\gamma_o, \omega)\sin\left[n\omega t + \delta_n(\gamma_o, \omega)\right] \quad (7)$$

Equations (7) and (2) are similar except that in addition to input frequency, the nonlinear stress has contributions from higher order frequencies too. Also, in Equation (7), stress is a function of strain amplitude while in Equation (2) stress is independent of strain amplitude. The fundamental challenge of delineating the elastic and viscous contributions and deriving physical meaning from the higher order contributions to the overall stress response. To overcome this, Cho et al. (2005) proposed two axioms based on a geometric interpretation of Lissajous Bowditch plots and noted that these plots are single closed loops as such stress is a continuous and differentiable function of x and y, and secondly, stress is an odd function of both x and y, that is $\sigma(-x, -y) = -\sigma(x, y)$. Using these two axioms and a corollary, $\sigma(-x, y) = -sigma(x, -y)$, stress under LAOS can be written as;

$$\sigma(x, y) = \left[\frac{\sigma(x, y) - \sigma(-x, y)}{2}\right]$$
$$+ \left[\frac{\sigma(x, y) - \sigma(x, -y)}{2}\right] \quad (8)$$

It can be seen that the first term on RHS of Equation (8) is an odd function of x and even function of y (denoted as σ_{OE}), and the second term on RHS of Equation (8) is an even function of x and an odd function of y (denoted as σ_{EO}). It can be shown that the circular integration of σ_{OE} and σ_{EO} with respect to x and y respectively

is zero, thereby showing that stress under LAOS is symmetric with respect to strain and strain rate. Cho et al (2005) used this definition to define elastic and viscous stress in nonlinear range as

$$\sigma' = \sigma_{OE} \quad \text{and} \quad \sigma'' = \sigma_{EO} \quad (9)$$

This nonlinear stress (7) can be written in a similar way as linear stress as shown in (6). Since elastic stress is odd for x and the viscous stress is odd for y, these two can be expressed as shown in (10)

$$\sigma' = \zeta'(x, y_o)x \quad \text{and} \quad \sigma'' = \zeta''(y, y_o)y \quad (10)$$

where ζ' and ζ'' are referred as generalized dynamic moduli. Cho et al. (2005) successfully decompose the stress into elastic and viscous components but the generalized dynamic moduli provide no physical meaning. The problem was simplified from describing shapes of L-B plots (shear stress vs strain or strain-rate) to lines as elastic stress and viscous stress produced single-valued functions in terms of strain and strain-rates respectively. A polynomial fit doesn't yield any information and depends on the order of the polynomial chosen. Ewoldt et al. (2008) proposed the use of Chebyshev polynomials of the first kind as a set of orthogonal basis to characterize these single-valued functions. Using these functions, the elastic and viscous contributions to the output shear stress is written as

$$\sigma'(x) = \gamma_o \sum_{n:odd} e_n(\omega, \gamma_o)T_n(x) \quad (11)$$

$$\sigma''(y) = \dot{\gamma}_o \sum_{n:odd} v_n(\omega, \gamma_o)T_n(y) \quad (12)$$

where $T_n(x)$ is the nth-order Chebyshev polynomial of the first kind, $x = \gamma/\gamma_o$; $y = \dot{\gamma}/\dot{\gamma}_o$ to ensure a domain of $[-1, +1]$ as these polynomials are orthogonal over a finite domain $[-1, +1]$. Also, $e_n(\omega, \gamma_o$ and $v_n(\omega, \gamma_o$ referred as the elastic and viscous Chebyshev coefficients respectively which reduce to storage and loss modulus in the linear regime as shown below

$$e_1 \to G'; v_1 \to \eta' = G''/\omega \forall e_3/e_1 \ll 1 \& v_3/v_1 \ll 1 \quad (13)$$

An advantage of these polynomials is that these are unique, orthogonal, provide a physical interpretation and deduce to LVE measures in the linear regime. Ewoldt et al. (2008) interpreted these elastic and viscous Chebyshev coefficients in terms of intracycle viscoelasticity using their algebraic sign. The physical interpretations are given Table 1.

4 RESULTS AND DISCUSSION

4.1 Elastic and viscous contributions

In this study, a terminally blended crumb rubber modified asphalt binder was tested at 0.5 Hz frequency

Table 1. Chebyshev elastic and viscous coefficients.

Elastic	$e_3 > 0$	Strain stiffening
	$e_3 < 0$	Strain softening
Viscous	$v_3 > 0$	Shear thickening
	$v_3 < 0$	Shear thinning

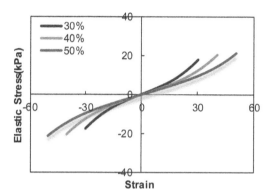

Figure 2. Elastic contribution as a function of strain.

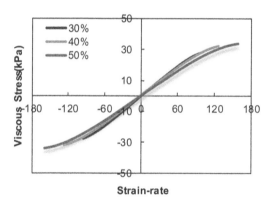

Figure 3. Viscous contribution as a function of strain-rate.

and a testing temperature of 30°C under large strains of 30%, 40%, and 50%. The entire stress response was recorded as a waveform and decomposed using the stress decomposition methodology described in Section 3.2. Firstly, it was observed that the stress response is non-sinusoidal, hence the classified as non-linear. The LVE measures lose their meaning in this region and thus, cannot be used to characterize the asphalt binders showing nonlinear behavior under large strains. Secondly, the elastic stress and viscous stress were obtained as a single-valued function of strain and strain-rate, and are plotted in Figure 2 and Figure 3 respectively.

It can be seen that these functions are non-linear and clearly deviate from LVE behavior. Further, the degree of nonlinearity increases with increasing amplitude indicating significant nonlinearity may exist under large strains, especially in case of modified asphalt binders.

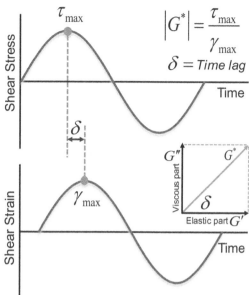

Figure 4. Third-order Chebyshev coefficients.

4.2 Chebyshev polynomial coefficients

Using the Chebyshev polynomial representation of elastic and viscous contributions to the total stress proposed by Ewoldt et al. (2008), the third-order Chebyshev coefficients were obtained and are plotted in Figure 4. It can be seen that all elastic third-order Chebyshev coefficients are have a positive sign while all viscous third-order Chebyshev coefficients have a negative sign. Interpretation it as per Table 1, it can be said that the elastic contribution is strain stiffening and viscous contribution is shear thinning at the applied strains of 30%,40% and 50% under the testing conditions.

5 CONCLUSION

In this study, the crumb rubber asphalt binder has been tested under large strains and the measured stress response is decomposed using orthogonal stress decomposition method. Further, elastic and viscous contributions were delineated, and Chebyshev coefficients were used for physical interpretations. It was found that the asphalt binder shows strain stiffening and shear thinning behavior under large strains. In this exploratory study, evidence for the presence of nonlinearity in the binder response has been reported and the need to characterize nonlinear asphalt binder response is further reiterated.

REFERENCES

Cho, K.S., Hyun, K., Ahn, K.H. and Lee, S.J., 2005. A geometrical interpretation of large amplitude oscillatory shear response. *Journal of rheology*, 49(3), pp.747-758.

Ewoldt, R.H., Hosoi, A.E. and McKinley, G.H. 2008. New measures for characterizing nonlinear viscoelasticity in large amplitude oscillatory shear. *Journal of Rheology*, 52(6), pp.1427–1458.

Gibson, N., Qi, X., Shenoy, A., Al-Khateeb, G., Kutay, M.E., Andriescu, A., Stuart, K., Youtcheff, J. and Harman, T. 2012. *Performance testing for superpave and structural validation.* (No. FHWA-HRT-11-045). United States. Federal Highway Administration.

Gulzar,S, Underwood, S. 2019a. Use of polymer nanocomposites in asphalt binder modification.In *Advanced Functional Textiles and Polymers* (Shahid-ul-Islam, B.S. Butola, ed.), John Wiley-Scrivener USA, pp. 405-431.

Gulzar,S, Underwood, S. 2019b. Nonlinear rheological behavior of asphalt binders.In *91st Annual Meeting of The Society of Rheology*, Raleigh, USA.

Gulzar, S, Underwood, S. 2020a. Nonlinear viscoelastic response of crumb rubber modified asphalt binder under large strains. *Transportation Research Record*, p.0361198120907097.

Gulzar, S, Underwood, S. 2020b. Fourier transform rheology of asphalt binder. Advances in Materials and Pavement Performance Prediction, May 2020, USA.

Gulzar, S, Underwood, S. 2020c. Large amplitude oscillatory shear of modified asphalt binder. Advances in Materials and Pavement Performance Prediction, May 2020, USA.

Hyun, K., Wilhelm, M., Klein, C.O., Cho, K.S., Nam, J.G., Ahn, K.H., Lee, S.J., Ewoldt, R.H. and McKinley, G.H. 2011. A review of nonlinear oscillatory shear tests: Analysis and application of large amplitude oscillatory shear (LAOS). *Progress in Polymer Science*: 36(12), pp.1697–1753.

Advances in Materials and Pavement Performance Prediction II – Kumar et al. (eds)
© 2021 Taylor & Francis Group, London, ISBN 978-0-367-46169-0

Effect of ageing on the molecular structure of bitumen

M. Guo, M.C. Liang & Y.B. Jiao
Beijing University of Technology, Beijing, China

Ye Fu
Beijing Technology and Business University, Beijing, China

ABSTRACT: Bitumen imparts most of its properties to the asphalt mixture. The chemical composition and molecular structure are the key factors affecting the ageing of bitumen. The objective of this study is to construct the average molecular structure of bitumen and study the effect of ageing on its micro characteristics. To achieve this goal, the chemical characteristics of bitumen was analyzed by Fourier Transform Infrared Spectroscopy (FTIR), Nuclear Magnetic Resonance (NMR), Gel Permeation Chromatography (GPC) and Elemental Analysis. Then the average molecular structure parameters were calculated by the improved B-L method. Results show that bitumen is a complex compound composed of aromatic rings, naphthenic and alkyl chains. After ageing, the molecular mass and polydispersity increased obviously. The oxygen reacted with bitumen to form polar functional groups. The aromatization and ring scission may occur during ageing. There were more aromatic rings, less naphthenic rings and the ring structure became more complicated.

1 INTRODUCTION AND BACKGROUND

Although bitumen accounts for only about 5% of the mass of asphalt mixtures, it has a significant impact on the performance of the pavement. (Zaumanis et al. 2014). Ageing, hardening and embrittlement of the bitumen lead to the disease (such as fatigue, rutting, cracking, etc.) of the pavement directly. (Eberhardsteiner et al. 2015). Ageing related properties of bitumen depend on the chemical composition of the molecules and especially on the structure of the molecules. (Liu et al. 2015; Weigel & Stephan 2017).

With the development of analytical technology, scholars have studied the ageing phenomenon of bitumen from chemical composition and molecular structure. For example, Menapace & Masad (2016) found that the conversion of light fractions to asphaltenes in ageing process resulted in the physical hardening of bitumen. Herrington et al. (2002) found that the dispersion and average molecular mass increased significantly during the ageing process. It is believed that the reasons of bitumen ageing also included the oxidation and the increase of strong polar compounds (containing carboxyl, carbonyl and other functional groups) (Lamontagne et al. 2001). However, there are no opinions on the reasons of such microscopic changes as the component migration, the process of oxygen action and the synthesis of polar compounds. The build of average molecular structure of bitumen may be a good way to understand the ageing process.

This study attempts to construct the average molecular structure of original and aged bitumen. It helps to understand the micro properties of bitumen and

the intrinsic reasons of bitumen ageing. In addition, the modification, recycling and substitute development of bitumen are highly dependent on its chemical composition and molecular structure.

2 MATERIALS AND TEST METHODS

2.1 Materials

An unmodified 70# bitumen produced in China was used in this study.

2.2 Laboratory ageing

The aged bitumen was conducted using the RTFOT and PAV. It was taken into RTFOT with a temperature of 163°C for 75 min first. Then it was taken into the PAV with a temperature of 100°C and an air pressure of 2.1 MPa for 20 h.

2.3 NMR

The ACSEndTM400 (AVANCE HD III) NMR spectrometer was used to infer the position of the hydrogen on the carbon chains. The deuterium substituted reagent was CDCl$_3$, and the chemical shift was 7.27 ppm.

2.4 FTIR

The SPECTRUM II FTIR with resolution of 0.5cm^{-1}, scanning times of 32 and test range of 4000-500cm^{-1} was used to study the functional group characteristics of bitumen.

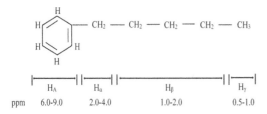

ppm   6.0-9.0   2.0-4.0   1.0-2.0   0.5-1.0

Figure 1.   Classification of ^{1}H-NMR spectral region.

Table 1.   The content of hydrogen in bitumen.

	H$_A$	Hα	H$_\beta$	H$_\gamma$
Bitumen	0.0582	0.1546	0.5863	0.2008
Aged bitumen	0.0642	0.1511	0.5776	0.2070

2.5   GPC

The Waters 515-717-2410 GPC System was used to get the molecular mass of bitumen. The column group consisted of three Waters Styragel (HT6E-HT5-HT3) columns connected in series. The column temperature was 35°, and the elution rate was 1 mL/min. The range of peak molecular weight of polystyrene standard sample (PS) was 162-2300000. The detector was Waters-2410. The mobile phase was THF.

2.6   Elemental analysis

The VarioEL cube Elemental Analyzer was used to analysis the element composition of bitumen. The test sample was wrapped in tin container, fully combusted and decomposed at 1200°.

3   RESULTS AND DISCUSSION

3.1   NMR analysis

Current studies generally classified the types of hydrogen in the spectrogram as H$_A$, H$_\alpha$, H$_\beta$, H$_\gamma$ according to the different chemical shift. The classification and chemical shift of ^{1}H-NMR spectral region were shown in Figure 1.

According to the integral curve of H$_A$, H$_\alpha$, H$_\beta$ and H$_\gamma$, the content of them was obtained. After normalization, it was shown in Table 1.

Among the four kinds of hydrogen, the content of H$_\beta$ was the largest, accounts for 50% ~ 60%. After ageing, the content of H$_A$ and H$_\gamma$ increased. The content of H$_\alpha$ and H$_\beta$ decreased a little. It indicated that the aromatic rings may gradually increase during ageing. However, the alkyl chains connected with the aromatic ring broken, and the number of branches may increase.

3.2   FTIR analysis

The results of FTIR of bitumen were shown in Figure 2. There was absorption peak at the wave numbers of 1600cm^{-1}, which was characterized by the C = C skeleton vibration of aromatic rings. It showed that

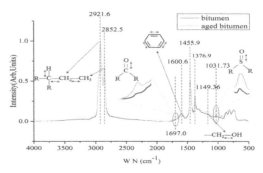

Figure 2.   FTIR spectra of bitumen.

there were aromatic rings in bitumen. The absorption peaks at 2852–2958cm^{-1} of bitumen indicated that there were alkane C-H, C-CH$_3$, C-CH$_2$-.

After ageing, there was a considerable difference between the two replicates at the carbonyl (C=O) and sulfoxide (S=O) functional groups (peaks at 1700 cm^{-1} and 1031 cm^{-1}). The changes in the carbonyl and sulfoxide area were often used to characterize ageing (Hofko et al. 2018). The carbonyl area is commonly defined around 1700 cm^{-1} peak and the sulfoxide area is around 1031 cm^{-1}. The aliphatic group (around 1460 and 1376 cm^{-1}) is commonly used as a reference group, since it is not affected by ageing. The carbonyl and sulfoxide indices were obtained as shown below:

$$CI = \frac{CA}{Cref.} \tag{1}$$

$$SI = \frac{SA}{Sref.} \tag{2}$$

where CI = carbonyl indices; CA = carbonyl area; $Cref.$ = $Sref.$ = reference area; SI = sulfoxide indices; SA = sulfoxide area.

After ageing, the CI changed from 0.0087 to 0.0183. The SI changed from 0.0349 to 0.1962. It meant that the oxygen reacted with bitumen to form polar functional groups such as carbonyl group and sulfoxide group, during the ageing process. And in the reaction, there were more sulfoxide group than carbonyl group. The SI may be a better estimate of the ageing.

3.3   GPC analysis

According to GPC, different average molecular mass can be obtained, such as number average molecular mass (M$_n$), weight average molecular mass (M$_w$) and Z-average molecular mass (M$_z$). As shown in Table 2, after ageing, the molecular mass increased obviously, the distribution index (M$_w$/M$_n$) was larger. The molecular structure became more complex.

3.4   Elemental analysis

The elements composition of bitumen was shown in Table 3. The carbon and hydrogen were main elements

Table 2. The molecular mass of bitumen.

	M_n	M_w	M_z	M_w/M_n
Bitumen	726	2332	7474	3.2121
Aged bitumen	804	3391	11360	4.2150

Table 3. The elements composition of bitumen.

	N W/%	C W/%	H W/%	S W/%	O W/%	H/C
Bitumen	1.24	85.73	9.038	2.964	1.028	1.27
Aged bitumen	1.04	85.98	9.005	2.055	1.92	1.26

Table 4. The average molecular formula of bitumen.

	Average Molecular Formula
Bitumen	$C_{166.60}H_{210.77}S_{2.16}N_{2.07}O_{1.50}$
Aged bitumen	$C_{242.97}H_{305.36}S_{2.18}N_{2.52}O_{4.07}$

Table 5. The number of different types of hydrogen in bitumen.

H_T	H_A	H_α	H_β	H_γ
Bitumen	210.77	13.13	32.08	121.93	43.65
Aged bitumen	96.49	1.84	5.76	63.57	25.32

Table 6. Symbols and calculation formulas of average structural parameters.

Symbol	Calculation formulas
f_A	$f_A = [C_T/H_T - (H_\alpha + H_\beta + H_\gamma)/(2H_T)]/(C_T/H_T)$
σ	$\sigma = C_\alpha/C_P = (H_\alpha/2)/(H_A + H_\alpha/2)$
H_{Au}/C_A	$H_{Au}/C_A = (H_A + H_\alpha/2)/C_A$
C_A	$C_A = f_A C_T$
R_A	$R_A = (C_A - 4)/3$ (Peri condensed aromatic rings)
R_T	$R_T = C_T + 1 - H_T/2 - C_A/2$
R_N	$R_N = R_T - R_A$
C_N	$C_N = 3R_N$ (Peri condensed aromatic rings)
C_S	$C_S = C_T - C_A$
C_P	$C_P = C_S - C_N = C_T - C_A - C_N$
f_N	$f_N = C_N/C_T$
f_P	$f_P = C_P/C_T = 1 - f_A - f_N$
f_S	$f_S = C_S/C_T$
C_A*	$C_A* = (2.503/H_{Au}/C_A)^2$ (Peri condensed aromatic rings)
n	$n = C_A/C_A*$
L	$L = C_P/(H_\gamma/3)$

in bitumen, accounting for about 95%. After ageing, the content of oxygen increased obviously, the content of carbon increased slightly, and the content of nitrogen, sulfur and hydrogen decreased slightly. It showed that the oxygen had a great effect on bitumen ageing. The ageing of bitumen was mainly thermo-oxidative ageing.

3.5 Analysis of average molecular structure

The weight average molecular mass (M_w) of bitumen was measured by GPC, and the content of carbon, hydrogen, oxygen, nitrogen and sulfur was obtained by Element Analysis. The average molecular formula was shown in Table 4.

Based on the hydrogen content of bitumen and the average molecular formula, the total number of hydrogen (H_T) and the total number of carbon (C_T) can be got. The number of different types of hydrogen were shown in Table 5.

The calculation of average structural parameters of bitumen mainly refers to the methods proposed by Brown and Ladner. (Brown & Ladner 1960). The basic assumptions of the calculation are:

1. The average molecules are all hydrocarbon structures;
2. The hydrogen carbon ratio of the saturated hydrocarbon are all 2, that is, $H_\alpha/C_\alpha = (H_\beta+H_\gamma)/(C_\beta+C_\gamma) = 2$;

3. There is no Quaternary Carbon in the average molecule.

The number of aromatic carbons (C_A), is obtained by the aromatic carbon fraction (f_A) and total number of carbons (C_T). The number of units (n) is the ratio of C_A to the aromatic carbon number per unit (C_A*). The H_{AU}/C_A is defined as the aromatic ring condensation degree parameter. Parameter σ represents the carbon substitution ratio around the aromatic ring. The aromatic ring number (R_A), total ring number (R_T), and naphthenic ring number (R_N) are deduced from the aromatic ring structure. The naphthenic carbon number (C_N) divided by the total carbon number gives the naphthenic fraction (f_N). The saturated carbon number (C_S) divided by the total carbon number gives the saturated carbon ratio (f_S). The paraffinic carbons (C_P) divided by the total carbon number gives the paraffinic carbon ratio (f_P). The average chain length (L) is related to the paraffinic carbons (C_P) in the chain. The average structural parameters of bitumen can be calculated by the formulas in Table 6. And the average structural parameters of bitumen were shown in Table 7.

Besides carbon and hydrogen, there are also nitrogen, sulfur and oxygen in bitumen. It is necessary to modify the average structural parameters. In general, it is considered that the sulfur exists as the sulfoxide group and thioethers, the nitrogen exists in the form of pyridine and the oxygen exists in the form of hydroxyl oxygen (-OH), ketone (C=O) in the molecular structure of bitumen. According to FTIR, there are hydroxyl oxygen (-OH) in bitumen before and after ageing. After ageing, the carbonyl (C=O) and sulfoxide (S=O) functional groups appeared.

After modification, the average molecular structure of bitumen was shown in Figure 3.

After ageing, the total ring number and aromatic ring number increased obviously. The naphthenic carbon number decreased. There may be aromatization

Table 7. Average structural parameters of bitumen.

	Bitumen	Aged bitumen
f_A	0.41	0.42
σ	0.55	0.54
H_{Au}/C_A	0.46	0.42
C_A	63.31	102.05
R_A	19.77	32.68
R_T	30.56	40.27
R_N	10.79	7.59
C_N	32.38	22.76
C_S	103.29	140.92
C_P	70.91	118.16
f_N	0.19	0.09
f_S	0.62	0.58
n	2.14	2.87
f_P	0.4	0.49
L	4.87	5.62

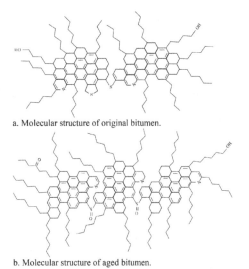

a. Molecular structure of original bitumen.

b. Molecular structure of aged bitumen.

Figure 3. Average molecular structure of bitumen.

of naphthenic rings and alkyl chains. The naphthenic carbon ratio decreased a lot but the aromatic carbon ratio and alkane carbon ratio increased, which showed the broken of naphthenic rings. The alkyl chains may break, and the number of branches increased. In addition, the oxygen reacted with bitumen to form polar functional groups. The thioethers was oxidized to sulfoxides and the hydroxyl oxygen was oxidized to ketone.

4 CONCLUSIONS

The objective of this study was to construct the average molecular structure of bitumen and studied the change between the original and aged bitumen from micro-level. The following is a summary of conclusions that can be drawn based on the aforementioned results and discussion:

1. The average molecular structure of bitumen obtained by this study was basically consistent with the molecular composition measured by FTIR and NMR. It showed that bitumen was a complex compound composed of aromatic rings, naphthenic and alkyl chains.
2. The carbon and hydrogen were the main elements in bitumen, accounting for about 95%. After ageing, the content of oxygen increased obviously and the oxygen reacted with bitumen to form polar functional groups. The oxidation of thioethers to sulfoxides had more increments. The sulfoxide index may be a better estimate of the ageing.
3. After ageing, the molecular mass and polydispersity increased obviously. The aromatization and ring scission may occur during ageing. There were more aromatic rings, little naphthenic rings and the ring structure became more complicated. The alkyl chains connected with the aromatic ring may break, and the number of branches increased.

REFERENCES

Brown, J. K. & Ladner, W. R. A. 1960. Study of the Hydrogen Distribution in Coal-like Materials by High-resolution Nuclear Magnetic Resonance Spectroscopy?——A Comparison with Infrared Measurement and the Conversion to Carbon Structure. *Fuel.* 39:87-96.

Eberhardsteiner. et al. 2015. Towards a microstructural model of bitumen ageing behavior. *International Journal of Pavement Engineering.* 16(10): 939–949.

Herrington, P. R. et al. 2002. Oxidation of Roading Asphalts. *Ind Eng Chem Res.* 33(11): 2801–2809.

Hofko, B. et al. 2018. FTIR spectral analysis of bituminous binders: reproducibility and impact of ageing temperature. Mater Struct. 51(2):45.

Lamontagne, J. et al. 2001. Comparison by fourier transform infrared (FTIR) spectroscopy of different ageing techniques: application to road bitumen, *Fuel.* 80:483–488

Liu, B. et al. 2015. Nanoscaled Mechanical Properties of Asphalt Binders Caused by Aging. *Proceeding of New Frontiers in Road and Airport Engineering:* 17–24.

Menapace, I. & Masad, E. 2016. Evolution of the microstructure of unmodified and polymer modified asphalt binders with aging in an accelerated weathering tester. *Journal of Microscopy,* 263(3):341–356.

Weigel, S. & Stephan, D. 2017. Modelling of rheological and ageing properties of bitumen based on its chemical structure. *Materials and Structures.* 50(1):83.

Zaumanis, M. et al. 2014. Influence of six rejuvenators on the performance properties of Reclaimed Asphalt Pavement (RAP) binder and 100% recycled asphalt mixtures. *Constr Build Mater:* 71:538–50.

Advances in Materials and Pavement Performance Prediction II – Kumar et al. (eds)
© 2021 Taylor & Francis Group, London, ISBN 978-0-367-46169-0

Importance of triaxial stress state on asphalt binder tensile failure

Ramez Hajj & Amit Bhasin
The University of Texas, Austin, TX, USA

ABSTRACT: The failure stress and strain, as well as ductility, depend greatly on the stress state to which the specimen is subjected. In previous studies of asphalt binders, the dependence of material behavior on temperature, loading rate, and aging has been well-examined. However, the influence of the triaxial state of stress that confines binders between mineral aggregates has not been studied extensively. In this work, a review of other materials indicates the importance of triaxiality on the failure of soft materials. Based on this review, the state of stress in asphalt materials was considered at three scales using a finite element simulation- a full depth asphalt pavement, a dense graded asphalt mixture, and a fine aggregate matrix (FAM) mix. Results indicated that the triaxiality increased as the scale became smaller. Further simulations also indicated that the poker chip test can serve as a "worst-case" scenario test where a very high triaxiality can be replicated in the lab for binder testing.

1 MOTIVATION, BACKGROUND, AND SCOPE

Fatigue cracking is typically thought of as a result of energy dissipated in the viscoelastic asphalt binder at high loading rates later in the pavement life. Because fatigue cracks typically form through the binder or as a result of adhesion issues between binder and aggregate, binder properties are of utmost importance to evaluate the fatigue cracking resistance of a flexible pavement. Over the years, many methods have been used to quantify the fatigue cracking resistance of asphalt binders. In summary, binder testing methods that are currently used in practice for evaluating fatigue and fracture resistance are largely focused on obtaining the linear viscoelastic properties of the bulk material. In addition, the tests that do evaluate actual fatigue and fracture of the binder, rather than linear properties, have not generally considered the state of stress that the binder is subjected to in a mix.

1.1 *Importance of asphalt binder stress state*

Asphalt binder researchers have largely focused on testing the binder in appropriate and various temperatures, loading rates, and strain levels, but has not often considered the state of stress that the binder experiences. However, a large number of researchers in other fields have successfully identified this problem of stress state and worked to solve it. Many of the previous studies on this topic considered particulate polymers (Kody & Lesser 1997; Wang et al. 2002) or composites (Asp et al. 1995). Among various materials, one finding that was consistent among a wide range of studies is that the stress state that a material experiences, whether artificially prescribed or observed naturally in a composite, is critical in determining the failure mechanism

and failure strain (Asp et al. 1995; ;ishi et al. 1998; Kody and Lesser 1997; Wang et al. 2002).

Asp et al. (1995) examined the state of stress in epoxy composites to better understand a phenomenon that had been observed in the material- namely that failure strains for the composite when transversely loaded were lower than those of the bulk material in uniaxial tension. A similar phenomenon is observed for asphalt binders, where the failure strain of the mix is much lower than that of the bulk binder at similar temperature and loading conditions. To understand the stress state inherent in their problem, Asp et al. used finite element modeling to examine the stress state at the most critical points in the composite. The metric used to quantify the stress state was simply to determine the triaxial stress state by comparing the stress components in the x, y, and z directions. They found a worst-case scenario of stresses 1:1:2, respectively, based on the highest local stress observed in the epoxy. One conclusion was that the poker chip test could be used to recreate this stress state in the lab by specifying a certain aspect (diameter/thickness) ratio, in this case 7.5.

Kishi et al. (1998) observed similar phenomena in epoxy resins. Their findings indicated, like other studies, that the deformation behavior of the material is dependent on its state of stress. In fact, they argued, the term ductility is only valid if a stress state is specified. Based on the observations in this study, the stress states were classified as uniaxial (in which high stresses on the material only exist in the direction of loading) or triaxial (where high stresses occur in all three directions, not only the loading direction). Uniaxial stresses were observed to be associated with high plastic strain before failure, or a ductile response. Triaxial stress states led to less plastic strain before their brittle failure. A relationship between high hydrostatic stresses and more brittle failure was also observed by Kody & Lesser (1997). This is similar to existing theory on soft solids in which the failure mode transitions from fracture to flow based on the film thickness (Shull and Creton 2004; Nase et al. 2008), which has also been theorized to apply to asphalt binders (Hajj et al. 2019).

A number of studies have used the term triaxiality to define the stress state and quantify its confinement (e.g. Wang et al. 2002). The triaxiality factor can be defined in many ways. Manjoine (1982) defined a triaxiality factor as, essentially, the sum of the principal stresses divided by the octahedral shear stress (Equation 1), which is a common method of denoting it in the literature. Note that this factor is essentially a statement of hydrostatic divided by von Mises stress. Multiple studies showed that with an increase in this factor, ductility decreased (Manjoine 1982).

$$TF = \frac{\sqrt{2}(\sigma_1 + \sigma_2 + \sigma_3)}{\sqrt{(\sigma_1 - \sigma_2)^2 + (\sigma_2 - \sigma_3)^2 + (\sigma_3 - \sigma_1)^2}} \quad (1)$$

Wang et al. (2002) used a similar approach to quantify triaxiality, which involved dividing the hydrostatic stress by the von Mises stress equivalent. Although it is outside the scope of this paper, it is very important to consider the effects of the stress state on binder cavitation based on a recent study indicating the cavitation mechanism that underlies asphalt binder fracture (Hajj et al. 2019).

1.2 Previous studies on asphalt mixture stresses using tomography and computational tools

Previous works have used finite element analysis to investigate the stresses that exist in asphalt mixtures. Some of the earliest studies regarding this issue (Bahia et al. 1999) used a digital scanner to obtain cross-sectional images of asphalt mixtures. These images were then meshed and analyzed in two dimensions using ABAQUS commercial finite element software. They found very high localized strains in the binder, roughly 85 times those found in the bulk mixture.

In later studies, x-ray computed tomography (CT) was introduced as a way to capture the microstructure of a mix by scanning 3D specimens of asphalt mixtures and generating a stack of 2D cross sectional images of the mix. In these studies, the images were subsequently stacked to form a 3D microstructure for use in computational modeling (e.g. Onifade et al. 2013). However, previous studies have mostly limited their scope to segmenting the asphalt mix into two fractions-coarse aggregates with a minimum size ranging from 1-3 mm, and asphalt mortar, a viscoelastic material which includes the binder and fine aggregates, or smaller than the minimum coarse aggregate size. In these types of models, mortar describes the composite of fine aggregate and binder that is typically considered to be a continuum of material with a singular material property.

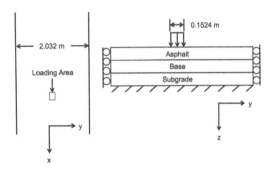

Figure 1. Schematic of virtual pavement structure and loading (adopted from Huang et al. 2011).

2 METHODS AND MATERIALS

Laboratory specimens were created to evaluate the mixes at two scales, a full mix dense graded specimen and a fine aggregate matrix (FAM) specimen. The FAM specimen is a composite of the binder and the fine fractions of aggregate less than 1.18 mm in size. Both mix designs including gradation and optimum binder content were based on commonly used designs for the state of Texas.

After compacting, cutting, and coring these specimens, it was necessary to scan them using x-ray CT to obtain images of the microstructure of the materials. A detailed description of the CT method has been provided by Ketcham and Carlson (2001), and a brief description of the method used is provided below. High-resolution x-ray CT directs a fan beam at a material specimen from all directions in a single two-dimensional (2D) plane. Detectors on the opposite side of the source measure the intensity of x-rays passing through the specimen, which is determined by the density of the materials that are passed through. When done from all angles, this can provide a 2D "slice" that consists of a density map of the phases present, with more dense materials appearing lighter in color in the final image. This is performed at various heights, with a known distance between slices, to generate a three-dimensional stack of slices of the material microstructure. For the laboratory specimens described above, the middle third of each specimen was captured via CT to avoid any artifacts of compaction near the ends of the specimen.

3 MULTISCALE FINITE ELEMENT ANALYSIS

3.1 Flexible pavement stress analysis

A three-dimensional model of a flexible pavement was developed to better understand stresses at the largest length scale possible. The pavement structure and loading geometry were adopted from a previous study by Huang et al. (2010) (Figure 1).

Elastic material properties were assigned to the 0.1524 m layers of the pavement model to consider

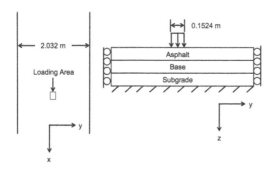

Figure 2.   Third principal stress in pavement layer.

a single snapshot in time. Since exact magnitudes and time dependent responses were not required, this simplification was deemed appropriate.

Using finite element (FE) analysis, the stresses in the asphalt pavement layer upon application of the load were examined. Based on the principal stresses, it was found that the asphalt layer remained in compression throughout the layer in the z-direction, but the bottom of the layer experienced high tension in the other two directions under the tire load. An example of one such analysis is shown in Figure 2. Based on these stresses, hydrostatic and von Mises stresses were computed, which resulted in triaxiality measurements at various points in the structure. The highest tensile triaxiality was observed at the bottom of the pavement layer, with a value of roughly 0.25.

3.2   Full mix and FAM finite element analysis

As mentioned above, CT scans were obtained of a full dense graded mix for use in constructing 3D FE models to examine the stress state of binder in the mix. The 2D slices were obtained from the test and stacked to form a 3D model, which was segmented considering two material phases- the coarse aggregates (larger than 1.18 mm) and the mortar or FAM, which was considered to be a continuum of material comprised of binder and fine aggregate. Based on the input parameters, the size of each voxel of material was 50.3 μm.

A single side of the specimen was held fixed, and loading was applied in a manner consistent with the principal stress findings in the above pavement section. The same ratio of x to y to z stresses as found there was applied.

The results indicated higher stresses at critical points, where two aggregates were in close proximity to each other with aggregate coating film of mortar/binder between them. Images of both scales' simulations can be found in other work (Hajj 2019). High triaxiality was also observed near these interactions, although points of high triaxiality were not exclusive to these locations. Based on a survey of approximately 20 points per slice of the specimen over four slices, the highest triaxiality noted was roughly 1.2. In some cases, local stresses observed were nearly

20 times higher than the bulk stresses applied on the specimen, which validated previous findings.

The FAM specimen indicated similar findings. For this specimen, a similar segmentation and meshing process was undertaken. The material constituents were segmented into two phases- the fine aggregates and the mastic, which was treated as a continuum. The mastic consists of the binder and aggregates smaller than 0.075 mm. Similar to the previous scales, linear elastic properties were assumed. Mastic modulus was 0.2 times of the aggregate modulus, and Poisson's ratio 0.4 was assumed for mastic, while 0.25 for aggregate.

Local stresses were again found to be much higher than the bulk applied stress, in this case as high as 10 times higher than the bulk stresses applied in the simulation. A composite effect between the FAM stresses and stresses in the full mix specimen indicates that local stresses could therefore be up to 200 times the bulk applied stress.

Considering triaxiality also led to findings of a higher confined stress state, in this case yielding a maximum triaxiality value of up to 2.2, compared to 1.2 for the full mix, based on a similar number of sampled points. Based on this rapidly increasing triaxiality, it is determined that it will be beneficial to develop a test which can properly represent the triaxial stress state in order to properly measure the binder's strength and ductility.

4   POKER CHIP TEST FOR MEASURING BINDER STRENGTH AND DUCTILITY

As described above, it is of utmost importance to test the binder's strength and ductility in a realistic state of stress. The above analyses indicated that a triaxial state of stress appropriately represents the stress state of binders confined between aggregates.

The poker chip geometry has previously been used (Hajj et al. 2019; Sultana et al. 2014) to represent the binder's state of stress, with the assumption that the high confinement of this test realistically represents the binder stress state. This test confines a thin film of binder between two steel plates and applies monotonic tension to get the full stress-strain curve of the material, theoretically applying a high hydrostatic stress state throughout. To determine how true this assumption is, an analysis was conducted of the poker chip test at various aspect ratios.

Previous results based on theoretical solutions (Sultana et al. 2014) indicated that the stress state between the two plates was only uniform at high aspect (diameter/thickness) ratios. Simulations of this geometry were conducted of using the FE software. 3D specimens were generated using steel plates bonding soft elastic material which had varying thicknesses that created aspect ratios ranging from 5 to 50. It was observed that the stresses were almost entirely uniform at aspect ratios 40 and higher but were subject to edge effects at lower aspect ratios. Figure 3 shows the first principal stress invariant (related to hydrostatic stress) for

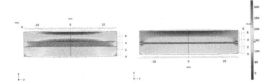

Figure 3. First invariant of stress distribution at (a) aspect ratio = 10 and (b) aspect ratio = 40.

specimens of aspect ratios 10 and 40 respectively to demonstrate this.

Similar results were observed based on the triaxiality of the specimen. A uniform distribution of triaxiality was observed at higher aspect ratios, again above 40. In this simulation, it was necessary to assume a value for Poisson's ratio. However, it should be noted that the Poisson's ratio can have a large effect (magnitudes) on the triaxiality. Therefore, it should be noted that a specific triaxiality cannot be prescribed in the specimen by the geometry unless the Poisson's ratio is known. This is further complicated by the fact that the true Poisson's ratio of asphalt binder is viscoelastic rather than elastic. However, at the typical range of Poisson's ratio for asphalt binder, the triaxiality was found to be generally at least twice as high as those found in the FAM specimen (2.2). Therefore, it is reasonable to consider the poker chip geometry an "extreme case" scenario for triaxial testing of asphalt binders.

5 CONCLUSIONS

At the critical points of high local stress in the mastic, a very high hydrostatic (confining) stress was observed, indicating that the failure mechanism of the binder will likely be in hydrostatic tension rather than pure tension. Previous works have shown hydrostatic tension to result in a significantly lower failure stress for a material. This can also be considered a high triaxiality, which was observed in the specimen to be higher than at larger scales.

The pavement layer's compressive stresses, focused at the top of the layer, are higher in magnitude than the tensile stresses at the most critical point (the bottom of the pavement layer in this case). However, this does not indicate that cracking will not be the most dominant pavement distress, due to both the location of critical points for crack formation (in the binder), and the high magnitude of local stresses at critical points in the mix.

It is desirable for the purpose of material selection and design to know the tensile strength of the material. For asphalt binders, one must focus on the tensile strength as exhibited in a realistic stress state compared to the one experienced in a mix. Based on the findings of this study, the binder should be tested in a confined stress state with high hydrostatic stress. The poker chip test creates this type of stress state.

The use of aspect ratios in a poker chip test should be greater than 40 based on the use of FE analysis to validate previous analytical solutions and experiments. This aspect ratio will guarantee uniform stress distribution and brittle failure.

REFERENCES

Asp, L. E., Berglund, L. A., & Gudmundson, P. (1995). Effects of a composite-like stress state on the fracture of epoxies. *Composites science and technology*, 53(1), 27–37.

Bahia, H. U., Zhai, H., Bonnetti, K., & Kose, S. (1999). Non-linear viscoelastic and fatigue properties of asphalt binders. *Journal of the Association of Asphalt Paving Technologists*, 68, 1–34.

Hajj, R. (2019). Origins and evolution of instabilities in asphalt binders. *Ph.D. Dissertation*.

Hajj, R., Ramm, A., Bhasin, A., & Downer, M. (2019). Real-time microscopic and rheometric observations of strain-driven cavitation instability underlying micro-crack formation in asphalt binders. *International Journal of Pavement Engineering*, 1–13.

Huang, C. W., Abu Al-Rub, R. K., Masad, E. A., & Little, D. N. (2010). Three-dimensional simulations of asphalt pavement permanent deformation using a nonlinear viscoelastic and viscoplastic model. *Journal of Materials in Civil Engineering*, 23(1), 56–68.

Ketcham, R. A., & Carlson, W. D. (2001). Acquisition, optimization and interpretation of X-ray computed tomographic imagery: applications to the geosciences. *Computers & Geosciences*, 27(4), 381–400.

Kishi, H., Shi, Y. B., Huang, J., & Yee, A. F. (1998). Ductility and toughenability study of epoxy resins under multi-axial stress states. *Journal of materials science*, 33(13), 3479–3488.

Kody, R. S., & Lesser, A. J. (1997). Deformation and yield of epoxy networks in constrained states of stress. *Journal of materials science*, 32(21), 5637–5643.

Manjoine, M. J. (1982). Creep-Rupture Behavior of Weldments. *Welding J.*, 61(2), 50.

Nase, J., Lindner, A., & Creton, C. (2008). Pattern formation during deformation of a confined viscoelastic layer: From a viscous liquid to a soft elastic solid. *Physical review letters*, 101(7), 074503.

Onifade, I., Jelagin, D., Guarin, A., Birgisson, B., & Kringos, N. (2013). Asphalt internal structure characterization with X-ray computed tomography and digital image processing. In *Multi-scale modeling and characterization of infrastructure materials* (pp. 139–158). Springer, Dordrecht.

Shull, K. R., & Creton, C. (2004). Deformation behavior of thin, compliant layers under tensile loading conditions. *Journal of Polymer Science Part B: Polymer Physics*, 42(22), 4023–4043.

Sultana, S., Bhasin, A., & Liechti, K. M. (2014). Rate and confinement effects on the tensile strength of asphalt binder. *Construction and Building Materials*, 53, 604–611.

Wang, T., Kishimoto, K., & Notomi, M. (2002). Effect of triaxial stress constraint on the deformation and fracture of polymers. *Acta Mechanica Sinica*, 18(5), 480.

Advances in Materials and Pavement Performance Prediction II – Kumar et al. (eds)
© 2021 Taylor & Francis Group, London, ISBN 978-0-367-46169-0

Micro-scale observations of fatigue damage mechanism in asphalt binder

Ramez Hajj
Department of Civil and Environmental Engineering, University of Illinois, Urbana-Champaign, IL, USA

Kiran Mohanraj & Amit Bhasin
Department of Civil, Environmental, and Architectural Engineering, University of Texas, Austin, TX, USA

Adam Ramm & Michael Downer
Department of Physics, University of Texas, Austin, TX, USA

ABSTRACT: The fatigue cracking mechanism of asphalt binders in asphalt mixtures is often described using mathematical modeling and theoretical diagrams but has yet to be observed using optical microscopy techniques. We use dark field optical microscopy to observe the bulk microstructure of asphalt binder subjected to high amplitude cyclic loading. This preliminary observation informs that once cavitation begins in asphalt binder, the cavities serve as nucleation sites for microcracks to form. In the first stage of material failure, the cavities grow until they collapse and form macrocracks connecting individual cavities, which serves as stage two of failure. Finally, the cavities are joined by microcracks to form a macrocrack in the specimen. As this process happens, bulk microstructures are pushed between the existing cracks to form high density areas of structure in the undamaged region.

1 INTRODUCTION

Fatigue cracking is a critical distress in flexible pavements that often leads to premature failure of a pavement structure. Fatigue cracking is thought to be the most common mechanism of load-related cracking in pavements, caused by repeated traffic loading on the pavement surface. Asphalt pavements are constructed of two main constituents-mineral aggregates and asphalt binder, which acts as an adhesive to bond the aggregates. Fatigue cracking is believed to typically originate in the binder of a mix, due to high local stresses and the binder having a lower tensile strength than the aggregates. Therefore, it is of utmost importance to characterize both the fatigue behavior of a binder and its ability to resist fatigue cracking before use in the field. The following section provides a review of previous works that have focused on this area.

1.1 *Previous studies on binder fatigue cracking*

Asphalt binders are complex, thermoviscoelastic materials, which can vary greatly in composition and physical properties based on their source and even when they are sampled from the same source. Currently, a binder's resistance to fatigue cracking is classified in the United States based on linear viscoelastic properties over a small number of loading cycles. However, in the last 20 years, researchers have proposed to use alternative methods to evaluate a

binder's ability to resist fatigue cracks. Although a thorough review of these attempts can be found elsewhere (Hajj & Bhasin 2018) the following paragraphs serve as a description of some of these works. The time sweep test was developed for use as an actual fatigue test of asphalt binders (Bahia et al. 2001) in the Dynamic Shear Rheometer (DSR), where the number of cycles to failure could be used as a metric to describe a binder's resistance to fatigue damage. This approach was also used to study the self-healing behavior that is inherent in asphalt binders (Stimilli et al. 2012). However, this test often took too long to perform, and also suffered from issues related to edge instabilities (Anderson et al. 2001) caused by nonuniform stress distribution in the DSR parallel plate geometry. Shan et al. (Shan et al. 2011) also noted that thixotropy must be considered in the observed reduction of modulus in this test. The timing issues with regard to the test were later addressed by the implementation of the Linear Amplitude Sweep (LAS) test (Wen & Bahia 2009). However, this test still suffered from some similar limitations in terms of specimen geometry to the time sweep test. The test was later reconfigured into the Binder Yield Energy Test (BYET), but at that point was considered by the authors to be more of a "damage tolerance" test rather than a true fatigue test (Hintz 2012).

To compensate for the specimen geometry issue, other researchers investigated different geometries that could be used to perform time sweep and amplitude sweep tests. Time sweep tests were performed in alternate geometries including the cone and plate (Motamed & Bahia 2011) and the Annular Shear Rheometer (ASR) (van Rompu et al. 2012).

1.2 Previous observations of binder microstructure

In recent years, understanding the microstructure of asphalt binders has become an area of increased interest. With regard to cracking, which is assumed to originate in the binder due to a lower tensile strength than aggregates, it is crucial to understand the material's microstructure and its relationship to damage. Loeber et al. (1996) were among the first to use Atomic Force Microscopy (AFM) to observe the binder's microstructure. Their observations included a structural formation at the surface of the binder which was called by the authors a "bee" structure, due to the fact that its shape is similar to that of the body of a bumblebee. These structures, which were a few micrometers in size, were originally believed by the authors who discovered them to be related to the asphaltene phase of the binder, which is the stiffest polar fraction. They proposed that the classic colloidal model of bitumen may be reinforced by these findings. Many subsequent studies have continued using AFM as a tool to observe bee structures and their relationship to chemical and physical properties of the binder (e.g. Masson et al. 2006).

However, AFM has some significant drawbacks when used for observing asphalt binder microstructure. For one thing, AFM can only be used at a specific temperature range; it is difficult to observe a very wide range of temperatures using this technique. In addition, the specimen preparation is often cumbersome. Most importantly, the AFM technique is only effective for observing structures that exist at the surface of the material. Therefore, researchers have pursued a method to better understand the structure of the material beyond the surface.

A dark field microscopy technique that utilizes scattered light from inclusions in the matrix of the material was applied (Ramm et al. 2016). This technique led to the observation of smaller microstructures in the bulk of the material, which were termed "ant" structures due to their smaller size than the "bee" structures.

With the newly found ability to observe bulk microstructures in the binder available, researchers have explored further applications of this method to explore the fundamental mechanisms of damage in the binder. A recent study (Hajj et al. 2019) observed the initiations of damage in the binder using this technique combined with the application of monotonic strain to the binder. The observations of this study indicated that in the case of monotonic loading, damage first appears in the form of cavitation at the point of a single microstructural feature in the bulk. However, this study contradicted studies in the past which postulated that damage would initiate near each microstructural feature, and then agglomerate to form cracks.

Rather, it appears that damage initiates at a single microstructural feature locally and pushes others as the dissipated energy grows the defect. The study by Hajj et al. (2019) also indicated that the damage observed in the binder matched strongly with the stress-strain curve of the material under similar testing conditions. Finally, a theoretical framework was presented for understanding the initiation of cavitation, which suggests that the "ant" structures act as stiff inclusions in a soft bulk matrix.

1.3 Scope of this paper

In summary, there is a lack, in the literature, of observation of the initiation and nucleation of direct fatigue damage in a binder. This paper presents recent advances toward understanding the mechanism that underlies fatigue cracking in asphalt binders. The goal of this paper is to study the origins of fatigue damage in the binder and how it grows before it nucleates into a macro-sized crack. This paper presents preliminary findings regarding the fatigue mechanism in asphalt binders.

2 METHODS AND MATERIALS

A Performance Graded (PG) binder was selected for evaluation in this study. The method used for observing the fatigue damage in these binders is based using the dark field optical microscopy technique that was developed in previous works to observe the binder's microstructure at a depth penetrating about 25 ìm into the binder's bulk. This allows for viewing of damage that is happening inside the bulk of the material, rather than at its surface.

The apparatus used for this study is similar to the one used in the previous study by this group (Hajj et al. 2019). A schematic of the apparatus can be found in Hajj et al. (2019), but the basic idea of this setup is that an objective is placed below the specimen, which is fixed on a thick plexiglass slide to a metal table. The asphalt binder specimen is sandwiched between the thick slide and a plexiglass cylinder which is attached to a piezoelectric actuator. Near infrared light is absorbed by the specimen and the inclusions, as well as damage, scatter back light which is captured by the objective to provide a view of the formation and nucleation of instabilities in the binder.

The specimen geometry used in this test was the poker chip geometry, originally developed by Gent and Lindley (1959). The poker chip geometry has been used in previous studies to understand the tensile behavior of the binder in a stress state that represents the actual state of stress a binder experiences in an asphalt mix (Hajj et al. 2019). Previous works (Sultana et al. 2014) have also indicated that aspect ratios (thickness/diameter) above 40 are required in this test to achieve a uniform distribution below the parallel plates confining the specimen.

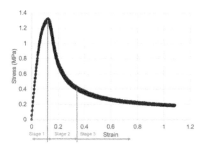

Figure 1. Three stage model for damage by monotonic loading.

Figure 2. Images of fatigue damage after (a) a few cycles, (b) about 9000 cycles, and (c) about 18000 cycles.

3 THREE STAGE DAMAGE MODEL FOR MONOTONIC LOADING

The recently published work from Hajj et al. (2019) examined the failure of bitumen specimens under monotonic loading. Based on the findings from this study, the failure process of the binder under monotonic tension can be classified into three stages. In stage 1, the material remains undamaged, with no visible instabilities in the specimen. Meanwhile, the engineering stress observed in the material continues to increase until the peak load is reached.

In the second stage, the load in the specimen begins to decrease. This results in a decrease in the effective material modulus due to initiation of damage in the material. A sharp decrease in load occurs, which is concurrent with new cavities initiating in the binder specimen. Finally, in the third stage of damage, the load continues to decrease but slows drastically. From the microscopic perspective, this phenomenon is defined by the growth of existing cavities, but without the formation of any new cavities. From the curve, it is observed at about 25% of the peak load. Figure 1 shows the three stages of damage on the stress-strain curve.

It is important to note that previous studies with regard to modeling damage in viscoelastic materials have observed that damage initiates before the peak load is reached. However, the recent study by Hajj et al. (2019) did not observe this. Rather, at very high rates of loading, or nearly glassy behavior, the binder displayed a response similar to plasticity in metals, in which the specimen can deform without observable damage, based on the microscopic observations.

4 MICROSCALE FATIGUE MECHANISM OF ASPHALT BINDERS

While the previous study examined the fracture mechanism of asphalt binders subjected to monotonic tension, it is also critical to consider the response of the material to fatigue or repeated loading. Fatigue loading is particularly important from a pavement materials perspective due to repeated loading of the pavement surface by vehicular traffic. Therefore, experiments focused on fatigue loading were performed on the b inder specimen.

Specimens were loaded using the actuator and microscope system described in the work by Hajj et al. (2019) and in Section 2 of this paper. Two specimens were prepared; in both cases, the aspect ratio varied above 40, although different aspect ratios were used to generate two different loading amplitudes based on the capabilities of the actuator used for loading. Initial testing was performed on both sets of binders with high amplitude (up to 100% strain), high frequency (10 Hz), cyclic loading applied. However, as this loading was applied, it was noted that no visible damage appeared in the specimen. Although at first this may seem counterintuitive, it is important to consider that the rate of loading is very high compared to the previous experiments by Hajj et al. (2019) and described in this paper. Note that at a loading rate of 5 im/s, Hajj et al. (2019) observed damage initiation only at loading rates of 75% strain amplitude or higher. In this section, the rates are magnitudes higher than this, so it is not unexpected to observe no visible damage in the specimen.

Therefore, to induce damage in the specimen, cavities were induced in the specimen using a monotonic, slower rate of loading similar to the previous study. This can be thought of as similar to creating a notch in a specimen, as is traditionally done for fatigue testing of materials.

The first specimen was prepared at an aspect ratio (diameter/thickness) of 70. Cavitation was induced in the specimen, then the specimen was allowed to recover for a few minutes. At this point, the cavity induced was very small and barely visible. Then, fatigue loading began at an amplitude of 50% strain and 10 Hz loading. Initiating cavitation using this method resulted in a rapid growth in the size of the cavities over a short number of cycles. After many cycles, the cavities stopped scattering as much as previously, and cracks began growing from existing cavities. These cracks grew toward cracks growing from other cavities, which provides some insight into the mechanism by which cracks form in asphalt pavement structures- first by cavitation, then microcracking near the cavity, then by the cracks connecting to form a macrocrack. Figure 2 shows images of three points over the 30 minutes at which the binder was loaded.

Next, a specimen of aspect ratio 125 was subjected to a similar loading with twice as high an amplitude for a total strain of 100% each cycle at 10 Hz loading frequency. Cavitation was first initiated as described for the previous specimen. At this level of loading,

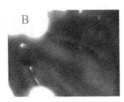

Figure 3. Images of the fatigue damage at higher amplitude after (a) a few cycles and (b) about 30,000 cycles.

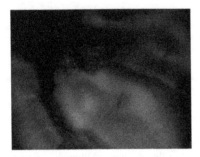

Figure 4. Persistent damage in the specimen from Figure 3 after 3 days at rest.

cracking occurred much more quickly in the specimen, which was tested for 1 hour. It is also important to note that in this specimen, the microstructures were observed to be pushed between cracks, similar to findings in the previous study (Hajj et al. 2019). Images of this test are shown in Figure 3. Finally, the specimen was permitted to rest after loading for three days at room temperature. In this time frame, the cracks did not heal, and were still visible (Figure 4), indicating "permanent" damage.

5 CONCLUSIONS AND FUTURE WORK

This study presents an insight into the mechanism by which cracking occurs in asphalt binders subjected to fatigue loading. The major findings in this study are summarized as follows:

1. Fatigue cracking occurs in asphalt binders from existing instabilities in the binder's microstructure. These cracks occur first by growing cavitation in the binder, then by forming microcracks peripheral to the cavities, and finally by joining these microcracks to form macro-sized cracks.
2. Fatigue damage does not appear to initiate in the binder from a visual perspective at very high loading rates, even at high strain levels, without a point of initiation from low rate loading. However, when low rate loading is initiated, fatigue cracks can grow rapidly from this point. It is critical to note that continuous high rate loading although classically used in fatigue testing, is not a realistic allegory to loading of the pavement
3. The overall mechanism where damage forms in fatigue is not very different from monotonic

loading, except that visible cracks connect instabilities in the binder.
4. Fatigue damage due to many cycles of high amplitude loading does not appear to heal, even after very long rest periods.

A more comprehensive study is currently underway to evaluate more binders and capture load measurements for the binder during testing. This will provide insight into the mechanism observed and its universal applicability to asphalt binders.

REFERENCES

Anderson, D. A., Le Hir, Y. M., Marasteanu, M. O., Planche, J. P., Martin, D., & Gauthier, G. (2001). Evaluation of fatigue criteria for asphalt binders. *Transportation Research Record*, 1766(1), 48–56.

Apostolidis, P., Kasbergen, C., Bhasin, A., Scarpas, A., & Erkens, S. (2018). Study of Asphalt binder fatigue with a new dynamic shear rheometer geometry. Transportation Research Record, 2672(28), 290–300.

Bahia, H. U., Zhai, H., Zeng, M., Hu, Y., & Turner, P. (2001). Development of binder specification parameters based on characterization of damage behavior. *Journal of the Association of Asphalt Paving Technologists*, 70.

Gent, A. N., & Lindley, P. B. (1959). Internal rupture of bonded rubber cylinders in tension. *Proceedings of the Royal Society of London. Series A. Mathematical and Physical Sciences*, 249(1257), 195–205.

Hajj, R., & Bhasin, A. (2018). The search for a measure of fatigue cracking in asphalt binders–a review of different approaches. *International Journal of Pavement Engineering*, 19(3), 205–219.

Hajj, R., Ramm, A., Bhasin, A., & Downer, M. (2019). Real-time microscopic and rheometric observations of strain-driven cavitation instability underlying micro-crack formation in asphalt binders. *International Journal of Pavement Engineering*, 1–13.

Hintz, C. (2012). *Understanding mechanisms leading to asphalt binder fatigue* (Doctoral dissertation, The University of Wisconsin-Madison).

Loeber, L., Sutton, O., Morel, J, V I M Valleton, J. M, & Müller, G. (1996). New direct observations of asphalts and asphalt binders by scanning electron microscopy and atomic force microscopy. *Journal of microscopy*, 182(1), 32–39.

Masson, J. F., Leblond, V., & Margeson, J. (2006). Bitumen morphologies by phase-detection atomic force microscopy. *Journal of Microscopy*, 221(1), 17–29.

Motamed, A., & Bahia, H. U. (2011). Influence of test geometry, temperature, stress level, and loading duration on binder properties measured using DSR. *Journal of Materials in Civil Engineering*, 23(10), 1422–1432.

Ramm, A., Sakib, N., Bhasin, A., & Downer, M. C. (2016). Optical characterization of temperature-and composition-dependent microstructure in asphalt binders. *Journal of microscopy*, 262(3), 216–225.

Shan, L., Tan, Y., Underwood, B. S., & Kim, Y. R. (2011). Separation of thixotropy from fatigue process of asphalt binder. *Transportation Research Record*, 2207(1), 89–98.

Stimilli, A., Hintz, C., Li, Z., Velasquez, R., & Bahia, H. U. (2012). Effect of healing on fatigue law parameters of asphalt binders. *Transportation Research Record*, 2293(1), 96–105.

Sultana, S., Bhasin, A., & Liechti, K. M. (2014). Rate and confinement effects on the tensile strength of asphalt binder. *Construction and Building Materials*, 53, 604–611.

Van Rompu, J., Di Benedetto, H., Buannic, M., Gallet, T., & Ruot, C. (2012). New fatigue test on bituminous binders: Experimental results and modeling. *Construction and Building Materials*, 37, 197–208.

Wen, H., & Bahia, H. (2009). Characterizing fatigue of asphalt binders with viscoelastic continuum damage mechanics. *Transportation Research Record*, 2126(1), 55–62.

Advances in Materials and Pavement Performance Prediction II – Kumar et al. (eds)
© 2021 Taylor & Francis Group, London, ISBN 978-0-367-46169-0

Laboratory and field ageing of asphalt mixtures

R. Jing, A. Varveri, D. van Lent & S. Erkens
Delft University of Technology, Delft, The Netherlands
Netherlands Organization for Applied Scientific Research (TNO), Delft, The Netherlands

ABSTRACT: Ageing of bituminous materials contributes to various forms of pavement failures, thus leading to the degradation of pavement performance. To understand the evolution of pavement performance, the development of a proper laboratory protocol to simulate long-term ageing process of asphalt mixtures is of uppermost importance. In this study, the porous and dense asphalt slabs with thickness of 5 cm were exposed to oven ageing at 85°C for 3 and 6 weeks in the laboratory. Cyclic Indirect Tensile tests were performed to investigate the effect of ageing on the mechanical properties of asphalt mixture. The results were used to correlate with the change in the mechanical properties of the porous and dense pavement in the field. Pavement test sections were constructed in 2014 and have been exposed to actual environmental conditions since then. To study the temporal changes in the mechanical properties of the pavements, asphalt cores were collected from the test sections annually. The results show that porous asphalt has a higher ageing rate than dense asphalt due to its high porosity. Porous asphalt aged at 85°C for 3 and 6 weeks in the laboratory have the same stiffness change as that aged 3 and 3.5 years in the field, respectively. Dense asphalt aged at 85°C for 3 weeks in the laboratory have the same stiffness change as that aged 4 years in the field.

1 INTRODUCTION

The majority of Dutch national highway and provincial road network is paved with a top layer of porous asphalt (PA) and dense asphalt (SMA, stone matrix asphalt), respectively. The performance of these pavements change over time due to ageing (Petersen & Glaser 2011). Though this may benefit the pavement through enhanced rutting resistance, ageing can cause or accelerate several distresses such as fatigue, low temperature cracking and moisture damage (Woo et al. 2008). A proper protocol to simulate the long-term ageing could contribute to the prediction of pavement performance in time and to the development of longer-lasting pavement materials.

The commonly used protocol to simulate long-term ageing process of bituminous materials is pressure ageing vessel (PAV). The test is performed on bitumen films with 3.2 mm thickness at temperatures between 90 and 110°C under pressurized conditions (Lu & Isacsson 2002). Even though the protocol is widely accepted by the pavement industry worldwide, it has been reported that the PAV protocol cannot accurately simulate the long-term ageing condition for porous mixtures (Jing et al. 2019; van Lent et al. 2016). In addition, the performance of bitumen aged in a multiphase system (mixture) is different than ageing bitumen by itself. Ageing susceptibility does not only depend on the physicochemical properties of bitumen, but also depends on the interaction with filler, and mixture morphology which is essentially a result of aggregate packing, porosity, air void distribution and their interconnectivity (Erkens et al. 2016; Huang & Zeng 2007).

The goal of this study is to evaluate the behavior of porous and dense asphalt mixtures ageing both in the laboratory and field. The main objectives are to (i) determine the changes in the mechanical properties of porous and dense asphalt due to laboratory and field ageing by means of Cyclic Indirect Tensile test, (ii) correlate the results of laboratory ageing with those of field ageing.

2 MATERIALS AND AGEING METHODS

2.1 Asphalt mixture design

The same asphalt mixture design was used for mixtures produced in the laboratory and laid in the field test sections for both porous (PA) and dense (SMA) asphalt. Moreover, the laboratory slabs and test sections were produced using the same bitumen, filler and aggregate types. Table 1 shows the specifications of the used materials. The bitumen content was 5.0 % and 6.4 %, and the target air void (AV) content was 20 % and 5 % for the PA and the SMA mixtures, respectively. The gradations of both mixtures are shown in Figure 1.

2.2 Laboratory ageing

Two asphalt slabs (namely one PA and one SMA) were prepared in the laboratory. The asphalt slabs were 5 cm in thickness, and 50 cm in length and width. The

Table 1. Materials specifications.

Name	Type	Properties
Bitumen	PEN 70/100	Penetration at 25°C (dmm) 70-100
		Softening point (°C) 43-51
Filler	Wigro 60K	Density (kg/m3) 2780
		Hydrated lime content (%) 25
Aggregate	Bestone	Density (kg/m3) 2740
		Nominal Max. size (mm) 16

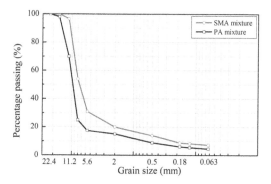

Figure 1. Gradations of PA and SMA mixtures.

asphalt slabs were compacted using a roller compactor to reach the design density and void content. Then the compacted slabs were cut into three smaller slabs (16×50×5 cm). One of these slabs was used as a reference sample at the fresh (unaged) condition. The other two slabs were aged in the oven at 85° C for 3 and 6 weeks, respectively. In order to prevent the slabs from deforming during ageing and simulate the field ageing situation, slabs were wrapped by duct tape on the four sides and the bottom. After laboratory ageing, three core samples (10 cm in diameter and 5 cm in thickness) were drilled from each slab.

2.3 Field ageing

The construction phase of the test sections started with the removal of the existing old pavement surface, which had 10 cm thickness. After the milling process, a bitumen emulsion tack coat layer was sprayed on the surface. Then the new stone asphalt concrete (STAC) layer of 6 cm thickness was laid first as the base layer, and the 5 cm thickness top porous (PA) and dense (SMA) asphalt layers were placed on the left and right lanes separately. The layers were compacted using a roller compactor. The construction of the test sections was done in October 2014. Since then the test sections have been continuously exposed to the environment and three core samples with a diameter of 10 cm and a thickness 5 cm are drilled from the PA and SMA layers every year.

3 EXPERIMENTAL METHODS

To verify that the volumetrics of the laboratory- and field-produced samples are comparable, the internal morphology of the cores (at unaged conditions) was determined by means of X-ray CT scanner. The samples were scanned along the height with a vertical resolution of 0.6 mm. Then the image processing software Simpleware was used to identify and quantify the different phases of the mixtures.

To study the temporal changes in the mechanical properties of the porous and dense asphalt mixture due to oxidative ageing, Cyclic Indirect Tensile tests (IT-CY) were performed according to NEN-EN 12697-26. The dynamic modulus of core samples was determined using the Universal Testing Machine (UTM) at five frequencies (i.e. 0.5, 1, 2, 5 and 10 Hz) and four testing temperature (i.e. 0, 10, 20 and 30° C). Three replicate samples for each ageing condition were tested. Each sample were conditioned at the testing temperature for 4 hours before testing to equilibrate sample's temperature.

4 RESULTS AND DISCUSSION

4.1 Volumetrics composition of the mixtures

On the basis of the CT scan images, the distribution of the voids, mortar and aggregates over the height of the samples (PA and SMA) was determined. Figure 2 shows the results for the laboratory and field samples. Overall, Figure 2 shows that PA has higher air void content as expected and lower mortar content in comparison with SMA. In addition, the laboratory-produced PA sample has slightly higher void content than the PA sample cored from the field. Considering the target air void content of 20%, it appears that the test sections were over-compacted resulting to a lower air void percentage. In contrast, SMA samples from the laboratory and field slight differences in the percentage of each component.

4.2 Mechanical properties of the mixtures

The Time-Temperature Superposition (TTS) principle was used to generate the master curves of the dynamic modulus at a reference temperature of 20 °C. Figure 3 illustrates the evolution of dynamic modulus for PA mixtures due to laboratory and field ageing. The black dashed lines denote the results of the laboratory samples and the colored lines denote the results of the field samples.

Figure 3 shows that that the field sample after production and construction (sample in 2014) is stiffer than the sample after production in the laboratory (sample 0W). This could be attributed to the differences in the air void content between the field PA sample (15% AV) and the laboratory PA sample (20% AV), with the latter being more porous (Figure 2) possible due to the application of higher compaction effort.

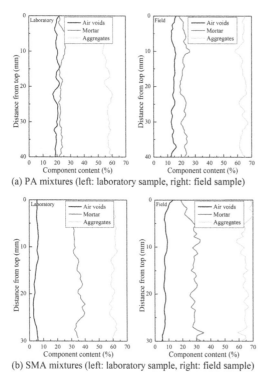

(a) PA mixtures (left: laboratory sample, right: field sample)

(b) SMA mixtures (left: laboratory sample, right: field sample)

Figure 2. Distribution of air voids, mortar and aggregates over the height of the core sample. (mortar is the mix of bitumen, filler and fine sand with particle size less than 2 mm).

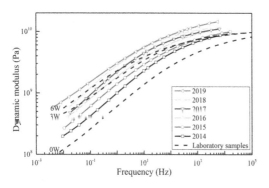

Figure 3. Dynamic modulus of PA samples at different ageing conditions.

Such difference in the air void content would result in differences in the ageing rates of the laboratory-aged and the field-aged mixtures. The laboratory-aged samples would essentially have a higher ageing rate than the field samples based on their volumetrics composition. However, it is still meaningful to correlate the rates of laboratory and field ageing. Considering that the absolute dynamic modulus values of PA samples from laboratory and field are significantly different at fresh condition, the change in the dynamic modulus

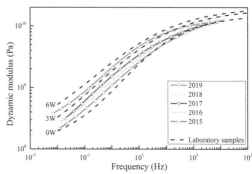

Figure 4. Dynamic modulus of SMA samples at different ageing conditions.

is used to correlate laboratory and field ageing. To be specific, the change in the dynamic modulus at 10 Hz (since it is a commonly used frequency in a dynamic test) and 20° C (room temperature) of PA samples is selected and plotted in Figure 5. The results show that after 3 weeks of ageing the laboratory-produced PA samples have approximately the same stiffness change as the PA cores after 3 years of ageing in the field. Moreover, the PA samples after 6 weeks of laboratory ageing have the same stiffness change as the PA sample that were aged 3.5 years in the field.

Figure 4 demonstrates the evolution of dynamic modulus for SMA mixtures due to laboratory and field ageing. Unfortunately, there were no cores taken from the SMA section right after laying in 2014. Since SMA samples from the laboratory and field have the similar percentage of each component, as shown in Figure 2, it is assumed that the initial stiffness (at fresh state) for the field sample is similar to the laboratory produced sample. Figure 4 shows that 3 weeks of laboratory ageing simulate 4 years of field ageing for the SMA sample. Based on the results, the stiffness levels of the SMA samples after 6 weeks of laboratory ageing exceed the stiffness of 5 years of field ageing.

Finally, on the basis of the results in Figure 3 and 4, the temporal evolution of dynamic modulus (at 10 Hz and 20° C) for both mixtures after field and laboratory ageing were calculated and plotted in Figure 5. The results clearly show that the rate of stiffness change is higher for the PA sample than for SMA samples for both laboratory and field ageing conditions. This is mainly because of the high void content of PA, which leads to an inherently high sensitivity of these mixtures to oxidative ageing. After a certain time, the stiffness of the PA mixtures tends to become stable and that occurs at an earlier time than for the SMA mixture. In addition, the difference in the stiffness changes between two mixtures is higher for field ageing than for the laboratory ageing. This could probably be related to other environmental factors; for instance moisture and ultraviolet radiation can contribute to pavement ageing and you would expect that their effects on the two mixtures to be different due to their distinct morphologies.

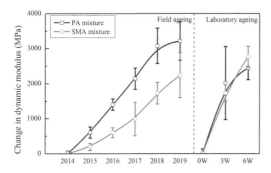

Figure 5. Temporal changes in the dynamic modulus (at 10 Hz and 20° C) of PA and SMA mixtures at different ageing conditions.

5 CONCLUSIONS

To better understand the long-term behavior of asphalt pavements and to select proper materials and structures that can delay potential pavement failure, the development of accelerated ageing protocols is necessary and crucial. In this study, the ageing susceptibility of porous and dense asphalt mixtures was investigated using field aged cores from test sections (constructed in 2014) and laboratory aged cores (oven ageing at 85° C for 3 and 6 weeks) taken from 5 cm thickness asphalt slabs. A series of stiffness tests were conducted on asphalt cores, which were taken from the laboratory-aged slabs and the pavement sections.

The CT scan results show that there is a difference between laboratory and field compaction, especially for the PA mixture. To be specific, the laboratory-produced PA sample has higher void content than the field PA sample, while for the SMA samples slight difference in the voids content were found. This differences in the voids content could higher the application of accelerated ageing protocols for the prediction of the ageing sensitivity of mixtures in the field. To overcome this issue, the ageing rate (namely the temporal change of stiffness due to ageing) was used as the ageing sensitivity index. The results show that change in stiffness for both laboratory and field ageing is greater for the porous than the dense mixture, because of the high porosity of porous asphalt that allows air (oxygen) to easily flow into the mixture. Due to its high ageing rate, the porous mixture is expected to reach a constant stiffness value faster than the dense mixture, as the rate of stiffness increase tends to decrease with time. Overall, the difference of the ageing rates between the two mixtures is larger when the mixtures are aged in the field. This could probably be related to other environmental factors; for instance moisture and ultraviolet radiation can contribute to pavement ageing. From a comparison between the results of laboratory- and

field-aged samples, it can be concluded that porous asphalt aged at 85° C for 3 and 6 weeks in the laboratory has the same stiffness change as the field aged 3 and 3.5 years, respectively. On the other hand, dense asphalt aged at 85° C for 3 weeks in the laboratory has the same stiffness change as the field-aged 4 years. The stiffness of the dense samples after 6 weeks of laboratory ageing would probably exceed the stiffness of samples after 5 years of field ageing.

As a continuation of this research, the core samples will be cut into three slices from top to bottom. Bitumen will be extracted from each slice. Chemical and rheological tests will be performed on the recovered bitumen to investigate the ageing profile of the mixtures. Results of the recovered bitumen from the laboratory- and field-aged samples will be used to verify the relationship between laboratory and field ageing.

ACKNOWLEDGMENTS

The authors gratefully acknowledge the Dutch Ministry of Infrastructure and Water Management (Rijkswaterstaat) for funding this project.

REFERENCES

Erkens, S., Porot, L., Glaser, R., & Glover, J. 2016. Aging of Bitumen and Asphalt Concrete: Comparing State of the Practice and Ongoing Development in the United States and Europe. *Transportation Research Board 95th Annual Meeting*, Washington DC, United States, 10–14 June.

Huang, S., & Zeng, M. 2007. Characterization of Aging Effect on Rheological Properties of Asphalt-filler System. *International Journal of Pavement Engineering* 8(3): 213–223.

Jing, R., Varveri, A., Liu, X., Scarpas, A., & Erkens, S. 2019. Laboratory and Field Ageing Effect on Bitumen Chemistry and Rheology in Porous Asphalt Mixture. *Transportation Research Record* 2673(3): 365–374.

Lu, X., & Isacsson, U. 2002. Effect of Ageing on Bitumen Chemistry and Rheology. *Construction and Building Materials* 16: 15–22.

Petersen, J., & Glaser, R. 2011. Asphalt Oxidation Mechanisms and the Role of Oxidation Products on Age Hardening Revisited. *Road Materials and Pavement Design* 12(4): 795–819.

van Lent, D., Mookhoek, S., van Vliet, D., Giezen, C., & Leegwater, G. 2016. Comparing Field Aging to Artificial Laboratorial Aging of Bituminous Binders for Porous Asphalt Concrete using Black Space Graph Analysis. *6th Eurasphalt & Eurobitumen Congress*, Prague, Czech Republic, 1–3 June.

Woo, W., Chowdhury, A., & Glover, C. 2008. Field Aging of Unmodified Asphalt Binder in Three Texas Long-Term Performance Pavements. *Transportation Research Record* 2051(1): 15–22.

Advances in Materials and Pavement Performance Prediction II – Kumar et al. (eds)
© 2021 Taylor & Francis Group, London, ISBN 978-0-367-46169-0

Asphalt complex modulus and phase angle by equilibrium molecular dynamics

M. Masoori & M.L. Greenfield
Department of Chemical Engineering, University of Rhode Island, Kingston, RI, USA

ABSTRACT: Relating changes in asphalt mechanics to asphalt chemistry would be useful for understanding pavement performance. Towards that end, stress relaxation modulus from prior molecular dynamics simulations of a model asphalt was converted to complex modulus, and random noise was decreased with tools from signal processing. Time-temperature superposition was found for magnitude of complex modulus over 400 to 533 K, while phase angle showed less good superposition. Time shift factors were in a similar range to those from rotation rates of individual molecules. Trends were described by the CAM asphalt rheology model.

1   INTRODUCTION

Rheology serves a predominant role in identifying suitable asphalts for specified environmental conditions. Laboratory experiments and field studies have refined Superpave guidelines (Kennedy et al. 1994) for properties that enable good road performance and are measurable over reasonable times.

Rheology models can describe material effects in characterization experiments. Fundamental rheology models are sometimes applied to asphalt data. This includes fits using a multicomponent Maxwell model, which employs contributions over multiple relaxation times. The Christensen-Anderson-Marasteanu model (CAM; Christensen & Anderson 1992; Christensen et al. 2017; Marasteanu & Anderson 1999) fits experimental data well for complex modulus frequency dependence. It correlates with relaxation time distribution yet is not connected to a direct molecular-scale interpretation. Molecular dynamics (MD) simulations of asphalt use a detailed composition model with explicit representations for molecules and chemical interactions.

It would be desirable if molecular dynamics could lead to predictions of mechanical properties for asphalt pavement design. Then the same systems and methods could model how well-defined chemical changes directly impact mechanics. Such "chemo-mechanics" has potential for understanding impacts of asphalt oxidation, aging, and modification on asphalt properties and pavement performance.

Model asphalt composition has been addressed in prior work. Zhang & Greenfield (2007a) proposed 3-component saturate/naphthene/asphaltene systems. A 6-component system included more diverse chemical structures, but molecular weights were too low. Hansen et al. (2013) proposed a 4-component model with better molecular weight and some chemical diversity. Li & Greenfield (2014a) proposed models representative of SHRP asphalts AAA-1, AAK-1, AAM-1 using 12 components. Differences among asphalts were obtained by composition variations, i.e. how many of each molecule type were present. Predictions for the model AAA-1 system (Li & Greenfield 2014b) were comparable to properties previously reported from experiments on AAA-1. Predictions were made of complex modulus at 260°C (Masoori & Greenfield 2014), and results were reinforced by longer simulations that used a different force field, (Khabaz & Khare 2015, 2018).

Obtaining MD predictions over relevant shear rates poses challenges. Specifications indicate a dynamic shear rheometry (DSR) oscillation frequency of 10 rad/s, representing stress and strain deformation frequencies induced by traffic. Corresponding time scales are ~ 0.1 s. Molecular dynamics simulations proceed with time steps of order 10^{-15} s, and a long simulation can be 10^6 to 10^8 steps. The total times (1 to 100 ns) are much shorter than a single oscillation in DSR. Relating MD and DSR requires time-temperature superposition with larger extrapolations than in bulk rheology, where frequency shifts of a few orders of magnitude are sufficient.

Another constraint on using molecular dynamics to estimate asphalt rheology is inherent noise. The small system size of $\sim 10^5$ atoms, compared to $\sim 10^{24}$ in a bulk sample, leads to larger random fluctuations and noise compared to average forces. While increasing system size and duration of averaging can help to reduce noise, estimates of stress relaxation at the longest achievable time and complex modulus at the lowest attainable frequency still inherently have noise. Previous work (Masoori & Greenfield 2017) demonstrated how numerical tools from signal processing may be applied to identify the underlying signal for complex modulus from within noisy simulation results. Those initial results were demonstrated at only one temperature. This work extends that analysis to multiple temperatures and demonstrates time-temperature superposition for complex modulus results on the nanoscale. An ultimate outcome is a quantified numerical asphalt rheology model for a system that was simulated by molecular dynamics.

2 SIMULATION METHODS

Simulation results were taken from prior calculations (Li & Greenfield 2014a, b). The system contains 12 molecule types of molecular weight 290 to 890 g/mol that are distributed among saturate, naphthene aromatic, polar aromatic, and asphaltene solubility classes. Molecular dynamics simulations of up to 6 ns at 533.15, 443.15, 400.15 K, and lower temperatures were run with LAMMPS2001 software and OPLS force field. Equilibrium densities were reached at constant temperature and pressure (1 atm) using the Nose-Hoover method. Sampling of equilibrium fluctuations was performed at constant volume with a temperature control parameter of 0.1 ps.

Instantaneous stress under zero-strain conditions was calculated using a molecular virial, which sums only forces between molecules. The time correlation function of stress was used to obtain stress relaxation modulus via (Mondello & Grest 1997)

$$G(t) = \frac{V}{10k_B T} \sum_i \sum_j \langle \sigma_{ij}^{s,t}(0)\sigma_{ij}^{s,t}(t) \rangle \quad (1)$$

as in earlier work (Zhang & Greenfield 2007b, Li & Greenfield 2014b). V is volume; $k_B T$ is Boltzmann constant and temperature. σ_{ij}^{st} indicates component (i,j) of a symmetric and traceless stress tensor, $\sigma_{ij}^{st} = (\sigma_{ij} + \sigma_{ji})/2 - \Sigma\sigma_{ii}/3$. All possible time origins were used in this average, so "t" is a time difference between two snapshots in the simulation. For a small time difference, the stress is not expected to change much: $\langle \sigma_{ij}^{st}(0)\sigma_{ij}^{st}(t) \rangle$ should be near $\langle \sigma_{ij}^{st2} \rangle$. For a longer time difference, the two stresses become less similar as molecules rearrange. Such molecular motions underlie the time-dependent relaxation of imposed stress during creep and DSR experiments.

The complex modulus was obtained via

$$G^*(\omega) = G' + iG''$$
$$= G_e + i\omega \int_0^\infty G(t)\exp(-i\omega t)dt \quad (2)$$

The zero frequency tensile modulus G_e equals zero for a viscoelastic liquid. Phase angle $\delta = \tan^{-1} G''/G'$.

Steps to improve the signal-to-noise ratio of G^* were described elsewhere (Masoori & Greenfield 2017). Briefly, a moving average of $G(t)$ smooths fluctuations over ±0.1 ps. During the Fourier transform (Eq 2), a window function with a decaying exponential scaled stress relaxation so it reached zero by an imposed maximum time, with a corresponding inverse process applied to the resulting frequency spectrum. Odd and even symmetries were applied to the sine and cosine components of the Fourier transform; this converts $G(t)$ into a periodic function over the transformation domain, which lessens numeric side effects. Next, components of complex modulus of low magnitude ($|G| < 1$ Pa) were removed. Finally, a moving average was applied to the $G^*(\omega)$ results.

Time-temperature superposition was applied by a time shift factor $a_T = t/t_{ref} = \omega_{ref}/\omega$. Changes in modulus with temperature and density were accounted for by $b_T = (\rho_{ref}T_{ref} / \rho T)$, which corresponds to a master curve for $(G^*/\rho T)$ rather than $|G^*|$. Use of this full time-temperature superposition was applied for interpreting high strain rate molecular simulations of hydrocarbon rheology (McCabe et al. 2002).

3 RESULTS AND DISCUSSION

Molecular dynamics simulations led directly to the stress relaxation $G(t)$, shown previously (Masoori & Greenfield 2014, 2017). Applying Fourier transformation and signal processing noise reduction leads to the $|G^*|$ and δ results in Figures 1 to 3 for temperatures 533, 443, and 400 K (260, 170, 127°C).

Figure 1 is a Black-van Gurp-Palmen plot of phase angle vs. complex modulus (log scale). Moduli at 443 and 400 K have been shifted to 533 K (reference) using density and temperature. Qualitative overlap is seen for more viscous results (δ near 90°). This suggests that some extent of time-temperature superposition can be achieved. Smooth lines show trends of the CAM model of asphalt rheology. The dot-dash curve (CAM) uses a parameter set for the simulation results of this work, described below. The curves labeled AC-1,2 use parameters reported by da Silva et al. (2004) for fits to their measurements on different asphalts. Figure 1 indicates that the simulation results show a similar range of complex modulus and phase angle as experimental data.

Figure 2 depicts frequency dependence of $|G^*|$ at 533 K. Time shift factors a_T were chosen to achieve a sufficient overlap at lower frequencies. Vertical shift factors b_T depend only on temperature. Numeric regression to obtain a "best" fit was not done because the extent of time-temperature superposition was only semi-quantitative for phase angle. Superposition is good for frequencies $\omega_{ref} < 2$ rad/ps. At higher frequencies, results show a loss peak for time scales that are shorter than bond relaxation. Efforts focused on less high frequencies. Higher temperatures allow more relaxation and thus equivalently longer times compared to measurements at colder temperatures; they correspond to smaller equivalent frequencies. Results in Figure 2 depict this: results from 533 K show the lowest $|G^*|$ and frequencies.

Figure 3 shows frequency dependence of phase angle. The same time shift factors a_T are applied as

459

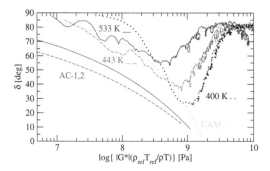

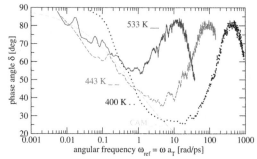

Figure 1. Black-van Gurp-Palmen plot for model AAA-1 asphalt at 533 (solid), 443 (dashed), 400 K (dotted). CAM indicates an approximate fit using the CAM model. AC-1,2 shows CAM fits from da Silva et al. (2004) for two asphalts.

Figure 3. Time-temperature superposition of phase angle for model AAA-1 asphalt at 533 (solid), 443 (dashed), 400 K (dotted). Dot-dash line shows the CAM fit.

Table 1. Shift factors from time-temperature superposition of complex modulus and from rotational relaxation time.

T	a_T from $|G^*|$	a_T range, rotational relaxation time
533.15	1.0	1.0 (reference state)
433.15	7.0	4.7 to 52
400.15	30.0	23 to 537

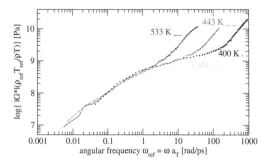

Figure 2. Time-temperature superposition of complex modulus magnitude for model AAA-1 asphalt at 533 (solid), 443 (dashed), 400 K (dotted). Dot-dash line shows the CAM fit.

in Figure 2. Time-temperature superposition is semi-quantitative between 533 and 443 K. Phase angles at 400 K show some similarity to 443 K results, though decreases in δ with frequency are more rapid at the lower temperature. The more poor display of time-temperature superposition is consistent with the differences in shape seen in Figure 1 at $|G^*| \sim 10^9$ Pa. The high frequency loss peak is clear in Figure 3. A plot of $\delta(\omega)$ without applying the shift factor shows that this peak arises at a temperature-independent frequency $\omega \sim 10$ rad/ps. It is anticipated this peak is an artifact of the thermostat time constant of 0.1 ps.

Approximate parameters of the CAM rheology model $|G^*| = G_g[1 + (\omega/\omega_0)^{-\nu}]^{\nu/w}$ that can describe the simulation results were obtained via the shifted δ and $|G^*|$ results. First, an approximate crossover frequency $\omega_0 \approx 0.9$ rad/ps was identified at which $\delta \approx 45° = w\pi/2/[1 + (\omega/\omega_0)^\nu]$. Next, the exponent $\nu = 0.5$ and parameter $w = 1$ were chosen so the model represented the general trend of the superposed phase angles. Finally, the plateau modulus log G_g [Pa] $\approx$ 9.25 was found to provide a good description of the $|G^*|$ data. As with the shift factors, these parameters are approximate rather than a numeric "best fit". They describe complex modulus results adequately, other than the potentially artificial loss peak.

The time shift factors a_T obtained from shifting the complex modulus data are listed in Table 1. Also listed are alternate ranges of shift factors that could be obtained by $a_T = \tau_{r,i}(T) / \tau_{r,i}(T_{ref})$, where $\tau_{r,i}$ is the average rotational relaxation time for molecule type i in the model asphalt. These were obtained previously (Li & Greenfield 2014b) from a time correlation function that quantified how slowly each molecule changed its orientation. Each shift factor falls within the range calculated for different molecule types. In that prior work, it was assumed that viscosity tracks the slowest relaxation time in the system. Here, we find instead that the complex modulus tracks a relaxation time that is typical for the range of relaxation times found in the multicomponent system, rather than only the fastest or slowest.

The CAM model with $w = 1$ is equivalent to the Christensen-Anderson model with $\nu = (\log 2)/R$. Rheological index R indicates the decrease in modulus (on a log scale) between the high frequency limit and the crossover frequency. For the CA data fit reported here, rheological index $R = 0.6$ indicates a more narrow relaxation time distribution compared to that found in a real asphalt. The crossover frequency $\omega_0 = 0.9$ rad/ps corresponds to $\omega_0 = 10^{6.2}$ or $10^{4.6}$ rad/s if shifted to 40 or 25°C using a shift factor a_T from rotational relaxation. Real AAA-1 has a rheological index of 1.5 and crossover frequency of 10^{-2} rad/s at a reference temperature of -19°C (Anderson et al. 1994). Parameters from this simulated asphalt were somewhat outside the ranges shown by real asphalts. This could be due to the high temperature; Anderson et al. (1994) noted that rheological index was smaller near regimes of viscous flow.

Others have computed asphalt complex modulus with molecular simulation. Lemarchand et al. (2018) simulated a system representative of asphalt using dissipative particle dynamics. A similar scaling of $|G^*|$ with ω was found. Khabaz & Khare (2018) used nonequilibrium MD to impose oscillatory shear on the Li 12-component AAA-1 over 260 to 500 K. They computed shift factors and master curves for viscosity, complex modulus, and creep compliance. Slightly smaller shift factors were found here for each temperature range. Khabaz & Khare (2018) depicted a much wider frequency range because their nonequilibrium simulations spanned much wider temperature ranges. They found smaller changes in $|G^*|$ as a function of frequency, via imposed oscillatory strain, than were computed here from spontaneous fluctuations in stress in the absence of strain. This corresponds to a larger rheological index, which is more consistent with data for real asphalts. Time-temperature superposition of phase angle master curves was not as good as for viscosity and complex modulus, similar to findings here.

4 CONCLUSIONS

The goal of this work was to demonstrate the extent that time-temperature superposition can be achieved for stress relaxation data that originate from equilibrium molecular dynamics simulations of an undeformed model asphalt. A new innovation was applying noise reduction techniques with a basis in signal processing to reduce the inherent statistical error so underlying asphalt rheology mechanics can emerge. A Black-van Gurp-Palmen plot showed that semi-quantitative time-temperature superposition is possible. Superposing the magnitude of complex modulus showed very good overlap. Phase angles showed qualitative superposition with the same shift factors. Dependences on angular frequency were consistent with the CAM model. Time shift factors were within a range exhibited by single molecule rotations in an earlier analysis of simulation results. Temperature changes in complex modulus and viscosity followed typical material time scales, rather than the fastest or slowest relaxation time. The CAM model parameterized for the simulated asphalt represented the general trends of the angular frequency dependence, though its parameters were somewhat outside the ranges exhibited by real asphalts. Improvements require yet more accurate model asphalt compositions.

ACKNOWLEDGMENTS

This material is based in part upon work supported by the Rhode Island Dept. of Transportation and the Federal Highway Administration (agreement # DTFH61-07-H-00009). Opinions, findings, and conclusions or recommendations expressed in this work are those of the Authors and do not necessarily reflect the view of the Fed. Highway Administration.

REFERENCES

Anderson, D.A., Christensen, D.W., Bahia, H.U., Dongre, R., Sharma, M.G., Antle, C.E., & Button, J. 1994. *Binder Characterization and Evaluation, Volume 3: Physical Characterization*. Technical Report SHRP-A-369. Strategic Highway Research Program.

Christensen, D.W. & Anderson, D.A. 1992. Interpretation of dynamic mechanical test data for paving grade asphalt cements. *J. Assoc. Asph. Paving Technol.* 61: 67–116.

Christensen, D.W., Anderson, D.A. & Rowe, G.M. 2017. Relaxation spectra of asphalt binders and the Christensen–Anderson rheological model. *Road Mater. Pavement Des.* 18(sup1): 382–403.

Hansen, J.S., Lemarchand, C.A., Nielsen, E., Dyre, J.C. & Schrøder, T. 2013. Four-component united-atom model of bitumen. *J. Chem. Phys.* 138: 094508.

Kennedy, T.W., Huber, G.A., Harrigan, E.T., Cominsky, R.J., Hughes, C.S., Von Quintus, H. & Moulthrop, J.S. 1994. *Superior Performing Asphalt Pavements (Superpave): The Product of the SHRP Asphalt Research Program*. Technical Report SHRP-A-410. Strategic Highway Research Prog.

Khabaz, F. & Khare, R. 2015. Glass transition and molecular mobility in styrene-butadiene rubber modified asphalt. *J. Phys. Chem. B* 119: 14261–14269.

Khabaz, F. & Khare, R. 2018. Molecular simulations of asphalt rheology: Application of time-temperature superposition principle. *J. Rheol.* 62: 941–954.

Lemarchand, C.A., Greenfield, M.L., Dyre, J.C. & Hansen, J.S. 2018. ROSE bitumen: Mesoscopic model of bitumen and bituminous mixtures. *J. Chem. Phys.* 149: 214901.

Li, D.D. & Greenfield, M.L. 2014a. Chemical compositions of improved model asphalt systems for molecular simulations. *Fuel* 115: 347–356.

Li, D.D. & Greenfield, M.L. 2014b. Viscosity, relaxation time, and dynamics within a model asphalt of larger molecules. *J. Chem. Phys.* 140: 034507.

Marasteanu, M.O. & Anderson, D.A. 1999. Improved model for bitumen rheological characterization. In *Proceedings of the Eurobitume Workshop 99, Luxembourg*. Paper no. 133. the Netherlands: European Asphalt Association.

Masoori, M. & Greenfield, M.L. 2014. Frequency analysis of stress relaxation dynamics in model asphalts. *J. Chem. Phys.* 141: 124504.

Masoori, M. & Greenfield, M.L. 2017. Reducing noise in computed correlation functions using techniques from signal processing. *Mol. Simul.* 43: 1485–1495.

McCabe, C., Manke, C.W. & Cummings, P.T. 2002. Predicting the Newtonian viscosity of complex fluids from high strain rate molecular simulations. *J. Chem. Phys.* 116: 3339–3342.

Mondello, M. & Grest, G.S. 1997. Viscosity calculations of *n*-alkanes by equilibrium molecular dynamics. *J. Chem. Phys.* 106: 9327–9336.

Silva, L.S., Camargo Forte, M.M., Alencastro Vignol, L. & Cardozo, N.S.M. 2004. Study of rheological properties of pure and polymer-modified Brazilian asphalt binders. *J. Mater. Sci.* 39: 539–546.

Zhang, L. & Greenfield, M.L. 2007a. Analyzing properties of model asphalts using molecular simulation. *Energy Fuels* 21: 1712–1716.

Zhang, L. & Greenfield, M.L. 2007b. Relaxation Time, Diffusion, and Viscosity Analysis of Model Asphalt Systems using Molecular Simulation. *J. Chem. Phys.* 127: 194502.

Advances in Materials and Pavement Performance Prediction II – Kumar et al. (eds)
© 2021 Taylor & Francis Group, London, ISBN 978-0-367-46169-0

Evaluation of models for binder dynamic shear modulus and phase angle

I. Onifade, K. Huang & B. Birgisson
TEES, Texas A&M University, College Station, TX, USA

ABSTRACT: An evaluation of the predictive capabilities of the Bari-Witczak and the Onifade-Birgisson models for the prediction of dynamic shear modulus and phase angle from conventional steady-state viscosity is performed. The study considers five different geographical locations with a wide range of temperature in order to account for the range of in-service pavement temperature conditions in the US. The Bari-Witczak models predicted accurate tendencies of dynamic shear modulus and phase angle for pavement sections in the warmer regions, while several irregularities in model predictions were observed for the pavement sections in the colder region. On the other hand, the Onifade-Birgisson model is sensitive to the variations in the pavement temperature with realistic tendencies predicted for the different pavement locations and seasons. The results of the model predictions can have serious implications for the reliable estimation of mixtures dynamic modulus, evaluation of susceptibility of binders to low-temperature cracking, and the evaluation of pavement long-term performance.

1 INTRODUCTION

The viscoelastic properties of asphalt binder are important material properties, which indicate the binder frequency- and temperature-dependent response. The viscoelastic properties is characterized by the dynamic shear modulus ($|G_b^*|$) obtained as the ratio of the peak values of the shear stress to the shear strain, and the phase angle which is the lag between the shear stress and the shear strain in one load cycle. The phase angle reflects the tendencies for viscous or elastic material response. The phase angle ranges between 0° to 90°, where materials with purely elastic material behavior exhibit a phase angle of 0°.

The viscoelastic properties of asphalt binder plays a significant role in the estimation of parameters used for differentiating between poor or high performance binders with respect to fatigue cracking and rutting. For instance, the Glover-Rowe parameter is used as an indicator for the evaluation of the cracking susceptibility of both modified and unmodified binders (Anderson et al. 2011). The Glover-Rowe parameter is estimated using the binder viscoelastic properties (complex shear modulus and phase angle) measured at a temperature of 15°C, and a reference frequency of 0.0005 radians/sec. Research effort during the Strategic Highway Research Program (SHRP) led to the development of Superpave rutting and fatigue parameters. A conceptual approach was adopted for the development of the Superpave parameters. Higher value of the ratio of the dynamic shear modulus and the phase angle ($|G_b^*| \cdot \sin(\delta_b)$) was recommended to resist permanent deformation at intermediate to high temperature, while minimal values of ($|G_b^*| \cdot \sin(\delta_b)$) was recommended to resist fatigue cracking at the lower temperature range. The viscoelastic properties are also required to construct the Black Space diagram, which is utilized to study different types of phenomena in viscoelastic materials. Thus, accurate measurement and estimation of the viscoelastic properties of asphalt binders is essential for the characterization of their performance with respect to pavement applications (Zhang et al. 2019).

The frequency dependent behavior of asphalt concrete mixtures due to the contribution of the interaction of the aggregate structure and the viscoelastic properties of the binder plays an important role in the characterization of asphalt concrete mixture dynamic modulus at different loading frequencies. A number of models for predicting asphalt mixture dynamic modulus have been proposed, which requires the binder dynamic modulus and phase angle as input parameter e.g., the Hirsch model (Christensen & Bonaquist 2015). The general trend in the development of new mixture dynamic modulus predictive models indicates the need for accurate models for the interconversion of conventional steady-state viscosity data into frequency-dependent viscoelastic binder properties to enable the use of existing viscosity data in new model calibration and validation.

In the paper, the prediction accuracy of new and existing models for the prediction of dynamic shear modulus and phase angle of asphalt binder is evaluated. Particular emphasis is placed on the prediction accuracy the models over the range of temperature conditions in five different geographical locations in the USA. The geographical locations are selected in order to evaluate the predictive capabilities of the models outside the range experimental conditions at which the model calibration was performed.

2 MODELS FOR ASPHALT BINDER DYNAMIC SHEAR MODULUS AND PHASE ANGLE PREDICTION

Two different models considered in this study are the Bari-Witczak model (Bari & Witczak 2007) and the Onifade-Birgisson model (Onifade & Birgisson 2020). Both models are used for the conversion of steady-state viscosity to dynamic shear modulus and phase angle, and are developed based on different underlying principles. However, both models are developed using the same binder dataset, with the binder properties obtained at a temperature range of 15°C to 115°C, and loading frequencies of 1, 10 and 100 rad/sec. Details about the models are presented below:

2.1 Bari and Witczak model

Bari and Witczak used the Cox-Merz rule to develop a new dynamic shear modulus G* predictive models for asphalt binders. Analogy between the dynamic viscosity and the frequency dependent steady-state viscosity was assumed with correction factors introduced to account for the discrepancies between the two measures of viscosity. The correction factor considers the effect of the dynamic shear loading frequency on the material response. The Witczak-Bari equation for the dynamic shear modulus is given as:

$$|G_b^*| = 0.0051 \cdot f_s \cdot \eta_{fs,T} \cdot (\sin(\delta_b))^{7.1542 - 0.4929 \cdot fs + 0.0211 \cdot f_s^2} \tag{1}$$

where $|G_b^*|$ is the dynamic shear modulus (Pa), f_s is the dynamic shear loading frequency (Hz), $\eta_{fs,T}$ is the viscosity of the asphalt binder as a function of both loading frequency and temperature (cP) and δ_b is the phase angle. The maximum value of the dynamic shear modulus is limited to a value of 1 GPa.

The expression for the phase angle is given as:

$$\delta_b = 90 + (b_1 + b_2 \cdot VTS) \cdot \log(f_s \cdot \eta_{fs,T})$$
$$+ (b_3 + b_4 \cdot VTS) \cdot \log(f_s \cdot \eta_{fs,T}) \tag{2}$$

where VTS' is the adjusted VTS value, f_s is the dynamic shear loading frequency (Hz), and b1, b2, b3 and b4 are models parameters with values of -7.3146, -2.6162, 0.1124, and 0.2029 respectively.

Bari and Witczak introduced equations to extend the conventional ASTM A-VTS relationship to account for the influence of the loading frequency on the viscosity measurements. They hypothesized in their work that the A-VTS values are frequency-dependent parameters with suitable factors introduced to account for the effect of the loading rate. The modified ASTM A-VTS relationship is expressed as:

$$\log\log(\eta_{fs,T}) = A' + VTS' \cdot \log(T_R)$$
$$= c \cdot A + d \cdot VTS \cdot \log(T_R) \tag{3}$$

where c and d are the frequency adjustment factor for the A and the VTS respectively.

Bari and Witczak expressed the parameters c and d as power law functions of the loading frequency expressed as:

$$c = c_o \cdot f_s^{c_1}$$
$$d = d_o \cdot f_s^{d_1} \tag{4}$$

where c_o, c_1, d_o and d_1 are model parameters.

Bari-Witczak model expressed the adjusted VTS value as:

$$VTS' = d \cdot VTS = do \cdot f_s^{d1} \cdot VTS \tag{5}$$

2.2 Onifade and Birgisson model

Onifade and Birgisson proposed new models for the prediction of asphalt binder dynamic shear modulus and phase angle from conventional steady-state viscosity measurements. The Onifade-Birgisson model employed generalized logistic functions in the model development to capture the asymptotic behavior of asphalt binders over wide range of temperature conditions. Onifade-Birgisson model expressed the dynamic shear modulus for unmodified binders as:

$$\log|G_b^*| = \left[-3.541 + \frac{5.879}{1 + 0.4575 \cdot e^{(-0.325 \cdot \log(\eta \cdot \omega))}}\right]$$
$$\cdot(-VTS)^{0.58} \tag{6}$$

where $|G_b^*|$ is the dynamic shear modulus (psi), η is the viscosity in (MegaPoise), VTS is the regression slope of the Viscosity Temperature Susceptibility plot, and ω is the angular frequency in (rad/sec).

The study by Onifade and Birgisson identified the existence of a new parameter termed the "reduced viscosity" which is a combination of the steady-state viscosity, frequency and a viscosity shift factor. The "reduced viscosity" obtained using a suitable viscosity shift factor results in a smooth continuous function of the binder phase angle for both modified and unmodified binders. The reduced viscosity is expressed as:

$$\eta_r = \eta \cdot \omega^{k_f} \tag{7}$$

where k_f is the single-valued viscosity shift factor. Onifade-Birgisson model expressed the phase angle for unmodified binders in terms of the reduced viscosity as:

$$\delta_b = \frac{\log|G_b^*|_{ref}}{0.0267 \cdot e^{(-0.08037 \cdot \log(\eta \cdot \omega^{0.7}))}}$$
$$+ 0.04144 \cdot e^{(0.2432 \cdot \log(\eta \cdot \omega^{0.7}))} \tag{8}$$

where δ_b is the phase angle, $|G_b^*|ref$ is the reference dynamic shear modulus (evaluated using Equation 6 at a reference temperature of 96°C), and the single-valued viscosity shift function is set to 0.7. The maximum value of the phase angle is set to 90°.

Figure 1. The locations of pavement sections with four different climatic zone distributions in the US.

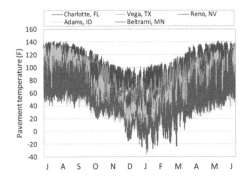

Figure 2. Annual pavement temperature for chosen locations.

3 EVALUATION OF ASPHALT BINDER FIELD RHEOLOGICAL PROPERTIES USING MODELS

In this section, the performance evaluation of the Bari-Witczak and the Onifade-Birgisson models is performed. The performance evaluation is based on the comparison of the predicted asphalt binder dynamic shear and phase angle under different temperature conditions in the United States using a standard PG binder (PG 67-22). Figure 1 shows the locations of pavement sections in different climatic zone in the United States. These locations were carefully chosen to cover a wide variety of temperature conditions in the United States with a gradually decrease manner. The pavement sections are located in four climate zones, which include Dry-Freeze, Dry-Nonfreeze, Wet-Freeze, and Wet-Nonfeeze. The one-dimensional heat transfer model (Han et al. 2011) is adopted to calculate the pavement temperature. The model comprehensively considered the heat sources including the solar radiation, atmospheric down-welling and outgoing longwave radiation and convection enhanced by wind. Figure 2 shows the calculated annual pavement temperature for the chosen pavement locations based on the local weather data. In this figure, the horizontal axis is the time starts from July 1st. More details of pavement temperature for these five location are shown in Table 1. The model

Table 1. The details of pavement temperature for the chosen locations.

Location	Climate zone	High temp. (F)	Low temp. (F)
Charlotte, FL	WNF	142.00	39.92
Vega, TX	DNF	137.06	13.45
Reno, NV	DF	137.48	10.62
Adams, ID	DF	138.46	−6.17
Beltrami, MN	WF	128.82	−35.79

prediction at the lower temperature range in the dry-freeze and wet-freeze regions is of particular interest in order to evaluate the model performance outside the range of temperature conditions at which the model calibration was performed.

4 RESULTS AND DISCUSSIONS

Figure 3 shows the phase angle predictions for the different pavement geographical location using both models. The comparison of the phase angle predictions shows that both models predict relatively close magnitude and tendencies of the phase angle in relatively warm locations like Charlotte, FL, Vega, TX and Reno, NV. However, the Bari model hits the lower bound of 2.5° prescribed in their model when during the winter of Adams, ID and Beltrami, MN (highlighted using the dashed boxes in Figure 3). In contrast, the Onifade-Birgisson model is sensitive to the variations in the pavement temperature with realistic tendencies in the phase angle predicted for the different pavement locations and seasons.

Figure 4 shows the dynamic shear modulus predictions for the chosen locations using both models. The dynamic shear modulus predictions using both models are almost identical in the warmer areas like Charlotte, FL, Vega, TX and Reno, NV for all year around, and in the warmer seasons in Adams, ID and Beltrami, MN. However, limitation of the Bari-Witczak model in the lower temperature region is observed and obvious in dynamic shear modulus predictions. A flat upper bound of dynamic shear modulus prediction is observed when the temperature is cold enough in the winter of Adams, ID and Beltrami, MN. Moreover, the results of the predictions using the Bari-Witczak model presents an irregular tendency in the dynamic shear modulus prediction with lower dynamic shear modulus values predicted in the winter seasons of Adams, ID and Beltrami, MN. This is accompanied by an irregular variation between extreme highs and lows in the winter of Adams, ID and Beltrami, MN, highlighted using the dashed boxes. Using these irregular variations and extremely low dynamic shear modulus might bring serious errors if they adopted to predict the performance of the pavement in these regions. The comparison of both models firstly indicates that the Onifade-Birgisson model predicts more accurate

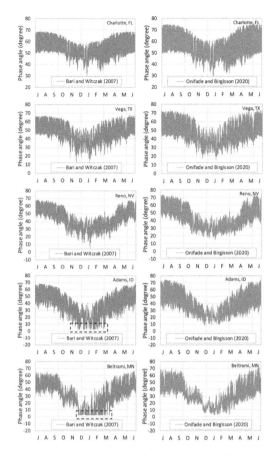

Figure 3. Phase angle predictions for the chosen locations using Bari-Witczak and Onifade-Birgisson model.

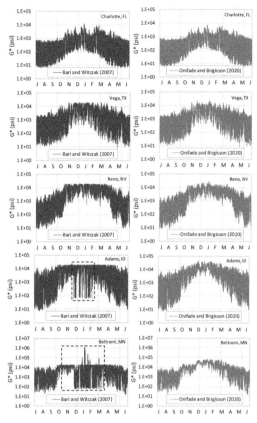

Figure 4. Shear modulus predictions for the chosen locations using Bari-Witczak and Onifade-Birgisson model.

phase angle and dynamic shear modulus tendencies compared to the Bari-Witczak model. Secondly, the limitation of the Bari-Witczak especially in the area where the pavement temperature is extremely low is highlighted in the results presented in Figure 3 and Figure 4.

5 CONCLUSIONS

The study evaluates the predictive capabilities of Bari-Witczak model and the Onifade-Birgisson model for the prediction of asphalt binder dynamic shear modulus and phase angle. A wide range of temperature conditions in different climate zones was considered to cover the temperature range of in-service pavement section in the US.

The results of the model evaluation show that both models predict reasonable tendencies of the binder dynamic shear modulus and phase angle in the warmer climate regions. However, shortcomings of the Bari-Witczak model were identified in the colder regions in Adams, ID and Beltrami, MN where inaccurate tendencies in the phase angle and dynamic shear modulus are observed in the winter season. A flat upper

bound of dynamic shear modulus is predicted by the Bari-Witczak model with irregular variations between extreme lows and highs of dynamic shear modulus. The Bari-Witczak model also predicts lower dynamic shear modulus values in the winter seasons of Adams, ID and Beltrami, MN compared to the warmer seasons.

The limitations of the Bari-Witczak model in terms of applicable temperature range is of significant importance especially in the area where the pavement temperature is extremely low. The irregular tendencies and extremely low dynamic shear modulus associated with the Bari-Witczak model in the cold temperature regions will result in significant errors in the estimation of asphalt mixture properties and evaluation of long-term pavement performance. The observed irregularities may also affect the prediction of the aged viscoelastic properties of the binder as well as the susceptibility of binder to low temperature cracking.

REFERENCES

Anderson, R.M., King, G.N., Hanson, D.I., Blankenship, P.B., 2011. Evaluation of the Relationship between Asphalt Binder Properties and Non-Load Related Cracking, in: Journal of the Association of Asphalt Paving Technologists. Presented at the Asphalt Paving Technology 2011.

Bari, J., Witczak, M., 2007. New Predictive Models for Viscosity and Complex Shear Modulus of Asphalt Binders: For Use with Mechanistic–Empirical Pavement Design Guide. Transportation Research Record: Journal of the Transportation Research Board.

Christensen, D.W., Bonaquist, R., 2015. Improved Hirsch model for estimating the modulus of hot-mix asphalt. Road Materials and Pavement Design 16, 254–274. https://doi.org/10.1080/14680629.2015.1077635

Onifade, I., Birgisson, B., 2020. Improved models for the prediction of asphalt binder dynamic shear modulus and phase angle. Construction and Building Materials 250, 118753. https://doi.org/10.1016/j.conbuildmat.2020.118753

Zhang, D., Birgisson, B., Luo, X., Onifade, I., 2019. A new short-term aging model for asphalt binders based on rheological activation energy. Mater Struct 52, 68. https://doi.org/10.1617/s11527-019-1364-7

Han, R., Jin, X., and Glover, C.J. Modeling pavement temperature for use in binder oxidation models and pavement performance prediction. Journal of Materials in Civil Engineering, 2011, 23 (4), 351–359.

Advances in Materials and Pavement Performance Prediction II – Kumar et al. (eds)
© 2021 Taylor & Francis Group, London, ISBN 978-0-367-46169-0

DSR geometry impact on asphalt binder linear viscoelastic properties

J.A. Rodrigues, G.S. Pinheiro, K.L. Vasconcelos & L.B. Bernucci
Department of Transportation Engineering, University of São Paulo, São Paulo, SP, Brazil

R.C.O. Romano & R.G. Pileggi
Department of Civil Construction Engineering, University of São Paulo, São Paulo, SP, Brazil

ABSTRACT: Asphalt binder is a rheological complex material with elastic and viscous components response, according to the temperature and loading conditions, being its rheological characterization an important tool to better predict the material performance. The Dynamic Shear Rheometer (DSR), using parallel-plates geometry, is so far the most common oscillatory testing carried out for rheological characterizations in bituminous materials. However, the rheological response needs to be characterized considering the stages of binder application and use, as the material have different behavior for each phase during hot asphalt mixture production, application and design life performance. So, the setup of rheological tests must be adequate to correctly represent the real conditions during each stage. The main purpose of this work was to compare the impact of different test configurations (using parallel-plate, Vane and DIN) in the response of a neat asphalt binder applying the strain sweep test at different temperatures.

1 INTRODUCTION

Rheology (from the Greek, "ρεω", translated literally as "to flow" and "λογοσ", that means "science", hence, "the study of the flow") consists in the science in which the relationship between an induced stress/strain on a material and the resulting flow or deformation is studied (Airey 2002; Tabilo-Munizaga & Barbosa-Cánovas 2005). Some materials are rheologically and structurally complex, i.e. suspensions of particles, paint, blood, foodstuffs and some asphalt binders, such as polymer solutions, presenting a variety of behaviors. Specifically, for asphalt binders, this behavior may vary from Newtonian to highly viscoelastic material (one that exhibits both elastic and viscous components response) according to temperature changes and loading conditions (Wood 1958; Phan-Thien & Mai-Duy 2017). Empirical properties are still used as an indicative of asphalt rheological characteristics, commonly obtained through softening point and penetration tests. However, it is well known that fundamental tests, based on known physics concepts and equations, can determine the true properties of the materials, enabling to define their behavior and predict performance (Hraiki 1974; Tabilo-Munizaga & Barbosa-Cánovas 2005).

1.1 Oscillatory test

Oscillatory tests conducted in a small-strain range, using a dynamic shear rheometer (DSR), are important tools to determine and better understand the asphalt viscoelastic properties. Those tests are operated within the region of linear viscoelastic (LVE) response and this response can be presented in different ways, through the rheological parameters (i.e. $G*$, G', G'', $\eta*$, δ, tan δ, etc.) versus loading time at specific temperatures (isothermal plot), or versus temperature at specific loading times (isochronal plot) (Airey 2002; Mezger 2014). Furthermore, the rheometers present different measuring systems, or geometries, that can be used to perform the rheological tests. The following sub-sections list distinct aspects of these measuring apparatus according to Mezger (2014).

1.1.1 Parallel-plate (PP) geometry

The PP is composed by a statically inferior and a motion superior cylindrical plate (Figure 1, on the left), which cause shearing in the sample, producing a deflection path and a deflection angle. Although, for a strict rheological property determination, sample adhesion to the plates (avoiding any wall-slip effect) and homogeneous deformation throughout the entire shear gap must be established. Some advantages of the apparatus can be mentioned: short preparation time, errors normally reduced compared to the cone-plate geometry, and simple cleaning process. But, melt fracture, inhomogeneous flow, surface effects can be pointed as disadvantages; further, larger gap heights, increase the temperature gradient in the sample (Mezger 2014; Hung et al. 2015).

1.1.2 DIN geometry

The DIN geometry consists of two concentric cylinders (CC) (Figure 1, in the center), also known as

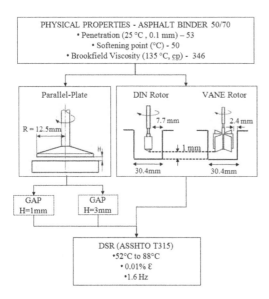

Figure 1. Experimental design flowchart.

"cup-and-bob". It has the capability to prevent the material to flow off the shear gap (even for high shear rates); can guarantee a better temperature control compared to parallel- and cone-plate geometries; and does not require trimming the sample (avoiding any unexpected disruption). Nonetheless, a large amount of material is necessary, and it is more timing consuming to change, prepare and stabilize the temperature (Mezger 2014).

1.1.3 Vane geometry

The Vane geometry is composed by several rectangular vanes connected to a cylinder (Figure 1, on the right). The vane insertion into the sample causes far less disturbance which is a significant factor with thixotropic systems, such as asphalt binders. However, the sample localized between the vane area might not be sheared, and there might occur inhomogeneous "plastic" deformation behavior, if the test is conducted at high rotational speeds (Mezger 2014).

There is a diversity of geometries configurations that may led to different viscoelastic responses, due to the type of flow produced. Various authors from distinguished areas studied geometry types that might better represent the material characteristics and application. Hung et al. (2015) used DIN geometry to investigate and compare traditional PP, at 40° C, 65° C and 90° C, for quality control on projects that used asphalt rubber binders. Their results indicated no significant difference between the two geometries. Baumgardner & D'Angelo (2012) investigated the ability of the DIN geometry to investigate rheological characteristics of conventional and modified binders. The authors concluded that the DIN geometry indicates similar results for Superpave parameters, as well as Multiple Stress Creep Recovery (MSCR) testing, performed at 64° C,

which exhibited the same peak strain in the creep portion of the curve, indicating that both geometries had the same shear rate. But differences in the compliance results were evident, reaching values up to 71,4% higher for the DIN geometry, when compared to the PP for one of the tested binders.

Polacco et al. (2003) tested the PP and DIN geometries, varying from 60°C up to 90° C using neat and modified asphalt binders. It was observed that PP geometry presents satisfactory responses until the material approaches the liquid state, when the tendency to flow out from the PP geometry and the low torque values appears. Hence, both problems can be solved with the DIN geometry, giving more reliable data than PP in this case. Puga and Williams (2015) investigated the influence of varying gap height and geometry types, PP and CC, on asphalt rubber binders. The authors detected different responses when compared the same material with higher G^* values from the DIN testing geometry compared to the ones obtained with PP. Also, Rønholt et al. (2013) observed that the elastic modulus (G'), obtained from small amplitude oscillatory shear rheology tests, has considerably dependence. For this tests, nine different geometry types where used (including PP, DIN and Vane) on the same material.

There is no consensus on the use of different geometries for the determination of asphalt binder rheological characterization, since several investigations have been conducted with different conclusions with respect to the effect of geometry. The objective of this study was to evaluate different DSR geometries and their accuracy for the rheological characterization of the viscoelastic properties of asphalt binders and then try to identify which one would be the most suitable geometry for this purpose.

2 EXPERIMENTAL DESIGN

The rheological tests were performed on a neat asphalt binder, CAP 50/70 (penetration grade between 50- and 70- 10^{-1}mm). Comparisons were made between PP geometry using 1- and 3-mm gaps, and CC with two different rotor geometries: DIN and Vane. The CC includes a cup with radius of 30.4 mm and both rotors have 42 mm height, with radius of 15mm (DIN) and 28 mm (Vane), and same 1 mm bottom gap. DSR tests were performed according to the ASSHTO T315 (2019) standard procedure varying the temperature from 52° C to 88° C, applying 20 min. of conditioning time for each one, with 0.01% oscillation strain, keeping the frequency at 1.6 Hz. Figure 1 shows the testing plan resume. Five samples per each gap in PP geometry and three samples per rotor type (DIN and Vane) in the CC geometry were tested. At first, limiting value of the linear viscoelastic region (LVR) material range in terms of the shear strain were empirically found by amplitude sweep test, according to ASTM D7175 (2015).

3 RESULTS AND DISCUSSIONS

In order to better evaluate the efficiency and accuracy of different testing geometries in rheological characterization of asphalt binders, results from DSR tests were analyzed using different graphs.

Figure 2 presents the dynamic shear modulus results as function of temperature and difference between DIN results and others can be observed. Although the exact cause of results divergence has not been conclusive identified, it's important to point that rheological response of the material depends on many factors. One of them is the gap between the cup inner surface and the rotor side, which can contribute for the occurrence of wall-slip effects. DIN has 7.7 mm lateral gap and Vane has only 2.2 mm, more like PP gaps, then during DIN tests the flux of material inside the cup are less influenced by the friction generated between the wall and the material inside the cup than VANE and PP. In addition, the second important factor is that each rheological instrument affects differently the flow imposed to the material. Once physical conditions vary for each rheological measuring system adopted, the sample properties may exhibit different responses. (Rønholt et al. 2013).

The black space graph (Figure 3), that correlates for each geometry with the average results of Dynamic Shear Modulus ($|G^*|$) and Phase Angle (δ), shows a considerable difference among the results of the different geometries. For the CC, using Vane or Din geometries, the correlation between $|G^*|$ and δ follows the same pattern (but with distinct absolute values), results followed the expected trend, where the phase angle tends to 90° with the temperature increase. A different pattern was observed in the results obtained with the PP geometry (even without influence of gap position during the test). PP results showed an unexpected material behavior above 64° C: there was an inversion on the phase angle with the decrease in the dynamic shear modulus.

Thus, the results obtained using the different geometry cannot be assumed as effective material response in a generalized way, since as already mentioned, the rheological response depends of the type of the flow generated along the test, even using the same sample of the binder.

The first hypothesis for the conflicted results was some error caused by the contact loss between binder sample and the plates of the geometry. However, according to the results presented in Figure 4, there was not an abrupt change in the axial force measured with the change of the temperature, indicating the accuracy during the test. Additionally, the axial force differences observed for 1mm and 3mm gap in the PP geometry indicate that the amount of material sample can also influence the axial displacement as the material response - expansion or retraction- during the test. Hence, the axial forces were up to 10 times higher in CC than for PP for the same range of temperature.

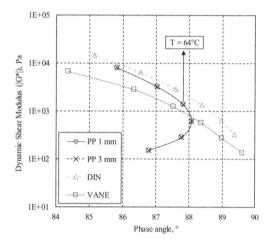

Figure 3. Black space diagram of the different geometries correlating the Dynamic Shear Modulus and phase angle.

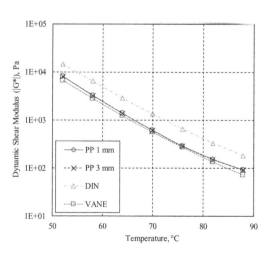

Figure 2. Dynamic Shear Modulus versus temperature.

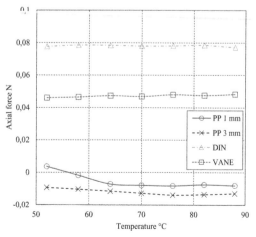

Figure 4. Axial force of the different geometries.

These results suggest better efficiency of CC geometry (considering the neat asphalt binder evaluated), especially at higher temperatures when asphalt binder tends to flow. So, the confinement guaranteed by the cup and the shear condition applied by the complete immersion of geometry inside the sample can prevent the asphalt binder from draining at high temperatures, assuring the effective material contact with the rotors (DIN or Vane) responsible for measuring the rheological response of material.

However, tests carried out using CC geometries need special care during sample conditioning time, which can interfere in the homogeneity of the sample temperature. As the amount of asphalt binder used in CC is around ten times greater than the PP samples, it requests longer conditioning times. In contrast, temperature control of the sample is guaranteed due to the relatively large contact area at the wall of the cup (Mezger 2014).

4 CONCLUSIONS

The influence of the geometry used during DSR tests and rheological characterization of asphalt binder materials was clearly identified for a neat asphalt binder. Even though an asphalt binder of the same origin and properties has been used for all the tests, rheological response of the material was considered dependent on the geometry of the test.

Although PP geometry is the standard and most commonly used worldwide in rheological characterization of asphalt materials, the results shown above identified important limitations of the geometry for certain test conditions (clearly associated with the higher temperatures) which compromises the results obtained. The rheological properties, $|G^*|$ and δ, are associated with the type of flow generated during the test. A representation of what the material will face during its application and service life would be the most recommended. Two other geometries were tested. DIN and Vane rotors combined with CC do not have the same limitation (material flow out of the geometry) identified in PP at temperatures above its softening point. Therefore, temperatures below 50° C were not evaluated and they might be an issue and limitation of

its use for asphalt binder characterization in a broader range of temperatures.

In summary, the study identified that the correct geometry selection depends on the temperature range and applicability of the material, as these variables yield the material to different stress state conditions, and consequently different rheological behaviors.

ACKNOWLEDGMENTS

Supported by São Paulo Research Foundation (FAPESP: 2017/25708-7; 2019/11354-4 and 2019/08415-1).

REFERENCES

Airey, G.D. 2002. Use of Black Diagrams to Identify Inconsistencies in Rheological Data *Road Materials and Pavement Design* (n. 4) vol. 3: 403–424.

Baumgardner G. & D'Angelo J.A. 2012. Evaluation of New Dynamic Shear Rheometer Testing Geometry for Performance Testing of Crumb Rubber–Modified Binder. *Journal of the Transportation Research Board* vol. 2293.

Hraiki S. 1974. Rheological properties of bituminous materials. *Rheologica Acta* vol. 13: 567–570.

Hung S. et al. 2015. Comparison of Concentric Cylinder and Parallel Plate Geometries for Asphalt Binder Testing with a Dynamic Shear Rheometer. *Journal of the Transportation Research Board* vol. 2505.

Mezger, T.G. (ed. 4) 2014. *The Rheology Handbook*. Hanover: Vincentz Network.

Phan-Thien N. & Mai-Duy N. (ed. 3) 2017 *Understanding Viscoelasticity an Introduction to Rheology*. Cham: Springer International Publishing.

Polacco G. et al. 2003. Dynamic Master Curves of Polymer Modified Asphalt from Three Different Geometries. *Applied Rheology* vol. 13 (3): 118–124.

Puga K. (dissertation) 2015. *Comprehensive Study on the Sustainable Technology of Asphalt Rubber for Hot Mix Asphalt Binders and Mixes*. Ames: Iowa State University

Rønholt S. et al. 2013. Small Deformation Rheology for Characterization of Anhydrous Milk Fat/Rapeseed Oil Samples. *Journal of Texture Studies* (45): 20–29.

Tabilo Munizaga G. & Barbosa-Cánovas G.V. 2005. Rheology for the food industry. *Journal of Food Engineering* (67) 147–156.

Wood P.R. 1958. Rheology of Asphalts and Its Relation to Behavior of Paving Mixtures. *Highway Research Board Bulletin* (n. 192): 20–25.

Advances in Materials and Pavement Performance Prediction II – Kumar et al. (eds)
© 2021 Taylor & Francis Group, London, ISBN 978-0-367-46169-0

Geometric nonlinearities of bituminous binder and mastic using Large Amplitude Oscillatory Shear (LAOS)

I.J.S. Sandeep & S. Sai Bhargava
Department of Civil Engineering, IIT Madras, India

A. Padmarekha
Department of Civil Engineering, SRM Institute of Science and Technology, India

J. Murali Krishnan
Department of Civil Engineering, IIT Madras, India

ABSTRACT: Fourier transformation rheology, geometric parameters from Lissajous-Bowditch plots, and Chebyshev polynomials are some of the techniques used for determining linear and nonlinear response while conducting Large Amplitude Oscillatory Shear (LAOS). In this investigation, LAOS technique is used to characterize the response of an unmodified bituminous binder and mastic. The bituminous binder and mastic are tested at 40°C and subjected to 1% and 5% strain amplitude at frequencies of 0.1, 1, and 10 Hz. Lissajous-Bowditch plots based geometric measures are used in this study. The difference in the response of binder and mastic is observed at 10 Hz frequency. The elastic and viscous linear limit is observed to be different for bituminous binder and mastic at 10 Hz frequency. Strain stiffening ratio (S) and shear thickening ratio (T) defined from the geometric measures follow no particular trend across the test conditions.

1 INTRODUCTION

The linear viscoelastic response of bituminous binder and mastic are characterized using a Dynamic Shear Rheometer (DSR) in small amplitude oscillatory shear (SAOS). The response from the SAOS is characterized in terms of the material functions, storage modulus (G'), loss modulus (G'') and phase angle (δ). According to ASTM D7175 (2015), the linear viscoelastic region is defined as the range in strains where the value of the dynamic modulus is 90% or more of the initial value. At large strains, when the material response is nonlinear, characterizing the response using linear viscoelastic material functions (G', G'', δ) from SAOS is no longer valid. Thus, there is a need for additional material functions to delineate the nonlinear behavior of the bituminous material in oscillatory shear.

In the recent past, large amplitude oscillatory shear (LAOS) is used to study the material response at large strains. LAOS differs from SAOS in terms of data collection. Only the peak values of the response are collected in SAOS, while full stress and strain waveform data are collected in LAOS. The full waveform obtained is further used to delineate the nonlinear response using different techniques such as Harmonic analysis using Fourier transformation rheology (Hyun et al. 2011), Lissajous-Bowditch plots (Ewoldt et al. 2008), and Chebyshev polynomial (Hyun et al. 2011).

Ewoldt et al. (2008) described new measures to study the intra/inter cycle nonlinearities and defined dimensionless indices such as strain stiffening ratio (S) and shear thickening ratio (T) using the Lissajous-Bowditch plots. Padmarekha et al. (2013) studied the response of unmodified and modified bituminous binders in LAOS mode and used geometric measures from the Lissajous plot to delineate the nonlinear response. This investigation explores the response of a bituminous binder and mastic using LAOS.

Many studies using the bituminous mastic with different types of fillers were characterized in linear viscoelastic regime using SAOS testing (Chen et al. 2019; Kim and Little 2004; Liao et al. 2015; Phan et al. 2016;. Underwood & Kim (2015), Diab & You (2018) studied the nonlinear viscoelastic behavior of the bituminous mastic in LAOS using different mineral fillers and different volumetric concentrations. Diab & You (2018) compared the extent of nonlinearity using S value, which is contradictory to the description of S as presented by Ewoldt et al. (2008).

Delineating the linear and nonlinear response for the binder and mastic is essential for comparing their response in oscillatory shear. As of now, no clear-cut measures exist for prescribing the nonlinearity of bituminous mastic. LAOS comes in handy here and can be used to delineate the nonlinearity for both binder and mastic. In this study, an attempt was made to characterize the nonlinear response of binder and mastic using geometric material functions from the Lissajous-Bowditch plots as defined by Ewoldt et al. (2008).

2 ANALYSIS OF LISSAJOUS-BOWDITCH PLOTS

From the elastic Lissajous-Bowditch plot, G'_M is defined as a minimum strain elastic modulus, and G'_L is defined as a large strain elastic modulus (Ewoldt et al. 2008). Equation 1 shows the mathematical form for the same. For a given test condition, a dimensionless index S, known as strain stiffening ratio, is defined as shown in Equation 2. $S=0$ indicates linear elastic behavior, whereas $S>0$ indicates strain stiffening, and $S<0$ indicates strain softening.

$$G'_M = \frac{d\sigma}{d\gamma}\Big|_{\gamma=0}, \quad G'_L = \frac{\sigma}{\gamma}\Big|_{\gamma=\pm\gamma_0} \quad (1)$$

$$S = \frac{G'_L - G'_M}{G'_L} \quad (2)$$

Similarly, from the viscous Lissajous-Bowditch plot, minimum rate dynamic viscosity (η'_M), and large rate dynamic viscosity (η'_L) are defined as shown in Equation 3 (Ewoldt et al. 2008). The dimensionless index T known as shear thickening ratio is shown in Equation 4. $T=0$ indicates linear viscous behavior, whereas $T>0$ indicates shear thickening, and $T<0$ indicates shear thinning.

$$\eta'_M = \frac{d\sigma}{d\gamma}\Big|_{\dot{\gamma}=0}, \quad \eta'_L = \frac{\sigma}{\gamma}\Big|_{\gamma=\pm\dot{\gamma}_0} \quad (3)$$

$$T = \frac{n'_L - n'_M}{n'_L} \quad (4)$$

Both S and T values indicate intracycle elastic and viscous nonlinearities, respectively.

3 EXPERIMENTAL INVESTIGATION

In this study, an unmodified VG30 grade of bitumen (IS 73 2018) is used as the base bituminous binder. Granite dust passing 75 μm sieve and retained on 30 μm sieve is used as a filler material to prepare the mastic. The specific gravity of the filler material is 2.777. The filler at 175 ° C was added to the bitumen at 165 ° C in the proportion of 1:1 by weight. The VG30 binder and the filler are blended at 165 ° C for 20 min using a mechanical blender. The produced mastic was tested within three days from the date of preparation to avoid phase separation within the mastic.

Both VG30 and mastic are subjected to oscillatory shear using the Dynamic Shear Rheometer (DSR). An Anton-Paar DSR, MCR 702, was used in this study. An 8 mm parallel plate geometry at a gap of 2 mm is used. The tests were conducted at a temperature of 40 ° C

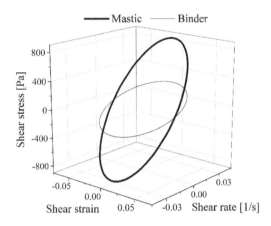

Figure 1a. 3D Lissajous-Bowditch plot for binder and mastic at 5 %, 0.1 Hz and 40 ° C

at 1 % and 5 % strain amplitude and at a frequency of 0.1, 1 and 10 Hz. The tests were performed using the LAOS testing protocol. 1000 cycles of loading were applied to the sample for all the conditions of loading. Torque, shear strain, shear rate, and stress waveform data were collected every 5 seconds with a total of 512 data points in each waveform. For further analysis, 1000th waveform data was considered.

4 RESULTS AND DISCUSSION

The responses from the LAOS can be visualized in the form of 3D space, as shown in Figure 1(a). The shear strain, shear rate, and the shear stress form the three orthogonal co-ordinate axes. The projection of the 3D space on to the stress vs. strain plane results in the elastic Lissajous-Bowditch plot and on to the stress vs. shear rate plane results in the viscous Lissajous-Bowditch plot.

In Figure 1(a), it is observed that the shear stress in the case of the mastic is higher than the binder indicating the higher stiffness of the mastic at 5 % strain amplitude and 0.1 Hz frequency. To study the variation in the behavior of the mastic with respect to the binder, the normalized shear stress with the peak value is plotted against shear strain and shear rate (Figure 1b, 1c, 2a, and 2b).

The overlapping plots in Figure 1(b) and 1(c) indicates similar behavior of bituminous binder and mastic at 0.1 Hz. Figure 2(a), 2(b) show the Lissajous-Bowditch plot for binder and mastic subjected to 5 % strain amplitude, and 10 Hz frequency. The difference in the response of binder and mastic is observed. With the increase in frequency from 0.1 to 10 Hz, the effect of granular material in the mastic can be observed.

Following Ewoldt et al. (2008), elastic characterization measures (G'_M, G'_L, S) and viscous characterization measures (η'_M, η'_L, T) are calculated and tabulated in Table 1 and Table 2. G' and η' in Table 1 and Table 2 are the first harmonics of the stress waveform

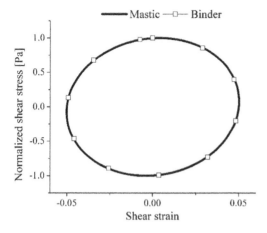

Figure 1b. Elastic Lissajous-Bowditch plot for binder and mastic at 5 %, 0.1 Hz and 40°C.

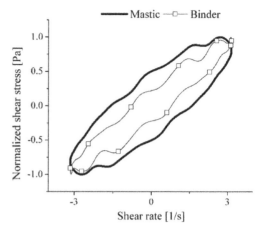

Figure 2b. Viscous Lissajous-Bowditch plot for binder and mastic at 5 %, 10 Hz and 40°C.

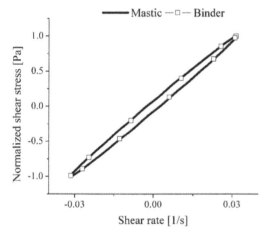

Figure 1c. Viscous Lissajous-Bowditch plot for binder and mastic at 5 %, 0.1 Hz and 40°C

Table 1. Strain stiffening ratio (*S*) for binder and mastic.

Sample	Strain amplitude %	G'_L Pa	G'_M Pa	G' Pa	S
VG30 at	1	398	822	397	−1.065
0.1 Hz	5	315	293	320	0.071
Mastic at	1	1613	2933	1578	−0.818
0.1 Hz	5	1324	1379	1352	−0.041
VG30 at	1	3268	2668	3229	0.183
1 Hz	5	6601	6780	6656	−0.027
Mastic at	1	9493	9067	9455	0.045
1 Hz	5	8319	8096	8312	0.027
VG30 at	1	279280	287050	274180	−0.028
10 Hz	5	80511	247420	81337	−2.073
Mastic at	1	417880	376460	418830	0.099
10 Hz	5	399250	511570	375140	−0.281

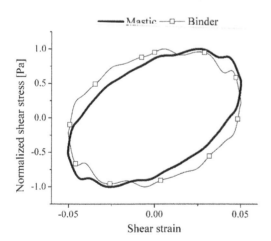

Figure 2a. Elastic Lissajous-Bowditch plot for binder and mastic at 5 %, 10 Hz and 40°C.

when written in terms of shear strain and shear rate, respectively.

None of the values of *S* and *T* converge to zero, as observed from Table 1 and Table 2. But *S* and *T* values are sensitive to the geometric material functions obtained from the experiments. Considering the experimental results of the geometric material functions to vary within ±5%, the corresponding value of *S* and *T* ranges from -0.095 to +0.095. The response of the material can be considered as linear elastic and linear viscous within the range. From Table 1 and Table 2, the values of *S* and *T* do not follow any specific trend with an increase in strain amplitude. A similar observation was reported by Diab & You (2018).

At 1% strain amplitude and 1 Hz frequency, the intracycle nonlinearity indicating strain stiffening is observed in the binder as against mastic exhibiting a linear elastic response. This indicates a difference in the linearity limit of binder and mastic at 1 Hz frequency. Similar behavior is observed at 10 Hz frequency (Table 1).

Table 2. Shear thickening ratio (T) for binder and mastic.

Sample	Strain amplitude %	η'_L Pa.s	η'_M Pa.s	η' Pa.s	T
VG30 at	1	8179	7945	8009	0.029
0.1 Hz	5	7897	8026	7895	−0.016
Mastic at	1	31628	31295	30831	0.011
0.1 Hz	5	27936	25318	27940	0.094
VG30 at	1	3254	3541	3346	−0.088
1 Hz	5	7532	7333	7533	0.026
Mastic at	1	12262	12232	12261	0.003
1 Hz	5	11208	11383	11220	−0.016
VG30 at	1	2330	6495	2885	−1.788
10 Hz	5	5494	1854	5366	0.663
Mastic at	1	13165	14377	13605	−0.092
10 Hz	5	10956	8098	10733	0.261

From Table 2, at 1 % strain amplitude and 10 Hz frequency, the intracycle viscous nonlinearity indicating shear softening is observed in the binder as against mastic exhibiting linear response. This indicates a difference in the linearity limit of binder and mastic at 10 Hz frequency.

From Table 1, both the binder and mastic at 0.1 Hz and the binder at 1 Hz exhibit nonlinear response at 1 % strain amplitude as against exhibiting linear response at 5 % strain amplitude. It is interesting to note such a counter-intuitive trend on the influence of strain amplitude on the extent of nonlinearity.

5 CONCLUSION

LAOS technique is used to study the difference in the response of bituminous binder and mastic. From the Lissajous-Bowditch plot, it is observed that the presence of granular material influences the behavior of mastic at 10 Hz frequency. The geometric measures of elastic and viscous Lissajous-Bowditch plots and the dimensionless indices, S and T, were used to delineate the linear/nonlinear response using the waveform data recorded. In addition, S and T values provided information on the behavior of binder and mastic in the intracycle elastic and viscous nonlinear regime. The elastic and viscous linear limit of binder and mastic are not identical at all the test frequencies.

The evaluation of material functions describing the intracycle nonlinearity qualitatively using the geometry of the Lissajous-Bowditch plot, as shown in this study, is the first step in analyzing the material response in the nonlinear regime. From the full waveform data obtained, one can perform different types of analysis, such as energy dissipation, temperature susceptibility, and fatigue damage, to mention a few.

REFERENCES

Chen, M., Javilla, B., Hong, W., Pan, C., Riara, M., Mo, L. & Guo, M. 2019. Rheological and interaction analysis of asphalt binder, mastic and mortar. *Materials*, 12(1), 128.

Diab, A. & You, Z. 2018. Linear and nonlinear rheological properties of bituminous mastics under large amplitude oscillatory shear testing. *Journal of Materials in Civil Engineering*, 30(3), 1–11.

Ewoldt, R. H., Hosoi, A. E. & McKinley, G. H. 2008. New measures for characterizing non-linear viscoelasticity in large amplitude oscillatory shear. *Journal of Rheology*, 526, 1427–1458.

Hyun, K., Wilhelm, M., Klein, C.O., Cho, K.S., Nam, J.G., Ahn, K.H., Lee, S.J., Ewoldt, R.H. & McKinley, G.H. 2011. A review of nonlinear oscillatory shear tests: Analysis and application of large amplitude oscillatory shear (LAOS). *Progress of Polymer Science*, 36, 1697–1753.

IS 73, 2018.: Paving bitumen – specification. *Bureau of Indian Standards*, 4th revision, New Delhi, India.

Kim, Y.R. & Little, D.N. 2004. Linear viscoelastic analysis of asphalt mastics. *Journal of Materials in Civil Engineering*, 16(2), 122–132.

Liao, M.C., Chen, J.S. & Airey, G. 2015. Characterization of viscoelastic properties of bitumen-filler mastics. *Asian Transport Studies*, 3(3), 312–327.

Padmarekha, A., Chockalingam, K., Saravanan, U., Deshpande, A.P. & Krishnan, J.M. 2013. Large amplitude oscillatory shear of unmodified and modified bitumen. *Road Materials and Pavement Design*, 14(1), 12–24.

Phan, C.V., Benedetto, H.D., Sauzéat, C. & Lesueur, D. 2016. Influence of hydrated lime on the linear viscoelastic properties of mastics. In proceedings, 6th Eurasphalt & Eurobitume Congress, 211, Prague, Czech Republic.

Underwood, B. S. & Kim, Y. R. 2015. Nonlinear viscoelastic analysis of asphalt cement and asphalt mastics. *International Journal of Pavement Engineering*, 16(6), 510–529.

Advances in Materials and Pavement Performance Prediction II – Kumar et al. (eds)
© 2021 Taylor & Francis Group, London, ISBN 978-0-367-46169-0

Molecular dynamics simulation of tensile failure of asphalt binder

W. Sun & H. Wang
Department of Civil and Environmental Engineering, Rutgers University, New Brunswick, USA

ABSTRACT: This study proposed an innovative computer modeling method to investigate tensile strength and cohesive crack of asphalt binder using Molecular Dynamics (MD) simulation. The tensile stress-strain curve was obtained from MD simulations and fracture properties were calculated using cohesive zone models. The result showed that the effect of loading rate on cohesive failure properties was not significant, which agreed with previous experiment findings. The increase of temperature reduced tensile strength but increased fracture toughness of asphalt binder. The simulation results suggest the potential of MD simulation in studying chemo-mechanical link of asphalt binder.

1 INTRODUCTION

Asphalt binder works as binding agent in asphalt mixture, providing cohesion and adhesion with aggregates. Under repetitive traffic loading and complicated environmental influences, microcracks may develop in asphalt binder and the fracture resistance of asphalt binder decrease due to aging. Damage occurs within asphalt binder under tensile or shear stress was regarded as cohesive failure. Previous researches have shown that cohesive failure of asphalt binder can occur along with the debonding of asphalt-aggregate system and results in fracture in asphalt mixtures (Fromm 1974; Zhang et al. 2016).

Experimental research usually employed tensile test on thin film of asphalt binder to measure tensile strength, considering the effects of loading rate, temperature, thickness of film, and aspect ratios (Harvey & Cebon 2003, 2005; Masad et al. 2010). Besides the influence of testing parameters, another important aspect that affect the cohesive behavior of asphalt binder is the property of material. It is well known that asphalt binder was an extremely complex compounds with four major components, asphaltene, resin, aromatic, and saturate (SARA). As the largest and most complicated components with highly polarity, asphaltene is normally assumed to contribute the majority part on the properties of asphalt. Sultana & Bhasin (2014) reported that the larger fraction of polarity component, such as asphaltene, would induce increase of stiffness and tensile strength. However, the observations in chemical structural and microscopic characteristics of asphalt binder remained to be controversial (Mullins 2010). Some research indicated that the asphaltene existed in the admixture as separated colloidal micelles (Murgich, et al. 1999), while others proposed that asphaltene was soluble to maltenes (Redelius 2010). These observations implied the importance of understanding molecular structure of asphalt binder.

Numerous researches have been conducted to quantify functional groups of asphalt binder using laboratory measurements, such as gel-permeation chromatography (GPC) or Fourier-transform infrared spectroscopy (FTIR) (Hofko et al. 2017). However, due to the large amount of molecules in asphalt binder with different molecular sizes and configurations, it is extremely difficult to obtain an accurate model of asphalt binder. Instead, molecular dynamics (MD) simulation method employed a concept of assuming molecular models of fractional components and assembling them in different ratios to simulate various asphalt binders. More importantly, MD can provide detailed molecular-level information that how the chemical composition and molecular structure of materials affect properties.

Recently MD simulation has been used to investigate the microscopic behavior of asphalt binder, such as adhesion with aggregate (Xu &Wang 2016; Gao et al. 2018), diffusion and self-healing (Bhasin et al. 2011; Sun et al. 2016; Xu & Wang 2018). However, the cohesive failure of asphalt binder has not been studied with MD simulation, which usually emerged simultaneously with debonding of asphalt-aggregate, and was the initial stage for the occurrence of self-healing.

This paper aims at using MD simulation as a computational tool to (1) conduct tensile tests for asphalt binder models, (2) investigate the effect of temperature and loading rate on cohesive failure behavior of asphalt binders.

2 MODELS AND SIMULATION METHODS

2.1 Molecular models for asphalt binder

Asphalt is a main byproduct from crude oil with highly complicated chemical composition. The chemical composition and molecular structure of asphalt binder are known to affect its rheological and mechanical

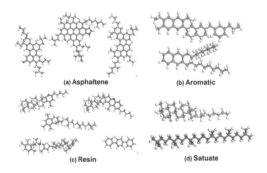

(a) Asphaltene   (b) Aromatic

(c) Resin   (d) Satuate

Figure 1. Molecular model illustration of 12 components for asphalt binder.

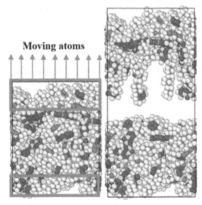

Moving atoms

Fixed atoms

Figure 2. Illustration of tensile test simulation on asphalt binder: before failure (left) and after failure (right).

properties and thus field performance of asphalt pavement. Asphalt is commonly divided as four components, which are known as SARA (Satuarate, Aromatic, Resin and Asphaltene). Asphaltenes were regarded as an important part in affecting the overall properties, and its molecular composition and structures can be assumed as polycyclic aromatic compounds with heteroatoms and alkyl side chains. Li and Greenfield (2014a) developed asphalt model system of SHRP AAA-1 asphalt constituted by 12 components as shown in Figure 1. The model was extensively employed in related research (Li & Greenfield 2014b, Khabaz & Khare 2015), which showed competence in revealing molecular interaction and behaviors of asphalt binder. The mass fraction of SHRP AAA-1 asphalt is 16.5% for asphaltene, 38.1% for resin, 30.6% for aromatic, and 10.7% for saturate, respectively. The model was relaxed in a 200ps NPT emsemble followed by a 500ps NVT emsemble.

2.2 Simulation of tensile test

The tensile failure and cohesive strength of asphalt binder can be determined from direct tension test in the traditional laboratory test. In this study, direct tension tests were simulated utilizing MD simulations in order to study the cohesive failure phenomenon of asphalt binder at the atomistic scale.

In the simulation of direct tension test, some constraints need to be applied in order to create microcracks within the asphalt binder to simulate the real experimental test. Therefore, a thin film of atoms at both the top and bottom of the asphalt model was controlled as rigid body to have no velocity and force throughout the simulation process to simulate the tensile grips. With the upward movement at a specific velocity of the fixed top atoms, the rest segments of atoms will be affected by the inter-molecular force from each other. During the tensile test, non-periodical boundary condition (shrink wrapping) was applied along z-direction. Because of constricting the movement of molecules at the bottom, the whole asphalt binder model behaves like to be pulled apart (as shown in Figure 2). By recording the traction force during the whole simulation, the tensile stress-strain curve can

be obtained to quantify cohesive failure behavior of asphalt binder. To investigate the effect of other impact factors, a baseline case of tensile test simulation was performed at temperature of $0°$ and loading rate of 10m/s. It should be noted that system's dimension was 38*38*50Å after equilibration, meaning that the tensile strain rate was very high as $2*10^9$/s compared with macroscopic experiment.

2.3 Molecular dynamics simulation details

The OPLS-AA force field (Robertson 2015), which was developed and adopted to simulate organic substances, was employed for asphalt binder model to describe the valence and non-bond interaction between atoms. To validate the molecular model, the density values of well-equilibrated asphalt model was calculated and the values was 0.996 g/cm^3 at 25°, which showed general agreement with the results obtained from experiments and previous MD simulations.

3 RESULTS AND DISCUSSIONS

3.1 Stress-strain relationship and exponential cohesive zone model

As shown in Figure 2, micro-crack formed with the proceeding of tensile test. The traction forces between the atoms in the separate crack faces were recorded. The results of stress-strain relationship in the tension-induced cohesive failure process are shown in Figure 3. The stress-strain curve showed that the tensile stress experienced initially quickly increase up to a peak value. Then the tensile stress began to decrease because the growing separation of atoms reduce the molecular interaction, which can be considered as the cohesive damage initiation. With the tensile strain increased further, micro-crack propagated continuously and the stress kept decreasing to nearly zero, which indicated the asphalt binder was fully cracked.

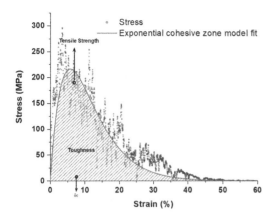

Figure 3. Illustration of tensile test simulation on asphalt binder.

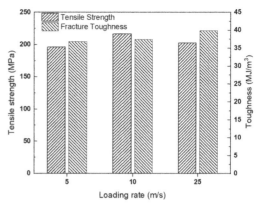

Figure 4. Effect of loading rate on tensile test indicators.

The tensile stress data were collected every 100 fs in the simulation, which showed certain variations due to the Brownian movements of atoms in the system. An exponential cohesive zone model was selected to describe the tensile stress-strain relationship as shown in Eq. (1).

$$\sigma = \sigma_c \left(\frac{\delta}{\delta_c} \right) e^{\left(1 - \frac{\delta}{\delta_c} \right)} \qquad (1)$$

Where, σ_c is peak stress value; and δ_c is corresponding strain to σ_c.

From the fitting of cohesive zone model, several mechanical properties regarding to tensile fracture failure of asphalt binder can be calculated, as shown in Figure 3. Tensile strength was defined as the peak stress in the failure process; while the area beneath the fitting curve was calculated as fracture toughness. The CZM fitting curve appeared insufficiently to capture the peak value due to the greater variation. However, as the temporal and special limitation of MD simulation, the value can hardly compare with experiments. Therefore, the model fitting employed was mainly focused on demonstrating the trend of influencing factors.

3.2 Effect of loading rate

The loading rates considered in the sensitivity analysis were 5m/s, 10m/s, and 25m/s. These rates were much greater than the ones used in macroscopic experiments due to the consideration of computational expense. The calculated results were shown in Figure 4. It should be first noted that the calculated values of tensile strength and fracture toughness were much higher than experimental values. This is expected due to the time and spatial scope used in MD simulations, which also existed in other similar research (Xu & Wang 2016; Wang et al. 2017). Although the simulated values at the nano-scale cannot be directly used to represent the mechanical properties measured at traditional experiment, MD simulation can provide a powerful and

convenient way to study the effects of influencing factors.

From Figure 4, it can be concluded that the increasing loading rate has minor effects on tensile strength and fracture toughness. Tensile strength showed variations with the change of loading rate, while fracture toughness exhibited slight increase with the increase of loading rate. The results were in agreement with previous experimental works, demonstrating that loading rate had little impact on tensile test results of asphalt thin films (Marek & Herrin 1968; Harvey & Cebon 2005).

3.3 Effect of temperature

The mechanical properties of asphalt binder are significantly affected by the temperature. In this study, three temperatures ($-10°C$, $0°C$, and $10°C$) were adopted to conduct the tensile test simulation. Figure 5 showed the calculated values of tensile stress and fracture toughness of asphalt binder at three temperatures. It can be seen that with the increase of temperature, the tensile strength exhibited reduction, especially from $0°$ to $10°$. However, fracture toughness gradually increased with the temperature. This is expected since viscoelastic materials tend to transit from brittle status to ductile status when temperature increases. From this perspective, it was reasonable that asphalt binder showed better post-failure resistance even though its tensile strength decreased as temperature increased, as the ductility offered the material to tolerate more deformation.

4 CONCLUSIONS

This study performed MD simulations to investigate the tensile strength and cohesive failure pattern of asphalt binder. During the MD simulation process of asphalt binder, the tensile stress-strain curves followed the trend observed from laboratory experiments conducted at the macroscopic sale. The tensile strength and fracture toughness can be calculated using an exponential CZM model.

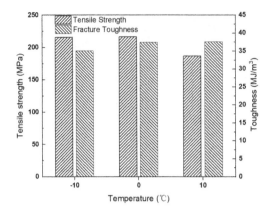

Figure 5.   Effect of temperature on tensile test indicators.

The loading rate was found not significantly influence tensile properties of asphalt binder, which agreed with experimental findings. The effect of temperature on cohesive failure of asphalt binder was observed. Although tensile strength decreased at high temperature, fracture toughness increased as the materials became more ductile.

The simulation results suggest the potential of MD simulation in studying chemo-mechanical link of asphalt binder. Future work will be conducted to study the effect of SARA fractions and aging on mechanical properties of asphalt binder.

REFERENCES

Bhasin, A., Bommavaram, R., Greenfield, M. L. & Little, D. N. 2011. Use of Molecular Dynamics to Investigate Self-Healing Mechanisms in Asphalt Binders. *Journal of Materials in Civil Engineering,* 23**,** 485–492.

Fromm, H. J. 1974. the mechanisms of asphalt stripping from aggregate surfaces. Proc., Assoc. of Asphalt Paving Technologist, 43, 191–223.

Gao, Y. M., Zhang, Y. Q., Gu, F., Xu, T. & Wang, H. 2018. Impact of minerals and water on bitumen-mineral adhesion and debonding behaviours using molecular dynamics simulations. Construction and Building Materials, 171, 214–222.

Harvey, J. A. F. & Cebon, D. 2003. Failure mechanisms in viscoelastic films. Journal of Materials Science, 38, 1021–1032.

Harvey, J. A. F. & Cebon, D. 2005. Fracture Tests on Bitumen Films. Journal of Materials in Civil Engineering, 17, 99–106.

Hofko, B., Alavi, M. Z., Grothe, H., Jones, D. & Harvey, J. 2017. Repeatability and sensitivity of FTIR ATR spectral analysis methods for bituminous binders. Materials and Structures, 50, 187.

Li, D. D. & Greenfield, M. L. 2014a. Chemical compositions of improved model asphalt systems for molecular simulations. Fuel, 115, 347–356.

Li, D. D. & Greenfield, M. L. 2014b. Viscosity, relaxation time, and dynamics within a model asphalt of larger molecules. J Chem Phys, 140, 034507.

Khabaz, F. & Khare, R. 2015. Glass Transition and Molecular Mobility in Styrene-Butadiene Rubber Modified Asphalt. J Phys Chem B, 119, 14261–14269.

Marek, C. R. & Herrin, M. 1968. Tensile behavior and failure characteristics of asphalt cements in thin films. Journal of Association of Asphalt Pavement Technologists, 37, 386–421.

Masad E, E. H. J., Bhasin A, Caro S, N Little D. 2010. Relationship of ideal work of fracture to practical work of fracture. J Assoc Asphalt Paving Technol, 79, 81–118.

Mullins, O. C. 2010. The Modified Yen Model. Energy & Fuels, 24, 2179–2207.

Murgich, J., Abanero, J. A. & Strausz, O. P. 1999. Molecular recognition in aggregates formed by asphaltene and resin molecules from the Athabasca oil sand. Energy & Fuels, 13, 278–286.

Redelius, P. 2004. Bitumen solubility model using Hansen solubility parameter. Energy & Fuels, 18, 1087–1092.

Robertson, M. J., Tirado-Rives, J. & Jorgensen, W. L. 2015. Improved Peptide and Protein Torsional Energetics with the OPLS-AA Force Field. Journal of Chemical Theory and Computation, 11, 3499–3509.

Sultana, S. & Bhasin, A. 2014. Effect of chemical composition on rheology and mechanical properties of asphalt binder. Construction and Building Materials, 72, 293–300.

Sun, D. Q., Lin, T. B., Zhu, X. Y., Tian, Y. & Liu, F. L. 2016. Indices for self-healing performance assessments based on molecular dynamics simulation of asphalt binders. Computational Materials Science, 114, 86–93.

Xu, G. J. & Wang, H. 2016. Molecular dynamics study of interfacial mechanical behavior between asphalt binder and mineral aggregate. Construction and Building Materials, 121, 246–254.

Xu, G.J. & Wang, H. 2018 Diffusion and Interaction Mechanism between Rejuvenating Agent and Virgin and Recycled Asphalt Binders A Molecular Dynamics study, Molecular Simulation, 44(17) 1433–1443

Wang, H., Lin, E.Q. & Xu, G.J. 2017, Molecular Dynamics Simulation of Asphalt-Aggregate Adhesion Strength with Moisture Effect, International Journal of Pavement Engineering, 18(5), 414–423

Zhang, J., Airey, G. D. & Grenfell, J. R. A. 2016. Experimental evaluation of cohesive and adhesive bond strength and fracture energy of bitumen-aggregate systems. Materials and Structures, 49, 2653–2667.

Advances in Materials and Pavement Performance Prediction II – Kumar et al. (eds)
© 2021 Taylor & Francis Group, London, ISBN 978-0-367-46169-0

Dissolution simulation of polymers in bitumen

H. Wang, P. Apostolidis, H. Zhang, X. Liu & S. Erkens
Section of Pavement Engineering, Faculty of Civil Engineering and Geosciences, Delft University of Technology, Delft, The Netherlands

A. Scarpas
Department of Civil Infrastructure and Environmental Engineering, Khalifa University of Science and Technology, Abu Dhabi, United Arab Emirates
Section of Pavement Engineering, Faculty of Civil Engineering and Geosciences, Delft University of Technology, Delft, The Netherlands

ABSTRACT: Fundamental models should be developed and utilized in order to facilitate the chemo-mechanical design of modified binder systems for paving applications but not only. Especially, the fact that the incorporation of new chemical substances used as bio-based modifiers or alternative binders is attracting great interest to replace traditional technologies, the development of tools able to provide insight into the various physio-chemical phenomena is crucial. Among other polymer-bitumen interaction phenomena, the dissolution mechanism of polymers in bitumen is a significant aspect that should be considering in order to enhance binder properties through polymer modification. The current research gives emphasis on modelling the mechanism of dissolution for rubbery polymers in bitumen.

1 INTRODUCTION

Building fundamental tools and insights into the chemo-mechanical design of polymer-bitumen systems remains crucial in the field of asphalt chemistry. Bio-based modifiers provide a unique opportunity for replacing petroleum derived chemicals in pavement systems and their success depends on selecting the suitable bitumen and polymers in order the latter to be disentangled at the desired rate and extent over time. For instance, lignin or bio-based waxes are naturally occurring glassy or crystal-like polymers with unique functionalities when they are dissolved in bitumen.

An appropriate understanding of the various controlling steps in the dissolution process is needed to tailor the bitumen modification techniques, especially when semi-crystalline or glassy bio-based polymers are incorporating in bitumen. Moreover, the control of disentanglement rate of polymeric chains in binders used for paving applications is associated with thermodynamics changes. Thus, the knowledge of the dissolution behaviour of certain polymers will enable material designers to develop well-dissolved modified binders of desired micro-structural and morphological features and ultimately mechanical properties. Herein, in order to provide guidance for the selection of suitable bitumen through the polymer modification, the dissolution mechanism of glassy polymers in binders is introduced and assessed as well by performing numerical simulations.

2 BACKGROUND

The mechanism of polymer dissolution into a solvent, such as bitumen, involves two phenomena; solvent diffusion and chain disentanglement. Once a polymer is in contact with a miscible solvent, species of the latter diffuses into the polymer and a gel-like layer is formed adjacent to the solvent-polymer interface due to the solvent induced-plasticization of polymer (Wang et al. 2019; Wang et al. 2020a; Wang et al. 2020b). Once semi-crystalline polymers are exposed to miscible solvents, unfolding of the crystalline regions is an additional step accompanying solvent diffusion and chain disentanglement. Hence, semi-crystalline polymer dissolution becomes equivalent to amorphous polymer dissolution after unfolding the crystals in polymer.

The polymer dissolution differs from dissolution of a non-polymeric material. Polymers require a time period before starting to dissolve, while non-polymeric materials dissolve instantaneously. Also, polymer dissolution can be controlled either by the disentanglement of the polymer chains or by chains diffusion through a boundary layer adjacent to the solvent-polymer interface. Moreover, swelling occurs and the polymer begins to release its contents to the surrounding solvent, either via solute diffusion or polymer dissolution, and hence, certain solvents exhibit an altered dissolution rate of the exposed areas of polymers.

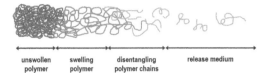

unswollen polymer | swelling polymer | disentangling polymer chains | release medium

Figure 1. Polymer dissolution process.

The entanglement process of glass polymers has been extensively assessed via modelling the concentration of acting species at polymer-liquid interface (Papanu et al. 1989) and the polymeric thickness (Devotta et al. 1994a, 1994b; Lee and Peppas 1987; Peppas et al. 1994) changes over time. Disentanglement kinetics related to the swelling rate have been modelled as well predicting thus the changing mobility of the disengaging macromolecule at the gel-liquid interface (Devotta et al. 1995). In this research, special emphasis was given on modelling the transport phenomena at the gel-liquid interface.

3 MUTUAL DIFFUSE-REACTION MODEL

As the solvent penetrates into the polymeric film, the solvent species diffuses into the material, the thickness of the latter may increase (swelling) depending on the natural of polymer, and a gel phase is formed at the polymer-solvent interface. Simultaneously, the polymeric chains at the gel-liquid interface disentangle themselves (see Figure 1). If the size of diffusing species of solvent is small, then the latter penetrates into the polymer quickly. Thus, the mobility of the gel-like chain increases faster (high diffusivity) and such highly mobile chains disengage at the gel-liquid interface rapidly.

By assuming the physical entanglement in the gel phase to chemical crosslinks, the chemical driving force of mutual diffusion process is determined by the chemical potential difference for the species of interacting phases p (polymer) and s (solvent) at the $s-p$ interface (see Figure 2) and expressed as

$$\Delta G_M = \sum_i^n \varphi_i^0 \left[\mu_{is}\left(\varphi_{is}^{\Gamma_{sp}}\right) - \mu_{ip}\left(\varphi_{ip}^{\Gamma_{sp}}\right) \right] \quad (1)$$

φ_i^0 where is the concentration of interacting species (i) transferred over the interface, n is the number of different interacting species (P and S), $\varphi_{is}^{\Gamma_{sp}}$ and $\varphi_{ip}^{\Gamma_{sp}}$ are the molecular fractions of interacting species of s and p phases at the s-p interface (denoted as Γ_{sp}), respectively, μ_{is} and μ_{ip} are the chemical potentials of the ith species in s and p phases.

In thermodynamic equilibrium, the chemical potential of solvent and polymeric species is equal at Γ_{sp} and it is expressed as

$$\mu_{Ps}(\varphi_{PS}, T, p) = \mu_{Pp}(\varphi_{Pp}, T, p) \text{ at } \Gamma_{sp} \quad (2)$$

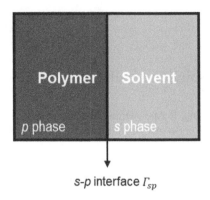

Polymer | Solvent

p phase | s phase

s-p interface Γ_{sp}

Figure 2. Schematic representation of the polymer and solvent phases.

In the complete binary case ($\varphi_{Ps} \neq 1$), there is also a relation similar to Eq. 2 for the other component.

For the case under consideration, both solvent phase and polymer phase are immobile. The mass balance equations for species P and S in the solvent phase are

$$\frac{\partial(\rho_s \varphi_{Ps})}{\partial t} - \nabla \cdot (\rho_s D_{Ps} \nabla \varphi_{Ps}) = 0 \quad (3)$$

$$\frac{\partial(\rho_s \varphi_{Ss})}{\partial t} - \nabla \cdot (\rho_s D_{Ps} \nabla \varphi_{Ss}) = 0 \quad (4)$$

where ρ_s is the density of solvent; D_{Ps} is the liquid binary diffusion coefficient; D_{Ss} is the self-diffusion coefficient of solvent.

The mutual diffuse-reaction process is controlled by thermodynamic equilibrium at the interface and is translated into a simple Dirichlet condition. Therefore, an equilibrium concentration for species P at the polymer-solvent interface is expressed as

$$\varphi_{Ps} = \varphi_{eq} \text{ at } \Gamma_{sp} \quad (5)$$

The mass balance equation for species P and S at the $s-p$ interface gives

$$(\rho_S D_{Ps} \nabla \varphi_{Ps}) \cdot n_{sp} = (\rho_p \varphi_{Pp} w) \cdot n_{sp} \text{ at } \Gamma_{sp} \quad (6)$$

$$(\rho_S D_{Ss} \nabla \varphi_{Ss}) \cdot n_{sp} = (\rho_p \varphi_{Sp} w) \cdot n_{sp} \text{ at } \Gamma_{sp} \quad (7)$$

The total mass balance at the $s-p$ interface gives

$$\rho_s w \cdot n_{sp} = \rho_p w \cdot n_{sp} \text{ at } \Gamma_{sp} \quad (8)$$

where w is the velocity of the s-p interface. The whole balance equations presented above are sufficient to solve the physical problem, provided that the overall surrounding boundary conditions are given as

$$n_{sp} \cdot w = n_{sp} \cdot \left(\frac{\rho_s}{\rho_p(1 - \varphi_{Ps})} D_{Ps} \nabla \varphi_{PS} \right) \text{ at } \Gamma_{sp} \quad (9)$$

$$0 = n_{sp} \cdot \left(\frac{\rho_s - \rho_p}{\rho_p(1 - \varphi_{Ps})} D_{PS} \nabla \varphi_{PS} \right) \text{ at } \Gamma_{sp} \quad (10)$$

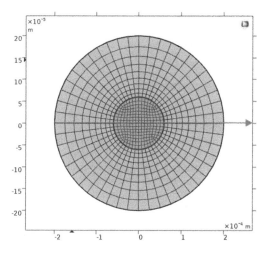

Figure 3. Finite element geometry and mesh (number of quad elements: 240 and 430 for polymer and solvent, respectively).

All the above expressions give the recession velocity and the s-phase velocity at the interface, which are necessary to implement the direct explicit numerical methods.

4 FINITE ELEMENT SIMULATIONS

The two-dimensional (2D) polymer dissolution in bitumen at $100°C$ was simulated herein by implementing the above described concurrent mutual diffuse-reaction model in a commercially available multiphysics tool. 2D mesh is demonstrated in Figure 3 where the blue part represents the polymer surrounding with the solvent. The disentanglement rate and the activation energy were selected as 2000 1/s and 30 kJ/mol at $100\,°C$. The diffusion coefficient assumed constant as 1e-13 m^2/s ($D_{PS} = D_{Sp}$), with the initial concentration of interacting species to be equal as 0.001 mol/m^3 ($\varphi_{PS} = \varphi_{Ss} = \varphi_{Pp} = \varphi_{Sp}$).

Figure 4a and 4b show the general prediction of the variation of surface concentration of the polymeric and dissolved species in the domains of both polymer and solvent phases after the diffusion and disentanglement, respectively. In particular, Figure 4a clearly demonstrates the continuous decrease of polymeric species concentration from the centre of polymer phase to solvent. From Figure 4b, it can be found the polymer-solvent interface zone shows higher concentration of dissolved species.

To further investigate the concentration profile, the concentration of dissolved species along the horizontal axis (shown as the red line in Figure 3) over time is plot in Figure 5. Obviously, the highest concentration of dissolved species is formed at the polymer-solvent interface where abundant polymer and solvent species are available to react resulting in the highest reaction/dissolution rate. Toward the inner layer of polymer

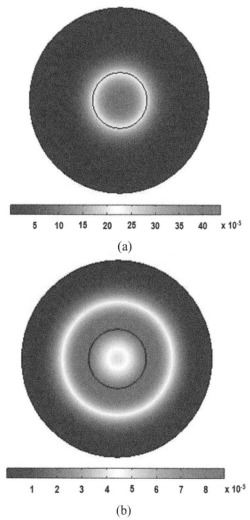

(a)

(b)

Figure 4. Predicted surface concentrations of (a) polymeric species and (b) dissolved species after diffusion disentanglement.

phase, the concentration decreases. In solvent, similar phenomenon is observed at the locations away from the interface. This is due to lack of either polymeric or solvent species. As reaction proceeds, more and more polymeric species are dissolved into solvent and the dissolved species diffuse to the surrounding space, causing an overall concentration increase of dissolved species.

Figure 6 summarizes the concentrations of solvent, polymer and dissolved species in the polymer phase at the $s - p$ interface over time. It is apparent that the concentration of solvent species gradually decreases while the concentration of dissolved species gradually increases as reaction proceeds. However, for the polymeric species in the solvent, it first increases due to the diffusion from the polymer phase and

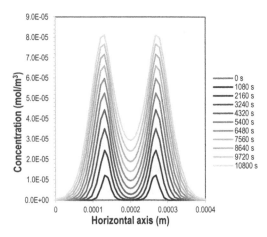

Figure 5. Predicted variation of concentration of dissolved species along the horizontal axis over time.

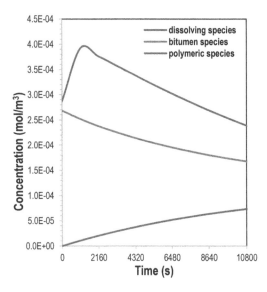

Figure 6. Predicted variation of concentration of solvent, polymer and dissolved species in the dissolution medium at the *s-p* interface over time.

then, their concentration starts decreasing. The dissolution rate is higher than the diffusion rate, which enables the continuous decrease of the amount of polymeric species. The dissolution rate of polymer increases with increased amount of solvent species until a maximum rate is reached. After reaching a maximum, the dissolution rate decreases resulting in reduced swelling.

5   CONCLUSIONS

The polymer-bitumen interaction plays critical role in controlling the properties of polymer modified binders. At high temperatures, the polymer-bitumen interaction is a polymer dissolution process, which consists of swelling and chain disentanglement, and a degradation process. Full control of dissolving species in bitumen will help on obtaining the desired micro-structural and morphological features and subsequently the goal mechanical properties. In this study, emphasis was given on modeling the mechanism of dissolution, and particularly of chain disentanglement, of an already swollen polymer in solvent, or bitumen. Predictions on mass transport of reacting species are provided assisting on understanding the controlling thermodynamic parameters of polymer-bitumen interaction phenomena at micro-scale.

For future studies, temperature effects and the impact of polymer size and structure on the dissolution process may be considered.

REFERENCES

Devotta, I, et al. 1995. Unusual retardation and enhancement in polymer dissolution: role of disengagement dynamics. *Chemical Engineering Science 50(16)*, pp. 2557–2569.

Devotta, I., & R.A. Mashelkar. 1996. Role of thermodynamic and kinetic factors in polymer dissolution in mixed solvents. *Chemical Engineering Communications 156(1)*, pp. 31–45.

Papanu, J.S., et al. 1989. Transport models for swelling and dissolution of thin polymer films. *Journal of Applied Polymer Science 38(5)*, pp. 859–85.

Ranade, V.V., & R.A. Mashelkar. 1995. Convective diffusion from a dissolving polymeric particle. *AIChE Journal 41(3)*, pp. 666–676.

Wang, H., et al. 2019. Numerical investigation of rubber swelling in bitumen. *Construction and Building Materials 214*, pp. 506–515.

Wang, H., et al. 2020a. Experimental investigation of rubber swelling in bitumen. *Transportation Research Record*, 0361198120906423.

Wang, H., et al. 2020b. The role of thermodynamics and kinetics in rubber–bitumen systems: a theoretical overview. *International Journal of Pavement Engineering*, 1–16.

Advances in Materials and Pavement Performance Prediction II – Kumar et al. (eds)
© 2021 Taylor & Francis Group, London, ISBN 978-0-367-46169-0

Preliminary study on using lignin as aging inhibitor in bitumen

Y. Zhang
School of Highway, Chang'an University, Xi'an, China
Section of Pavement Engineering, Delft University of Technology, Delft, The Netherlands

X. Liu, R. Jing, P. Apostolidis & S. Erkens
Section of Pavement Engineering, Delft University of Technology, Delft, The Netherlands

A. Scarpas
Department of Civil Infrastructure and Environmental Engineering, Khalifa University of Science and Technology,
Abu Dhabi, United Arab Emirates
Section of Pavement Engineering, Delft University of Technology, Delft, The Netherlands

ABSTRACT: During oxidative aging, oxygen reacts with active molecules present in bitumen producing polar compounds, principally ketones and sulfoxides, and increasing in the portion of asphaltenes. In general, oxidation reactions in bitumen yields to change its generic chemical composition and finally its colloidal structure deteriorating the physico-mechanical properties. Lignin is a natural polymer, which has been used in this study as an aging inhibitor to bitumen. Particularly, the effect of aging on the microstructure morphology, surface properties, chemical composition and rheological changes of lignin and the impact of latter as anti-oxidant in bitumen were evaluated. For the purposes of this study, Environmental Scanning Electron Microscope, Helium Pycnometer, Dynamic Vapor Sorption devices and were used to analyze the microstructure, density and specific surface area, respectively. Moreover, Fourier Transform Infrared spectroscopy was used to track the compositional changes in lignin-modified bitumen after PAV aging. Dynamic Shear Rheometer was used to analyze the rheological properties. Overall, decreasing in the carbonyl and sulfoxide compounds were tracked in lignin-modified binders confirm that lignin act as an aging inhibitor in bitumen.

1 INTRODUCTION

Bitumen is a complex petroleum-based material which is the most widely used binder for paving applications. However, considering the uncertainty in crude oil supply, alternative binders are encouraged to be used as a replacement of bituminous binders or performance modifiers. Especially, lignin, among others, has attracted considerable attention as a partial substitute or modifier of bitumen (Xu et al. 2017). Lignin is the most abundant natural polymer on the Earth, with the total amount of lignin present in the biosphere to exceed 300 billion tons and increase by approximately 20 billion tons every year (Bruijnincx et al. 2016). Lignin can be found as well in co-products of timber production, or in byproducts of paper and hydrolytic industries. Thus, the utilization of lignin in binders specially designed for pavements may bring large economic benefits to sustainable development.

Nevertheless, the use of lignin is not limited only as a replacement of bitumen or modifier but it can be incorporated as an aging inhibitor as well in order to minimize the aging potential and to improve the durability of binders (Apostolidis et al. 2017). Particularly, during oxidative aging, oxygen reacts with molecules present in bitumen producing polar compounds, principally ketones and sulfoxides, and secondary causing increase in the portion of asphaltenes. In general, oxidation reactions in bitumen yields to change its generic chemical composition and finally its colloidal structure deteriorating the physicomechanical properties. Lignin has found to retard bitumen oxidation (Batista et al. 2018; Gosselink et al. 2011; Sundstrom et al. 1983; Williams et al. 2008) mainly due to the lignin radical scavenging activity, its polyphenolic structure (Boeriu et al. 2004; Dizhbite 2004; Pan 2012), and the physical interaction between bitumen and lignin (Zhang 2019). However, lignin itself also has a large number of carbonyl and sulfoxide functional groups, which should be taken into account separately from the bituminous functional groups that evolve due to bitumen aging (Petersen 2009). Thus, a preliminary study was performed to evaluate the anti-aging effect of lignin in bitumen by differentiating the aging functional groups of bitumen and lignin.

2 MATERIALS AND METHODS

2.1 Material preparation

In this study, a 70/100 pen grade bitumen with 47.5°C softening point was used. The organsolv lignin (brownish powder) was provided by Chemical Point UG (Germany), having a purity of above 87%. On the basis of previous studies (Van Vliet et al. 2016), various contents of lignin were added into the bitumen in this study. The abbreviations Bref, BL05, BL10, BL20, BL30 stood for the added lignin by 0, 5, 10, 20 and 30% of the mass of bitumen.

The high shear mixer was used to mix lignin with bitumen. Lignin was gradually added in bitumen, the mixing temperature and rate were 163°C and 3000 rpm, respectively. The whole mixing process took about 30 minutes. Pressure Aging Vessel (PAV) was carried out to simulate the long-term aging of bitumen based on the standard testing procedure (NEN-EN 14769). To be specific, 50 ± 0.5 g bitumen was poured into a PAV pan to form a film with 3.2-mm thickness. Then, the PAV test was performed at a temperature of 100°C at 2.10 MPa for 20 hours.

2.2 Morphology, density and specific surface area characterization

An Environmental Scanning Electron Microscope (ESEM) was used to observe the surface morphology of lignin at ambient temperature. The magnification was varied by $\times 125$ and $\times 1000$. To ensure drying of lignin before scanning, lignin was oven-conditioned at 80°C for 24 hours. Further, approximately 20 mg of lignin was poured onto an 8-mm diameter sample holder to prepare the ESEM samples. It was important to prevent the samples from dust or other impurities before ESEM scanning.

Density measurements of lignin particles were conducted in a Helium Pycnometer (HP). A high precision electronic scale was used to weigh a small amount of sample. Helium was gradually filled in the chambers containing the lignin when the equilibrium was reached, the pressure of the chamber was measured. Then, expansion valve opened automatically and the gas flowed into the expansion chamber. Once the equilibrium was reached, the pressure of the current condition was measured. By comparing twice equilibrium pressures and volume of chamber, the volume of lignin calculated and density was determined.

The specific surface area was calculated by Braunauer-Emmett-Teller (BET) method after conducting surface measurements in a Dynamic Vapor Sorption (DVS) device. The amount of lignin was about 50 mg necessary due to the high uptakes of vapors on the high surface area. The ambient temperature 25°C was selected, and a 100-sccm flow rate was used in the test. The typical partial pressure range for BET experiments was 0.05 to 0.30% (P/P_0). It was therefore ideal to collect an isotherm between these limits using 5% P/P_0 steps. The BET method assumed a physical adsorption mechanism in monolayer coverage was obtained, followed by multilayer adsorption, it could be applied to any adsorption system. The total surface area and the specific surface area were determined.

2.3 Chemical characterization

FTIR spectroscopy of a single-point Attenuated Total Reflectance (ATR) fixture was used to collect spectra data of lignin and bitumen samples and to track chemical compositional changes. The wavenumber ranged from 600 to 4000 cm^{-1} with a resolution of 4 cm^{-1}. Nine replications for each sample were analyzed. Different functional groups have a different light-absorption spectrum. The functional group aging index (AI) was used for the main absorption bands of lignin to compare the changes of functional groups with the changes of spectra. The range of chemical functional groups of lignin and bitumen were calculated and considered has been summarized in (Zhang et al. 2019).

The aging effect of lignin itself is not obvious or even negligible but still contains the aging functional groups. In order to eliminate the effect of lignin itself, the increase ratio index (IRI) was calculated by the changes in the aging index under different aging conditions and the proportion of bitumen in the mix. The equation is listed below.

$$IRI = \frac{AI_{Aged} - AI_{Fresh}}{W} \qquad (1)$$

where AI_{Aged} is the aging index of samples in aged condition; AI_{Fresh} is the aging index of samples in unaged condition; W is the mass fraction of bitumen in the mix, $W = m_{bitumen}/m_{total}$.

2.4 Rheological characterization

To characterize the rheological properties, complex shear modulus (G^*) of bitumen over a wide range of temperatures and frequencies were performed by the Dynamic Shear Rheometer (DSR). In this study, the parallel-plates geometry with an 8-mm diameter and 2-mm gap were used at the range of temperature from -10 to 30°C with 10°C increment, and for relatively elevated temperature range from 30 to 60°C with 10°C increments used the 25-mm diameter plates and 1-mm gap. The samples were tested with a frequency sweep from 100 to 0.1 rad/s at each temperature. The master curves were generated by the results at a reference temperature of 20°C. The effect of the different content of lignin and aged conditions were depicted by master-curves.

3 RESULTS AND DISCUSSION

3.1 Microstructural morphology

The microstructure of lignin particles in different magnification (x125 and x1000) was shown in Figure 1. 10–200 μm was the range of the size of lignin particles, smaller fractions of particles that seem to be crushed from larger ones. The lignin particles were observed some angularities in fresh condition. There was an agglomeration phenomenon after aging, the

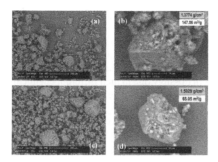

Figure 1. ESEM images; Magnification x125 of (a) fresh, (c) aged lignin; Magnification x1000 of (b) fresh and (d) aged lignin.

agglomeration phenomenon was confirmed as the density of aged particles increased. And the surface of the particles became smoother and color appearance was darker.

The density of the lignin was 1.3774 g/cm^3, which was measured by HP. The specific surface area was 147.0593 m^2/g, which was calculated by the BET method in DVS. The physical properties of lignin were measured after aging as well. The overall color of lignin particles became darker, the density increased to 1.5029 g/cm^3 and the specific surface area decreased to 65.0475 m^2/g. The specific surface area of fresh (unaged) lignin was two times more than that of aged lignin. The larger the powder area, the greater the friction between the particles. After aging, the density of lignin increased and its specific surface area decreased. In addition, its surface became smoother, possibly because it was in contact with oxygen during aging.

3.2 *Chemical compositional changes*

IR spectra results of various lignin modified bitumen are shown in Figure 2. Each IR spectra was the average result of the nine replications. The peaks of certain functional groups could be found in both base materials. They corresponded one by one, it showed that the lignin added in bitumen did not produce new chemical functional group peaks. This figure illustrated that lignin itself contains carbonyl and sulfoxide functional groups. The absorbance indices of fresh and aged conditions are shown in Figure 3. The values of indices were also the average of the nine measurements.

Figure 3 gives an overview of the aging indicators for different materials at fresh and aged conditions. The aging indices of various samples at the fresh state increased with the increase of lignin content. Aging indices of samples with 5% content of lignin in bitumen were relatively small, so the increase was very slight. Obviously, all aging indices increased with aging. To evaluate the anti-aging effect of different lignin modified binders, the aging indices were used to calculate the IRI described by Eq. 1. The results are shown in Figure 4.

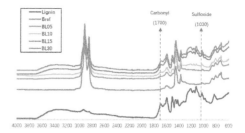

Figure 2. IR spectra of lignin-modified bitumen.

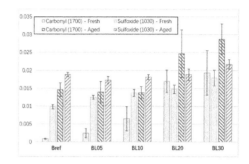

Figure 3. Carbonyl and sulfoxide indices of lignin-modified bitumen.

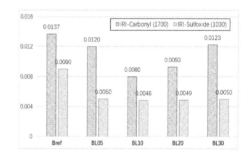

Figure 4. IRI of carbonyl and sulfoxide indices.

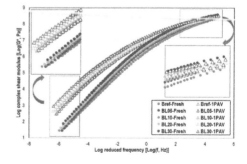

Figure 5. Master curves of complex shear modulus.

IRI had eliminated the effect of lignin. The higher the IRI value is, the faster the aging functional groups increase at the same aging process, vice versa. Apparently, the addition of lignin declined the growth rate of both carbonyl and sulfoxide compounds comparing to neat bitumen (Bref). It showed that lignin shows to an anti-aging effect since it reduced the production of related functional groups. The changes in the two functional groups with the content of lignin were different. The increase rate of the carbonyl group decreased first and then increased with the content of lignin increasing. IRI reached the lowest when the content of lignin was 10%. However, the increase rate of sulfoxide basically did not change with lignin content. The IRI of sulfoxide was slightly different from each other compared with that of the carbonyl. Compared to neat bitumen (Bref), it was still significantly lower.

The top surface of bitumen which is exposed to oxygen, ages first. As time proceeds, the internal species of bitumen starts to be oxidized due to the diffusion of oxygen into the bitumen. In the case of bitumen with organosolv lignin particles, the diffusion of oxygen is potentially prohibited to flow and react with the sensitivity carbon and sulfur species of bitumen and to produce carbonyl and sulfoxide compounds. In other words, it would take a longer time for the oxygen to diffuse and react with these bituminous species. Here, on the basis of spectroscopic results, not new reaction products are observed with the addition of lignin in bitumen, and thus it is believed that lignin possibly is able to inactivate oxidation promotors and oxygen receptors that might be in bitumen. The mechanism of inhibition of oxidation is thus based on prohibiting the susceptible to oxygen species to be transformed to carbonyl or sulfoxide compounds and thus decelerating the overall oxidation process of bitumen.

3.3 Rheological changes

The master-curves of complex modulus of different materials are shown in Figure 5. High modulus of the material indicated increase resistance to deformation. Each tested material had three replicates. Obviously, the modulus increased with the increase of (fresh or aged) lignin content in bitumen and over different aging time periods. Over the range of high frequencies, the properties between all samples were very close. Although modulus increase improved the resistance to deformation, it may be prone to brittle fracture, which affected the low-temperature performance. However, the effect of lignin on bitumen mainly reflected at high temperatures showing improved high resistance to deformation.

4 CONCLUSION

The research presented in this paper has shown the preliminary conclusions on the use of lignin as an aging inhibitor in bitumen. After aging, the lignin particles agglomerated, the density of lignin increased and its specific surface area decreased. In addition, its surface became smoother, the color of appearance became darker. Moreover, the IRI of aging functional groups was designed to accurately evaluate the anti-aging performance of lignin. The addition of lignin in bitumen reduced production and inhibited the formation rate of carbonyl and sulfoxide compounds. The changes in IRI in these two compounds were different possibly due to their different chemical properties. Finally, the master-curves of complex modulus have demonstrated that lignin in bitumen improved high-temperature resistance to deformation and potentially led to the enhancement of brittle fracture at low temperatures.

ACKNOWLEDGMENTS

The first author would like to acknowledge the scholarship from the China Scholarship Council and the financial support by Nedvang.

REFERENCES

Apostolidis, P., et al. 2017. Synthesis of asphalt binder aging and the state of the art of antiaging technologies. *Transportation Research Record 2633*, 147–153.

Arafat, S., et al. 2019. Sustainable lignin to enhance asphalt binder oxidative aging properties and mix properties. *Journal of Cleaner Production 217*, 456–468.

Batista, K.B., et al. 2018. High-temperature, low-temperature and weathering aging performance of lignin modified asphalt binders. *Industrial Crops and Products 111*, 107–116.

Boeriu, C.G., et al. 2004. Characterization of structure-dependent functional properties of lignin-natural antioxidants. *Industrial Crops and Products 20*, 205–218.

Bruijnincx, P., et al. 2016. *Lignin valorisation: The importance of a full value chain approach*. Utrecht University.

Dizhbite, T., et al. 2004. Characterization of the radical scavenging activity of lignins-natural antioxidants. *Bioresource Technology 95*, 309–317.

Gosselink, R. 2011. *Lignin as a renewable aromatic resource for the chemical industry*. PhD Thesis, UWageningen.

Pan, T. 2012. A first-principles based chemophysical environment for studying lignins as an asphalt antioxidant. *Construction and Building Materials 36*, 654–664.

Peterson, J.C. 2009. *A review of the fundamentals of asphalt oxidation: chemical, physicochemical property, and durability relationship*. Transportation Research Circular, E-C140, Transportation Research Board of National Academies, Washington, D.C.

Sundstrom, D.W., et al 1983. Use of byproduct lignins as extenders in asphalt. *Ind. Eng. Chem. Prod. Res. Dev. 22*, 496–500.

van Vliet, D., et al. 2016. Lignin as a green alternative for bitumen. In *Proceedings of E&E Congress*.

Williams, R.C. et al 2008. *The utilization of agriculturally derived lignin as an antioxidant in asphalt binder*. InTrans Project Reports 14.

Xu, G., et al. 2017. Rheological properties and anti-aging performance of asphalt binder modified with wood lignin. *Construction and Building Materials 151*, 801–808.

Zhang, Y., et al. 2019. Chemical and rheological evaluation of aged lignin-modified bitumen. *Materials 12(24)*, 4176.

Author index

Printed and bound by CPI Group (UK) Ltd, Croydon, CR0 4YY

24/10/2024

01778293-0011